T0201762

PROBABILITY CONCEPTS AND THEORY FOR ENGINEERS

PROBABILITY CONCEPTS AND THEORY FOR ENGINEERS

Harry Schwarzlander
Formerly
Department of Electrical and Computer Engineering
Syracuse University, NY, USA

A John Wiley and Sons, Ltd, Publication

This edition first published 2011
© 2011 John Wiley & Sons, Ltd

Registered office
John Wiley & Sons Ltd, The Atrium, Southern Gate, Chichester, West Sussex, PO19 8SQ, United Kingdom

For details of our global editorial offices, for customer services and for information about how to apply for permission to reuse the copyright material in this book please see our website at www.wiley.com.

Library of Congress Cataloging-in-Publication Data

Schwarzlander, Harry.
  Probability concepts and theory for engineers / Harry Schwarzlander.
      p. cm.
  Includes bibliographical references and index.
  ISBN 978-0-470-74855-8 (hardback)
  1. Probabilities. 2. Electrical engineering–Mathematics. I. Title.
  TK7864.S39 2011
  519.202′462–dc22

                                        2010033582

Print ISBN: 9780470748558 (hb)
ePDF ISBN: 9780470976470
oBook ISBN: 9781119990895
ePub ISBN: 9780470976463

A catalogue record for this book is available from the British Library.

Set in 9.5/11.5 pt Times by Thomson Digital, Noida, India
Printed in the UK by CPI Antony Rowe, Chippenham, Wiltshire

This book is dedicated to
students young and old
whose thinking will
help shape the future.

Contents

Preface   **xi**
Introduction   **xiii**

Part I   The Basic Model
Part I Introduction   2
Section 1   Dealing with 'Real-World' Problems   3
Section 2   The Probabilistic Experiment   6
Section 3   Outcome   11
Section 4   Events   14
Section 5   The Connection to the Mathematical World   17
Section 6   Elements and Sets   20
Section 7   Classes of Sets   23
Section 8   Elementary Set Operations   26
Section 9   Additional Set Operations   30
Section 10   Functions   33
Section 11   The Size of a Set   36
Section 12   Multiple and Infinite Set Operations   40
Section 13   More About Additive Classes   44
Section 14   Additive Set Functions   49
Section 15   More about Probabilistic Experiments   53
Section 16   The Probability Function   58
Section 17   Probability Space   62
Section 18   Simple Probability Arithmetic   65
Part I Summary   71

Part II   The Approach to Elementary Probability Problems
Part II Introduction   74
Section 19   About Probability Problems   75
Section 20   Equally Likely Possible Outcomes   81
Section 21   Conditional Probability   86
Section 22   Conditional Probability Distributions   91
Section 23   Independent Events   99
Section 24   Classes of Independent Events   104
Section 25   Possible Outcomes Represented as Ordered k-Tuples   109
Section 26   Product Experiments and Product Spaces   114
Section 27   Product Probability Spaces   120

Section 28   Dependence Between the Components in an Ordered k-Tuple   125
Section 29   Multiple Observations Without Regard to Order   128
Section 30   Unordered Sampling with Replacement   132
Section 31   More Complicated Discrete Probability Problems   135
Section 32   Uncertainty and Randomness   140
Section 33   Fuzziness   146
Part II Summary   152

Part III   Introduction to Random Variables
Part III Introduction   154
Section 34   Numerical-Valued Outcomes   155
Section 35   The Binomial Distribution   161
Section 36   The Real Numbers   165
Section 37   General Definition of a Random Variable   169
Section 38   The Cumulative Distribution Function   173
Section 39   The Probability Density Function   180
Section 40   The Gaussian Distribution   186
Section 41   Two Discrete Random Variables   191
Section 42   Two Arbitrary Random Variables   197
Section 43   Two-Dimensional Distribution Functions   202
Section 44   Two-Dimensional Density Functions   208
Section 45   Two Statistically Independent Random Variables   216
Section 46   Two Statistically Independent Random Variables—Absolutely
              Continuous Case   221
Part III Summary   226

Part IV   Transformations and Multiple Random Variables
Part IV Introduction   228
Section 47   Transformation of a Random Variable   229
              a)   Transformation of a discrete random variable   229
              b)   Transformation of an arbitrary random variable   231
              c)   Transformation of an absolutely continuous random variable   235
Section 48   Transformation of a Two-Dimensional Random Variable   238
Section 49   The Sum of Two Discrete Random Variables   243
Section 50   The Sum of Two Arbitrary Random Variables   247
Section 51 n-Dimensional Random Variables   253
Section 52   Absolutely Continuous n-Dimensional R.V.'s   259
Section 53   Coordinate Transformations   263
Section 54   Rotations and the Bivariate Gaussian Distribution   268
Section 55   Several Statistically Independent Random Variables   274
Section 56   Singular Distributions in One Dimension   279
Section 57   Conditional Induced Distribution, Given an Event   284
Section 58   Resolving a Distribution into Components of Pure Type   290
Section 59   Conditional Distribution Given the Value of a Random Variable   293
Section 60   Random Occurrences in Time   298
Part IV Summary   304

Part V   Parameters for Describing Random Variables and Induced Distributions

Part V Introduction		306
Section 61	Some Properties of a Random Variable	307
Section 62	Higher Moments	314
Section 63	Expectation of a Function of a Random Variable	320
	a) Scale change and shift of origin	320
	b) General formulation	320
	c) Sum of random variables	322
	d) Powers of a random variable	323
	e) Product of random variables	325
Section 64	The Variance of a Function of a Random Variable	328
Section 65	Bounds on the Induced Distribution	332
Section 66	Test Sampling	336
	a) A Simple random sample	336
	b) Unbiased estimators	338
	c) Variance of the sample average	339
	d) Estimating the population variance	341
	e) Sampling with replacement	342
Section 67	Conditional Expectation with Respect to an Event	345
Section 68	Covariance and Correlation Coefficient	350
Section 69	The Correlation Coefficient as Parameter in a Joint Distribution	356
Section 70	More General Kinds of Dependence Between Random Variables	362
Section 71	The Covariance Matrix	367
Section 72	Random Variables as the Elements of a Vector Space	374
Section 73	Estimation	379
	a) The concept of estimating a random variable	379
	b) Optimum constant estimates	379
	c) Mean-square estimation using random variables	381
	d) Linear mean-square estimation	382
Section 74	The Stieltjes Integral	386
Part V Summary		393

Part VI   Further Topics in Random Variables

Part VI Introduction		396
Section 75	Complex Random Variables	397
Section 76	The Characteristic Function	402
Section 77	Characteristic Function of a Transformed Random Variable	408
Section 78	Characteristic Function of a Multidimensional Random Variable	412
Section 79	The Generating Function	417
Section 80	Several Jointly Gaussian Random Variables	422
Section 81	Spherically Symmetric Vector Random Variables	428
Section 82	Entropy Associated with Random Variables	435
	a) Discrete random variables	435
	b) Absolutely continuous random variables	438
Section 83	Copulas	443
Section 84	Sequences of Random Variables	454
	a) Preliminaries	454

	b)	Simple gambling schemes	455
	c)	Operations on sequences	458
Section 85		Convergent Sequences and Laws of Large Numbers	461
	a)	Convergence of sequences	461
	b)	Laws of large numbers	464
	c)	Connection with statistical regularity	468
Section 86		Convergence of Probability Distributions and the Central Limit Theorem	470
Part VI Summary			477

Appendices   **479**

Answers to Queries   479

Table of the Gaussian Integral   482

Part I Problems   483

Part II Problems   500

Part III Problems   521

Part IV Problems   537

Part V Problems   556

Part VI Problems   574

Notation and Abbreviations   **587**

References   **595**

Subject Index   **597**

Preface

This book had its earliest beginnings as a gradually expanding set of supplementary class notes while I was in the Department of Electrical Engineering (later renamed Department of Electrical and Computer Engineering, and now Department of Electrical Engineering and Computer Science) at Syracuse University. Our graduate course in 'Probabilistic Methods' was one of the courses I taught quite a few times—in the day and evening programs on campus as well as at the University's off-campus Graduate Centers and via satellite transmission.

Early on I found that existing textbooks, while providing a good coverage of probability mathematics, seemed to lack a consistent and systematic approach for applying that mathematics to problems in the 'real world.' My notes therefore focused initially on that aspect. I began to understand that mathematics *cannot* be 'applied to the real world.' One can only apply the mathematics to one's *thoughts* about the external world. This makes it important that these thoughts are appropriately structured—that a real-world problem gets conceptualized in a manner that allows the correct and consistent application of mathematical principles. This conceptualizing is of little concern in most applications of mathematics since it occurs rather automatically. But the application of Probability Theory calls for much more attention to be devoted to formulating a suitable conceptual model.

Each time I taught the course I tried to improve and add to my notes. After my retirement from the Department I happened to look at these notes again, and they struck me as sufficiently interesting to make it worthwhile to expand and rework them into a textbook. Now, many years later, after incorporation of much more material and a great deal of editing and revising, as well as the creation of a large number of problems and review exercises, here is that book.

Naturally, many individuals and sources have helped me to move forward with this project and to bring it to completion. First, I am indebted to Profs. C. Goffman, H. Teicher, and M. Golomb, among others, in the Department of Mathematics at Purdue University, for helping me strengthen my mathematical thinking. Of course, I derived inspiration and deepened my understanding through the Probability textbooks I used in my course at different times—books by Gnedenko, Meyer, Papoulis, Parzen, Pfeiffer, and Reza—as well as through interactions with my students. Discussions with colleagues at Syracuse University have also been helpful. Of these, I want to specifically acknowledge Prof. D. D. Weiner and Dr. M. Rangaswamy, who introduced me to spherically symmetric random variables; as well as Prof. F. Schlereth, who directed my attention to copulas.

However, the writing of this book could not have come to completion without the loving care and encouragement I received over all these years from my wife, Patricia Carey Schwarzlander. I also appreciate the understanding and support of my children, as well as my grandchildren, over the stressful period of nearly two years during which I converted the manuscript into publishable form and worked my way through the page proofs. I am also grateful for the assistance and encouragement provided by the staff of Wiley, Chichester. And I am indebted to the Department of Electrical Engineering and Computer Science at Syracuse University for having continued to provide me with an office, without which this project would have been very difficult to carry out.

Harry Schwarzlander
Syracuse, NY
November 2010

Introduction

Motivation and General Approach

The intent of this book is to give readers with an engineering perspective a good grounding in the basic machinery of Probability Theory and its application. The level of presentation and the organization of the material make the book suitable as the text for a one-semester course at the beginning graduate level, for students who are likely to have had an introductory undergraduate course in Probability. Nevertheless, the basic approach to applying probability mathematics to practical problems is not short-changed. Many years of teaching graduate students in Electrical Engineering have made it clear to me that it is very helpful to begin with a thorough treatment of the basic model, rather than gloss over this important material in favor of rushing on to a greater variety of advanced topics. This is the strategy pursued here.

Although the material is developed from first principles, a greater mathematical maturity is called for than is usually expected from most undergraduates, and intuitive motivational arguments are minimized. Nevertheless, portions of the book can be used for an undergraduate course, although it would differ significantly from a typical undergraduate 'Probability and Statistics' course. Furthermore, the material is presented in sufficient detail and organized in such a way that the book is also a very suitable text for self-study.

The book is unusual in a number of respects. It is intended for the student who applies probability theory. Yet there is not much discussion of specific applications, which would require much space to be devoted to establishing the specific problem contexts. Instead, it has been my intent to provide a thorough understanding (1) of probability theory as a mathematical tool, emphasizing its structure, and (2) of the underlying conceptual model that is needed for the correct application of that tool to practical problems. The manner in which a connection is made between a real-world problem and its mathematical representation is stressed from the beginning. On the other hand, not much emphasis is placed on combinatorics and on the properties of special distributions.

The degree of sophistication in probability expected of engineering students at the graduate level has changed over the years. In preparation for further study and research in stochastic processes, information theory, automata theory, detection theory, radar, and many other areas, familiarity with some of the more abstract topics, such as transformations of random variables, singular distributions, and sequences of random variables, is desirable. Such material is introduced with care at appropriate points in the book.

Throughout, my aim has been to help the reader achieve a solid understanding and feel comfortable and confident in applying probability theory in engineering research as well as in practical problems. Since most graduate students in engineering do not have the opportunity to go through two or three theoretical mathematics courses before starting into probability, this has meant placing more emphasis on introducing or reviewing various mathematical ideas needed in the development of the theory.

Organization of Material

The basic unit of presentation is the *Section*. Sections are numbered consecutively but are grouped into six major subdivisions—Parts I through VI—which could also be thought of as chapters. Each Section introduces and develops one or several related new concepts. There are *problems* associated with each Section and these are presented in the Appendix. Sections are somewhat self-contained, and the *numbering* of equations, definitions, theorems and problems begins anew in each Section, with the Section number as prefix.

To help the student reflect on the various concepts and become more familiar with them, each Section ends with one or more 'queries,' which are short review exercises. These also facilitate self-study. An answer table to the queries appears at the beginning of the Appendix.

Part I covers the development of the basic conceptual and mathematical models: the probabilistic experiment and the probability space. Devoting all of Part I to this introductory development was motivated by (1) my experience indicating that even students who do have an undergraduate probability background tend nevertheless to have a weak grasp of this material, and (2) the fact that it is an essential foundation on which to build a good understanding of more advanced concepts. For further emphasis, the first few Sections have intentionally been kept short.

In Part II there is brought together a mixture of topics pertaining to 'elementary probability problems'— that is, to problems formulated in terms of a basic probability space without requiring the introduction of random variables. This portion of the book provides a bridge between Part I and the introduction of random variables in Part III. Some of the material is relatively simple, such as sampling with and without replacement. It is included for the sake of completeness, and some of this might be skipped in a graduate-level course. On the other hand conditional probability, independence, and product spaces are of course essential.

Part III serves to introduce the basics of random variables and an induced probability space. Discrete and absolutely continuous random variables are clearly distinguished. Discussion of multidimensional random variables has been divided into two steps. First, only two-dimensional random variables are introduced, whereas higher-dimensional random variables are discussed in Part IV. The reasons for this are that (1) it allows the student to first get a good grasp of the two-dimensional case, which is more easily visualized, and (2) it allows pursuit of a somewhat simplified syllabus from which higher-dimensional random variables are omitted. From Part III onward, some special attention is devoted to Gaussian distributions and Gaussian random variables because of their mathematical importance as well as their significance in engineering applications.

Transformations of random variables are addressed in considerable detail in Part IV. In addition to higher-dimensional random variables, Part IV also covers a number of other topics that do not require reference to expectations and moments, which are dealt with extensively in Part V. Delaying this important material to a later portion of the book is a choice I have made for pedagogic reasons. I believe it is helpful to gain a thorough grasp of the basic mechanics of random variables before getting involved in the various descriptors of random variables. Naturally, an instructor using this book can bring in some of that material earlier, if desired. Part V concludes with an introduction to the Riemann–Stieltjes integral, which allows some of the definitions in Part V to be extended to singular random variables. This has made it possible to bypass a discussion of the Lebesgue integral.

Part VI begins with a Section on complex random variables and then introduces the characteristic function and the generating function. There is a discussion of multidimensional Gaussian and Gaussian-like ('spherically symmetric') random variables. And following up on the Section on 'entropy' in Part II, this measure of randomness is now applied to random variables. Also introduced is the relatively new topic of 'copulas.' Then, sequences of random variables are treated in some detail, leading to the laws of large numbers and the Central Limit Theorem.

Notation and Terminology

Specialized notation and abbreviations have been avoided as much as possible. Frequent reference to a 'probabilistic experiment' has made the abbreviation 'p.e.' convenient. For *intervals* I have found Feller's [Fe2] overbar notation useful in order to avoid a proliferation of parentheses and brackets (see Section 36). It has been my experience that students readily accept this. A table summarizing all abbreviations and notation is provided in the Appendix.

Representative Syllabi

For a one-semester graduate course, the following is one way of selecting material from this book and grouping it into twelve weekly assignments:

Week:	Sections:	Subject matter:
1	1–8	The probabilistic experiment and set theory concepts.
2	9–16	More set theory, functions, statistical regularity, probability.
3	17–22	Probability space, simple probability problems, conditional probability.
4	23–27	Independence, product spaces.
5	34–39	Discrete and absolutely continuous random variables.
6	41–46	Two-dimensional random variables, statistical independence.
7	40, 47–50	Gaussian random variables, transformations.
8	51–55, 57	n-dimensional random variables, coordinate transformations, conditional distributions.
9	59–63	Conditional distributions, random occurrences in time, expectations.
10	64, 67–71	Variance of a function of a random variable, conditional expectations, correlation coefficient, covariance matrix.
11	75–80	Characteristic functions, n-dimensional Gaussian random variables.
12	84–86	Sequences, laws of large numbers, Central Limit Theorem.

For an undergraduate course, the book would be used in a different way, with greater emphasis on the development of the basic conceptual and mathematical ideas and on discrete probability. In this case, the following path into the book might be found useful:

Week:	Sections:	Subject matter:
1	1–7	The probabilistic experiment and set theory concepts.
2	8–13	Set operations, functions.
3	14–18	Statistical regularity, the probability function and probability space
4	19–22	Simple probability problems, conditional probability.
5	23–26	Independence, product experiments.
6	27–30	Product probability spaces, sampling experiments.
7	34–37	Random variable defined.
8	38, 39, 41, 42	C.d.f. and p.d.f., two random variables.
9	43, 44, 46	2-dimensional c.d.f. and p.d.f., statistical independence.
10	61–63, 66	Expectation and variance, sampling.
11	67–70	Conditional expectations, correlation coefficient.
12		Selections from the remainder of the book.

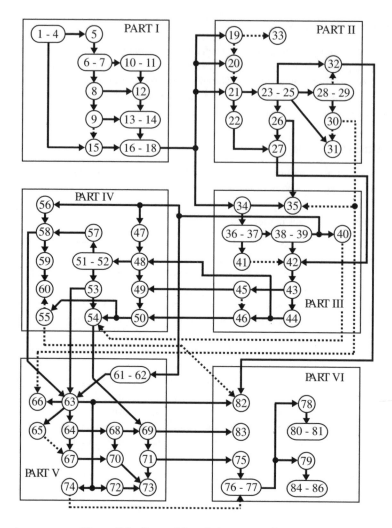

Figure F.1 Prerequisite relations among Sections

A flow diagram depicting the prerequisite relationships among Sections is shown in Figure F.1. A reader who is interested primarily in studying a particular Section can see from this diagram with what background material he or she should be acquainted.

Part I

The Basic Model

Part I Introduction

This First Part is devoted to the development of the basic conceptual and mathematical formulation of a probability problem. The conceptual formulation serves to build up a clear and consistent way of thinking about a problem situation to which probability analysis is to be applied. The mathematical formulation is then constructed, step by step, by connecting set-theoretic concepts to the conceptual building blocks, and the relevant notation is introduced. Also, those principles of Set Theory that are needed for the mathematical formulation are presented in detail. The result is the 'basic model', which consists of a conceptual part (the 'probabilistic experiment') and a mathematical part (the 'probability space'). Part I ends with some preliminary exercising of this model, and the establishment of several simple rules of probability arithmetic.

For the sake of emphasis, the individual Sections of Part I have intentionally been kept short. This is also intended to help ease the reader into a somewhat faster pace in the remaining Parts. Readers who have some familiarity with Probability principles should not skip Part I, but may find that they can assimilate the material presented here more quickly than those without that background.

About Queries:
Queries appearing at the end of a Section are short-answer review exercises that are intended to be worked out, pencil in hand, after studying that Section. A table of answers to queries appears in the Appendix. The number in brackets at the end of a query is the key for locating the answer in the answer table.

About Problems:
There are problems associated with each Section. These are located in the Appendix, arranged by Section.

SECTION *1*

Dealing with 'Real-World' Problems

Probability theory is a branch of mathematics that has been developed for more effective mathematical treatment of those situations in the real world that involve uncertainty, or randomness, in some sense. Most nontrivial real-world problems incorporate just such situations, so that the practical importance of probability theory hardly needs to be stressed. It turns out, however, that intensive study of the theory alone can still leave the would-be practitioner peculiarly inept in its application. Why should this be so?

It is not inappropriate to ask this question right at the outset. After all, when commencing the exploration of new territory, it helps to be alerted about the nature of the obstacles that lie ahead. A few reflections can lead to an answer to our question.

Example 1.1

> A young man working as a sales clerk in a furniture store is asked to count all the chairs that are on display in the store. These chairs may differ from each other in design and material. Nevertheless, the sales clerk automatically adds to his count each time his gaze comes to rest on an object that, in fact, anyone else would also regard as an instance of 'chair'. This is because the *concept* 'chair' is firmly established in his mind. Possession of the concept 'chair' permits the clerk in this situation to carry out the simple mathematical operation of *counting*. Furthermore, the widely shared agreement regarding the concept 'chair' assures that the count will be the same as that obtained by anyone else, barring oversights or counting errors. We may say that with the aid of the concept 'chair', the clerk avoids *conceptual errors* in his task, although not necessarily mathematical errors, or procedural errors.

We can imagine how the clerk in this Example has his store environment modeled in terms of a large variety of concepts, some as simple and common as 'chair'. These concepts serve to classify the environment, and act as discriminators in simple tasks such as discussed in the Example. On the other hand, it is important to realize that the applicability of a concept is not always clear-cut; concepts have a certain 'fuzziness'. Surely we can devise an object that neither obviously is a chair, nor obviously isn't a chair.

Probability Concepts and Theory for Engineers, First Edition. Harry Schwarzlander.
© 2011 John Wiley & Sons, Ltd. Published 2011 by John Wiley & Sons, Ltd.

In Example 1.1, the sales clerk is performing a '*real-world*' task. He is concerned with the immediate experience of his store environment, to which he is applying his counting ability. This is the sense in which we will use 'real world'. A person who picks up two dice and throws them acts in the real world, in contrast to another person who merely talks about throwing two dice.

Whenever mathematics is applied to a real-world problem, an important step is the establishment of a suitable model, or idealization, of the physical problem under consideration. This makes possible the application of the logical rules of mathematics to the real world. When we have a very simple problem, the modeling seems to take place rather automatically, so that we may not even be aware of it, as is surely the case with the sales clerk in Example 1.1. In more sophisticated situations, more effort generally goes into the modeling process. Because this modeling process always involves 'conceptualization'—fitting the real-world problem at hand into our way of thinking about things—we will call such models *conceptual models*.

Example 1.2

An electronic engineer is handed a small cylindrical capsule with a short wire protruding from each end. He tries to 'identify' this object as a particular kind of electronic component, such as a resistor, or a capacitor. Actually, 'resistor', 'capacitor', etc., are conceptual models for real-world objects—the engineer's way of viewing the great variety of electronic components he encounters. His task is, therefore, to determine which of these models is most appropriate for the object at hand.

The model called 'resistor', for instance, pertains to real-world objects having two (or more) distinct places for electrical connections. Furthermore, the applicability of this model depends on the effect produced by the component in question when incorporated into an electric circuit. This effect depends on the internal construction of the component.

We noted before that a conceptual model makes possible the application of mathematics to the real world. In the situation just described it is the mathematics of electrical network theory. For instance, the conceptual model 'resistor' implies to the engineer that under suitable conditions the relationship

$$i = ke \tag{1.1}$$

adequately characterizes the current i through the device in question, at an instant when the voltage across the terminals is e, where k is a proportionality constant. The conceptual model 'resistor' can therefore be considered as a bridge between a real-world object and the *mathematical model* that is expressed by the relation (1.1).

Of course, for a given real-world problem, the choice of a suitable model need not be unique, and the mathematical result obtained may depend on the particular model used. The appropriateness of any one model can be judged by how well the results obtained with it agree with, or predict, the actual physical nature or behavior of the situation being considered.

'Networks' are the conceptual models to which Network Theory is applied. Network Theory is not used in *building* the model, when some real-world electrical circuit is analyzed. But, once a model has been decided on, then Network Theory allows various conclusions to be drawn because it provides rules for applying abstract mathematical techniques to the model. The same process can be discerned whenever mathematics is applied to real-world problems. *The application of Probability Theory to situations of randomness is also made through suitable conceptual models.* However, as we will see in Section 2, the conceptual models required for probability problems are considerably more complicated than those needed in connection with most other branches of applied mathematics.

Just as Network Theory cannot be applied where there are no network models, so is Probability Theory useless without appropriate models. The establishment of an adequate model is therefore an essential ingredient in tackling any real-world problem by means of Probability Theory. Neglect of this important point easily results in confused and erroneous applications of the theory. Here lies the answer to the question posed at the beginning of this Section.[1]

As we pursue our study of Probability in this book, we will try to pay attention to the connection between real-world problems and their mathematical formulation from the start. This will be possible with the help of a standard scheme for building conceptual models—a scheme that can be applied to all probability problems.

---

[1] For further reading on the role of models in Applied Mathematics and the Sciences, see for example [Fr1], [Ri1], [MM1].

SECTION *2*

The Probabilistic Experiment

To begin with, a name should be given to situations of uncertainty in the real world to which we will apply Probability Theory. The word 'experiment' suggests itself when thinking of various real-world situations that involve uncertainty:

> the throwing of dice;
> the measuring of a physical parameter, such as length, temperature,
>   or magnetic field strength;
> sampling a batch of manufactured items.

We want to be able to express quantitatively the likelihood with which particular results will arise in situations such as these. But before we can even begin to apply any mathematics to such problems, we must have a clear picture of what it is we are actually dealing with. Now, 'experiment' is a very widely used word, and therefore imprecise in meaning. We will qualify it and will refer to any real-world 'experiment' as a '*probabilistic experiment*' if it has been modeled in a manner that allows probability mathematics to be applied. It is also worth noting that, when trying to envision real-world situations such as referred to in the above list, we find that in each case there appears an observer—the *experimenter*—as an explicit or implicit part of the setting. This should not be surprising, since 'uncertainty' can only exist as the result of some sort of contemplation, and thus is really a state of mind—the state of mind of the experimenter.

Thus, the first question to be addressed is: what are the essential features of a suitable conceptual model—a probabilistic experiment? These are stated in the form of four distinct requirements in the Definition below, and are then illustrated by means of an Example. These requirements will assure that the model brings into focus those aspects of real-world 'experiments' that are essential to the correct application of the mathematical theory. In other words, we shall 'talk probability' (ask probability questions, and compute probabilities) only in those real-world contexts that we are able to model in the manner specified below. When this is not possible we do not have a legitimate probability problem. (We note that definitions within the conceptual realm lack the precision of a mathematical definition. As a reminder of this, the word 'Definition' appears in parentheses.)

> **(Definition) 2.1**: Probabilistic experiment.
> A *probabilistic experiment* is a conceptual model that consists of four distinct parts, as follows:
>
> 1. *Statement of a purpose.* By this is meant an expression of the intent to make specific real-world observations. It must specify a particular real-world context, configuration, or

Probability Concepts and Theory for Engineers, First Edition. Harry Schwarzlander.
© 2011 John Wiley & Sons, Ltd. Published 2011 by John Wiley & Sons, Ltd.

environment, which is pertinent to these observations. It must also include a listing or precise delineation of the various distinct 'possibilities' (properties, alternatives, facts, values, or observable conditions) that are considered to be of interest and that are to be watched for when the intended real-world observations are actually carried out.

2. *Description of an experimental procedure.* By experimental procedure is meant the specification of an unambiguous sequence of actions to be carried out by the experimenter, leaving no choice to the experimenter, and leading to the observations that are called for in the *purpose*.

3. *Execution of the experimental procedure.* This may only take place after the purpose has been stated and the description of the experimental procedure is completed.

4. *Noting of the results.* By this is meant the identification of those 'possibilities', listed as part of the *purpose*, which are actually observed to exist upon complete execution of the experimental procedure.

The following Example provides a first step toward understanding the significance of this Definition. Here, we model the throwing of a die as a probabilistic experiment.

Example 2.1: A die-throwing experiment.

1. *The purpose of the experiment is* to determine one of the following six possibilities that are observable upon throwing a given die, whose faces are marked in the standard way, onto a firm and reasonably smooth horizontal surface of adequate size to allow the die to roll freely and come to rest:

one dot faces up

two dots face up

. . . . . .

. . . . . .

six dots face up

2. *The procedure is*: The experimenter is to throw the die in the customary manner onto a surface such as specified in the purpose, with enough force so as to cause the die to roll. When the die comes to rest, the experimenter is to observe the number of dots facing up.

3. *This procedure is carried out*: An appropriate surface and a die are available. The experimenter throws the die in the specified manner. It comes to rest at a certain spot on the surface with various numbers of dots facing in various directions and, in particular, five dots facing upward.

4. Of all that is perceived by the experimenter upon carrying out the procedure—the path of the die, the final location or orientation of the die, the lighting conditions, the sound accompanying the throw of the die, etc.—only the property 'five dots face up' is one of the possibilities that was initially stated to be of interest. *This property is noted.*

This example is, of course, rather trivial. Nevertheless, it serves to illustrate the importance of each of the four steps making up the probabilistic experiment; for, knowledge of only one (or even two or three) of the four parts making up the model does not guarantee a unique description of the complete experiment. Thus, suppose that item 1 in the description of the die-throwing experiment were missing. From a knowledge of items 2, 3, and 4 it is not certain that 1 should read as it actually appears above. For instance, the purpose might actually have been to observe one of the properties

five dots face up

five dots face sideways

neither of the above.

It turns out that if this were the purpose, the probability problem arising from the experiment would be quite different.

Now suppose the description of the experimental procedure were missing. Even if item 3 is known to us—that is, we have seen the experimental procedure being carried out—we cannot usually be sure what the procedure really was. Just consider the possibility of the following sentence added to item 2 in the above Example: 'If on the first throw the number of dots facing up is six, the die is thrown again.' There would be no way of inferring this from a knowledge of items 3 and 4 alone.

Without knowledge of item 3, on the other hand, it is not clear whether properties noted under item 4 are observed as a result of carrying out the procedure as specified in item 2, or some other sequence of actions. It might also be proposed that items 2 and 3 be combined; however, there is a definite reason for maintaining them as separate items: The description of the experimental procedure must not change during the course of the experiment—that is, while the procedure is being carried out. Another way of stating this is to say that item 2 is strictly 'deterministic', whereas item 3 brings into evidence occurrences and properties about which there is initial uncertainty.

Finally, it is clear that an essential part of the experiment is missing if knowledge about the observed properties of interest, as specified in item 4, is not available. This situation will be considered again in Section 3.

Thus, we have seen that even as innocent an activity as throwing a die calls for a careful description if it is to be modeled as a probabilistic experiment. Even more care will be needed when dealing with more complicated experiments, but this can only be appreciated after our study has progressed.

Particular attention has to be given to the *purpose* of a probabilistic experiment. The purpose cannot be in the form of a question, for instance. Consider: 'Will hydrochloric acid produce a precipitate in the unknown liquid sample?' This sentence does not conform with item 1 of our Definition and therefore cannot be the statement of purpose of a probabilistic experiment; it gives no clear account of the possible alternatives that are to be looked for in this experiment. Also, a probabilistic experiment cannot have as its purpose the determination of a probability. At this point we have not yet assigned a technical meaning to the word 'probability'. We are not yet concerned with probability, only with experiments. But if we stick to the common language sense of 'probability', which we might paraphrase as 'likelihood', or 'chance', this surely is not a property that is observable in the real world as a result of performing an experiment. Similarly, the purpose of a probabilistic experiment cannot be to decide something. Decisions might be made on the basis of the results of the experiment, but this circumstance does not enter into our model.

Probability Theory is actually not concerned with completed probabilistic experiments—only with probabilistic experiments prior to their execution. Nevertheless, a picture of the complete probabilistic experiment must be in our mind, as will become clearer as we proceed. We will sometimes refer to a probabilistic experiment *prior to execution*—i.e., to parts 1 and 2 of the Definition—as the '*experimental plan*'.

The discussion so far leads us to view our involvement in a real-world problem as taking place in three different domains, as illustrated in Figure 2.1. In the external world or 'Real World', a person experiences a profusion of perceptions. The 'Conceptual World' provides the opportunity to clarify, organize and prioritize these perceptions. Furthermore, it forms the bridge between real-world experience and mathematical analysis. In the Conceptual World, as mentioned earlier, 'definitions' are of a different kind from those in the Mathematical World. They generally are rules for relating the real world to the conceptual world, and thus are susceptible to interpretation. There is usually a fuzziness to these definitions, that is, their application to any given problem is not always entirely

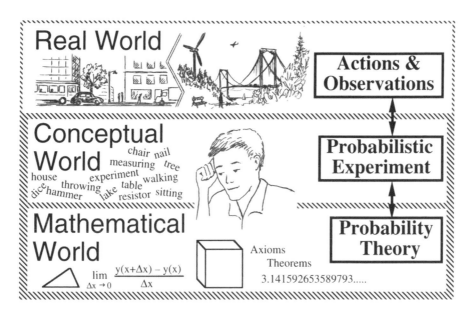

Figure 2.1 Applying mathematics to real-world problems

clear-cut. It is then necessary to use judgment based on experience, or to proceed with caution and be prepared to modify one's approach.

The notion of probability evolved in the conceptual world. It does not exist in the real world. Designating our conceptual experimental model a 'probabilistic experiment' therefore seems fitting. The definition of a probabilistic experiment is our first encounter with a 'definition' in the Conceptual World. We may think of it also as a *rule* for constructing the right kind of conceptual model for any given real-world experiment—a rule telling us how to organize our thinking about the experiment. A probabilistic experiment will always serve as our conceptual model when we try to apply Probability Theory to real-world problems, as indicated in Figure 2.1. Probability Theory may not lead to meaningful results when dealing with a real-world problem that cannot be modeled in accordance with Definition 1 or some other suitable conceptual model.[2]

Real-world problems that can be modeled as probabilistic experiments have a peculiar feature that is not shared by most other types of problems to which Mathematics can be applied. It is the fact that a human being, an observer—the experimenter—is an essential part of the model. *The experimenter's* mind harbors the uncertainty (or certainty) about the properties that will be observed in an experiment. Thus, Probability Theory is unique, since it permits the application of mathematical techniques to a class of problems that in a certain sense include the interaction of a human being with his/her environment.

---

[2] The Definition of this Section, and the additional requirements put on the conceptual model in Section 15, follow closely the specifications presented in [AR1]. Other 'systems' of probability exist, which are not built on the particular kind of conceptual model introduced here. This fact can be a source of confusion, because nearly the same terminology and mathematics appears in all of them. Our approach here is along those lines that are now most widely accepted, especially in the physical and natural sciences and in engineering.

Queries

Note: Queries appearing at the end of a section are to be considered part of the material of that section. The reader should answer them, pencil and paper at hand, in order to assure a correct understanding of the ideas presented. The number in brackets following each Query identifies the answer in the Answer Table (see Appendix).

2.1 Which of the following are correct statements, which incorrect, based on a reasonable interpretation of the definition of a probabilistic experiment? If incorrect, which part(s) of Definition 2.1 are violated?

a) A person playing a game of checkers is conducting a probabilistic experiment.

b) A person, prior to playing a game of checkers, is deciding on a strategy. This constitutes a probabilistic experiment.

c) A person decides to take a walk along a certain stretch of seashore to see if she might find something interesting, such as driftwood, shells, etc. She is embarking on a probabilistic experiment.

d) A person is waiting at Kennedy airport for a friend who is to arrive on a particular flight. She wonders whether or not the friend will arrive safely. She is engaged in a probabilistic experiment.

e) A person with high blood pressure is about to fly from New York to London. He wonders whether he will get there or whether he will die from a heart attack or perish in a crash along the way. He is engaging in a probabilistic experiment.

f) A mail clerk puts a letter on a letter scale to see whether it weighs more than an ounce or not. He is performing a probabilistic experiment.   [184]

SECTION *3*

Outcome

The 'probabilistic experiment' (henceforth abbreviated p.e.) is the conceptual model through which we will view any real world problem to which we want to apply Probability Theory. As we proceed, we will become quite accustomed to working with p.e.'s. We will also explore some of the difficulties that can arise when the conceptual model is bypassed. But before establishing connections to mathematics, a little more needs to be said about the model.

(Definition) 3.1
Of the distinct possibilities of interest in a particular p.e., all those that are actually observed after the procedure has been carried out are collectively called the *actual outcome* (or simply, *outcome*) of the p.e.

For instance, the die-throwing experiment in Example 2.1 has as the actual outcome the single property 'five dots face up'. But an outcome can also consist of several properties, all of which are observed upon executing the procedure. On the other hand it follows from the above Definition that an experiment cannot have more than one outcome. An experimental plan leads to one actual outcome from among a variety of candidates, upon completion of the experiment. These various candidates are referred to as the *possible outcomes* of the p.e., or of the experimental plan. Thus, the die-throwing experiment of Section **2** has six possible outcomes, and each happens to be associated with exactly one of the six properties that are of interest.

It is often convenient to incorporate the notion of 'possible outcomes' in the description of a p.e.: In the statement of purpose of a p.e., we can replace the list of 'possibilities' or 'properties', etc., by a listing of all the possible outcomes. In our die-throwing experiment this happens to cause no change, since the possible outcomes coincide with the various properties of interest. This is not always the case.

Example 3.1: Throw of two dice.
An experiment similar to the one modeled in Example 2.1 is to be performed with *two* dice—one red and one white die. There are now 12 properties of interest:

one dot faces up on red die	one dot faces up on white die
. . .	. . .
. . .	. . .
six dots face up on red die	six dots face up on white die

Probability Concepts and Theory for Engineers, First Edition. Harry Schwarzlander.
© 2011 John Wiley & Sons, Ltd. Published 2011 by John Wiley & Sons, Ltd.

On the other hand, this p.e. has $6 \times 6 = 36$ possible outcomes, covering all possible combinations of results on the red and the white die. Thus, each possible outcome is made up of *two* 'properties of interest'. In the statement of purpose for this p.e., it is simpler to list all properties of interest than all possible outcomes.

Now let us suppose the die-throwing experiment is performed on a sidewalk and the die falls down a drain. How does this situation fit into the schema of Definition 2.1? None of the properties of interest in the experiment can be observed in this case, so that there is *no outcome*! Although this experiment may have been 'properly' performed, in the sense that the experimenter carried out all the required actions, it nevertheless turned out in an undesirable manner. We say that such an experiment, which has no outcome, was *unsatisfactorily performed*.

Henceforth, unsatisfactorily performed experiments will be *excluded from our considerations*. The theory to be developed, which will allow us to treat mathematically those situations in which observables arise with some degree of uncertainty, presupposes experiments which are modeled according to Definition 2.1—i.e., as probabilistic experiments—and furthermore, these experiments must be satisfactorily performed. This should not be regarded as restricting the applicability of the theory; it merely refines our conceptual model. For, we have in fact two ways of accommodating the situation described above, where a die falls down a drain:

a) As an unsatisfactorily performed experiment it does not exist in our conceptual model world—we simply do not think of the actions that lead to the loss of the die as belonging to a p.e. Or:
b) As an experiment whose possible outcomes include 'die is lost' or some such property, it does have an acceptable representation in our conceptual model world. In this way it can be arranged that an otherwise unsatisfactorily performed experiment gets treated as a satisfactorily performed one.

In practice, any situation involving uncertainties can be framed within some kind of hypothetical experiment to which the scheme of Definition 2.1 can be applied. Often, several different possible experiments suggest themselves. Each of these different formulations may result in a different mathematical treatment of the problem at hand.

Queries

3.1 Consider various p.e.'s whose procedure calls for one throw of an ordinary die onto a table top. For each of the following lists, decide whether it can be a listing of the *possible outcomes* of such an experiment.

a) 'one dot faces up'
   'two dots face up'
   'three dots face up'
   'four dots face up'
   'five dots face up'
b) 'three or fewer dots face up'
   'an even number of dots faces up'
   'four or more dots face up'
c) 'one dot faces up'
d) 'an even number of dots faces up'
   'one dot faces up'
e) 'die comes to rest on table top'
   'die falls off table'

f) 'the experiment is satisfactorily performed'
   'the experiment is unsatisfactorily performed'.   [42]

3.2 In the statement of purpose of a certain p.e. there appears a list of exactly *three* observable properties, or features, which are to be looked for upon executing the procedure. Allowing only satisfactorily performed experiments, what is the maximum number of possible outcomes that might exist for this experiment? The minimum number?   [226]

3.3 The purpose of a particular p.e. includes a listing of various observable features, or properties, that are of interest in this experiment and that are to be looked for upon executing the procedure. What is the least number of observable features that must be in this listing if it is known that the p.e. has five possible outcomes?   [195]

3.4 An experimenter who is about to perform a particular p.e. claims that she knows which one of the possible outcomes will be the actual outcome, prior to performing the experiment. This implies which of these:
a) She will see to it that the experiment will be unsatisfactorily performed
b) The experiment is 'rigged'
c) The experiment has only one possible outcome
d) She has performed the experiment previously.   [213]

SECTION *4*

Events

Often, we will be especially interested in some group or collection of possible outcomes from among all the various possible outcomes of a p.e. We may then wish to express the fact that the actual outcome, when the experiment is performed, belongs to this particular collection of possible outcomes. Consider again the die-throwing experiment with possible outcomes 'one dot faces up', 'two dots face up', . . . 'six dots face up'. We might be particularly interested to see whether the actual outcome of this experiment is characterized by the description

$$\text{'an even number of dots face up'} \tag{4.1}$$

This description encompasses several possible outcomes, namely:

'two dots face up'
'four dots face up'
'six dots face up'.

(Definition) 4.1
Suppose that an experimental plan and a listing of the possible outcomes have been specified. Any characterization of results that might be observed upon performing the experiment is called an *event*. An event therefore represents some collection of possible outcomes. If the experiment is performed and its actual outcome belongs to an event that has been specified for this experiment, then that event is said to have *occurred*.

Statement 4.1, above, is an example of an event that can be defined for the die-throwing experiment. We see that the description of an event need not explicitly identify all the possible outcomes that it encompasses.

For a given p.e. or experimental plan we may find it convenient to define *several* events, in order to simplify various statements about the experiment. Using again the die-throwing experiment as an example, we may be interested in stating whether, upon performing the experiment, more than three dots face up. Thus, we define for this experiment another event

'more than three dots face up'

Probability Concepts and Theory for Engineers, First Edition. Harry Schwarzlander.
© 2011 John Wiley & Sons, Ltd. Published 2011 by John Wiley & Sons, Ltd.

which consists of the possible outcomes

'four dots face up'
'five dots face up'
'six dots face up'.

In the die-throwing experiment described in Example 2.1, the actual outcome was 'five dots face up', so that the event 'more than three dots face up' did occur, but not the event 'an even number of dots is showing'.

There is, of course, another way of expressing whether a throw of the die results in more than three dots facing up. That is to revise the model by defining the probabilistic experiment in such a way that 'more than three dots face up' is one of the *possible outcomes*. Then we would be able to say that, in Example 2.1, the *outcome* was 'more than three dots face up'. In typical applications, however, this approach is cumbersome: once a fairly straightforward model has been established, it is usually preferable to stick with it.

It should be realized that the notion of 'event', as we have defined it, exists only in our conceptual model world. It has no counterpart in the real world in the absence of a conceptual model. It arises from the way in which we have decided to think about situations of uncertainty in the real world. Naturally, our specific use of the word 'event' must not be confused with its use in ordinary language, where it is synonymous with 'happening', or 'occurrence'.

Before concluding this Section we note some relationships between experiments, possible outcomes, and events.

a) Whenever we speak of events and possible outcomes, a particular experimental plan is assumed to underlie the discussion, even if it is not explicitly stated.
b) If two events are defined in such a way that both represent the same collection of possible outcomes, then they are considered to be the same event, even if their verbal descriptions differ.
c) An event can be defined in such a way that it comprises only *one* of the possible outcomes. Such an event is called an *elementary event*.
d) An event can be defined in such a way that it comprises *none* of the possible outcomes (the particular collection of outcomes that is empty). Such an event cannot occur when the experiment in question is performed (since we do not allow an experiment to have *no* outcome), and is therefore called an *impossible event*.
e) The event that comprises *all* possible outcomes of a particular p.e. is called the *certain event*, since it *must* occur when the experiment is (satisfactorily) performed.
f) *Several* different events may occur when an experiment is performed, even though an experiment always has only one outcome.
g) There may be uncertainty as to which of several possible outcomes will be the actual outcome when an experiment is about to be executed. But after the experiment has been performed, all uncertainty has been removed. It must then be perfectly clear what the actual outcome is; and whether or not a particular event has occurred.

Queries

4.1 Consider two different die-throwing experiments:

p.e. no. 1 has these possible outcomes:
'one or two dots face up'
'three or four dots face up'
'five or six dots face up'.

p.e. no. 2 has these possible outcomes:
  'an odd number of dots faces up'
  'an even number of dots faces up'.
For each of the following statements, indicate whether it characterizes an event in no. 1, an event in no. 2, in both, or in neither:

  a) more than two dots face up
  b) an odd number of dots faces up
  c) one dot faces up
  d) one or more dots face up
  e) no dots face up.   [117]

4.2 For a particular p.e., three events are specified, no two of which are the same.
   a) What is the minimum number of possible outcomes this p.e. must have?
   b) Same question, if it is known that all three events can occur together?   [19]

The Connection to the Mathematical World

Our next concern is the *connecting link* between the conceptual model and the mathematics of probability. We developed our conceptual model to the point where we talked about collections of outcomes—namely, events. It is therefore natural to turn to the branch of mathematics called Set Theory, which is a completely abstract theory applicable to problems that involve *collections of objects*. Two basic features of Set Theory are:

a) It requires *well-defined* collections of objects: A collection of objects is well defined if, given any object, this object either is a member of the collection, or it is not a member of the collection—it cannot be both.
b) There exists no uncertainty, vagueness or ambiguity. It is a strictly *deterministic* theory.

We may best grasp the significance of these two features by trying to think of examples of collections that lack them.

Example 5.1

Here is a collection that is *not* well-defined: 'Of all the children registered this year in a particular school, the collection of all those who are nine years old.' This collection is not well-defined because at different times of the year, not all the same children will be nine years old. Besides, some children may transfer into or away from the school during the year.

Example 5.2

We often make reference to collections of objects that are governed by some degree of uncertainty. Consider a 'bridge hand'. In the game of bridge, 13 cards are dealt to each of the four players, from a shuffled deck of 52 different cards. When we speak of a 'bridge hand', we are referring to a collection of 13 cards such as might be dealt to one of the players, without specifying which particular cards they are.

Looking now at the conceptual model we have developed so far, it should be clear that we have carefully constructed it in such a way that properties (a) and (b), listed above, are satisfied if the 'objects' in question are taken to be the *possible outcomes* of a probabilistic experiment.

Probability Concepts and Theory for Engineers, First Edition. Harry Schwarzlander.
© 2011 John Wiley & Sons, Ltd. Published 2011 by John Wiley & Sons, Ltd.

We note the following set-theoretic terms:

> **(Definition) 5.1**
> a) A *set* is a well-defined collection of abstract objects.
> b) *Elements* are the basic objects we are dealing with.
> c) A *space* is a nonempty set; that is, a set containing at least one element. Usually we mean by *space* that particular set that contains *all* the elements that are defined in a given problem or discussion.

The *elements* of a set, as abstract mathematical objects, are considered devoid of attributes, but are distinct. This seems contradictory since the absence of attributes would seem to make elements indistinguishable. We get around this by thinking of each of the elements of a set as *representing* a conceptual object so that, in particular, we can label and identify the individual elements of a set.

A set of elements, therefore, is the mathematical model for a collection of conceptual objects (or for the corresponding real-world objects), which abstracts the distinctness of the objects and their belonging to the collection, nothing else. A set of elements may also serve as the model for a collection of other mathematical objects (with mathematical attributes)—for instance, the collection of integers from 1 to 6. In such a case it becomes tedious and also unnecessary to maintain the distinction that a 'set of six elements' is a *model* for the collection of integers from 1 to 6. We speak then simply of 'the set of integers from 1 to 6', or the like.

It should now be clear that our earlier discussion of the conceptual model needs to connect with the mathematical entities set, element and space in the manner that is illustrated in Figure 5.1.

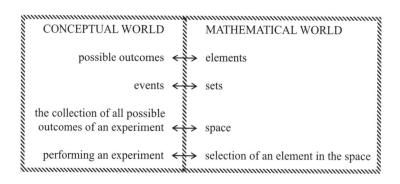

Figure 5.1   Making connections between the conceptual and mathematical worlds

The naturalness of this correspondence accounts for an easy diffusion of terminology across the boundary separating the conceptual and the mathematical world. Also, the possible outcomes are often called *samples*, and the elements *points*; these terms are often combined to give *sample points*, and it has become customary to refer to the collection of all possible outcomes of an experiment as the space of outcomes, or *sample space*. On the other hand, 'sample space' and 'event' are also used in Set Theory when this theory is applied to probability problems.

If the correspondence between the conceptual and mathematical worlds is made carelessly, inconsistencies can arise that will result in paradoxical results. Such difficulties have existed throughout much of the history of Probability. Establishing a firm correspondence between the conceptual and mathematical world is no trivial matter. Besides permitting the use of mathematics for the solution of real-world problems, it can lead to a better understanding of the conceptual model.

It may also lead to the introduction of additional constraints into the conceptual model that are inherent in the mathematical structure. For example, we have spoken of impossible events. Conceptually it may seem appropriate to consider a variety of different impossible events, in connection with a given probabilistic experiment. In the next Section we see that, as a result of the particular way in which we make the connection to the mathematical world, these various impossible events become indistinguishable. Our mathematical model allows only a single impossible event.

In the next few Sections we will explore those elementary notions of Set Theory that pertain to the mathematics of Probability.

Queries

5.1 Which of the following collections can be modeled by a set of elements?
  a) All the chairs on display in a particular furniture store.
  b) Consider a die-throwing experiment that has not yet been performed. The collection of those attributes of interest in this experiment that will *not* be observed upon performing the experiment.
  c) The collection of all mathematical theorems that have not yet been discovered (i.e., never stated, not even as conjectures).
  d) The collection of all the different sets that can be formed from elements in a given set.
  e) The collection of all aqueous solutions in which a drop of concentrated hydrochloric acid produces a precipitate at $20\,°C$.
  f) The collection of all the meanings attributable to some word that is given.   [36]

5.2 Consider the die-throwing experiment of Example 2.1. Let the collection of all possible outcomes be represented by a space of elements. For each of the following, state whether it is represented by an element, a set, or neither.
  a) 'the experiment has not been performed'
  b) 'the experiment has been performed'
  c) 'more dots face up than the next time the experiment will be performed'
  d) 'one dot faces up'
  e) 'more than one dot faces up'
  f) 'more than six dots face up'.   [122]

SECTION *6*

Elements and Sets

It is worth repeating that Set Theory is concerned with *well-defined* collections of objects. In any problem involving Set Theory, one well-defined collection of objects must always be the collection of *all* the elements (objects) that can arise in that problem. In other words, the problem statement must give enough information to define the set of *all* the elements that are under consideration in the particular problem. As mentioned in Section **5**, we call this set the *space*, or *sample space*, of the problem; and we denote it S.[3] If the space is not defined, Set Theory can lead to inconsistent results.

Once the space S is defined for a given problem, then it must be true that any element ξ arising in that problem *belongs to* (or 'is an element of') the particular set that serves as the space S. This relation is expressed symbolically by

$$\xi \in \mathsf{S}.$$

Henceforth, a space S will always be assumed to have at least two elements.

> **(Definition) 6.1**
> Given a space S and sets A, B. Then A is a *subset* of B, denoted
>
> $$\mathsf{A} \subset \mathsf{B}, \text{ or } \mathsf{B} \supset \mathsf{A}$$
>
> if and only if $\xi \in \mathsf{A}$ implies $\xi \in \mathsf{B}$, for all elements ξ (i.e., for all elements $\xi \in \mathsf{S}$).

For example, if a space S is defined for a given problem, then for any set A that arises in the problem the statement $\mathsf{A} \subset \mathsf{S}$ must be true. The statements $\mathsf{A} \subset \mathsf{B}$, $\mathsf{B} \supset \mathsf{A}$ are also read 'A is contained (or included) in B', 'B contains (or includes) A'.

The relationships expressed by $\in$ and $\subset$ are easily visualized in pictorial form by means of a so-called 'Venn diagram', as shown in Figure 6.1. In this diagram the large rectangular outline encloses a region that symbolizes the collection of all the points making up the space S for a given problem. Any point in this region then represents an element in S, that is, an element that can be talked about. Collections of such elements—that is, sets—are then represented by suitable regions within the area allotted to S. Individual elements are sometimes indicated by dots. When the region representing a set A is drawn inside the region representing another set B, as in Figure 6.1, this is meant to indicate that $\mathsf{A} \subset \mathsf{B}$.

---

[3] Other symbols commonly used to designate the space of elements are X, Ω, S, U. The term 'universal set' is also used.

Probability Concepts and Theory for Engineers, First Edition. Harry Schwarzlander.
© 2011 John Wiley & Sons, Ltd. Published 2011 by John Wiley & Sons, Ltd.

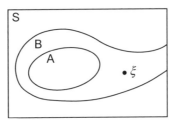

Figure 6.1   Venn diagram

The above notions relate readily to our conceptual model. The actual outcome of a p.e. must always be one of the possible outcomes. In set theoretic terms this means: Whenever we are modeling a p.e., every sample point that pertains to this p.e. must belong to the sample space for the p.e. Also, all the events that can be talked about consist of possible outcomes that belong to the sample space. Therefore, we may bring some Set Theory terminology into our conceptual model and say that these events are 'contained in' the collection of all possible outcomes, the sample space. For any given problem, we can think of the sample space as specifying a kind of 'universe'. All manipulations we might perform on its contents must keep us within its domain of definition.

The difference in the meanings of the two symbols $\in$, $\subset$ must be kept in mind. The symbol $\in$ relates an object to a *collection* of such objects. The symbol $\subset$ relates two things of like category, namely, *two collections* of similar objects. This is the way the symbols $\in$, $\subset$ are defined, so that their use in any other way is meaningless. The following statements, implied by Figure 6.1, illustrate the correct usage (where $\notin$ and $\not\subset$ are negations of $\in$ and $\subset$, respectively):

$$\xi \in S, \quad \xi \notin A, \quad \xi \in B, \quad A \subset B, \quad A \subset S, \quad S \not\subset B$$

We have used letter symbols A, B, C to identify sets. A set may also be identified by listing all its elements and enclosing this listing between braces. If ξ_a, ξ_b denote two distinct elements, then $\{\xi_a, \xi_b\}$ is *the set whose elements are ξ_a and ξ_b*. Braces will always be read 'the set whose elements are (contents of brace)'. Rather than enumerating all the elements that make up a set, we may characterize them in some way. Thus, the expression $\{\xi : \xi \notin A\}$ stands for 'the set of all ξ's *such that* ξ is not an element of A'. Another way of saying it is 'the set of those ξ's that satisfy the condition 'ξ is not an element of A'.'

It is important to distinguish between an *element* ξ and the *set whose only element is* ξ, denoted $\{\xi\}$. Thus, $\xi \in \{\xi\}$ and $\{\xi\} \subset S$; but the following statements are meaningless: $\xi \subset \{\xi\}$, $\{\xi\} \in S$. This illustrates how the details of the mathematical model can force additional constraints onto the conceptual model. Because of the manner in which we established the correspondence between the conceptual model and Set Theory, we must make a careful distinction between a possible outcome and the event consisting of only that one possible outcome. The former corresponds to an *element* in the space S, the latter to a *subset* of the space S.

Finally, it is possible to speak of a collection that is entirely devoid of objects. Thus, we have an *empty set*, which is denoted ϕ. Since there are no elements ξ belonging to ϕ, then in Definition 6.1 the requirement stated there is satisfied 'vacuously' if A stands for ϕ. Thus, $\phi \subset B$ is true for every set $B \subset S$. Note that corresponding to the empty set of Set Theory we have in our conceptual model an 'impossible event'.

Above we defined the relation '$\subset$' between sets, also called the 'inclusion relation'. Another important relation between sets is *equality*.

(Definition) 6.2
Given a space S and sets A, B. Then A and B *are the same set*, or *are equal*, denoted

$$A = B$$

if and only if A ⊂ B and also A ⊃ B.

It follows that if A is an empty set and B is an empty set, then A = B; i e., there is only one empty set φ ⊂ S. Here again, the mathematical model influences the conceptual model. While we might envision many different kinds of impossible events in connection with a particular p.e., we will speak of only *one* impossible event (as noted in Section **5**) since there is only one empty set in the mathematical model.

If two sets A, B are *not* the same set, i.e., are not equal, then they are called *distinct* and we write A ≠ B.

Queries

6.1 A space S is given, with subsets A, B. Also, let ξ_a, ξ_b be elements in S such that ξ_a, $\xi_b \in A$ and $\xi_b \in B$, but $\xi_a \notin B$. For each of the following statements, determine whether it is true, possibly true (not enough information is available to decide), or false:

a) A ⊂ B

b) B ⊂ A

c) A = B

d) A ≠ B

e) φ ⊂ A

f) A = S

g) B ⊂ B.

[108]

6.2 A space S consists of four elements, and the set B(B ⊂ S) consists of two elements. How many elements can the set A consist of, if it is known that:

a) A ⊂ B and A ≠ B

b) A⊄B

c) A ⊂ φ

d) A ⊃ B and A ≠ B.

[179]

Classes of Sets

So far, we have considered only those collections of objects in which the objects were elements in S, and we called these collections 'sets'. But sets are also 'objects', and thus we can also talk about *collections of sets*. It would be confusing to refer to such a collection again as a set; it is a 'set of sets', and the word 'class' is commonly used to describe such a collection. We will use script capital letters to signify classes.

Example 7.1
 Consider a space S and sets A and B that are subsets of S (we write $A, B \subset S$). Let

$$\mathscr{C} = \{A, B, S\}$$

be the *class* that consists of the sets A, B and S. Note that although S is the space, and $A \subset S$ as well as $B \subset S$, S is, nevertheless, a *set* and can be one of the sets making up a class. The relationship between $\mathscr{C}$ and S is then

$$S \in \mathscr{C}$$

since S is one of the sets that belong to the class $\mathscr{C}$. Furthermore, if $\mathscr{D}$ denotes the class $\{A, B\}$, then $\mathscr{D} \subset \mathscr{C}$. And if we specify $B = S$, then $\mathscr{D} = \mathscr{C}$, since a particular object cannot occur more than once in a given collection.

 This Example shows how the symbols $\in$, $\subset$ are used with sets and classes of sets, in order to express relationships analogous to those that we discussed in connection with elements and sets. There is an important difference, however: For A and $\mathscr{C}$ as defined in Example 7.1, for instance, we have $A \in \mathscr{C}$ and $S \in \mathscr{C}$, but also $A \subset S$. When dealing with elements and sets, we can say that $\xi_a \in S$ and $\xi_b \in S$, but there exists no inclusion relation between ξ_a and ξ_b. The 'elements', or sample points, remain the basic building blocks; they must not be thought of as other kinds of 'sets' made up of even more fundamental units.

 We can also speak of a class that is entirely devoid of sets, an *empty class*, which is denoted by $\{\ \}$. This notation cannot be mistaken for an empty set, since we reserve the symbol ϕ for the latter. Also, note that $\{\phi\}$ does not represent an empty class but the class whose only set is the empty set!

 Certain kinds of classes of sets that will be of particular interest to us are called *partitions*.

Probability Concepts and Theory for Engineers, First Edition. Harry Schwarzlander.
© 2011 John Wiley & Sons, Ltd. Published 2011 by John Wiley & Sons, Ltd.

(Definition) 7.1
Given a space S and a nonempty set $A \subset S$. A nonempty class $\mathscr{C}$ of nonempty subsets of A is called a *partition of* A if each element of the set A belongs to one and only one of the sets in $\mathscr{C}$. In case $A = S$, such a class of sets is simply called a *partition*.

A partition is easily represented in a Venn diagram by a set of nonoverlapping regions that completely cover S. On the other hand, arbitrary classes of sets are not always easily displayed or perceived on a Venn diagram, because the points in the diagram represent elements, not sets of elements.

Example 7.2
Consider a space of four elements, $S = \{\xi_a, \xi_b, \xi_c, \xi_d\}$. Let $\mathscr{C}$ be the class made up of the three sets $\{\xi_a\}, \{\xi_b\}, \{\xi_c, \xi_d\}$. Then $\mathscr{C}$ is a partition (Figure 7.1a). Let $\mathscr{D}$ be the class made up of the sets $\{\xi_a\}, \{\xi_b\}$. Then $\mathscr{D}$ is a partition of the set $A = \{\xi_a, \xi_b\}$ (Figure 7.1b).

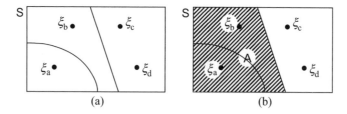

(a)   (b)

Figure 7.1   Venn diagram representations of (a) the partition $\mathscr{C}$, and (b) the set A and its partition $\mathscr{D}$

Note that a class such as $\mathscr{E} = \{A, \{\xi_a\}, \{\xi_b\}\}$ is not easily visualized on a single Venn diagram. It is shown in Figure 7.2 by using separate Venn diagrams to depict each of the member sets.

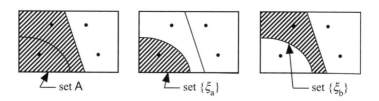

set A   set $\{\xi_a\}$ set $\{\xi_b\}$

Figure 7.2   The class $\mathscr{E}$

A partition is a particular case of a more general type of class called a disjoint class:

(Definition) 7.2
Two sets having no elements in common are called *disjoint*. A class of sets such that each of the member sets has no elements in common with any other member set is called a *disjoint class*.

Note that any set and ϕ constitute a pair of disjoint sets. Therefore, ϕ can be a member of a disjoint class. Furthermore, according to Definition 7.2, an empty class is a disjoint class, and so is any class consisting of a single set. According to Definition 7.1, a class consisting of a single nonempty set can also be regarded as a partition of that set. Disjoint classes and partitions are related in the following way.

Theorem 7.1
Given a space S and a disjoint class $\mathscr{C}$ of nonempty subsets of S. Then either $\mathscr{C}$ is a partition, or it can be made into a partition by incorporating in it one additional set.

Proof
Every element $\xi \in S$ belongs to at most one of the sets in $\mathscr{C}$, since $\mathscr{C}$ is a disjoint class. If every element belongs to exactly one of the sets in $\mathscr{C}$, then $\mathscr{C}$ is a partition. If this is not the case, let

$$A = \{\xi : \xi \notin C, \text{ for every set } C \in \mathscr{C}\},$$

the set of all those elements that belong to none of the sets in $\mathscr{C}$. The sets in $\mathscr{C}$, together with the set A, satisfy the definition of a partition.

Two sets that are disjoint correspond, in the conceptual model, to two events having no possible outcomes in common. These events are then said to be *mutually exclusive*, because the occurrence of one implies the nonoccurrence of the other. In other words, the actual outcome that arises when the experiment is performed can only belong to at most *one* of the two events.

Queries

7.1 Given a space S and sets A, B $\subset$ S where A $\neq$ B. Consider the two classes $\mathscr{C}_1 = \{S, \phi\}$ and $\mathscr{C}_2 = \{A, B\}$. If it is known that $\mathscr{C}_1 \neq \mathscr{C}_2$ then for each of the following state whether it is true, possibly true, false, or meaningless:

a) A $\neq$ S   d) $\phi \subset$ B   g) S $\subset \mathscr{C}_1$
b) $\mathscr{C}_1 \subset \mathscr{C}_2$ e) $\phi \in \mathscr{C}_2$ h) $\mathscr{C}_1 \neq \{\phi\}$.
c) A $\subset$ B   f) B $\in$ S   [99]

7.2 Given a space S and two nonempty sets A, B $\subset$ S. Let $\mathscr{C}$ be a partition of the set A, and let $\mathscr{D}$ be a partition of the set B. For each of the following statements, decide whether it is true, possibly true, or false.

a) If A $=$ B, then $\mathscr{C} = \mathscr{D}$ d) If A $\subset$ B, then $\mathscr{C} \subset \mathscr{D}$
b) If $\mathscr{C} \subset \mathscr{D}$, then A $\subset$ B   e) If A $=$ S, then $\mathscr{C} \subset \mathscr{D}$.
c) If $\mathscr{C} = \mathscr{D}$, then A $=$ B   [240]

SECTION **8**

Elementary Set Operations

Sets can be manipulated in various ways. The rules for these manipulations, and the properties that characterize them, constitute the so-called 'set algebra'. The various set manipulations all derive from the three elementary set operations introduced in Definition 8.1 below. In each case, performing a set operation on one or two sets results again in a set.[4]

(Definition) 8.1
Let some space S be given, with $A, B \in S$.

Operation	Performed on	Notation for result	Meaning	Venn diagram (result is shown shaded)
a) Union	A, B	$A \cup B$ or $A + B$	$\{\xi: \xi \in A$ or $\xi \in B$ or both$\}$	
b) Intersection	A, B	$A \cap B$ or AB	$\{\xi: \xi \in A$ and also $\xi \in B\}$	
c) Complementation	A	A^c or A'	$\{\xi: \xi \notin A\}$	

In words, the union of A and B is the set of all those elements belonging to *at least one* of the sets A, B. Similarly, the intersection of A and B is the set of all those elements belonging to *both* of the sets A, B. And the complement of A is the set of all those elements *not* belonging to A.

---

[4] It is possible to define a single set operation that suffices to express all possible set manipulations. An operation called a 'stroke' or a 'Scheffer stroke', defined as the complement of an intersection, has this property. It is not convenient for our purposes, however.

Probability Concepts and Theory for Engineers, First Edition. Harry Schwarzlander.
© 2011 John Wiley & Sons, Ltd. Published 2011 by John Wiley & Sons, Ltd.

Part (c) of Definition 8.1 states that $A^c = \{\xi : \xi \notin A\}$, so that we also have

$$(A^c)^c = \{\xi : \xi \notin A^c\} = \{\xi : \xi \in A\} = A;$$

double complementation changes nothing. Furthermore, since S is the set of all elements ξ that are defined in a given problem, we see that $S^c = \{\xi : \xi \notin S\}$—the set of those ξ's not among all possible ξ's—so that $S^c = \phi$. Then it also follows that $\phi^c = S$.

Suppose that two sets $A, B \subset S$ are such that $A \cap B = \phi$. This means that there exist no elements ξ such that $\xi \in A$ and also $\xi \in B$; so that A and B have no elements in common and therefore are disjoint. Thus, the statement '$A \cap B = \phi$' is equivalent to the statement 'A and B are disjoint'.

In expressions involving more than one set operation, parentheses are utilized in the usual way to indicate the order in which the operations are to be performed.

From Definition 8.1 follow various basic rules of set algebra. Given any three sets A, B, C (subsets of a space S), we have the following:

	For unions	For intersections	
Idempotent law	$A \cup A = A$	$AA = A$	(8.1a, 8.1b)
Commutative law	$A \cup B = B \cup A$	$AB = BA$	(8.2a, 8.2b)
Associative law	$A \cup (B \cup C) = (A \cup B) \cup C$	$A(BC) = (AB)C$	(8.3a, 8.3b)
Distributive laws	$A \cup (BC) = (A \cup B)(A \cup C)$	$A(B \cup C) = (AB) \cup (AC)$	(8.4a, 8.4b)
These specialize to the			
Reduction rule	$A \cup (AB) = A$	$A(A \cup B) = A$	(8.5a, 8.5b)
Operations performed on a set and …			
its complement	$A \cup A^c = S$	$A \cap A^c = \phi$	(8.6a, 8.6b)
the empty set	$A \cup \phi = A$	$A \cap \phi = \phi$	(8.7a, 8.7b)
the space	$A \cup S = S$	$A \cap S = A$	(8.8a, 8.8b)
DeMorgan's laws	$(A \cup B)^c = A^c \cap B^c$	$(A \cap B)^c = A^c \cup B^c$	(8.9a, 8.9b)
Inclusion	$A \subset A \cup B$	$A \cap B \subset A$	(8.10a, 8.10b)

The verification of these rules is left as an exercise (Problem 8.5). For example, in the case of Equation (8.9a), the complement of $A \cup B$ is the set of elements that do *not* belong to A or B or both, that is, $\{\xi : \xi \notin A$ and also $\xi \notin B\}$. But this is also the set expressed by $A^c \cap B^c$.

Equations (8.1) through (8.10) can be applied whenever it is necessary to simplify or manipulate a complicated set-algebraic expression. This eliminates the need for detailed examination of such an expression in order to clarify the conditions it imposes on elements $\xi \in S$.

From Equations (8.3a) and (8.3b) it follows that parentheses are not needed in expressions such as $A \cup B \cup C$ and ABC, because the order in which the indicated operations are carried out does not affect the result. The order *is* important, however, in an expression involving both $\cup$ and $\cap$ (as in Equations (8.4a) and (8.4b)). In such cases, by convention, if no parentheses are used then the intersections are meant to be performed first. Thus, we can write $A \cup BC$ for the left side of Equation (8.4a), but parentheses must be used on the right side of Equation (8.4a). On the other hand, when complementation is indicated, it always gets performed before other set operations so that, for example, the expression AB^c is equivalent to $A \cap (B^c)$.

More complicated identities can be derived from rules (8.1) through (8.10).

Example 8.1: Extension of DeMorgan's laws to n sets.

Suppose a space S is given and some number n of subsets $A_1, \ldots, A_n$ of S ($n > 2$). Consider the set $A_1 \cup A_2 \cup A_3 \cup \ldots \cup A_n$. Repeated application of rules (8.9a) and (8.9b) give the following expression for the complement of this set:

$$
\begin{aligned}
(A_1 \cup A_2 \cup A_3 \cup \ldots \cup A_n)^c &= (A_1 \cup A_2 \cup \ldots \cup A_{n-1})^c \cap A_n^c \\
&= (A_1 \cup A_2 \cup \ldots \cup A_{n-2})^c \cap A_{n-1}^c \cap A_n^c \\
&\quad \cdot \quad \cdot \quad \cdot \quad \cdot \quad \cdot \quad \cdot \\
&= A_1^c \cap A_2^c \cap A_3^c \cap \ldots \cap A_n^c,
\end{aligned}
$$

and we have extended rule (8.9a) to a larger number of sets. Taking complements of both sides and renaming the sets as follows,

$$
A_1 = B_1^c, A_2 = B_2^c, \ldots, A_n = B_n^c,
$$

gives the extension of rule (8.9b) to n sets; namely,

$$
B_1^c \cup B_2^c \cup \ldots \cup B_n^c = (B_1 \cap B_2 \cap \ldots \cap B_n)^c.
$$

As another illustration in the use of the various set operations we derive the following useful property:

Theorem 8.1

Let A, B $\subset$ S. Then AB and AB^c are disjoint sets, and

$$
AB \cup AB^c = A. \tag{8.1}
$$

Proof

Application of Equations (8.2b), (8.1b), (8.6b), and (8.7b) gives

$$
AB \cap AB^c = ABAB^c = AABB^c = ABB^c = B\phi = \phi,
$$

which demonstrates the disjointness of AB and AB^c. Next, using Equation (8.4b) with C replaced by B^c, and applying Equations (8.6a) and (8.8b), results in

$$
AB \cup AB^c = A(B \cup B^c) = AS = A.
$$

The theorem is illustrated by the Venn diagrams in Figure 8.1. The shaded region on the left represents the set A, and the shaded region on the right the set $AB \cup AB^c$, which equals A. If both AB and AB^c are nonempty, then $\{AB, AB^c\}$ is a partition of A.

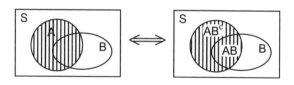

Figure 8.1   Illustration of Theorem 8.1

Since classes are also sets, set operations can be applied to classes as well. Thus, given a space S, then $\{\ \}^c = \{C: C \notin \{\ \}\}$, which is the class of *all* subsets of S.

Queries

8.1 Let A, B, C be three subsets of S about which it is known that $A = B \cup C$ and also $A = B \cap C$. For each of the following, state whether it is true, possibly true, or false:

a) $B = A$ c) $A \cap B = C$

b) $B = \phi$ d) $B \neq C$.   [132]

8.2 Starting with the sets ϕ and S, how many new sets (distinct from ϕ and S) can be generated by applying the operations $\cup$, $\cap$, and ^{c} in various ways? [196]

8.3 Given a space S and a partition $\mathscr{P} = \{A, B, C\}$; i.e., A, B, C are nonempty, disjoint, and their union is S. Consider the class $\mathscr{Q} = \{A \cup B, B \cup C, A \cup C\}$ and determine:

a) How many sets are in the class $\mathscr{P} \cap \mathscr{Q}$?

b) How many sets are in the class $\mathscr{P} \cup \mathscr{Q}$? [23]

Additional Set Operations

In addition to the elementary set operations defined in the previous Section, there are two other operations that are represented by distinct symbols.

(Definition) 9.1
Let some space S be given, with A, B $\subset$ S.

Operation	Performed on	Notation for the result	Meaning	Venn diagram (result is shown shaded)
a) *Directed difference*	A, B (in that order)	A–B	$\{\xi: \xi \in A \text{ and also } \xi \notin B\}$	
b) *Symmetric difference*	A, B	AΔB	$\{\xi: \xi \in A \text{ and also } \xi \notin B, \text{ or } \xi \in B \text{ and also } \xi \notin A\}$	

In words, A–B is the set of elements belonging to A but not to B. AΔB is the set of elements belonging to exactly one of the sets A, B.[5]

Derivation of the following basic rules from Definition 9.1 and from the rules of Section 8 is left as an exercise (see Problem 9.8).

	Directed difference	Symmetric difference	
Annihilation law	$A - A = \phi$	$A \Delta A = \phi$	(9.1a, 9.1b)
Commutative law	(not applicable)	$A \Delta B = B \Delta A$	(9.2)
Associative law	(not applicable)	$A\Delta(B \Delta C) = (A \Delta B)\Delta C$	(9.3)
Distributive laws	$A(B - C) = AB - AC$	$A(B \Delta C) = AB \Delta AC$	(9.4a, 9.4b)
	$(AB) - C = (A - C)(B - C)$		(9.4c)

---

[5] The symmetric difference operation can be seen to be analogous to the 'exclusive or' operation of Abstract Algebra or Logic.

Probability Concepts and Theory for Engineers, First Edition. Harry Schwarzlander.
© 2011 John Wiley & Sons, Ltd. Published 2011 by John Wiley & Sons, Ltd.

Operations performed on a set and . . .

its complement	$A - A^c = A$	$A \triangle A^c = S$	(9.5a, 9.5b)
the empty set	$\phi - A = \phi$	$A - \phi = A \triangle \phi = A$	(9.6a, 9.6b)
the space	$A - S = \phi$	$S - A = S \triangle A = A^c$	(9.7a, 9.7b)
Complementation	$(A - B)^c = A^c \cup B$	$(A^2 B)^c = AB \cup A^c B^c$	(9.8a, 9.8b)

In Equations (9.4a) and (9.4b), parentheses around the intersections have been omitted. Here, as in Section 8, it is to be understood that intersections are performed first.

The two operations '$-$' and '$\triangle$' do not extend the manipulative power of the three elementary set operations, since both $A - B$ and $A \triangle B$ are also expressible in terms of the elementary operations of Section 8. This is shown by the following identities that derive from the definitions of the various set operations:

$$A - B = A \cap B^c \tag{9.9}$$

$$A \triangle B = (A - B) \cup (B - A) = AB^c \cup BA^c \tag{9.10}$$

Actually, not even all three of the elementary set operations of Section 8 need to be utilized in order to perform any desired set manipulations. According to Equation (8.9a), a union can always be expressed in terms of intersection and complementation. Similarly, according to Equation (8.9b), an intersection is expressible in terms of union and complements. Of the three elementary set operations of Section 8, only complementation cannot be expressed in terms of the other two.

Beginning with some class $\mathscr{C}$ of sets contained in a given space S, we can see that the application of the various set operations to the members of this class may cause new sets to be generated that are not in $\mathscr{C}$. For example, let $\mathscr{C} = \{A, S\}$, where $A \neq \phi$. Applying complementation to A then results in the new set $A^c \notin \mathscr{C}$. However, it is possible to define a class $\mathscr{A}$ of sets in such a way that any set obtained by performing set operations on members of $\mathscr{A}$ is again a member of $\mathscr{A}$. Such a class is said to be 'closed' under the various set operations and is called an *additive class*. An additive class is assumed to be nonempty. Since two of the elementary set operations (unions and complements; or, intersections and complements) suffice to carry out all the various set manipulations, an additive class is specified more concisely in the following way:

(Definition) 9.2
Given a space S and a nonempty class $\mathscr{A}$ of subsets of S. $\mathscr{A}$ is an *additive class* (or *algebra*) of sets if
a) for every set $A \in \mathscr{A}$, the set A^c also belongs to $\mathscr{A}$; and
b) for every pair of sets $A, B \in \mathscr{A}$, the set $A \cup B$ also belongs to $\mathscr{A}$.

Additive classes will be discussed in more detail in Section 13. At this point, we merely note two simple examples of additive classes:

a) the class of *all* subsets of S;
b) the class consisting of only the two sets ϕ and S.

Verify that these two classes are additive classes!

Queries

9.1 Given a space S and *nonempty* subsets A, B. If $A \cup B^c = S$, state for each of the following, whether it is necessarily true, possibly true, or false:

a) $S - A \supset B^c$

b) $S - A \supset B$

c) $S - A = \phi$

d) $A - B = A^c B$

e) $A^c - B = B^c - A$

f) $B \cup A^c = S$.   [101]

9.2 Given four sets A, B, C, $D \subset S$ about which the following is known:

none of the sets are equal;

$A \subset B \subset C$; and

$\{A, B, C, D\}$ is an additive class.

Express D in terms of one or more of the other three sets.   [78]

9.3 Given a space S and a partition $\mathscr{P} = \{A, B, C\}$.

a) A new class $\mathscr{Q}$ is formed such that:

A, B, C belong to $\mathscr{Q}$, and

$\mathscr{Q}$ is closed with respect to symmetric difference.

What is the smallest number of sets that must belong to $\mathscr{Q}$?

b) Same question if, instead, $\mathscr{Q}$ is closed with respect to the directed difference operation.   [229]

SECTION *10*

Functions

Engineers and other applied scientists are accustomed to thinking of a *function* primarily as a correspondence between real numbers (values of real variables). 'A function of x', $G(x)$, usually means that to any real number x (or, to any point on the x-axis, which is the *domain* of the function G), gets assigned a real value $y = G(x)$ (a point on the y-axis). The collection of all such y-values (not necessarily the whole y-axis) is the *range* of the function G.

Example 10.1

When no further specifications are given or implied, then the statement $y = x^2$ is taken to mean that for every real number x there is specified a corresponding real number y, which is obtained according to the computational rule $y = x^2$. Thus, we have here a function whose domain consists of all the real numbers (the x-axis) and whose range consists of all non-negative real numbers (the non-negative y-axis). This happens to be a very simple function because

  i) it is specified by a simple computational rule; and
  ii) the same rule applies over the whole domain, that is, for all x.

In Set Theory, functions occur in a more general way, namely, as correspondences between elements or subsets of a space, and element or subsets of another space or of the same space. These functions include as a special case our ordinary real functions of a real variable, which assign elements of the space of real numbers to elements of the same space. We distinguish between two types of functions in Set Theory:

(Definition) 10.1

a) A *point function* is a function whose domain is a set of elements.
b) A *set function* is a function whose domain is a class of sets.

Functions of interest to us will primarily be those that assign *real numbers* to the elements of a given sample space, or to the subsets of a space; i.e., *real-valued functions*. We always assume such a function to be *single-valued*; only *one* real number is assigned to each object (element or set) in the domain of the function. A simple example of a real-valued point function is a 'counting function', or 'index'.

Probability Concepts and Theory for Engineers, First Edition. Harry Schwarzlander.
© 2011 John Wiley & Sons, Ltd. Published 2011 by John Wiley & Sons, Ltd.

Example 10.2

In the die-throwing experiment of Section **2**, the possible outcomes are:

'one dot faces up'

'two dots face up'

$\ldots\ldots$

'six dots face up'

The Set Theory model corresponding to this experiment is a space S consisting of six elements. Let these elements be denoted

$$\xi_a, \xi_b, \xi_c, \xi_d, \xi_e, \xi_f.$$

Now consider the real-valued point function G that assigns the integer 1 to ξ_a, 2 to ξ_b, ..., 6 to ξ_f; so that we write $G(\xi_a) = 1, G(\xi_b) = 2, \ldots, G(\xi_f) = 6$. This function G can be pictured or visualized in the manner shown in Figure 10.1. In this representation, the domain of G is depicted as a Venn diagram with all elements shown that belong to S, but no sets other than S. The range of G is the set of integers from 1 to 6, shown in set notation. An arrow originates at each element of S and points to the integer being assigned to that element. These arrows represent the function.

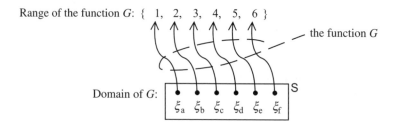

Figure 10.1   The point function G

This particular function G happens to fulfill two different roles simultaneously: It *counts*, or *indexes*, the elements in S; and it *assigns* to each element ξ a real number that relates to the conceptual model, namely, the number of dots facing up in the corresponding outcome. If we are only concerned with the first of these two roles—the indexing of the elements in S—then we can incorporate the function G into our notation by *renaming* the elements $\xi_1, \xi_2, \ldots, \xi_6$.

A simple example of a real-valued *set function* is again a counting function—that is, a function that assigns a counting integer, or index, to each *set* in a given class.

A property of interest primarily in connection with point functions is 'one-to-one'. We call a (single valued) point function G *one-to-one* if for every pair of distinct elements ξ_1, ξ_2 in the domain of G we have $G(\xi_1) \neq G(\xi_2)$. To every element in the range of such a function there corresponds a unique element in the domain of the function. Thus, any counting function is one-to-one. On the other hand, the function described in Example 10.1 is not one-to-one.

If G is a one-to-one function, it has a (unique) inverse, G^{-1}, otherwise not. The function described in Example 10.1 does not have an inverse. However, if the domain of that function is restricted to, say, all non-negative x, then the function is one-to-one and has an inverse.

Suppose G_1 is a function defined on a particular domain, and G_2 is a function defined on the *range* of G_1. Let x denote one of the objects making up the domain of G_1. Then it is possible to first operate on x with G_1 and then on the result obtained with G_2 This is written $G_2G_1(x)$, where G_2G_1 is a new function that assigns to x the value $G_2[G_1(x)]$. For example, if G_1 is a one-to-one function, we have $G_1^{-1}G_1(x) = x$.

Queries

10.1 Consider a set $A = \{\xi_a, \xi_b, \xi_c\}$.

    a) How many different point functions can be defined whose domain is A and whose range is a subset of A? (In other words, each such function is a rule that assigns to each element in A one of the elements in A. Keep in mind that A itself is also a subset of A.)

    b) How many different *one-to-one* point functions can be defined whose domain is A and whose range is A? [20]

10.2 Given a space S and the sets $A, B, C \subset S$. Suppose a set function K is defined on the class $\{A, B, C\}$, with $K(A) = 1$, $K(B) = 5$, and $K(C) = -1/3$. Under each of the following conditions, determine whether $K(A \cup B)$ is defined, and if so, evaluate:

    a) $A \subset B$ d) $A \cup B = C$

    b) $B = A \cap C$ e) $A = B \, \Delta \, C$.

    c) $\{A, B, C\}$ is a partition   [167]

The Size of a Set

It is now possible to define the size of a set in terms of a function. This may at first glance appear devious, but isn't it a common approach to express the size of a collection in terms of a *count* of its elements? And keeping a count of the elements by means of an index amounts to introducing a function—in particular, a one-to-one function.[6]

But what about an infinite set? It is not satisfactory to say that a set has infinite size if its elements result in an infinite count, because such a statement does not make precise the notion of infinity. Besides, it would not be very convenient to apply such a criterion in order to test whether or not a set is infinite. Instead we have:

(Definition) 11.1

a) A set A is *infinite* if it has a subset B, $B \neq A$, such that there exists a one-to-one point function whose domain contains A and whose range is B.

b) A set is called *finite* if it is not infinite.

The elements of a *finite* set A can be put in one-to-one correspondence with the set of integers $\{1, \ldots, n\}$, for some n; and we say that 'A has size n', which is denoted $|A| = n$. And if $B \subset A$, then it follows from Definition 6.1 that B is also finite, with $|B| \leq n$ (see also Problem 11.3).

As an example of part (a) of Definition 11.1, consider the set I_+ of all positive integers. Does it have a subset J (with $J \neq I_+$) such that the elements of I_+ can be put into one-to-one correspondence with the elements of J? Indeed, there are many such subsets—for instance, the set of all positive integers greater than 1. A one-to-one function G satisfying the Definition would then be $G(i) = i + 1$, for $i = 1, 2, 3, \ldots$.

Some reflection should convince the reader that the notion of an 'infinite' set as introduced in part (a) of Definition 11.1 is in exact agreement with the way 'infinite' is generally understood in Mathematics. The advantage of the formulation in (a) is that the definition is 'constructive', that is, a set can be demonstrated to be infinite by carrying out the simple construction that is specified there.

It is not always sufficient to classify a set as infinite. We need to be able to distinguish between two types of infinite sets:

---

[6] Instead of 'size,' the term 'cardinality' is commonly used in Mathematics.

Probability Concepts and Theory for Engineers, First Edition. Harry Schwarzlander.
© 2011 John Wiley & Sons, Ltd. Published 2011 by John Wiley & Sons, Ltd.

(Definition) 11.2

a) A set A is *countable* if it is possible to specify a one-to-one function whose domain is A and whose range is some subset of the set of positive integers.

b) A set is said to be *uncountable* if it is not countable.

Note that all finite sets are countable. A set that is countable and infinite is also called *countably infinite*. Thus, the set of all positive integers is a countably infinite set: the one-to-one function that is the *identity function*, defined on the elements of this set, has as its range again the set of the positive integers which, of course, is a subset of itself.

It is a bit more difficult to demonstrate that a given set is uncountable.

Example 11.1: Coin-toss sequences.

a) Consider the tossing of a coin n times in succession. The following experimental plan might model this situation for, say, $n = 10$:

1) The purpose is to construct a randomly selected sequence of T (tails) and H (heads) of length 10. (We call such a sequence a 'coin-toss sequence'.)

2) The procedure is to select an ordinary coin and toss it ten times in succession, recording the observed *result* (tail or head) after each toss by means of a T or an H.

Suppose this experimental plan is executed. Then:

3) The procedure is carried out, leading to (say)

4) The following record: T T H H T H H H H T.

The actual outcome in this case is described by the sequence T T H H T H H H H T, and this outcome is representative of the $2^{10} = 1024$ different possible outcomes for this p.e. To this p.e., then, corresponds a sample space of size 2^{10}. Similarly, for any other (finite) value of n, the sample space is finite, and has size 2^n.

b) If the coin is to be tossed a very large number of times, then it may seem convenient not to worry about the exact value of n and to simply postulate $n \to \infty$ in our conceptual model. The possible outcomes are then coin-toss sequences of infinite length. The sample space S for this p.e. is made up of elements ξ each of which represents a distinct infinite-length coin-toss sequence. This sample space turns out to be *uncountable*, as will now be shown.

Since the procedure suggested by Definition 11.2 calls for disproving countability, we will try to set up a counting function G by listing infinite-length coin-toss sequences in succession, so that for the nth sequence in the list the count is $G(\xi) = n$, for $n = 1, 2, 3, \ldots$, as shown in Figure 11.1. (Clearly, it should make no difference in which order we list the sequences.) Uncountability is demonstrated if, no matter how we order the various coin-toss sequences in the list, it can be shown that there exists at least one sequence that is not contained in the list. Such a sequence would have to differ from each sequence in our list in at least one letter position. In fact, many such sequences can be found. For instance, for the listing indicated in Figure 11.1, consider the 'diagonal sequence'

$$T \quad H \quad H \quad T \ldots \ldots$$

and change each of the letters, to give

$$H \quad T \quad T \quad H \ldots \ldots$$

This sequence differs from the first sequence in Figure 11.1 in at least the first letter, it differs from the second sequence in at least the second letter, it differs from the third sequence in at least the

$G(\xi)$	ξ
1	T T H H T H H H H . . . . .
2	T H T T H H H T T . . . . .
3	H H H T H H H T T T . . . . .
4	T T T T H T T H H . . . . .
etc.	. . . . . .

Figure 11.1 A table of infinite-length coin-toss sequences

third letter, and so on. Thus, it cannot be identical to any of the sequences contained in the list, whatever these are. The list, therefore, cannot be complete, so that there are sequences that do not belong to the domain of G.

It must be noted that 'infinities' do not arise in real-world experiments, only in the conceptual models by means of which we describe some of those experiments. Therefore we should not expect 'physical insight', or 'physical intuition' to guide us in problems involving infinite sets. For instance, in connection with the above example it may be appealing to assume (incorrectly) that since the sample space for any arbitrary number n of coin tosses is countable (namely, of size 2^n), it must be countably infinite for an infinity of tosses. Intuition fails here because an *infinity* of coin tosses is not physically meaningful. Nevertheless, it is often convenient to use infinite sets in the conceptual model. But the notion of 'infinity' must then be treated strictly according to the mathematical rules.[7]

A few basic properties of infinite sets are brought together in the following Theorem.

Theorem 11.1
Given a space S, with A, B $\subset$ S. Let C $=$ A $\cup$ B.
a) If A and B are both finite, then C is finite.
b) If A is infinite, then C is infinite.
c) If A and B are both countable, then C is countable.
d) If A is uncountable, then C is uncountable.

Proof
a) If $|A| = n$ and $|B| = m$, then $|C| \leq n + m < \infty$
b) Suppose C is finite, $|C| = n$. According to Equation (**8**.10a), A $\subset$ C; from which it follows that A is also finite. But A is infinite; therefore C cannot be finite.
c) Let G_A, G_B be one-to-one point functions satisfying Definition **11**.2a for sets A and B, respectively, and define

$$G_C = \begin{cases} 2G_A(\xi) - 1, & \text{for } \xi \in A \\ 2G_B(\xi), & \text{for } \xi \in B - A. \end{cases}$$

---

[7] For further discussion of infinite sets, countable sets and uncountable sets, see any book on Set Theory, such as [Fr2] or [HK1].

In other words, G_C associates odd integers with the elements of A, and even integers with the elements of B − A. Then G_C is one-to-one with domain C, and its range is some subset of the positive integers.

d) Let A be uncountable and suppose that C is countable. Then, a one-to-one function G_C exists whose domain is C and whose range is some subset of the positive integers. This function also maps any subset of C into some subset of the positive integers. Since $A \subset C$, this implies that A is countable, which is a contradiction.

Queries

11.1 Given two sets A, $B \subset S$ where A is finite and B is infinite. For each of the following statements, determine whether it is true, possibly true, or false.

    a) S is infinite   c) A^c is infinite

    b) $A \cap B$ is infinite   d) B^c is infinite.   [128]

11.2 Consider the collection of all the different infinite-length coin-toss sequences. The sequences in this collection are modified as follows, and only distinct sequences are retained. In which case(s) is the resulting collection countable?

    a) Only the first 1000 letters (H or T) of each sequence are retained

    b) The first 1000 letters of each sequence are deleted

    c) Alternate letters (the second, fourth, sixth, . . .) are deleted from each sequence

    d) Only the (2^n)th letters of each sequence are retained, where $n = 1, 2, 3, \ldots$

    e) Only the sequences that start with H are retained

    f) Only the sequences that contain a single H are retained.   [49]

Multiple and Infinite Set Operations

In Example 8.1 we considered a set $A_1 \cup A_2 \cup A_3 \cup \ldots \cup A_n$. The following is a more compact notation for representing this union of n sets:

$$\bigcup_{i=1}^{n} A_i \qquad (12.1a)$$

In words, this is the set of all those elements (in the space) that belong to *at least one* of the sets $A_1, \ldots, A_n$. If we denote this collection of sets by $\mathscr{C}$,

$$\mathscr{C} = \{A_1, \ldots, A_n\}$$

then another form for the expression (12.1a) is

$$\bigcup_{A \in \mathscr{C}} A \qquad (12.1b)$$

where 'A' serves as generic symbol for a set. Analogously, for the set

$$A_1 \cap A_2 \cap A_3 \cap \ldots \cap A_n$$

we can write

$$\bigcap_{i=1}^{n} A_i, \quad \text{or} \quad \bigcap_{A \in \mathscr{C}} A \qquad (12.2)$$

In words, the expressions (12.2) stand for the set of all those elements that belong to *each* of the sets $A_1, \ldots, A_n$.

In the following Definition the union and the intersection of multiple sets are extended to include the possibility of $\mathscr{C}$ being infinite.

(Definition) 12.1

Let $\mathscr{C}$ be any countable class (finite or infinite) of subsets of a space S. Then:

Probability Concepts and Theory for Engineers, First Edition. Harry Schwarzlander.
© 2011 John Wiley & Sons, Ltd. Published 2011 by John Wiley & Sons, Ltd.

a) $\bigcup_{A\in\mathscr{C}} A$ denotes the set of all those elements that belong to at least one of the sets in $\mathscr{C}$, and is called a *countable union*.

b) $\bigcap_{A\in\mathscr{C}} A$ denotes the set of all those elements that belong to each of the sets in $\mathscr{C}$, and it is called a *countable intersection*.

If, in particular, $\mathscr{C}$ is countably infinite, and the sets in $\mathscr{C}$ are assumed indexed so that we can write

$$\mathscr{C} = \{A_1, A_2, A_3, \ldots\} \tag{12.3}$$

then $\bigcup_{A\in\mathscr{C}} A$ is an *infinite union*, which can also be written $\bigcup_{i=1}^{\infty} A_i$; and $\bigcap_{A\in\mathscr{C}} A$ is an *infinite intersection*, which can also be written $\bigcap_{i=1}^{\infty} A_i$. It should be noted, however, that an infinite union differs from a finite union in that we cannot simply write it in the form

$$A_1 \cup A_2 \cup A_3 \cup \ldots$$

since this would require developing the definition of an infinite union as a limit, namely,

$$\lim_{n\to\infty} A_1 \cup A_2 \cup A_3 \cup \ldots \cup A_n.$$

The same comment applies to an infinite intersection.[8]

The complementary nature of the operations $\cup$ and $\cap$, as expressed in Equation (8.9), holds also for infinite set operations. This is stated in Theorem 12.1, which covers finite as well as countably infinite set operations. The proof is left as an exercise (Problem 12.3).

Theorem 12.1: DeMorgan's Laws for countable set operations.
Given a countable class $\mathscr{C}$ of subsets of a space S, then

$$\left(\bigcup_{A\in\mathscr{C}} A\right)^c = \bigcap_{A\in\mathscr{C}} A^c \tag{12.4}$$

and

$$\left(\bigcap_{A\in\mathscr{C}} A\right)^c = \bigcup_{A\in\mathscr{C}} A^c \tag{12.5}$$

If the sets in the countably infinite class $\mathscr{C}$ are indexed in the manner indicated in (12.3), then they form an *infinite sequence* of sets,

$$A_1, A_2, A_3, \ldots$$

However, an infinite sequence of sets should not be regarded synonymous with an infinite class. In the previous Section we dealt with infinite coin-toss sequences—such a sequence is made up of but

---

[8] This would have to get addressed in a more rigorous development of Set Theory.

two distinct letters, T and H. In the same way, an infinite sequence of sets can be defined from but a finite number of distinct sets. Thus, using only two distinct sets A and B, we can specify the infinite sequence

$$\text{A, B, A, B, A, B, A, B, } \ldots \tag{12.6}$$

On the other hand, in a class, no meaning attaches to mentioning any set more than once: A set either belongs to the class or it does not. The sets making up the sequence (12.6) are the members of the class {A, B}.

These considerations lead to an alternate method for defining countable unions and countable intersections.

(Definition) 12.1′

Let $A_1, A_2, A_3, \ldots$ be an infinite sequence of subsets of a space S. Then:

a) $\bigcup_{i=1}^{\infty} A_i$ denotes the set of all those elements that belong to at least one of the sets making up the sequence, and is called a countable union.

b) $\bigcap_{i=1}^{\infty} A_i$ denotes the set of all those elements that belong to each of the sets making up the sequence, and is called a countable intersection.

The following Theorem answers a question that often arises in connection with countable unions.

Theorem 12.2

A countable union of countable sets is countable.

Proof

Let $A_1, A_2, A_3, \ldots$ be an infinite sequence of countable subsets of a space S. Let each set in this sequence be expressed in terms of its elements as follows:

$$A_i = \{\xi_{i,1}, \xi_{i,2}, \xi_{i,3}, \xi_{i,4}, \ldots\}, \quad \text{for} \quad i = 1, 2, 3, \ldots$$

Let $B = \bigcup_{i=1}^{\infty} A_i$. B is countable if a one-to-one function can be defined whose domain is B and whose range is some subset of the positive integers (Definition 11.2). We envision a table containing all the elements making up the various sets in the sequence $A_1, A_2, A_3, \ldots$, arranged as shown in the array (12.7).

$$\left. \begin{array}{llllll} \xi_{1,1} & \xi_{1,2} & \xi_{1,3} & \xi_{1,4} & \xi_{1,5} & \cdots \\ \xi_{2,1} & \xi_{2,2} & \xi_{2,3} & \xi_{2,4} & \xi_{2,5} & \cdots \\ \xi_{3,1} & \xi_{3,2} & \xi_{3,3} & \cdots \\ \xi_{4,1} & \xi_{4,2} & \xi_{4,3} & \cdots \\ \cdots \cdots \cdots \cdots \\ \cdots \cdots \cdots \cdots \end{array} \right\} \tag{12.7}$$

Every element of B appears at least once in this array. Unless the sets $A_1, A_2, A_3, \ldots$ constitute a disjoint class, some elements appear more than once; in other words, there are distinct entries in the array that stand for the same element. Now we superimpose onto the array (12.7) the pattern of integers shown in Figure 12.1, which can be extended indefinitely. This yields the required function. It assigns to each element in B the integer appearing in the location occupied by that element in the array (12.7). To an element that appears more than once in the array (12.7), only the smallest applicable integer is assigned. Thus, a rule has been found for assigning to every element in B a distinct positive integer, i.e., a one-to-one function, which proves that B is countable.

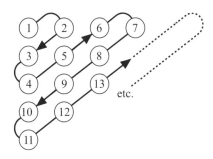

Figure 12.1   Counting the entries in a doubly infinite array

---

Queries

12.1 If the size of a class $\mathscr{C}$ is 10, what are the minimum and maximum possible sizes of the set $\bigcup_{A \in \mathscr{C}} A$? Of the set $\bigcap_{A \in \mathscr{C}} A$?   [175]

12.2 In the proof of Theorem 12.2, what integer in the pattern of Figure 12.1 gets assigned to $\xi_{1,100}$? To $\xi_{99,99}$?   [141]

12.3 Given a sequence $A_1, A_2, A_3, \ldots$ of nonempty subsets of a space S, such that $\bigcup_{i=1}^{\infty} A_i = A_1$, and $\bigcap_{i=1}^{\infty} A_i = A_2$.

  a) What is $A_1 \cup A_2$?
  b) What is $A_1 \cap A_2$?
  c) Is it true, possibly true, or false that $A_1 - A_2$ belongs to the sequence?   [94]

SECTION *13*

More About Additive Classes

In Section 9 we spoke of classes that are closed with respect to the various set operations, and named such a class an *additive class*. It turns out that there is an important connection between additive classes and partitions. A partition is frequently the starting point in constructing or describing an additive class. Beginning with some partition $\mathscr{P}$ of a space S, an additive class can be formed by starting with the sets in $\mathscr{P}$ and then including all those sets that can be obtained by using the union and complementation operations. The resulting additive class is referred to as the *additive class generated by the partition $\mathscr{P}$*, which we denote $\mathscr{A}(\mathscr{P})$. More generally, given an arbitrary class $\mathscr{C}$, by 'the additive class generated by $\mathscr{C}$' will be meant the class that contains the sets in $\mathscr{C}$ and all sets that can be obtained from these by finite set operations.

As we proceed, the term 'proper subset' will be useful, and it appears in the following Example. The meaning is: If $A \subset B$ but $A \neq B$, then A is a *proper subset* of B.

Example 13.1

Let S be some space of elements and let A be a nonempty proper subset of S. Then $\mathscr{P} = \{A, A^c\}$ is a partition. The additive class $\mathscr{A}(\mathscr{P})$ generated by $\mathscr{P}$ is then the class $\{\phi, A, A^c, S\}$.

Clearly, every additive class has subclasses which are partitions; for instance, the (degenerate) partition $\{S\}$ is a subclass of *every* additive class. Furthermore:

Theorem 13.1

Every finite additive class $\mathscr{A}$ contains a unique *largest* partition $\mathscr{P}$, and this is the partition that generates $\mathscr{A}$; i.e., $\mathscr{P}$ is such that $\mathscr{A} = \mathscr{A}(\mathscr{P})$.

Proof

Consider a finite additive class $\mathscr{A}$ of subsets of a given space S.

First, we show that there is a unique largest partition. Let $\mathscr{P}_1$, $\mathscr{P}_2$ be two distinct partitions (i.e., $\mathscr{P}_1 \neq \mathscr{P}_2$), both of which are subclasses of $\mathscr{A}$, and assume that both have the same size. Then there must be at least one element $\xi_a \in S$ such that in $\mathscr{P}_1$ it belongs to a set A_1 and in $\mathscr{P}_2$ it belongs to a different set $A_2 \neq A_1$ (see Figure 13.1); otherwise, $\mathscr{P}_1 = \mathscr{P}_2$.

Probability Concepts and Theory for Engineers, First Edition. Harry Schwarzlander.
© 2011 John Wiley & Sons, Ltd. Published 2011 by John Wiley & Sons, Ltd.

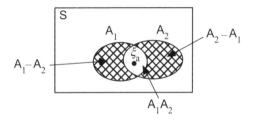

Figure 13.1   The sets $A_1 \in \mathscr{P}_1$ and $A_2 \in \mathscr{P}_2$

From $A_2 \neq A_1$ it follows that at least one of the two sets $A_1 - A_2$, $A_2 - A_1$ is nonempty; without loss of generality, say that $A_1 - A_2$ is nonempty. Then there exists a partition $\mathscr{P}_3 \subset \mathscr{A}$ that is larger than $\mathscr{P}_1$ and $\mathscr{P}_2$ and that is obtained from $\mathscr{P}_1$ through replacement of A_1 by the two sets $A_1 - A_2$ and $A_1 \cap A_2$. Therefore, if $\mathscr{P}$ is the largest partition contained in A it must be unique.

To show that $\mathscr{P}$ generates $\mathscr{A}$, we first note that $\mathscr{P}$ generates an additive class $\mathscr{A}_1 \subset \mathscr{A}$, since $\mathscr{P} \subset \mathscr{A}$ and $\mathscr{A}$ is an additive class. If $\mathscr{A}_1 \neq \mathscr{A}$, then there must be some set B that belongs to $\mathscr{A}$ but not to $\mathscr{A}_1$. Thus, B is not expressible as a union of sets in $\mathscr{P}$, and there must be a larger partition $\mathscr{P}_3 \subset \mathscr{A}$, consisting of all intersections of the sets belonging to $\mathscr{P}$ with the sets B and B^c. Therefore, $\mathscr{A}_1 \neq \mathscr{A}$ is inconsistent with the hypothesis that $\mathscr{P}$ is the largest partition contained in $\mathscr{A}$.

Given a partition $\mathscr{P}$ of size n, $\mathscr{P} = \{A_1, A_2, \ldots, A_n\}$, then the size of $\mathscr{A}(\mathscr{P})$ can be determined by listing the sets making up $\mathscr{A}(\mathscr{P})$. Such a listing appears in Table 13.1.[9]

The expressions on the right can be recognized as the *binomial coefficients*, that is, the coefficients of successive terms in the expansion of $(a + b)^n$. Using the notation

$$\binom{n}{k} = \frac{n!}{(n-k)!k!},$$

we find that summing the number of sets in the above listing gives, according to the binomial theorem,

$$\binom{n}{0} + \binom{n}{1} + \binom{n}{2} + \ldots + \binom{n}{n} = (1+1)^n = 2^n$$

Thus, a partition of size n generates an additive class of size 2^n. From Theorem 13.1 it follows that the size of a finite additive class must be a power of 2, and the size of the largest partition contained therein must be equal to the exponent.

There is another way to arrive at the size of the additive class generated by a partition $\mathscr{P} = \{A_1, \ldots, A_n\}$. Every set $B \in \mathscr{A}(\mathscr{P})$ can be characterized by identifying those members of

---

[9] To arrive at the expressions for the number of sets, consider the general case of unions of k sets ($k \leq n$). How many different ways can k sets be chosen out of a total of n sets? The first of the k sets can be any one of the n sets; the second set can be any one of the remaining $n-1$ sets, etc.; giving a total of $n(n-1) \ldots (n-k+1)$. But this procedure has assigned a separate count to each *ordering* of a particular complement of k sets. Because of commutativity and associativity of the union operation, these different orderings result in the same union. Therefore it is necessary to divide the above total by the number of distinct orderings. For any given k sets, each one can be the first set, any of the remaining $(k-1)$ can be the second set, etc.; giving the divisor $k(k-1)(k-2) \ldots 1$.

Table 13.1   Determining the size of $\mathscr{A}(\mathscr{P})$

Sets in $\mathscr{A}(\mathscr{P})$:	Number of sets:
The empty set: ϕ	1
The sets in $\mathscr{P}$: $A_1, A_2, \ldots, A_n$	n
Pairwise unions: $A_1 \cup A_2, A_1 \cup A_3, \ldots, A_{n-1} \cup A_n$	$\dfrac{n(n-1)}{2}$
Triple unions: $A_1 \cup A_2 \cup A_3$, etc.	$\dfrac{n(n-1)(n-2)}{2 \cdot 3}$
......	......
......	......
n-fold union: $\displaystyle\bigcup_{i=1}^{n} A_i$	1

$\mathscr{P}$ which are contained in B. This can be done by assigning to every $B \in \mathscr{A}(\mathscr{P})$ an n-digit binary number or code, where a 'zero' in the ith position means $A_i \not\subset B$ and a 'one' in the ith position means $A_i \subset B$. Each n-digit binary number identifies a different set in $\mathscr{A}(\mathscr{P})$. Since there are 2^n different n-digit binary numbers, the size of must be 2^n. (Thus, two different ways of counting to 2^n have been illustrated.)

Example 13.2
The Venn diagram in Figure 13.2 shows a space S partitioned by $\mathscr{P} = \{A_1, A_2, A_3, A_4\}$. The binary number representation for each member of $\mathscr{P}$ is also shown. Each set in $\mathscr{A}(\mathscr{P})$ has its unique four-digit binary number. For instance, 1010 represents the set $A_1 \cup A_3$, and S is identified by 1111.

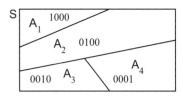

Figure 13.2   Venn diagram with binary code

Now let an arbitrary finite class $\mathscr{C} = \{A_1, A_2, \ldots, A_n\}$ of subsets of a space S be given. Let $\mathscr{P}$ be the largest partition contained in the additive class generated by $\mathscr{C}$. How are the sets in $\mathscr{P}$ related to the sets in $\mathscr{C}$?

This can be answered by beginning with the simpler class $\mathscr{C}_1 = \{A_1\}$, and then building up to $\mathscr{C}$. First we note that the additive class generated by $\mathscr{C}_1$ is as in Example 13.1 (with A replaced by A_1), and $\mathscr{P} = \{A_1, A_1^c\}$. An exception has to be made if $A_1 = \phi$ or S. In that case, ϕ has to be dropped from the specification of $\mathscr{P}$, leaving the degenerate partition $\{S\}$.

Next, consider $\mathscr{C}_2 = \{A_1, A_2\}$. From Theorem 8.1 it follows that $\mathscr{P}$ is now made up of the sets $A_1 A_2$, $A_1 A_2^c$, $A_1^c A_2$, $A_1^c A_2^c$, with any duplicate or empty set deleted. Continuing in this manner, let $\mathscr{C}_{n-1} = \{A_1, \ldots, A_{n-1}\}$ and let $\mathscr{P}$ consist of all the distinct nonempty sets defined by the expressions

$$\bigcap_{i=1}^{n-1} A_i^x$$

where the superscript x denotes here, for each of the A_i's, either complementation or noncomplementation. Then, for $\mathscr{C}_n = \mathscr{C} = \{A_1, \ldots, A_n\}$, Theorem 8.1 yields a partition $\mathscr{P}$ that is made up of all the distinct nonempty sets defined by the expressions

$$\bigcap_{i=1}^{n} A_i^x.$$

If it turns out that all these expressions represent distinct nonempty sets, then $|\mathscr{P}| = 2^n$; otherwise $|\mathscr{P}| < 2^n$. We thus have:

Theorem 13.2
The largest partition in the additive class generated by a finite class $\mathscr{C} = \{A_1, \ldots, A_n\}$ of subsets of a space S consists of all distinct nonempty sets given by $\bigcap_{i=1}^{n} A_i^x$.

It must be understood that an additive class need not be closed with respect to *countable* unions and *countable* intersections. Therefore we introduce the following:

(Definition) 13.1
A *completely additive class* (or *σ-algebra*) of sets is a nonempty class $\mathscr{A}$ such that
a) for every countable subclass $\mathscr{C} \subset \mathscr{A}$, the set $\bigcup_{A \in \mathscr{C}} A$ also belongs to $\mathscr{A}$; and
b) the complement of every set belonging to $\mathscr{A}$ also belongs to $\mathscr{A}$.

Since all other set operations can be expressed in terms of union and complementation, a completely additive class can be described simply as a nonempty class that is closed with respect to countable set operations.

Example 13.3
Let S be the space of all positive integers, $S = I_+$. Let $\mathscr{C}$ be the class made up of all the one-element subsets of S, as well as all those sets that can be generated by performing *finite* set operations on these one-element sets. Thus, $\mathscr{C}$ is an additive class. Now consider the set E of all even positive integers. E is an *infinite* (countable) union of one-element sets,

$$E = \bigcup_{i=1}^{\infty} \{2i\}$$

and according to Theorem 13.3, stated below, cannot be obtained by performing finite set operations on one-element sets. Therefore, E does not belong to $\mathscr{C}$. Thus, $\mathscr{C}$ is *not* a completely additive class.

Theorem 13.3

Let K be an infinite subset of a space S, and such that K^c is also infinite. Then K cannot be obtained by performing finite set operations on finite subsets of S.

Proof

Since the Theorem involves finite set operations, it is sufficient to consider a finite class $\mathscr{C} = \{A_1, A_2, \ldots, A_n\}$ of finite subsets of an infinite space S. Let

$$B = \bigcup_{i=1}^{n} A_i$$

which (by repeated application of Theorem 11.1(a)) is finite. Then, from Example 8.1,

$$B^c = \bigcap_{i=1}^{n} A_i^c$$

and since S is infinite, it follows from Theorem 11.1(a) that B^c is infinite. Furthermore, according to Theorem 13.2, B^c belongs to the largest partition in the additive class generated by $\mathscr{C}$. That is, no proper nonempty subset of B^c can be obtained by performing set operations on members of $\mathscr{C}$.

Now let K be a set that is obtained by performing finite set operations on members of $\mathscr{C}$. If K is infinite, then $K \supset B^c$ since B is finite and since K cannot be a proper subset of B^c. Therefore $K^c \subset B$ (Problem 8.8a), which implies that K^c is finite. Thus, it is not possible for both K and K^c to be infinite.

It should be noted that a finite class that is additive is also completely additive. Of course, every completely additive class is also additive. In other words, the distinction between 'additive' and 'completely additive' is of significance only for infinite classes.

Queries

13.1 On a space S we define two distinct nonempty sets A and B, where A, B ≠ S. How many (distinct) sets belong to the additive class generated by $\mathscr{C} = \{A, B\}$? [26]

13.2 Let S be the space of all positive integers, and let $\mathscr{C}$ be the additive class generated by all the one-element subsets of S, as in Example 13.3. For each of the following statements, decide whether it is true, possibly true, or false
a) Every subset of S belongs to C
b) If $\mathscr{D}$ is the additive class generated by all the two-element subsets of S, then $\mathscr{D} = \mathscr{C}$
c) No infinite sets belong to $\mathscr{C}$
d) Let A, B, C be subsets of S such that $A = B \cup C$. If $A \in \mathscr{C}$, then $B \in \mathscr{C}$, and $C \in \mathscr{C}$. [130]

13.3 Given a space S and two partitions, $\mathscr{P}_1$ and $\mathscr{P}_2$, both having size 10 and such that $\mathscr{P}_1 \neq \mathscr{P}_2$. For each of the following statements, decide whether it is true, possibly true, or false.
a) $\mathscr{P}_1 \cup \mathscr{P}_2$ is a partition
b) $\mathscr{P}_1 \cap \mathscr{P}_2$ is a partition
c) Every set belonging to $\mathscr{P}_1$ is a subset of a set belonging to $\mathscr{P}_2$. [248]

SECTION *14*

Additive Set Functions

Now we return again to functions, in particular, to real-valued *set* functions. There is another property that such functions are often required to exhibit, in addition to being single-valued, in order to be useful.

Let $\mathscr{A}$ be some additive class of sets and H a real-valued set function defined on this class. Thus, if $A \in \mathscr{A}$, then the function H assigns to A a real number $H(A)$. We are familiar with ways of manipulating real numbers, that is, the rules of ordinary arithmetic. We are now also familiar with set operations. Can the two go hand-in-hand? For instance, if we take a *union* of sets in $\mathscr{A}$, their H-values might *add* to give the H-value of the union. A property of this sort would be convenient.

It is helpful at this point to introduce the notation

$$A \uplus B$$

for the union of two sets A, B, if it is to be noted that A and B are *disjoint*. We refer to $A \uplus B$ as a *disjoint union*. Also, if $\mathscr{C}$ is a disjoint class, then for $A_1, \ldots, A_n \in \mathscr{C}$ we have the disjoint union

$$\biguplus_{i=1}^{n} A_i = A_1 \uplus A_2 \uplus \ldots \uplus A_n$$

where no parentheses are needed on the right since the disjointness of the unions is not dependent on the order in which they are performed. It should also be noted that every union can be expressed in terms of a disjoint union; for instance,

$$A \cup B = A \uplus (B - A). \tag{14.1}$$

The following Definition involves disjoint unions:

(Definition) 14.1
Given an additive class $\mathscr{A}$ of sets, and a set function H on $\mathscr{A}$.[10] H is said to be an *additive* set function if, for every pair of disjoint sets A, $B \in \mathscr{A}$, the following equation holds:

$$H(A \uplus B) = H(A) + H(B).$$

---

[10] An additive set function can also be defined without requiring the class on which it is defined to be additive. This is a little more cumbersome, however, and provides no advantage for our purposes.

Probability Concepts and Theory for Engineers, First Edition. Harry Schwarzlander.
© 2011 John Wiley & Sons, Ltd. Published 2011 by John Wiley & Sons, Ltd.

A simple example of an additive set function over an additive class of *finite* sets is the function that assigns to each set the number which expresses the size of the set. Suppose $S = \{\xi_a, \xi_b, \xi_c, \xi_d\}$, and let H be the set function expressing the size of any subset of S; that is, for $A \subset S$, $H(A) = |A|$. Now consider the disjoint union

$$\{\xi_a\} \cup \{\xi_a\}^c = S.$$

Here, $H(\{\xi_a\}) = 1$, $H(\{\xi_a\}^c) = 3$, and $H(S) = 4$ so that, in agreement with Definition 14.1,

$$H(S) = H(\{\xi_a\}) + H(\{\xi_a\}^c).$$

Any additive set function can be thought of as expressing a property analogous to 'size', such as 'mass', or 'weight'—the allotment of some noncompressible commodity to various sets. For this reason, an additive set function is also called a *distribution*; it expresses how the total available amount of the commodity is *distributed* over the various sets. Thus, the property of additivity that is introduced in Definition 14.1 does not merely provide a mathematical convenience, it also models the manner in which we think about various physical phenomena.

Example 14.1

Let the space $S = \{\xi_a, \xi_b, \xi_c, \xi_d\}$ represent the collection of four weights available with a balance, as shown in Figure 14.1. Subsets of S then represent combinations of weights that can be placed on the tray of the balance. Let W be the set function expressing the weight. Thus, $W(\{\xi_a\}) = 1$, $W(\{\xi_b\}) = 2$, $W(\{\xi_a, \xi_b\}) = 3$, etc. Clearly, W turns out to be an *additive* set function, but why? Why do we assign $W(\{\xi_a, \xi_b\}) = 3$, for instance? It is because in the conceptual world we automatically apply the principle of conservation of mass in order to say that the collection of weights made up of the 1 gram weight and the 2 gram weight represents a total of 3 grams. Thus, the additivity of W in this case is arrived at through physical reasoning.

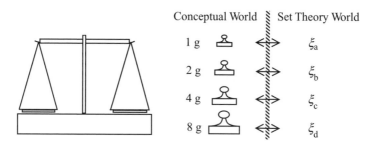

Figure 14.1   A balance and four weights

It should be noted that Definition 14.1 requires that if H is an additive set function, then $H(\phi) = 0$. This is because $H(A) = H(A \cup \phi) = H(A) + H(\phi)$.

(Definition) 14.2

Given a completely additive class of sets, and a set function H on $\mathcal{A}$. H is said to be a *countably additive* set function if, for every infinite sequence $A_1, A_2, \ldots$ of disjoint sets in $\mathcal{A}$ the following equation holds:

$$H\left(\biguplus_{i=1}^{\infty} A_i\right) = \sum_{i=1}^{\infty} H(A_i) \tag{14.2}$$

Example 14.2
Consider the sample space $S = \{\xi_a, \xi_b\}$ consisting of two elements, and the class $\mathscr{A}$ of all subsets of S, namely, $\mathscr{A} = \{\phi, S, \{\xi_a\}, \{\xi_b\}\}$. For any set A in $\mathscr{A}$ we again denote $|A|$, the size of A, by $H(A)$ in order to emphasize that for finite sets, 'size' is a set function. Thus, $H(\phi) = 0$; $H(\{\xi_a\}) = H(\{\xi_b\}) = 1$; $H(S) = 2$.

Additivity of the set function H is easily checked by considering all possible disjoint pairs of sets in $\mathscr{A}$; for instance,

$$H(\{\xi_a\} \uplus \{\xi_b\}) = H(S) = 2$$

and also

$$H(\{\xi_a\}) + H(\{\xi_b\}) = 1 + 1 = 2.$$

The set function H in this case is also *countably additive*. Any infinite sequence of sets satisfying Definition 14.2 must be one of the following, or a reordering of one of the first four:

a) $S, \phi, \phi, \phi, \phi, \phi, \ldots$
b) $\{\xi_a\}, \phi, \phi, \phi, \phi, \phi, \ldots$
c) $\{\xi_b\}, \phi, \phi, \phi, \phi, \phi, \ldots$
d) $\{\xi_a\}, \{\xi_b\}, \phi, \phi, \phi, \phi, \ldots$
e) $\phi, \phi, \phi, \phi, \phi, \ldots$

The empty set is the only set that may appear repeatedly in a sequence of disjoint sets. The definition for countable additivity therefore reduces here to that of simple (pairwise) additivity.

Queries

14.1 Given $S = \{\xi_1, \ldots, \xi_6\}$, and sets $A = \{\xi_1, \xi_2\}$, $B = \{\xi_1, \xi_3, \xi_5\}$.
Let $\mathscr{C} = \{AB, A - B, B - \{\xi_3\}, B - A, B^c A^c\}$. Exactly one set is to be removed from $\mathscr{C}$ so that the remaining sets form a disjoint class. Which one must be removed?
[81]

14.2 Given sets $A, B \subset S$, where $A \cup B \neq S$ and $A \cap B \neq \phi$. Let $\mathscr{C}$ be an additive class, with $A, B \in \mathscr{C}$. An additive set function H, defined on $\mathscr{C}$, assigns the following values: $H(A) = 18$, $H(A \cup B) = 19$, and $H(A \cap B) = 0.7$. Find $H(B)$.
[11]

14.3 Consider the experiment of throwing a die and noting the number k of dots facing up. Suppose we assign values to various events as shown:

Event: 'k is even'   Value: $+10$
         '$k > 3$' -10
         '$k > 4$' $+7$

What value should be assigned to the event '$k = 2$ or 6' if the assignment is to be an additive set function?   [224]

More about Probabilistic Experiments

We have now discussed sets and their manipulation, classes of sets, and set functions. This completes the coverage of those elements of Set Theory that provide the basis for the mathematics of Probability. We return now to our conceptual model in order to establish additional links between the conceptual and mathematical worlds.

So far we have been concerned with only a single p.e. at a time. In such a context, the various *possible* outcomes would seem to be of rather little importance compared to the one outcome that is actually the result of the experiment. In this Section we expand our conceptual model so that it can encompass a multiplicity of p.e.'s, which is essential to the definition of 'probability' and its applications. For this purpose, three properties will be introduced.

> **(Definition) 15.1**
> We say that a number of probabilistic experiments are *identical* if the following three requirements are satisfied:
>
> a) The experimental procedures are identical.
> b) The sets of possible outcomes are identical.
> c) The relevant environmental conditions that exist during the execution of the experiments are, to the best of our knowledge, the same.

The three requirements listed in this Definition are obviously subject to interpretation, especially the third one. This illustrates again the inherent lack of precision in our conceptual model world. It is easy to imagine, for instance, that in some real-world experiment it is disputable whether a particular value of ambient temperature is a relevant environmental condition or not; and if it is relevant, how precise a temperature measurement is required so that the ambient temperature can be considered 'the same' during different executions of this experiment. But it must be remembered that we are here concerned with identical *probabilistic experiments*; the identity occurs in the conceptual world, not in the physical world. In other words, when we speak of two identical experiments we mean two real-world experiments whose *conceptual models* we take to be identical p.e.'s. The appropriateness

Probability Concepts and Theory for Engineers, First Edition. Harry Schwarzlander.
© 2011 John Wiley & Sons, Ltd. Published 2011 by John Wiley & Sons, Ltd.

of the models (and of their presumed identity) is ultimately decided by the success with which they lead to answers that agree with real-world observations.

(Definition) 15.2
Consider a number of real-world experiments, each of which is modeled by a separate probabilistic experiment. If there is reason to believe that for each of the experiments, the results obtained in that experiment do not affect the results that are observed upon performing the remaining experiments, then we say that the corresponding probabilistic experiments are *independent*.

The first sentence of this Definition makes it clear that here we are considering some physical situations that are modeled by a multiplicity of probabilistic experiments, rather than by a single probabilistic experiment. The remainder of the Definition makes it clear that, given several probabilistic experiments, these can only be *assumed*, or *judged*, to be independent, not proven to be so. In many cases, the independence assumption is readily made. For instance, several die-throwing experiments, identical to the one described in Section **2**, can be considered independent experiments: There is no reason to suspect that, under normal circumstances, the result of one die throw influences the result of another die throw. Even if the same die is used in each of the experiments, it is assumed that the die is well shaken between throws, and allowed to roll.

(Definition) 15.3
Suppose that in a probabilistic experiment $\mathfrak{E}$ an event A is of interest. Suppose further that real-world activities take place which can be modeled as a large number of independent probabilistic experiments, all of which are identical to $\mathfrak{E}$. If the event A is found to occur in a definite fraction of these experiments, then we say there is *statistical regularity relative to the experiment $\mathfrak{E}$ and the event* A.

By a 'definite fraction' is meant the following. Suppose the large number of experiments is divided arbitrarily into several groups, then in each of these groups approximately the same fraction of experiments gives rise to event A. The possible *absence* of statistical regularity becomes particularly plausible if the large number of experiments in question take place sequentially in time: It is then conceivable that the event A would gradually occur more and more often, (or less and less often) as, say, the experimenter gains experience, or as the apparatus gradually wears (although ostensibly remaining unchanged so that the experiments are still deemed identical).

Suppose that in a p.e. $\mathfrak{E}$ an event A is of interest. Suppose that n independent p.e.'s are carried out, all of which are identical to $\mathfrak{E}$. If in these n experiments, A is found to occur n_A times, then the ratio n_A/n expresses how frequently the event A occurred in the n experiments. Accordingly, we call this ratio the *relative frequency of occurrence of A* (for these particular n experiments) and denote it rf(A). For another such group of n experiments, a different value of rf(A) is likely to be obtained. Now, if statistical regularity exists relative to the p.e. $\mathfrak{E}$ and the event A, then for large n, rf(A) will tend toward a definite fixed value. This value is the 'definite fraction of the experiments giving rise to A', referred to in Definition 15.3. This definite value is also referred to as the *chance*, or *probability*, that A occurs.

The importance of statistical regularity should now be clear: In the absence of statistical regularity it would be meaningless to speak of 'probability' in the sense in which we are using this term here.

We note, incidentally, that if statistical regularity exists relative to a p.e. $\mathfrak{E}$ and an event A and also an event B, where A and B are mutually exclusive, then statistical regularity must also exist with respect to the event 'A or B' consisting of the possible outcomes of both A and B. Continuing with this thought, it should become clear that, for a given probabilistic experiment, the events that exhibit statistical regularity correspond in the mathematical model to the various sets in some *additive class*.

Lack of statistical regularity with respect to an experiment and an event A does not necessarily imply lack of statistical regularity with respect to every other event that can be defined for this experiment, as is illustrated in the following Example. Of course, the impossible event and the certain event always exhibit statistical regularity—even in an experiment where some events lack statistical regularity.

Example 15.1

Consider practice arrow shooting on a particular field that is subject to strong sporadic crosswinds. The shooting of arrows by an experienced bowman can be modeled by a sequence of probabilistic experiments that may reasonably be considered identical and independent. All experiments in the sequence have in common:

a) Procedure: Experienced bowman shoots from a standard position, whereupon position of arrow in target is noted.
b) Sample Space: The set of distinguishable positions on the target.

Missing the target altogether might be regarded as an unsatisfactorily performed experiment. The experiments also have in common the relevant environmental condition 'strong sporadic cross-winds'. Suppose the events of interest are defined by an arrow entering one of the three disjoint regions A, B, C on the target as shown in Figure 15.1. If the wind is totally unpredictable then it is possible that statistical regularity is found to hold, and the wind is one source of randomness in the experiment—perhaps the major source. But it is also possible that *no* statistical regularity is observed with respect to events A, B, and C because the wind characteristics may vary with the time of day, the weather, or the seasons.

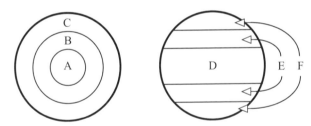

Figure 15.1 Two different partitionings of the target surface

On the other hand, the alternate partitioning D, E, F of the target surface shown in Figure 15.1 very likely would not be so sensitive to the variations and irregularities of the crosswind. Upon conducting a large sequence of these experiments one would be able to check the appropriateness of the assumption of statistical regularity with respect to events D, E and F, even if there is a lack of statistical regularity with respect to A, B and C.

Just as the properties of *identity* and *independence* of probabilistic experiments, so is statistical regularity a property of our *conceptual model*, the probabilistic experiment. We assign this property to our model whenever the real-world context indicates that it is *reasonable*. For, even observing a very large number of independent, identical, experiments will never yield with complete certainty a 'definite fraction' of experiments in which some specified event does occur. Frequently, it is possible to infer statistical regularity from the physical nature of a given real-world problem, without actually performing measurements. That is, we fall back on our accumulated experience, as embodied in our conceptualization of various aspects of the real world, in order to decide whether statistical regularity is an appropriate characteristic to ascribe to a particular event in a particular p.e. For example, when considering the repeated throwing of a die, we readily tend to assume statistical regularity.

Since statistical regularity requires independent, identical experiments, we will from now on assume that a probabilistic experiment is of such a nature that it is meaningful to think of the experiment as one instance among a large number of independent, identical experiments. The implication is that the experimental procedure must bring the experimenter into contact with some combination of uncontrollable and unpredictable influences that constitute a definite 'random mechanism', and the nature of this random mechanism remains undisturbed from one p.e. to the next.

Of course, the environmental conditions relevant to any given p.e. cannot remain 'the same' forever. The time frame over which repetitions of a p.e. are considered or performed has an effect on whether statistical regularity can be considered to hold. If environmental conditions change gradually, the repetitions cannot be spread out over too long a time. In that case, also, the results have to be understood as representing the situation under conditions that prevailed at the time. On the other hand, if some environmental conditions fluctuate within a fairly well-established fixed range (as in Example 15.1), and the relative frequencies are to reflect as much as possible the full range of environmental influences, then it may be necessary to *enlarge* the time frame. In that way, short-term variations get incorporated in the underlying random mechanism, and will reflect in whatever relative frequencies are of interest.

Example 15.2

A friend of mine has moved to a second-floor apartment. I am about to visit him for the first time and as I walk up the stairs, taking two steps at a time, I wonder whether the stairs have an even number of steps. This can be modeled in the form of a probabilistic experiment, the experimental plan being:

a) Purpose: to determine whether or not the stairs can be negotiated taking only double steps;
b) Procedure: climb the stairs two steps at a time, and note whether or not a single step needs to be taken at the top.

However, my uncertainty about the outcome is removed once the observation has been made. It is not meaningful in this case to consider repetitions of the experiment. Once I am familiar with the stairs, I will not again set out to determine whether there is an even number. In this case, no 'random mechanism' is affecting the result obtained.

Queries

15.1 For a given p.e., five mutually exclusive events are defined that together account for all the possible outcomes. In other words, on the sample space for this experiment, the five sets that correspond to these events form a partition. Is it possible for statistical regularity to exist
a) for all except one of the events?
b) for exactly one of the events?   [233]

15.2 In each of the following experiments, decide whether any of the Definitions 15.1, 15.2 or 15.3 are violated, and if so, which ones.

a) Of three persons, each buys a lottery ticket in a lottery. Each is performing an experiment. (Are these experiments independent, identical experiments, with statistical regularity?)

b) Two persons play a game of chess. Each move can be considered an experiment. (What is the experiment?)

c) A die-throwing experiment is performed in which the die is thrown and the number facing up is recorded. However, if this throw results in a 'six', the die is thrown again and the number obtained on the second throw is recorded.

d) A die-throwing experiment is performed many times in succession. In this experiment, the die is thrown once and the number facing up is added to the number observed in the previous throw, and this sum constitutes the result of the experiment.   [116]

SECTION *16*

The Probability Function

Suppose that, in a given p.e. $\mathfrak{E}$, some event A is of interest. Suppose further that n independent, identical versions of $\mathfrak{E}$ are performed, and statistical regularity holds relative to A. For the n experiments, a certain value of *relative frequency* of A, rf(A), is obtained. Clearly, rf(A) must be a non-negative number not exceeding 1. Furthermore, if we consider two mutually exclusive events A and B whose relative frequencies of occurrence are rf(A) and rf(B), respectively, then the relative frequency of occurrence of the event 'A or B' must be rf(A) + rf(B).

These considerations will now be carried over into the abstract domain of Set Theory, in order to establish the basis for an 'objective' or 'axiomatic' Probability Theory that can be applied to real-world situations. This is done by extending the Set Theory model by means of a suitable *set function*. This set function assigns to any given set a numerical value that represents an *idealization of the relative frequency of the corresponding event*.

> **(Definition) 16.1**
> Given a space S. A *probability function* (or *probability distribution*) is a non-negative, countably additive set function P defined on a completely additive class of subsets of S, with $P(S) = 1$.[11]

If the space S in Definition 16.1 models the collection of possible outcomes of the p.e. $\mathfrak{E}$ considered above, then S has a subset A which represents, in the mathematical model, the event A. What numerical value should the probability function P assign to the set A? The relative frequency of occurrence of event A in the p.e. $\mathfrak{E}$ is a number that is based on actual observation. It is the fraction of experiments giving rise to the event A when a number of independent, identical experiments are performed. It is not a constant. However, if statistical regularity holds, rf(A) exhibits 'stability'—that is, for a large number of experiments, the rf(A) tends to be close to a definite value. This definite value is the 'idealization of relative frequency' mentioned above, and is expressed by $P(A)$.

Much later in our study a mathematical justification will be established for the notion that, given statistical regularity, the relative frequency of an event tends toward a definite value as the number of experiments increases. For now we will simply assume that, if statistical regularity holds in our

---

[11] The definition of probability in terms of a countably additive set function was first introduced by A. N. Kolmogorov in his famous work *Grundbegriffe der Wahrscheinlichkeitsrechnung* [Ko1]. A non-negative, countably additive set function defined on a completely additive class is also called a *measure*. A set function P satisfying this Definition is therefore also sometimes called a *probability measure*. The development and study of the abstract properties of measures is called *Measure Theory*. Cf. [Ha1].

Probability Concepts and Theory for Engineers, First Edition. Harry Schwarzlander.
© 2011 John Wiley & Sons, Ltd. Published 2011 by John Wiley & Sons, Ltd.

conceptual model world, then there is inherent in our model an idealized relative frequency, of which the actually observed relative frequencies are more or less crude approximations. It is this idealized relative frequency that is expressed by P in the mathematical model.

The chart in Figure 5.1 can now be continued as shown in Figure 16.1. In the mathematical world, $P(A)$ is a unique and fixed value. The idealized relative frequency of event A, on the other hand, if it cannot be inferred form the physical nature of the experiment, must be estimated from the results that are observed in a sequence of experiments.

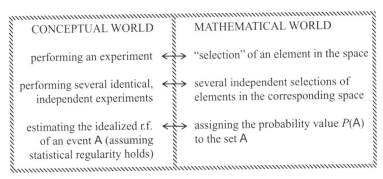

Figure 16.1   More connections between the conceptual and mathematical worlds

A comment is in order about the distinction between a probability function P being 'defined' on a particular completely additive class, and this function P being 'known'. Suppose we have a completely additive class $\mathscr{A}$ of subsets of a space S. When we *define* a set function P on this class, we are merely saying: there shall be such a function, and its name shall be P. If, furthermore, we say that this set function shall be non-negative, countably additive, and that it assigns the value 1 to S, then P is a function of the type that satisfies Definition 16.1, so that we call it a probability function. On the other hand, if the *values* that P assigns to the various members of $\mathscr{A}$ are given or known as well, then P is *known*.

It might be asked why, in Section 15, we spoke only of relative frequencies of *events*, not of possible outcomes. This can now be answered by again considering Definition 16.1. The probability function is defined as a *set function*, in order to correspond to relative frequencies of *events*. Mathematical inconsistencies could arise if the probability function were to be regarded also as a point function. Thus, in order to harmonize with the mathematical model, we avoid thinking about the relative frequency of occurrence of a particular possible outcome. Instead, we consider the relative frequency of occurrence of the *event* consisting of that particular possible outcome.

All the requirements spelt out in Definition 16.1, except countable additivity (but including simple additivity) of P are based directly on the conceptual model we have developed. The requirement of *countable* additivity, while compatible with the conceptual model, represents a refinement of the mathematical model over physical reasoning. This is because physical arguments, by means of which the 'relative frequency' notion was developed in the previous Section, do not extend to 'infinities'.

Example 16.1: Random Integers.

Consider the experiment of picking 'at random', without bias, an integer from the set of integers $\{1, 2, 3, \ldots, 20\}$, and let A_i ($i = 1, \ldots, 20$) denote the event 'integer i was selected'. The mathematical model consists of a space $S = \{\xi_1, \ldots, \xi_{20}\}$, where, for $i = 1, \ldots, 20$, ξ_i represents the possible outcome 'integer i gets selected'. Then $|S| = 20$. Also, let the one-element sets A_1,

$A_2, \ldots, A_{20}$ represent the events $A_1, \ldots, A_{20}$, respectively. The set $\biguplus_{i=1}^{20} A_i$ then corresponds to the *certain event*, so that

$$P\left(\biguplus_{i=1}^{20} A_i\right) = P(S) = 1.$$

Since $A_1, \ldots, A_{20}$ are disjoint sets that represent equally likely events, we specify $P(A_i) = 0.05$ for $i = 1, \ldots, 20$, so as to satisfy

$$\sum_{i=1}^{20} P(A_i) = 1.$$

On the other hand, suppose 20 is replaced by ∞. That is, the experiment now involves picking an integer 'at random and without bias' from among *all* positive integers, in the sense that no group of n integers is favored over any other group of n integers, for any n. This is not a *physically meaningful* experiment, but we can talk about such an experiment at least in terms of the conceptual model. We must have

$$P\left(\biguplus_{i=1}^{\infty} A_i\right) = 1$$

But $P(A_i) = 0$. Is then

$$\sum_{i=1}^{\infty} P(A_i) = \sum_{i=1}^{\infty} 0 = P\left(\biguplus_{i=1}^{\infty} A_i\right) = 1? \tag{16.1}$$

No! The trouble lies in the definition of P. When 20 is replaced by ∞, the set function P is changed and no longer satisfies the definition of a probability function: Countable additivity is violated, as can be seen in Equation (16.1).

Now suppose, instead, that Definition 16.1 did *not* require countable additivity. Then the set function P in this example *would* be a probability function, but we simply could not expect the relation

$$\sum_{i=1}^{\infty} P(A_i) = P\left(\biguplus_{i=1}^{\infty} A_i\right)$$

to hold. We conclude that the p.e. of picking an integer 'at random and without bias' (in the sense in which we interpreted that phrase) not only lacks an exact real-world counterpart, but it is also mathematically untenable within the theory presented here.[12]

In order to maintain the correspondence between the conceptual and mathematical worlds we therefore must refrain from formulating any experiment that calls for the *unbiased* selection of an

---

[12] It is possible to overcome the limitation of the mathematical model that is exhibited by Example 16.1. This can be done by resorting to a more elaborate mathematical model, as described in [Re1]. For purposes of applications there seems to be no need to do this, however.

integer from the set of all positive integers—or more generally, an experiment requiring the unbiased selection of an element from some countably infinite collection of elements.

Queries

16.1 A die is loaded in such a way that the probability of k dots coming to face up is proportional to k ($k = 1, \ldots, 6$). What is the probability of obtaining a 'six' with this die?   [214]

16.2 Consider again the experiment of picking an integer 'at random' from among *all* the positive integers, as discussed in Example 16.1. We have seen that the set function P as defined there is not a probability function, since it is not countably additive. Suppose instead that for every set A, $P(A)$ is defined as

$$P(A) = \lim_{n \to \infty} P_n(A)$$

where $P_n(A)$ is the probability of the set $A \cap S_n$ in the experiment of picking at random, without bias, an integer from the set of integers $S_n = \{1, 2, \ldots, n\}$. What values does the set function P defined in this manner assign to the following sets (where i denotes a positive integer):
a) $\{1, 2, 3, 4, 5\}$
b) $\{i: i > 1000\}$
c) $\{i: i \text{ even}\}$
d) $\{i: i \text{ odd}\}$
e) The set of all integers whose last digit (in the decimal representation) is a 5.   [186]

16.3 Given a sample space S and a countably infinite partition $\mathscr{P} = \{A_1, A_2, A_3, \ldots\}$. If for each i ($i = 1, 2, 3, \ldots$), the event A_i is exactly twice as likely as A_{i+1}, what is the probability of A_3?   [17]

Probability Space

In Section 16, *two* ingredients have been added to the mathematical model that we are building in order to deal analytically with a p.e. These new ingredients are: a completely additive class of sets and a probability function. Previously we have already seen that the *space* stands for the Set Theory model of an experimental plan. The *Probability Theory model* of a p.e. is not merely the sample space; it is called a *probability space* and consists of three items: the sample space S, a completely additive class of subsets of S, and a probability function P defined on $\mathscr{A}$. The notation used to designate a probability space therefore has the form of a *triple*:

$$(\mathsf{S},\ \mathscr{A},\ P)$$

Given a probability space, if any one of the three items S, $\mathscr{A}$, P is changed, a new probability space results that corresponds to a different p.e. On the other hand, several different p.e.'s may lead to the same probability space; such experiments are 'probabilistically equivalent'.

Example 17.1: Throw of two dice.

Consider again the experiment of throwing two distinguishable dice in order to determine the number of dots coming to face up on each. Say they are a red and a white die. For this experiment we have a sample space S consisting of 36 elements. This is the first ingredient of the probability space for this experiment. When drawing a Venn diagram for this space, it is convenient to exhibit the connection of the 36 elements to the possible outcomes of the underlying p.e. by indicating the properties of interest that characterize each possible outcome, as shown in Figure 17.1.

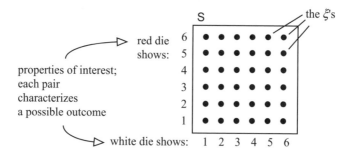

Figure 17.1 Sample space for the throw of two dice

Probability Concepts and Theory for Engineers, First Edition. Harry Schwarzlander.
© 2011 John Wiley & Sons, Ltd. Published 2011 by John Wiley & Sons, Ltd.

Next, we have to specify a completely additive class $\mathscr{A}$ of subsets of S. Since S is finite, it is sufficient to specify $\mathscr{A}$ as an additive class. The additive class $\mathscr{A}$, in the case of a finite sample space, is usually taken to be the class of *all subsets of* S, assuming there is statistical regularity with respect to all the corresponding events. To completely specify P it is not necessary to indicate the value assigned by P to every single member of $\mathscr{A}$. One way in which P can be completely specified is by specifying it for the partition that generates $\mathscr{A}$. Thus, if $\mathscr{A}$ is the class of all subsets of S, we might specify P only on the one-element subsets of S; additivity of P takes care of the rest. Assuming the dice are 'fair', then $P(\{\xi\}) = 1/36$, for all $\xi \in$ S. This completes the specification of the probability space (S, $\mathscr{A}$, P) for the throw of two distinguishable dice.

Another probabilistic experiment that is represented by the very same probability space as the one described in Example 17.1 would be the 'random' (i.e., 'blind' or unbiased) selection of a card from a deck of 36 distinguishable cards. That experiment is therefore probabilistically equivalent to the throw of two distinguishable dice, and its mathematical treatment differs in no way from that of the dice experiment.

(Definition) 17.1
Given a probability space (S, $\mathscr{A}$, P). The sets that belong to the partition that generates $\mathscr{A}$ will be called the *elementary sets* of the probability space.

Normally, the elementary sets of a probability space (S, $\mathscr{A}$, P) are all the one-element subsets of S, as in Example 17.1. But if (S, $\mathscr{A}$, P) is such that not all one-element sets have a probability assigned to them, then some elementary sets have a size greater than 1.

The *mathematics* of probability begins with the probability space, and thus it is divorced from conceptual modeling and from questions of applications. A mathematical probability problem, stated in terms of a particular probability space, applies at once to all probabilistically equivalent experiments that are represented by that probability space. A great economy of thought can therefore be achieved by stating probability problems directly in terms of the mathematical model. Nevertheless, making the connection to the real world via a suitable conceptual model must not be lost track of.

In developing our conceptual model we provided for an impossible event that corresponds, in the mathematical model, to the *empty set*. We also saw that the empty set ϕ is unique, and $P(\phi) = 0$ always. On the other hand, if an event is specified that has probability zero, this does not necessarily mean it is the impossible event—i.e., the empty set. For, when defining a probability space, it is not necessary that only the empty set be assigned zero probability. One can easily partition a space S into two nonempty subsets A, A^c and set $P(A) = 0$, $P(A^c) = 1$. This would be a legitimate probability distribution. But what is the significance that attaches to the specification $P(A) = 0$, if it does not necessarily imply that A is the impossible event?

In order to clarify this, we refer again to Example 11.1. Every conceivable coin-toss sequence (of infinite length) is a possible outcome of the experiment in which a coin is tossed repeatedly, without end. The collection of possible outcomes was found to be uncountable. Here, the probability that any *particular* (initially specified) sequence will occur is

$$\lim_{n \to \infty} \left(\frac{1}{2}\right)^n = 0.$$

Every countable subset of this space, therefore, has probability zero as well.

This may not appeal to intuition. Here again, the 'infinity' in our *model* plays tricks with our intuition. There is no infinity in the real world. Nevertheless, it is often convenient to construct the mathematical model in terms of an infinite space. We may think of a set with zero probability, then, as representing not necessarily an impossible event, but an event whose occurrences remain finite in number as the number of repetitions of the experiment grows without bound.

Queries

17.1 Which of the following experiments are probabilistically equivalent:
  a) It is to be observed whether one, two, three, four, five or six dots will come face up when a die is tossed. The procedure is to throw a die in the usual manner and note the number of dots facing up.
  b) It is to be observed whether or not a single dot will come to face up when a die is tossed. The procedure is to throw a die in the usual manner and note whether or not a single dot faces up.
  c) It is to be observed whether 'head' or 'tail' comes to face up when a coin is tossed. The procedure is to toss a single coin and note whether the result is 'head' or 'tail'.
  d) It is to be observed how many 'heads' are obtained when a coin is tossed five times in succession. The procedure is to begin with a count of zero. The coin is then tossed and if the result is 'head', the count is increased by one. This step is repeated until the coin has been tossed five times. The final count constitutes the result of the experiment.
  e) It is to be observed whether an even or an odd number of dots will come to face up when a die is tossed. The procedure is to throw a die in the usual manner, to observe the number of dots facing up. It is only noted whether this is an even number or an odd number.   [64]

Simple Probability Arithmetic

When a probability space has been specified for some p.e., then the additivity of the probability function permits us to perform probability calculations in the manner discussed in Section 14 for arbitrary additive set functions. We examine now some of the computational rules that apply to a probability function, and thus complete the first part of our study—the development of the basic conceptual and mathematical models for probability problems.

Let there be defined some probability space $(S, \mathscr{A}, P)$, with sets $A, B, C \in \mathscr{A}$. We have already considered instances in which the probability of a *disjoint union* is found (Section 16). The simplest example of two disjoint sets is a set A and its complement. Since $A \cup A^c = S$, we have $P(A) + P(A^c) = P(S) = 1$, giving

$$P(A^c) = 1 - P(A) \tag{18.1}$$

The special case where $A = S$ results in

$$P(\phi) = 0 \tag{18.2}$$

Thus, if a set function is a probability function, it must assign the value 0 to the empty set. On the other hand, as noted in Section 17, a set to which zero probability is assigned is not necessarily the empty set.

From the identity $A = AB \uplus AB^c$ of Theorem 8.1 follows that $P(A) = P(AB \uplus AB^c) = P(AB) + P(AB^c)$. Writing $AB^c = A - B$, this becomes

$$P(A - B) = P(A) - P(AB) \tag{18.3a}$$

In the special case where $A \supset B$, Equation (18.3a) simplifies to

$$P(A - B) = P(A) - P(B) \tag{18.3b}$$

Probability Concepts and Theory for Engineers, First Edition. Harry Schwarzlander.
© 2011 John Wiley & Sons, Ltd. Published 2011 by John Wiley & Sons, Ltd.

Since AB^c can always be written $A - AB$, and $A \supset AB$ always holds, Equation (18.3a) also follows from Equation (18.3b). If $P(A) = 0$ in Equation (18.3b), since the left side cannot be negative, it follows that $P(B) = 0$; similarly, if $P(B) = 1$, then $P(A) = 1$.

In the above we have used the additivity of P to express the probability of a disjoint union. What is to be done in the case of a union of sets that are not disjoint? The answer is that, in order to be able to use additivity for computing probabilities, such a union must first be expressed as a disjoint union. According to Equation (14.1), $A \cup B = A \uplus (B - A)$. Use of Equation (18.3a) then gives

$$P(A \cup B) = P(A) + P(B) - P(AB) \tag{18.4}$$

This can be visualized by means of the Venn diagram in Figure 18.1. Thus, the expression $P(A) + P(B)$ incorporates the probability of the set AB *twice*, which accounts for the last term on the right side of Equation (18.4).

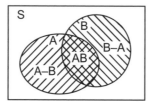

Figure 18.1   Expressing a union as a disjoint union

Equation (18.4) can be extended to triple and higher unions. In the case of a triple union we have (see Problem 14.4):

$$P(A \cup B \cup C) = P(A) + P(B) + P(C) - P(AB) - P(AC) - P(BC) + P(ABC) \tag{18.5}$$

Equations (18.1)–(18.5) express simple rules of probability arithmetic. When a probability space $(S, \mathscr{A}, P)$ is defined for a given probabilistic experiment, then the probability function P is also defined, but usually *not all* its values (i.e., not the probabilities of *all* sets in $\mathscr{A}$) are initially specified. The values that P assigns to other sets in $\mathscr{A}$ can then be found using the above and other rules.

We are now able to completely work through a simple probability problem.

Example 18.1

A manufacturer uses an automatic machine M to produce a particular type of fastener in large quantities. Certain dimensions of the completed fastener are critical to its correct functioning— we will call them dimension a and dimension b. In order to assure correct functioning of every fastener sold, the manufacturer passes the output of M through an automatic testing arrangement T that discards every fastener that does not satisfy the tolerance requirement. It is determined that out of a large sample of fasteners produced, 2.0% are discarded. An engineer examining the discarded fasteners finds that some of these actually function correctly. Upon carefully measuring a large number of discards, he finds that 75% of them exceed tolerance on dimension a and 55% of them exceed tolerance on dimension b. Then, he also determines that the *faulty* fasteners are exactly those for which dimension *a* and dimension *b* *both* exceed their respective tolerances. What percentage of the fasteners produced by the machine are actually faulty?

Here the *probabilistic experiment* involves the production and testing of a single fastener. The purpose is to determine whether the fastener is faulty and, more specifically, whether or not dimension *a* and dimension *b* lie within their respective tolerances. The procedure is to accept a single fastener from M during its normal operation, pass it through T, and, if it is rejected, measure dimensions *a* and *b*. (Without more detailed information available, it has to be assumed that repetitions of this experiment are identical and independent, and that there is statistical regularity with respect to the way in which dimensions *a* and *b* turn out.) The following are the possible outcomes of the p.e., along with their designations as elements of the sample space:

$$\xi_a \leftrightarrow \text{'fastener is rejected and only dimension } a \text{ exceeds tolerance'}$$
$$\xi_b \leftrightarrow \text{'fastener is rejected and only dimension } b \text{ exceeds tolerance'}$$
$$\xi_c \leftrightarrow \text{'fastener is rejected and both } a \text{ and } b \text{ exceed tolerance'}$$
$$\xi_p \leftrightarrow \text{'fastener is passed by T'}$$

In the mathematical model, therefore, $S = \{\xi_a, \xi_b, \xi_c, \xi_p\}$, and we take $\mathscr{A}$ to be the class of all subsets of S. Labels get assigned to those members of $\mathscr{A}$ that are of particular interest, as follows:

$$A = \{\xi_a, \xi_c\} \leftrightarrow \text{'}a \text{ exceeds tolerance'}$$
$$B = \{\xi_b, \xi_c\} \leftrightarrow \text{'}b \text{ exceeds tolerance'}$$
$$F = \{\xi_c\} \leftrightarrow \text{'fastener is faulty'}$$

These sets are identified in the Venn diagram in Figure 18.2. Rejection of the fastener by T corresponds to $A \cup B$.

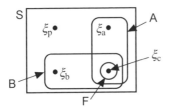

Figure 18.2   Venn diagram

Since no other information is available, the probability function in this case can only express the *actually observed* relative frequencies, rather than 'idealized' relative frequencies. So we write:

$$P(A \cup B) = 0.02$$
$$P(A) = 0.02 \times 0.75 = 0.015$$
$$P(B) = 0.02 \times 0.55 = 0.011$$

The probability function P has now been specified for some of the subsets of S. Of course, we also know that $P(S) = 1$ and $P(\phi) = 0$. What is to be found is $P(F)$. Since $F = AB$, application of Equation (18.4) gives

$$P(F) = P(A) + P(B) - P(A \cup B) = 0.015 + 0.011 - 0.02 = 0.006.$$

This completes the use of the mathematical model. Returning now to the conceptual domain, we can say that, in a large sample of fasteners produced by the machine under the same conditions, approximately 0.6% will actually be faulty.

It can happen that the probability of a particular event is desired but cannot be determined from the available information. In such a case it may be possible to establish a range within which the desired value of probability must lie.

Example 18.2

Consider the throw of a *loaded* die, for which it is known that an odd number of dots occurs with probability 0.6, less than four dots with probability 0.64, and four dots with probability 0.15. What is the probability of getting a 'six' with this die?

In this case the p.e. is familiar and we can move right into the mathematical model, with $|S| = 6$ and $\mathscr{A}$ the class of all subsets of S. Some of the members of $\mathscr{A}$ can be labeled as follows (see Figure 18.3):

$$A \leftrightarrow \text{'odd number of dots'}$$
$$B \leftrightarrow \text{'less than four dots'}$$
$$C \leftrightarrow \text{'four dots'}$$

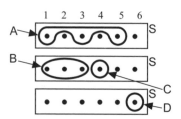

Figure 18.3   The sets A, B, C and D

We can now specify P by stating that $P(A) = 0.6$, $P(B) = 0.64$, and $P(C) = 0.15$. In this way, all the available information has become incorporated into the probability space $(S, \mathscr{A}, P)$ for this problem. Now let

$$D \leftrightarrow \text{'six dots'}.$$

By noting which elements make up each of the sets, it is easy to express D in terms of the other sets, namely, $D = (A \cup B \cup C)^c$. Using Equations (18.1) and (18.5), we have

$$P(D) = 1 - P(A \cup B \cup C)$$
$$= 1 - [P(A) + P(B) + P(C) - P(AB) - P(AC) - P(BC) + P(ABC)].$$

Since $AC = \phi$ and $BC = \phi$, it follows that $P(AC) = P(BC) = P(ABC) = 0$ and

$$P(D) = 1 - [P(A) + P(B) + P(C) - P(AB)]$$
$$= P(AB) - 0.39. \tag{18.6}$$

But $P(AB)$ is not known, so that it is not possible to completely evaluate $P(D)$.

Another approach is to express D as follows:

$$D = A^c - (B \cup C).$$

Application of Equation (18.3a) gives

$$P(D) = P(A^c) - P[A^c(B \cup C)].$$

With the help of Equations (18.1) and (**8**.4b) this becomes

$$P(D) = 1 - P(A) - P(A^cB \cup A^cC)$$
$$= 1 - P(A) - P(A^cB) - P(C) \qquad (18.7)$$
$$= 0.25 - P(A^cB).$$

Again, we cannot complete the calculation—in this case because $P(A^cB)$, the probability that two dots occur, is not known.

It is possible, however, to go a little further and determine the *range of possible values* for $P(D)$. From Equation (18.6), the largest possible value for $P(D)$ is obtained if the probability of joint occurrence of A and B is as large as possible (see Figure 18.4). This is the case if $A \subset B$ (since $P(A) < P(B)$). In that case $P(AB) = P(A)$ and Equation (18.6) becomes

$$P(D) = 0.6 - 0.39 = 0.21.$$

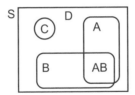

Figure 18.4   Single Venn diagram for the sets A, B, C and D

The other extreme is obtained if $P(AB)$ has the smallest possible value, which from Equation (18.6) can be seen to be 0.39, giving $P(D) = 0$. Thus, we have been able to obtain the following information about $P(D)$:

$$0 \leq P(D) \leq 0.21.$$

The same range for $P(D)$ can also be deduced from Equation (18.7).

Whereas this result might not be of much help to a gambler interested in using this particular loaded die, the derivation does illustrate the application of basic probability arithmetic.

Queries

18.1 Given a probability space $(S, \mathscr{A}, P)$, with A, B $\in \mathscr{A}$. If $P(A) = 0.6$ and $P(B) = 0.7$, what is the minimum possible value for $P(A \cup B)$? The maximum possible value?   [5]

18.2 Given a probability space $(S, \mathscr{A}, P)$ in which $\mathscr{A}$ consists of 32 sets. What is the smallest number of these sets whose probability (that is, value assigned by P) must be specified in order for all other probabilities to be obtainable by calculation?   [191]

18.3 Given a probability space $(S, \mathscr{A}, P)$, with $A, B \in \mathscr{A}$, where neither of the sets A, B is a subset of the other, and both are distinct from ϕ, S. Suppose only $P(A)$, $P(A^c)$, $P(\phi)$, $P(S)$, and $P(AB)$ are initially known. How many other probabilities can be found by means of simple probability arithmetic?   [221]

Part I Summary

Part I introduced the reader to two basic types of mental models to be used for treating practical problems involving uncertainty. The first of these is a *conceptual model*—the 'probabilistic experiment'. This model guides our thinking about a problem in such a way that it becomes easy to correctly apply probability mathematics to the problem. The second type of model is a *mathematical model*—the 'probability space', where mathematical analysis takes place.

It was shown how to connect the two types of models for a given problem, so that the appropriate mathematical analysis gets carried out and the mathematical results get interpreted correctly. This interplay between a probabilistic experiment and a probability space will be explored further in Part II and will remain a background consideration throughout the book.

Before defining a probability space, it was necessary to introduce a number of mathematical notions with which the reader may not have had adequate prior familiarity. These mathematical notions are elements of *Set Theory*, including functions defined on elements and on sets (point functions and set functions). Actually, point functions are not required to set up a probability space, but they will be used extensively later in this study. The probability space was then defined as a combination of three entities: a space of elements, a class of sets (subsets of the space), and a set function (the probability function) defined over the sets in this class. Because of the property of additivity, probability arithmetic can be carried out in the probability space, and simple probability problems can be solved with this machinery as developed so far.

Upon completing Part I, the reader should begin to feel comfortable with formulating basic probability problems—first in terms of the conceptual model, then in terms of the mathematical model—and in simple cases obtaining a solution. Also, skill should have been acquired in manipulating sets and working with classes of sets, including partitions and additive classes, as well as with set functions. All of this material provides the foundation for the remainder of this study.

Part II

The Approach to Elementary Probability Problems

Part II Introduction

This Part is concerned with applying the 'basic model' developed in Part I to a range of discrete probability problems. These are here referred to as 'elementary probability problems'—not because they are necessarily simple, but because they can be treated without requiring major extensions to the basic model. Those extensions get introduced in Part III and are used in the remaining Parts.

Part II begins with further examination of how to apply the basic model to real-world situations. The basic model then gets enhanced through the introduction of additional principles: conditional probability and independence. These not only significantly extend the manipulative power of the mathematical model, but also give more flexibility to the ways we can conceptualize a probability problem.

Then we enlarge the mathematical model through the introduction of product probability spaces. This allows the treatment of multiple executions of a p.e. or 'trial', which is then applied to various categories of sampling experiments. No great emphasis is placed, however, on problems involving complex combinatorics. As an approach to less-tractable problems, the use of flow and state diagrams is illustrated.

Part II ends with a brief introduction to two ancillary topics. One of these is the mathematical notion of 'entropy' as a measure of uncertainty or randomness, which is revisited and developed further in Section 82. The other, the principle of 'fuzziness', is actually not a probability topic at all. It is included because it illustrates a formal approach for dealing with a type of uncertainty that is of a quite different nature than the one to which we apply probability mathematics.

About Probability Problems

In Part I we have delineated the kinds of real-world phenomena to which we might address ourselves when we use the notion of probability. These phenomena involve not merely 'things', but include the thoughts and actions of a human being—the 'experimenter'. We have tried to systematize our view of such a complicated situation by means of the 'probabilistic experiment', which guides us to view the real world in accordance with a particular kind of conceptual model. This conceptual model turns out to be easily brought into correspondence with a suitable mathematical model—the probability space.

This basic mathematical model for probability problems—the probability space—is built upon the branch of Mathematics called Set Theory, and we said just enough about Set Theory in order to be able to define a probability space and be prepared to work with it. By way of review we recall that a probability space combines three set-theoretic notions, namely: a space of elements, a completely additive class of sets, and a countably additive set function. When a probability space gets defined for a particular probabilistic experiment, then

a) the elements of the space represent the various possible outcomes;
b) the completely additive class of sets is generated by a class of sets that correspond to events that may be of interest and with respect to which there exists statistical regularity; and
c) the countably additive set function defined on this class expresses idealized relative frequencies of the various events, approximately as they would be observed in a large number of identical, independent repetitions of the given probabilistic experiment.

These various considerations go into 'setting up' a probability problem for solution.

Generally, a distinction is made between the task of modeling a particular experiment and then computing various probabilities from other probabilities which are known, and the task of determining probabilities from measurements. Problems concerned with measurement (i.e., estimation) of probabilities through experimentation are classified under the name of *Statistics*. In a Probability problem, the probabilities of some events are generally known. These may either be assumed values, or they have been obtained through suitable measurements, or they can be inferred from the physical nature of the problem. Frequently we can choose our conceptual model in such a way that probabilities are easily inferred. This is particularly true if we can arrange for the possible outcomes to be equally likely, as discussed in Section 20.

Thus, probability problems are basically of the following type: *Given the probabilities associated with some of the events in a probabilistic experiment, the probabilities of various other events are to be found, or some related questions are to be answered.* Although this may seem rather straightforward, we will see that the implications of this statement are far-reaching and lead us further into an involvement with a broad range of mathematical concepts and techniques.

As our attention focuses more on the *mathematics* of probability, we must not forget that much care is also needed to assure proper and consistent modeling of the real-world problem. In this we are aided by the rules developed in Part I for building up the conceptual model. These rules also help discriminate between problems that are legitimate probability problems and those that are not. For, the mere appearance of the word 'probability' in the statement of a problem does not automatically make it a probability problem, as the following example illustrates.

Example 19.1

a) Of 1000 patients tested in a certain clinic, 30 are found to exhibit both symptom x and symptom y. None of the patients exhibited either symptom x alone or symptom y alone. What is the probability that symptoms x and y must always occur together?

We begin by assembling the conceptual model. We envision a p.e. that consists of testing a patient for the purpose of determining whether he or she exhibits symptoms x, y. Because the problem is stated in an abstract form and we do not know what the symptoms x and y are, nor what observations need to be made, we can only *assume* that there is a well-defined procedure for observing the existence of these properties of interest:

> 'x is present'
> 'x is not present'
> 'y is present'
> 'y is not present'

The possible outcomes are therefore:

> 'both x and y are present'
> 'only x is present'
> 'only y is present'
> 'neither x nor y is present'

No information is available for judging statistical regularity; it will be assumed to hold. Also, it has to be assumed that the 1000 patients represent distinct individuals and are representative of the larger public.

Then a four-element sample space gets defined, in which the following sets or events can be identified (see Figure 19.1):

> A ↔ 'symptom x is present'
> B ↔ 'symptom y is present'

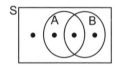

Figure 19.1   Venn diagram

We have available the observed relative frequencies from a sequence of 1000 experiments that presumably are independent and identical. Therefore, we use the available relative frequency information to define probabilities:

$$P(\mathsf{AB}) = 0.03, \quad P(\mathsf{A}\Delta\mathsf{B}) = 0$$

Other probabilities can be computed, such as $P(\mathsf{A}^c \cup \mathsf{B}^c) = 0.97$; etc. If we now try to answer the question being posed, we realize that '*symptoms x and y always occur together*' is not an event in our p.e. Instead, it is the *specification* that $\mathsf{A}\Delta\mathsf{B} = \phi$. Therefore, we cannot determine its probability. It is part of the description of the physical circumstances—a restriction on the 'random mechanism'. As such, it either holds or does not hold; it is not something that 'occurs' with statistical regularity. We conclude that the question *as stated* does not constitute a probability problem.

b) Now suppose the problem is restated in a slightly modified form, as follows: 'How certain is it that the probability of A and B not occurring separately is greater than (say) 0.99?' Here, 'how certain is it that . . .' may be interpreted as 'what is the probability that . . .' or 'how close to 1 is the probability that . . .'.

This question *does* constitute a legitimate probability problem. We are now concerned with the event $(\mathsf{A}\Delta\mathsf{B})^c$ in the sample space of Figure 19.1. The probability of this event is assumed to lie between 0.99 and 1.0. We are asked with what probability this assumption leads to the actually observed relative frequency of $\mathsf{A}\Delta\mathsf{B}$ in 1000 executions of the experiment (the testing of 1000 patients). However, that probability is not defined on the four-element sample space of Figure 19.1. Instead, an enlarged probability space called a 'product space' needs to be considered. This will be explored further on, in Section 27.

From this Example we see how adherence to the rules for building the model prevents us from 'getting lost' in the problem. In practical applications it can easily happen that a problem statement gets formulated in such a way that no solution is possible. It may then be necessary to further clarify the circumstances that gave rise to the problem, and also to clarify what information is really wanted.

Example 19.2

You are told that two balls have been removed 'blindly' from a bag containing a mixture of black and white balls, ten balls in all. The balls that were drawn were one black and one white ball. You are asked: What is the probability that the mixture in the bag consisted of an equal number of black and white balls?

You specify the p.e., which involves determining the color of two balls drawn from the given bag. A suitable procedure is assumed, which allows balls to be withdrawn without bias. The possible outcomes are:

'two white balls'
'two black balls'
'a white and a back ball'

As you begin to specify events, you realize that your p.e. has no event 'bag contains an equal number of black and white balls', so that the question as it was posed cannot be answered by solving a probability problem based on the p.e. you have constructed—and no other plausible p.e. can be found.

On the other hand, a probability that *can* be computed is the probability of drawing a black and a white ball when two balls are drawn from a bag containing five black and five white balls. In the p.e. that is implied there, the mixture of five black and five white balls is part of the experimental setup; it does not 'occur'. But this is not what was asked for.

Now suppose you investigate the matter further and you find out the following from the person who asked the question: Originally there had actually been *two* bags, one containing five white and five black balls, and the other containing two white and eight black balls. Without preference, one of the bags had been removed, and the two balls were drawn from the remaining bag.

This additional information changes the problem considerably. You are now able to set up the p.e. so that it includes the observable properties 'bag A remains' (the bag with 5 white and 5 black balls) and 'bag B remains' (the other bag). The experimenter in this case is not chosen to be the person who withdraws the balls from the bag, but a hypothetical behind-the-scene supervisor and observer. You have to assume that a valid procedure is used that leads to the following possible outcomes:

'two white balls from bag A'	'two white balls from bag B'
'two black balls from bag A'	'two black balls from bag B'
'a white and a back ball from bag A'	'a white and a back ball from bag B'

The sample space is described by the Venn diagram in Figure 19.2. You need to find the probability of the event 'bag A remains' in an experiment in which the event 'a white and a black ball drawn' is known to occur. This is most conveniently done using 'conditional probability', which is discussed in Sections 21 and 22 (see Example 22.3).

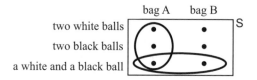

Figure 19.2   Venn diagram for the drawing of two balls

A valid probability problem must lead to the formulation of a p.e. that puts the experimenter into contact with the workings of some kind of *random mechanism*. The possible outcomes of the p.e. must be able to come about through the operation of this random mechanism. Thus, the problem specification must provide some information about the effects that the random mechanism generates.

In Example 19.1, for instance, the appearance of symptoms in patients is due to a random mechanism. We don't know how or why it operates, but we do have some relative frequency data—namely, about the joint occurrence of symptoms x and y. But, 'x and y always occur together' is not a possible result of this random mechanism. In the modified problem statement, an additional random mechanism is invoked which accounts for the uncertainty inherent in relative frequency measurements. In Example 19.2, the 'blind' removal of two balls is the random mechanism. It does not control the number of white and black balls in the bag. Again, in the revised version of the problem, an additional random mechanism gets introduced—the arbitrary removal of one of the two bags. It governs the number of black and white balls in the remaining bag.

Here is another example illustrating the importance that attaches to careful construction of the conceptual model, the probabilistic experiment.

Example 19.3

In a television quiz show, the contestant is shown three doors and is asked to choose one of them. Behind one of the doors, the contestant is told, there has been placed a large prize, while behind each of the other two doors is tethered a goat, signifying 'no prize'. After the contestant has made known his choice of a door, the quiz master opens up one of the two *other* doors—one that has a goat behind it. The quiz master then offers to the contestant the option of switching his/her choice to the remaining door which is still closed. Should the contestant accept this option—that is, is the contestant's chance of winning the prize improved, diminished, or unaffected by switching from the original choice to the other door that is still closed?[1]

We resist the impulse to jump to a conclusion and instead begin by developing the appropriate conceptual model. What sort of model is required here? The first question to resolve is: Who is the experimenter—the contestant, the quiz master, or a hypothetical neutral observer? It is the contestant who is facing the uncertainty about winning the prize and who has to decide whether to switch the original choice of door; so we can cast the contestant as the experimenter. Now we recall that the experimental procedure of a p.e. does not allow any choice to the experimenter. The experimenter is able, however, to *choose between two different probabilistic experiments*. Therefore, we model the situation at hand in terms of two different p.e.'s.

In one experiment the contestant's procedure is to simply choose one of the three doors, thereby making a selection from three equally likely possible outcomes (a door with the prize and two doors with a goat) and this selection is unaffected by later actions of the quiz master. The probability of winning the prize is 1/3.

In the other experiment the procedure is to choose one door, have the quiz master open another door with a goat behind it, and then select the remaining door. We note that if the prize is behind the door first chosen, then behind the last door must be a goat. Similarly, if one of the two doors with a goat behind it is first chosen, then the prize will be found behind the last door. Therefore, in this case the procedure again boils down to a selection from among three equally likely outcomes, *two* of them resulting in the prize and one in a goat. The probability of winning the prize is then 2/3. The contestant is thus better off with the second experiment—i.e., switching the choice of door.

In the next few Sections we will concern ourselves with some basic techniques that are very useful in dealing with many kinds of probability problems. In all of Part II we will restrict our attention to problems in which the possible outcomes of the probabilistic experiment are distinguished primarily in terms of *qualitative* attributes. The sample space is then usually finite. We refer to such problems as 'elementary probability problems', although this is not meant to imply that they are necessarily simple.

Thereafter, in Part III, we will expand our study in order to treat probability problems in which the possible outcomes are described primarily *quantitatively*, and where this quantitative information is to be incorporated into the mathematical model.

---

[1] This problem has been referred to as the 'Monty Hall Paradox,' named after the moderator of the TV quiz show in which it arose. It is discussed in [Se1].

Queries

19.1 Which of the following are legitimate probability problems:
a) A coin is tossed and a 'head' is obtained. What is the probability that the coin is double-headed (a head on both sides)?
b) A coin is tossed ten times. A 'head' is obtained each time. What is the probability that the coin is double-headed?
c) Ten coins are tossed, resulting in ten 'heads'. What is the probability that at least one of the coins is double-headed?
d) A coin is blindly chosen from ten coins, one of which is known to be double-headed. What is the probability that the chosen coin, when tossed, results in a head?
e) Two coins are blindly chosen from ten coins, one of which is known to be double-headed. The two chosen coins are tossed, resulting in two heads. What is the probability that one of the chosen coins is the double-headed coin?   [50]

Equally Likely Possible Outcomes

Historically, problems that led to the earliest intuitive notions about probability involved experiments with a finite number of possible outcomes that could reasonably be assumed to be *equally likely*. This is a feature that is approximated in many game-playing and gambling situations, such as the throw of a die or a coin, picking a card from a standard deck, etc. In fact, the assumption of equally likely outcomes for a long time formed the basis for solving probability problems. With the mathematical model we have developed in Part I, this is of course not the case, and we should be on the alert for those situations in which experiments would *seem* to lead to equally likely possible outcomes, but actually don't.

This brings us to the important question: How can we decide, in a given probabilistic experiment for which probabilities are not initially stated, whether or not the outcomes are equally likely? This question must be resolved by considering the *procedure* of the probabilistic experiment. The procedure must bring into evidence the particular random mechanism with which the experimenter comes into contact when he or she performs the experiment. If the procedure is such that the random mechanism cannot be expected to favor any one possible outcome over some other possible outcome, then all possible outcomes are equally likely. Or, we may say that the possible outcomes are equally likely if the observable properties represented by the various possible outcomes can be freely interchanged without affecting the nature of the p.e. Thus, in the case of rolling a die (of good quality) or tossing a coin, the unpredictable motion of a physical object, coupled with its symmetry, readily leads us to assume that no possible outcome (face of the die or side of the coin) is favored over the other possible outcomes. Alternately, if in the case of the die we renumber the faces, we would not think of this as a modification of the p.e. Therefore, we accept the equal likelihood of the possible outcomes as reasonable. On the other hand, under some circumstances, such as gambling, the possibility of skill and deception being used to produce biased outcomes must not be overlooked (cf. [Mc1]).

In other instances, the nature of the random mechanism underlying a probabilistic experiment may be more difficult to discern. Particular care must be taken in real-world situations involving the 'random selection' of an object from a collection of objects.

Example 20.1

We are told that a ball is to be picked from a container in which there are four balls, marked 1, 2, 3, and 4, respectively, and it is to be noted which particular ball is picked. We seem to have an experiment with four possible outcomes. But let us make sure the experimental plan is completely specified. We have a *purpose:* to determine whether ball 1, ball 2, ball 3, or ball 4 is selected, when a ball is picked from the container. What is the procedure? It is not clear from what has been said so far. The one sentence that has been used to describe a possible real-world situation does not allow us to proceed without some reservations. The following factors may affect the experiment significantly and need to be clarified:

a) *Distinguishability of the objects during the selection process.* If the objects are distinguishable during selection, then this may influence, or bias, the selection process. (For example, some balls may be larger than others.)

b) *The physical arrangement of the objects.* The physical arrangement of the objects may be such as to favor certain ones over others. (As an extreme example, we may visualize the container as narrow and deep, so that the balls are stacked in essentially a single column, with only one ball directly accessible at any one time. Many card tricks depend on creating the impression of equally likely selection when in fact a very particular card is selected.)

c) *The actual manner in which the selection is made.* This, when combined with (a) and (b), determines whether all the possible outcomes are equally likely. 'Manual' selection of a ball by the experimenter usually does not lead to equally likely outcomes if either the balls are distinguishable during the selection process or their physical arrangement is significant.

In the probability literature the drawing of a ball from an urn containing various balls is frequently taken as the stereotypical model for an experiment having equally likely outcomes. The thinking there is different, however. Drawing a ball from an urn in that case is merely a way of 'visualizing' or assigning a conceptual meaning to *an abstract experiment*: an experiment that has been *defined* to have equally likely outcomes. The possibility of a deviation from equal likelihood is therefore categorically excluded. In Example 20.1, on the other hand, we are alerted to the care with which facts about a *real-world* situation must be assembled, when the probability model is to be constructed.

Often it is considerably more difficult to *devise* a real-world experiment in such a way that outcomes can truly be assumed equally likely, than to analyze it in terms of Probability Theory. The task of designing suitable experiments is encountered in statistical quality control, population surveys, and many other applications of Statistics.

Example 20.2

A particular book is opened by a reader to an arbitrary page and closed again. Another person then randomly opens the book. What is the probability that she will open it to the same page as the first person?

Here we are easily led to assume equal likelihood of opening to any page. We may then say that the p.e. need only be concerned with the actions of the second person, since the first person merely 'specifies' a particular page, and it makes no difference which one it is. If the book can be opened to n different places (pairs of pages, including title pages, etc.) then the desired probability is obtained at once as $1/n$.

However, we have arrived at this answer rather hastily, bypassing the conceptual model almost entirely. Instead, we must look carefully at the procedure, and ask what is meant by the second person opening the book 'randomly'. Suppose she uses a suitable random mechanism for generating page numbers with equal likelihood, such as might be done with dice or a roulette wheel, and then opens the book to the page selected in this manner. In that case we can reasonably say that all possible outcomes are equally likely, and the above answer is correct.

On the other hand, if the second person opens the book 'blindly' (or even with some deliberation) then our simple model is no longer applicable, since she is more likely to open it to a place close to the center than near the beginning or end of the book. In addition, it is a familiar fact that a book opens more readily to some pages than to others; and if it has been opened to a particular place it will more readily be opened to that place again. Thus, we cannot assume equally likely possible outcomes, nor can we neglect the role of the first person. His or her activity must be included in the p.e. Furthermore, it becomes necessary to examine carefully whether the assumption of statistical regularity is justified; if not, no probability space can be defined.

A p.e. in which all possible outcomes are indeed equally likely has associated with it a particularly simple probability space $(S, \mathscr{A}, P)$. First, the sample space S must be finite, as we saw in Example 16.1; and we can write $|S| = n$ $(n < \infty)$. Since there is statistical regularity with respect to all events that represent single possible outcomes, every one-element subsets of S belongs to the class $\mathscr{A}$. The largest partition $\mathscr{P}_0$ contained in $\mathscr{A}$ must therefore be the class of all one-element subsets of S. To each member of this class $\mathscr{P}_0$ the probability function P assigns the value $1/n$, since $|\mathscr{P}_0| = n$; and of course

$$P(S) = n\left(\frac{1}{n}\right) = 1.$$

In Example 17.1 we saw that specification of P on all the elementary sets $\{\xi\}$ was sufficient for a complete specification of P over the whole class $\mathscr{A}$. Additivity makes it possible to compute the probabilities of all the remaining sets. In the case at hand, this computation is very simple. For any set $A \subset S$, $P(A)$ is the number of elements in A divided by the number of elements in S:

$$P(A) = \frac{|A|}{|S|} = \frac{|A|}{n}. \tag{20.1}$$

These comments can be summarized as follows:

Theorem 20.1
Let $(S, \mathscr{A}, P)$ denote the probability space for a p.e. with n equally likely possible outcomes. Then

S consists of n elements, $|S| = n$;
$\mathscr{A}$ consists of all the 2^n subsets of S; and
$P(A) = |A|/n$, for every set $A \subset S$.

Clearly, if two different p.e.'s with equally likely possible outcomes have *the same number* of possible outcomes, then the two p.e.'s are probabilistically equivalent.

We have now seen one way in which the physical circumstances of an experiment lead us to infer a particular probability distribution, namely, equal probability of all the elementary events. Further on we will become acquainted with other kinds of probability distributions which may be inferred from the physical circumstances of an experiment.

If the possible outcomes of an experiment are equally likely, then finding the probability of an event generally reduces to a problem of counting elements, or possible outcomes. Since even something as simple as counting can get rather involved, it is worthwhile to devote some attention to methods of counting, which we do in later Sections.

In the discussion leading up to Theorem 20.1, the role of the partition $\mathscr{P}_o$ in specifying the probability distribution P was emphasized. This significance of the largest partition of S applies not only when the possible outcomes are equally likely, and we have the following:

Theorem 20.2
Given a countable probability space $(S, \mathscr{A}, P)$. P is completely specified if $P(A)$ is given for every set A belonging to the largest partition contained in $\mathscr{A}$.

Example 20.3: Throw of two dice.
We return to the experiment considered in Example 17.1, which has 36 equally likely possible outcomes. Suppose that the *sum* of the dots showing on the two dice is of interest. Then, for $i = 2, \ldots, 12$, we consider the events

$$A_i \leftrightarrow \text{'sum of the dots showing equals } i\text{'}.$$

Application of Theorem 20.1 yields the following probabilities for these events:

$$P(A_i) = \begin{cases} \dfrac{i-1}{36}, & i = 2, \ldots, 7 \\[2mm] \dfrac{13-i}{36}, & i = 8, \ldots, 12 \end{cases} \tag{20.2}$$

On the other hand, if we are only concerned about the probabilities associated with various sums of dots, then we might simplify the probability space. In the probability space $(S, \mathscr{A}, P)$ of Example 17.1, $\mathscr{A}$ is the additive class generated by the partition $\mathscr{P}_o$ consisting of all 36 distinct one-element subsets of S, $\mathscr{A} = \mathscr{A}(\mathscr{P}_o)$. Instead, we can use a smaller (or 'coarser') partition $\mathscr{P}_1$ as the defining partition, where $\mathscr{P}_1 = \{A_2, A_3, \ldots, A_{12}\}$, and let $\mathscr{A}_1 = \mathscr{A}(\mathscr{P}_1)$. The two partitions $\mathscr{P}_o$ and $\mathscr{P}_1$ are illustrated in Figure 20.1.

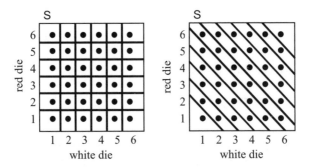

Figure 20.1   The two partitions $\mathscr{P}_o$ and $\mathscr{P}_1$

We have then a different probability space, $(S, \mathscr{A}_1, P_1)$, whose elementary sets are $A_2, \ldots, A_{12}$. The space of elements S remains the same. $\mathscr{A}_1$ is the additive class generated by $\mathscr{P}_1$, and P_1 is the probability distribution over $\mathscr{A}_1$. In accordance with Theorem 20.2, P_1 is fully specified by the requirement that $P_1(A_i) = P(A_i)$, $i = 2, \ldots, 12$, as given in Equation (20.2).

Probability distributions are often characterized by the probabilities that get assigned to some partition—usually, to the partition that generates the completely additive class that defines the probability space. In the above example, the distribution P in the probability space $(S, \mathscr{A}, P)$ is a *uniform* or *equal-probability* distribution, because it assigns the same probability to every set in $\mathscr{P}_o$. This is not true for the distribution P_1 in $(S, \mathscr{A}_1, P_1)$. Example 20.3 illustrates how a probability space in which P is a particular distribution can also exhibit a quite different distribution (over a different partition).

Queries

20.1 In each of the following, consider the appropriate p.e. and determine whether the possible outcomes can be considered equally likely:

  a) A stroboscopic (high-speed) photograph is taken of the rim of a rapidly turning wheel. The wheel has six spokes that are numbered, and the photograph covers an area which cannot contain more than one spoke. The possible outcomes are that spoke 1, 2, 3, 4, 5, or 6 show in the photograph.

  b) A checker is dropped from a six-foot height onto a checkerboard of 64 equal-sized squares. Of interest is on which square the major part of the checker comes to rest.

  c) A witness who claims to be able to identify the perpetrator of a particular crime is brought to a lineup of five suspects. Of interest is which of the suspects will be identified by the witness.

  d) A machine will fail if any one of three different subsystems fails. Of interest is which of the three subsystems fails first and thus causes the machine to fail.

[44]

20.2 A p.e. $\mathfrak{E}_1$ has ten equally likely possible outcomes. Another p.e., $\mathfrak{E}_2$, has 11 equally likely possible outcomes. For each of the following, determine whether it is true, possibly true, or false.

  a) $\mathfrak{E}_1$ and $\mathfrak{E}_2$ are probabilistically equivalent.

  b) $\mathfrak{E}_2$ can be redefined (by suppressing one of the possible outcomes) to make it probabilistically equivalent to $\mathfrak{E}_1$.

  c) $\mathfrak{E}_1$ can be redefined (by introducing an additional possible outcome) to make it probabilistically equivalent to $\mathfrak{E}_2$.

  d) $\mathfrak{E}_1$ can be redefined by adding a coin toss to the experimental procedure and appropriately revising the possible outcomes, so as to make the resulting p.e. probabilistically equivalent to $\mathfrak{E}_2$.

[250]

Conditional Probability

Once a particular probability problem has been developed to the point where a conceptual model has been constructed and the probability space has been defined, then this p.e. and this probability space should be retained while any additional steps are carried out that may be required to solve the problem. But, after the probability space has been defined, it can happen that the need arises to introduce some restrictions into the model. For instance, there may be some event in the initial model, the occurrence of which is later to be taken as a precondition for various further considerations. This would seem to call for a revision of the probabilistic experiment in order to take into account such a new specification.

Example 21.1

For the experiment of throwing a red and a white die in order to determine the number of dots that will be showing on each, we have a clearly developed conceptual model, and the associated probability space was discussed in Example 17.1. Suppose now that we roll two dice in the context of a board game that has this special rule:

A six must show on one of the dice in order for a player to be able to start, and his starting move is then determined by the number of dots showing on the other die.

We are interested in what particular starting move a player will make. Do we have to define a new experiment in order to treat this case? After all, a throw in which no six occurs is now not of interest—we might regard it an unsatisfactorily performed experiment. It turns out, however, that we can describe this new state of affairs in terms of our previously discussed model and sample space. We merely need a way of specifying that we now *require a particular event to occur with certainty*, namely the event 'at least one six is thrown'.

Generalizing the situation described in Example 21.1, consider a p.e. 𝕰 with associated probability space (S, $\mathscr{A}$, P). Suppose our interest in 𝕰 changes: The results of this p.e. are now of interest only if a particular event 'A' occurs. In a sequence of independent, identical repetitions of this experiment, this means that we disregard all those executions of the experiment in which A does not occur. The relative frequencies of occurrence of various events can still be observed, but these are now 'conditional' frequencies, *subject to the condition that the event A occurs*. For instance, the relative frequency of the event A itself, under this condition, clearly is 1. It can also be readily seen

Probability Concepts and Theory for Engineers, First Edition. Harry Schwarzlander.
© 2011 John Wiley & Sons, Ltd. Published 2011 by John Wiley & Sons, Ltd.

that the relative frequency of any event that is mutually exclusive of A is zero, under the condition that A occurs.

Thus, the specification that event A must occur does not necessitate a revision of $\mathfrak{E}$ as initially formulated—all that is needed is to augment the conceptual model by introducing the notion of '*conditional relative frequency*'. In the same way, rather than revising the probability space (S, $\mathscr{A}$, P), it is possible to augment it by defining additional probability values on the class $\mathscr{A}$, called *conditional probabilities*.

Suppose that the *conditional* relative frequency of some event B of $\mathfrak{E}$ is of interest, under the condition that A occurs. Then this is the relative frequency of the joint occurrence of A and B, among those executions of $\mathfrak{E}$ in which A occurs. In terms of ordinary (or 'unconditional') relative frequencies, it can therefore be expressed as rf(AB)/rf(A). Transferring this result into the mathematical model we have:

(Definition) 21.1
Given a probability space (S, $\mathscr{A}$, P), with A $\in \mathscr{A}$ such that $P(A) \neq 0$. For every B $\in \mathscr{A}$, the *conditional probability of* B *given* A is

$$P(B|A) \equiv \frac{P(B \cap A)}{P(A)}$$

$P(B \mid A)$ remains undefined if $P(A) = 0$.

Assuming $P(A) \neq 0$, it follows immediately that $P(AB|A) = P(B|A)$.

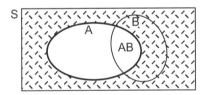

Figure 21.1   The condition 'event A must occur'

A Venn diagram as shown in Figure 21.1 can be used to visualize the significance of Definition 21.1. It should be clear that the conditional probability $P(B|A)$ represents an idealization of the relative frequency of occurrence of B in many repeated executions of the experiment, but counting only those executions in which A occurs. We noted above that, in $\mathfrak{E}$, the relative frequency of A, under the condition that A must occur, is 1. In terms of conditional probability, assuming $P(A) \neq 0$, we now have $P(A|A) = P(A)/P(A) = 1$. Also,

$$P(A^c|A) = P(A^c A|A) = \frac{P(\phi)}{P(A)} = 0.$$

Example 21.1 (Continued)
In the probability space (S, $\mathscr{A}$, P) described in Example 17.1, let

$$A \leftrightarrow \text{'at least one die shows six dots'}.$$

The set A is shown shaded in the Venn diagram of Figure 21.2. Now we can introduce the conditional probabilities $P(|A)$ that express probabilities of various events under the condition that A occurs. For instance, if $\{\xi_a\}$ is some one-element subset of A, then the conditional probability of $\{\xi_a\}$, given A, is

$$P(\{\xi_a\}|A) = \frac{P(\{\xi_a\} \cap A)}{P(A)} = \frac{P(\{\xi_a\})}{P(A)} = \frac{1/36}{11/36} = \frac{1}{11}.$$

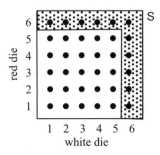

Figure 21.2   Venn diagram

In terms of the conceptual model, suppose $\{\xi_a\}$ represents the event 'red six, white two'. Our result means that the relative frequency of this event, in a large number of throws, is approximately 1/11 if we count only those throws in which at least one 'six' appears.

Here is another illustration of the use of conditional probability.

Example 21.2

Two coins, a penny and a nickel, are tossed simultaneously, and it is then observed which side faces up on each of the coins. We ask: What is the probability of obtaining two heads, under the condition that the penny shows head? This experiment has four possible outcomes:

$$\xi_1 \leftrightarrow \text{'head on both'}$$
$$\xi_2 \leftrightarrow \text{'head on penny, tail on nickel'}$$
$$\xi_3 \leftrightarrow \text{'tail on penny, head on nickel'}$$
$$\xi_4 \leftrightarrow \text{'tail on both'}$$

From the nature of the experiment we can assume equally likely outcomes. Let

$$H_p \leftrightarrow \text{'penny shows head'}$$

that is, $H_p = \{\xi_1, \xi_2\}$. Then

$$P(\{\xi_1\}|H_p) = \frac{P(\{\xi_1\} \cap H_p)}{P(H_p)} = \frac{P(\{\xi_1\})}{P(H_p)} = \frac{1/4}{1/2} = \frac{1}{2}.$$

Next, we obtain the probability of obtaining two heads, under the condition that at least one of the coins shows head. Let

$$H \leftrightarrow \text{'at least one coin shows head'}$$

that is, $H = \{\xi_1, \xi_2, \xi_3\}$. Then,

$$P(\{\xi_1\}|H) = \frac{P(\{\xi_1\} \cap H)}{P(H)} = \frac{P(\{\xi_1\})}{P(H)} = \frac{1/4}{3/4} = \frac{1}{3}.$$

The distinction between these two cases can be illustrated by means of the Venn diagrams shown in Figure 21.3. In the first case, we obtain the probability of $\{\xi_1\}$ under the restriction that we will consider only those experiments in which H_p occurs. In the second case, we obtain the probability of $\{\xi_1\}$ under the restriction that we will consider only those experiments in which H occurs.

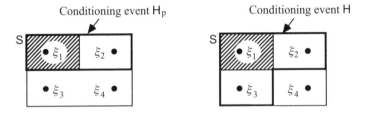

Figure 21.3 Two different conditioning events

At times, a problem may be stated in such a form that the conceptual basis for conditional probability is not clearly evident. For instance, the two questions considered in Example 21.2 might be stated in the following way:

(a) If it is known that the penny shows 'head,' what is the probability that both coins show 'head'?
(b) If it is known that one of the two coins shows 'head', what is the probability of a 'head' on both?

Noting that our results were 1/2 and 1/3, respectively, it may at first glance seem surprising that there should be any difference in the two cases. If at least one coin shows 'head', why should the additional knowledge that it is the penny affect the probability of both coins showing heads? Such incorrect intuition arises from an incorrect view of the problem. We must first bring the problem into the format required by our conceptual model, or we do not have a legitimate probability problem.

There are two distinct types of conceptual errors possible. First, we might be tempted to limit our thinking to only a single experiment, that is, a single and isolated execution: A pair of coins is tossed *once*. If one of the coins shows 'head', why should knowing that it is the penny make any difference? This way of thinking is incorrect because 'probability' corresponds to a relative frequency. The single toss of two coins must be thought of as a representative case out of many tosses.

Secondly, if we say that the experiment has been performed but we know only a *partial* result— that at least one of the coins shows a 'head', or that the penny shows a 'head'—we are violating the definition of a p.e., which specifies that upon execution of the experiment, *all* the observable features that are of interest are noted, and therefore known. In other words, the possibility of partial knowledge is specifically excluded. Statement (a), above, suggests that upon performing the experiment only the penny is observed initially. We must reinterpret such a problem statement (that is suggestive of a partially known result): Instead of a partially known result, we stipulate an *event that is required to occur*. Conditional probability then makes sense and the significance of the result that is obtained from the probability calculation is easily understood. In Example 21.2, for instance, we find that two heads occur more frequently if we restrict ourselves to those tosses in which the penny shows 'head', than if we restrict ourselves only to those tosses in which *at least* one head occurs.

Queries

21.1 Given a probability space (S, $\mathscr{A}$, P) with A, B $\in \mathscr{A}$, and such that $P(A)$, $P(B)$, $P(AB) \neq 0$. For each of the following statements, determine whether it is true, possibly true, or false:

(a) $P(S|A) = P(A|S)$

(b) $P(A \cup B|AB) = \dfrac{1}{2}[P(S|A) + P(S|B)]$

(c) $P(A - B|B) + P(B - A|A) = 1$

(d) $P(A \Delta B|A) = P(B^c|A)$

(e) $\dfrac{1}{P(A|B)} + \dfrac{1}{P(B|A)} = \dfrac{1}{P(AB|A \cup B)} + 1.$ [238]

21.2 I have three coins in my pocket that are indistinguishable to the touch but only two of them are normal coins, the other having a 'head' on each side. I take a coin out of my pocket, toss it, and obtain a 'head'. What is the probability that the coin I have tossed is in fact the double-headed one?   [13]

Conditional Probability Distributions

The mathematical significance of Definition 21.1 needs to be examined further. We note that if $P(A) \neq 0$ in Definition 21.1, then the 'conditional probability given A', $P(\cdot|A)$, is defined for every set belonging to $\mathscr{A}$.

Theorem 22.1
Given a probability space $(S, \mathscr{A}, P)$, with $A \in \mathscr{A}$ such that $P(A) \neq 0$. Then the conditional probabilities $P(\cdot|A)$ constitute a probability distribution over $\mathscr{A}$.

Proof
We must show that $P(\cdot|A)$ satisfies Definition 16.1. $P(\cdot|A)$ is clearly non-negative, with

$$P(S|A) = \frac{P(S \cap A)}{P(A)} = \frac{P(A)}{P(A)} = 1.$$

In order to show that $P(\cdot|A)$ is countably additive, we note that if $B_1, B_2, \ldots$ are disjoint sets belonging to $\mathscr{A}$, then so do $B_1 \cap A, B_2 \cap A, \ldots$ belong to $\mathscr{A}$ and are disjoint. Since P is countably additive,

$$P\left(\biguplus_i (B_i \cap A)\right) = \sum_i P(B_i \cap A)$$

and

$$P\left(\biguplus_i (B_i \cap A)|A\right) = \frac{P\left(\biguplus_i (B_i \cap A)\right)}{P(A)} = \frac{\sum_i P(B_i \cap A)}{P(A)}$$

$$= \sum_i P((B_i \cap A)|A).$$

Probability Concepts and Theory for Engineers, First Edition. Harry Schwarzlander.
© 2011 John Wiley & Sons, Ltd. Published 2011 by John Wiley & Sons, Ltd.

Thus, we may think of $P(\,|A)$ as the probability measure for a modified probability space, in which the basic space of elements is merely A. It is called a *conditional probability function*, or *conditional probability distribution*. On a Venn diagram this can be visualized by temporarily making the set A serve as the space of the problem so that the boundary of S is in effect brought inward to coincide with the boundary of A (see Figure 22.1). The probabilities of all subsets of A have to be normalized with respect to $P(A)$. Points outside A can then be thought of as representing unsatisfactory performances of the experiment. Under the condition that A occurs, the probability of the event AB (see Figure 22.1) becomes $P(AB|A) = P(AB)/P(A)$. This is the same as $P(B|A)$ since points of B that do not belong to A now represent outcomes that cannot arise. However, they do remain possible outcomes of the probability space, and belong to the set A^c to which $P(\,|A)$ assigns zero probability.

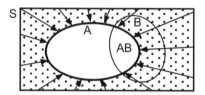

Figure 22.1   Sample space and Venn diagram

For a given probability space (S, $\mathscr{A}$, P), therefore, we can have many probability distributions—the ordinary (unconditional) distribution P, and various conditional probability distributions. Nevertheless, we retain the notation (S, $\mathscr{A}$, P) for a probability space, since all conditional distributions that can arise are implicit in that space. Henceforth, any statement made about a probability distribution also applies to conditional probability distributions. Besides, an unconditional probability function P can also be thought of as $P(\,|S)$.

Frequently, some of the probabilities that are initially known in a given problem are actually conditional probabilities. This must be recognized when the problem is begun. Oversights in this respect are a common source of error.

Example 22.1

The probability of obtaining a 'six' upon throwing a die is to be determined. The procedure is to select the die blindly from a group of three dice, two of which are 'fair', the third is loaded. For the loaded die, it is known that the relative frequency of occurrence of a 'six' is 1/4. The dice are indistinguishable during selection. The selected die is thrown and it is observed whether or not a 'six' appears.

For each die the possibilities are: 'a six', and 'not a six'. Since it is only of interest whether or not a six is thrown, we might be tempted to define a sample space of two elements, corresponding to the possible outcomes 'six' and 'not six'. But this will not bring into consideration both of the random mechanisms that are involved here—the random selection of a die, and the throwing of a die—and therefore will not permit the necessary probability calculation to be carried out. We have to adjust the conceptual model such that 'die number one (two, three) selected' each becomes a property of interest. The sample space then consists of six points, $S = \{\xi_a, \xi_b, \xi_c, \xi_d, \xi_e, \xi_f\}$, corresponding to the six possible outcomes, as shown in Figure 22.2.

The probabilities of the following events are 1/3:

$$D_1 = \{\xi_a, \xi_d\}; \quad D_2 = \{\xi_b, \xi_e\}; \quad D_3 = \{\xi_c, \xi_f\}$$

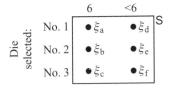

Figure 22.2   The sample space

These are the only values of the probability function P that are initially known, and they do not permit determination of the probability of the event of interest,

$$A = \{\xi_a, \xi_b, \xi_c\}.$$

The probability of getting a 'six' with a specified die, say No. 1, is *not* $P(\{\xi_a\})$; it is the conditional probability $P(\{\xi_a\}|D_1)$. Under the condition D_1 (see Figure 22.3) our experiment reduces to the single throw of an ordinary die (die number 1), each of whose faces is assumed to come up with equal likelihood. Therefore, under condition D_1, the probability of a 'six' is 1/6.

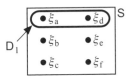

Figure 22.3   The condition D_1

Whereas the description of P is initially incomplete, we are completely informed about the conditional probability functions $P(\ |D_1)$, $P(\ |D_2)$, and $P(\ |D_3)$:

$$P(\{\xi_a\}|D_1) = 1/6, \quad P(\{\xi_d\}|D_1) = 5/6$$
$$P(\{\xi_b\}|D_2) = 1/6, \quad P(\{\xi_e\}|D_2) = 5/6$$
$$P(\{\xi_c\}|D_3) = 1/4, \quad P(\{\xi_f\}|D_3) = 3/4.$$

This information can be used to fill in the missing values of the function P. Thus, from Definition 21.1 and (8.5b),

$$P(\{\xi_a\}|D_1)P(D_1) = P(\{\xi_a\} \cap D_1) = P(\{\xi_a\}).$$

In a similar way $P(\{\xi_b\}|D_2)\, P(D_2) = P(\{\xi_b\})$, and $P(\{\xi_c\}|D_3)\, P(D_3) = P(\{\xi_c\})$. Then,

$$P(A) = P(\{\xi_a\}) + P(\{\xi_b\}) + P(\{\xi_c\})$$
$$= \left(\frac{1}{6} \cdot \frac{1}{3}\right) + \left(\frac{1}{6} \cdot \frac{1}{3}\right) + \left(\frac{1}{4} \cdot \frac{1}{3}\right) = \frac{7}{36}.$$

The significant features of the above example are:

a) The values of P are known for the members of a partition $\mathscr{P}$.
b) The probability of an event A is to be found, where A is not a union of members of $\mathscr{P}$.
c) For every set B $\in \mathscr{P}$ with nonzero probability, the conditional probability $P(A|B)$ is known.

The solution to such a problem can be expressed by a general formula as follows.

Theorem 22.2: 'Rule of Total Probability'.
Given a probability space (S, $\mathscr{A}$, P), a set A $\in \mathscr{A}$ and a partition $\mathscr{P} \subset \mathscr{A}$. Then

$$P(A) = \sum_{\substack{B \in \mathscr{P} \\ P(B) \neq 0}} P(A|B)P(B)$$

(22.1)

Proof
The class $\{AB: B \in \mathscr{P}\}$ is a partition of A (possibly augmented by ϕ); so that

$$P(A) = \sum_{B \in \mathscr{P}} P(AB)$$

But it may not be necessary to sum over *all* sets B $\in \mathscr{P}$. If any set B has $P(B) = 0$, then $P(AB) = 0$ also, and the corresponding term can be deleted from the summation. Therefore,

$$P(A) = \sum_{B \in \mathscr{P}} P(AB) = \sum_{\substack{B \in \mathscr{P} \\ P(B) \neq 0}} P(AB) = \sum_{\substack{B \in \mathscr{P} \\ P(B) \neq 0}} P(A|B)P(B).$$

We have seen that a conditional probability function $P(\,|A)$ has the same properties as an unconditional probability function P. An identity involving an arbitrary probability function P can therefore also be expressed in terms of $P(\,|A)$. Thus, from $P(B^c) = 1 - P(B)$ it follows that $P(B^c|A) = 1 - P(B|A)$. Or, consider the conditional probabilities of two *disjoint* sets B, C. These can be *added* to give

$$P(B \uplus C|A) = P(B|A) + P(C|A).$$

(22.2)

Another example is Equation (18.4), which can be expressed in terms of conditional probabilities to give

$$P(B \cup C|A) = P(B|A) + P(C|A) - P(BC|A).$$

From this follows a slight generalization of Equation (22.2), namely:

$$P(B \cup C|A) = P(B|A) + P(C|A) \quad \text{iff } P(ABC) = 0.$$

(22.3)

Or, Definition 21.1 can be conditioned by some event C $\in \mathscr{A}$, $P(C) \neq 0$, giving:

$$P(B|AC) = \frac{P(AB|C)}{P(A|C)}$$

(22.4a)

If Equation (22.1) is similarly modified, it becomes

$$P(A|C) = \sum_{\substack{B \in \mathscr{P} \\ P(BC) \neq 0}} P(A|BC)P(B|C). \tag{22.4b}$$

An expression such as $P(B|AC)$ can be read, 'the conditional probability of B, given A *and* C'.

On the other hand, there is usually no immediately recognizable significance to a sum of conditional probabilities if the conditioning events are not the same throughout. Note that in Equation (22.1) we do have a sum involving conditional probabilities where the conditioning events are all different; however, each of the conditional probabilities is weighted by the unconditional probability of the conditioning event.

Conditional probabilities are often useful for expressing the probability of an intersection of events. Consider two events A_1, A_2 with nonzero probability. First, we have, from the definition of conditional probability,

$$P(A_1 A_2) = P(A_1|A_2)P(A_2) = P(A_2|A_1)P(A_1). \tag{22.5}$$

This can now be extended to multiple intersections to give the following general *product rule* or 'chain rule':

$$P(A_1 A_2 \ \ldots \ A_n) = P(A_1|A_2 \ \ldots \ A_n) \cdot P(A_2|A_3 \ldots A_n) \cdot \ \ldots \ \cdot P(A_{n-1}|A_n) \cdot P(A_n). \tag{22.6}$$

Furthermore, in Equation (22.5) we have a relation between $P(A_1 \mid A_2)$ and $P(A_2 \mid A_1)$. This can be combined with Equation (22.1) to give

$$P(A_2|A_1) = \frac{P(A_1|A_2)P(A_2)}{\displaystyle\sum_{\substack{B \in \mathscr{P} \\ P(B) \neq 0}} P(A_1|B)P(B)} \tag{22.7}$$

for any partition $\mathscr{P}$. This relation is often called *Bayes' Rule*.

Example 22.2

Consider data transmission between two computers in different geographical locations. Data is transmitted in 'blocks', that is, in fixed-length sequences of digits. Suppose that the transmission may be routed over any one of three different routes, the selection being made automatically by a route selector according to existing traffic conditions, every time transmission of a block is initiated. Transmission takes place over the three routes various fractions of the time, as follows:

Route A: 60% of the time
Route B: 30% of the time
Route C: 10% of the time

Transmission errors can occur that render a block completely useless. The probability of a block experiencing such an error depends on the route. It is

0.001 when route A is used
0.002 when route B is used
0.003 when route C is used.

What is the probability that an error, when it does occur, arises on route C?

Solution:
The first step is the clarification of the p.e. that underlies this problem. As the purpose of the experiment we may take the determination, for a given data block, how it is routed (A, B, or C) and whether or not it was received in error. We skip the details of the procedure, assuming merely that the necessary observations can be made and that statistical regularity holds with respect to the events of interest (at least, during periods of similar traffic conditions). There are six possible outcomes, as indicated in Figure 22.4. Let

$$A \leftrightarrow \text{'Route A is used'}$$
$$B \leftrightarrow \text{'Route B is used'}$$
$$C \leftrightarrow \text{'Route C is used'}$$
$$E \leftrightarrow \text{'block is received in error'.}$$

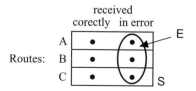

Figure 22.4   Sample space and Venn diagram

Since {A, B, C} is a partition, Equation (22.7) can be applied directly to give

$$P(C|E) = \frac{P(E|C)P(C)}{P(E|A)P(A) + P(E|B)P(B) + P(E|C)P(C)}$$

$$= \frac{0.003 \times 0.1}{0.001 \times 0.6 + 0.002 \times 0.3 + 0.003 \times 0.1} = \frac{0.0003}{0.0015} = 0.2.$$

We see that in the long run 1/5 of all transmission errors are caused on route C.

For another application of conditional probability we return to Example 19.2.

Example 22.3
In the second part of Example 19.2, let

$$A \leftrightarrow \text{'bag A remains'}$$
$$B \leftrightarrow \text{'bag B remains'}$$
$$C \leftrightarrow \text{'a white and a black ball are drawn'}$$

It is assumed that $P(A) = P(B) = 1/2$. What is wanted is $P(A|C)$, which, using Equation (22.7), can be written

$$\frac{P(C|A)P(A)}{P(C|A)P(A) + P(C|B)P(B)}. \tag{22.8}$$

For convenience, assume the two balls are drawn in succession, and let

$$W_i \leftrightarrow \text{'}i\text{th ball drawn is white' } (i = 1, 2)$$
$$L_i \leftrightarrow \text{'}i\text{th ball drawn is black' } (i = 1, 2).$$

Then $C = W_1 L_2 \uplus L_1 W_2$, and from Equation (22.2),

$$P(C|A) = P(W_1 L_2|A) + P(L_1 W_2|A)$$
$$P(C|B) = P(W_1 L_2|B) + P(L_1 W_2|W)$$

and application of Equation (22.6) to $P(W_1 L_2|A)$ gives $P(W_1 L_2|A) = P(L_2|W_1 A)P(W_1|A)$.

Now, we are told that balls are removed from the bag 'blindly', i.e., without bias. Since five out of the ten balls in bag A are white, $P(W_1|A) = 5/10 = 1/2$. With a white ball removed, five of the remaining nine balls are black, so that $P(L_2|W_1 A) = 5/9$. Then $P(W_1 L_2|A) = (5/9) \cdot (1/2) = 5/18$; and by symmetry, $P(L_1 W_2|A) = 5/18$ also, giving $P(C|A) = 5/9$.

An analogous calculation yields $P(C|B) = 16/45$. Substitution into Equation (22.8) then gives the desired result:

$$P(A|C) = \frac{(5/9)(1/2)}{(5/9)(1/2) + (16/45)(1/2)} = \frac{25}{41}.$$

As was to be expected, $P(A|C)$ is somewhat larger than $P(B|C) = 1 - P(A|C) = 16/41$.

Queries

22.1 Three urns are filled with a mixture of green and white balls that are indistinguishable to the touch, as follows:

urn no. 1:	three green balls,	nine white balls;
urn no. 2:	four green balls,	eight white balls;
urn no. 3:	six green balls,	six white balls.

By a suitable procedure that does not favor any one urn, the experimenter chooses one of the urns. Then she blindly picks a ball from the chosen urn.
a) What is the probability that the ball obtained in this manner is green?
b) If a green ball is picked, what is the probability that it came from urn no. 1?
[150]

22.2 Given an experiment $\mathfrak{E}$ with associated probability space $(S, \mathscr{A}, P)$, and some conditional probability distribution $P(\ |A)$ over $\mathscr{A}$. For each of the following statements, determine whether it is true, possibly true, or false:
a) The conditioning event can be an event with respect to which there is no statistical regularity.
b) Suppose there is some event B in the p.e. $\mathfrak{E}$ with respect to which there is no statistical regularity. Then, for some $A \in \mathscr{A}$, $P(B|A)$ is defined.
c) If there is a set $C \in \mathscr{A}$ whose conditional probability $P(C|A)$ equals its unconditional probability $P(C)$, then $P(C^c|A) = P(C^c)$.

d) The conditional probability of the empty set is zero.

e) Assume that $P(A^c) \neq 0$. Then for every $B \in \mathscr{A}$, $P(B|A) + P(B|A^c) = 1$.   [135]

22.3 Given a probability space $(S, \mathscr{A}, P)$ with $A, B, C \in \mathscr{A}$, where $A \subset B \subset C$ and $P(A) \neq 0$. For each of the following, determine whether it is true, possibly true, or false:

a) $\dfrac{P(A)}{P(C)} = P(A|B)P(B|C)$

b) $\dfrac{P(C)}{P(A)} = P(C|A)P(B|A)$

c) $\dfrac{P(C)}{P(B)} = \dfrac{P(A|B) + P(C|B) - 1}{P(A|C)}$.   [244]

Independent Events

So far we have used the word 'independence' in connection with distinct probabilistic experiments. In that context, 'independence' expresses a relationship within the conceptual world. A similar notion arises in the mathematical model, where we introduce the following precise criterion for declaring two *events* to be independent.

> **(Definition) 23.1**
> Given a probability space $(S, \mathscr{A}, P)$. Two events $A_1, A_2 \in \mathscr{A}$ are termed *independent* if and only if
>
> $$P(A_1 \cap A_2) = P(A_1)P(A_2) \tag{23.1}$$

At first glance, the motivation for calling two events independent if they satisfy relation (23.1) may escape the reader. It becomes clearer through the use of conditional probability. Assume $P(A_1) \neq 0$, then Equation (23.1) can be written

$$P(A_2|A_1) = P(A_2) \tag{23.2}$$

In words, the specification 'the event A_1 must occur' does not affect the probability of event A_2—the probability of event A_2 *does not depend* on whether or not event A_1 is required to occur. We may view this in terms of the conceptual model. Given a large number of independent executions of the same underlying p.e., we are saying that the relative frequency of a particular event A_2 is not affected if we restrict our attention to only those experiments in which the event A_1 occurs. Use of the word 'independent' for the property (23.1) is therefore, consistent with the concept of independence introduced earlier. If $P(A_2) \neq 0$, the positions of A_1 and A_2 in Equation (23.2) can of course be interchanged.

Definition 23.1 also permits A_1 to stand for ϕ or S, and the same is true for A_2. From the mathematical point of view there is no reason to exclude these possibilities; the Definition is simple as it stands, and it does not create mathematical inconsistencies. However, these special cases have little intuitive appeal when viewed in the conceptual model world. For, we find that the Definition tells us that any event A is independent of the certain event, and also of the impossible event. Even more surprising is the implication of the relations

$$P(\phi \cap \phi) = P(\phi)P(\phi), \quad \text{and} \quad P(S \cap S) = P(S)\,P(S).$$

Probability Concepts and Theory for Engineers, First Edition. Harry Schwarzlander.
© 2011 John Wiley & Sons, Ltd. Published 2011 by John Wiley & Sons, Ltd.

We see that the impossible event (as well as any other event having zero probability) is 'independent of itself', and also the certain event (as well as any other event having probability 1) is 'independent of itself'. Here, we have again an example of the way in which some aspect of the mathematical model can influence the way we think or speak about a situation in the conceptual world.

Independent events must not be confused with disjoint events.

Example 23.1

In the probability space for the throw of two dice (Example 17.1), consider the following two events (see Figure 23.1):

$$W_1 \leftrightarrow \text{'one dot shows on the white die'}$$

$$R_1 \leftrightarrow \text{'one dot shows on the red die'}$$

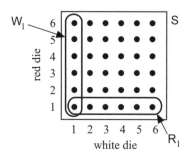

Figure 23.1   Throw of two dice

These two events are *independent*, since

$$P(W_1 \, R_1) = \frac{1}{36} = P(W_1) \, P(R_1).$$

Clearly, W_1 and R_1 are not disjoint.

Note that independence of two events is a property that arises from the way the probability function is defined, but disjointness is not. In the above Example, for instance, *any* two six-element events are independent if they have exactly one element in common. More specifically, we have:

Theorem 23.1

Given two sets A, B in a probability space $(S, \mathscr{A}, P)$, such that $P(A)$, $P(B) \neq 0$. Then A and B cannot be both disjoint and independent.

Proof

Suppose A, B are independent. Then $P(AB) = P(A) \, P(B) \neq 0$, so that $AB \neq \phi$; i.e., A, B are not disjoint.

Definition 23.1 specifies the *mathematical property of independence*, and we have seen how it corresponds readily to an intuitive idea of independence of events in the conceptual model (except in those cases where ϕ or S is one of the events in question). How is it used? It turns out that independence is an important consideration in many probability problems. There are two distinct ways in which this property can enter into a problem: *by assumption*, or *by computation*.

Example 23.2

Consider again the probability space (S, $\mathscr{A}$, P) for the throw of two distinguishable dice, as described in Example 17.1. Let us examine in detail how one arrives at the probability distribution $P(\{\xi\}) = 1/36$, for all ξ. The first assumption is that the two dice are 'fair' and are shaken well before they are thrown, so that for each die, when it comes to rest, the six faces are equally likely to end up on top. Using the notation of Example 23.1, this gives $P(W_1) = 1/6$, $P(R_1) = 1/6$, etc. Then we *assume independence* between the number that comes up on one die and the number on the other die. The justification for this assumption is that it 'seems reasonable'—it corresponds to our understanding of the physical nature of the underlying p.e. Now, if the element ξ_1 corresponds to the possible outcome 'both dice show one dot', then

$$P(\{\xi_1\}) = P(W_1\,R_1) = P(W_1)P(R_1) = \frac{1}{6}\cdot\frac{1}{6} = \frac{1}{36}.$$

In the same manner, 1/36 is obtained for the probability of *every* elementary set in this probability space.

Thus, an awareness of some pertinent features of a probabilistic experiment can lead us to recognize that particular events may reasonably be said to be independent. In this way, additional facts can be introduced into the mathematical model. These can help in formulating the probability space, or they can simplify the problem at hand.

Example 23.2 (continued)

Assume now that the probability distribution P is completely defined and consider the following two events:

A $\leftrightarrow$ 'the sum of the dots on both dice is 7'

B $\leftrightarrow$ 'the difference in the number of dots on the two dice is either 0 or 3'

It is not immediately clear from the conceptual model whether or not these events are independent. Suppose this question is to be resolved. This can be done by *computation*:

$$P(A) = 1/6, \quad P(B) = 1/3, \quad P(AB) = 1/18 = P(A)P(B)$$

Therefore, A and B are independent events.

An important property associated with independent events is the following.

Theorem 23.2

If A and B are independent events, then

A, B^c is a pair of independent events;

A^c, B is a pair of independent events;

and A^c, B^c is a pair of independent events.

Proof

Each case is proven in analogous fashion. Consider A and B^c. The disjoint union $AB \uplus AB^c = A$ makes it possible to write

$$P(AB^c) = P(A) - P(AB) = P(A) - P(A)P(B)$$
$$= P(A)[1 - P(B)] = P(A)P(B^c).$$

This result is extended in Problem 23.9.

Example 23.3

Suppose that within 10 000 hours of operation, the probability of failure of a particular type of transistor when operating under certain specified conditions is 0.05. What is the probability that an electronic device containing two such transistors, operating under these same conditions, will operate for at least 10 000 hours without failure?

The p.e. involves operating the device for 10 000 hours and observing whether failure has occurred by that time. We make two assumptions:

1) Failure of the device occurs only due to failure of one of the transistors, not due to failure of any other component. In other words, if a different type of failure occurs, the experiment is considered unsatisfactorily performed. Therefore, the procedure must include suitable monitoring of the device so that the reason for a failure can be established.
2) The two transistors are subject to independent failure—the failure mechanisms affecting the two transistors are independent. Thus, we are, for instance, excluding the possibility that, as one of the transistors degrades slightly, the operating conditions of the other transistor changes, thus affecting its probability of failure.

The sample space is then as shown in Figure 23.2, where

$$F_1 \leftrightarrow \text{'Transistor \#1 fails within 10 000 hours'}$$
$$F_2 \leftrightarrow \text{'Transistor \#2 fails within 10 000 hours'}$$

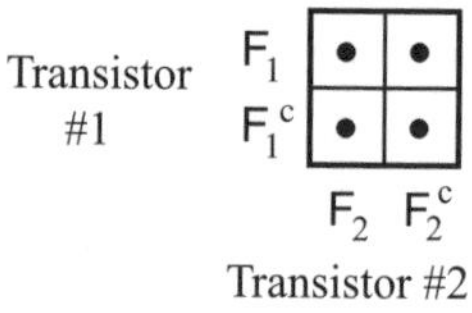

Figure 23.2   Sample space

In order to translate the independence of the failure mechanisms into the mathematical model we specify F_1 and F_2 to be independent events. From Theorem 23.2, F_1^c and F_2^c are also independent events and we have

$$P(F_1^c F_2^c) = P(F_1^c)P(F_2^c) = (1 - 0.05)(1 - 0.05) = 0.9025.$$

We could have tried to set this up somewhat differently. Suppose we specify the p.e. in such a way that the procedure simply calls for observation of the device until failure of a transistor occurs, but no longer than 10 000 hours. In that case, there would be only *two* possible outcomes: 'failure' or 'no failure' of the device within the first 10 000 hours (due to transistor failure). But this model would not bring into play the information that is available about the random mechanism involved here: the probability of failure of transistors #1 and #2 in 10 000 hours, and the independence of these failures. Thus, the desired probability cannot be obtained with this formulation.

Queries

23.1 A red and a white die are thrown. Which of the following are pairs of independent events:

   a) A ↔ '*one* dot faces up on white die'
   B ↔ 'red *one* and white *one*, or red *not one* and white *two*';

   b) A ↔ 'no number greater than *two* on either die'
   B ↔ 'each die shows a *one* or a *six*';

   c) A ↔ 'no number greater than *two* shows on either die'
   B ↔ 'the sum of the dots is either 2, 5 or 9';

   d) A ↔ 'white die shows no more than *two* dots'
   B ↔ 'red die shows no more than *two* dots';

   e) A ↔ 'an unequal number of dots shows on the two dice'
   B ↔ 'the sum of the dots is *seven*'.   [38]

23.2 For each of the following cases, state whether it is necessarily true, possibly true, or false, that the two events A and B, where $B \neq \phi$, are independent as well as disjoint:

   a) $P(A) = P(B)$ b) $P(A) = 1 - P(B)$

   c) $A = S$ d) $A = \phi$.   [129]

23.3 Given a probability space $(S, \mathscr{A}, P)$ with $A, B \in \mathscr{A}$. If A, B are independent and $P(A \triangle B) = P(A)$, evaluate $P(A)$.   [198]

Classes of Independent Events

In the previous Section, independence was defined as a relation between two sets or events. If more than two sets are under consideration, this relation may hold for *each pair* of sets. We are therefore led to the following criterion for calling more than two sets 'independent'.

(Definition) 24.1

Given a probability space $(S, \mathscr{A}, P)$ and a class $\mathscr{C} = \{A_1, \ldots, A_n\} \subset \mathscr{A}, n \geq 2$. The sets in $\mathscr{C}$ are termed *pairwise independent*, or $\mathscr{C}$ is termed *a class of pairwise independent events*, if and only if every pair of sets in $\mathscr{C}$ is a pair of independent sets.

In this connection it should be noted, for given subsets A, B, C of a space S, that if A, B are independent sets, and A, C are independent sets, then B, C are *not* necessarily independent. For instance, B and C might be identical sets with probability other than 0 or 1.

There is, however, another way of extending the property of independence from two sets to several sets, resulting in a stronger requirement than expressed in Definition 24.1. This type of independence comes about in the following way. Given a class $\mathscr{C} = \{A_1, \ldots, A_n\}$, suppose that $\mathscr{C}$ is not only a class of pairwise independent events, but that the following extensions of Equation (23.1) are satisfied as well:

$$\left.\begin{aligned} P(A_1 A_2 A_3) &= P(A_1)P(A_2)P(A_3) \\ P(A_1 A_2 A_3 A_4) &= P(A_1)P(A_2)P(A_3)P(A_4) \\ & \cdot \ \cdot \ \cdot \ \cdot \ \cdot \ \cdot \ \cdot \ \cdot \ \cdot \ \cdot \ \cdot \\ P\left(\bigcap_{i=1}^{n} A_i\right) &= \prod_{i=1}^{n} P(A_i) \end{aligned}\right\} \tag{24.1}$$

Furthermore, suppose that the first of these relations holds for *every* choice of three sets in $\mathscr{C}$, the second relation for every choice of four sets in $\mathscr{C}$, and so on. If $\mathscr{C}$ satisfies all these requirements, it is called an *independent class,* and the sets $A_1, \ldots, A_n$ are termed *mutually independent*. Stated more compactly, we have:

(Definition) 24.2

Given a probability space $(S, \mathscr{A}, P)$ and a class $\mathscr{C} = \{A_1, \ldots, A_n\} \subset \mathscr{A}, n \geq 2$. Then $\mathscr{C}$ is termed an *independent class,* and the sets $A_1, \ldots, A_n$ are termed *mutually independent*, if and only if

Probability Concepts and Theory for Engineers, First Edition. Harry Schwarzlander.
© 2011 John Wiley & Sons, Ltd. Published 2011 by John Wiley & Sons, Ltd.

$$P\left(\bigcap_{i\in K}A_i\right) = \prod_{i\in K}P(A_i)$$

holds for *every subset* K of the set of integers $\{1, \ldots, n\}$ satisfying $2 \leq |K| \leq n$.

Example 24.1

Consider again the experiment of throwing two dice (Example 17.1). Let

$A_1 \leftrightarrow$ 'six dots on red die'

$A_2 \leftrightarrow$ 'five dots on white die'

$A_3 \leftrightarrow$ 'equal number of dots on both dice'

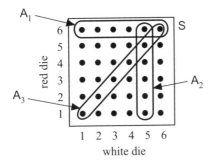

Figure 24.1   Pairwise independence

These sets, identified in Figure 24.1, are *pairwise independent*, as is easily verified, but they are not mutually independent, since

$$P(A_1 A_2 A_3) = P(\phi) = 0 \neq P(A_1)P(A_2)P(A_3).$$

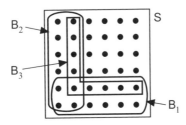

Figure 24.2   No pairwise independence

Now let (see Figure 24.2)

$B_1 \leftrightarrow$ 'one or two dots on red die'

$B_2 \leftrightarrow$ 'one or two dots on white die'

$B_3 \leftrightarrow$ 'neither die shows one dot, and at least one die shows two dots'.

Here we have

$$P(\mathsf{B}_1 \mathsf{B}_2 \mathsf{B}_3) = \frac{1}{36}, \quad \text{and also} \quad P(\mathsf{B}_1)P(\mathsf{B}_2)P(\mathsf{B}_3) = \frac{1}{36}.$$

However, $P(\mathsf{B}_1 \mathsf{B}_3) = \frac{5}{36} \neq P(\mathsf{B}_1)P(\mathsf{B}_3) = \frac{1}{3} \cdot \frac{1}{4} = \frac{1}{12}$; so that the sets B_1, B_2, B_3 are not mutually independent, and not even pairwise independent. On the other hand, consider B_1 and B_2 together with

$$\mathsf{B}_4 \leftrightarrow \text{'neither one nor five nor six dots show on either die'.}$$

This is illustrated in Figure 24.3. Here,

$$P(\mathsf{B}_1 \mathsf{B}_2 \mathsf{B}_4) = \frac{1}{36}, \quad \text{and} \quad P(\mathsf{B}_1)P(\mathsf{B}_2)P(\mathsf{B}_4) = \frac{1}{3} \cdot \frac{1}{3} \cdot \frac{1}{4} = \frac{1}{36}.$$

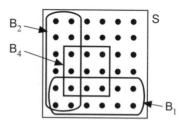

Figure 24.3   Mutual independence

Also,

$$P(\mathsf{B}_1 \mathsf{B}_2) = \frac{1}{9} = P(\mathsf{B}_1)P(\mathsf{B}_2)$$

$$P(\mathsf{B}_1 \mathsf{B}_4) = \frac{1}{12} = P(\mathsf{B}_1)P(\mathsf{B}_4)$$

$$P(\mathsf{B}_2 \mathsf{B}_4) = \frac{1}{12} = P(\mathsf{B}_2)P(\mathsf{B}_4).$$

Therefore, the sets B_1, B_2, B_4 are mutually independent.

The significance of mutual independence of several sets is best explained using conditional probabilities. Considering again a class $\mathscr{C} = \{\mathsf{A}_1, \ldots, \mathsf{A}_n\}$, let A_i be any set in $\mathscr{C}$, and let $\mathscr{D}$ be any nonempty subclass of $\mathscr{C} - \{\mathsf{A}_i\}$ satisfying the requirement that $P\left(\bigcap_{\mathscr{D}} \mathsf{A}_j\right) \neq 0$. If $\mathscr{C}$ is an independent class then Definition 24.2 implies that

$$P\left(\mathsf{A}_i \Big| \bigcap_{\mathscr{D}} \mathsf{A}_j\right) = P(\mathsf{A}_i).$$

For instance, if $\mathsf{A}_i = \mathsf{A}_1$ and $\mathscr{D} = \{\mathsf{A}_2, \mathsf{A}_3\}$ (where $P(\mathsf{A}_2\mathsf{A}_3) \neq 0$), we have

$$P(\mathsf{A}_1|\mathsf{A}_2\mathsf{A}_3) = \frac{P(\mathsf{A}_1\mathsf{A}_2\mathsf{A}_3)}{P(\mathsf{A}_2\mathsf{A}_3)} = \frac{P(\mathsf{A}_1)P(\mathsf{A}_2)P(\mathsf{A}_3)}{P(\mathsf{A}_2)P(\mathsf{A}_3)} = P(\mathsf{A}_1). \tag{24.2}$$

Therefore, for any two or more sets in $\mathscr{C}$, the right side of Equation (22.6) reduces to a product of unconditional probabilities. On the other hand, if $\mathscr{C}$ is not an independent class but merely a class of pairwise independent events, Equation (24.2) does not necessarily hold because $P(A_1 A_2 A_3)$ can then not be assumed to equal $P(A_1) P(A_2) P(A_3)$. Every set in an independent class, therefore, is not only independent of every other set (which is pairwise independence), but is also independent of all possible intersections (two at a time, three at a time, etc.) of the other sets in the class.

Clearly, if $\mathscr{C}$ is an independent class, then a subclass of $\mathscr{C}$ having two or more sets is also an independent class. An independent class of size 2 consists simply of a pair of independent sets.

In Example 24.1 we used the *computations* called for in Definition 24.2 in order to establish that the sets B_1, B_2, B_4 are mutually independent. Sometimes it is justified to introduce mutual independence of a class $\mathscr{C}$ by *assumption*. This is possible, for instance, if all those events in the underlying experiment that correspond to the sets in $\mathscr{C}$ are governed by distinct random mechanisms that can be assumed to operate independently.

Example 24.2

Three dice (red, white, and green) are thrown simultaneously. Assuming that the dice are well shaken before throwing, it is justified to consider that the numbers of dots that will show on the three dice are the results of independent random mechanisms—each die is released in a random manner and follows an unpredictable path. Collisions between the dice are also random and therefore cannot introduce a dependence between the results obtained. The events

$$R_1 \leftrightarrow \text{'one dot on red die'}$$
$$W_4 \leftrightarrow \text{'four dots on white die'}$$
$$G_2 \leftrightarrow \text{'two dots on the green die'}$$

are therefore taken to be mutually independent *by assumption*, or *by specification*. On the other hand, consider the following three events:

$$A \leftrightarrow \text{'the red and the white die show the same number of dots'}$$
$$B \leftrightarrow \text{'the red and the green die show the same number of dots'}$$
$$C \leftrightarrow \text{'the white die shows two, three, or four dots and the green die one or two'.}$$

It is not apparent from the physical nature of the experiment that these should be mutually independent events, because each die affects two of the events. However, mutual independence of these three events can be established *by calculation* (Problem 24.2).

We conclude this Section with the following useful property of independent classes, proof of which is left as an exercise (Problem 24.11).

Theorem 24.1

In a probability space $(S, \mathscr{A}, P)$, let $\mathscr{C} \subset \mathscr{A}$ be an independent class. Consider two finite classes $\mathscr{C}_1$ and $\mathscr{C}_2$ such that $\mathscr{C}_1, \mathscr{C}_2 \subset \mathscr{C}$ and $\mathscr{C}_1 \cap \mathscr{C}_2 = \{ \}$. Then any set in $\mathscr{A}(\mathscr{C}_1)$, the additive class generated by $\mathscr{C}_1$, is independent of any set in $\mathscr{A}(\mathscr{C}_2)$.

Queries

24.1 It is known that A, B are a pair of independent events, with $P(A)$, $P(B) \neq 0, 1$. Which of the following are independent classes?

a) $\{A^c, B^c\}$ b) $\{A, B, A^c, B^c\}$

c) $\{\phi, A, B, S\}$ d) $\{\phi, A, A^c\}$.   [68]

24.2 An additive class $\mathscr{A}$ consists of eight sets, of which only ϕ has zero probability. What is the largest possible size of a subclass of $\mathscr{A}$ if it is an independent class?   [192]

24.3 Let $\{A, B, C\}$ be an independent class of sets, with $P(ABC) = 1/2$. Evaluate the product $P(AB) \, P(AC) \, P(BC)$.   [216]

24.4 At least one of three *equiprobable* events A, B, C *must* occur. If, furthermore, $P(ABC) = P(A) \, P(B) \, P(C) = 1/8$, what is the probability that exactly two of the events A, B, C occur?   [18]

Possible Outcomes Represented as Ordered *k*-Tuples

In Section 20 we saw that in a p.e. with possible outcomes that are equally likely, the *size* of the sample space completely determines the probability function. In such a case, therefore, *counting* all possible outcomes of the p.e. becomes a basic part of the probability problem. This can be an awkward task if the sample space is large, but it is often possible to visualize the p.e. in a particular way that will simplify the task of counting. In this and the next few Sections we examine particular types of p.e.'s—and the associated sample spaces—for which this is possible.

Basic to our discussion is the familiar notion of an *ordered k-tuple* (or simply, *k-tuple*), which is a consecutive arrangement, or sequence, of *k* entities or symbols, where *k* is an integer greater than 1. If $k = 2$, we also call it an *ordered pair*, if $k = 3$, an *ordered triple*. The notation consists of a listing of the *k* entities or symbols, separated by commas, with the listing enclosed in parentheses. Thus,

$$(u, b, r, r, s, b, a, t) \tag{25.1}$$

denotes an 8-tuple of *letters*. We refer to the symbol appearing in the *i*th position as the *i*th *component*, for $i = 1, \ldots, k$. In expression (25.1), for instance, the first component is 'u', the eighth component is 't'. In a *k*-tuple the order of the components is considered significant and needs to be preserved.

When *k*-tuples are used, it must be clear in what context they are being used and what they represent. An understanding of what a particular *k*-tuple represents should also make it clear from what collection of entities or symbols each of the components can be chosen. For instance, the *k*-tuple (25.1) suggests that all components are taken from the same collection—the letters of the English alphabet. However, without additional information, the mere inspection of a single *k*-tuple does not yield the rules governing its components. Thus, each component in expression (25.1) might have been selected from a different subset of the alphabet.

The context can arise in the conceptual domain or in the mathematical domain. Our characterization of a probability space is an example of an ordered triple in the mathematical domain, in which the first component is a set, the second component a class, and the third component a set function.

In this Section we utilize *k*-tuples in the conceptual domain, where they can be useful in characterizing the possible outcomes of a probabilistic experiment.

Probability Concepts and Theory for Engineers, First Edition. Harry Schwarzlander.
© 2011 John Wiley & Sons, Ltd. Published 2011 by John Wiley & Sons, Ltd.

From Section 3 we recall that a p.e. may be of such a nature that each of the possible outcomes represents a *multiplicity* of properties of interest. In an experiment involving the throw of a red and a white die, for instance, one of the possible outcomes is characterized by the two properties: 'two dots face up on the red die' and 'five dots face up on the white die'. Each of the other possible outcomes of that experiment also consists of two properties, one property of the type '*x* dots on the red die' and the other property of the type '*y* dots on the white die'. It is, therefore, possible to represent each possible outcome in this experiment by an ordered pair of symbols, such as:

$$(\boxdot, \boxed{\vdots})$$

Here, the first component is meant to identify a property 'of the first type' (a face of the *red* die) and the second component a property 'of the second type' (a face of the *white* die). Together, the two components identify a particular possible outcome.

Now, there are six different 'properties of the first type'. There are also six different 'properties of the second type', irrespective of the nature of the first component. Therefore, there exist $6 \times 6 = 36$ distinct ordered pairs, each representing one of the possible outcomes in this p.e. It follows that in this case the sample space has size $|S| = 36$.

This idea is generally applicable in any p.e. in which each of the possible outcomes encompasses *several* observed properties—one property from each of k *different types* of observed properties. In that case, each possible outcome can be characterized by means of an ordered k-tuple $(a_1, a_2, \ldots, a_k)$. Here a_1 is a parameter that identifies a particular observed property of the first type, a_2 of the second type, and so on. Each element of the sample space is then identified by a distinct k-tuple from the collection of all such k-tuples. If n_1 is the number of different observable properties of the first type (the number of values that can be assumed by the parameter a_1), and in general, n_i is the number of properties of type i (where $i = 1, \ldots, k$), and all possible combinations can arise, then the size of the sample space is

$$|S| = n_1 \, n_2 \cdots n_k. \tag{25.2}$$

This does not apply, of course, if there is a dependence between the components of the k-tuple so that, for instance, n_2 depends on n_1.

Example 25.1: Picking a card.

Consider the following experiment. Six decks of playing cards are on hand. One is a full deck of 52 cards, and the others are partial decks containing 48, 42, 32, 22, and 12 cards, respectively. One deck is chosen according to the result of a die throw, and one card is picked blindly from the chosen deck. Of interest is the identity of the card that is picked, and the deck from which it is chosen.

Each possible outcome consists of two observable properties—a die throw result and the identity of a playing card—and therefore can be represented by an ordered pair of the type (d, c) where d is a die throw result and c a playing card. Although there are six possibilities for the first component and 52 possibilities for the second component, Equation (25.2) does not apply. This is because the picking of a card can result in 52 possibilities only if the first deck is chosen.

Instead, in this case, $|S| = 52 + 48 + 42 + 32 + 22 + 12 = 208$.

Consider now a p.e. whose procedure calls for the same action to be carried out *several times*, with each repetition to be carried out independently and under essentially the same initial conditions. If each of the actions results in one observed property, then carrying out the action k times results in k observed properties, each from the same collection of observable properties. Typical of such an experiment are k successive 'blind' or 'unbiased' selections of an object from a pool of similar objects, with the selected object each time getting thrown back into the pool prior to the next selection. Such an experiment is frequently referred to as *ordered sampling with replacement*. A single one of the blind selections is often called a 'trial', so that we are thinking here about an experiment that consists of k trials. (It is assumed that the *order* in which the observed properties occur is significant.) The possible outcomes, therefore, may again be represented by k-tuples, and the expression for the size of the sample space is a special case of Equation (25.2), since now $n_1 = n_2 = \ldots = n_k$:

$$|\mathsf{S}| = n_1^k \qquad (25.3)$$

Example 25.2: Ordered sampling with replacement.

Beginning with a shuffled standard deck of 52 playing cards,[2] the following procedure is carried out five times: A card is drawn blindly, examined, then returned to the deck and the deck is thoroughly reshuffled. Thus, each possible outcome represents a particular sequence of five cards, which may include duplicates. What is the probability that the first three draws will yield at least one face card (Jack, Queen, or King) and the last two draws will not yield any face card?

Each element of the sample space can be characterized by a 5-tuple, such as the one shown in Figure 25.1. Each of the five components in the 5-tuple is taken from a collection of 52 playing cards. The selections are independent and equally likely, so that the possible outcomes are equally likely, and Theorem 20.1 applies. From Equation (25.3), $|\mathsf{S}| = 52^5$. But finding the size of the event of interest seems to require some elaborate counting. To simplify things it helps to express that event as an intersection, $\mathsf{A} \cap \mathsf{B}$, where

$\mathsf{A} \leftrightarrow$ 'at least one face card in the first three draws'

$\mathsf{B} \leftrightarrow$ 'no face card in the last two draws'.

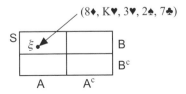

Figure 25.1 A card is drawn five times

Since the events A and B are determined by independent physical mechanisms (selection of the first three cards, and selection of the last two cards, respectively) they are independent events, so that the desired probability is

$$P(\mathsf{AB}) = P(\mathsf{A})P(\mathsf{B}).$$

---

[2] A standard deck of 52 playing cards (also called a 'Bridge deck') is made up of four 'suits' (spades, hearts, diamonds and clubs, represented by the symbols ♠, ♥, ◇, and ♣, respectively) of 13 cards each. Each suit consists of ten 'number cards' ('one' to 'ten', where the 'one' card is called 'Ace') and three 'face cards' (Jack, Queen and King).

We note that A is made up of possible outcomes having one, two, or three face cards in the first three draws. It is easier here to first find the probability of the event

$$A^c \leftrightarrow \text{'}no \text{ face card in the first three draws'}.$$

The deck contains 12 face cards, leaving 40 other cards. The number of 5-tuples that can be formed using

a) selections from the 40 nonface cards in each of the first three positions, and
b) selections from the full deck for each of the remaining two positions, is

$$|A|^c = 40 \cdot 40 \cdot 40 \cdot 52 \cdot 52 = 40^3 \cdot 52^2.$$

Thus, $P(A^c) = \frac{40^3 \cdot 52^2}{52^5} \doteq 0.455$, and $P(A) = 1 - P(A^c) \doteq 0.545$. (The notation $\doteq$ is used to indicate that the equality is not exact due to round-off error.) Similarly,

$$|B| = 52 \cdot 52 \cdot 52 \cdot 40 \cdot 40 = 52^3 \cdot 40^2$$

giving $P(B) = \frac{52^3 \cdot 40^2}{52^5} \doteq 0.592$. The desired result is then

$$P(AB) = P(A)P(B) \doteq 0.545 \times 0.592 \doteq 0.322.$$

The simplest experiments involving 'ordered sampling with replacement' are those where each trial is *binary*; that is, in each trial the observed property is one of just two possibilities. If the binary trials are identical, they are also called 'Bernoulli trials'.

Example 25.3: Multiple coin toss.
A coin is tossed ten times in succession, as in Example 11.1a. The experiment, therefore, consists of ten independent trials, with each trial yielding one of the observable properties H (heads) or T (tails). Each possible outcome can then be represented as a 10-tuple having components that are either H or T. For instance, if the execution of the p.e. yields the record TTHHTHHHHT as in Example 11.1a, then this outcome is described by the 10-tuple (T, T, H, H, T, H, H, H, H, T).
There is no 'sampling' here, but the p.e. is probabilistically equivalent to an experiment of, say, blindly drawing ten times, with replacement, a ball from an urn containing a mixture of an equal number of red and green balls.
All possible outcomes are equally likely. This can be seen by considering two 10-tuples (i.e., possible outcomes) that differ in exactly one component, such as

$$(T, T, H, H, T, \underline{H}, H, H, H, T) \quad \text{and} \quad (T, T, H, H, T, \underline{T}, H, H, H, T).$$

Clearly, neither of these two possible outcomes is more likely than the other. Repeated application of this argument shows that any two possible outcomes are equally likely, and therefore all possible outcomes are equally likely.
From Equation (25.3), the space S for this p.e. has size $|S| = 2^{10} = 1024$. The probability distribution is uniform; that is, every one-element subset A of S has $P(A) = 2^{-10}$.
Now suppose that what is really of interest is just the *number of heads* obtained in the ten tosses. Let $H_i \leftrightarrow \text{'}i \text{ heads'}$ $(i = 0, 1, \ldots, 10)$. Then, $|H_i| = \binom{10}{i}$, for each i. These numbers are binomial

coefficients, arrived at in the same manner as in Section 13. On the partition $\mathscr{P}_H = \{H_0, H_1, H_2, \ldots, H_{10}\}$, therefore, probabilities are *not* assigned uniformly. Instead, the probabilities that get assigned to the members of the partition $\mathscr{P}_H$ are

$$P(H_i) = \binom{10}{i} 2^{-10}, \quad i = 0, 1, \ldots, 10. \tag{25.4}$$

These probabilities constitute a particular form of the so-called *binomial distribution*. If we specify an additive class $\mathscr{A}_H$ of subsets of S as $\mathscr{A}_H = \mathscr{A}(\mathscr{P}_H)$, the additive class generated by $\mathscr{P}_H$, then a different probability space $(S, \mathscr{A}_H, P_H)$ can be defined for our coin toss p.e., where P_H is then also called a binomial probability distribution.

Queries

25.1 A fair coin is tossed ten times.
   a) What is the probability that all ten tosses produce the same result?
   b) What is the probability that the results alternate, that is, Tail is followed by Head, and Head followed by Tail?
   c) What is the probability that the first five tosses produce identical results?   [153]

25.2 A die is thrown n times in succession. If A denotes the event

$$A \leftrightarrow \text{a 'one' is obtained at least once}$$

find the smallest n for which $P(A) \geq 1/2$.   [8]

25.3 Which of the following experiments have possible outcomes described by equally likely k-tuples, and such that all combinations of components are allowed:
   a) Five cards are dealt from a shuffled deck of cards and the identities of the cards are noted, in the order dealt.
   b) A bridge deck is dealt to four players (13 cards to each player). It is noted how many aces are held by each of the players.
   c) At a square dance there are 20 men and 20 women. A chance device is used to assign partners. It is noted which woman gets partnered with which man.
   d) Four persons participate in a board game. Each throws a die and it is noted who has the highest throw. If there is a tie, the tie is broken by repeated throws by those players involved in the tie.
   e) Three dice—white, green and red—are thrown. If any of the dice show a 'three', the throw is repeated (until none of the dice show a three). The number of dots on each of the dice is noted.   [46]

SECTION 26

Product Experiments and Product Spaces

Another way to regard ordered sampling with replacement, and any other experiment involving two or more *independent* trials, is as a 'product experiment'.

> **(Definition) 26.1**
> Given a real-world situation that can be modeled in terms of k *independent* p.e.'s $\mathfrak{E}_1, \ldots, \mathfrak{E}_k$. If, instead, this real-world situation is regarded as a *single* p.e. $\mathfrak{E}$, then $\mathfrak{E}$ is a *product experiment*, denoted symbolically
>
> $$\mathfrak{E} = \mathfrak{E}_1 \times \mathfrak{E}_2 \times \ldots \times \mathfrak{E}_k.$$
>
> The experiments $\mathfrak{E}_1, \ldots, \mathfrak{E}_k$ then become the *component experiments* of the product experiment $\mathfrak{E}$.

If a product experiment $\mathfrak{E}$ has k component experiments, then it follows from Section 25 that each possible outcome of $\mathfrak{E}$ is representable as a k-tuple, the ith component being a possible outcome from the ith component experiment ($i = 1, \ldots, k$). Note that a product experiment involving ordered sampling with replacement has *identical* component experiments, but this is not required in Definition 26.1.

Example 26.1

Consider an experiment where ten urns are available, as well as a box filled with blue, green and orange ping-pong balls. A ball is to be drawn blindly and deposited in a randomly selected urn. It is of interest what color ball is deposited in which urn.

It is easy to envision this experiment in terms of two independent p.e.'s, that is, as a product experiment $\mathfrak{E} = \mathfrak{E}_1 \times \mathfrak{E}_2$, where $\mathfrak{E}_1$ is the experiment of selecting a ball, and $\mathfrak{E}_2$ is the experiment of selecting an urn.

Now consider this slightly different experiment: Ten urns are available, each filled with a different mixture of blue, green and orange ping-pong balls. An urn is to be selected randomly and a ball drawn from it. It is of interest what color ball is drawn. Here, one may be tempted to think in terms of a product experiment analogous to the above. But $\mathfrak{E}_1$ and $\mathfrak{E}_2$ are now not independent experiments! Since the ball is to be drawn *from the urn that has been chosen*, this part of the experiment is clearly dependent upon the choice of the urn. This is *not* a product experiment.

The term 'product experiment' is motivated by the corresponding Set Theory model. For each of k independent p.e.'s $\mathfrak{E}_1, \dots \mathfrak{E}_k$ there is a sample space. Thus, if we want to consider all k p.e.'s at once, the mathematical model would involve k different sample spaces. But this is not a useful approach—in a given probability problem we can have only a *single* basic space of elements. Then it is necessary to combine the k different sample spaces in an appropriate way. Following up on Definition 26.1 we therefore have:

(Definition) 26.2
Given k separate spaces of elements $S_1, \dots, S_k$. The *Cartesian product* of these k spaces is the space

$$S = S_1 \times S_2 \times \dots \times S_k$$

whose elements correspond to all the distinct *ordered k-tuples* that have as their ith component an element of S_i (for $i = 1, \dots, k$). That is,

$$S = \{(\xi_1, \dots, \xi_k) : \xi_1 \in S_1, \xi_2 \in S_2, \dots, \xi_k \in S_k\}.$$

The space S in Definition 26.2 is called a *product space*; it models a product experiment. The size of S is given by Equation (25.2), where now $n_i = |S_i|$, for $i = 1, \dots, k$; that is,

$$|S| = |S_1| \cdot |S_2| \cdot \ \dots \ \cdot |S_k|.$$

Some or all of the spaces $S_1, \dots, S_k$ in Definition 26.2 may, of course, be identical. But they still are distinguished by their position in the Cartesian product. We also note that each ordered k-tuple $(\xi_1, \dots, \xi_k)$ in the mathematical model corresponds to an ordered k-tuple of possible outcomes—one from each of the k component experiments.

Example 26.2: Throw of two dice.
Although the p.e. described in Example 17.1 is stated as a single experiment, we can envision it modified into a product experiment $\mathfrak{E}$ with two component experiments: $\mathfrak{E}_1$ that involves the throwing of the white die, and $\mathfrak{E}_2$ the red die. Our interpretation of the physical situation leads us to accept either of the conceptual models as appropriate. In other words, we feel no need to make a distinction between throwing the two dice together and throwing them separately.
    The two component experiments of $\mathfrak{E}$, although not identical, are probabilistically equivalent and have the sample spaces

$$S_1 = \{\xi_a, \xi_b, \xi_c, \xi_d, \xi_e, \xi_f\}, \quad \text{and} \quad S_2 = \{\xi_a, \xi_b, \xi_c, \xi_d, \xi_e, \xi_f\}.$$

Use of the same symbols for elements of S_1 and elements of S_2 is not meant to imply any relationship. S_1 and S_2 are completely separate universes of elements.
    The sample space for the product experiment is the product space $S = S_1 \times S_2$. Every element $\xi \in S$ corresponds to an ordered pair consisting of one element from S_1 followed by one element from S_2, as illustrated in Figure 26.1.

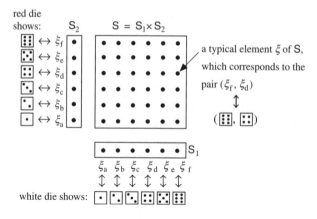

Figure 26.1   Throw of two dice viewed as a product experiment

When a product space has only two component spaces, its product nature is conveniently depicted by drawing the sample space in the form of a rectangular array of elements, as in Figure 26.1. For higher-dimensional product spaces, of course, such a visual device no longer works.

Since a space is a set, Definition 26.2 can also be taken as the definition of a *product set*. More generally, if $A_1 \subset S_1, \ldots, A_k \subset S_k$, then the Cartesian product

$$
\begin{aligned}
A &= A_1 \times A_2 \times \ldots \times A_k \\
&= \{(\xi_1, \ldots, \xi_k) : \xi_1 \in A_1, \xi_2 \in A_2, \ldots, \xi_k \in A_k\}
\end{aligned}
\tag{26.1}
$$

is a *product set* that is a subset of $S = S_1 \times \ldots \times S_k$, with $|A| = |A_1| \cdot |A_2| \cdot \ldots \cdot |A_k|$. In the product space, a component set such as A_1 no longer appears; instead, it becomes a product set. For instance, A_1 becomes $A_1 \times S_2 \times \ldots \times S_k$; it is the set of all k-tuples such that the first component belongs to A_1 while each of the remaining components is any element chosen from $S_2, S_3, \ldots, S_k$, respectively.

It follows from Equation (26.1) that a *one-element* product set is the Cartesian product of one-element sets: $\{(\xi_1, \ldots, \xi_k)\} = \{\xi_1\} \times \ldots \times \{\xi_k\}$. Furthermore, if any of the component sets in Equation (26.1) is empty, then $A = \phi$.

Example 26.3

In Example 26.2, consider an event in component experiment $\mathfrak{E}_1$,

   $A_1 \leftrightarrow$ 'more than three dots on white die'

and an event in component experiment $\mathfrak{E}_2$,

   $A_2 \leftrightarrow$ 'two dots on the red die'

In the product experiment $\mathfrak{E} = \mathfrak{E}_1 \times \mathfrak{E}_2$ in which both a red die and a white die are thrown, these events become $A_1 \times S_2$ and $S_1 \times A_2$ (see Figure 26.2).

The intersection of these two product sets is the product set

$$
(A_1 \times S_2) \cap (S_1 \times A_2) = A_1 \times A_2
$$

where

$A_1 \times A_2 \leftrightarrow$ 'more than three dots on white die, and two dots on red die'.

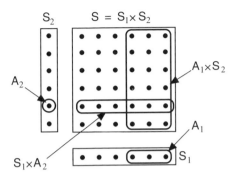

Figure 26.2   Product sets

Of course, not all subsets of a product space $S = S_1 \times \ldots \times S_k$, $k \geq 2$, are product sets. In the above Example, for instance, the set $B \leftrightarrow$ 'sum of dots is three' cannot be represented in the form of Equation (26.1) and is not a product set. Furthermore, since the various elementary sets whose union is B are product sets, we see that a union of product sets is not necessarily a product set.

It is possible to take the Cartesian product of two product sets. If $A = A_1 \times \ldots \times A_k$ and $B = B_1 \times \ldots \times B_l$, then $A \times B = A_1 \times \ldots \times A_k \times B_1 \times \ldots \times B_l$. We also note the following basic rule:

Theorem 26.1
Given two spaces S, T (where S and T could be the same space), and sets A, $B \subset S$ and C, $D \subset T$. If $A \subset B$ and also $C \subset D$, then $A \times C \subset B \times D$.

Proof
Suppose $A \subset B \subset S$, and $C \subset D \subset T$. Let ξ_1 be any element of S, and ξ_2 any element of T. Then for every $\xi_1 \in A$, it is true that $\xi_1 \in B$. Similarly, for every $\xi_2 \in C$, it is true that $\xi_2 \in D$.

Now consider the element $(\xi_1, \xi_2) \in S \times T$, and suppose $(\xi_1, \xi_2) \in A \times C$. This means $\xi_1 \in A$ and $\xi_2 \in C$, and therefore also $\xi_1 \in B$ and $\xi_2 \in D$; i.e., $(\xi_1, \xi_2) \in B \times D$. Thus, $A \times C \subset B \times D$.

Our conceptual world should also allow a product experiment made up of an infinity, or unending sequence, of component experiments. In fact, we encountered such a case in Example 11.1. In the p.e. described there, each component experiment consists of a single coin toss, i.e., it has two possible outcomes; and the possible outcomes of the product experiment are *infinite sequences* of coin-toss results. We saw that the sample space for that product experiment is uncountable. Of course, such a model strains our notion of reality: No experimenter can perform an infinite, i.e., unending, sequence of experiments. In the physical world, nothing can go on 'forever'. Conceptually, however, we can view this as an idealization that corresponds to a limiting case in the mathematical world. It is even harder to conceive of a large number of repetitions of such an experiment. But, rather than being consecutive, these 'repetitions' could be thought of as taking place simultaneously, perhaps in separate locations.

A product experiment is made up of *independent* component experiments, as stated in Definition 26.1. On the other hand, for a product *space* it is not necessarily true that its component spaces represent independent trials. A product space arises whenever the possible outcomes of a p.e. are representable as k-tuples (with k fixed), such that all available combinations of components are allowed. Dependence can take a special form, as illustrated in the next Example.

Example 26.4

In a certain factory two different machines produce the same kinds of dice. It is discovered that all the dice coming from one of the machines are biased in the same way. Dice from the other machine are fair. Furthermore, the biased and unbiased dice cannot be distinguished by physical inspection. Suppose that a batch of dice from this factory contains a mixture of dice produced by the two machines, with 40% of the dice being of the biased kind. A technician picks a die from this batch and throws it three times, observing what number of dots is obtained on each throw.

Even though the dice are indistinguishable, when setting up the p.e. we must take into account which kind of die is picked, since this influences the results of the die throws. An experimenter other than the technician has to be specified, who in principle can distinguish the biased from the unbiased dice. Hence, each possible outcome consists of four properties of interest:

a) what kind of die is picked (biased or unbiased)
b) result of the first die throw (1, 2, 3, 4, 5, or 6 dots)
c)   ”   second   ”   ”
d)   ”   third   ”   ”

Each possible outcome is therefore representable as a 4-tuple, such as (b, two dots, one dot, four dots), where 'b' designates 'biased'. The sample space is a product space S, each of whose component spaces is associated with one component of the 4-tuple. Thus, $S = S_1 \times S_2 \times S_3 \times S_4$, where $|S_1| = 2$, and $|S_2| = |S_3| = |S_4| = 6$, so that $|S| = 2 \cdot 6^3 = 432$.

The experiment, however, is not a product experiment. As we have noted, the die throw results depend on what kind of die is chosen. Yet, after the die has been chosen, the remainder of the experiment consists of independent trials. The independence of the trials, therefore, is *conditional*, conditioned on the first component in the 4-tuples that represent the possible outcomes. Accordingly, we can regard this kind of experiment a '*conditional product experiment*'.

Queries

26.1 Let $\mathfrak{E}_1$ denote the experiment of tossing a single coin.
   a) What is the number of possible outcomes of $\mathfrak{E} = \mathfrak{E}_1 \times \mathfrak{E}_1$?
   b) Suppose the coin is tossed 11 times. How many executions of the experiment $\mathfrak{E}$ does this constitute?   [230]

26.2 For each of the following, determine whether it can be regarded as a product experiment. If it can, what is the size of the sample space?
   a) A card is picked from a shuffled standard deck of bridge cards, and its suit is noted; then a card is picked from another shuffled standard deck and it is noted whether or not it is a 'face card'.
   b) A set of 100 light bulbs is undergoing life tests; i.e., they are operated until they fail.
      The time till the first failure occurs is measured.
      The time till the second failure occurs is measured.
          . . .
          . . .
      The time till the last bulb fails is measured.
   c) A card is picked from a shuffled deck to determine whether or not it is the ace of hearts. If it is not, the card that has been selected is put aside and another card is picked for the same purpose; and so on, until the ace of hearts is found.

d) Refer to Example 24.1. A red and a white die are thrown and it is determined
   i) whether or not event B_1 occurred;
   ii) whether or not event B_2 occurred;
   iii) whether or not event B_4 occurred.
e) A count is kept for one hour of the number of vehicles passing a particular spot along
   a highway. 1000 yards downstream from the observation point is an intersection.
   1000 yards beyond the intersection another observation point is established and a
   count of vehicles is made over the same one-hour period.   [183]

26.3 Given a space S and a nonempty set $A \subset S$, with $A \neq S$. Now consider all 16 expressions
of the form $\alpha \times \beta$, where both α and β can be any one of the sets ϕ, A, A^c, S. How many of
these expressions represent the empty set?   [193]

SECTION 27

Product Probability Spaces

In Section 26 it was seen how a number of independent trials can be represented as a *product experiment*, with a sample space that is a product space. We examine now what kind of *probability space* is to be associated with a product experiment. For the sake of simplicity, $k = 2$ will be used in this discussion; that is, we consider a product experiment $\mathfrak{E} = \mathfrak{E}_1 \times \mathfrak{E}_2$ with sample space $S = S_1 \times S_2$. Extension to arbitrary k is immediate.

Let $(S_1, \mathscr{A}_1, P_1)$ and $(S_2, \mathscr{A}_2, P_2)$ be the probability spaces associated with the component experiments $\mathfrak{E}_1$ and $\mathfrak{E}_2$, respectively. To be described is the probability space $(S, \mathscr{A}, P)$ of the product experiment $\mathfrak{E}$. It has already been established that $S = S_1 \times S_2$, but $\mathscr{A}$ and P remain to be specified in terms of $\mathscr{A}_1$, $\mathscr{A}_2$, and P_1, P_2, respectively.

First we note the following.

(Definition) 27.1

For each $i = 1, \ldots, k$, let $\mathscr{C}_i$ be a class of subsets of a space S_i. By the *product class*

$$\mathscr{C} = \mathscr{C}_1 \times \mathscr{C}_2 \times \ldots \times \mathscr{C}_k$$

is meant the class of all those subsets of $S = S_1 \times S_2 \times \ldots \times S_k$ that are product sets of the form $A = A_1 \times A_2 \times \ldots \times A_k$, where $A_i \in \mathscr{C}_i$, for $i = 1, \ldots, k$.

Now it might be tempting to designate $\mathscr{A}_1 \times \mathscr{A}_2$ as the completely additive class of subsets of $S = S_1 \times S_2$. However, $\mathscr{A}_1 \times \mathscr{A}_2$ is not an additive class! It is a class of product sets and, as was observed in Section 26, the union of two product sets is not necessarily a product set. Instead, a completely additive class of subsets of S will be specified as $\mathscr{A} = \mathscr{A}(\mathscr{A}_1 \times \mathscr{A}_2)$, the *completely additive class generated by $\mathscr{A}_1 \times \mathscr{A}_2$*.

This still leaves the probability function P to be defined over all of $\mathscr{A}(\mathscr{A}_1 \times \mathscr{A}_2)$. But first an Example.

Example 27.1

Referring to Example 26.3, let $(S_1, \mathscr{A}_1, P_1)$ denote the probability space associated with $\mathfrak{E}_1$, the throw of the white die. Similarly, let $(S_2, \mathscr{A}_2, P_2)$ denote the probability space associated with $\mathfrak{E}_2$, the throw of the red die. We see that in $(S_1, \mathscr{A}_1, P_1)$ we have $P_1(A_1) = 1/2$. This should carry over to the probability space $(S, \mathscr{A}, P)$ of the product experiment $\mathfrak{E}$, so that $P(A_1 \times S_2) = 1/2$.

Probability Concepts and Theory for Engineers, First Edition. Harry Schwarzlander.
© 2011 John Wiley & Sons, Ltd. Published 2011 by John Wiley & Sons, Ltd.

Furthermore, $P_2(A_2) = 1/6$, so that we expect $P(S_1 \times A_2) = 1/6$. The important argument is now the following:

i) $\mathfrak{E}_1$ and $\mathfrak{E}_2$ are independent experiments.

ii) $A_1 \times S_2$ is concerned only with what happens in $\mathfrak{E}_1$; $S_1 \times A_2$ is concerned only with what happens in $\mathfrak{E}_2$. Then $A_1 \times S_2$ and $S_1 \times A_2$ must be independent events in S.

iii) Therefore, $P(A_1 \times A_2) = P(A_1 \times S_2) \cdot P(S_1 \times A_2) = P_1(A_1) \cdot P_2(A_2)$, giving $P(A_1 \times A_2) = 1/12$.

The probability of every other product set, that is, of every set in $\mathscr{A}_1 \times \mathscr{A}_2$, is determined in an analogous manner.

Example 27.1 suggests how independence of the component experiments allows the probability distribution P to be defined on $\mathscr{A}(\mathscr{A}_1 \times \mathscr{A}_2)$ in a consistent manner. First, to every set $A_1 \times A_2 \in \mathscr{A}_1 \times \mathscr{A}_2$ gets assigned the probability $P(A_1 \times A_2) = P_1(A_1) P_2(A_2)$. The function P is then *extended* to all of $\mathscr{A}(\mathscr{A}_1 \times \mathscr{A}_2)$ via countable additivity. We refer to the distribution arrived at in this manner as a *product distribution* and denote it $P_1 \bullet P_2$.

In the general case of k component experiments we have:

(Definition) 27.2

For $i = 1, \ldots, k$, let $(S_i, \mathscr{A}_i, P_i)$ be the probability space that corresponds to component experiment $\mathfrak{E}_i$. Let $\mathscr{A}$ be the completely additive class of subsets of $S = S_1 \times S_2 \times \ldots \times S_k$ that is generated by $\mathscr{A}_1 \times \mathscr{A}_2 \times \ldots \times \mathscr{A}_k$. Then the *product probability distribution* $P = P_1 \bullet P_2 \bullet \ldots \bullet P_k$ is that probability distribution on $\mathscr{A}$ that assigns to every product set $A_1 \times A_2 \times \ldots \times A_k \in \mathscr{A}_1 \times \mathscr{A}_2 \times \ldots \times \mathscr{A}_k$ the value

$$P(A_1 \times A_2 \times \ldots \times A_k) = P_1(A_1)P_2(A_2) \ldots P_k(A_k)$$

and that is defined on the remaining sets of $\mathscr{A}$ through countable additivity.

The following compact definition of a product probability space can now be given:

(Definition) 27.3

For $i = 1, \ldots, k$, let $(S_i, \mathscr{A}_i, P_i)$ be the probability space corresponding to p.e. $\mathfrak{E}_i$. The *product probability space* $(S, \mathscr{A}, P)$ that corresponds to the product experiment $\mathfrak{E} = \mathfrak{E}_1 \times \ldots \times \mathfrak{E}_k$ is defined by:

i) $S = S_1 \times \ldots \times S_k$

ii) $\mathscr{A} = \mathscr{A}(\mathscr{A}_1 \times \ldots \times \mathscr{A}_k)$

iii) $P = P_1 \bullet \ldots \bullet P_k$.

Example 27.2

In a specified number n of throws of an ordinary and fair die, it is of interest how many sixes are obtained. Thus, beginning with the single die-throw experiment and denoting it $\mathfrak{E}_1$, we define an n-fold product experiment

$$\mathfrak{E} = \underbrace{\mathfrak{E}_1 \times \ldots \times \mathfrak{E}_1}_{n \text{ times}}.$$

In this product experiment, then, what is the probability that i sixes are obtained ($i = 0, 1, \ldots, n$)?

$\mathfrak{E}_1$ could be formulated as a *binary* experiment, with the two possible outcomes 'six dots face up' and 'other than six dots face up'. However, we proceed according to the usual way of modeling a single die throw. Then, the probability space for $\mathfrak{E}_1$ is $(S_1, \mathscr{A}_1, P_1)$, where $|S_1| = 6$, and P_1 distributes probability equally over the one-element sets of S_1. As in Example 26.2, let $\xi_f \leftrightarrow$ 'a six is thrown'. Abbreviating $P_1(\{\xi_f\})$ by p, we have $p = 1/6$. The only other event of interest in $\mathfrak{E}_1$ is 'a six is not thrown', i.e., $\{\xi_f\}^c$, which arises with probability $P_1(\{\xi_f\}^c) = 1 - p = 5/6$. $\{\xi_f\}$ and $\{\xi_f\}^c$ partition S_1, but their probabilities are unequal.

The probability space for $\mathfrak{E}$ is the product probability space $(S, \mathscr{A}, P)$ as given in Definition 27.3, but with all component experiments now being identical and with k replaced by n. It is convenient to carry over to S the partition $\{\{\xi_f\}, \{\xi_f\}^c\}$ of the component spaces. In other words, we will concern ourselves only with sets of the form

$$\underbrace{\{\xi_f\}^x \times \{\xi_f\}^x \times \ldots \times \{\xi_f\}^x}_{n \text{ times}}$$

where each exponent x indicates either complementation or no complementation. There are 2^n such sets, which together constitute a partition $\mathscr{P}$ of S. A similarity to Example 25.3 may be noted; there we had $|S| = 2^{10}$, here $|\mathscr{P}| = 2^n$. But an important difference is that the sets in $\mathscr{P}$ are not equiprobable. What are their probabilities?

That depends on how many of the components making up a given product set $A \in \mathscr{P}$ are complements. Suppose A has $\{\xi_f\}$ as i of its components, and therefore $\{\xi_f\}^c$ as the remaining $n - i$ of its components ($i = 0, 1, \ldots, n$). Then, it follows from Definition 27.3, item (c), that

$$P(A) = p^i (1 - p)^{n - i} = \frac{5^{n - i}}{6^n}.$$

The remaining step proceeds as in Example 25.3. Let $H_i \leftrightarrow$ 'i sixes' (or i 'Hits'), where $i = 0, 1, \ldots, n$. Instead of $|H_i|$, the number of elements in H_i, we now want the number of sets in $\mathscr{P}$ that are subsets of H_i. This is $\binom{n}{i}$, and since all these sets have the same probability $p^i (1 - p)^{n - i}$, we obtain

$$P(H_i) = \binom{n}{i} p^i (1 - p)^{n - i} = \frac{n(n - 1) \cdots (n - i + 1)}{i(i - 1) \cdots 1} \cdot \frac{5^{n - i}}{6^n}. \tag{27.1}$$

In the center we have the general form of the *binomial distribution*[3] that, in this case, expresses the probabilities over the partition $\{H_0, H_1, \ldots, H_n\} \subset \mathscr{A}$. Note that we cannot simply say that the product probability distribution P is a binomial distribution. Clearly, P is uniform, since it assigns equal probability 6^{-n} to every one-element set. However, P becomes a binomial distribution *over a particular partition* of S.

Even in some simple problems it can be expedient to introduce a model where a product experiment consists of an *unending sequence* of component experiments, as noted in Section 26.

---

[3] To explain the name 'binomial distribution,' let $q = 1 - p$ in Equation (27.1). Then the expression in the center of Equation (27.1) represents the ith term in the binomial expansion of $(p + q)^n$.

Example 27.3: Runs.

How many times must a fair die be thrown till the first 'six' appears? Unless there is a reason to limit the number of throws in the conceptual model, it is simpler here to assume an unlimited number of throws. That will provide the answer to the question, whatever practical limit there may be. So we postulate a p.e. $\mathfrak{E}$ in which the possible outcomes are

first 'six' appears on first throw,

first 'six' appears on second throw,

first 'six' appears on third throw,

. . . .

. . . .

The procedure can be to throw the die repeatedly, keeping a tally of the number of throws, until a six is obtained, then stop. Thus, execution of the experiment $\mathfrak{E}$ need not continue indefinitely. Nevertheless, the sample space for $\mathfrak{E}$ is countably infinite.

However, with the procedure specified in this way, $\mathfrak{E}$ is not a product experiment. Instead, this situation can be modeled as a product experiment $\mathfrak{E}'$ with an infinite set of component experiments,

$$\mathfrak{E}' = \mathfrak{E}_1 \times \mathfrak{E}_1 \times \mathfrak{E}_1 \times \ldots,$$

where $\mathfrak{E}_1$ is the single die–throw experiment. The possible outcomes of $\mathfrak{E}'$ are infinite sequences of die-throw results, and the possible outcomes of $\mathfrak{E}$ become events in $\mathfrak{E}'$. Although this may seem like a more complex formulation, it allows the rules for probabilities of product events to be applied. Of course, this change in the model does not affect the nature of the p.e.; the execution can still be terminated once one of the events of interest has been observed.

We have, then, the product probability space $(S, \mathscr{A}, P)$ made up of identical component probability spaces $(S_1, \mathscr{A}_1, P_1)$, each representing a single die throw.

In $\mathfrak{E}_1$, let 'a six is thrown' $\leftrightarrow D_6 \in \mathscr{A}_1$.

In $\mathfrak{E}'$, let 'first six occurs on the ith throw' $(i = 1, 2, 3, \ldots) \leftrightarrow A_i \in \mathscr{A}$;

and 'no six occurs ever' $\leftrightarrow A_\infty \in \mathscr{A}$.

$\{A_1, A_2, A_3, \ldots, A_\infty\}$ is a partition of the sample space for $\mathfrak{E}'$, and the sets in this partition correspond to the possible outcomes identified earlier for $\mathfrak{E}$. Then,

$$P(A_i) = \underbrace{\left[P_1(D_6{}^c) \ldots P_1(D_6{}^c)\right]}_{i-1 \text{ times}} P_1(D_6) \cdot 1 \cdot 1 \cdot 1 \cdot \ldots$$

$$= P_1(D_6{}^c)^{i-1} P_1(D_6) = \frac{5^{i-1}}{6^i}.$$

This is a so-called *geometric* probability distribution over the partition $\{A_1, A_2, \ldots, A_\infty\}$.

The development of product probability spaces given in this Section can also be applied to problems where the experiment is a *conditional product experiment*, as introduced in Example 26.4. This is illustrated below.

Example 27.4

Consider again the situation described in Example 26.4. Also, suppose that the throw of a biased die results in one, two, three, four, five, or six dots with probabilities 0.05, 0.1, 0.15, 0.2, 0.2 and 0.3, respectively.

We formulate the probability space $(S, \mathscr{A}, P)$ for this experiment:

a) S is the product space given in Example 26.4, which can also be written $S_1 \times S_2 \times S_2 \times S_2$, since the last three component spaces are identical.

b) Let $\mathscr{A}_1$ be the additive class of all subsets of S_1, and $\mathscr{A}_2$ the additive class of all subsets of S_2. Then, $\mathscr{A} = \mathscr{A}(\mathscr{A}_1 \times \mathscr{A}_2 \times \mathscr{A}_2 \times \mathscr{A}_2)$, as in Definition 27.3.

c) To be able to specify P, we first write $S_1 = \{\xi_x, \xi_y\}$, where

$$\xi_x \leftrightarrow \text{'biased die is chosen', and } \xi_y \leftrightarrow \text{'unbiased die is chosen'}$$

The component spaces S_2, like the component spaces in Example 26.2, can be described by $\{\xi_a, \xi_b, \xi_c, \xi_d, \xi_e, \xi_f\}$. Over $\mathscr{A}_2$, the *conditional* distributions $P_2(\,|\{\xi_x\})$ and $P_2(\,|\{\xi_y\})$ are known. It should be noted that the same conditioning event applies to all three die throws.

We can now express the probability of any elementary set in $\mathscr{A}$, such as $\{(\xi_x, \xi_b, \xi_a, \xi_d)\}$; that is, a biased die is chosen and in three successive throws two, one, and four dots come up:

$$P(\{(\xi_x, \xi_b, \xi_a, \xi_d)\}) = P_1(\{\xi_x\}) \cdot P_2 \bullet P_2 \bullet P_2(\{(\xi_b, \xi_a, \xi_d)\}|\{\xi_x\})$$
$$= P_1(\{\xi_x\}) \cdot P_2(\{\xi_b\}|\{\xi_x\}) \cdot P_2(\{\xi_a\}|\{\xi_x\}) \cdot P_2(\{\xi_d\}|\{\xi_x\}).$$

Suppose we wish to find the (unconditional) probability of the event 'more than three dots are obtained on each of the three throws' $\leftrightarrow A \in \mathscr{A}$, where $A = S_1 \times A_2 \times A_2 \times A_2$, and $A_2 \leftrightarrow$ 'more than three dots are obtained in a given throw'. From the data given above we have $P_2(A_2|\{\xi_x\}) = 0.7$, whereas $P_2(A_2|\{\xi_y\}) = 0.5$. Then $P(A) = 0.4 \cdot 0.7^3 + 0.6 \cdot 0.5^3 = \underline{0.2122}$.

Queries

27.1 Given a space S and a nonempty set $A \subset S$, with $A \neq S$. Consider the partition $\mathscr{P} = \{A, A^c\}$.

a) What is the size of the product class $\mathscr{P} \times \mathscr{P}$?
b) What is the size of $\mathscr{A}(\mathscr{P} \times \mathscr{P})$?
c) What is the size of $\mathscr{A}(\mathscr{P}) \times \mathscr{A}(\mathscr{P})$?
d) Is $\mathscr{P} \times \mathscr{P}$ a partition of $S \times S$? [29]

27.2 Given a p.e. $\mathfrak{E}$ with probability space $(S, \mathscr{A}, P)$, where $S = \{\xi_1, \xi_2\}$, and $P(\{\xi_1\}) = 0.1$, $P(\{\xi_2\}) = 0.9$. In the probability space for $\mathfrak{E} \times \mathfrak{E} \times \mathfrak{E}$, is there an event whose probability is

a) 0.009? b) 0.018?
c) 0.18? d) 0.9? [55]

27.3 In the conditional product experiment of Example 27.4, which of the following three events have the same probability?

a) $\{(\xi_x, \xi_a, \xi_c, \xi_f), (\xi_y, \xi_a, \xi_c, \xi_f)\}$
b) $\{(\xi_x, \xi_a, \xi_c, \xi_f), (\xi_y, \xi_b, \xi_d, \xi_e)\}$
c) $\{(\xi_x, \xi_b, \xi_d, \xi_e)\}, (\xi_y, \xi_b, \xi_d, \xi_e)\}$. [41]

Dependence Between the Components in an Ordered k-Tuple

In Section 25 we considered experiments where the possible outcomes are representable as ordered k-tuples, and where a component in one position of the k-tuple does not influence the occurrence of a component in another position. There are other kinds of experiments where this independence between components does not apply—where there are constraints or dependencies between some or all components in the k-tuples that describe possible outcomes. Such experiments can usually be modeled as conditional product experiments such as encountered in Examples 26.4 and 27.4.

One simple kind of dependence between components of a k-tuple, and therefore between component experiments, is typified by the situation described in the following Example.

Example 28.1: Dealing a card to each of four players.
An experiment calls for dealing a card to each of four players from a shuffled deck (a standard deck of 52 Bridge cards). It is of interest *which* card gets dealt to each player. The possible outcomes are therefore representable as quadruples (or 4-tuples),

$$(a_1, a_2, a_3, a_4)$$

where a_i identifies the card dealt to player i ($i = 1, 2, 3, 4$).

The card dealt to player 1 is any one of 52 cards; i.e., component a_1 gets chosen from among $n = 52$ possibilities. Player 2 can also receive any one of 52 cards; however, it cannot be the same card as the one dealt to player 1. Once a particular component a_1 is specified in our quadruple, therefore, there remain only $n - 1 = 51$ different possibilities for a_2. Similarly, once a_1 and a_2 are specified there remain $n - 2$ different possibilities for a_3; and once a_1, a_2, a_3 are specified, then $n - 3$ different possibilities remain for a_4. Accordingly, $|S| = 52 \cdot 51 \cdot 50 \cdot 49 = 6\,497\,400$.

If the deck of cards is assumed to be well shuffled prior to dealing, then equally likely possible outcomes (4-tuples) can be assumed. Now consider the event

$$A \leftrightarrow \text{'each player receives a spade'}.$$

To find $P(A)$ we need $|A|$. Player 1 can be dealt any one of 13 spade cards, so that there are 13 different possibilities for the first component, a_1, among the 4-tuples belonging to A. For given a_1

there are only 12 possibilities for a_2; and so on. In this way we obtain $|A| = 13 \cdot 12 \cdot 11 \cdot 10 = 17\,160$, giving $P(A) = \frac{17\,160}{6\,497\,400} \doteq 0.00264$.

Experiments of this kind, where each possible outcome is characterized by an ordered k-tuple of distinguishable objects or 'properties', drawn from a common pool of such objects, are frequently referred to as *ordered sampling without replacement*. In the card-dealing situation of Example 28.1, the deck can also be reshuffled prior to the selection of each successive card, which would not change the essential nature of the experiment. When regarded in this way, it is easy to see how each selection of a card can be viewed as a separate component experiment. But these component experiments don't combine into a product experiment! Instead, they become a *conditional* product experiment, since the second, third and fourth selection is made from diminished decks each of whose compositions are determined by what cards have already been dealt.

From Example 28.1 it can be seen that in experiments involving the selection, from n possibilities, of an *ordered sample of size k without replacement*, the size of the sample space is given by the general expression

$$|S| = n(n-1)(n-2)\ldots(n-k+1) = \frac{n!}{(n-k)!}. \tag{28.1}$$

n is here the total number of distinguishable objects (or 'properties') from which the selection is made. From Equation (28.1) it can be seen that if n is large and $k \ll n$, then $|S| \approx n^k$; i.e., the situation approaches ordered sampling *with* replacement.

A special case of formula (28.1) arises when $k = n$. In that case,

$$|S| = n \cdot (n-1) \cdot (n-2) \cdot \ldots \cdot 3 \cdot 2 \cdot 1 = n! \tag{28.2}$$

This expresses the number of *permutations* which are possible for n distinguishable objects. It is the number of distinct n-tuples differing only by a rearrangement of their components, provided that all these components are distinct.

We have now considered experiments that can be regarded as 'ordered sampling *with* replacement' (Sections 25–27), and those which can be regarded as 'ordered sampling *without* replacement' (in this Section). In addition, there is 'unordered sampling', which we address in the next two Sections. The following chart can help orient the reader:

SAMPLING:	ordered	unordered
with replacement	Sections 25–27	Section 30
without replacement	Section 28	Section 29

When more general kinds of dependence exist between the components of a k-tuple, the probability of an event may be expressible in terms of the product rule (22.6). Thus, suppose the possible outcomes of an experiment are characterized by k-tuples

$$(a_1, a_2, a_3, \ldots, a_k).$$

For $i = 1, \ldots, k$, let A_i be an event governed solely by component a_i. Then we know from the chain rule (22.6) that

$$P(A_1 A_2 \ldots A_k) = P(A_1) \cdot P(A_2|A_1) \cdot P(A_3|A_1 A_2) \cdot \ldots \cdot P(A_k|A_1 \ldots A_{k-1}). \tag{28.3}$$

Example 28.2

A digital device generates random binary sequences of length 10, made up of the binary digits **0** and **1**, according to the following rule. The first digit in the sequence is **0** or **1** with equal probability. Furthermore, whenever a **0** occurs, the next digit is again **0** with probability 0.9. Also, whenever a **1** occurs, the next digit is again **1** with probability 0.9. What is the probability of the sequence

$$(0, 0, 0, 0, 0, 1, 1, 1, 1, 1)?$$

For $i = 1, \ldots, 10$, let $A_i \leftrightarrow$ the ith digit in a sequence of length 10 is as specified. Since the probability of each digit depends only on the immediately preceding digit, formula 28.3 simplifies to

$$P(A_1 A_2 \ldots A_{10}) = P(A_1) \cdot P(A_2|A_1) \cdot P(A_3|A_2) \cdot \ldots \cdot P(A_{10}|A_9)$$

and the desired probability is $0.5 \times 0.9^4 \times 0.1 \times 0.9^4 \doteq \underline{0.0215}$.

It is also possible that some or all components of a k-tuple are constrained in a more complicated way by the remaining entries in the k-tuple. Problems of that kind may require special approaches, such as discussed in Section 31.

Queries

28.1 From a pool of 25 job applicants, five individuals are picked arbitrarily and their names are written down in a list. What is the probability that the resulting list will be in alphabetic order? (Assume there are no identical names.)   [21]

28.2 The numbers $1, 2, \ldots, n$ (where $n \geq 3$) are randomly arranged in a sequence, such that all possible arrangements are equally likely. What is the probability that the subsequence 1, 2, 3 appears somewhere within the random sequence?   [161]

28.3 An urn contains six red balls, four white balls, and four green balls. Three balls are withdrawn blindly, in succession, without replacement, and their colors are observed. However, each time a ball is withdrawn, *two* balls of that same color are added to the urn and the contents mixed. Which of the following results is more likely:

a) (red, green, green), or   b) (white, white, white)?   [95]

Multiple Observations Without Regard to Order

In the last few Sections we have been concerned with experiments whose possible outcomes are made up of a multiplicity of observations, or properties. Furthermore, each such observation or property belonged to a different 'type'. Observations might differ according to where they occur within a sequence of observations — the *first* observation, the *second* observation, and so on—or they might differ in nature, as would be the case with the result of a die throw and the result of a coin toss. But this feature, where each observation belongs to a different type, does not always apply.

Example 29.1

Consider the situation described in Example 28.1, but this time let the purpose of the experiment be merely to determine *which* four cards were dealt, no matter to whom. If we did keep track of the recipients as well, we would have 52·51·50·49 possible outcomes, as in Example 28.1. But that is too large a number—each possible outcome of this new experiment is counted many times: as many times as there are ways of arranging the four (distinct) components in a 4-tuple, which according to Equation (28.2) is $4! = 24$. Therefore, it is necessary to divide the above number by the multiplicity of the count, giving $\dfrac{52 \cdot 51 \cdot 50 \cdot 49}{4!} = 270\,725.$

In this Example each possible outcome encompasses k *distinct* properties, but all these properties are of the same 'type'; i.e., the k properties (cards) are not associated with particular recipients. The order in which they are listed is therefore of no significance. Such an experiment is sometimes referred to as *unordered sampling without replacement* (or simply, sampling without replacement). We can again represent the possible outcomes of such an experiment as k-tuples. Each component in the k-tuple is selected from the same collection of properties or objects, which is initially of size n. However, the following constraints apply:

(a) No two components can be the same.
(b) No two k-tuples are allowed that differ only by a rearrangement of their components.

Taking into account only constraint (a), then from Equation (28.1) we obtain $\dfrac{n!}{(n-k)!}$ as the number of distinct k-tuples. But this means that k-tuples that differ from each other only by a rearrangement of their components are included in the count. For a given complement of components there are $k!$ distinct arrangements possible, so that the count is too high by a factor of $k!$. The size of the sample space for an experiment involving the selection of an unordered sample of size k without replacement (from n distinguishable objects or properties) is therefore:

$$|\mathsf{S}| = \frac{n!}{(n-k)!} \cdot \frac{1}{k!} = \binom{n}{k}. \tag{29.1}$$

This is the binomial coefficient first encountered in Section **13**. It expresses 'the number of distinct combinations of n things taken k at a time'.

If a p.e. consists of 'unordered sampling without replacement' from a collection of distinct objects or properties, are the possible outcomes equally likely? Let $\mathfrak{E}_1$ be such an experiment. Assuming that all the properties or objects from which samples are taken are *distinct*, then it is possible to specify an experiment $\mathfrak{E}_2$ that differs from $\mathfrak{E}_1$ only in that the *order* of the samples (the components of the k-tuples) is observed and retained. In other words, $\mathfrak{E}_2$ consists of *ordered* sampling without replacement. It should be clear from the above discussion that to every possible outcome of $\mathfrak{E}_1$ corresponds an event of size $k!$ in $\mathfrak{E}_2$. Since $\mathfrak{E}_2$ has equally likely outcomes (see Section **28**), we find that all events of size $k!$ in $\mathfrak{E}_2$ have the same probability, and this is the probability of the elementary events in $\mathfrak{E}_1$. The possible outcomes of $\mathfrak{E}_1$ are thus equally likely.

In this discussion we have made use of a particular relationship between two experiments, $\mathfrak{E}_1$ and $\mathfrak{E}_2$. The following definition makes this relationship more explicit.

(Definition) 29.1
Given a p.e. $\mathfrak{E}_1$ with sample space S_1, and another p.e. $\mathfrak{E}_2$ with sample space S_2.
(a) If with each one-element set $\{\xi\} \subset \mathsf{S}_1$ can be associated a set $\mathsf{A}_\xi \subset \mathsf{S}_2$ having the same probability, and such that the collection of all the sets A_ξ is a partition of S_2, then we say that $\mathfrak{E}_1$ is a *contraction* of $\mathfrak{E}_2$; and the probability space of $\mathfrak{E}_1$ is a contraction of the probability space of $\mathfrak{E}_2$. Also, $\mathfrak{E}_2$ is then called a *refinement* or *dilation* of $\mathfrak{E}_1$.
(b) If all the sets A_ξ have the same size, then $\mathfrak{E}_1$ is a *uniform contraction* of $\mathfrak{E}_2$, or $\mathfrak{E}_2$ is a *uniform refinement* or *uniform dilation* of $\mathfrak{E}_1$.

The discussion that precedes Definition 29.1 considered unordered sampling without replacement where all the properties or objects that are sampled are *distinct*. We see that such an experiment is a *uniform contraction* of *ordered* sampling without replacement, for the same real-world context. Therefore, the possible outcomes are equally likely. However, this is generally not true if the properties or objects that are sampled are *not* all distinct, as is illustrated in Part b of Example 29.2.

Example 29.2:
(a) Consider the p.e. $\mathfrak{E}_1$ in which a card is drawn from a standard deck in order to observe only the *value* of the card, irrespective of the suit. $\mathfrak{E}_1$ has a sample space S_1 of size $|\mathsf{S}_1| = 13$ (i.e., each $\xi \in \mathsf{S}_1$ represents a possible *value*). This p.e. is a uniform contraction of the experiment $\mathfrak{E}_2$ in which both the suit and the value are observed. To each elementary event in $\mathfrak{E}_1$ corresponds a 4-element event in $\mathfrak{E}_2$. In both experiments the possible outcomes are equally likely.
(b) Now modify both $\mathfrak{E}_1$ and $\mathfrak{E}_2$ by requiring *two* cards to be drawn (without replacement), without regard to order. Then $\mathfrak{E}_1$ is a *nonuniform contraction* of $\mathfrak{E}_2$. For instance, to the

possible outcome (K, Q) in $\mathfrak{E}_1$—'a king and a queen'—corresponds a 16-element event in $\mathfrak{E}_2$; but to the possible outcome (K, K) in $\mathfrak{E}_1$ corresponds a 6-element event in $\mathfrak{E}_2$. The possible outcomes of $\mathfrak{E}_2$ are equally likely, but not those of $\mathfrak{E}_1$. We also note that $\mathfrak{E}_2$ in this case is a uniform contraction of a third experiment, $\mathfrak{E}_3$, which differs from $\mathfrak{E}_2$ only in that the order of appearance of the two cards is taken into account.

An instance of nonuniform refinement has been encountered previously in Example 27.3, where the experiment $\mathfrak{E}$ is replaced by $\mathfrak{E}'$.

In order to compute probabilities of an event in which not all the observed properties or objects are distinguishable, it is easy to begin with the refined experiment in which all observable properties or objects are assumed distinguishable, and then remove the distinguishability.

Example 29.3: Removal of distinguishability.
A container holds four red and two white balls, which are indistinguishable to the touch. Three balls are removed blindly. What is the probability that two red balls and one white ball are drawn (irrespective of order)?

We define a p.e. $\mathfrak{E}_3$ in which the balls are numbered—let them be designated r_1, r_2, r_3, r_4 and w_1, w_2—and in which *ordered* sampling is used. With all balls distinguishable, ordered samples of size 3 without replacement are equally likely and occur with probability (see Equation (28.1))

$$\frac{1}{|\mathsf{S}|} = \frac{(n-k)!}{n!} = \frac{3!}{6!} = \frac{1}{120}.$$

Going to *unordered* samples (but with balls still numbered) implies a uniform contraction of $\mathfrak{E}_3$ to an experiment $\mathfrak{E}_2$. 3! possible outcomes of $\mathfrak{E}_3$ map into a single possible outcome in $\mathfrak{E}_2$. The probability of an elementary event in $\mathfrak{E}_2$ is therefore

$$\frac{3! \cdot 3!}{6!} = \frac{1}{20}.$$

Now we want to remove the distinguishability among the red balls and among the white balls. This suggests a *nonuniform contraction* from $\mathfrak{E}_2$ to another experiment, $\mathfrak{E}_1$, in which the event whose probability is to be found is an elementary event. $\mathfrak{E}_1$ has three possible outcomes, corresponding to one, two, or three red balls drawn. (At least one red ball must get drawn.)

We ask: How many possible outcomes of $\mathfrak{E}_2$ correspond to 'two red balls and one white ball' in $\mathfrak{E}_1$? Since *any* two red balls and *any* one white ball is allowed, the answer is the number of ways of choosing two red balls out of four, multiplied by the number of ways of choosing one white ball out of two:

$$\binom{4}{2}\binom{2}{1} = \frac{4!}{2!\,2!} \cdot \frac{2!}{1!\,1!} = 3 \cdot 2 \cdot 2 = 12.$$

The desired probability is therefore $\dfrac{12}{20} = \underline{0.6}$.

This result can easily be generalized as follows. Suppose the container holds a total of n balls of which r are red and the others white. If k $(1 \le k \le n)$ balls are withdrawn, what is the probability that j $(0 \le j \le k,\ r)$ of these are red? In the experiment $\mathfrak{E}_1$ (reformulated with these new parameters), let $\mathsf{R}_j \leftrightarrow$ 'j red balls are drawn'. Then

$$P(\mathsf{R}_j) = \frac{\dbinom{r}{j}\dbinom{n-r}{k-j}}{\dbinom{n}{k}} \tag{29.2}$$

For given n and k, in the probability space for $\mathfrak{E}_1$, the sets R_j $(0 \leq j \leq k, r)$ constitute a partition, and the probabilities $P(\mathsf{R}_j)$ represent a 'hypergeometric' distribution over that partition.

Queries

29.1 In each of the following, determine whether:
  (a) $\mathfrak{E}_1$ is a uniform contraction of $\mathfrak{E}_2$;
  (b) $\mathfrak{E}_1$ is a nonuniform contraction of $\mathfrak{E}_2$; or
  (c) $\mathfrak{E}_1$ is not a contraction of $\mathfrak{E}_2$.

  i) $\mathfrak{E}_1$: A coin is tossed five times, and the result of each toss is recorded
     $\mathfrak{E}_2$: A coin is tossed six times, and the result of each toss is recorded.
  ii) $\mathfrak{E}_1$: A coin is tossed five times, and the number of heads is noted
     $\mathfrak{E}_2$: A coin is tossed five times, and the result of each toss is noted.
  iii) $\mathfrak{E}_1$: A die is tossed five times, and the result of each toss is recorded, but in
     each toss only the numbers 1 through 5 are of interest
     $\mathfrak{E}_2$: A die is tossed five times, and the result of each toss is recorded, where in
     each toss all six faces are of interest. [59]

29.2 An urn contains six balls, marked with the numbers 1 to 6. Four balls are drawn blindly. What is the probability that the numbers 1 through 4 are drawn (in any order)? [215]

Unordered Sampling with Replacement

'Unordered sampling with replacement' refers to a category of experiments where

a) there are identical trials;
b) the initial experimental conditions are restored after each trial; and
c) there is no interest in the order in which the results of the trials are observed.

What gets recorded is only *how many times* each possible outcome of a trial is observed.

A p.e. involving unordered sampling with replacement generally has possible outcomes that are *not* equally likely. To develop the probability space for such an experiment, it is helpful to begin with a refined experiment—namely, *ordered* sampling with replacement. This refined experiment may have equally likely outcomes. Such an approach was illustrated in Example 25.3. We consider now a different example, where the refined experiment does not have equally likely outcomes and where the trials are not binary.

Example 30.1

A large container is filled with a mixture of ping-pong balls of different colors. There are n_b blue balls, n_g green balls, n_r red balls, and n_y yellow balls, the total number of balls being $n = n_b + n_g + n_r + n_y$, where $n_b, n_g, n_r, n_y \geq 1$. Ten times a ball is withdrawn blindly, and returned each time and mixed in among the other balls. Of interest is how many balls of each of the four colors are drawn.

This situation can be modeled as a p.e. $\mathfrak{E}$ whose procedure calls for ten independent executions of the basic trial of blindly withdrawing a ball, noting its color, returning it and mixing it in with the other balls. The experimenter keeps a tally. On a sheet with four columns headed 'blue,' 'green,' 'red,' and 'yellow,' she puts a mark in the appropriate column every time the color of a ball is observed, until the ten withdrawals have been completed—i.e., until ten marks have been placed on the tally sheet.

The size of the sample space equals the number of ways in which ten marks can be distributed over four columns, which is 286. This can be determined as follows: Let n_2 be the number of marks in the first two columns. For a given value of n_2 there are $n_2 + 1$ ways of distributing this number

---

Probability Concepts and Theory for Engineers, First Edition. Harry Schwarzlander.
© 2011 John Wiley & Sons, Ltd. Published 2011 by John Wiley & Sons, Ltd.

of marks over the first two columns, and $10 - n_2 + 1$ ways of distributing the remaining marks over the last two columns, giving $(n_2 + 1)(10 - n_2 + 1)$ possible arrangements. The total number of possible arrangements is therefore

$$\sum_{n_2=0}^{10} (n_2 + 1)(11 - n_2) = 286.$$

But these elementary events are not equally probable, not even if $n_b = n_g = n_r = n_y$. To find their probabilities, we can refine $\mathfrak{E}$ into a product experiment

$$\mathfrak{F} = \underbrace{\mathfrak{T} \times \mathfrak{T} \times \ldots \times \mathfrak{T}}_{10 \text{ times}}$$

where each trial $\mathfrak{T}$ consists of one selection of a ball. Each trial has four possible outcomes, representing the four different colors that can be drawn. In $\mathfrak{T}$, the probability of a blue ball is $p_b = n_b/n$, of a green ball $p_g = n_g/n$, of a red ball $p_r = n_r/n$, of a yellow ball $p_y = n_y/n$. The probability space $(S, \mathscr{A}, P)$ for $\mathfrak{F}$ has size $|S| = 4^{10} = 1\,048\,576$. A given elementary set in S, corresponding to k_b blue, k_g green, k_r red and k_y yellow balls drawn *in a particular order* (where $k_b + k_g + k_r + k_y = 10$), has probability

$$p_b^{k_b} \cdot p_g^{k_g} \cdot p_r^{k_r} \cdot p_y^{k_y}. \tag{30.1}$$

This completely specifies the probability distribution P for the refined probability space. (We note that in the special case where $n_b = n_g = n_r = n_y$, so that $p_b = p_g = p_r = p_y$, the distribution P is uniform, assigning a probability of 4^{-10} to every elementary set in $\mathscr{A}$.)

The p.e. $\mathfrak{E}$ in which we are interested is a contraction of $\mathfrak{F}$; but not a uniform contraction. Let $\mathscr{P}$ be a partition of S such that each set of $\mathscr{P}$ is made up of all those elements of S that correspond to a particular 4-tuple (k_b, k_g, k_r, k_y). To obtain the probability of such a set, the expression (30.1) needs to get multiplied by the size of the set. This is the number of ordered 10-tuples made up of k_b b's (blue balls), k_g g's (green balls), k_r r's (red balls), and k_y y's (yellow balls). There are $\binom{10}{k_b}$ ways to arrange the k_b b's. For a given arrangement of b's there are $\binom{10 - k_b}{k_g}$ ways to arrange the k_g g's; then $\binom{10 - k_b - k_g}{k_r}$ ways to arrange the k_r r's. This leaves the k_y remaining positions for the y's. For $k_b + k_g + k_r + k_y = 10$, the size of a set in $\mathscr{P}$ is then

$$\binom{10}{k_b} \cdot \binom{10 - k_b}{k_g} \cdot \binom{10 - k_b - k_g}{k_r} \cdot 1 = \frac{10!}{k_b! k_g! k_r! k_y!}.$$

We therefore associate with $\mathfrak{E}$ the probability space $(S_1, \mathscr{A}_1, P_1)$ in which each element of S_1 identifies one of the sets in $\mathscr{P}$, where $\mathscr{A}_1$ is the class of all subsets of S_1, and where P_1 assigns to the elementary sets contained in S_1 the probabilities

$$\frac{10!}{k_b! k_g! k_r! k_y!} (p_b^{k_b} \cdot p_g^{k_g} \cdot p_r^{k_r} \cdot p_y^{k_y}). \tag{30.2}$$

This completely specifies the distribution P_1, which is called a *multinomial distribution*.

The possible outcomes of the p.e. 𝕰 in this Example can still be characterized by 10-tuples, but a mere reordering of the components within any one 10-tuple is of no consequence. Therefore, each 10-tuple might as well be arranged in alphabetical order with all the b's first, then the g's, then the r's and last the y's. But this is equivalent to an ordered *4-tuple* indicating the number of b's, g's, r's and y's, respectively; with the constraint that the sum of the components must equal 10.

The discussion in Example 30.1 can be generalized to yield the following:

Theorem 30.1
An experiment 𝕰 consists of k identical, independent trials. Each trial leads to the observation of exactly one out of m properties of interest. To be determined is the number of times each of the properties is observed in the k trials. Then 𝕰 has associated with it a probability space with a *multinomial distribution* over its elementary sets. Specifically, in a single trial, let p_i be the probability of the ith property being observed; $i = 1, \ldots, m$. Then, in 𝕰, the probability that

the first property occurs k_1 times
the second property occurs k_2 times
....
the mth property occurs k_m times

where $k_1 + k_2 + \ldots + k_m = k$, is given by

$$\frac{k!}{k_1! k_2! \ldots k_m!} \cdot p_1^{k_1} \cdot p_2^{k_2} \cdot \ldots \cdot p_m^{k_m} \tag{30.3}$$

The summing of multinomial probabilities can be awkward. Suppose that, in an experiment 𝕰 of the kind specified in Theorem 30.1, an event is of interest whose probability is not given directly by Equation (30.3). It may then be easier to redefine the experiment and develop the appropriate distribution in a manner analogous to Example 30.1. For instance, suppose that in Example 30.1, we seek the probability of obtaining k_b blue balls (irrespective of what other colors are drawn). This can be answered by replacing 𝕰 by a simpler experiment $𝕰_b$ in which only the observation of either a blue ball or not a blue ball gets tallied. Theorem 30.1 applies to $𝕰_b$ with $m = 2$, and we immediately obtain the probabilities

$$\binom{10}{k_b} p_b^{k_b} (1 - p_b)^{10 - k_b} \tag{30.4}$$

describing a *binomial* probability distribution. Thus, we see that the binomial distribution is a special case of the multinomial distribution; Equation (30.3) becomes the expression for a binomial distribution when $m = 2$.

Queries

30.1 In the throw of five dice (e.g., the game of Yahtsee), what is the probability that each of the dice shows a different number of dots?   [77]

30.2 Is the probability of getting exactly two 'heads' in ten coin tosses
  a) greater than
  b) equal to, or
  c) less than
  the probability of getting exactly one 'head' in five coin tosses?   [204]

More Complicated Discrete Probability Problems

There are a great variety of discrete probability problems that involve the counting of possible outcomes but do not lend themselves to the approaches discussed in the last few Sections. Such problems usually require the judicious application of conditional probability, combinatorics, and careful accounting, and can be quite tedious. As an illustration we consider an example of an *assignment problem*.

Example 31.1

At a dance, five couples draw lots to decide which man is to dance with which woman. Later they wish to change partners, and draw lots again. One man ends up with the same partner as before and complains that the drawing must have been biased. If the drawing is in fact unbiased, what is the probability of at least one pair being the same after the second drawing?

To formulate the experiment, we denote the five women by $w_1, \ldots, w_5$ and the five men by $m_1, \ldots, m_5$, where matching subscripts identify partners after the first drawing. The p.e. involves generating an assignment of m's to w's. Every distinct assignment is a possible outcome, and it is the purpose of the p.e. to determine which assignment gets generated as a result of the second drawing. The possible outcomes, therefore, can be characterized by 5-tuples such as $(m_5, m_1, m_3, m_2, m_4)$, where the first component indicates assignment to w_1, the second component to w_2, and so on. The procedure requires use of an appropriate unbiased random mechanism for generating this assignment in such a way that all possible assignments are equally likely. We see that this is ordered sampling without replacement; but the discussion of Section 28 does not provide a convenient approach for finding the desired probability.

The size of the sample space for this p.e. is given by Equation (28.2): $|S| = 5! = 120$. Of interest is the event

$$A \leftrightarrow \text{'at least one } (w, m)\text{-pair has matching subscripts'}$$

and $P(A) = \dfrac{|A|}{|S|}$. As is usual when considering an event that involves 'one or more' of something, we work with the complementary event. Thus, $P(A) = 1 - P(A^c) = 1 - \dfrac{|A^c|}{|S|}$. What is $|A^c|$?

In order to compute the number of possible outcomes making up A^c, suppose we try to use the approach of Example 28.1. To w_1 can be assigned m_2, m_3, m_4, or m_5, so there are four possibilities.

Probability Concepts and Theory for Engineers, First Edition. Harry Schwarzlander.
© 2011 John Wiley & Sons, Ltd. Published 2011 by John Wiley & Sons, Ltd.

But it gets more complicated when we come to w_2. If m_2 is assigned to w_1, then there are four possibilities for w_2; otherwise there are only three. Similar distinctions must be made when counting possible assignments to w_3 and w_4.

Instead, in order to keep track of the conditions that affect the count, we resort to a *tree diagram* (see Figure 31.1). The top node (circle) in the tree represents assignments to w_1; i.e., the first component in our 5-tuple. The number of possibilities is identified by the number 4 next to the circle for this node. The assignment that is made is denoted m_a, as indicated inside the circle. Thus, m_a stands for m_2, m_3, m_4 or m_5; i.e., $a = 2, 3, 4,$ or 5.

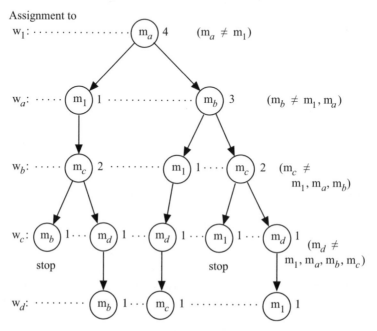

Figure 31.1   Tree diagram

In the next row of nodes we consider possible assignments to w_a, i.e., the ath component in our 5-tuple. This assignment can have either the subscript 1 or a different subscript b ($b \neq 1, a$). In each case, the number of possibilities is indicated by the numeral next to the node. The third row contains possible assignments to w_b. For the left branch of the tree, b has not been defined as yet and can be chosen arbitrarily (but again $b \neq 1, a$). Proceeding in this manner, the tree shown in Figure 31.1 is obtained. Branches marked 'stop' cannot be completed; i.e., they result in assignments that do not belong to A^c.

Each path reaching the lowest level, the level of w_d, represents a collection of valid assignments. The number of distinct assignments (5-tuples) represented by such a path is obtained by beginning at the lowest level node and tracing the path to the top, multiplying the numbers encountered on the way. Doing this for each path that reaches the lowest level yields

$$|A^c| = 1 \cdot 1 \cdot 2 \cdot 1 \cdot 4 + 1 \cdot 1 \cdot 1 \cdot 3 \cdot 4 + 1 \cdot 1 \cdot 2 \cdot 3 \cdot 4 = 8 + 12 + 24 = 44.$$

Therefore, $P(A) = 1 - \dfrac{|A^c|}{|S|} = 1 - \dfrac{44}{120} = \dfrac{19}{30} \doteq \underline{0.633}$. Thus, it is more likely than not that at least one of the pairs will be the same.

On the other hand, the probability that a *particular* man, say m_1, will be dancing with the same woman a second time is easily found. Since each woman has the same likelihood of being paired with m_1, this probability is 1/5.

Whenever there is a need to trace one or more sequences of steps that arise either in the experimental procedure, or in the situation to which the procedure is being applied, or in the analysis of the problem, a *state diagram* can be useful. A tree diagram such as utilized in Example 31.1 is a special kind of state diagram which has no closed loops. A state diagram can be static or dynamic. The tree diagram in Example 31.1 is a static diagram—nothing is 'happening'; it serves to visualize all the various possible assignments. In Example 31.2 a dynamic state diagram is used to help in the determination of probabilities.

Example 31.2

A machine processes jobs that arrive at arbitrary times. For the purpose of describing the operation of the machine, time is partitioned into successive intervals of equal duration. During any given interval, a new job arrives with probability p_a. No more than one job can arrive in one interval. If a newly arrived job cannot be accepted by the machine because it is busy with another job, the newly arrived job enters a queue (waiting line). If the machine is processing a job at the beginning of an interval, then there is a fixed probability p_c ($p_c \neq 0, 1$) of completing the job during that interval, irrespective of how long the job has been running. (The possibility of an arrival or completion at the exact start or end of an interval is ignored.) If the machine completes a job during a particular interval, the machine immediately accepts the next job waiting in the queue or, if the queue is empty, accepts a job that may be newly arriving during that interval. If the queue is full when a new job arrives, and remains full till the end of the interval, the new job gets rerouted for processing elsewhere. Suppose that the queue cannot hold more than two jobs. What is the probability that a job arrives during an interval in which the queue remains full, so that the job must be turned away?

Here is a dynamic situation or 'system' that, at the beginning of any time unit, can be in one of several possible states, as follows:

S_0: The machine is idle and no jobs are waiting
S_1: The machine is busy but no jobs are waiting
S_2: The machine is busy and one job is waiting
S_3: The machine is busy and two jobs are waiting

When operation commences, the system is presumably in state S_0. In successive intervals the other states become possible, so that the effect of the system starting out in state S_0 gradually disappears and the state probabilities approach 'steady-state' values, i.e., values that no longer change with time. We will assume that a *steady state* has been reached.

The experiment, therefore, consists of observing the system during an arbitrarily chosen interval well beyond the start-up period. What is observed is the state of the system at the beginning of the interval, and what happens during the interval: Does a new job arrive? And, does the machine complete a job it is processing? This leads to a sample space S as shown in Figure 31.2. But more convenient is a state diagram, indicating the various possible transitions from states at the beginning of the time interval being observed to states at the end of that interval. This is shown in Figure 31.3.

The probability of an arrival is given. Also known is the conditional probability of a completion, given states S_1, S_2, or S_3. What is asked for is the probability of the event A shown shaded in the Venn diagram: the system being in state S_3 at the beginning of the interval, a job arriving, and the

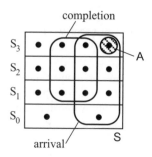

Figure 31.2   Sample space

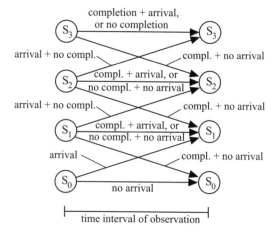

Figure 31.3   State diagram

processor not completing the job it is processing. But we first need to determine the probability that the system is in S_3.

Because of the steady-state assumption, the probabilities associated with the various states at the beginning of the observation interval are the same as at the end of this interval. This makes it possible to simplify the state diagram so that only one set of states is shown. Each arrow is labeled with the conditional probability of the respective state transition, given the state from which the arrow originates. Then the state diagram takes on the form of Figure 31.4, where q_a denotes $1 - p_a$

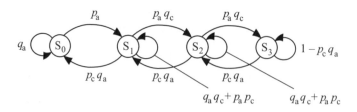

Figure 31.4   Simplified state diagram

and q_c denotes $1 - p_c$. As a check, we observe that the conditional probabilities of all the arrows leaving any given state sum to 1.

Let p_i denote the probability that the system is in state S_i ($i = 0, 1, 2, 3$). We are seeking p_3. From the state diagram we see that S_3 can arise in two different ways—two arrows lead to it. Summing the probabilities with which each of these two state transitions can occur yields p_3:

$$p_3 = (1 - q_a\, p_c)p_3 + p_a\, q_c\, p_2$$

This can be written $p_3 = \alpha\, p_2$, where $\alpha = \dfrac{p_a q_c}{q_a p_c}$. In a similar manner the following equation for p_2 is obtained:

$$p_2 = p_c\, q_a\, p_3 + (q_c\, q_a + p_a\, p_c)p_2 + p_a\, q_c\, p_1$$

Substitution for p_3 and simplification yields $p_2 = \alpha\, p_1$. Similarly, the equation for p_1 results in $p_1 = \alpha \dfrac{p_0}{q_c}$.

p_0 needs to be found also, but a fourth equation analogous to the above, based on the state diagram, cannot be used because it would be redundant. Instead, we now make use of the requirement $p_0 + p_1 + p_2 + p_3 = 1$. This gives $p_0 = \dfrac{q_c}{\alpha + \alpha^2 + \alpha^3 + q_c}$, so that

$$p_3 = \frac{\alpha^3}{\alpha + \alpha^2 + \alpha^3 + q_c}.$$

The probability that an arriving job cannot be accepted is then

$$P(\mathsf{A}) = p_3\, p_a\, q_c = \frac{\alpha^3 p_a\, q_c}{\alpha + \alpha^2 + \alpha^3 + q_c}. \qquad (31.1)$$

Queries

31.1 In Example 31.1, what is the probability that
    a) each pair is the same after the second drawing?
    b) exactly one man ends up with a different partner?   [166]

31.2 In Example 31.2, suppose that $p_a = 0.1$. How large must p_c be so as to assure that no job will get turned away?   [197]

Uncertainty and Randomness

All along we have used our basic model, the probabilistic experiment, to address situations containing elements of unpredictability that satisfy the requirements of Section 15. The unpredictability inherent in such a situation gives rise to a vague sense of uncertainty in the mind of a beholder. However, once the beholder has modeled the situation as a p.e. and regards himself or herself as the experimenter, then the uncertainty has been made precise, so to speak. In other words, the beholder then has made a decision which defines exactly what it is that he or she is uncertain about—namely, *which of various possible outcomes* will be observed the next time this p.e. is executed.

This last statement can be reformulated slightly by placing it in the context of the probability space that gets defined for the p.e. in question, Thus, bringing it into the mathematical model. Accordingly, we can say that the uncertainty of the experimenter consists in not knowing *which elementary event* will occur when the p.e. is next executed. *All* of the experimenter's uncertainty (pertaining to this p.e.) is removed when the p.e. has been satisfactorily performed; i.e., when the experimenter has observed and noted which of the possible outcomes is the actual outcome. For, the uncertainty concerning the occurrence or nonoccurrence of *every* event that can be defined has thereby been removed.

The uncertainty experienced by the experimenter, furthermore, is tied to a particular 'random mechanism' which comes into play when the p.e. is executed—or to several random mechanisms. Random mechanisms reside in the physical world, and are the source of the unpredictability associated with a p.e.[4] So we have the following progression: A random mechanism produces unpredictability in some situation. The unpredictability generates uncertainty in the mind of a beholder. This motivates the beholder to model the situation as a p.e. In relation to this p.e., the influence of the random mechanism has then been made precise, and also the uncertainty experienced by the beholder. This uncertainty is eliminated upon executing the p.e.

The question arises how the randomness created by some random mechanism, or the uncertainty experienced by an experimenter, can be quantified. Clearly, some situations exhibit more randomness than others.

---

[4] More specifically, a random mechanism is *a way of conceptualizing* some aspect of the physical world which defies detailed analysis, measurement or understanding, and that, therefore, exhibits a behavior that is deemed unpredictable. Thus, we 'believe' that when a good die is thrown in a manner that would seem to preclude the imparting of any bias by the thrower, then the detailed motion of the die and the rest position it attains is unknown and unknowable to the experimenter until actually observed. This means that we do *not* believe, or put out of our mind the possibility, that the experimenter or the die itself or some extraneous observer or some spirit entity is able to influence the result of the throw.

Probability Concepts and Theory for Engineers, First Edition. Harry Schwarzlander.
© 2011 John Wiley & Sons, Ltd. Published 2011 by John Wiley & Sons, Ltd.

Example 32.1

Consider the simplest situation of uncertainty. This is a p.e. with *two* possible outcomes since, if there is only one possible outcome, there is no uncertainty. Assume statistical regularity to hold, so that the situation can be modeled in terms of a binary (two-element) probability space. Denote the two elementary sets A and B. P remains to be specified. Certainly there is some randomness in this situation—the experimenter has some uncertainty regarding what will happen when the experiment is performed. Now we ask, how does this randomness depend on P?

Consider the extreme case where $P(A) = 1$ and $P(B) = 0$. Then there really is no uncertainty, and the randomness produced by the underlying random mechanism is zero. But the same is true if $P(B) = 1$ and $P(A) = 0$. In general, though, as we have noted, there is uncertainty associated with such a p.e. Therefore, the randomness and uncertainty must increase as $P(A)$ is increased from 0, and it must decrease as $P(A)$ is increased toward 1. (Symmetry would suggest that the maximum occurs at $P(A) = P(B) = 0.5$, since there is no reason to expect other inflection points, beside a simple maximum, in the manner in which uncertainty depends on $P(A)$.)

From this Example it can be concluded that randomness and uncertainty depend on P. The question we will consider now is *how* it depends on P. Consider first some p.e. $\mathfrak{E}$ with *equally likely* possible outcomes. Then P is fixed once the size of the sample space has been specified. If $|S| = n$, then all elementary events have probability $1/n$. The uncertainty associated with $\mathfrak{E}$, therefore, can only depend on n. In fact, we would expect it to *increase* with n. (It has already been noted in Example 32.1 that, if $n = 1$, then there is no uncertainty.) Letting $H_{\mathfrak{E}}$ denote the uncertainty experienced by the experimenter prior to executing $\mathfrak{E}$, we therefore have

$$H_{\mathfrak{E}} = f(n) \tag{32.1}$$

where f is the function to be determined.

Suppose $\mathfrak{E}$ is to be performed again. Assuming independent experiments, the same amount of uncertainty, $H_{\mathfrak{E}}$, is again experienced by the experimenter prior to executing the experiment a second time. Altogether, in connection with performing the experiment twice, the experimenter therefore experiences (and is then relieved of) an amount of uncertainty $2H_{\mathfrak{E}}$. But we can also model the two executions of $\mathfrak{E}$ as a single execution of the product experiment $\mathfrak{E}^2 = \mathfrak{E} \times \mathfrak{E}$, without affecting the associated uncertainty or randomness of $2H_{\mathfrak{E}}$. Since $\mathfrak{E}^2$ has n^2 equally likely outcomes, we see that

$$2H_{\mathfrak{E}} = f(n^2). \tag{32.2}$$

Combining Equations (32.1) and (32.2) yields $2f(n) = f(n^2)$. The function that satisfies this relationship for arbitrary positive n is the logarithm function. The uncertainty or randomness associated with a probabilistic experiment $\mathfrak{E}$ having n equally likely possible outcomes can therefore be expressed as

$$H_{\mathfrak{E}} = c \log n \tag{32.3}$$

where c is an arbitrary scale factor.

We note that for $n = 1$, Equation (32.3) evaluates to zero as it should. The smallest amount of randomness that is possible (with equally likely possible outcomes) arises when $n = 2$. If this amount is chosen as the basic unit of randomness, or uncertainty, we require $c \log 2 = 1$, which is satisfied if $c = 1$ and the base of the logarithm is chosen as 2. This gives the randomness in *binary units* or *bits*:

$$H_{\mathfrak{E}} = \log_2 n \text{ bits.} \tag{32.4}$$

Thus, a simple coin toss has $\log_2 2 = 1$ bit of randomness, whereas a die throw has $\log_2 6 \doteq 2.585$ bits, and the throw of two dice has $\log_2 36 \doteq 2 \times 2.585 = 5.170$ bits. This last result demonstrates that if the throw of two dice is to be simulated using coin tosses, then at least six coin tosses will be required since the amount of randomness associated with five tosses is only $5 \times 1 = 5.000$ bits.

In the above derivation we have restricted ourselves to experiments where the possible outcomes are equally likely. As a result, $H_\mathbb{E}$ depended only on n. Nevertheless, as noted earlier, $H_\mathbb{E}$ must depend on P. Since P is completely specified by the values it assigns to all the elementary events, we can make the dependence of H on P explicit by writing the probabilities of the elementary events as arguments of the function H. So we have, in the case of equally likely possible outcomes,

$$H\left(\frac{1}{2}, \frac{1}{2}\right) = \log_2 2 = 1 \text{ bit}$$

$$H\left(\frac{1}{3}, \frac{1}{3}, \frac{1}{3}\right) = \log_2 3 \doteq 1.585 \text{ bits, etc.}$$

and in general,

$$H\left(\frac{1}{n}, \frac{1}{n}, \ldots, \frac{1}{n}\right) = \log_2 n \text{ bits.} \tag{32.5}$$

Examination of these expressions leads to the question whether it is possible to express the function H as a *sum* of terms, where each term is a function of the probability of one of the elementary events. Indeed, Equation (32.5) will be satisfied if each elementary event A_i ($i = 1, \ldots, n$, where $n = |S|$) contributes to such a summation an amount $P(A_i) \log_2 \frac{1}{P(A_i)}$, or $-P(A_i) \log_2 P(A_i)$. Then

$$\sum_{i=1}^{n} [-P(A_i) \log_2 P(A_i)] = \sum_{i=1}^{n} \frac{1}{n} \log_2 n = \log_2 n$$

as required by Equation (32.5). This formulation generalizes to arbitrary probability distributions P, as can be substantiated by other considerations (cf. [Kh1]), and we have:

(Definition) 32.1
Given a countable probability space $(S, \mathscr{A}, P)$, let $p_1, p_2, p_3, \ldots$ be the probabilities of the sets belonging to the partition that generates $\mathscr{A}$ (the elementary sets). Then the randomness, uncertainty or *entropy* associated with this space, and with any p.e. modeled by this space, is

$$H(p_1, p_2, p_3, \ldots) = -\sum p_i \log_2 p_i \text{ bits}$$

where $-p_i \log_2 p_i = 0$ if $p_i = 0$.

Clearly, experiments that are probabilistically equivalent have the same entropy. It also follows from Definition 32.1 that entropy is a non-negative quantity, which is necessary if it is to express uncertainty or randomness. Furthermore, zero entropy can arise if and only if one of the probabilities p_i equals 1.

We see from Definition 32.1 that the entropy associated with a probability space $(S, \mathscr{A}, P)$ is computed from the probabilities of the elementary sets; i.e., the sets belonging to the partition $\mathscr{P}$ that generates $\mathscr{A}$. The entropy of $(S, \mathscr{A}, P)$ can therefore be thought of as associated with $\mathscr{P}$, so that the notation $H(\mathscr{P})$ can also be used. In a similar way, an entropy can be computed for some other partition $\mathscr{P}_1$. After all, $\mathscr{P}_1$ generates an additive class $\mathscr{A}_1 \subset \mathscr{A}$ and therefore a new probability space $(S, \mathscr{A}_1, P_1)$, which is like a contraction of the original space except that $|S|$ is unchanged. We call the entropy $H(\mathscr{P}_1)$ associated with $(S, \mathscr{A}_1, P_1)$ a *partial entropy* of the original space, because it is predicated on a partial observation of the result of the underlying p.e. The motivation for computing a partial entropy might arise if, in the execution of the underlying p.e., not all the properties of interest are immediately discernible or adequately recorded, and therefore not all of the experimenter's uncertainty gets removed at once.

Example 32.2

In the probability space $(S, \mathscr{A}, P)$ that models the throw of a red and a white die, the partition $\mathscr{P}$ which generates $\mathscr{A}$ has size $|\mathscr{P}| = 36$, and since the elementary sets are equally probable the entropy is $\log_2 36 \doteq 5.170$ bits, as noted above.

Case A. Suppose now that in the execution of the experiment the number of dots showing on the red die is noted and recorded first. In the mathematical model this can be represented by introducing the partition $\mathscr{P}_r$ made up of the six equiprobable events

$$R_i \leftrightarrow \text{'}i \text{ dots on the red die'} \ (i = 1, \ldots, 6),$$

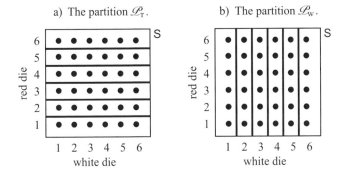

a) The partition $\mathscr{P}_r$.   b) The partition $\mathscr{P}_w$.

Figure 32.1   Two partitions

as shown in Figure 32.1a. $\mathscr{P}_r$ generates the new additive class $\mathscr{A}_r$ and thus defines the new space $(S, \mathscr{A}_r, P_r)$ whose entropy $H(\mathscr{P}_r)$ is simply that of a single die throw, $\log_2 6 = 2.585$ bits. Thus, upon observing the red die result, the experimenter's initial uncertainty of 5.170 bits has been reduced to 2.585 bits, one-half the original amount. To become relieved of the remaining uncertainty, the experimenter's must of course observe the white die, which can be modeled in terms of the partition $\mathscr{P}_w$ illustrated in Figure 32.1(b). $\mathscr{P}_w$ defines the probability space $(S, \mathscr{A}_w, P_w)$ whose entropy $H(\mathscr{P}_w)$ is also 2.585 bits. So we see that the entropy $H(\mathscr{P})$ of the throw of both dice can be decomposed into $H(\mathscr{P}_r) + H(\mathscr{P}_w)$, just as the entropies of two separate throws of a die (or of two different dice) add up to the entropy of the throw of two dice.

Case B. Now we consider a slight variation of the above and suppose that for some reason the experimenter begins by noting the *sum* of the dots showing on the red and the white die. This is modeled by the partition $\mathscr{P}_1$ specified in Example 20.3, and illustrated in Figure 20.1. $\mathscr{P}_1$ leads to the modified space $(S, \mathscr{A}_1, P_1)$ described in Example 20.3. Using Definition 32.1, the partial entropy due to observing only the sum of the dots is then

$$H(\mathscr{P}_1) = \frac{1}{36}\log_2 36 + \frac{1}{18}\log_2 18 + \ldots + \frac{1}{36}\log_2 36 \doteq 3.274 \text{ bits.}$$

Here, the first observation reduces the uncertainty of the experimenter to 1.896 bits. One way to eliminate all remaining uncertainty is to make the second observation the same as in Case A; i.e., observe how many dots show on the white die. But $H(\mathscr{P}_w) = 2.585$ bits, which is more than the 1.864 bits needed to remove all uncertainty. In other words we now find that $H(\mathscr{P}_1) + H(\mathscr{P}_w)$ is greater than the entropy of the two-dice experiment. Why?

The answer is that there is a dependence between the two observations made in Case B, whereas there is no dependence in Case A. Thus, in Case A, whatever result is noted during the first observation (the number of dots showing on the red die) has no effect on the probabilities with which different numbers can arise on the white die. This is not true in Case B since, for instance, a sum of 'two' immediately fixes the number of dots that can appear on the white die. To accommodate situations as in Case B, a *conditional entropy* needs to get defined.

Let some p.e. be given, with probability space (S, $\mathscr{A}$, P) and entropy $H(\mathscr{P})$, where $\mathscr{P}$ is the partition that generates $\mathscr{A}$. If Part 4 of the p.e. (see Definition 2.1) is accomplished in two consecutive distinct observational steps, then each step defines a partition. Let the partition defined by the first observational step be denoted $\mathscr{P}_1$. $\mathscr{P}_1$ generates the new space (S, $\mathscr{A}_1$, P_1) whose entropy $H(\mathscr{P}_1)$ is a partial entropy of (S, $\mathscr{A}$, P). If $\mathscr{P}_2$ is the partition defined by the second observational step, we define the *conditional entropy* $H(\mathscr{P}_2|\mathscr{P}_1)$ in the following way. Let $\mathscr{P}_1 = \{A_1, A_2, A_3, \ldots, A_i, \ldots\}$, and $\mathscr{P}_2 = \{B_1, B_2, B_3, \ldots, B_j, \ldots\}$. Then

$$H(\mathscr{P}_2|\mathscr{P}_1) = -\sum_i \sum_j P(A_i\, B_j)\log_2 P(B_j|A_i). \qquad (32.6)$$

It is important to observe that the conditional probability $P(B_j|A_i)$ appears only as the argument of the logarithm, not as the multiplier.

It can now be stated that

$$H(\mathscr{P}) = H(\mathscr{P}_1) + H(\mathscr{P}_2|\mathscr{P}_1) \qquad (32.7)$$

provided $\mathscr{P} = \{A_iB_j: A_i \in \mathscr{P}_1, B_j \in \mathscr{P}_2\}$. This is easily verified from the definitions of the various entropies (Problem 32.2). On the other hand, if there is no dependence between the successive observations, i.e., the sets in $\mathscr{P}_2$ are independent of the sets in $\mathscr{P}_1$, then $H(\mathscr{P}_2|\mathscr{P}_1)$ reduces to $H(\mathscr{P}_2)$. Then the partial entropies $H(\mathscr{P}_1)$ and $H(\mathscr{P}_2)$ sum to $H(\mathscr{P})$, as in Case A of Example 32.2.

An important property of the entropy of a finite probability space is expressed by the following Theorem.

Theorem 32.1

Given a finite probability space (S, $\mathscr{A}$, P). Let $\mathscr{P}$ be the partition which generates $\mathscr{A}$, where $|\mathscr{P}| = n$; and let $p_1, p_2, p_3, \ldots, p_n$ be the probabilities of the elementary sets. Then the entropy is maximized if and only if

$$p_1 = p_2 = p_3 = \ldots = p_n = \frac{1}{n}.$$

In other words, among all p.e.'s with a given number of possible outcomes, those with equally likely possible outcomes give rise to the greatest uncertainty, and no others.

Proof

Let $p_1, p_2, p_3, \ldots, p_n$ (where $\sum_i p_i = 1$) be an arbitrary probability assignment. It is to be shown that

$$H\left(\frac{1}{n}, \frac{1}{n}, \frac{1}{n}, \ldots, \frac{1}{n}\right) - H(p_1, p_2, p_3, \ldots, p_n) \geq 0. \tag{32.8}$$

Multiplying through by $\ln 2$ converts all logarithms to base e. The first term is then changed into

$$(\ln 2) \cdot H\left(\frac{1}{n}, \frac{1}{n}, \frac{1}{n}, \ldots, \frac{1}{n}\right) = \ln n.$$

which can also be written $\sum p_i \ln n$, and the second term is similarly modified. The left side of Equation (32.8) then becomes

$$\sum p_i \ln n + \sum p_i \ln p_i = \sum p_i \ln n p_i = -\sum p_i \ln \frac{1}{n p_i}.$$

Using the inequality $\ln x \leq x - 1$ which holds for all $x > 0$, and noting that

$$\lim_{p_i \to 0} \left(p_i \ln \frac{1}{n p_i}\right) = 0$$

we obtain

$$-\sum p_i \ln \frac{1}{n p_i} \geq -\sum p_i \left(\frac{1}{n p_i} - 1\right) = -\sum \left(\frac{1}{n} - p_i\right) = -1 + 1 = 0.$$

Furthermore, the inequality becomes an equality only if equality holds for every term of the summation, i.e., $\ln \frac{1}{n p_i} = \frac{1}{n p_i} - 1$, or $p_i = \frac{1}{n}$, for $i = 1, \ldots, n$.

We return to this topic in Section 82. Entropy, in the sense of Definition 32.1, plays a fundamental role in Information Theory. Further development and applications may be found in any basic textbook on Information Theory.

Queries

32.1 The entropy associated with a particular probability space is 3.75 bits. What is the smallest possible size of the sample space? [220]

32.2 Given two experiments $\mathfrak{E}_1, \mathfrak{E}_2$, with probability spaces $(S_1, \mathscr{A}_1, P_1)$ and $(S_2, \mathscr{A}_2, P_2)$, respectively, and with entropies H_1 and H_2, respectively. Also, let $|S_1| = n$, and $|S_2| = m$. For each of the following statements, determine whether it is true, possibly true, or false:
a) If $n > m$, then $H_1 > H_2$.
b) If $H_1 = H_2$, then $n = m$.
c) If $\log_2 n \geq H_1 > \log_2 (n - 1)$ bits, then one of the elementary sets in $\mathscr{A}_1$ has zero probability.
d) If $H_1 = 1$ bit, then $\mathfrak{E}_1$ is probabilistically equivalent to the experiment of tossing a single coin.
e) Let $\mathfrak{E}_2$ be a uniform refinement of $\mathfrak{E}_1$, obtained by replacing each elementary event A_i of $\mathfrak{E}_1$ by two elementary events, each having half the probability of A_i. Then $H_2 = H_1 + 1$ bits. [134]

SECTION 33

Fuzziness

The conceptual and theoretical development presented up to now, and to be pursued further in the remainder of this book, makes possible the application of mathematical techniques and numerical evaluation to many kinds of situations of *uncertainty*. However, they do not permit the treatment of every possible situation of uncertainty. We have been careful to formulate a conceptual model that leads to correct application of the theory—and that delimits the kinds of situations of uncertainty to which this theory is applicable. These are situations where uncertainty arises from the unpredictability of what will happen. Careful attention to the conceptual model allows us to sort out what kind of questions can be appropriately answered by means of probability theory, and which ones cannot.

Other types of 'uncertainty' arise that might be dealt with through various modifications of the theory presented here (cf. [Fi1]), while still others call for a fundamentally different approach. In this Section we examine the kind of 'uncertainty' that is called *fuzziness*, as well as a few principles of the *theory of fuzzy sets* that can be applied in situations of fuzziness.

Example 33.1

Suppose I am searching for a certain Mr. Lavec who I know lives in a town somewhere in France. Furthermore, I am told that he lives in *a town near Paris*. Assuming that this information has come from a dependable source, I am now able to confine my search to towns near Paris. But which towns are the ones 'near Paris', and which ones are not?

I have been given imprecise descriptive information. Descriptive information serves to narrow down the number or variety of possibilities. Imprecision, on the other hand, is a source of uncertainty—in the case at hand, uncertainty as to whether a particular town is to be regarded as being 'near Paris' or not.

Since this situation involves a collection of things—of towns in France—a set-theoretic model would seem to be applicable. Assuming I have a register of all the towns in France, then I can define a space S of elements ξ where each element represents a particular town in France. The additional information I have obtained suggests a subset A of S, whose elements are all the 'towns near Paris'. But in order to specify this set A I need to know how close to Paris a town must be in order for its representative element ξ to belong to A.

Now suppose that I later get additional information: I find out that Mr. L. lives in *a large town*. Here again, I can think in terms of some subset B of S, consisting of all those elements ξ that represent 'a large town'. The towns in which Mr. L. might be residing are then those whose

Probability Concepts and Theory for Engineers, First Edition. Harry Schwarzlander.
© 2011 John Wiley & Sons, Ltd. Published 2011 by John Wiley & Sons, Ltd.

representative elements make up the intersection $A \cap B$. But first I need to know which towns are considered 'large', so that I can define the set B.

The important aspects of Example 33.1 are: A collection of 'things' or 'possibilities' are given. This collection gets narrowed down, but in an *imprecise* way, so that it is not necessarily clear whether any given thing or possibility belongs to the reduced collection or not; there remains some 'uncertainty' regarding this belonging. But it is not uncertainty due to unpredictability of what will happen; and of course, there is no underlying random mechanism, and no statistical regularity. What we have here is *fuzziness*.

It is possible to continue with the set-theoretic description begun in Example 33.1 by making use of the following new idea.

(Definition) 33.1
Given a space S of elements ξ. A *fuzzy set* $A \subset S$ is defined by a point function μ_A, called a *membership function*, which assigns to every $\xi \in S$ a weight $\mu_A(\xi)$, where $0 \leq \mu_A(\xi) \leq 1$.

The weight $\mu_A(\xi)$ that the membership function μ_A assigns to an element ξ expresses the 'degree of membership of ξ in the fuzzy set A'—i.e., the degree or definiteness with which ξ is a member of A. We can draw the Venn diagram for a fuzzy set A by means of a graduated or diffuse boundary, as shown in Figure 33.1. This diagram suggests that the degree of membership for points in the dark area in the center is 1 or close to 1, and it is 0 for points outside the shaded area. Intermediate values apply in the lighter shaded region.

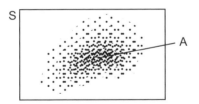

Figure 33.1   A fuzzy set A

A special case of Definition 33.1 applies when a set $A \subset S$ is an ordinary nonfuzzy or '*crisp*' set. In that case, $\mu_A(\xi) = 0$ or 1. S itself has the membership function $\mu_S(\xi) = 1$, for all $\xi \in S$, and the empty set is characterized by $\mu_\phi(\xi) = 0$, for all $\xi \in S$.

In a sense, all verbal descriptions—i.e., descriptions in the conceptual domain—are fuzzy. Take colors. Suppose you have a can of red paint into which you mix a very small amount of yellow paint; then a little more yellow, and a little more. At what point does the mixture become orange? There is no such 'point'. The color of the paint changes gradually, so that the descriptor 'red' gradually becomes less appropriate while the descriptor 'orange' gradually becomes more appropriate. Fuzziness can arise in other ways as well. For instance, numerical properties of a physical object (length, mass, temperature, etc.) cannot be established with arbitrary precision. This is because any physical measurement is limited by noise and inaccuracies inherent in the measurement process, and by the precision with which scales can be read. Therefore, a numerical descriptor is also fuzzy if it is meant to express a numerical value on a continuous scale.

An important question is: How are the values $\mu_A(\xi)$ of the membership function of a fuzzy set A determined? We know that the *probability* of an *event* A is a number that represents the ideal value of relative frequency of occurrence of A, which can be estimated experimentally. With fuzzy sets it is different. Generally, the membership function of a fuzzy set is not uniquely determined or derivable, and cannot be measured. This feature has caused the theory of fuzzy sets to be received at times with skepticism. We return to Example 33.1 to examine this question.

Example 33.1 (Continued)

Consider the fuzzy set $A \leftrightarrow$ 'a town near Paris'. If ξ_1 represents some particular town—say, Metz—there is no way to immediately specify the value of $\mu_A(\xi_1)$. Rather, the relevant argument is this: The further from Paris a town is, the smaller should be its degree of membership in A. Similarly, the closer to Paris a town is, the larger should be its degree of membership in A. Except, of course, the degree of membership cannot be less than 0 or larger than 1.

To begin with, then, it is necessary to determine for each town its distance from Paris. This might be done in terms of the geographical distance between the town's center and the center of Paris, for instance, or in some other appropriate way. Then, if d_i denotes the distance between Paris and town ξ_i, and $d_i > d_j$, then μ_A should satisfy $\mu_A(\xi_i) < \mu_A(\xi_j)$. (This assumes $\mu_A(\xi_i) \neq 1$, $\mu_A(\xi_j) \neq 0$; otherwise $\mu_A(\xi_i) = \mu_A(\xi_j)$.) Furthermore, μ_A must satisfy $\mu_A(\xi_i) = \mu_A(\xi_j)$ if $\xi_i = \xi_j$. The important requirement, in other words, is that μ_A be nonincreasing with increasing d. The exact form of the function μ_A, however, remains arbitrary and is chosen on the basis of reasonableness and convenience. This is permitted in Fuzzy Set Theory because it is not so much concerned about the exact value of a membership function μ_A at a point ξ_1, as about whether $\mu_A(\xi_1)$ is smaller or larger than the value of μ_A at some other point ξ_2.

Then, for instance, if d_{max} denotes the greatest distance a French town can have from Paris, we might specify

$$\mu_A(\xi_i) = 1 - \frac{d_i}{d_{max}}. \tag{33.1}$$

This membership function assigns to a town that is $d_{max}/2$ miles from Paris a degree of membership of 1/2. Subjectively, this may be deemed too large, so that a concave-upward function may be preferable to the linear function (33.1). Thus, a better choice might be

$$\mu_A(\xi_i) = \frac{(d_{max} - d_i)^2}{d_{max}^2}. \tag{33.2}$$

It could also be argued that μ_A should level off near 1, but this added complication in the functional form is not likely to yield a significant advantage.[5]

The membership function of B, on the other hand, has to be chosen more carefully. An expression somewhat analogous to (1) or (2), with d_i replaced by π_i, the population size of town i, would give a membership of 1 to Paris but a considerably smaller membership to all other towns. Instead, the logarithm of the population size might be used, so that

$$\mu_B(\xi_i) = \frac{\log_{10} \pi_i}{\log_{10} \pi_{max}}. \tag{33.3}$$

In this way, a town gets assigned a degree of membership of 0.5 in B if its population size is the square root of the size of Paris.

---

[5] One could imagine a language in which different words identify specific types of fuzziness. For instance, there might then be two different ways of saying 'a town near Paris' that would make it clear whether Equation (33.1) or Equation (33.2) is intended to be used.

Set inclusion is expressed in terms of membership functions as follows:

(Definition) 33.2
Given a space of elements S and fuzzy sets A, B $\subset$ S with membership functions μ_A and μ_B, respectively. Then A is a *subset* of B, denoted

$$A \subset B, \text{ or } B \supset A$$

if and only if $\mu_A(\xi) \leq \mu_B(\xi)$ for all elements $\xi \in$ S.

This can be seen to be consistent with Definition 6.1 in the case of crisp sets. We turn now to set operations involving fuzzy sets. They are defined in the following way [Za1]:

(Definition) 33.3
Given a space of elements S and fuzzy sets A, B $\subset$ S with membership functions μ_A and μ_B, respectively.
a) A$\cup$B is the (fuzzy) subset of S whose membership function is max(μ_A, μ_B).
b) A$\cap$B is the (fuzzy) subset of S whose membership function is min(μ_A, μ_B).
c) A^c is the (fuzzy) subset of S whose membership function is $1 - \mu_A$.

It should be recognized that Definition 33.3 is consistent with the set operations for crisp sets. Thus, Definition 8.1 can be considered a special case of Definition 33.3 where the membership functions μ_A and μ_B take on only the values 0 and 1. From Definitions 33.2 and 33.3 it follows immediately that if A $\subset$ B, then A$\cap$B = A and A$\cup$B = B.
The following Example illustrates the use of the fuzzy set operations of Definition 33.3.

Example 33.2
Let S be the space of real numbers. The set of those real numbers x that are larger than a given number y is an ordinary set of numbers, namely, the subset of the real numbers specified by $\{x: x > y\}$. However, if the symbol > is replaced by $\gg$, it is no longer an ordinary set but a fuzzy set: 'the set of numbers x that are much larger than y'. Complete specification of this set requires an appropriate membership function to be given. Let the set be denoted A$_y$, then a suitable choice of membership function is [Za1]

$$\mu_{A_y}(x) = \begin{cases} 0, & x \leq y \\ 1 - \dfrac{y}{x}, & x > y. \end{cases}$$

This function is sketched in Figure 33.2.

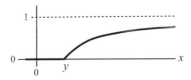

Figure 33.2   The membership function $\mu_{A_y}(x)$

From Definition 33.3 we now find that the set A_y^c of numbers that are not much larger than y has the membership function (see Figure 33.3)

$$\mu_{A_y^c}(x) = \begin{cases} 1, & x \le y \\ \dfrac{y}{x}, & x > y. \end{cases}$$

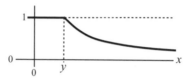

Figure 33.3   The membership function $\mu_{A_y^c}(x)$

We see that numbers less than or equal to y have full membership in this set; while numbers x that are larger than y have fractional membership that decreases with increasing x.

The set of numbers much larger than 1 and also much larger than 2 ought to be the same as the set of numbers much larger than 2. Fuzzy set theory agrees with this intuitive result and yields $A_2 \subset A_1$, so that $A_1 \cap A_2 = A_2$. Since

$$\mu_{A_1}(x) = \begin{cases} 0, & x \le 1 \\ 1 - \dfrac{1}{x}, & x > 1 \end{cases} \quad \text{and} \quad \mu_{A_2}(x) = \begin{cases} 0, & x \le 2 \\ 1 - \dfrac{2}{x}, & x > 2 \end{cases}$$

we obtain $\mu_{A_1 \cap A_2}(x) = \min[\mu_{A_1}(x), \mu_{A_2}(x)] = \mu_{A_2}(x)$, for all x.

However, the union $A_1 \cup A_1^c$ leads to a surprising departure from ordinary Set Theory. Since for a *crisp* set A, $A \cup A^c = S$, we would expect every fuzzy set A to satisfy $\mu_{A \cup A^c}(x) = \mu_S(x) = 1$ for all x. Instead, in the case of A_1 we obtain

$$\mu_{A_1 \cup A_1^c}(x) = \max[\mu_{A_1}(x), \mu_{A_1^c}(x)]$$

$$= \begin{cases} 1, & x \le 1 \\ \dfrac{y}{x}, & 1 < x \le 2y \\ 1 - \dfrac{y}{x}, & x > 2y \end{cases}$$

which is sketched in Figure 33.4. But some reflection on the meaning of the fuzzy set $A_1 \cup A_1^c$—'the set of numbers x that are either much larger than 1 or *not* much larger than 1 (or both)'—it becomes plausible that there should be numbers that do not fully correspond to this specification: numbers that might be considered to be 'somewhat larger than 1'. In this case, it should be noted, the exact form of the membership function has a significant effect; for instance, it determines at what x-value $\mu_{A_1 \cup A_1^c}(x)$ is smallest.

There is no further discussion of fuzzy sets in this book.[6]

---

[6] Applications of fuzzy set theory range from fuzzy logic and fuzzy control to fuzzy system analysis and many other topics. Textbooks are available which present a detailed development of Fuzzy Set Theory, such as [KF1].

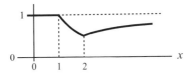

Figure 33.4   The membership function $\mu_{A_i \cup A_i^c}(x)$

Queries

33.1 On a space S there are defined three fuzzy sets A, B, C with membership functions μ_A, μ_B, μ_C, respectively. Suppose that $\mu_B = (1/2)\mu_A$, and $\mu_C = 1 - (1/2)\mu_A$. For each of the following statements, determine whether it is true, possibly true, false, or meaningless.

a) There exists a fuzzy set D such that $B = A \cap D$

b) $B = (1/2)A$

c) $B \cup C = S$

d) $B \cap C = \phi$

e) $B = C$.   [105]

33.2 For any given number y, use the membership functions of Example 33.2 to characterize the set of numbers much larger than y, and the set of numbers not much larger than y. Now define the fuzzy set A as the set of numbers x that are much larger than 2 but not much larger than 1. What numerical value x has the greatest degree of membership in this set?   [10]

Part II Summary

Part II has served to expand the conceptual and mathematical formulations that can be applied to probability problems. Furthermore, it provided the reader with opportunities to set up and solve a variety of elementary probability problems. By 'elementary probability problems' we mean those problems where the sample space is countable and where the possible outcomes are either non-numerical or, if they are numerical, this attribute need not be taken into account in formulating the problem.

We began with some basic issues, and saw how formulation of an elementary probability problem is particularly simple if the possible outcomes can be regarded as equally likely. The simple but important new concepts of conditional probability and independence were then introduced. They greatly extend the power of the basic mathematical model and will be of use throughout the remainder of the book. Thereafter, the emphasis was on approaches to a variety of 'counting' problems, that is, problems where the major effort that is required involves determining the sizes of various sets. These were problems categorized as ordered or unordered sampling, with or without replacement. In addition, some more complicated counting problems were considered.

While the idea of 'sampling' as addressed here will not be of significant interest in the remainder of the book, one important development that was undertaken along the way will continue to play a role—namely, the notions of product spaces and product experiments.

Part II concluded with an examination of two ideas that are somewhat peripheral to the main thrust of this book. These are the expression of uncertainty in a quantitative manner, and a mathematical approach to fuzziness.

Part III

Introduction to Random Variables

Part III Introduction

We move now into new territory as we extend the mathematical model in order to be able to represent possible outcomes having numerical values. The full development of this extension, and the exploration of its ramifications, will be the concern of the remainder of the book, beginning with Part III. The basic new tool is a point function defined on a probability space, which is called a 'random variable'. We begin this exploration by considering the simplest kinds of random variables, namely, 'discrete random variables'.

Some basic properties of the real numbers are discussed next. Familiarity with this material is necessary for a full understanding of the mathematical implications of different types of random variables. Then comes the general definition of a random variable, which is followed by a discussion of ways in which a probability distribution over the real numbers can be described. Because of its theoretical significance as well as its importance in a variety of applications, the Gaussian distribution is considered in some detail. It will receive further attention in later portions of the book as well.

The remainder of Part III is concerned with defining *two* random variables on a probability space, and various topics that arise from this. Those topics include ways of describing probability distributions in two dimensions, the application of two random variables in a given problem situation and computing probabilities of events of interest, as well as the notion of statistical independence of two random variables.

Numerical-Valued Outcomes

When constructing the mathematical model for a given probability problem, we begin with a space S of elements, chosen to be of the correct size so that we can think of each of the elements as representing one of the possible outcomes of the probabilistic experiment. Aside from this, the elements have no attributes whatsoever, numerical or qualitative, but are purely abstract entities. Each possible outcome does, of course, represent one or more observable properties or attributes. It frequently happens that such properties are expressed numerically; if so, it is often desirable to bring these numerical values into the mathematical model. In fact, this is what occurred in Example 10.2. In that Example, a *number* got associated with each element in the sample space, so that a *real-valued point function* became defined on the space. This point function served to *count* the elements of the space, but at the same time it associated with each element a numerical value that expressed 'the number of dots facing up' in the corresponding possible outcome.

Thus, we recall that a mathematical device for associating numerical values with the elements of a space S is a real-valued point function. But now we have a slightly different situation: We are not merely concerned with a space S, but with a probability space (S, $\mathscr{A}$, P). What difference is there between assigning numerical values to elements of S and assigning numerical values to the elements of a probability space?

The difference is that, in the case of the probability space, *the probability assignment should carry over* into the realm of the real numbers which become associated with the elements of the space. We shall want to speak of 'the probability of a five', or 'the probability of all numbers greater than 100', etc. Therefore, a point function that is meant to associate numerical values with the elements of a probability space *must also behave like a set function*; it must map sets in $\mathscr{A}$ uniquely into sets of real numbers. In this way, *the probability* of a given set of real numbers becomes defined as the probability of the set in $\mathscr{A}$ that maps into that given set of real numbers.

Throughout Part III, we explore the ramifications of this requirement as we extend our mathematical model to accommodate *numerical-valued* outcomes and events. This extension plays a central role in the remainder of the book.

Example 34.1
The familiar experiment of throwing a single die leads to the probability space (S, $\mathscr{A}$, P) described in Figure 34.1. We have a space S of six elements, the class of all subsets of S, and a probability function that assigns to each set a value proportional to its size.

Probability Concepts and Theory for Engineers, First Edition. Harry Schwarzlander.
© 2011 John Wiley & Sons, Ltd. Published 2011 by John Wiley & Sons, Ltd.

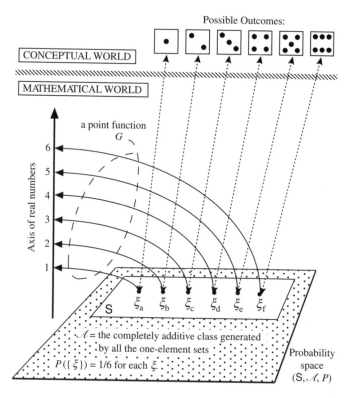

Figure 34.1   A real-valued point function G defined on the probability space for the throw of a single die

Also indicated in the figure is the same point function G that was defined in Example 10.2. After defining this function G on our probability space $(S, \mathscr{A}, P)$, we want to be able to make probability statements *about sets of real numbers*. We want to be able to manipulate, or compute with, probabilities of sets of real numbers in the same way as probabilities of subsets of S.

We recall that in order to do this we need a completely additive class. This class must now consist of subsets of R, the *space of real numbers*. But, does not the function G specify such a class? For each subset of S—i.e., for each set in $\mathscr{A}$—the function G identifies a corresponding subset of R. Thus, the set $\{\xi_c, \xi_d\} \in \mathscr{A}$ has as its 'image', via G, the set $\{3, 4\} \subset R$. In a similar way each of the subsets of S has an image that is a subset of R. Also, the largest partition in $\mathscr{A}$ has as its image, via G, a disjoint class of subsets of R, consisting of the six one-element sets $\{1\}, \{2\}, \{3\}, \{4\}, \{5\}, \{6\}$. This class can be enlarged into a partition, $\mathscr{P}_G$, by including in it the set $Q = R - \{1, 2, 3, 4, 5, 6\}$, that is, the set of all real numbers other than the integers from 1 to 6. It is therefore possible to define an additive class of subsets of R, namely, the additive class generated by $\mathscr{P}_G$, which we denote $\mathscr{A}_G$. Since it is finite, $\mathscr{A}_G$ is also *completely additive*.

Finally we ask, what is the probability function that becomes defined on the members of $\mathscr{A}_G$ and that we denote P_G. Clearly, we have $P_G(\{1\}) = P_G(\{2\}) = \ldots = P_G(\{6\}) = 1/6$ and $P_G(Q) = 0$ for the probabilities on the partition $\mathscr{P}_G$. The remaining values of P_G are taken care of automatically by additivity. Thus, $P_G(\{3, 4\}) = P_G(\{3\}) + P_G(\{4\})$, etc.

Upon reflecting on what has been done here, it should be clear that the point function G has been used to establish a *new probability space*, $(\mathsf{R},\ \mathscr{A}_G,\ P_G)$. The elements of this space are the real numbers, the completely additive class is a very specific finite class of subsets of R, and a probability function P_G has been prescribed on the class $\mathscr{A}_G$. We call P_G the *induced probability distribution*, or the *probability distribution induced by the function G*.[1]

An important convenience results from the fact that the basic elements of this new probability space are the real numbers: It is now easy to describe the probability distribution graphically. This is because the real numbers can be represented by points on a line, the 'real line'. Thus, for the induced probability distribution P_G specified in Example 34.1, the bar graph in Figure 34.2 can be drawn. The horizontal axis represents the real numbers and is labeled x. The letter x will often be used as the generic symbol for an arbitrary real number, just as ξ has served as the generic symbol for elements of the basic space S. A vertical 'bar' is positioned, in Figure 34.2, at every x for which $\{x\} \in \mathscr{A}_G$. The height of the bar expresses $P_G(\{x\})$.

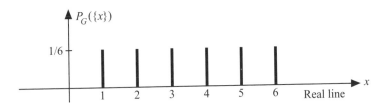

Figure 34.2   A discrete probability distribution

The point function G described above is an example of a 'discrete random variable'.

(Definition) 34.1
A real-valued point function G on a probability space $(\mathsf{S},\ \mathscr{A},\ P)$ is called a *discrete random variable* if and only if
a) its range is a countable subset of R, and
b) for each number x in the range of G, the set $\{\xi: G(\xi) = x\}$ is a member of the class $\mathscr{A}$.

The requirement (a) is automatically satisfied if S is countable. Requirement (b) is automatically satisfied if all the elementary sets are one-element sets, as was the case in Example 34.1. The following Example illustrates how it is possible for a point function to violate this requirement.

Example 34.2
Consider a probability space $(\mathsf{S},\ \mathscr{A},\ P)$ where

$$\mathsf{S} = \{\xi_a, \xi_b, \xi_c\},$$
$$\mathscr{A} = \{\phi, \{\xi_a\}, \{\xi_a\}^c, \mathsf{S}\},\ \text{and}$$
$$P\ \text{is arbitrary}.$$

Such a probability space might arise, for instance, in connection with the following game: A coin is tossed. If the result is 'Tail', nothing further happens. If it is 'Head', the player must guess the

---

[1] Notation must be watched carefully: If we write $P(\mathsf{A})$, then A must be a subset of S and belong to $\mathscr{A}$, since P is defined over $\mathscr{A}$. If we write $P_G(\mathsf{A})$, then A must be a set of numbers and belong to $\mathscr{A}_G$, since P_G is defined over $\mathscr{A}_G$.

age (in years) of a person unknown to the player. If the guess is correct, the player receives $10; if not, he or she must pay $1.

To the elements of the space correspond three possible outcomes:

$\xi_a \leftrightarrow$ 'coin toss results in Tail'

$\xi_b \leftrightarrow$ 'coin toss results in Head and player guesses correctly'

$\xi_c \leftrightarrow$ 'coin toss results in Head and player guesses incorrectly'

and we have $P(\{\xi_a\}) = P(\{\xi_a\}^c) = 1/2$. However, it cannot be assumed that statistical regularity holds with respect to the guessing, so that $\{\xi_b\}, \{\xi_c\} \notin \mathscr{A}$.

Now, let the amount won be expressed by a point function H defined on S as follows: $H(\xi_a) = 0$, $H(\xi_b) = 10$, $H(\xi_c) = -1$ (see Figure 34.3). We note that, whereas 10 is a number in the range of H, the set $\{\xi: H(\xi) = 10\} = \{\xi_b\}$ is not a member of $\mathscr{A}$ and thus no probability is defined for it. Accordingly, the one-element set $\{10\} \subset \mathsf{R}$ is a set to which H assigns no probability value. Thus, requirement (b) of Definition 34.1 is not satisfied, and we conclude that the point function H is *not* a discrete random variable.

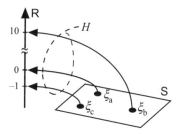

Figure 34.3   The point function H

The probability distribution induced by a discrete random variable assigns probability 1 to some *countable* subset of R. Such an induced probability distribution is called a *discrete probability distribution*. In principle, discrete probability distributions can always be described graphically in the form of a bar graph such as shown in Figure 34.2.

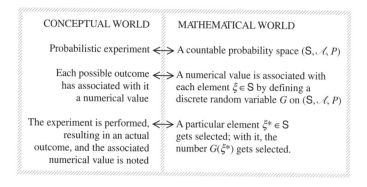

Figure 34.4   Relating a discrete random variable to the conceptual model

(Definition) 34.2
Let G be a real-valued point function on a space S, and A some subset of the real numbers, $A \subset R$. The set $\{\xi: G(\xi) \in A\}$ is called the *inverse image of* A *under the function* G, or simply, the *inverse image of* A, and we denote it $G^{-1}(A)$.

It is to be noted that G is a point function over the elements of S, and it maps into real numbers, not sets of real numbers. G^{-1} would then normally also be regarded as a point function, assigning elements in S to individual real numbers. Instead, Definition 34.2 says that G^{-1} is to be regarded as a *set function* that maps subsets of R into subsets of S. In Definition 34.1, for instance, it is now possible to replace $\{\xi: G(\xi) = x\}$ by $G^{-1}(\{x\})$.

If G is a function such as specified in Definition 34.2, then $G^{-1}(A)$ exists for every $A \subset R$. For instance, consider the function G of Example 34.1, and let $A = \{x : x \geq 6\}$. Then, according to Definition 34.2, $G^{-1}(A) = \{\xi: G(\xi) \in A\} = \{\xi_f\}$; that is, only the element ξ_f gets mapped by G into the set A. On the other hand, $G^{-1}(\{x: x > 6\}) = \phi$. However, with P_G as defined in Example 34.1, $P_G(\{x: x \geq 6\})$ and $P_G(\{x: x > 6\})$ are meaningless expressions because the sets $\{x: x \geq 6\}$ and $\{x: x > 6\}$ are not members of $\mathscr{A}_G$. (We will remove this limitation in Section 37.)

The relation of a discrete random variable to the conceptual model can be summarized as shown in Figure 34.4.

Here is another illustration of a discrete random variable.

Example 34.3:   Runs.
The probability space $(S, \mathscr{A}, P)$ for the p.e. $\mathfrak{E}$ in Example 27.3, in which a die is thrown repeatedly, has an uncountable sample space. The events $A_1, A_2, A_3, \ldots, A_\infty$ defined in that Example serve to partition S. It is possible to define a random variable, N, on the elements of this probability space which expresses the number of throws till the first 'six' appears. Thus, if an element ξ belongs to the set A_n (where $n = 1, 2, 3, \ldots$), then $N(\xi) = n$. Strictly speaking, an exception must be made for $\xi \in A_\infty$: It is not possible to assign $N(\xi) = \infty$, because '∞' is not a real number. But since $P(A_\infty) = 0$, the induced probability distribution is not affected by the value that N assigns to A_∞. For instance, $N(\xi)$ can arbitrarily be set to 0 in that case. (An alternative approach would be to regard an infinite-length die throw sequence that is totally devoid of sixes as the result of an unsatisfactorily performed experiment.)

The new probability space resulting from the discrete random variable N is $(R, \mathscr{A}_N, P_N)$, where $\mathscr{A}_N$ is generated by the partition $\mathscr{P}_N = \{\{1\}, \{2\}, \{3\}, \ldots, Q\}$, Q being the set $R - \{1, 2, 3, \ldots\}$. N induces a *geometric distribution*, which is given in its general form by

$$P_N(\{x\}) = (1-p)^{x-1} p, \ x = 1, 2, 3, \ldots$$

and is described by the bar graph in Figure 34.5. This distribution expresses the probabilities associated with different run lengths, in the case of product experiments made up of Bernoulli

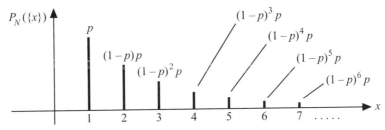

Figure 34.5   Bar graph of a geometric distribution

trials. p and $1 - p$ are the probabilities associated with the individual Bernoulli trial. In the case of Example 27.3, $p = 1/6$, giving the run length probabilities found in that Example.

We also note that N induces zero probability on the set Q, since $\sum_x P_N(\{x\}) = 1$.

Queries

34.1 What is the size of the class $\mathcal{A}_G$ defined in Example 34.1?   [222]

34.2 A probability space (S, $\mathcal{A}$, P) is defined as follows:

$$S = \{\xi_a, \xi_b, \xi_c, \xi_d\}$$
$$\mathcal{A} = \{\phi, \{\xi_a, \xi_b\}, \{\xi_c, \xi_d\}, S\}$$
$$P(\{\xi_a, \xi_b\}) = 1/3.$$

For each of the following specifications of a point function on S, determine which ones define a discrete random variable on (S, $\mathcal{A}$, P).

a) $G(\xi_a) = 1$, $G(\xi_b) = 2$, $G(\xi_c) = 3$, $G(\xi_d) = 4$
b) $G(\xi_a) = 1$, $G(\xi_b) = -1$, $G(\xi_c) = 1$, $G(\xi_d) = -1$
c) $G(\xi_a) = 0$, $G(\xi_b) = 0$, $G(\xi_c) = 1/2$, $G(\xi_d) = 1/2$
d) $G(\xi_a) = G(\xi_b) = G(\xi_c) = G(\xi_d) = 3$
e) $G(\xi_a) = \xi_c$, $G(\xi_b) = \xi_d$, $G(\xi_c) = \xi_a$, $G(\xi_d) = \xi_b$.   [62]

34.3 Consider again the probability space defined in the preceding Query, and the different point functions specified there. In each of the five cases determine whether the inverse image of the set of real numbers $\{-1, 0, 1\}$ exists. In which cases is this inverse image a set in $\mathcal{A}$?   [35]

The Binomial Distribution

A variety of discrete probability distributions arise frequently in practical problems. One of these is the binomial distribution, which we have previously encountered in Sections 25 and 30, but not as an *induced* distribution.

Example 35.1

Consider again the coin-toss experiment of Example 11.1, part (a). The possible outcomes of this experiment can be represented as 10-tuples, where each entry in any 10-tuple can be either 'T' or 'H'. There are 2^{10} equally likely possible outcomes, so that the probability space $(S, \mathscr{A}, P)$ for this p.e. has $|S| = 2^{10}$ and $P(\{\xi\}) = 2^{-10}$ for each $\xi \in S$.

Suppose we define a discrete random variable B on this space that expresses 'the number of heads obtained in the ten tosses'. B then simply associates one of the integers from 0 to 10 with each of the elementary sets $\{\xi\}$, and the probability distribution P_B assigns a positive probability to each of these eleven integers, regarded as one-element sets in R. From Equation (30.1) we see that the values of these probabilities are

$$P_B(\{m\}) = \frac{\binom{10}{m}}{2^{10}}, \quad m = 0, 1, 2, \ldots, 10$$

and these values completely specify the induced probability distribution P_B.

The above expression gives the probability values assigned by P_B to the sets $\{0\}, \{1\}, \{2\}, \ldots, \{10\}$, and those values are plotted as a bar graph in Figure 35.1. From these, the probabilities assigned by P_B to other sets of real numbers can be obtained by additivity.

Because the expressions for the probabilities $P_B(\{m\})$ involve binomial coefficients, the distribution obtained in this Example is called a *binomial distribution* and B is a *binomial random variable*. A binomial random variable arises if a p.e. involves k ($k \geq 1$) independent, identical trials; if some event associated with a trial is of interest; and if the random variable is to express *how many times* this event occurs during those k trials.

Suppose the p.e. consists of independent, identical *binary* trials, i.e., Bernoulli trials. The possible outcomes in that case are all the 2^k different possible sequences of results from the k trials. If the

Probability Concepts and Theory for Engineers, First Edition. Harry Schwarzlander.
© 2011 John Wiley & Sons, Ltd. Published 2011 by John Wiley & Sons, Ltd.

letters H and T are used to identify the two possible results of a single Bernoulli trial, then a typical possible outcome of the p.e. is represented by a k-tuple such as

$$\underbrace{(\text{T, T, H, T, H, H}, \dots, \text{T})}_{k \text{ components}}.$$

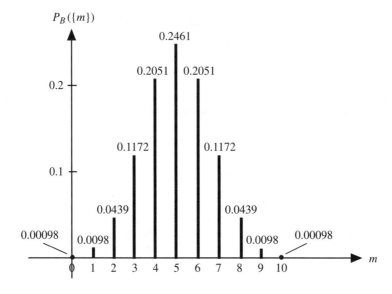

Figure 35.1   The probability of m heads in ten tosses of a coin

To each possible outcome gets assigned a numerical value, namely, a count of the H's (or of the T's) appearing in the k-tuple. In the mathematical model, this corresponds to defining a real-valued point function on the probability space for this p.e. This function is a binomial random variable.

In Example 35.1, the two possible results of each trial (H and T) were equally likely. To remove that restriction and to allow trials other than binary trials, let some product experiment $\mathfrak{E}$ consist of k identical trials, $\mathfrak{E} = \underbrace{\mathfrak{E}_1 \times \mathfrak{E}_1 \times \dots \times \mathfrak{E}_1}_{k \text{ times}}$. The sample space for $\mathfrak{E}_1$ contains an event H that is of interest. Denote $P_1(\text{H}) = p$, so that $P_1(\text{H}^c) = 1 - p$. On the probability space for $\mathfrak{E}$ we introduce the random variable B that expresses, for any sequence of k trials, the number of times H occurs. There are $\binom{k}{m}$ sequences of length k in which H occurs exactly m times, and each of them has probability $p^m (1-p)^{k-m}$. Summing over the $\binom{k}{m}$ sequences gives[2]

---

[2] The experiment $\mathfrak{E}$ either consists of ordered sampling with replacement, or $\mathfrak{E}$ (or a suitable contraction of $\mathfrak{E}$) is probabilistically equivalent to ordered sampling with replacement. We saw in Section 25 that, for such an experiment, probabilities are distributed binomially over a certain partition of the probability space. The random variable B maps these same probabilities into the induced space.

$$P_B(\{m\}) = \binom{k}{m} p^m (1-p)^{k-m}, \quad m = 0, 1, 2, \ldots, k \qquad (35.1)$$

This is the general form of the binomial distribution. For $p = 1/2$, the symmetrical form results, as obtained in Example 35.1:

$$P_B(\{m\}) = \binom{k}{m} \left(\frac{1}{2}\right)^k, \quad m = 0, 1, 2, \ldots, k. \qquad (35.2)$$

If $p \neq 1/2$, the binomial distribution is not symmetric about the midpoint of its range.

Evaluation of the binomial formula (35.1) is awkward if k is large. A simpler formula, approximating Equation (35.1) for large k and small p, can be derived as follows. Comparing the two expansions

$$(1-p)^k = 1 - kp + \frac{k(k-1)}{2!} p^2 - \ldots, \text{ and}$$

$$e^{-x} = 1 - x + \frac{x^2}{2!} - \frac{x^3}{3!} + \ldots$$

it can be seen that *for large k*, $(1-p)^k \approx e^{-kp}$. If, in addition, p is very small, then for values of m much smaller than k, Equation (35.1) becomes

$$P_B(\{m\}) = \frac{k(k-1)\ldots(k-m+1)}{m!} \frac{p^m}{(1-p)^m} (1-p)^k$$

$$\approx \boxed{\frac{(kp)^m}{m!} e^{-kp}} \qquad (35.3)$$

This result is called the *Poisson approximation* to the binomial distribution.

Example 35.2

A 'random deck' of 52 playing cards is to be assembled by blindly picking one card from each of 52 standard decks of playing cards. We are interested in the number of aces of hearts in the random deck.

It is convenient to specify the p.e. in such a way that the order is noted in which cards enter the random deck. Keeping track only of whether a card is an ace of hearts ('A') or not ('N'), each possible outcome is identified by a binary 52-tuple such as

$$(N, N, N, N, A, N, \ldots\ldots\ldots, N).$$

This makes a total of 2^{52} possible outcomes.

On the corresponding probability space we can define a discrete random variable Y that expresses for each element (52-tuple) the number of A's. The procedure in this p.e. consists of 52 identical, independent trials, and the probability of an A in any one trial is 1/52. The discrete distribution induced by the random variable Y is therefore binomial, with $k = 52$, $p = 1/52$:

$$P_Y(\{m\}) = \binom{52}{m} \left(\frac{1}{52}\right)^m \left(1 - \frac{1}{52}\right)^{52-m}$$

Intuition may suggest that it is highly likely that the random deck will contain a single ace of hearts. Using the Poisson approximation to find the probability of this event, we have immediately

$$P_Y(\{1\}) \approx e^{-1} = 0.3679\ldots.$$

On the other hand, the probability of no ace of hearts at all is approximately the same:

$$P_Y(\{0\}) \approx e^{-1} = 0.3679\ldots.$$

For comparison, direct computation yields $P_Y(\{1\}) \doteq 0.37146$, $P_Y(\{0\}) \doteq 0.36431$.

Formula (35.3) not only *approximates* a probability distribution, but is itself a probability distribution, called the *Poisson distribution*. To verify that Equation (35.3) represents a probability distribution we check whether it assigns only non-negative values, and whether it assigns the value 1 to the whole space, R. The first requirement is clearly satisfied; and the second holds if the formula is defined to hold for *all non-negative* integers m, giving

$$\sum_{m=0}^{\infty} \frac{(kp)^m}{m!} e^{-kp} = e^{-kp} \sum_{m=0}^{\infty} \frac{(kp)^m}{m!} = e^{-kp} e^{kp} = 1.$$

A discrete random variable that induces a Poisson distribution is called a *Poisson random variable*. It assigns the positive probabilities given by Equation (35.3) to the members of some countably infinite class of disjoint subsets of R—typically, the class consisting of the sets $\{0\}$, $\{1\}$, $\{2\}$, …. Poisson random variables arise in many practical problems, such as discussed in Section 60.

Queries

35.1 A coin is tossed ten times. What is the probability of getting
a) either exactly three 'heads' or exactly three 'tails'?
b) more 'heads' than 'tails'? [147]

35.2 Consider repeated throws of a fair die. Determine the probability of
a) getting a 'six' exactly once in six throws
b) getting a 'six' exactly ten times in 60 throws. [4]

35.3 Which of the following discrete random variables induce a binomial distribution:
a) The random variable that expresses the number of heads obtained (0 or 1) in a single coin toss.
b) The random variable that expresses the number of times a coin must be tossed till the first 'head' is obtained.
c) The random variable that expresses the number of people born in January among the first 35 visitors to enter Disney World in Florida on a given morning.
d) In ten throws of a die, the random variable that expresses the number of occurrences of that face of the die that occurs most frequently. [47]

The Real Numbers

We have become aware of the need for defining real-valued point functions on a probability space when a problem involves numerical observations. We have also seen how such a function G, if it satisfies the requirements of Definition 34.1, leads to a new probability space $(R, \mathscr{A}_G, P_G)$ for the given problem. The new space of elements is the space of real numbers, R. And we saw that, by proceeding as in Example 34.1, it is not difficult to define a completely additive class of subsets of R onto which G induces a legitimate probability distribution. Nevertheless, a great simplification would be achieved if a *single* completely additive class of subsets of R could be found that will at once be acceptable for every problem. Such a class exists; but prior to specifying it (in the next Section) we examine some properties of the real numbers.

In Part I it was noted that a 'space', in the sense of Set Theory, is made up of abstract elements having no attributes beyond their individual identity. The space R of real numbers is different in that its elements have other attributes. These arise from the existence of an 'order relation' that is defined for every pair of distinct elements a, b in R, namely, $a < b$ or $a > b$. This permits the elements of R to be ordered, resulting in the geometrical picture of points on a line, the '*real line*'.

A fundamental property of R is the following.

Theorem 36.1
R is an uncountable set.

Proof
This is easily established by the method of Section 11. Replace T and H in the infinite coin-toss sequences of Example 11.1 by 0 and 1, respectively, then each coin toss sequence becomes the binary expansion of a real number in the range from $0.0000\ldots$ to $0.11111\ldots = 1.000\ldots$. Since the set of infinite-length coin-toss sequences is uncountable, the real numbers between 0 and 1 are also uncountable. This set of numbers is a subset of R, so that by Theorem 11.1d, R must be uncountable.

A set of numbers $\{x: a \leq x \leq b\}$, where $a < b$, is called an *interval*, and a, b are the *endpoints* of the interval. In particular, since both endpoints belong to the set, it is a *closed interval*. If the endpoints do not belong to the set, it is an *open interval*, and if only one of the endpoints belongs to the set, it is a *semi-closed* or *semi-open interval*.

Probability Concepts and Theory for Engineers, First Edition. Harry Schwarzlander.
© 2011 John Wiley & Sons, Ltd. Published 2011 by John Wiley & Sons, Ltd.

The interval $\{x: a \le x \le b\}$ can be put into one-to-one correspondence with the real numbers from 0 to 1 by means of the function

$$f(x) = \frac{x-a}{b-a}, \quad a \le x \le b$$

which leads to the following:

Corollary 36.1
Every interval is uncountable.

It is helpful to use a special notation for intervals.[3] For $a, b \in \mathsf{R}$ and $a < b$:

$\overline{a,\ b}$ denotes the open interval $\{x: a < x < b\}$;

$\overline{\vphantom{b}a,\ b}$ denotes the closed interval $\{x: a \le x \le b\}$; and

$\overline{a,\ b}$ and $\overline{a,\ b}$ denote the semi-closed (or semi-open) intervals $\{x: a < x \le b\}$ and $\{x: a \le x < b\}$, respectively.

Set operations apply to intervals in the usual way.

Example 36.1
Consider the real numbers a, b, c, where $a < b < c$. Then

$$\overline{a,\ b} \cup \overline{b,\ c} = \overline{a,\ c}.$$

This must not be mistaken for a disjoint union. The two sets $\overline{a,\ b}$ and $\overline{b,\ c}$ have one element in common—the element b. On the other hand, the intervals $\overline{a,\ b}$ and $\overline{b,\ c}$ are disjoint, and so are the intervals $\overline{a,\ b}$ and $\overline{b,\ c}$. Thus, it is correct to write

$$\overline{a,\ b} \uplus \overline{b,\ c} = \overline{a,\ c}.$$

However, $\overline{a,\ b} \uplus \overline{b,\ c} \neq \overline{a,\ c}$; instead, $\overline{a,\ c} = \overline{a,\ b} \uplus \{b\} \uplus \overline{b,\ c}$.

Two important subsets of R, which partition R, are the set of *rational numbers* and the set of *irrational numbers*. We recall that a real number expressible as the ratio of two integers is a rational number, so that a rational number can also be thought of as an *ordered pair* of integers—a numerator integer and a denominator integer. The set of all ordered pairs of integers is countable, which is demonstrated by the counting rule shown in Figure 36.1. This figure contains part of a two-dimensional infinite array that assigns to every ordered pair of integers (i, j) a count. The first pair is $(0, 0)$, the second pair is $(0, 1)$, the third pair $(1, 0)$, etc. The pattern according to which the count proceeds is clear from the figure and analogous to the one used in the proof of Theorem 12.2. It can be continued uniquely to a table of arbitrary size; that is, for every pair of integers, positive or negative, a count is defined. For each rational number r there are of course many pairs of integers whose ratio

---

[3] The notation for intervals introduced in this Section is that of Feller [Fe2]. A different notation commonly used for open, closed, and semi-closed intervals is (a, b), $[a, b]$, and $(a, b]$ or $[a, b)$. We avoid this notation to prevent confusion with pairs of numbers, arguments of functions, and vectors.

is r, and only one of these pairs is needed to represent r. Furthermore, the first column of the array gets ignored since an integer divided by zero is not a rational number. Therefore, Figure 36.1 defines a one-to-one correspondence between the rational numbers and a subset of the set of all positive integers. We conclude:

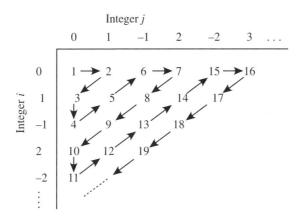

Figure 36.1 A counting rule for pairs of integers (i, j)

Theorem 36.2
The set of all rational numbers is countable.

Since the set of rational numbers has the infinite set of all integers as a subset, the rationals are also an infinite set.

Whereas the arrangement in Figure 36.1 results in an ordering of the rational numbers—by writing i/j for each pair (i, j) in the array and deleting all duplications—this is not the 'natural' ordering of the rational numbers. In fact, we cannot express two rational numbers that according to their natural ordering are adjacent: If i, j, k, l are integers and the rational numbers $i/j < k/l$ are given, there always exists another rational number x that lies between them; for instance, the number

$$x = \frac{(i/j) + (k/l)}{2} = \frac{il + jk}{2jl}.$$

Also, since $\overline{i/j, \ k/l}$ is an interval and therefore an uncountable set, we see that the set of *real numbers* lying between any two distinct rationals is uncountable.

The *size* of a *finite* set is simply the number of elements in the set. This notion of size is of no use for comparing *intervals* since all intervals are uncountable. Yet, using the geometric interpretation of the space of real numbers, it is possible to view an interval as some increment along the real line and characterize it by its 'length'. Thus, let L be a set function that expresses the *length* of an interval; that is, for $a, b \in \mathsf{R}$ and $a < b$,

$$L(\overline{a, b}) = L(\overline{\ a, b}) = L(\overline{a, b}) = L(\overline{a, b}) = b - a. \tag{36.1}$$

The symbol ∞ is used in the customary way. However, ∞ and $-\infty$ do not represent elements of R. Since above we required that $a, b \in \mathsf{R}$, it follows that the right side of Equation (36.1) is a *finite* positive number; i.e., $0 < b - a < \infty$. The intervals that appear on the left side of Equation (36.1) are therefore called *finite intervals*, although as sets they are uncountable. On the other hand, $\overline{-\infty, \, a}$ denotes the set of all real numbers less than or equal to a, $\{x : x \le a\}$. We call this a *semi-infinite interval* (or *infinite interval*), with $L(\overline{-\infty, \, a}) = \infty$. Other types of semi-infinite intervals are $\overline{-\infty, \, a}, \underline{a, \, +\infty}, \underline{a, \, +\infty}$. We can also write $\mathsf{R} = \overline{-\infty, \infty}$. Since there is no 'endpoint' at $-\infty$ and ∞, the intervals $\overline{-\infty, \, a}$ and $\overline{a, \, \infty}$ are considered *open* semi-infinite intervals; whereas $\overline{-\infty, \, a}$ and $\underline{a, \, \infty}$ are *closed* semi-infinite intervals.

The definition of the set function L can be extended via countable additivity to other subsets of R. This extended version of L is called the *Lebesgue measure*.

Example 36.2

In order to determine the value that L assigns to a one-element set $\{a\}$, we note that $\{a\} \uplus \underline{a, \, b} = \underline{a, \, b}$. Then $L(\{a\}) + L(\underline{a, \, b}) = L(\underline{a, \, b})$, giving

$$L(\{a\}) = (b - a) - (b - a) = 0.$$

Now let I denote the set of all integers. Then, we have

$$L(\mathrm{I}) = \sum_{i=-\infty}^{\infty} L(\{i\}) = \sum_{i=-\infty}^{\infty} 0 = 0. \tag{36.2}$$

The result obtained in Equation (36.2) extends to every countable set. Thus, we see, for instance, that the set of all rational numbers has measure 0.

Queries

36.1 How many distinct rational numbers are represented by those first 19 entries identified in the array of Figure 36.1? [15]

36.2 For each of the following statements, determine whether it is true, false, or meaningless:

a) $0 \cup \overline{0, \, 1} = \underline{0, \, 1}$

b) $\underline{-1, \, 0} \cup \overline{0, \, 1} = \underline{-1, \, 1}$

c) $\{0\} - \{1\} = \{0\}$

d) $\{\{0\}, \{1\}\} \subset \mathsf{R}$

e) $\overline{-\infty, \, 1} \cup \underline{1, \, \infty} = \mathsf{R}$. [103]

36.3 Determine the Lebesgue measure of the following sets:

a) $\overline{-1, \, 1} \cup \overline{0, \, 2}$

b) $\overline{-1, \, 1} \cap \overline{0, \, 2}$

c) $\underline{-1, \, 1} - \overline{-1, \, 1}$

d) $\underline{-1, \, 1} - \mathsf{R}_a$, where R_a denotes the set of all rational numbers. [188]

General Definition of a Random Variable

As stated at the beginning of the preceding Section, a completely additive class of subsets of R is desired that is sufficiently general to be used in all problem situations. That is, we would like *one* completely additive class that can be used in the definition of *any* probability space whose elements are real numbers—any probability space that might get induced by a random variable.

First, consider the class $\mathscr{V}$ of all closed semi-infinite intervals of the type $\overline{-\infty, a}$, in other words,

$$\mathscr{V} = \{\overline{-\infty, a} : a \in R\}. \tag{37.1}$$

This is an uncountable class: For each real number a, there is a distinct set in this class. The completely additive class that we seek can now be obtained from the class $\mathscr{V}$ as follows.

(Definition) 37.1

Let $\mathscr{B}$ denote the completely additive class of subsets of R that is generated from the class $\mathscr{V}$ of all semi-infinite intervals $\overline{-\infty, a}$, through countable set operations. $\mathscr{B}$ is called the *class of Borel sets*, and every set in $\mathscr{B}$ is called a *Borel set*.

The class $\mathscr{B}$ is a sufficiently comprehensive class of subsets of R to be able to serve as the completely additive class whenever a probability space is defined on R. In fact, to $\mathscr{B}$ belongs just about every subset of R that we might reasonably think of; and certainly any subset of R that has intuitive appeal and that therefore can have some connection to a conceptual model.

Example 37.1

To verify that one-element subsets of R belong to $\mathscr{B}$, we try to express such a set in terms of semi-infinite intervals of the kind specified in Definition 37.1, using countable set operations. Thus, an arbitrary real number $a_\text{o} \in R$ can first of all be expressed as an intersection,

$$\{a_\text{o}\} = \overline{-\infty, a_\text{o}} \cap \overline{a_\text{o}, +\infty}$$

The first set on the right belongs to $\mathscr{V}$. For the other set we can write $\overline{a_\text{o}, +\infty} = \overline{-\infty, a_\text{o}}^{\text{c}}$, but we still need to express $\overline{-\infty, a_\text{o}}$ in terms of closed sets belonging to $\mathscr{V}$. This is possible as follows:

$$\overline{-\infty, a_0} = \bigcup_{n=1}^{\infty} \overline{-\infty, a_0 - \frac{1}{n}} \tag{37.2}$$

In order to verify that a_0 is the smallest real number *not* belonging to the set on the right side, we note that for any $\varepsilon > 0$, $a_0 - \varepsilon$ is an element in the set because there is an n large enough to make $1/n < \varepsilon$. Therefore, a number smaller than a_0 belongs to the set. But there is no n such that $a_0 - 1/n = a_0$, so we conclude that a_0 does not belong to the set. Thus, it is correct to say that

$$\{a_0\} = \overline{-\infty, a_0} \cap \left(\bigcup_{n=1}^{\infty} \overline{-\infty, a_0 - \frac{1}{n}} \right)^c .$$

Accordingly, $\{a_0\}$ has been expressed in terms of countable set operations performed on sets in $\mathcal{V}$, and therefore $\{a_0\} \in \mathcal{B}$. An alternative expression for $\{a_0\}$ can be obtained using Equation (12.4):

$$\{a_0\} = \overline{-\infty, a_0} \cap \left(\bigcap_{n=1}^{\infty} \overline{-\infty, a_0 - \frac{1}{n}}^c \right) .$$

The significance of the class $\mathcal{V}$ and the class $\mathcal{B}$ is the following. Suppose a random variable X is defined on a probability space $(S, \mathcal{A}, P)$, and suppose that it induces on every set $\overline{-\infty, a}$ in $\mathcal{V}$ a probability $P_X(\overline{-\infty, a})$. Then the probability of every Borel set $B \in \mathcal{B}$ is also defined and can be obtained from the probabilities assigned to sets in $\mathcal{V}$ via countable additivity. $\mathcal{B}$ will serve as the completely additive class of subsets of R for the induced probability space.

We are now in a position to introduce a general definition of a random variable. If a real-valued point function X is defined on a probability space $(S, \mathcal{A}, P)$, then X is called a random variable if it induces a well-defined probability assignment P_X over $\mathcal{B}$, the class of Borel sets. Furthermore, in order to test whether a given random variable assigns probabilities to all the Borel sets, it is sufficient to check whether it assigns a probability to every set in $\mathcal{V}$.

Thus, given some real-valued point function X defined on a probability space, it is necessary to check whether X is such that the inverse image of every set in $\mathcal{V}$ belongs to $\mathcal{A}$:

$$X^{-1}(\overline{-\infty, a}) \in \mathcal{A}, \quad \text{for every } a \in R. \tag{37.3}$$

This step is usually easy to carry out. And if Equation (37.3) is satisfied, then the members of $\mathcal{V}$ have induced on them the probabilities P_X defined by

$$P_X(\overline{-\infty, a}) = P\left(X^{-1}(\overline{-\infty, a}) \right), \quad a \in R. \tag{37.4}$$

In other words, for each real number a, the probability assigned by P_X to the interval $\overline{-\infty, a}$ is the probability of that subset of S that is the inverse image of $\overline{-\infty, a}$.

If the class $\mathcal{V}$ of semi-infinite intervals is now extended, by means of countable set operations, to the class $\mathcal{B}$ of Borel sets, we find that most of the work is done. This is because the probability assignment P_X extends uniquely to all of $\mathcal{B}$ merely by the requirement that P_X is countably additive.

It can be shown that this procedure assures that, for every Borel set B, $P_X(B) = P(X^{-1}(B))$. The set function P_X thus defined on $\mathscr{B}$ is truly a probability function, since:

a) it is defined on a completely additive class, $\mathscr{B}$;
b) it is countably additive;
c) the value assigned to the empty Borel set ϕ is $P_X(\phi) = P(X^{-1}(\phi)) = P(\phi) = 0$ (which also follows from (b)); and
d) $P_X(R) = P(\{\xi: X(\xi) \text{ is any real number}\}) = P(S) = 1$.

On the basis of this discussion, we can now restate the above general definition of a random variable as follows.

(Definition) 37.2
Given a probability space (S, $\mathscr{A}$, P). A real-valued point function X defined on this space is a *random variable* (abbreviated r.v.) if and only if, for each real number a, the set $X^{-1}(-\infty, a^{|})$ belongs to $\mathscr{A}$.

There are several types of r.v.'s. One type is a discrete random variable, which was introduced in Section 34. In other words, Definition 34.1 is a special case of the above Definition. Henceforth, we will use the above Definition also for discrete r.v.'s.

Example 37.2
In Example 34.2 a real-valued point function H was defined on a probability space (S, $\mathscr{A}$, P) in such a way that it is not a discrete r.v. How does this function fail to satisfy Definition 37.2? To answer this question we consider various real numbers a and determine whether $H^{-1}(-\infty, a^{|}) \in \mathscr{A}$.

i) If a is any number less than -1 we see that $H^{-1}(-\infty, a^{|}) = \phi \in \mathscr{A}$.
ii) For $a \in \overline{-1, 0}$, we have $H^{-1}(-\infty, a^{|}) = \{\xi_c\} \notin \mathscr{A}$.

Here are real numbers a such that $H^{-1}(-\infty, a^{|}) \notin \mathscr{A}$. Therefore, H is not a r.v.

The *name* 'random variable', which is traditional, is misleading. It originated from a more intuitive approach to probability. In the context of our mathematical model, a random variable is neither random nor is it variable. It is a real-valued point function defined on a probability space, satisfying the condition stated in Definition 37.2. Figure 34.4, which summarizes the relation of a r.v. to the conceptual model, may be worth revisiting at this point.

If a r.v. X is defined on a probability space (S, $\mathscr{A}$, P), another probability space is automatically established, or 'induced', by X. This is the probability space (R, $\mathscr{B}$, P_X). The elements of this space are the real numbers, and the completely additive class is the class of Borel sets. P_X is the probability function induced on the class $\mathscr{B}$ by the r.v. X according to the rule $P_X(B) = P(X^{-1}(B))$, for every $B \in \mathscr{B}$. Henceforth, whenever we encounter the notation P_X, P_Y, etc., we will understand it as denoting an 'induced' probability function on $\mathscr{B}$, and associated with an 'induced' probability space (R, $\mathscr{B}$, P_X), (R, $\mathscr{B}$, P_Y), etc.

An important rule to keep in mind is this: If the symbol P is followed by an argument, this argument must be a subset of S; if the symbol P_X, P_Y, etc. is followed by an argument, this argument must be a set of real numbers.

Queries

37.1 For each of the following statements determine whether it is true, possibly true, or false.
  a) The class $\{\overline{-\infty, a} : a \in \mathsf{R}\}$ is a subclass of $\mathscr{V}$
  b) The Lebesgue measure of every set in $\mathscr{V}$ is ∞
  c) $\mathscr{V}$ is a disjoint class
  d) All sets in $\mathscr{V}$ are Borel sets.   [124]

37.2 A r.v. X is defined on the probability space $(\mathsf{S}, \mathscr{A}, P)$ for a single die throw. In other words, $|\mathsf{S}| = 6, |\mathscr{A}| = 64$, and $P(\{\xi\}) = 1/6$, for each $\xi \in \mathsf{S}$. Is it possible that there exists a set $\mathsf{A} \subset \overline{0, 1}$ on which X induces a probability $P_X(\mathsf{A})$ equal to

  a) 0.1   c) 0.5
  b) 1   d) 2/3 ?   [43]

37.3 A r.v. X is defined on a probability space $(\mathsf{S}, \mathscr{A}, P)$. For each of the following statements, determine whether it is true, possibly true, false, or meaningless.
  a) $P(\mathsf{S}) = 1$
  b) $P(X) = 0.5$
  c) $P_X(\{0\}) = 0$
  d) $P(\mathsf{R}) = 1$
  e) $P(X^{-1}(\mathsf{R})) = 0$
  f) $P_X(\phi) = P(\phi)$
  g) $P_X(\overline{-\infty, 1}) = P_X(\overline{1, \infty}) = 1$.   [98]

The Cumulative Distribution Function

In Section 34 we saw that the probability distribution induced by a discrete r.v. can be described by listing the probabilities induced on one-element subsets of R, or by displaying them in a bar graph. Another way to describe an induced probability distribution is by means of the *cumulative distribution function*, abbreviated c.d.f. The c.d.f. expresses, for each point along the real axis, 'how much probability lies to the left of it and directly on it'. Expressed concisely we have:

(Definition) 38.1
Given a probability space (S, $\mathscr{A}$, P) and a discrete random variable X defined on this space. The *cumulative distribution function* F_X associated with the r.v. X is defined, for every $x \in$ R, by

$$F_X(x) \equiv P_X(\overline{-\infty, x}) = P(\{\xi : X(\xi) \le x\})$$

In other words, a c.d.f. is an ordinary real-valued function of a real variable, and it expresses the probabilities of semi-infinite intervals. As the argument of a c.d.f. increases, this function 'accumulates' more probability. We use the notation $F_X(\infty)$ to express the limiting value $\lim_{x \to \infty} P_X(\overline{-\infty, x})$, so that

$$F_X(\infty) = P_X(\mathsf{R}) = P(\mathsf{S}) = 1.$$

Similarly, $F_X(-\infty)$ means $\lim_{x \to -\infty} P_X(\overline{-\infty, x})$, giving

$$F_X(-\infty) = \lim_{x \to -\infty} P_X(\overline{\mathsf{R} - x, \infty}) = P_X(\phi) = 0.$$

The discrete distribution illustrated in Figure 34.2, for instance, has a c.d.f. that rises from 0 to 1 in six equal steps, as shown in Figure 38.1. At the points of discontinuity, the value of the c.d.f. is the upper value; in Figure 38.1, $F_G(1) = 1/6$, for instance. This illustrates an important property of a c.d.f.—it is a *right-continuous* function: A function of a real variable, $F(x)$, is 'right-continuous' if for each real number x and for $\varepsilon > 0$, $F(x) = \lim_{\varepsilon \to 0} F_X(x + \varepsilon)$; whereas this need not be true for $\varepsilon < 0$.

We summarize the properties of a c.d.f. as follows.

Probability Concepts and Theory for Engineers, First Edition. Harry Schwarzlander.
© 2011 John Wiley & Sons, Ltd. Published 2011 by John Wiley & Sons, Ltd.

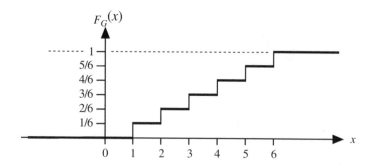

Figure 38.1   A cumulative distribution function

Theorem 38.1
A cumulative distribution function F_X

a) is a real-valued function whose domain is R and whose range is $\overline{[0, 1]}$;
b) is nondecreasing and right-continuous; and
c) satisfies the requirements $F_X(-\infty) = 0$ and $F_X(\infty) = 1$.

Conversely, any function satisfying Theorem 38.1 can be regarded as a c.d.f. and therefore defines a probability distribution over $\mathscr{B}$.

If a random variable X is not discrete, it induces a probability distribution P_X on R that cannot be described by a bar graph such as used in Section 34. However, the c.d.f. of an induced probability distribution always exists and it completely describes the induced distribution. This can be seen in the following way. According to Definition 38.1, a c.d.f. expresses the probabilities induced on all of the members of the class $\mathscr{V}$ (Equation (37.4)). These probabilities are defined since the members of $\mathscr{V}$ are Borel sets, so that the c.d.f. is defined. Furthermore, from Section 37, the class of Borel sets, $\mathscr{B}$, is the completely additive class generated by $\mathscr{V}$. The probability of *any* Borel set, therefore, can be obtained from probabilities of members of $\mathscr{V}$ (and therefore, from values of the c.d.f.) via countable additivity.

Example 38.1
Suppose the probability induced by a r.v. X on an open semi-infinite interval $\overline{-\infty, a_o}$ is to be expressed in terms of the c.d.f. F_X. In Equation (37.2) we saw how an open interval can be expressed in terms of sets in $\mathscr{V}$ using infinite set operations. To make use of countable additivity, the right side of Equation (37.2) needs to be expressed as a disjoint union:

$$\bigcup_{n=1}^{\infty} \overline{-\infty, a_o - \frac{1}{n}} = \overline{-\infty, a_o - 1} \uplus \left[\bigcup_{n=1}^{\infty} \overline{a_o - \frac{1}{n}, a_o - \frac{1}{n+1}} \right].$$

Thus, the nth approximation of $P_X(\overline{-\infty, a_o})$ is $F_X\left(a_o - \frac{1}{n+1}\right)$, so that

$$P_X(\overline{-\infty, a_o}) = \lim_{n \to \infty} F_X\left(a_o - \frac{1}{n}\right).$$

Of course, if the probability induced on the one-element set $\{a_o\}$ is already known, we have simply

$$P_X\overline{(-\infty, a_0)} = F_X(a_0) - P_X(\{a_0\}).$$

We examine now how a r.v. might arise that is not discrete.

Example 38.2: Chance selection of a real number between 0 and 1.

Consider a p.e. in which a coin is tossed an unending number of times (as in Example 11.1). Each possible outcome of this p.e. is an infinite coin-toss sequence—an infinite sequence of H's and T's. From Example 11.1 we know that the probability space (S, $\mathscr{A}$, P) for this p.e. is uncountable. Suppose that to each of the possible outcomes is to get assigned a real number in the manner of the Proof of Theorem 36.1. That is, let H = 0 and T = 1, so that each coin-toss sequence becomes the binary expansion of a real number between 0 and 1, and this is the number that is to be assigned to the coin-toss sequence. For instance, consider one of the possible outcomes of our p.e.—the outcome

$$T\ H\ T\ H\ H\ H\ H\ H\ H\ H\ H\ ....(no\ more\ T's).$$

With this outcome we associate the real number whose binary expansion is $0.10100000000.... = 0.101$. Changing to decimal form, this is the number 0.625.

So far, we have operated only in the conceptual model world, where we have specified a rule for associating real numbers with possible outcomes. In order to carry this assignment over into the mathematical model, we define a r.v. X on (S, $\mathscr{A}$, P) in such a way that it matches the conceptual rule just discussed. For instance, to the element

$$\xi_0 \leftrightarrow T\ H\ T\ H\ H\ H\ H\ H\ ....$$

X is to assign the number $X(\xi_0) = 0.625$. Since every real number from 0 to 1 gets assigned, the range of X is the interval $\overline{0, 1}$ (see Figure 38.2).

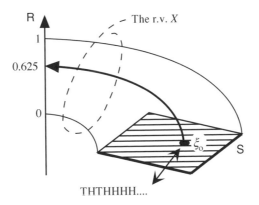

Figure 38.2   The random variable X

In order to specify the probability distribution P_X induced by this r.v. it is not sufficient to note that all possible outcomes are equally likely. For, since each outcome has zero probability, taking unions of one-element sets will never give more than zero. Instead, P_X can be described in terms of the probabilities that are associated with various intervals, such as the semi-infinite intervals $\overline{-\infty, x}$, for $x \in \mathbb{R}$. But by Definition 38.1 this is simply the c.d.f. $F_X(x)$ associated with X. What is this c.d.f.?

For any $x \leq 0$, the probability assigned to $\overline{-\infty, x}$ is zero, therefore $F_X(x) = 0$ for $x \leq 0$. Also, for $x \geq 1$, $F_X(x) = 1$ since all possible outcomes give rise to numbers no greater than 1. Now, let $0 \leq x \leq 1$; say, $x = 0.625$. What probability does X assign to the interval $\overline{-\infty, 0.625}$? We have

$$P_X(\overline{-\infty, 0.625}) = P(\{\xi : X(\xi) \leq 0.625\}).$$

The set $\{\xi : X(\xi) \leq 0.625\} \subset \mathsf{S}$ represents the collection of coin-toss sequences that lead to binary expansions less than or equal to 0.10100000. . . .

Here, it helps to define the following subsets of S:

$\mathsf{A} \leftrightarrow$ 'all coin-toss sequences starting with H'
$\mathsf{B} \leftrightarrow$ 'all coin-toss sequences starting with T H H'
$\mathsf{C} \leftrightarrow$ 'the single sequence T H T H H H H H H H H . . . . ' (a one-element set).

Then, $\{\xi : X(\xi) \leq 0.625\} = \mathsf{A} \uplus \mathsf{B} \uplus \mathsf{C}$, and

$$P(\{\xi : X(\xi) \leq 0.625\}) = P(\mathsf{A}) + P(\mathsf{B}) + P(\mathsf{C}) = \frac{1}{2} + \left(\frac{1}{2}\right)^3 + 0 = 0.625.$$

This calculation can be generalized to all $x \in \overline{0, 1}$, showing that $F_X(x) = x$ in the range $0 \leq x \leq 1$. $F_X(x)$ therefore has the graph shown in Figure 38.3.

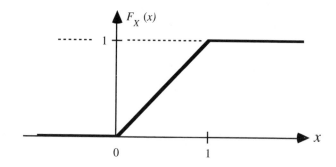

Figure 38.3   C.d.f. for the unbiased chance selection of a real number from the closed unit interval

We thus have an example of a random variable with a c.d.f. that is *continuous*, instead of consisting of a number of steps as would be the case for a discrete r.v.

(Definition) 38.2
A r.v. X is said to be *absolutely continuous* if it does not induce positive probability on any set of zero measure. The induced probability distribution P_X is then likewise called absolutely continuous.

An absolutely continuous r.v. spreads all probability over one or more *intervals*. Its c.d.f. therefore increases (with increasing argument) not in accordance with how much probability is located at specific points, as in Figure 38.1, but according to the *rate* at which it encounters or 'accumulates' probability.

Theorem 38.2
The c.d.f. $F_X(x)$ associated with an absolutely continuous r.v. X is a continuous function of x.

Proof
Let X be an absolutely continuous r.v. and suppose $F_X(x)$ is discontinuous at some point $x = x_0$. According to Theorem 38.1 this implies that

$$P_X\left(\overline{-\infty, x_0^\rceil}\right) - P_X\left(\overline{-\infty, x_0}\right) = P_X\left(\overline{-\infty, x_0^\rceil} - \overline{-\infty, x_0}\right) = P_X(\{x_0\}) > 0$$

in other words, X induces positive probability on a set of zero measure. But this contradicts the specification that X is absolutely continuous.

The r.v. X specified in Example 38.2 is absolutely continuous. However, the p.e. considered there, which gives rise to this r.v., seems rather artificial. In fact, it has no exact counterpart in the real world, since it is not possible to toss a coin an unending number of times. Is there not a simpler experiment that will give rise to an absolutely continuous r.v.?

Example 38.3: Spinning-pointer experiment.
Consider a pointer fastened to a horizontal surface and free to spin with perfect symmetry. The surface is marked with a scale of angular position, reading from 0 to 1 (Figure 38.4). An experiment is performed in which the pointer is spun and the angular position is observed at which the pointer comes to rest.

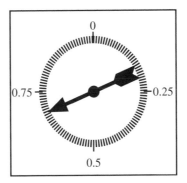

Figure 38.4   Spinning pointer

The possible outcomes of this experiment, therefore, are various angular position readings, given by numbers between 0 and 1. Are all the real numbers in this range represented? That depends on how the experiment is modeled. If we specify a p.e. $\mathfrak{E}$ that is to exactly model the real-world experiment, then there can be only a finite number of possible outcomes: The angular position scale can be read only to within some tolerance; for instance, to the nearest 0.01, or 1/360, or 0.001.

On the other hand, if we specify a p.e. $\mathfrak{F}$ in such a way that *any* real number (between 0 and 1) can arise, expressing an angular position, then the conceptual model ceases to be physically realizable and becomes just as 'artificial' as the infinite coin-toss experiment. Not only would the angular position scale of our apparatus require an infinite hierarchy of decimal markings (the tenths, the hundredths, the thousandths, etc.), which is physically impossible. But the experimenter would also have to make an unending number of readings in order to arrive at the infinite string of decimal positions that identifies the actual outcome.

It can be seen that $\mathfrak{F}$ is probabilistically equivalent to the p.e. in Example 38.2.

A r.v. and induced probability distribution that is absolutely continuous can arise only in connection with a p.e. that is 'artificial' in the same sense as in the above Example. Nevertheless, it is often convenient to formulate a conceptual model in this way. There are two reasons for this:

a) Many probability calculations are more conveniently carried out in terms of absolutely continuous distributions than discrete distributions.
b) In a given problem, the precision with which some numerical scale can be read may not be specified.

Reason (b) applies in Example 38.3. There, the p.e. $\mathfrak{F}$, which has a continuum of possible outcomes, serves as the ultimate refinement for *any* discrete 'spinning-pointer' experiment, no matter with what precision the scale is to be read.

A r.v. that induces positive probability on a countable subset of R, but does not induce probability 1 on any countable subset of R, satisfies neither the definition of a discrete r.v. nor that of an absolutely continuous r.v. It is called a *random variable of mixed type*. The c.d.f. of such a r.v. has step discontinuities as well as intervals with positive slope.

Queries

38.1 For each of the functions $g(x)$, $x \in$ R, specified below, determine whether it can be a c.d.f.

a) $g(x) = 1$, all x

b) $g(x) = \begin{cases} 0, & x < 0 \\ 1, & x \geq 0 \end{cases}$

c) $g(x) = \begin{cases} 0, & x < 0 \\ 1/2, & x = 0 \\ 1, & x > 0 \end{cases}$

d) $g(x) = \begin{cases} 0, & x < 0 \\ x, & 0 \leq x \leq 1/2 \\ 1, & x > 1/2. \end{cases}$ [56]

38.2 Which of the distributions specified below are absolutely continuous:

a) $F_X(x) = \begin{cases} 0, & x < 0 \\ \sin x, & 0 \leq x \leq \pi/2 \\ 1, & x > \pi/2 \end{cases}$

b) $F_X(x) = \begin{cases} 0, & x < 0 \\ 1 - \dfrac{1}{n}, & (n-1) \le x \le n, \quad n = 1, 2, 3, \dots \end{cases}$

c) $F_X(x) = \begin{cases} 0, & x < 0 \\ x, & 0 \le x \le 1/2 \\ 1, & x > 1/2. \end{cases}$ [212]

38.3 Let $F_X(x)$, $x \in \mathbb{R}$, be the c.d.f. associated with some r.v. X. For each of the following functions determine whether it is also a c.d.f.

a) $F_X(x/2)$ b) $F_X(x) + 1$

c) $F_X(1 - x)$ d) $F_X(2x + 1)$

e) $[F_X(x) + F_Y(x)]/2$, where F_Y is some other c.d.f.   [231]

The Probability Density Function

It is often convenient to describe the induced probability distribution of an absolutely continuous random variable not by means of the c.d.f. but by another function called the *probability density function*, abbreviated p.d.f.

(Definition) 39.1

If X is an absolutely continuous random variable with c.d.f. F_X, then the probability density function f_X is defined as the derivative of F_X,

$$\boxed{f_X(x) \equiv \frac{d}{dx} F_X(x)}$$

for all $x \in \mathsf{R}$ at which the derivative of F_X exists.

The p.d.f. expresses how 'densely' probability is distributed over any small increment of the real line.

Since for any given x, $F_X(x)$ is a probability, Definition 39.1 shows that the p.d.f. has to get *integrated* in order to yield a probability. Specifically, in order to find the probability that an absolutely continuous random variable X assigns to some *interval* $\overline{a, b^|}$, we have

$$P_X(\overline{a, b^|}) = \int_a^b f_X(x)\, dx = F_X(b) - F_X(a). \tag{39.1}$$

In particular,

$$\int_{-\infty}^{\infty} f_X(x)\, dx = F_X(\infty) - F_X(-\infty) = 1.$$

Since $F_X(x)$ is nondecreasing, $f_X(x) \geq 0$ must hold for all x. Also, *any* non-negative function of a real variable that integrates to 1 can be interpreted as a p.d.f.

Probability density functions that describe some commonly encountered distributions are specified and sketched in Figure 39.1. In each of the functions there are one or two unspecified

---

Probability Concepts and Theory for Engineers, First Edition. Harry Schwarzlander.
© 2011 John Wiley & Sons, Ltd. Published 2011 by John Wiley & Sons, Ltd.

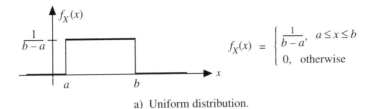

a) Uniform distribution.

$$f_X(x) = \begin{cases} \dfrac{1}{b-a}, & a \le x \le b \\ 0, & \text{otherwise} \end{cases}$$

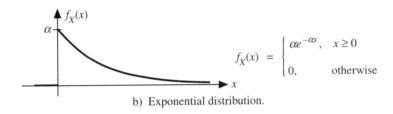

b) Exponential distribution.

$$f_X(x) = \begin{cases} \alpha e^{-\alpha x}, & x \ge 0 \\ 0, & \text{otherwise} \end{cases}$$

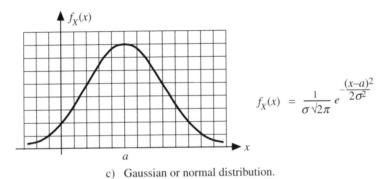

c) Gaussian or normal distribution.

$$f_X(x) = \frac{1}{\sigma\sqrt{2\pi}} e^{-\frac{(x-a)^2}{2\sigma^2}}$$

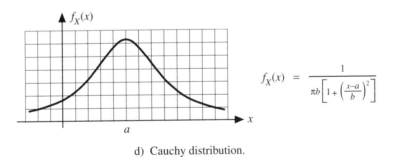

d) Cauchy distribution.

$$f_X(x) = \frac{1}{\pi b \left[1 + \left(\dfrac{x-a}{b} \right)^2 \right]}$$

Figure 39.1   P.d.f.'s of some common absolutely continuous probability distributions

parameters, so that actually a whole *family* of p.d.f.'s is represented in each part of the figure, each corresponding to a family of induced distributions. We see, for instance, that the induced probability distribution obtained in Example 38.2 is a *uniform distribution*—a member of the family of uniform distributions. Experiments giving rise to exponential distributions and to Gaussian distributions will be discussed in later Sections.

Example 39.1: Gaussian density function.

We verify that $f_X(x)$ as given in Figure 39.1c is a density function.

1. It is clearly non-negative.
2. Does it integrate to 1? The Gaussian p.d.f. does not lend itself to integration in terms of elementary functions. However, the integral $\int_{-\infty}^{\infty} f_X(x)\, dx$ can be evaluated by first applying the change of variable $w = (x-a)/\sigma$, giving

$$\int_{-\infty}^{\infty} f_X(x)\, dx = \frac{1}{\sigma\sqrt{2\pi}} \int_{-\infty}^{\infty} \exp\left[-\frac{(x-a)^2}{2\sigma^2} \right] dx = \frac{1}{\sqrt{2\pi}} \int_{-\infty}^{\infty} e^{-w^2/2}\, dw.$$

The square of this quantity can be written

$$\left[\int_{-\infty}^{\infty} f_X(x)\, dx \right]^2 = \frac{1}{2\pi} \int_{-\infty}^{\infty} \int_{-\infty}^{\infty} \exp\left[-\frac{w^2 + v^2}{2} \right] dw\, dv.$$

Conversion to polar coordinates, using $r^2 = w^2 + v^2$ and $\theta = \arctan\dfrac{v}{w}$, yields

$$\frac{1}{2\pi} \int_0^{2\pi} \int_0^{\infty} r e^{-r^2/2} dr\, d\theta = 1.$$

Therefore $\int_{-\infty}^{\infty} f_X(x)dx = \pm 1$, and since the result cannot be negative, it is $+1$, as required.

From Definition 38.2 it follows that an absolutely continuous probability distribution assigns *zero probability* to every one-element set $\{x\}$—irrespective of the value of the probability density function at x.[4] As a result, Equation (39.1) also equals $P_X(\overline{a, b}) = P_X(\overline{a, b}) = P_X(\overline{a, b})$. Furthermore, the value of a p.d.f. at any one point is not significant, since it does not affect the result of an integration. In Definition 39.1, $f_X(x)$ is left undefined at x-values where the derivative of $F_X(x)$ does not exist. We see now that an arbitrary value can be assigned to f_X at such an x-value. Typically, if $F_X(x)$ has a discontinuity in slope at $x = x_0$, then $f_X(x_0)$ is specified as either $f_X(x_0^-)$ or $f_X(x_0^+)$; that is, as either left-continuous or right-continuous at $x = x_0$. In Figure 39.1a, for instance, the probability density function of a uniform distribution is specified as being nonzero over a *closed* interval. However, the specification $a \le x \le b$ can be replaced by $a < x < b$ without affecting the probability distribution that is being described.

More generally, the value of a p.d.f. at isolated points can be changed arbitrarily without affecting the induced distribution that is being described. For instance, arbitrarily setting the Gaussian density function in Figure 39.1c to zero at the origin changes nothing. However, unless specifically noted otherwise, we will always express a p.d.f. $f_X(x)$ in the manner specified by Definition 39.1, with discontinuities only where the derivative of $F_X(x)$ does not exist.

As we saw earlier, the knowledge that all one-element sets have zero probability contributes little toward describing a probability distribution on the uncountable space $\mathbb{R}$—it merely indicates that the

---

[4] As noted in Section 17, a zero-probability event can be thought of as an event that will have only a finite number of occurrences in an infinite sequence of trials. But in practice there can be no infinity of trials. So, in order to avoid reference to an infinite sequence of trials, we might say that a zero-probability event is one that is 'never expected' to occur, rather than 'impossible.' Making such a distinction is motivated strictly by the conceptual model, where zero probability events readily arise in connection with an infinite sample space while impossible events are excluded from consideration in a p.e. and therefore *cannot* occur. On the other hand, physical reasoning does not really warrant such a distinction. In the real world, anything deemed impossible is simply something that is expected to never occur.

distribution at hand is not a discrete distribution. One must look at intervals. Thus, the constant density of the uniform distribution in Figure 39.1a does not signify that all one-element events $\{x\} \subset \overline{a, b}$ are 'equally likely'. They do all have zero probability, but this is also true if the distribution is Gaussian, or is any other absolutely continuous distribution. Instead, it signifies that *intervals of equal length*, lying wholly within $\overline{a, b}$, have equal probability.

Many absolutely continuous probability distributions encountered in practice, and their p.d.f.'s, can be thought of as consisting of a central 'body' where most of the probability is concentrated (which is not necessarily centered at the origin) and 'tails' where the remaining probability is dispersed. In both Gaussian and Cauchy distributions, the body corresponds to a single hump in the p.d.f. and the tails extend outward to each side. A uniform distribution, on the other hand, has only a body and no tails. Still different is an exponential distribution, which consists only of a tail.

As can be seen in Figure 39.1, the density functions of both Gaussian and Cauchy distributions are 'bell-shaped'. Yet, although the graphs of a Gaussian and a Cauchy p.d.f. appear similar, the natures of these two families are very different. In particular, even though both a Gaussian r.v. and a Cauchy r.v. distributes probability over all of R, the tails of a Gaussian distribution decay extremely rapidly, and those of a Cauchy distribution quite slowly. The tails of a Cauchy p.d.f. are of order x^{-2}, while a function with slightly shallower tails decaying only as $|x|^{-1}$ cannot be a p.d.f. (see Problem 39.10).

So far, a p.d.f. has been defined only for absolutely continuous distributions. However, the definition of the probability density function can be extended to those distributions that assign *positive* probability to various isolated points (one-element sets). This is done by means of the familiar δ-function or 'unit impulse'—which is not actually a *function* (a real-valued function defined over all of R) but is a convenient device for expressing the derivative of a step discontinuity. It is defined operationally by the following two rules:

$$\delta(x) = 0, \quad \text{for } x \neq 0 \tag{39.2}$$

$$\int_a^b \delta(x)\, dx = 1, \quad \text{for } a < 0 < b. \tag{39.3}$$

Formally, $\delta(x)$ stands for the derivative of a unit step at the origin.[5] A coefficient multiplying $\delta(x)$ expresses the 'weight' of the δ-function, since the 'magnitude' at $x = 0$ is undefined and can be thought of as infinite. The graphical representation of $\delta(x)$ is an upward arrow at $x = 0$.

Example 39.2

The probability density function f_G of the discrete r.v. G that induces the distribution described in Figure 38.1 is given by

$$f_G(x) = \frac{d}{dx} F_G(x) = \sum_{k=1}^{6} \frac{1}{6} \delta(x - k).$$

---

[5] More precisely, δ-functions have come about as a result of inserting into the mathematical domain the *conceptual* notion of the derivative of a step discontinuity. This has resulted in a symbolism that allows δ-functions to be treated in many ways like an ordinary real function, which they are not. However, a mathematically consistent and rigorous development of δ-functions is available in the theory of 'distributions' or 'generalized functions' (see [Li1], [Ze1]).

Example 39.3

A r.v. X induces probability 1/2 at the origin and distributes the remaining probability uniformly over the interval $\overline{0.5,\ 1}$. This is a r.v. of mixed type. Its c.d.f. is

$$F_X(x) = \begin{cases} 0, & x < 0 \\ \dfrac{1}{2}, & 0 \le x < 0.5 \\ x, & 0.5 \le x < 1 \\ 1, & x \ge 1 \end{cases}$$

and its p.d.f. is (see Figure 39.2).

$$f_X(x) = \frac{d}{dx}F_X(x) = \frac{1}{2}\delta(x) + \begin{cases} 1, & x \in \overline{0.5,\ 1} \\ 0, & \text{otherwise.} \end{cases}$$

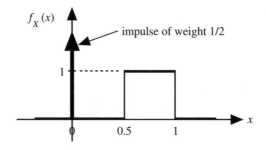

Figure 39.2   P.d.f. for a r.v. of mixed type

Any δ-function appearing in the specification of a p.d.f. must have positive weight. This follows from the fact that a c.d.f. is nondecreasing.

When a probability density function is used to describe a distribution that has discrete components, as in Example 39.3, more care is needed in carrying out probability calculations. Specifically, if the probability of an interval is to be computed from the p.d.f. (Equation (39.1)), it is necessary to specify for each endpoint whether or not it is to be included, since a δ-function might be located there. This is illustrated below.

Example 39.3   (continued)

Computing $P_X(\overline{0,\ 0.5})$, we obtain

$$P_X(\overline{0,0.5}) = \int_{0^+}^{0.5} f_X(x)\,dx = \lim_{\varepsilon \to 0}\int_{\varepsilon}^{0.5}\frac{1}{2}\delta(x)\,dx = 0.$$

In this calculation the discrete component of P_X at the origin, represented by the term $1/2\delta(x)$, is ignored because the origin does not belong to the interval $\overline{0,\ 0.5}$. On the other hand,

$$P_X(\overline{0,0.5}) = \int_{0^-}^{0.5} f_X(x)\,dx = \lim_{\varepsilon \to 0}\int_{-\varepsilon}^{0.5}\frac{1}{2}\delta(x)\,dx = \frac{1}{2}.$$

The probability that this r.v. X assigns to any one-element set $\{x\}$ is zero unless the element x is the point at which $f(x)$ has an impulse, namely, $x = 0$. Thus,

$$P_X(\{0\}) = \int_{0^-}^{0^+} f_X(x)\,dx = \lim_{\varepsilon \to 0} \int_{-\varepsilon}^{+\varepsilon} \frac{1}{2}\delta(x)\,dx = \frac{1}{2}.$$

Queries

39.1 Let $f(x)$, $x \in \mathbb{R}$, be an absolutely continuous probability density function. For each of the functions specified below, determine whether it is also a p.d.f.:

a) $f(x/2)$
b) $2\,f(2x)$
c) $f(1-x)$
d) $1 - f(x)$
e) $(1/2)[f(x) + \delta(x)]$. [52]

39.2 The c.d.f. associated with a r.v. X is described in the graph shown in Figure 39.3. What is the largest value taken on by the probability density function $f_X(x)$? [87]

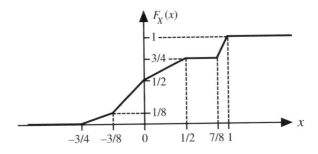

Figure 39.3 C.d.f. for the r.v. X in Query 39.2

39.3 A random variable X induces a uniform distribution over the interval $\overline{-1,\,3}$ and zero probability elsewhere. What is the probability that $|X(\xi)|$ is greater than $f_X(0)$? In other words, what is the probability that the value assigned by X, when the underlying experiment is performed, will have an absolute value greater than $f_X(0)$? [14]

39.4 The r.v. X induces a probability distribution characterized by the p.d.f.

$$f_X(x) = \begin{cases} \dfrac{k}{\sqrt{|x|}}, & -1 \le x \le 1 \\ 0, & \text{otherwise.} \end{cases}$$

With what probability does X take on the value 0; i.e., what is $P_X(\{0\})$? [199]

SECTION *40*

The Gaussian Distribution

A r.v. X is a *Gaussian* or *normal* r.v. if it induces a Gaussian or normal distribution, so that the p.d.f. f_X is of the form (see Figure 39.1)

$$f_X(x) = \frac{1}{\sqrt{2\pi}\,\sigma} \exp\left[-\frac{(x-a)^2}{2\sigma^2}\right]$$

(40.1)

As noted in the preceding Section, Equation (40.1) characterizes a two-parameter family of distributions. The parameter a can be any real number, and the parameter σ is a positive real number. Since $f_X(x)$ is a function of $(x-a)^2$, the parameter a identifies the point of symmetry of the p.d.f., $x = a$, which is also the point where f_X has its maximum. From Equation (40.1) it can also be deduced that the parameter σ affects the broadness of the hump of f_X.

A Gaussian distribution typically arises in experiments where the underlying random mechanism consists of the superposition of a large number of small disturbances. The mathematical principle involved here is the Central Limit Theorem, which will be discussed in Section 86. For now, we note that such an experiment might involve measurements that are subject to 'random errors'—as distinct from systematic errors. Or it might be concerned with measuring certain kinds of random noise.

Example 40.1

Suppose a noisy measurement of a physical parameter is such that with probability 0.9, the measurement result is known to lie within $\pm 10\%$ of the true parameter value. What induced distribution characterizes these measurements?

A physical measurement always has finite precision, implying a discrete set of possible outcomes. However, in order to simplify the mathematical model, and to obviate the need to know exactly which numerical values can be obtained with the measurement procedure, we idealize the situation and assume a continuum of possible outcomes. Thus, the measurement procedure is viewed as if it can yield any real number.

We have to assume that an appropriate p.e. can be specified, along with a corresponding uncountable probability space. We will also assume that the measurements are such that they are not subject to systematic errors (such as instrument calibration error), and that the random error is such that it is appropriately modeled by a Gaussian distribution. The measurement result is then characterized by a random variable X that is Gaussian. That is, X induces a Gaussian distribution, described by the p.d.f. (40.1). The center point of the distribution, a, represents the 'true parameter

Probability Concepts and Theory for Engineers, First Edition. Harry Schwarzlander.
© 2011 John Wiley & Sons, Ltd. Published 2011 by John Wiley & Sons, Ltd.

value', and the width of the 'hump' of the function (40.1) needs to be adjusted so that 90% of the area under the curve lies in $\overline{0.9a, 1.1a}$. This adjustment must involve the only other parameter that is available, which is σ. We will make that adjustment shortly.

Since a Gaussian distribution can be any distribution in the family that is characterized by Equation (40.1), it is convenient to agree on one particular Gaussian distribution that will serve as a preferred representative or 'spokesperson' for the whole family. This is the Gaussian distribution with $a=0$ and $\sigma=1$, which is called the *standardized* Gaussian distribution. It is induced by a *standardized Gaussian r.v.* If X_s is such a standardized Gaussian r.v., then

$$\boxed{f_{X_s}(x) = \frac{1}{\sqrt{2\pi}} \exp\left[-\frac{x^2}{2}\right]} \tag{40.2}$$

It can be seen that Equation (40.1) is obtained from Equation (40.2) by writing

$$f_X(x) = \frac{1}{\sigma} f_{X_s}\left(\frac{x-a}{\sigma}\right) \tag{40.3}$$

that is, by applying the shift a and scale factor σ. Therefore, if a graph of $f_{X_s}(x)$ has been plotted, it can be changed into a graph of $f_X(x)$ simply by designating the origin as the point $x=a$, along the x-axis replacing the integer markings ± 1, ± 2, etc., by $a \pm \sigma$, $a \pm 2\sigma$, ..., and dividing the vertical scale by σ.

As noted in Example 39.1, the c.d.f. of a Gaussian r.v. cannot be expressed in closed form in terms of elementary functions. The c.d.f. of a standardized Gaussian r.v. X_s is the function

$$F_{X_s}(x) = \int_{-\infty}^{x} f_{X_s}(x)dx = \int_{-\infty}^{x} \frac{1}{\sqrt{2\pi}} \exp\left(-\frac{u^2}{2}\right) du = \Phi(x) \tag{40.4}$$

where $\Phi(x)$ is also called the *Gaussian integral*, for which tables are available (see Appendix).

When using a table for values of $\Phi(x)$, it is important to be clear about what the table actually provides. Typically, tables of the Gaussian integral either list values of $\Phi(x) - 1/2$ for $x \geq 0$, that is,

$$\int_{0}^{x} \frac{1}{\sqrt{2\pi}} \exp\left(-\frac{u^2}{2}\right) du$$

or they list values of $1 - \Phi(x)$ for $x \geq 0$, that is,

$$\int_{x}^{\infty} \frac{1}{\sqrt{2\pi}} \exp\left(-\frac{u^2}{2}\right) du.$$

There are also tables of the so-called 'error function', defined for $x \geq 0$ as

$$\text{erf}(x) \equiv \frac{2}{\sqrt{\pi}} \int_{0}^{x} \exp(-t^2)dt$$

and there are tables of the 'complementary error function',

$$\text{cerf}(x) \equiv 1 - \text{erf}(x) = \frac{2}{\sqrt{\pi}} \int_{x}^{\infty} \exp(-t^2)dt.$$

Thus, an appropriate conversion must be applied to values obtained from a table, depending on the specific function that is tabulated.

Example 40.1 (continued)

We are now able to proceed with determination of the correct value of σ. A portion of the standardized Gaussian density function is plotted in Figure 40.1. The origin in this plot corresponds to the point $x = a$ in the Gaussian distribution we want to specify.

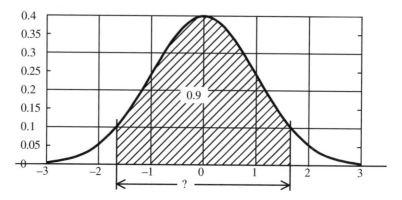

Figure 40.1   Standardized Gaussian p.d.f

Now we ask: How large must a symmetrical interval be (i.e., an interval centered at the origin), so that the standardized Gaussian distribution assigns a probability of 0.9 to that interval? In other words, we want the shaded region in the figure to have an area of 0.9. The right half of the shaded region then has an area of 0.45. Using the table of the Gaussian integral given in the Appendix, we see that an x-value of 1.645 yields the desired result of 0.45. The symmetrical interval with probability 0.9 therefore is $-1.645, 1.645$.

Making use of Equation (40.3), Figure 40.1 can thus be changed to the desired density function by replacing each abscissa value x by $a + \sigma x$. Then

$$a + 1.645\sigma = 1.1a, \text{ or}$$

$$1.645\sigma = 0.1a, \text{ giving}$$

$$\sigma \doteq 0.0608a.$$

Therefore, the desired distribution is described by the density function

$$f_X(x) \doteq \frac{1}{\sqrt{2\pi}\, 0.0608\, a} \exp\left(\frac{(x-a)^2}{2 \times (0.0608\, a)^2}\right) \doteq \frac{6.56}{a} \exp\left(\frac{(x-a)^2}{7.39 \times 10^{-3} a^2}\right)$$

where a is the true (but unknown) value of the parameter being measured.

There is another matter to be considered, however. We are dealing with the measurement of a 'physical parameter', which usually means that measurement results are positive. But the range of a Gaussian r.v. is all of $\mathbb{R}$. Isn't it necessary, then, to modify the result we have obtained and restrict f_X to positive x? In principle, that is true. But it turns that with $P_X(\overline{0.9a, 1.1a}) = 0.9$, the probability of X taking on a negative value is vanishingly small and for practical purposes can be ignored (see also Problem 40.2).

In order to investigate how probability declines in the tails of the Gaussian distribution, we turn to the function

$$1 - \Phi(x) = \int_x^\infty \frac{1}{\sqrt{2\pi}} \exp\left(-\frac{u^2}{2}\right) du, \quad \text{for} \quad x \geq 0.$$

It expresses the area under one of the tails of the standardized Gaussian p.d.f. from x upward. These tails decrease very rapidly as $|x|$ increases. For example,

if x equals	then $f_{X_s}(x)$ equals
2.0	0.0540
3.0	0.00443
4.0	0.000134
5.0	0.0000015

Since $\Phi(x)$ cannot be expressed in terms of elementary functions, an approximation to $1 - \Phi(x)$ is desirable that is easily manipulated and evaluated. One such approximation is the function

$$b(x) = \frac{1}{2} \exp\left(-\frac{x^2}{2}\right) = \int_x^\infty \frac{u}{2} \exp\left(-\frac{u^2}{2}\right) du$$

which is an upper bound on $1 - \Phi(x)$, for $x \geq 0$. To verify this, $b(x)$ can be compared with $1 - \Phi(x)$:

$$\Delta(x) = b(x) - [1 - \Phi(x)] = \int_x^\infty \left(\frac{u}{2} - \frac{1}{\sqrt{2\pi}}\right) \exp\left(-\frac{u^2}{2}\right) du. \tag{40.5}$$

The integrand on the right is positive for $u > \sqrt{2/\pi}$, so that $\Delta(x) > 0$ for $x \geq \sqrt{2/\pi}$, at least. Furthermore, since for $0 < x < 1$, $b(x)$ is convex upward ($b''(x) < 0$) and $1 - \Phi(x)$ is concave upward (second derivative is positive), and since $b(0) = 1 - \Phi(0) = 0.5$, it follows that $\Delta(x) \geq 0$ for $x \geq 0$.

$b(x)$ is a convenient bound for $1 - \Phi(x)$ because it is a simple function. A tighter bound can be obtained, however. For x greater than about 1, $\Delta(x)$ can be reduced by noting that the major contribution to the integral in Equation (40.5) is near the lower limit. Replacing $b(x)$ by $\gamma\, b(x)$ changes Equation (40.5) to

$$\Delta_\gamma(x) = \gamma b(x) - [1 - \Phi(x)] = \int_x^\infty \left(\gamma\frac{u}{2} - \frac{1}{\sqrt{2\pi}}\right) \exp\left(-\frac{u^2}{2}\right) du.$$

The integrand can be made zero at the lower limit by setting $\gamma\frac{u}{2} - \frac{1}{\sqrt{2\pi}} = 0$ for $u = x$. Then $\gamma = \frac{\sqrt{2/\pi}}{x}$, giving the new bound

$$b_1(x) = \frac{1}{\sqrt{2\pi}x} \exp\left(-\frac{x^2}{2}\right), \quad \text{for } x > 1. \tag{40.6}$$

$b_1(x)$ is still an upper bound, since $\Delta_{(\sqrt{2/\pi})/x}(x) =$

$$\int_x^\infty \left(\frac{1}{\sqrt{2\pi}} \frac{u}{x} - \frac{1}{\sqrt{2\pi}} \right) \exp\left(-\frac{u^2}{2} \right) du = \frac{1}{\sqrt{2\pi}} \int_x^\infty \left(\frac{u}{x} - 1 \right) \exp\left(-\frac{u^2}{2} \right) du > 0.$$

Example 40.2

We wish to estimate the probability that a standardized Gaussian r.v. X_s assigns a numerical value whose absolute value is greater than 3. Using the bound $b(x)$ we find $b(3) \doteq 5.5545 \times 10^{-3}$, indicating that the desired probability is no larger than $2 \times b(3) \doteq 0.0111$. The bound $b_1(x)$, on the other hand, gives $b_1(3) \doteq 1.4773 \times 10^{-3}$, showing that the desired probability is no larger than 0.002955. This is close to the probability that can be found from a table (which is based on a series expansion), namely, 0.00270.

Queries

40.1 For each of the following statements, determine whether it is true or false:
In the Gaussian p.d.f. (Equation (40.1)), the maximum value of the function
a) is always less than 1
b) can occur at a negative x-value
c) is increased if σ is increased
d) is increased if a is increased
e) is increased if the 'hump' of the p.d.f. is made narrower.   [139]

40.2 What are the x-coordinates of the points of inflection of the standardized Gaussian density function (40.2)?   [12]

40.3 For each of the following, determine whether it is true or false:
a) $\Phi(x) = 1 - \Phi(-x)$, $x \in \mathbb{R}$
b) $\Phi(x) = 0.5 + 0.5\ \mathrm{erf}(x)$, $x \geq 0$
c) $\frac{1}{2}\mathrm{cerf}(x) = 1 - \Phi(\sqrt{2x})$, $x \geq 0$.   [249]

Two Discrete Random Variables

So far, we have considered only those situations where a single numerical value gets associated with each possible outcome of a p.e. Sometimes, *two* numerical values are to be associated with each possible outcome. In this Section we consider such situations for the case of *discrete* random variables, using an approach analogous to that of Section 34. In other words, an induced distribution will not be defined over all Borel sets, as in Section 37.

Example 41.1

Consider again the throw of a die as in Example 34.1, with the random variable G and the induced distribution P_G defined in the same way as in that Example. Suppose you use that experiment to play a game with the following *pay-off rule*:

$$\text{one dot faces up} \Rightarrow \text{you lose \$1}$$

$$\text{two, three, four or five dots face up} \Rightarrow \text{no loss, no winning}$$

$$\text{six dots face up} \Rightarrow \text{you win \$1.}$$

In order to incorporate the pay-off rule in the mathematical model, a different function, H, can be specified on the probability space for the die throw in Example 34.1, which expresses your winnings in dollars:

$$H(\xi_a) = -1$$

$$H(\xi_b) = H(\xi_c) = H(\xi_d) = H(\xi_e) = 0$$

$$H(\xi_f) = +1.$$

H is also a r.v. It induces a probability distribution P_H that can be described graphically by the bar graph in Figure 41.1.

So far, what we have done is to first define one discrete random variable on the probability space for a die throw, and then another one. However, to define two discrete random variables on a probability space $(\mathsf{S}, \mathscr{A}, P)$ really means to associate a *pair of numbers* (x, y) with each element $\xi \in \mathsf{S}$. For instance, to the element ξ_a in Example 41.1 gets assigned the pair $(x, y) =$

Probability Concepts and Theory for Engineers, First Edition. Harry Schwarzlander.
© 2011 John Wiley & Sons, Ltd. Published 2011 by John Wiley & Sons, Ltd.

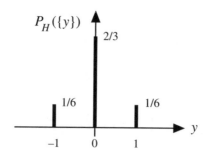

Figure 41.1   Bar graph for the distribution P_H

$(G(\xi_a), H(\xi_a)) = (1, -1)$; to the element ξ_b gets assigned $(G(\xi_b), H(\xi_b)) = (2, 0)$; and so on. Taken *jointly*, the functions G and H have as their range not real numbers, but *pairs of real numbers*; that is, they map into the product space $R \times R = R^2$, the Cartesian plane, whose elements are *pairs* of real numbers.

It is possible to arrange things so that we can talk about the probabilities of various subsets *of* R^2—subsets whose elements are *pairs* of numbers. This can be done by proceeding in a manner analogous to Example 34.1.

Example 41.1   (continued)

Consider some set in $\mathscr{A}$, such as the set $\{\xi_a, \xi_c\}$. The function pair (G, H) maps each element of this set into a pair of real numbers: $(1, -1)$ and $(3, 0)$. The collection of these two pairs is a subset of R^2; in other words, $\{(1, -1), (3, 0)\} \subset R^2$. In the same way, for each of the remaining sets in $\mathscr{A}$, the function pair (G, H) identifies a corresponding subset of R^2. The largest partition in $\mathscr{A}$ has as its image, via (G, H), a disjoint class of subsets of R^2 consisting of the six one-element sets

$$\{(1, -1)\}, \{(2, 0)\}, \{(3, 0)\}, \{(4, 0)\}, \{(5, 0)\}, \{(6, 1)\}. \qquad (41.1)$$

This class can be enlarged into a partition, $\mathscr{P}_{G,H}$, by including the set

$$Q_2 = R^2 - \{(1, -1), (2, 0), (3, 0), (4, 0), (5, 0), (6, 1)\}$$

It is therefore possible to define an additive class of subsets of R^2, namely, the additive class generated by $\mathscr{P}_{G,H}$. We denote this class $\mathscr{A}_{G,H}$. Since it is finite, it is also completely additive. The probability function $P_{G,H}$ that gets induced on $\mathscr{A}_{G,H}$ assigns probability $1/6$ to each of the sets listed in Equation (41.1), and probability 0 to Q_2. From these probabilities, the values that $P_{G,H}$ assigns to all other sets in $\mathscr{A}_{G,H}$ can be computed.

We see that the two functions G and H, taken jointly, serve to establish a new probability space, $(R^2, \mathscr{A}_{G,H}, P_{G,H})$. The induced probability distribution $P_{G,H}$ is a *two-dimensional* or *joint probability distribution*. It can be described by a bar graph as shown in Figure 41.2. In the induced probability space $(R^2, \mathscr{A}_{G,H}, P_{G,H})$, events are described by numbers of dots in a throw of the die, and/or in terms of specific amounts of pay-off. Questions about probabilities of these events can be answered using only the induced space, without recourse to the basic space $(S, \mathscr{A}, P)$.

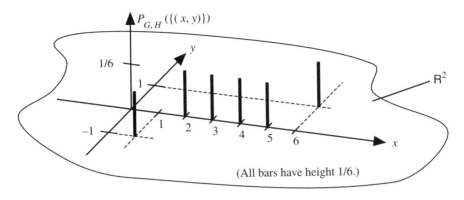

Figure 41.2   A joint probability distribution

Analogous to Definition 34.1, two discrete r.v.'s must satisfy the following:

(Definition) 41.1
Two real-valued point functions G and H defined on a probability space $(S, \mathscr{A}, P)$ are a pair of discrete random variables if and only if
a) their joint range is a countable subset of R^2, and
b) for each number pair (g, h) in their joint range, the set $\{\xi: G(\xi) = g \text{ and } H(\xi) = h\}$ is a member of the class $\mathscr{A}$.

We see that the definition of two discrete random variables on a probability space is a simple extension of a single random variable. The relationship to the conceptual model can be summarized as shown in Figure 41.3 (see also Figure 34.4).

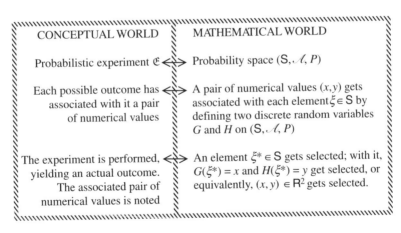

Figure 41.3   Relating a pair of discrete r.v.'s to the conceptual model

The two random variables G and H in Example 41.1 are related in a very definite way: Once we know what value the r.v. G assigns upon a particular execution of the underlying experiment (die throw), then we also know what value H assigns. The next Example illustrates the opposite extreme: neither random variable provides any information about the other random variable.

Example 41.2

Suppose two distinguishable dice—say, a red die and a white die—are thrown. We are familiar with the probability space for this experiment (Example 17.1). Suppose we want to bring into our mathematical model the number of dots facing up on each die. That is, we wish to associate with each element of the probability space two numbers,

$$r \leftrightarrow \text{number of dots showing on red die}$$
$$w \leftrightarrow \text{number of dots showing on white die}.$$

Thus, we define two functions on the probability space: the function G_r that assigns to each element the number of dots on the red die in the corresponding outcome; and the function G_w that assigns to each element the number of dots on the white die in the corresponding outcome. Considered jointly, the two functions then assign a *pair* of real numbers (r, w) to each element in the probability space. These number pairs are elements of the product space R^2; and G_r and G_w are two r.v.'s that, considered jointly, induce a new probability space $(\mathsf{R}^2, \mathscr{A}_{G_r,G_w}, P_{G_r,G_w})$ in the manner already discussed in Example 41.1. The joint probability distribution P_{G_r,G_w} is described by the bar graph in Figure 41.4.

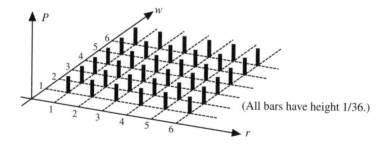

(All bars have height 1/36.)

Figure 41.4   Joint distribution of the two discrete r.v.'s G_r, G_w

The induced probability distribution P_{G_r,G_w} is completely defined once it is specified at all the points (r, w) that can arise:

$$P_{G_r,G_w}(\{(r, w)\}) = \frac{1}{36}, \quad \text{for } r, w \in \{1, 2, \ldots, 6\}.$$

For instance, probability zero gets assigned to the complement of the set $\{(r, w): r, w \in \{1, 2, \ldots, 6\}\}$.

In Example 41.2, suppose the experiment has been performed and we are told merely what value has been assigned by *one* of the two r.v.'s. This provides no clue whatsoever regarding the value assigned by the other r.v. Whenever two random variables are defined on a probability space in such a way that this is the case, then those r.v.'s are called *statistically independent*. We will explore this property further in Section 45. Statistical independence of two discrete random variables is expressed mathematically in the following way.

(Definition) 41.2

Let G and H be two discrete r.v.'s defined on a probability space $(S, \mathscr{A}, P)$. The probability space induced by G and H is $(R^2, \mathscr{A}_{G,H}, P_{G,H})$. G and H are *statistically independent* if and only if, for every elementary set $\{(g, h)\} \in \mathscr{A}_{G,H}$,

$$P_{G,H}(\{(g, h)\}) = P_G(\{g\})P_H(\{h\}). \tag{41.2}$$

In connection with this Definition it needs to be pointed out that if the real numbers g, h are such that $\{(g, h)\} \in \mathscr{A}_{G,H}$, then it follows that $\{g\} \in \mathscr{A}_G$ and $\{h\} \in \mathscr{A}_H$ (see Problem 41.5).

Applying Definition 41.2 to Example 41.1 shows that r.v.'s G and H are *not* statistically independent since, for instance, $P_{G,H}(\{(1, -1)\}) = 1/6 \neq P_G(\{1\})P_H(\{-1\}) = (1/6) \cdot (1/6) = 1/36$. As a matter of fact, the two r.v.'s in Example 41.1 have a very definite dependence, called a *functional dependence*, which can be expressed by writing

$$H(\xi) = \nu[G(\xi)], \quad \xi \in S. \tag{41.3}$$

The function ν in Equation (41.3) is the rule which uniquely indicates what value H assigns to any element in S, once the value assigned by G is known. In other words, it is the pay-off rule of Example 41.1:

For $\xi \in S$,	if $G(\xi) =$	then $H(\xi) =$
	1	-1
	2	0
	3	0
	4	0
	5	0
	6	1

It is not necessary to specify the function ν for other values $G(\xi)$, since none can arise.

We observe that in Example 41.1, H is functionally dependent on G, but G is not functionally dependent on H since $G(\xi)$ cannot be uniquely determined from $H(\xi)$ when $H(\xi) = 0$. It is also possible, of course, for two random variables to be defined in such a way that each is functionally dependent on the other.

If the *joint* probability distribution $P_{G,H}$ induced by two discrete random variables G and H is known, it is always possible to obtain the distributions P_G and P_H induced by the individual random variables. Let x be a number in the range of G, i.e., $\{x\} \in \mathscr{A}_G$. In the joint distribution, the probability $P_G(\{x\})$ may be spread over a multiplicity of points $\{(x, y_1)\}, \ldots, \{(x, y_n)\}$, which could also be a countably infinite collection of points. Therefore, $P_G(\{x\})$ is expressed by the sum $P_{G,H}(\{(x, y_1)\}) + P_{G,H}(\{(x, y_2)\}) + \ldots + P_{G,H}(\{(x, y_n)\})$, or a corresponding infinite summation; and we write

$$P_G(\{x\}) = \sum_y P_{G,H}(\{(x, y)\}).$$

Here, the summation is over all y such that, for given x, $\{(x, y)\} \in \mathscr{A}_{G,H}$. Similarly, for any given y in the range of H,

$$P_H(\{y\}) = \sum_x P_{G,H}(\{(x, y)\})$$

where the summation is over all x such that $\{(x, y)\} \in \mathcal{A}_{G,H}$. In Example 41.1, for instance,

$$P_H(\{0\}) = P_{G,H}(\{(2, 0)\}) + P_{G,H}(\{(3, 0)\}) + P_{G,H}(\{(4, 0)\}) + P_{G,H}(\{(5, 0)\})$$

$$= \frac{1}{6} + \frac{1}{6} + \frac{1}{6} + \frac{1}{6} = \frac{2}{3}.$$

On the other hand, $P_{G,H}$ cannot be obtained from P_G and P_H, unless either G and H are a pair of statistically independent random variables, in which case Definition 41.2 applies, or G and H are related by a functional dependence that is known.

Queries

41.1 Two discrete r.v.'s, G and H, are defined on a probability space in such a way that they jointly induce the probabilities indicated in Figure 41.5.

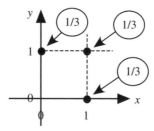

Figure 41.5   Joint probabilities for G and H

The induced space is $(\mathsf{R}^2, \mathcal{A}_{G,H}, P_{G,H})$, where $\mathcal{A}_{G,H}$ is defined in the manner described in this Section. For each of the following statements, determine whether it is true, possibly true, false, or meaningless.
a) G and H are functionally dependent
b) G and H are statistically independent
c) $P_{G,H}(\{(0, 1), (1, 0)\}) = 2/3$
d) $P_{G,H}(\{(0, 0)\}) = 0$
e) $\mathcal{A}_{G,H}$ is made up of 16 sets
f) The underlying space S consists of three elements.   [107]

41.2 The discrete r.v. X takes on values $x_1 = 1/6$, $x_2 = 1/3$, and $x_3 = 1/2$ with probabilities 3/6, 2/6 and 1/6, respectively. A r.v. Y is defined in such a way that it assigns to each element ξ of the underlying space a value

$$Y(\xi) = P_X(\{X(\xi)\}), \ \xi \in \mathsf{S}.$$

What is the probability of the event that X and Y take on the same value, that is, $P(\{\xi: X(\xi) = Y(\xi)\})$?   [73]

Two Arbitrary Random Variables

We will now consider the general case of two random variables (that are not necessarily discrete). This involves formulating a completely additive class of subsets of R^2 that is sufficiently general so it can be used in *every* probability space induced by two r.v.'s.

Let X_1 and X_2 be two r.v.'s defined on a probability space $(S, \mathscr{A}, P)$. They can be visualized in two different ways. The two r.v.'s can be thought of as two separate mappings, as shown in Figure 42.1a. Then, by considering the two mappings jointly, a *pair* of numbers, (x_1, x_2), gets assigned to each element $\xi \in (S, \mathscr{A}, P)$.

On the other hand, X_1 and X_2 can also be thought of as the 'components' of a *single* mapping that is accomplished by a *two-dimensional r.v.* $X = (X_1, X_2)$. Furthermore, it is sometimes convenient to regard a two-dimensional r.v. as a *vector random variable* (or *vector-valued random variable*), which can be written $X = [X_1, X_2]$ to indicate a two-component row-vector, or $X = \begin{bmatrix} X_1 \\ X_2 \end{bmatrix}$, to indicate a two-component column-vector.[6] X is considered to map each element ξ into a point in two-dimensional Euclidian space (see Figure 42.1b). Each component X_i $(i = 1, 2)$ of X is itself an ordinary (scalar) random variable.[7] Actually, even when thinking in terms of two separate mappings it is convenient to express the range of the two random variables along the coordinate axes of the Euclidian plane R^2 (as in Example 41.2), rather than in terms of two separate real lines.

We extend now our concept of an induced probability distribution to the general two-dimensional case. In R^2, each of the two 'axes' constitutes a replica of the space of real numbers; and a 'point' in R^2 is a vector, or ordered pair of numbers—one number from each of the two component spaces. Let the axes be denoted the x_1-axis and x_2-axis, respectively, and suppose that on each of them some set is defined—the set A_1 on the x_1-axis, and the set A_2 on the x_2-axis. Consider the set of all points in R^2 whose x_1-coordinates lie in A_1 and whose x_2-coordinates lie in A_2. This set of points in R^2 is the *Cartesian product* of the sets A_1 and A_2, denoted $A_1 \times A_2$ (see Definition 26.2 and Equation (26.1)). If A_1 or A_2 is empty, then $A_1 \times A_2 = \phi$.

A Cartesian product set in two-dimensional Euclidian space is also called a *rectangle set*. The region inside an ordinary rectangle whose sides are parallel to the coordinate axis is an example of a rectangle set. Such a set is obtained if both A_1 and A_2 are single intervals. Example 42.1 shows that a rectangle set can be more complicated.

---

[6] Vector r.v.s will be used in portions of Parts V and VI.

[7] In print, a bold-face letter is used to denote a vector r.v., whereas in written work an underline is simpler.

Probability Concepts and Theory for Engineers, First Edition. Harry Schwarzlander.
© 2011 John Wiley & Sons, Ltd. Published 2011 by John Wiley & Sons, Ltd.

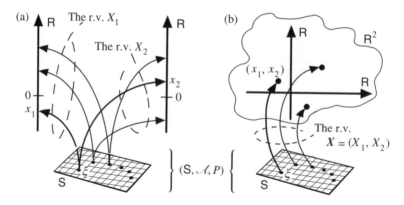

Figure 42.1   Two random variables viewed (a) as two separate mappings, and (b) as a two-dimensional or vector r.v. mapping into R^2

Example 42.1:   Rectangle set.
Let two subsets of R be defined as follows:

$$A_1 = \overline{-\infty, -1} \cup \{1\} \cup \overline{2.9, 5}; \quad \text{and } A_2 = \overline{1, 3} \cup \{\sqrt{12}\}$$

The Cartesian product $A_1 \times A_2$ is illustrated in Figure 42.2. It is a rectangle set consisting of six disconnected regions, each of which is itself a rectangle set: an ordinary rectangle, a semi-infinite strip, two finite line segments, a semi-infinite line segment, and an isolated point. Boundary lines that are not part of $A_1 \times A_2$ are shown as dashed lines. The isolated points and intervals making up the sets A_1 and A_2 are marked on the x_1-axis and x_2-axis, respectively. The intervals are identified by their endpoints, with a parenthesis indicating an endpoint not included in the interval, and a square bracket indicating an endpoint that is included.

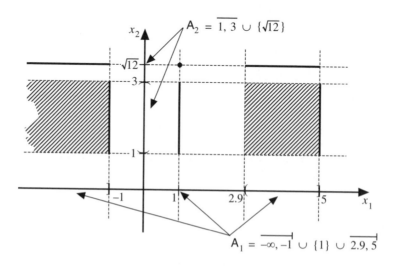

Figure 42.2   A rectangle set

While any rectangle set consisting of at least two points can be expressed as a disjoint union of rectangle sets, it is important to note that a union of rectangle sets, even a disjoint union, need not be a rectangle set.

We are interested particularly in those subsets of R^2 that are Cartesian products of (one-dimensional) Borel sets. Using the notation introduced in Section 27, we can denote the class of all such product sets by $\mathscr{B} \times \mathscr{B}$. This is not an additive class; but we can enlarge it into a completely additive class by closing it with respect to countable set operations, giving $\mathscr{A}(\mathscr{B} \times \mathscr{B})$, *the completely additive class generated by $\mathscr{B} \times \mathscr{B}$.*

(Definition) 42.1
The class $\mathscr{A}(\mathscr{B} \times \mathscr{B})$ of subsets of R^2 is called the *class of two-dimensional Borel sets*, and is denoted $\mathscr{B}^2$. Any set in $\mathscr{B}^2$ is called a *two-dimensional Borel set.*

Example 42.2
Consider the set $A = \{(x, y): 0 \leq x \leq 1, 0 \leq y \leq 1 - x\} \subset R^2$ described in Figure 42.3.

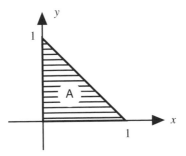

Figure 42.3   A set that is not a product set

The set A is not a product set; i.e., $A \notin \mathscr{B} \times \mathscr{B}$. However, $A \in \mathscr{A}(\mathscr{B} \times \mathscr{B})$, since it can be generated from sets in $\mathscr{B} \times \mathscr{B}$ via countable set operations. For instance, A can be expressed in the following way. Let $A_2, A_3, A_4, A_5, \ldots$ be successive approximations to A, defined as follows, each the result of set operations performed on rectangle sets:

$$A_n = \overline{[0, 1]} \times \overline{[0, 1]} - \bigcup_{i=2}^{n} \left(\overline{\left[\frac{i-1}{n}, \frac{i}{n}\right]} \times \overline{\left[1 - \frac{i-1}{n}, 1\right]} \right), \quad n = 2, 3, 4, \ldots$$

The sets A_2, A_3 and A_4 are depicted in Figure 42.4.

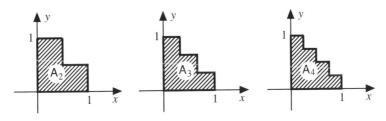

Figure 42.4   Successive approximations to the set A in Figure 42.3

Then $A = \bigcap_{n=2}^{\infty} A_n$, as the reader may verify (see Problem 42.3).

In two-dimensional space the class $\mathcal{B}^2$ is again so comprehensive that it covers all subsets of R^2 that we might be interested in. Therefore, $\mathcal{B}^2$ is generally accepted as the completely additive class on which a two-dimensional random variable must be able to induce a legitimate (2-dimensional) probability distribution. Thus, we have:

(Definition) 42.2
A point function $X = (X_1, X_2)$ defined on a probability space $(S, \mathcal{A}, P)$, which assigns a pair of real numbers to each element $\xi \in S$, is called a *two-dimensional random variable* if for every two-dimensional Borel set B, the set $X^{-1}(B) = \{\xi: X(\xi) \in B\}$ belongs to $\mathcal{A}$.

If a two-dimensional r.v. X is defined on a probability space $(S, \mathcal{A}, P)$, another probability space is automatically established. This is the induced probability space $(R^2, \mathcal{B}^2, P_X)$ consisting of the space R^2 of pairs of real numbers, the completely additive class $\mathcal{B}^2$, and the probability function P_X induced on this class by the r.v. X according to the rule $P_X(B) = P(X^{-1}(B))$, for each $B \in \mathcal{B}^2$.

From Definitions 42.1 and 42.2 it follows that a function that assigns a *pair* of real numbers to every element of a probability space is a two-dimensional r.v. if and only if each of its components is a (one-dimensional or scalar) r.v. Thus, if we define two r.v.'s X_1 and X_2 on a probability space, then the two-dimensional function X whose components are X_1 and X_2 is automatically a two-dimensional r.v. Similarly, given a two-dimensional r.v. $X = (X_1, X_2)$, we can immediately regard X_1 and X_2 as r.v.'s.

A rule already mentioned in Section 41 for a pair of discrete r.v.'s applies also in the general case: Given the joint probability distribution P_X induced by a two-dimensional random variable $X = (X_1, X_2)$, it is always possible to obtain the 'marginal' distributions—that is, the distributions induced by the component r.v.'s X_1 and X_2. Thus, for any Borel set $B \subset R$,

$$P_{X_1}(B) = P_X(B \times R) \tag{42.1a}$$

$$P_{X_2}(B) = P_X(R \times B). \tag{42.1b}$$

When applied to an arbitrary set B, these equations amount to collapsing P_X onto the x_1-axis and onto the x_2-axis, respectively. In simple cases this process can be carried out by inspection, as illustrated in the following Example.

Example 42.3
Consider a two-dimensional r.v. (X, Y) that induces the joint distribution shown in Figure 42.5, where probability 0.5 is spread uniformly over each of the two line segments in R^2. By collapsing the y-dimension and thus accumulating onto the x-axis the probability that is spread over two y-values, we see that X induces a uniform distribution over $\overline{0, 1}$. On the other hand, collapsing the x-dimension and thus accumulating onto the y-axis the probability that is spread over a range of x-values, we find that Y is a discrete r.v., assigning probability 0.5 to each of the point sets $\{0\}$ and $\{1\}$.

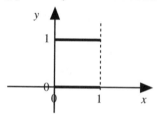

Figure 42.5   Probability 0.5 distributed uniformly over two line segments

If only the individual distributions induced by two r.v.'s X, Y are known, it is generally not possible to specify the two-dimensional distribution induced by the two r.v.'s jointly without some additional information.

Queries

42.1 Which of the sets (a), (b) and (c) described in Figure 42.6 and the additional three sets specified below in set notation are rectangle sets? (Boundaries and endpoints are included in the sets depicted in Figure 42.6.)

(d) $\{(0, 1)\}$ (e) $\{(x, y): x < 1\}$ (f) $\{(x, y): x \geq y\}$.   [67]

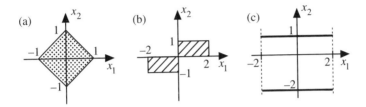

Figure 42.6   Three sets in R^2

42.2 Refer to Query 41.1. Which of the answers will be different if the induced space is now $(\mathsf{R}^2, \mathscr{B}^2, P_{G,H})$?   [207]

42.3 Two r.v.'s X, Y jointly induce a probability distribution $P_{X,Y}$ on R^2 as shown in Figure 42.7, with probability $1/2$ spread uniformly over each of the two line segments.

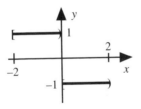

Figure 42.7   Two line segments in R^2

For each of the following statements, determine whether it is true, possibly true, false, or meaningless.

a) X is an absolutely continuous r.v., and Y is a discrete r.v.
b) X and Y are functionally related
c) $P_{X,Y}(\mathsf{R}) = 1$
d) $P_{X,Y}(\{(x, y): y = 1\}) = 1/2$
e) The underlying space S is finite.   [96]

Two-Dimensional Distribution Functions

When discussing a r.v. it is not always necessary to describe the probability space on which the r.v. is defined—the existence of an appropriate space is assumed. The same is true when we consider *two* r.v.'s. However, in the case of two r.v.'s, it is always assumed that these r.v.'s are defined on *one and the same* probability space, since in any given probability problem there is only one basic probability space $(S, \mathscr{A}, P)$. Thus, two random variables X and Y are two point functions, each of which assigns a real number to every element ξ of the one underlying space or, considered jointly, they assign a *pair* of numbers (x, y) to each element ξ of the space. In the last Section we saw that this is also equivalent to considering X and Y as the components of a two-dimensional r.v. or of a vector r.v.

In a similar way to the one-dimensional case, the two-dimensional probability distribution $P_{X,Y}$ induced by two random variables X and Y jointly on $\mathscr{B}^2$ is completely described by a two-dimensional c.d.f.

(Definition) 43.1

The *two-dimensional* (or *joint*) *cumulative distribution function* $F_{X,Y}(x, y)$ that describes the distribution induced jointly by two random variables X and Y is defined as

$$F_{X,Y}(x, y) \equiv P_{X,Y}(\overline{-\infty, x} \times \overline{-\infty, y}) = P(\{\xi: X(\xi) \le x,\ Y(\xi) \le y\})$$

If, instead, the two r.v.'s are regarded as the components of a two-dimensional or vector r.v. X and are denoted X_1, X_2, then the notation for the two-dimensional c.d.f. is $F_X(x_1, x_2)$.

Definition 43.1 suggests the following approach for evaluating a two-dimensional c.d.f. $F_{X,Y}$ at a point (x_o, y_o) in $\mathbb{R}^2$, given $P_{X,Y}$:

a) Draw a horizontal line going left from (x_o, y_o), and a vertical line downward from (x_o, y_o). (See Figure 43.1a.)
b) The semi-infinite region shown shaded in Figure 43.1b, together with the boundary lines themselves, represents the set of points whose probability is the value of $F_{X,Y}$ *at the point*

Probability Concepts and Theory for Engineers, First Edition. Harry Schwarzlander.
© 2011 John Wiley & Sons, Ltd. Published 2011 by John Wiley & Sons, Ltd.

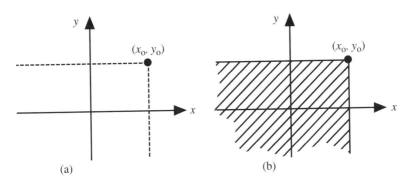

Figure 43.1   Evaluating $F_{X,Y}$ at the point $(x_\mathrm{o}, y_\mathrm{o})$

$(x_\mathrm{o}, y_\mathrm{o})$. In other words,

$$F_{X,Y}(x_\mathrm{o}, y_\mathrm{o}) = P_{X,Y}(\{(x, y): x \leq x_\mathrm{o}, y \leq y_\mathrm{o}\}),$$

the probability that 'X assigns a value less than or equal to x_o and also Y assigns a value less than or equal to y_o'.

Thus, at any point in the plane, the joint c.d.f. *accumulates* all probability lying to the left and below that point (and on the boundary of this region). From this fact follows immediately that a joint c.d.f. is *nondecreasing* and *right-continuous* in both arguments. In other words, if $x_\mathrm{a} \leq x_\mathrm{b}$ and $y_\mathrm{a} \leq y_\mathrm{b}$, then

$$F_{X,Y}(x_\mathrm{a}, y_\mathrm{a}) \leq F_{X,Y}(x_\mathrm{b}, y_\mathrm{b}).$$

It also follows that

$$F_{X,Y}(-\infty, -\infty) = \lim_{x,y \to \infty} F_{X,Y}(x, y) = 0$$

and $F_{X,Y}(+\infty, +\infty) = 1$. Now suppose that $P_{X,Y}$ is such that $F_{X,Y}(x, y) = 1$ over some region $\mathsf{A} \subset \mathsf{R}^2$. This means that for every point $(x_\mathrm{o}, y_\mathrm{o}) \in \mathsf{A}$, $P_{X,Y}$ induces probability 1 in the semi-infinite rectangle $\{(x, y): x \leq x_\mathrm{o}, y \leq y_\mathrm{o}\}$ (see Figure 43.2). But then there must be a smallest x_o and a smallest y_o in A—denote them x_o^* and y_o^*. The region A therefore cannot be as depicted in Figure 43.2, but must be a *semi-infinite rectangle*,

$$\mathsf{A} = \{(x, y): x \geq x_\mathrm{o}^*, y \geq y_\mathrm{o}^*\}.$$

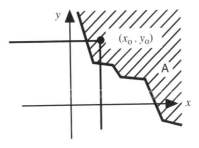

Figure 43.2   A region A where it is assumed that $F_{X,Y}(x, y) = 1$

The evaluation of a joint c.d.f. is most easily illustrated by considering two discrete r.v.'s, in which case the c.d.f. takes on a constant value over various rectangular regions.

Example 43.1

Consider a discrete two-dimensional r.v. $X = (X_1, X_2)$ that assigns the following probabilities to the points $(x_1, x_2) = (0, 2)$, $(1, 1)$, and $(2, 0)$:

$$P_X(\{(0, 2)\}) = 1/6, \; P_X(\{(1, 1)\}) = 1/3, \; P_X(\{(2, 0)\}) = 1/2.$$

Applying the steps described above to points in the various rectangular regions identified in Figure 43.3 will make it clear that F_X is constant in each of these regions, and is zero in the unshaded region. The various rectangular regions (which can be finite or semi-infinite) are obtained by drawing a vertical line upward and a horizontal line to the right from each point to which X assigns positive probability. After evaluation, a graphical representation of F_X can be drawn using an oblique view as shown. The value of F_X on a boundary of any of the rectangles is equal to its value above or to the right of that boundary, since F_X is right-continuous.

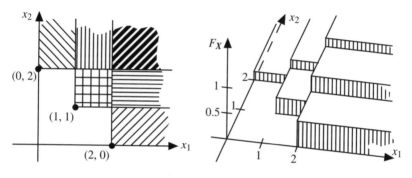

Figure 43.3   A c.d.f. for a discrete two-dimensional induced distribution

We examine now why a two-dimensional c.d.f. completely describes a two-dimensional probability distribution. According to Definition 43.1, a two-dimensional c.d.f. F_X expresses, at every point in R^2, the probability associated with a Cartesian product of semi-infinite intervals. In other words, it expresses the probabilities induced by X on the members of the class $\mathcal{V} \times \mathcal{V}$ (with $\mathcal{V}$ as defined in Section 37). Recall that in each of the component spaces, the class $\mathcal{V}$ generates $\mathcal{B}$ via countable set operations. Also, $\mathcal{B} \times \mathcal{B}$ generates $\mathcal{B}^2$. Therefore, *any set* in $\mathcal{B}^2$ can be expressed in terms of members of $\mathcal{V} \times \mathcal{V}$ using countable set operations. Accordingly, the *probability* of any set in $\mathcal{B}^2$ can be expressed in terms of a countable set of values of F_X—that is, in terms of an arithmetic expression involving a finite number of such values, or an infinite series of such values.

Example 43.2

The joint c.d.f. F_X of a two-dimensional random variable $X = (X_1, X_2)$ is given. We wish to find an expression for the probability $P_X(A)$, where A is the set of points lying in a square region centered at the origin as illustrated in the figure, that is,

$$A = \{(x_1, x_2): -1 < x_1 \leq 1, \; -1 < x_2 \leq 1\}.$$

Defining the additional sets (see Figure 43.4)

$$B = \{(x_1, x_2): x_1 \le -1, \ -1 < x_2 \le 1\}$$
$$C = \{(x_1, x_2): x_1 \le -1, \ x_2 \le -1\}$$
$$D = \{(x_1, x_2): -1 < x_1 \le 1, \ x_2 \le -1\}$$

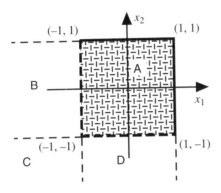

Figure 43.4   The sets A, B, C, and D

we see that $F_X(1, 1)$ is the probability of the set $A \uplus B \uplus C \uplus D$ that encompasses all points that do not lie to the right of, or above, the point $(1, 1)$. Similarly,

$$F_X(-1, 1) = P_X(B \uplus C), \ F_X(1, \ -1) = P_X(C \uplus D), \ F_X(-1, \ -1) = P_X(C).$$

Because of the disjoint unions, and from Equation (18.3b), we have:

$$
\begin{aligned}
P_X(A) &= F_X(1, 1) - P_X(B \uplus C \uplus D) \\
&= F_X(1, 1) - [F_X(-1, 1) + P_X(D)] \\
&= F_X(1, 1) - F_X(-1, 1) - [P_X(C \uplus D) - P_X(C)] \\
&= F_X(1, 1) - F_X(-1, 1) - F_X(1, \ -1) + F_X(-1, \ -1).
\end{aligned}
$$

The last line in the above Example generalizes in an obvious way to the probability assigned by X to an arbitrary rectangle with sides parallel to the coordinate axes:

Theorem 43.1
Given the joint c.d.f. F_X of a two-dimensional r.v. $X = (X_1, X_2)$ and a rectangle

$$A = \overline{a, b} \times \overline{c, d} \in \mathbb{R}^2.$$

Then $P_X(A) = F_X(b, d) - F_X(a, d) - F_X(b, c) + F_X(a, c)$.

The derivation in Example 43.2 is simple because A is a rectangle set. Example 43.3 illustrates a different case.

Example 43.3

With X as in Example 43.2, let B be the set of points lying on and below the line $x_2 = -x_1$. In order to express $P_X(B)$, we first approximate B by a sequence of rectangle sets as shown in Figure 43.5. The probabilities associated with these narrow strips are

$$F_X((n+1)\Delta, -n\Delta) - F_X(n\Delta, -n\Delta), \; n = 0, \; \pm 1, \; \pm 2, \; \ldots$$

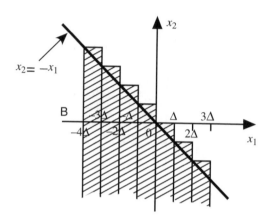

Figure 43.5   A set B with a diagonal boundary approximated by rectangle sets

Since F_X is right-continuous, we have

$$P_X(B) = \lim_{\Delta \to 0} \sum_{n=-\infty}^{\infty} [F_X((n+1)\Delta, -n\Delta) - F_X(n\Delta, -n\Delta)].$$

On the other hand, since F_X in general is not left-continuous everywhere (but right-continuous), it would not be correct to take the limit from below. That is,

$$\lim_{\Delta \to 0} \sum_{n=-\infty}^{\infty} [F_X(n\Delta, -n\Delta) - F_X((n-1)\Delta, -n\Delta)]$$

may not converge to $P_X(B)$.

The two-dimensional c.d.f. $F_{X,Y}$ for two random variables X, Y (considered jointly, or as components of a single vector r.v.) completely describes the distributions induced by X and Y individually. To see this, we note that the expression for $F_{X,Y}$ given in Definition 43.1 can be rewritten

$$F_{X,Y}(x, y) = P[X^{-1}(\overline{-\infty, x}) \cap Y^{-1}(\overline{-\infty, y})]. \tag{43.1}$$

In other words, at each point $(x, y) \in R^2$ the function $F_{X,Y}$ gives the probability of a joint event in the underlying space S—namely, the intersection of the events $X^{-1}(\overline{-\infty, x})$ and $Y^{-1}(\overline{-\infty, y})$. To determine the c.d.f. of X it is necessary to find $P[X^{-1}(\overline{-\infty, x})]$ for all x. From Equation (8.8b) it follows, for each x, that y must be chosen in such a way that $Y^{-1}(\overline{-\infty, y})$ equals S. This is achieved by letting $y = +\infty$. Thus,

$$F_{X,Y}(x, +\infty) = P[X^{-1}(\overline{-\infty, x}) \cap S] = P[X^{-1}(\overline{-\infty, x})] = F_X(x). \qquad (43.2)$$

$F_Y(y)$ is obtained in an analogous way. We have:

Theorem 43.2
Given two r.v.'s X, Y defined on a probability space, with joint c.d.f. $F_{X,Y}(x, y)$. Then

$$F_X(x) = F_{X,Y}(x, \infty)$$

and

$$F_Y(y) = F_{X,Y}(\infty, y).$$

On the other hand, if only F_X and F_Y are known, $F_{X,Y}$ cannot be obtained without additional information.

Queries

43.1 A pair of discrete r.v.'s X, Y have the joint c.d.f. described in Figure 43.6, where the encircled numbers are the values of $F_{X,Y}$ in the respective regions. $F_{X,Y} = 0$ in the shaded area.
  a) What is the probability that Y takes on a value no greater than 2.5?
  b) What is the probability that Y takes on a larger value than does X?   [89]

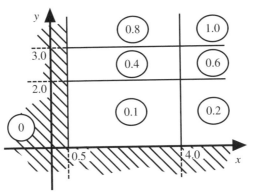

Figure 43.6   A two-dimensional c.d.f

43.2 Given a two-dimensional r.v. $X = (X_1, X_2)$ with $F_X(a, b) = F_X(b, a)$ for all pairs of real numbers (a, b). Which of the following statements are true, which are possibly true, and which are false:
  a) $F_X(a, a) \geq F_X(a, b)$ b) $F_{X_1}(a) = F_{X_2}(a)$
  c) $F_{X_1}(a) \geq F_X(a, b)$ d) $X_1(\xi) = X_2(\xi)$, for all $\xi \in S$,
                                                          the underlying space.   [245]

43.3 Which of the following are legitimate c.d.f.'s? $F_{X,Y}(x, y) =$

a) $\begin{cases} 0, & x+y < 0 \\ x+y, & 0 \leq x+y < 1 \\ 1, & x+y \geq 1 \end{cases}$ b) $\begin{cases} 1, & x, y \geq 1 \\ 0, & \text{otherwise} \end{cases}$ c) $\begin{cases} 0, & y < 0 \\ y, & 0 \leq y < 1 \\ 1, & y \geq 1. \end{cases}$ [208]

Two-Dimensional Density Functions

We saw in Section 36 how the Lebesgue measure of subsets of R is a generalization of the notion of 'length'. For subsets of R^2, Lebesgue measure represents a generalization of 'area'. The measure associated with a particular set therefore depends on the space in which the set is defined: A line segment has positive measure in R^1, where it is an 'interval'; whereas it has zero measure in R^2. For $A, B \subset R$, the rectangle set $A \times B$ has measure $L(A \times B) = L(A) \cdot L(B)$.

Example 44.1

a) The unit square: $L(\overline{0, 1} \times \overline{0, 1}) = L(\overline{0, 1} \times \overline{0, 1}) = L(\overline{0, 1}) \cdot L(\overline{0, 1}) = 1$.

b) The unit circle: $L(\{(x, y): x_2 + y_2 \le 1\}) = \pi$; but
$$L(\{(x, y): x_2 + y_2 = 1\}) = 0.$$

c) If $\{A_1, A_2, A_3, \ldots\}$ is a disjoint class of subsets of R^2, with

$$L(A_n) = 2^{-n}, \ n = 1, 2, 3, \ldots, \text{ then}$$

$$L\left(\biguplus_{n=1}^{\infty} A_n\right) = \sum_{n=1}^{\infty} L(A_n) = 1.$$

With the meaning of Lebesgue measure thus extended to two-dimensional Euclidian space, Definition 38.2 applies also to two-dimensional random variables:

(Definition) 44.1

A two-dimensional random variable X is said to be *absolutely continuous* if the induced probability distribution P_X does not assign positive probability to any subset of R^2 that has zero measure. P_X is then likewise called absolutely continuous.

Probability Concepts and Theory for Engineers, First Edition. Harry Schwarzlander.
© 2011 John Wiley & Sons, Ltd. Published 2011 by John Wiley & Sons, Ltd.

For $X = (X_1, X_2)$ to be absolutely continuous, it is not sufficient that the component random variables X_1, X_2 are absolutely continuous, as illustrated in the following Example.

Example 44.2

Consider again the 'spinning pointer' experiment of Example 38.3, with the final position of the pointer expressed on a *continuous* scale. Let a r.v. X be defined that represents the angular position at which the pointer comes to rest, measured clockwise. Suppose we are also interested in the angular position measured in the counterclockwise direction. Then we can define another r.v. Y on the same probability space in such a way that Y expresses the angular position at which the pointer comes to rest, measured counterclockwise. It is easily seen that the induced probability distributions P_X and P_Y are *identical* so that, for every real number $a, f_Y(a) = f_X(a)$. *Both* X and Y are absolutely continuous r.v.'s.

Now we examine the joint distribution $P_{X,Y}$. To each element ξ of the probability space, X assigns a real number $X(\xi)$ and Y a real number $Y(\xi)$. To be specific about how the endpoints of the ranges are to be treated, we will say that $X(\xi) \in \overline{0, 1}$ and $Y(\xi) \in \overline{0, 1}$. Then $Y(\xi) = 1 - X(\xi)$, and to each element ξ gets assigned a pair of numbers $(x, y) = (x, 1 - x), 0 \leq x < 1$. When viewed as the coordinates of points in a plane, the set of all such number pairs comprises a *line segment*, $y = 1 - x$, for $0 \leq x < 1$ (Figure 44.1a). This is the *range* of the two-dimensional random variable (X, Y). Probability 1 is therefore induced on a line segment that, considered as a subset of R^2, is a set of measure zero. The induced probability distribution, therefore, is *not* absolutely continuous.

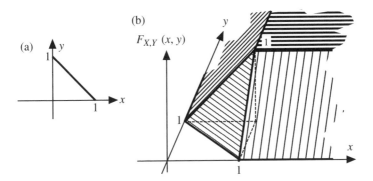

Figure 44.1 Probability distributed uniformly over an inclined line segment

The reader should verify that the two-dimensional c.d.f. $F_{X,Y}(x, y)$ has the form depicted in Figure 44.1b. This can be done using the procedure outlined in connection with Figure 43.1.

In this Example, the r.v.'s X and Y both induce the same probability distribution on subsets of R. Their c.d.f.'s are identical; their p.d.f.'s are identical. We refer to such r.v.'s as *identically distributed* r.v.'s. This must not be confused with *identical* r.v.'s.

(Definition) 44.2

Two r.v.'s X, Y are *identical*, denoted $X = Y$, if and only if $X(\xi) = Y(\xi)$ for each element ξ in the probability space on which X and Y are defined.

Of course, if two r.v.'s are identical, they are also identically distributed.

Theorem 44.1
If $X = (X_1, X_2)$ is absolutely continuous, then X_1 and X_2 are absolutely continuous (one-dimensional) r.v.'s. X_1 and X_2 are then also called *jointly absolutely continuous*.

Proof
For any zero-measure set $Z \subset R$ it must be shown that $P_{X_1}(Z) = P_{X_2}(Z) = 0$. We note that in R^2, $L(Z \times R) = 0$. Since $X = (X_1, X_2)$ is absolutely continuous, $P_X(Z \times R) = 0$. Use of Equation (42.1) then yields $P_{X_1}(Z) = 0$. Similarly, $P_{X_2}(Z) = 0$.

An absolutely continuous two-dimensional r.v. $X = (X_1, X_2)$ spreads probability over *regions* in R^2. $F_X(x_1, x_2)$ therefore increases (with increasing x_1 and/or x_2) not according to probability encountered at specific points (as in Example 43.1) or along certain line segments (as in Example 44.2), but according to how densely probability is spread over a region. In other words, F_X must be differentiable with respect to x_1 and x_2, except possibly on some set of zero probability that constitutes boundary lines across which the density of the induced distribution is discontinuous. Accordingly, we have:

(Definition) 44.3
The *joint* or *two-dimensional (probability) density function* of an absolutely continuous two-dimensional r.v. $X = (X_1, X_2)$ is the function

$$f_X(x_1, x_2) \equiv \frac{\partial^2}{\partial x_1 \, \partial x_2} F_X(x_1, x_2)$$

If X is absolutely continuous, then f_X is defined everywhere in R^2, except possibly on a set of zero measure (i.e. a set of zero probability) where it can be specified arbitrarily.

f_X is a non-negative function that integrates to 1 over R^2:

$$\int_{-\infty}^{\infty} \int_{-\infty}^{\infty} f_X(x_1, x_2) \, dx_1 \, dx_2 = \int_{-\infty}^{\infty} \int_{-\infty}^{\infty} \frac{\partial^2}{\partial x_1 \, \partial x_2} F_X(x_1, x_2) \, dx_1 \, dx_2$$

$$= F_X(+\infty, +\infty) - F_X(-\infty, -\infty) = 1 \tag{44.1}$$

In two dimensions, the p.d.f. has considerably more intuitive appeal than the cumulative distribution function. Nevertheless, we will *not* introduce δ-functions to extend the use of density functions to two-dimensional r.v.'s that are not absolutely continuous.[8] Thus, a two-dimensional p.d.f. will only be used to describe an absolutely continuous induced distribution.

We observe that the joint distribution of Example 44.2 cannot be described by a two-dimensional density function since the distribution is not absolutely continuous.

---

[8] In two dimensions, this should not be tried without recourse to a rigorous definition of the δ-function within an appropriate theoretical framework such as provided by Distribution Theory (cf. [Li1], [Ze1]).

An absolutely continuous two-dimensional probability distribution P_X, induced by a r.v. $X = (X_1, X_2)$, is completely described by the corresponding two-dimensional density function $f_X(x_1, x_2)$. The p.d.f. of one of the component r.v.'s is obtained from f_X by 'integrating out' the unwanted dimension. For instance, analogous to (43.2) we have

$$\int_{-\infty}^{\infty} f_X(x_1, x_2)\, dx_2 = \frac{d}{dx_1} F_X(x_1, +\infty) - \frac{d}{dx_1} F_X(x_1, -\infty)$$

$$= \frac{d}{dx_1} F_{X_1}(x_1) = f_{X_1}(x_1) \tag{44.2}$$

where use is made of Problem 43.4 and Definition 39.1. In this way, the second dimension is integrated out, leaving the one-dimensional or *marginal* p.d.f. f_{X_1}.

The probability induced by a two-dimensional absolutely continuous r.v. X on a set $\mathsf{B} \subset \mathsf{R}^2$ is obtained by integrating f_X over B. If $L(\mathsf{B}) = 0$, then $P_X(\mathsf{B})$ must be 0.

Example 44.3

Consider the two-dimensional r.v. (X, Y) with joint density function

$$f_{X,Y}(x, y) = \begin{cases} \dfrac{1}{4} e^{-x/2}, & x \geq 0,\ 0 \leq y \leq 2 \\[2mm] 0, & \text{otherwise} \end{cases}$$

as sketched in Figure 44.2. First, we verify that the given function is indeed a density function: Clearly, $f_{X,Y}(x, y) \geq 0$ is satisfied. Also,

$$\int_{-\infty}^{\infty} \int_{-\infty}^{\infty} f_{X,Y}(x, y)\, dx\, dy = \int_{0}^{2} \int_{0}^{\infty} \frac{1}{4} e^{-x/2} dx\, dy = \int_{0}^{2} \frac{1}{2} dy = 1$$

as required.

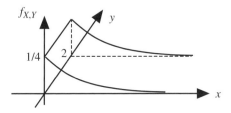

Figure 44.2   A joint p.d.f

In order to obtain the marginal density function $f_X(x)$, the y-coordinate gets integrated out, giving

$$f_X(x) = \int_{-\infty}^{\infty} f_{X,Y}(x, y)\, dy = \begin{cases} \displaystyle\int_{0}^{2} \frac{1}{4} e^{-x/2} dy = \frac{1}{2} e^{-x/2}, & x \geq 0 \\[2mm] 0, & x < 0. \end{cases}$$

Also,

$$f_Y(y) = \int_{-\infty}^{\infty} f_{X,Y}(x,y)dx = \begin{cases} \int_0^{\infty} \frac{1}{4}e^{-x/2}dx = \frac{1}{2}, & 0 \le y \le 2 \\ 0, \text{ otherwise.} \end{cases}$$

Thus, X is an exponential r.v., while Y is uniform.

What is the probability of the event 'the value assigned by X is greater than the value assigned by Y,

$$P(\{\xi : X(\xi) > Y(\xi)\})?$$

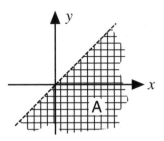

Figure 44.3   The set A

Let A denote the subset of R^2 that corresponds to this event. Integrating $f_{X,Y}$ over A (see Figure 44.3) gives:

$$P_X(\mathsf{A}) = \iint_A f_{X,Y}(x,\,y)\,dxdy = \int_0^2 \int_y^{\infty} \frac{1}{4}e^{-x/2}dxdy$$

$$= \int_0^2 \frac{1}{2}e^{-y/2}dy = 1 - e^{-1} \doteq 0.6321.$$

The specification of a two-dimensional density function often involves different analytic expressions that are applicable in different regions of R^2. In Example 44.3, for instance, $f_{X,Y}$ is specified by $(1/4)e^{-x/2}$ in a semi-infinite rectangular strip and is given by $f_{X,Y}=0$ in the remainder of the plane. The *boundary* of any such region of definition can be arbitrarily taken as belonging to the region or not. This is because the boundary is a set of measure zero, which has no effect on probability calculations *if* the induced distribution is absolutely continuous. However, at a point *within* a region where a two-dimensional p.d.f. is continuous, we always specify the value of the p.d.f. so as to maintain the continuity.

In Example 44.2 we encountered a two-dimensional induced distribution $P_{X,Y}$ that is neither discrete nor absolutely continuous, nor is it of the 'mixed type', in the sense of Section 38. Therefore, there must be yet another type of distribution, which we now define.

(Definition) 44.4
A two-dimensional r.v. X that induces probability 1 on a set of Lebesgue measure zero in R^2, but induces zero probability on every countable set, is said to be *singular*. Its induced probability distribution is then also called *singular*.

It should be kept in mind that the Lebesgue measure of a set depends on the dimensionality of the space. In Example 44.2, probability 1 is induced on a line segment in two-dimensional space—the space in which Lebesgue measure is a generalization of 'area'. A line segment is a set with zero area, hence its Lebesgue measure is zero. The two-dimensional r.v. (X, Y) in Example 44.2, therefore, is a singular r.v., although both X and Y are absolutely continuous.

Sometimes a two-dimensional induced probability distribution needs to be described in terms of *polar coordinates*. This is illustrated in the following Example.

Example 44.4: Polar coordinates.
Suppose two r.v.'s induce a distribution in R^2 in such a way that probability gets distributed uniformly over the region A consisting of the upper half of the unit circle centered at the origin, as shown in Figure 44.4. However, we want these two r.v.'s to map into each point in R^2 in accordance with the polar coordinates of that point. Let the r.v.'s be denoted R and Θ.

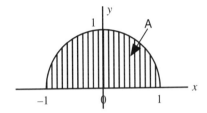

Figure 44.4   Probability distributed uniformly over a semi-circle

To obtain the joint p.d.f. $f_{R,\Theta}$, one approach is to first find the joint c.d.f. in the manner of Section 43. Definition 43.1 gives

$$F_{R,\Theta}(r, \theta) = P_{R,\Theta}(\overline{-\infty, r} \times \overline{-\infty, \theta}) = P(\{\xi : R(\xi) \leq r, \Theta(\xi) \leq \theta\}).$$

But the radial coordinate value, r, cannot be negative, and the angular coordinate value needs to be restricted to an interval of length 2π. We take that interval as $\overline{0, 2\pi}$. Then the above can be rewritten as follows:

$$F_{R,\Theta}(r, \theta) = P_{R,\Theta}(\overline{0, r} \times \overline{0, \theta}) = P(\{\xi : 0 \leq R(\xi) \leq r, 0 \leq \Theta(\xi) \leq \theta < 2\pi\}).$$

Now, at a point $(r, \theta) \in A$ (see Figure 44.5),

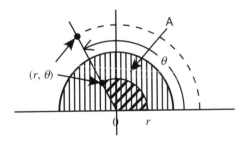

Figure 44.5   Use of polar coordinates

$$F_{R,\Theta}(r,\,\theta) = \frac{2}{\pi}\left(\frac{\pi r^2}{2}\right)\frac{\theta}{\pi} = \frac{r^2\theta}{\pi};$$

whereas for $(r,\,\theta)$ outside the region A and in the upper half-plane,

$$F_{R,\Theta}(r,\,\theta) = \frac{2}{\pi}\left(\frac{\pi}{2}\right)\frac{\theta}{\pi} = \frac{\theta}{\pi}.$$

Altogether, the joint c.d.f. is then:

$$F_{R,\Theta}(r,\,\theta) = \begin{cases} 0, & r < 0 \text{ or } \theta < 0 \\ r^2\theta/\pi, & 0 \le r \le 1 \text{ and } 0 \le \theta \le \pi \\ \theta/\pi, & r > 1 \text{ and } 0 \le \theta \le \pi \\ r^2, & 0 \le r \le 1 \text{ and } \theta > \pi \\ 1, & r > 1 \text{ and } \theta > \pi. \end{cases}$$

Now we obtain

$$f_{R,\Theta}(r,\,\theta) = \frac{\partial^2}{\partial r\,\partial\theta}\,F_{R,\Theta}(r,\,\theta) = \begin{cases} \dfrac{2r}{\pi}, & 0 \le r \le 1, \, 0 \le \theta \le \pi \\ 0, & \text{otherwise.} \end{cases}$$

The polar coordinate r.v.'s R and Θ, although restricted in their range, are just another pair of r.v.'s such as considered earlier in this Section. For instance, $f_{R,\Theta}$ must integrate to 1 (see Figure 44.6):

$$\int_0^\pi \int_0^1 f_{R,\Theta}(r,\,\theta)\,dr d\theta = \int_0^\pi \frac{r^2}{\pi}\bigg|_0^1 d\theta = \frac{\theta}{\pi}\bigg|_0^\pi = 1. \qquad (44.3)$$

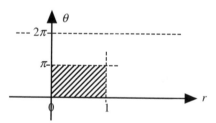

Figure 44.6   Polar coordinate variables in the $(r,\,\theta)$-plane

The *significance* of these r.v.'s—the fact that they are meant to express polar coordinates—is something we attach to them, and as a reminder of that significance we call them R and Θ.

The integration (44.3) should not be confused with using polar coordinates to integrate the *uniform* density $f_{X,Y}$ over A. In that case the differential is $r\,dr\,d\theta$, giving the integral

$$\int_0^\pi \int_0^1 \left(\frac{2}{\pi}\right)r\,dr\,d\theta$$

which again integrates to 1, as in Equation (44.3).

Queries

44.1 Two r.v.'s X, Y jointly induce a probability distribution which is *uniform* over the region shown shaded in Figure 44.7. That is, $f_{X,Y}$ is constant over this region, and zero everywhere else. Determine:

a) $f_{X,Y}(0, 0)$ b) $f_X(0)$ c) $P(\{\xi: X(\xi) = Y(\xi)\})$.   [170]

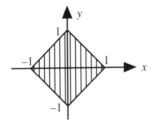

Figure 44.7 A square in $\mathbb{R}^2$

44.2 Given two r.v.'s X, Y defined on a probability space $(S, \mathcal{A}, P)$, such that X induces a uniform distribution over $\overline{0, 1}$, and Y induces a uniform distribution over $\overline{-1, 0}$. For each of the following statements, determine whether it is true, possibly true, or false:

a) $Y(\xi) < X(\xi)$, for all $\xi \in S$
b) Their joint distribution is singular
c) $P_{X,Y}(\{(1/2, -1/2)\}) = 0$
d) $F_X(x) \geq F_Y(x)$, for all $x \in \mathbb{R}$
e) $f_Y(y) = f_X(-y)$.   [136]

44.3 Consider sets A, B, C $\subset \mathbb{R}$ such that $L(A) = 0$, $L(B) = 1$, $L(C) = \infty$. For each of the following statements, determine whether it is true, possibly true, false, or meaningless:

a) $A \subset B \subset C$ b) $(A \times B) \cap (B \times B) = \phi$
c) $A \times B \times C \subset B \times C$ d) $A \times B = B \times A$
e) $L(BC) = 1$ f) $L(B \times B) = L(B)$.   [102]

Two Statistically Independent Random Variables

Section 23 dealt with the property of independence for pairs of events in a probability space. We examine now how this concept applies to random variables. The result will be a generalization of Definition 41.2 to two arbitrary r.v.'s.

First, we explore how independence of events in the induced space is related to independence of events in the basic probability space. Suppose X is a r.v. defined on a probability space $(S, \mathscr{A}, P)$, and B_1, B_2 are subsets of R (Borel sets) whose inverse images are the events $A_1 = X^{-1}(B_1) \in \mathscr{A}$ and $A_2 = X^{-1}(B_2) \in \mathscr{A}$. Then the induced probability distribution P_X causes B_1, B_2 to be independent if and only if their inverse images, A_1 and A_2, are independent events. This can be verified by considering the identity

$$X^{-1}(B_1 B_2) = X^{-1}(B_1) \cap X^{-1}(B_2) = A_1 A_2. \tag{45.1}$$

(For justification of the first equality in Equation (45.1), see Problem 45.6.) Now we observe that

$$P(A_1 A_2) = P[X^{-1}(B_1 B_2)] = P_X(B_1 B_2) \tag{45.2}$$

and furthermore,

$$P(A_1) P(A_2) = P_X(B_1) P_X(B_2) \tag{45.3}$$

Independence of either A_1 and A_2, or of B_1 and B_2, assures equality between Equation (45.2) and Equation (45.3).

This result generalizes to independent classes of arbitrary size as follows.

Theorem 45.1
Let a r.v. X be defined on a probability space $(S, \mathscr{A}, P)$, and let $(R, \mathscr{B}, P_X)$ be the induced probability space. A class $\{B_1, B_2, \ldots, B_n\} \subset \mathscr{B}$ is an independent class if and only if the class $\{A_1, A_2, \ldots, A_n\} \subset \mathscr{A}$ is an independent class, where the A_i's are the inverse images of the B_i's, that is,

$$A_i = X^{-1}(B_i), \ i = 1, \ldots, n.$$

Probability Concepts and Theory for Engineers, First Edition. Harry Schwarzlander.
© 2011 John Wiley & Sons, Ltd. Published 2011 by John Wiley & Sons, Ltd.

Thus, we may say that, in going from the basic space $(S, \mathscr{A}, P)$ to an induced space $(R, \mathscr{B}, P_X)$, independence of sets is preserved—provided that the *set function* X^{-1} (the inverse mapping) is used to specify which sets in $\mathscr{A}$ correspond to which Borel Sets. That this correspondence cannot be specified in terms of the r.v. itself is illustrated in the following Example.

Example 45.1

In the probability space $(S, \mathscr{A}, P)$ for the ordinary single die-throw experiment (Example 17.1) consider the two *independent* events

$$A_1 \leftrightarrow \text{'an odd number of dots shows'}$$
$$A_2 \leftrightarrow \text{'a number less than three shows'}.$$

Let X be a r.v. that assigns these values to the elements of S (see Figure 45.1):

$$X$$
$$\xi_a \rightarrow 0$$
$$\xi_b \rightarrow 1$$
$$\xi_c \rightarrow 1$$
$$\xi_d \rightarrow 2$$
$$\xi_e \rightarrow 3$$
$$\xi_f \rightarrow 4$$

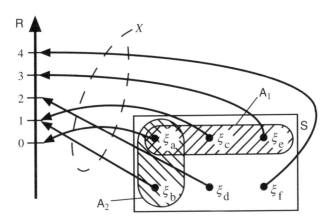

Figure 45.1   Two independent events

Then the elements of A_1 map into 0, 1 and 3, and the elements of A_2 map into 0 and 1. But the sets $\{0, 1, 3\}$ and $\{0, 1\}$ are *not* independent.

On the other hand, the induced distribution P_X does result in some pairs of independent sets in $\mathscr{B}$ (other than sets with probability 0 or 1), for instance the two sets $\{0, 1\}$ and $\{0, 3\}$: We have $P_X(\{0, 1\}) = 1/2$, $P_X(\{0, 3\}) = 1/3$, and

$$P_X(\{0, 1\} \cap \{0, 3\}) = 1/6 = P_X(\{0, 1\}) \, P_X(\{0, 3\}).$$

Furthermore, $X^{-1}(\{0, 1\}) = \{\xi_a, \xi_b, \xi_c\}$ and $X^{-1}(\{0, 3\}) = \{\xi_a, \xi_e\}$, which are independent sets in $\mathscr{A}$.

Next, consider *two* r.v.'s X, Y defined on $(S, \mathscr{A}, P)$, and again let B_1, $B_2 \in \mathscr{B}$. Also, let $A_1 = X^{-1}(B_1) \in \mathscr{A}$ and $A_2 = Y^{-1}(B_2) \in \mathscr{A}$. Figure 45.2 pictures the individual mappings X and Y. A_1 is the set of all those elements that get mapped by X into the set $B_1 \subset R$. The set B_1 is identified along the x-axis. A_2 is the set of all those elements that get mapped by Y into the set $B_2 \subset R$. The set B_2 is identified along the y-axis. (In Figure 45.2, for simplicity, B_1 and B_2 are shown as intervals.) $A_1 A_2$ is the set of elements that get mapped by X into B_1 and also by Y into B_2; i.e., taken jointly, the r.v.'s X, Y map $A_1 \cap A_2$ into some subset of $B_1 \times B_2 \subset R_2$. And since no other elements $\xi \in S$ get mapped into $B_1 \times B_2$, it follows that $A_1 A_2 = (X, Y)^{-1}(B_1 \times B_2)$. By proceeding as in Equations (45.1)–(45.3), it is easily verified that

$$P_{X,Y}(B_1 \times B_2) = P_X(B_1)\, P_Y(B_2) \tag{45.4}$$

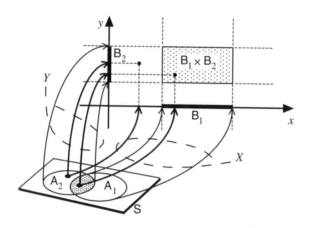

Figure 45.2   Mapping of a set A_1 by X and of a set A_2 by Y, where A_1, A_2 are independent events

if and only if the sets A_1 and A_2 are independent.

Now, it is possible for two r.v.'s X, Y to be defined in such a way that the relation (45.4) holds for *every* pair of Borel sets B_1, B_2. In that case, X and Y are said to be *statistically independent* random variables.

(Definition) 45.1

Two r.v.'s X, Y defined on a probability space $(S, \mathscr{A}, P)$ are called *statistically independent* (abbreviated s.i.) if and only if Equation (45.4) holds for every pair of Borel sets B_1, $B_2 \subset R$.

Example 45.2: Indicator r.v.'s of two independent events.

Consider a probability space $(S, \mathscr{A}, P)$. The *indicator r.v.* of an event $M \in \mathscr{A}$, denoted I_M, is a binary r.v. defined by

$$I_M(\xi) = \begin{cases} 1, & \xi \in M \\ 0, & \xi \notin M. \end{cases}$$

Let $M, N \in \mathscr{A}$ be two *independent* events with $P(M) = p$, $P(N) = r$, and let I_M, I_N be their indicator r.v.'s. The induced joint distribution is depicted in Figure 45.3, with I_M mapping into the x-axis

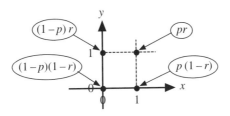

Figure 45.3   Joint distribution of I_M and I_N

and I_N into the y-axis. By applying Definition 45.1, we can verify that I_M and I_N are two statistically independent r.v.'s. That requires checking whether for every pair of Borel sets B_1, B_2,

$$P_{I_M, I_N}(B_1 \times B_2) = P_{I_M}(B_1)\, P_{I_N}(B_2). \tag{45.5}$$

But the only thing that affects the left-hand and right-hand sides of the equation is which of the points 0, 1 belong to B_1, and to B_2. Thus, only four different types of Borel sets need to be considered, which we denote B_a, B_b, B_c and B_d:

Let $B_a \in \mathscr{B}$ be such that $0 \in B_a$, $1 \in B_a$; then $P_{I_M}(B_a) = 1 = P_{I_N}(B_a)$.

Let $B_b \in \mathscr{B}$ be such that $0 \in B_b$, $1 \notin B_b$; then $P_{I_M}(B_b) = 1 - p$, $P_{I_N}(B_b) = 1 - r$.

Let $B_c \in \mathscr{B}$ be such that $0 \notin B_c$, $1 \in B_c$; then $P_{I_M}(B_c) = p$, $P_{I_N}(B_c) = r$. Finally,

let $B_d \in \mathscr{B}$ be such that $0 \notin B_d$, $1 \notin B_d$; then $P_{I_M}(B_d) = P_{I_N}(B_d) = 0$.

Since B_1 and B_2 can each be one of the four sets B_a, B_b, B_c, B_d, Equation (45.5) now has to be checked for 16 distinct cases, and can be found to hold in each of these. Thus,

$$P_{I_M, I_N}(B_a \times B_a) = P(S \cap S) = 1, \text{ and } P_{I_M}(B_a)\, P_{I_N}(B_a) = P(S)\, P(S) = 1$$
$$P_{I_M, I_N}(B_a \times B_b) = P(S \cap N^c) = 1 - r, \text{ and } P_{I_M}(B_a)\, P_{I_N}(B_b) = P(S)\, P(N^c) = 1 - r$$
etc.

The significance of Definition 45.1 can be clarified by rewriting the left side of Equation (45.4) in terms of conditional probability. Assuming X, Y statistically independent, this gives

$$P_{X,Y}(B_1 \times B_2) = P(A_1\, A_2) = P(A_1 | A_2)\, P(A_2)$$

while on the right side, $P_X(B_1)\, P_Y(B_2) = P(A_1)\, P(A_2)$. Thus,

$$P(A_1 | A_2) = P(\{\xi : X(\xi) \in B_1\} | \{\xi : Y(\xi) \in B_2\}) = P(A_1) = P(\{\xi : X(\xi) \in B_1\}).$$

Since this holds for *all* possible B_1, B_2, we conclude the following. Let the underlying experiment be performed with the requirement that the values that Y can assign are restricted to some Borel set B_2 (in other words, when this requirement is not satisfied, the experiment is considered unsatisfactorily performed). Then the probability with which X assigns a value in any given Borel set B_1 is not affected by this restriction. The roles of X and Y can, of course, be interchanged in this statement.

A r.v. that maps probability 1 onto a single point on the real line is a *unary random variable*. Such a r.v. can be regarded as 'degenerate' since there is no uncertainty or randomness associated with the distribution induced by this r.v. It follows from Definition 45.1 that a unary r.v. is statistically independent of any other r.v. defined on the same probability space (see Problem 45.4). While this may at first glance appear strange, it can be seen to be consistent with the interpretation given in the preceding paragraph.

Definition 45.1 is not usually convenient to apply when it is necessary to check whether two random variables are indeed statistically independent (s.i.). The following property of s.i. random variables is more useful for that purpose:

Theorem 45.2
Two random variables X, Y are statistically independent if and only if

$$F_{X,Y}(x, y) = F_X(x) F_Y(y)$$

(45.6)

for all real numbers x, y.

It is clear that Equation (45.6) must hold if X and Y are s.i., since for each (x, y), Equation (45.6) is a particular case of Equation (45.4). On the other hand, if Equation (45.6) holds for all (x, y), then Equation (45.4) can be derived for every choice of B_1, $B_2 \in \mathscr{B}$. This is possible because any Borel set can be expressed in terms of countable set operations applied to members of $\mathscr{V}$, the class of semi-infinite intervals defined in Section 37.

Theorem 45.2 also states that, if only the marginal distributions of two s.i. r.v.'s are known, then their joint distribution is completely specified via Equation (45.6).

Queries

45.1 In each of the following cases, determine whether it is true, possibly true, or false, that X and Y are statistically independent r.v.'s.

a) $F_X(x) = F_Y(x)$.

b) $P_{X,Y}(\overline{-\infty, a} \times \overline{-\infty, a}) = P_X(\overline{-\infty, a}) P_Y(\overline{-\infty, a})$, for every $a \in \mathbb{R}$.

c) $P_{X,Y}(A) = 1$, where A is the square $\overline{-1, 1} \times \overline{-1, 1}$.

d) X expresses the number of dots obtained in an ordinary die throw; Y expresses the number of dots obtained in a second throw of the same die.

e) X expresses the number of dots obtained in an ordinary die throw. Y assigns to each element ξ a value that is the inverse of the number assigned by X.   [241]

45.2 Consider two s.i. r.v.'s X and Y defined on a probability space $(S, \mathscr{A}, P)$. X has a probability density function as described in Figure 39.2. It is also known that $P(\{\xi: X(\xi) \leq z \text{ and } Y(\xi) \leq z\}) = z$, for $0 \leq z \leq 1$, $\xi \in S$. Evaluate $F_Y(0.2)$.   [201]

Two Statistically Independent Random Variables—Absolutely Continuous Case

If two r.v.'s X, Y are *jointly absolutely continuous* then Equation (45.6) can be differentiated with respect to x and y, and we have:

Theorem 46.1
Two jointly absolutely continuous random variables X, Y are s.i. if and only if

$$\boxed{f_{X,Y}(x,\ y) = f_X(x) f_Y(y)}$$

(46.1)

for all real numbers x, y (except possibly on a set of zero measure).

We can also say:

Corollary 46.1
If two absolutely continuous scalar r.v.'s X, Y are s.i., they are jointly absolutely continuous, with joint p.d.f. $f_{X,Y}(x,\ y) = f_X(x)\, f_Y(y)$.

Proof
If X, Y are s.i., then according to Theorem 45.2, $F_{X,Y}(x,\ y) = F_X(x) F_Y(y)$. Since X and Y are absolutely continuous, this can be written

$$F_{X,Y}(x,\ y) = \int_{-\infty}^{x} f_X(x)\, dx \int_{-\infty}^{y} f_Y(y)\, dy = \int_{-\infty}^{y} \int_{-\infty}^{x} f_X(x) f_Y(y)\, dxdy.$$

Thus, $f_{X,Y}(x, y)$ exists and equals $f_X(x) f_Y(y)$. Since $P_{X,Y}$ is described by a two-dimensional p.d.f., it cannot assign positive probability to a set of zero measure, so that X, Y are jointly absolutely continuous.

Probability Concepts and Theory for Engineers, First Edition. Harry Schwarzlander.
© 2011 John Wiley & Sons, Ltd. Published 2011 by John Wiley & Sons, Ltd.

To allow for situations where the marginal density functions are not initially known, the following more general form of Theorem 46.1 is useful:

Theorem 46.2
Two jointly absolutely continuous random variables X, Y are s.i. if and only if their joint density function can be written in the form

$$f_{X,Y}(x, y) = g_1(x) g_2(y) \tag{46.2}$$

for all real numbers x, y, where g_1 and g_2 are two non-negative functions. Then $f_X(x) = k\, g_1(x)$ and $f_Y(y) = (1/k)\, g_2(y)$, for some positive constant k.

In other words, in order to establish the statistical independence of two r.v.'s X and Y it is sufficient to show that the joint density function $f_{X,Y}(x, y)$ is factorable into a function of x and a function of y. The reader may supply a proof of Theorem 46.2 (see Problem 46.4). It should be noted that this Theorem must be applied with care, as the following Example illustrates.

Example 46.1
Consider two r.v.'s X, Y with joint density function (see Figure 46.1)

$$f_{X,Y}(x, y) = \begin{cases} 12\,xy, & x+y \le 1 \text{ and } x,\, y > 0 \\ 12\,(1-x)(1-y), & x+y > 1 \text{ and } x,\, y \le 1 \\ 0, & \text{otherwise.} \end{cases}$$

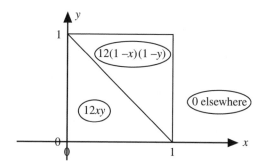

Figure 46.1   A two-dimensional c.d.f.

Application of Theorem 46.2 leads to the observation that, indeed, $12xy$ factors into a function of x and a function of y; for instance, $(12x)(y)$. Thus, it is possible to specify

$$g_1(x) = \begin{cases} 12\,x, & 0 \le x \le 1 \\ 0, & \text{otherwise} \end{cases} \qquad g_2(y) = \begin{cases} y, & 0 \le y \le 1 \\ 0, & \text{otherwise.} \end{cases}$$

Also, $12(1-x)(1-y)$ can be factored into

$$g_1(x) = \begin{cases} 12(1-x), & 0 \le x \le 1 \\ 0, & \text{otherwise} \end{cases} \qquad g_2(y) = \begin{cases} 1-y, & 0 \le y \le 1 \\ 0, & \text{otherwise.} \end{cases}$$

However, this function pair $g_1(x)$, $g_2(y)$ is different from the one specified first. Neither product $g_1(x) g_2(y)$ equals $f_{X,Y}$ over the whole plane. $f_{X,Y}$ cannot be expressed *over all x and y* in the form required by Theorem 46.2. We conclude that X and Y are *not* s.i. r.v.'s.

Significant analytic and computational simplification can result from applications of Theorems 46.1 and 46.2. This is one reason why statistical independence of random variables is a property of great interest.

The following property of statistically independent random variables is also very useful. While it does not establish the independence of two r.v.'s, it sometimes makes it possible to quickly recognize that two random variables are *not* s.i.:

Theorem 46.3
Let two jointly absolutely continuous random variables X, Y be s.i., and let $f_{X,Y} = 0$ over some portion of R^2. Then the subset of R^2 on which $f_{X,Y}$ does *not* vanish is a rectangle set, or it differs from a rectangle set by some set of measure zero.

Proof
Let X, Y be s.i., so that Equation (46.1) holds. Now let $B_1 \subset R$ be the set on which $f_X \neq 0$, and $B_2 \subset R$ the set on which $f_Y \neq 0$. Then $B_1 \times B_2$ is the set on which $f_{X,Y} \neq 0$, and this is a rectangle set. The possibility of $f_{X,Y}$ differing from a rectangle set by a set of zero measure is mentioned because the definition of a joint density function can be changed over a set of measure zero without affecting the distribution described by that density function.

Example 46.2
Two jointly absolutely continuous r.v.'s X, Y induce a distribution $P_{X,Y}$ described by the following p.d.f. (see Figure 46.2):

$$f_{X,Y}(x, y) = \begin{cases} \dfrac{1}{\pi}, & x^2 + y^2 \leq 1 \\ 0, & \text{otherwise} \end{cases}$$

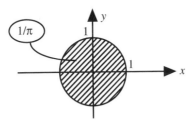

Figure 46.2   A joint p.d.f.

The subset of R^2 on which $f_{X,Y}$ does not vanish is the set

$$\{(x, y) : x^2 + y^2 \leq 1\}$$

which is not a rectangle set. Therefore X and Y are not s.i.

With appropriate interpretation, the property expressed in Theorem 46.3 also holds for other than absolutely continuous r.v.'s. For instance, if two discrete r.v.'s X, Y are s.i., then the set of all points in R^2 on which they induce positive probability must be a rectangle set.

We can conclude from Example 46.2 that two random variables whose joint distribution has circular symmetry cannot be s.i. if this joint distribution is restricted to the interior of some circle (and is not entirely concentrated at the origin). For instance, two s.i. *Gaussian* r.v.'s may have a joint density function with circular symmetry (see Problem 46.2), but this p.d.f. does not vanish anywhere in R^2.

On the other hand, suppose a probability distribution on R^2 has circular symmetry about the origin. If this distribution is induced by the *polar* r.v.'s R, Θ (see also Example 44.4), then R and Θ are s.i. This is the case since, for any $r_0 > 0$ such that $F_R(r_0) > 0$, $F_{R,\Theta}(r_0, \theta)$ increases *linearly* with θ over $0 \le \theta \le 2\pi$. Thus,

$$F_{R,\Theta}(r, \theta) = \frac{\theta}{2\pi} F_{R,\Theta}(r, 2\pi) = \frac{\theta}{2\pi} F_{R,\Theta}(r, \infty)$$

$$= \frac{\theta}{2\pi} F_R(r) = F_R(r) F_\Theta(\theta), \quad 0 \le \theta \le 2\pi. \tag{46.3}$$

Queries

46.1 Listed below are some functions $f(x, y)$, each involving a parameter a. In each case, what value(s) of a, if any, will make f the joint p.d.f. of two s.i. r.v.'s?

a) $f(x, y) = \begin{cases} a, & (x, y) \in [0, 1] \times [0, 1] \\ a + 0.5, & (x, y) \in [0, 1] \times [2, 3] \\ 0, & \text{otherwise} \end{cases}$

b) $f(x, y) = \begin{cases} 1, & 0 \le x \le 1,\ ax \le y \le 1 + ax \\ 0, & \text{otherwise} \end{cases}$

c) $f(x, y) = \begin{cases} a e^{-(x + ay)}, & x, y \ge 0 \\ 0, & \text{otherwise} \end{cases}$

d) $f(x, y) = \begin{cases} \dfrac{xy + 2y - 3x + a}{150 + 10a}, & (x, y) \in [-1, 1] \times [5, 10] \\ 0, & \text{otherwise.} \end{cases}$ [173]

46.2 Two absolutely continuous random variables X and Y are identically distributed and s.i. It is known that:

  a) Their joint p.d.f. $f_{X,Y}(x, y)$ vanishes at least in those regions shown shaded in Figure 46.3. It may also vanish in portions of the unshaded region. Also,

  b) $F_X(0.6) = F_Y(0.6) = 0.6$.

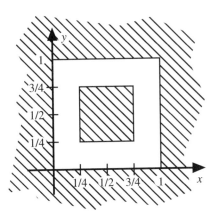

Figure 46.3   Region where $f_{X,Y}$ in Query 46.2 definitely vanishes

What is the probability that X and Y both take on values less than or equal to 1/4?   [84]

Part III Summary

In Part III the basic mathematical model developed in Part I and elaborated in Part II was extended in order to treat p.e.'s where the possible outcomes have numerical attributes. This involved a mapping to a supplementary probability space, the 'induced space', where the basic space of elements is the space of real numbers. The mapping function is called a 'random variable'.

We began by considering discrete random variables, where a simple-minded approach can be used to formulate the induced space—with the completely additive class depending on the particular real numbers that constitute the range of the random variable. However, in order to be able to introduce a general definition of a random variable, it was useful to introduce a completely additive class so comprehensive that it can be used in every induced space. This is the class of Borel sets.

Because the induced probability distribution is defined on sets of real numbers, it can be described by a simple function of a real variable, the cumulative distribution function. In the absolutely continuous case its derivative, the probability density function, is of greater intuitive appeal. It can be extended through the use of δ-functions to random variables with discrete components, or to random variables that are entirely discrete. A frequently encountered absolutely continuous distribution is the Gaussian distribution. This was discussed in some detail.

In the remainder of Part III, these considerations were extended to situations where *two* random variables arise in a given probability problem. In such a case, the two random variables can be regarded as a 'two-dimensional random variable' or 'vector random variable', which maps into the Euclidean plane, R^2. Again, the cumulative distribution function completely describes the induced distribution, but is not as easy to work with in two dimensions as in one dimension. If the two random variables are absolutely continuous, then the two-dimensional density function is much more convenient.

When dealing with two random variables, the property of statistical independence is of importance. It is an extension of the concept of independence of sets to random variables. Where it applies, it can simplify the approach to a problem. Statistical independence was discussed in the last two Sections of Part III.

Part IV

Transformations and Multiple Random Variables

Part IV Introduction

In many applications there arises a situation where a random variable gets defined on a probability space, and a second random variable is then specified as a function of the first one. This is referred to as a 'transformation' of a random variable. What actually gets transformed is the induced distribution. Transformations are considered in detail in this Part, beginning with the transformation of one- and two-dimensional r.v.'s in Sections 47 and 48. The special case of the sum of two r.v.'s is treated in Sections 49 and 50—for discrete r.v.'s and the general case, respectively. Coordinate transformations are examined in Section 53, and are applied to bivariate Gaussian r.v.'s in Section 54.

The treatment of multidimensional random variables, which in Part III was restricted to the two-dimensional case, is expanded in Sections 51 and 52 to more than two dimensions. This two-stage approach is being used in order to give the student the opportunity to first become comfortable with two-dimensional r.v.'s, for which it is still possible to visualize the c.d.f. or the p.d.f. It is then an easy step to understand the higher-dimensional case. Also, it is helpful for the reader to have a good grasp of the statistical independence of two r.v.'s before dealing with statistical independence of more than two r.v.'s, as presented in Section 55. Besides, this approach allows pursuit of a more streamlined syllabus that bypasses the higher-dimensional case.

In the remainder of Part IV, first of all, singular r.v.'s and distributions in one dimension are considered in Section 56. As shown there, such distributions can arise in straightforward engineering models. Conditional induced distributions are also discussed. Finally, Section 60 introduces experiments involving 'random occurrences in time', and various r.v.'s arising in such situations are examined. That Section, in a sense, is also a precursor to the study of sequences of r.v.'s in Part VI.

Transformation of a Random Variable

In Part III we have encountered situations in which *two* r.v.'s were defined on some underlying probability space (S, $\mathscr{A}$, P). We found that the two r.v.'s can be defined separately (as two separate mappings from the underlying space to two distinct real axes), or jointly as a two-dimensional random variable. Now we will consider a third possibility: One r.v. is defined as a *function*, or *transformation*, of the other r.v.

In this and the following Sections, we study methods for finding probability distributions when a transformation is applied to a r.v. We use the word 'transformation', rather than 'function', because when a new r.v. Y is specified as a function of a given r.v. X, then the induced distribution P_X gets transformed into the new induced distribution P_Y. Thus, the induced space (R, $\mathscr{B}$, P_X) gets modified by having an additional distribution, P_Y, defined on it. Alternatively, the transformation can be viewed as resulting in a *new* induced space, (R, $\mathscr{B}$, P_Y). We begin with the discrete case, for which the basic procedure is easiest to visualize.

a) *Transformation of a discrete random variable*

Transformation of a *discrete* r.v. presents no difficulty and can easily be carried out without following specific rules. However, we will—by means of an example—go through the essential steps in detail in order to establish a pattern of thinking that applies in general.

Example 47.1

The possible outcomes of the familiar experiment of throwing a die are the six different faces of the die pointing up, as shown in Figure 47.1.

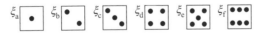

Figure 47.1   Possible outcomes for the throwing of a single die

Probability Concepts and Theory for Engineers, First Edition. Harry Schwarzlander.
© 2011 John Wiley & Sons, Ltd. Published 2011 by John Wiley & Sons, Ltd.

In Section 34 an appropriate probability space $(S, \mathcal{A}, P)$ was described, and it was seen how it is natural to define a discrete r.v. which assigns to each possible outcome an integer representing the number of dots facing up:

$$X(\xi_a) = 1,\; X(\xi_b) = 2, \ldots,\; X(\xi_f) = 6$$

Example 41.1 associated a *pay-off* with the throwing of the die according to the following rule:

> One dot $\rightarrow$ I pay \$1
>
> two, three, four or five dots $\rightarrow$ no payment, no winning
>
> six dots $\rightarrow$ I win \$1

This rule can be brought into the mathematical model by defining on $(S, \mathcal{A}, P)$ another r.v., Y, representing 'the amount I win'. But the amount I win depends on the number of dots facing up when the die is thrown, so that Y depends on X, or is a function of X, and we write

$$Y = g(X)$$

where g is a (single-valued) function. This says that for each element $\xi \in S$, $Y(\xi) = g[X(\xi)]$. Thus, Y is the name for the r.v. that assigns to each element ξ the value $g[X(\xi)]$.

The function g is defined by the above pay-off rule and can be represented by a graph as shown in Figure 47.2. The function g needs to be specified only for $x = 1, 2, 3, 4, 5, 6$. It is arbitrary for other values of x since these cannot arise.

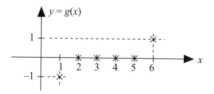

Figure 47.2   A pay-off rule

If a discrete r.v. Y is specified in terms of another discrete r.v. X, so that $Y = g(X)$, how can P_Y be obtained from P_X? In Example 47.1 this can be accomplished by inspection; but a little thought suggests the following general procedure:

Rule 47.1: Transformation of a discrete r.v.
Given a discrete r.v. X defined on a probability space $(S, \mathcal{A}, P)$, with induced distribution P_X; and another discrete r.v. Y defined by $Y = g(X)$. To find P_Y:
a) Consider some point y in the *range* of the r.v. Y.
b) Find the set of all x-values that get mapped into this value y by g. Denote this set K_y; in other words, $K_y = \{x : g(x) = y\}$.
c) Then $P_Y(\{y\}) = P_X(K_y)$, for the particular value y being considered.
d) Do this for all y-values in the range of the r.v. Y.

Example 47.1 (continued)

The application of Rule 47.1 is carried out in Table 47.1. The result, depicted as a bar graph in Figure 47.3, is of course the same distribution as P_H in Example 41.1.

Table 47.1   The transformation $Y = g(X)$.

a) Consider $y =$	1	0	-1	any other value
b) Then $K_y =$	$\{6\}$	$\{2, 3, 4, 5\}$	$\{1\}$	ϕ
c) Giving $P_Y(\{y\}) =$	$P_X(\{6\})$ $= 1/6$	$P_X(\{2, 3, 4, 5\})$ $= 2/3$	$P_X(\{1\})$ $= 1/6$	$P_X(\phi) = 0$

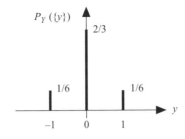

Figure 47.3   Bar graph of P_Y

It is worth noting that each of the sets K_y can be enlarged by a set of zero probability without affecting the results that are obtained. For instance, g could be defined as a continuous function for all x in the following way (with a graph as shown in Figure 47.4):

$$g(x) = \begin{cases} x - 2, & x < 2 \\ 0, & 2 \leq x \leq 5 \\ x - 5, & x > 5. \end{cases}$$

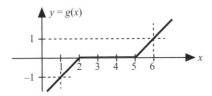

Figure 47.4 g defined as a continuous function

Then $K_0 = \overline{2, 5}$. However, $P_Y(\{0\})$ is still 2/3.

b) *Transformation of an arbitrary random variable*

We consider next the transformation of an arbitrary one-dimensional r.v. X, defined on some probability space $(S, \mathcal{A}, P)$, into another one-dimensional r.v. Y,

$$Y = g(X) \tag{47.1}$$

where g is some (single-valued) real function of a real variable. Equation (47.1) states that, for each element $\xi \in S$, $Y(\xi) = g[X(\xi)]$. Thus, the real number x that is $X(\xi)$ and the real number y that is $Y(\xi)$ are related by $y = g(x)$.

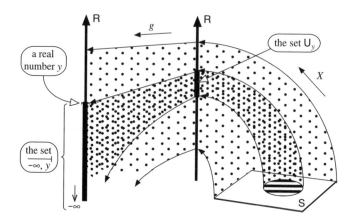

Figure 47.5   The transformation $Y = g(X)$, where X is an arbitrary r.v

Only a minor modification of Rule 47.1 is required so that it will apply in the general case. This modification is needed because we must now use *c.d.f.'s* to describe the induced probability distributions. Thus, we must ask, what probability gets assigned to a semi-infinite interval $\overline{-\infty, y}$ as a result of performing the transformation g in Equation (47.1). This can be visualized as in Figure 47.5. The r.v. X maps the elements of S into elements of R. The function g is a mapping from R to R. We need to determine the probability that gets assigned to intervals $\overline{-\infty, y}$, for different y. Therefore, we trace each of these intervals back to its inverse (on the first R-axis) and find the probability of that inverse image. We have:

Rule 47.2: Transformation of an arbitrary r.v.
Given a r.v. X defined on a probability space $(S, \mathscr{A}, P)$, with c.d.f. F_X; and another r.v. Y defined by $Y = g(X)$. To find the c.d.f. F_Y of the r.v. Y:
a) Consider some $y \in R$.
b) Determine the inverse image of the set $\overline{-\infty, y}$. Denote that inverse image U_y. Thus, $U_y = \{x: g(x) \le y\}$.
c) Then $F_Y(y) = P_X(U_y)$, for the particular value y being considered.
d) Do this for all $y \in R$.

This rule, of course, covers the transformation of a discrete r.v. as well, but Rule 47.1 is easier to use in that case.

One of the simplest transformations is multiplication by some constant a: $Y = aX$. The effect, if $|a| > 1$, is for probability to get spread more widely over the real axis; and if $0 < |a| < 1$, it gets more compressed. In addition, if $a < 0$, then also a reversal occurs with respect to the origin. If $a = 1$, Y is identical to X, whereas $a = -1$ merely reverses the induced distribution on the real axis.

Specifically, if $a > 0$, then for any choice of y, Rule 47.2 gives

$$F_Y(y) = P_X(\mathsf{U}_y) = P_X\left(\left\{x : x \leq \frac{y}{a}\right\}\right) = F_X\left(\frac{y}{a}\right).$$

If X is absolutely continuous, so is Y and $f_Y(y) = \frac{1}{a}f_X\left(\frac{y}{a}\right)$. On the other hand, if $a < 0$, then for any choice of y, Rule 47.2 gives

$$F_Y(y) = P_X(\mathsf{U}_y) = P_X\left(\left\{x : x \geq \frac{y}{a}\right\}\right) = 1 - F_X\left(\frac{y}{a}\right) + P_X\left(\left\{\frac{y}{a}\right\}\right);$$

and if X is absolutely continuous, then $f_Y(y) = -\frac{1}{a}f_X\left(\frac{y}{a}\right) = \frac{1}{|a|}f_X\left(\frac{y}{a}\right)$. For $a = -1$, this becomes simply $f_Y(y) = f_X(-y)$. In that case, if X induces a symmetrical distribution such as Gaussian, then $f_X(-y) = f_X(y)$; i.e., X and Y are identically distributed (i.d.). But they are *not* identical (unless $f_X(x) = \delta(x)$). Finally, if $a = 0$, then the induced distribution collapses into the origin, making Y a unary r.v. with $f_Y(y) = \delta(y)$.

Now we consider a less trivial application of Rule 47.2.

Example 47.2

A r.v. X is given, with known c.d.f. $F_X(x)$. A new r.v. Y is defined by $Y = X^2$. What is $F_Y(y)$, expressed in terms of $F_X(x)$? We proceed in accordance with Rule 47.2 (as illustrated in Figure 47.6):

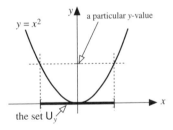

Figure 47.6   Application of Rule 47.2 to the transformation $Y = X^2$

a) Pick some $y < 0$.
b) The set $\mathsf{U}_y = \phi$.
c) Then $F_Y(y) = P_X(\mathsf{U}_y) = 0$.
d) This holds for every $y < 0$.

Continuing:

a) Pick some $y \geq 0$.
b) $\mathsf{U}_y = \{x : -\sqrt{y} \leq x \leq \sqrt{y}\}$.
c) Therefore,

$$F_Y(y) = P_X(\mathsf{U}_y) = F_X(\sqrt{y}) - F_X(-\sqrt{y}) + P_X(\{-\sqrt{y}\}). \tag{47.2}$$

d) This holds for every $y \geq 0$.

If X is absolutely continuous, this result can be expressed in terms of f_X:

$$F_Y(y) = \begin{cases} \int_{-\sqrt{y}}^{\sqrt{y}} f_X(x)dx & y \geq 0 \\ 0, & y < 0. \end{cases} \tag{47.3}$$

Differentiation with respect to y then results in[1]

$$f_Y(y) = \begin{cases} \dfrac{1}{2\sqrt{y}}\left[f_X(\sqrt{y}) + f_X(-\sqrt{y}) \right], & y \geq 0 \\ 0, & y < 0. \end{cases} \tag{47.4}$$

For instance, if X is 'unif. $\overline{-2,2}$', that is, X induces a uniform distribution with $f_X(x) = \begin{cases} \dfrac{1}{4}, & |x| \leq 2 \\ 0, & \text{otherwise} \end{cases}$, then $f_Y(y) = \begin{cases} \dfrac{1}{4\sqrt{y}}, & 0 < y \leq 4 \\ 0, & \text{otherwise} \end{cases}$, which is sketched in Figure 47.7.

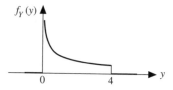

$f_Y(y)$

0   4 y

Figure 47.7 P.d.f. of $Y = X^2$ where X is unif. $\overline{-2, 2}$

Great care is needed if a δ-function appears in the expression for f_X. In Example 47.2, suppose X is a binary r.v. taking on the values ± 2 with equal probability, so that

$$f_X(x) = \frac{1}{2}\delta(x-2) + \frac{1}{2}\delta(x+2).$$

Substituting into Equation (47.4), the result for $y \geq 0$ is

$$f_Y(y) = \frac{1}{2\sqrt{y}}\delta(\sqrt{y} - 2). \tag{47.5}$$

In order to express this in more familiar form, a change of variable must be applied to the δ-function. The argument of the δ-function is zero if $y = 4$. But it is not correct to simply replace $\delta(\sqrt{y} - 2)$ in

---

[1] Leibnitz's rule for differentiating an integral with respect to a parameter t that appears in the integrand as well as in the limits of integration is:

$$\frac{d}{dt}\int_{a(t)}^{b(t)} c(x,t)dx = \frac{db(t)}{dt}c(b(t),t) - \frac{da(t)}{dt}c(a(t),t) + \int_{a(t)}^{b(t)} \frac{dc(x,t)}{dt}dx,$$

where $dc(x,t)/dt$ must exist and be continuous for $x \in \overline{a(t), b(t)}$ and $t \in \overline{t_1, t_2}$, and $da(t)/dt$ and $db(t)/dt$ must exist and be continuous for $t \in \overline{t_1, t_2}$; in which case the formula applies for $t \in \overline{t_1, t_2}$.

Equation (47.5) by $\delta(y-4)$. Instead, making use of Equation (39.3) and letting $v = \sqrt{y}$, we note that

$$\int_{4-\varepsilon}^{4+\varepsilon} \frac{1}{2\sqrt{y}} \delta(\sqrt{y}-2)\, dy = \int_{2-\varepsilon'}^{2+\varepsilon'} \delta(v-2)dv = 1.$$

Thus, $\frac{1}{2\sqrt{y}}\delta(\sqrt{y}-2)$ is a *unit* impulse at $y=4$, and Equation (47.5) becomes $f_Y(y) = \delta(y-4)$.

In general, if the variable appearing in the argument of the δ-function is changed from u to v, then the δ-function has to get multiplied by dv/du. In the above, the need for this can be avoided by starting with Equation (47.3) instead of Equation (47.4) (see Problem 47.1).

A comment about the function g in Equation (47.1) is appropriate. In order for Y to be a r.v., g cannot be completely arbitrary. We have already specified g as a single-valued real function of a real variable; but it must also be such that the requirement of Definition 37.2 is satisfied by $g(X)$. Thus, g must be such that the inverse image of every Borel set B is also a Borel set:

$$\{x : g(x) \in \mathsf{B}\} \in \mathscr{B}$$

A function g that satisfies this requirement is called a *Borel function* (or Borel-measurable function). However, the collection of Borel functions is so broad that it is difficult to construct a function that is not a Borel function. For this reason we will not be concerned about verifying, for any given transformation g, that it is indeed a Borel function.

c) *Transformation of an absolutely continuous random variable*

There exists a simpler approach to the transformation $Y = g(X)$ if X is absolutely continuous, and if g is monotonic and has a piecewise continuous derivative. In order to develop this approach, we consider the joint distribution of X and Y, which is known since f_X is assumed known and since all probability must be concentrated on the graph of the function g.

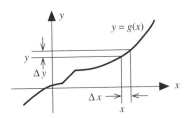

Figure 47.8   Locus of the joint distribution of two r.v.'s X and $Y = g(X)$

Assume for the moment that g is monotonically increasing. A small interval of size Δx at point x corresponds to an interval of size $\Delta y \approx \frac{dy}{dx}\Delta x$ at point $y = g(x)$, as depicted in Figure 47.8. Therefore,

$$f_Y(y) = \frac{f_X(x)}{\frac{dy}{dx}}, \quad \text{where } \frac{dy}{dx} > 0 \text{ and } x = g^{-1}(y).$$

If g is monotonically *decreasing*, dy/dx is negative but this does not affect f_Y. In order to take this into account, we must write

$$f_Y(y) = \frac{f_X(x)}{\left|\frac{dy}{dx}\right|}, \quad \text{where } \frac{dy}{dx} \neq 0 \text{ and } x = g^{-1}(y). \tag{47.6}$$

On the other hand, if g is constant over some x-interval I_x, the following δ-function gets introduced into f_Y:

$$P_X(I_x)\delta(y - g(x_0)), \quad x_0 \in I_x;$$

while over the increasing or decreasing portions of g, Equation (47.6) applies.

The requirement of g being monotonic can be circumvented by partitioning the x-axis into intervals over which g is monotonically increasing, monotonically decreasing, or constant, and summing the contributions from all these 'branches' of g.

Example 47.3
Let $Y = X^2$ as in Example 47.2, so that $\frac{dy}{dx} = \frac{d}{dx}x^2 = 2x$. Two branches of g have to be considered: $x > 0$ and $x < 0$. Thus,

$$f_Y(y) = \frac{f_X(x)}{2x}\bigg|_{x>0} + \frac{f_X(x)}{-2x}\bigg|_{x<0} = \frac{1}{2\sqrt{y}}[f_X(\sqrt{y}) + f_X(-\sqrt{y})], \quad y > 0.$$

The joint distribution does not extend to negative y-values, so that $f_Y(y) = 0$ for $y < 0$, and we have the same result as in Example 47.2.

Queries

47.1 Consider a binomial r.v. B that induces the probabilities

$$P_B(\{k\}) = \binom{10}{k}2^{-10}, \quad k = 0, 1, 2, \ldots, 10.$$

For each of the following transformations, what is the set K_1?

a) $D = 8 - B$ b) $D = \log_{10}(1 + B)$

c) $D = (B - 5)^2$ d) $D = \min(1, B)$.   [32]

47.2 A r.v. X has a probability density function as described in Figure 39.2. Which of the transformations of X listed below define a r.v. Y such that $f_Y \neq f_X$?

a) $Y = \begin{cases} X, & x < 1 \\ 1, & x \geq 1 \end{cases}$ b) $Y = \begin{cases} X, & x \neq 0 \\ 2, & x = 0 \end{cases}$

c) $Y = |X|$ d) $Y = \begin{cases} 0, & x = 0.1, 0.2, 0.3, \ldots \\ X, & \text{otherwise} \end{cases}$

e) $Y = 0.5X + 0.5|X|$.   [70]

47.3 Given a r.v. X with c.d.f.

$$F_X(x) = \begin{cases} 1, & x > 2 \\ x/4 + 1/2, & -2 \leq x \leq 2 \\ 0, & x < -2. \end{cases}$$

In each of the following cases, specify a function $g(x)$ in the range $-2 \leq x \leq 2$ so that $Y = g(X)$. (Answers are not unique.)

a) $F_Y(y) = \begin{cases} 1, & y > 1 \\ y/2 + 1/2, & -1 \leq y \leq 1 \\ 0, & y < -1 \end{cases}$

b) $F_Y(y) = \begin{cases} 1, & y > 4 \\ y/4, & 0 \leq y \leq 4 \\ 0, & y < 0 \end{cases}$

c) $F_Y(y) = \begin{cases} 1, & y > 1 \\ y^2, & 0 \leq y \leq 1 \\ 0, & y < 0. \end{cases}$ [158]

Transformation of a Two-Dimensional Random Variable

The Transformation Rules 47.1 and 47.2 easily extend to two-dimensional or vector r.v.'s. Let $X = (X_1, X_2)$ be some two-dimensional r.v. defined on a probability space $(S, \mathscr{A}, P)$, and let $Y = (Y_1, Y_2)$ be a two-dimensional r.v. that is defined by a transformation of X:

$$Y = g(X). \tag{48.1}$$

This says that, for each element $\xi \in S$, the pair of real numbers $x = (x_1, x_2) = X(\xi)$, and the pair of real numbers $y = (y_1, y_2) = Y(\xi)$, are related by $y = g(x)$. In other words, the function g assigns a (single) pair of numbers $y = (y_1, y_2)$ to each pair of numbers x. Here is another way of expressing Equation (48.1):

$$Y = (g_1(X), g_2(X)) \tag{48.2}$$

where $g_1(X) = Y_1$, $g_2(X) = Y_2$, and $(g_1, g_2) = g$.

First, consider X to be a discrete r.v., so that Y is also discrete.

Example 48.1

We return once again to the familiar experiment of throwing two distinguishable dice (see Example 41.2). Let x_1 denote the number of dots observed on the red die, x_2 on the white die. The joint distribution of X_1 and X_2 was described in Example 41.2 (where the r.v.'s were named G_r and G_w, respectively). What is now wanted is the distribution associated with the smaller of the two numbers thrown, and the larger of the two numbers thrown, per throw. Then we let $Y_1 = \min(X_1, X_2)$, and $Y_2 = \max(X_1, X_2)$; or, in the notation (48.2),

$$Y = (Y_1, Y_2) = (\min(X_1, X_2), \max(X_1, X_2)).$$

Both Y_1 and Y_2 can take on any one of the values 1, 2, 3, 4, 5, or 6, but the r.v. Y cannot take on all 36 combinations of these values. The *range* of Y is easily seen to be as indicated in Figure 48.1. What is the probability distribution P_Y?

Probability Concepts and Theory for Engineers, First Edition. Harry Schwarzlander.
© 2011 John Wiley & Sons, Ltd. Published 2011 by John Wiley & Sons, Ltd.

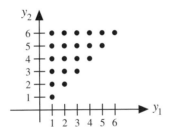

Figure 48.1   Range of the r.v. Y

Rule 47.1 can be rewritten in the following way to fit this two-dimensional case:

Rule 48.1
Given a two-dimensional discrete r.v. X defined on a probability space $(S, \mathscr{A}, P)$, with induced distribution P_X; and another two-dimensional discrete r.v. $Y = g(X)$. To find P_Y:
a) Consider some point y in the range of the r.v. Y.
b) Find the set of all those points x that the function g maps into this point y. Denote this set K_y;
   i.e., $K_y = \{x: g(x) = y\} \subset R^2$.
c) Then $P_Y(\{y\}) = P_X(K_y)$ for the particular point y being considered.
d) Do this for all y in the range of Y.

Example 48.1 (continued)
The steps of Rule 48.1 are applied in Table 48.1. The resulting joint distribution P_Y assigns probability 1/36 to those points shown in Figure 48.1 for which $y_1 = y_2$, and probability 1/18 to all the remaining points.

Table 48.1   The transformation $Y = (\min(X_1, X_2), \max(X_1, X_2))$

a) Consider $y =$	(1, 1)	(1, 2)	(1, 3)	etc.
b) Then $K_y =$	$\{(1, 1)\}$	$\{(1, 2), (2, 1)\}$	$\{(1, 3), (3, 1)\}$	$\ldots$
c) Giving $P_Y(\{y\}) =$	$P_X(\{(1, 1)\}) = 1/36$	1/18	1/18	$\ldots$

Now let X and Y in Equation (48.1) be arbitrary two-dimensional random variables. As in the one-dimensional case (Section 47), g must then be a Borel function. But here, g is a mapping from a plane into a plane. Therefore, specifying g to be a Borel function now means that the inverse image of every two-dimensional Borel set must be a two-dimensional Borel set. That is, for every $B \in \mathscr{B}^2$ there is the requirement that

$$\{x : g(x) \in B\} \in \mathscr{B}^2.$$

This requirement will always be assumed satisfied.

The c.d.f. of the r.v. Y in Equation (48.1) is obtained by means of a straightforward extension of Rule 47.2. We have:

Rule 48.2
Given a two-dimensional r.v. X defined on a probability space $(S, \mathcal{A}, P)$, with c.d.f. F_X. To find the c.d.f. F_Y of the r.v. $Y = (Y_1, Y_2)$, where $Y = g(X) = (g_1(X), g_2(X))$:
a) Pick some $y = (y_1, y_2) \in \mathbb{R}^2$.
b) Determine the inverse image of the set $\overline{-\infty, y_1}^| \times \overline{-\infty, y_2}^|$. Denote this inverse image U_y.
   Thus, $U_y = \{x: g_1(x) \le y_1 \text{ and } g_2(x) \le y_2\}$.
c) Then $F_Y(y) = P_X(U_y)$ for the particular point y being considered.
d) Do this for all points $y \in \mathbb{R}^2$.

The procedure specified by this rule can be visualized with the help of Figure 48.2, where arrows suggest the function X and the function g. The following example illustrates the application of this rule.

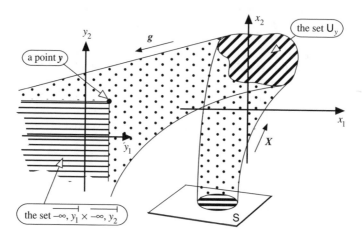

Figure 48.2   Transformation of (X_1, X_2) into (Y_1, Y_2)

Example 48.2: Conversion from rectangular to polar coordinates.
Polar coordinate r.v.'s were first considered in Example 44.4. Now we begin with two r.v.'s X and Y, and apply a transformation that yields polar coordinate r.v.'s R and Θ. In other words, for each ξ, the number pair $(R(\xi), \Theta(\xi))$ is to represent the polar coordinates of the rectangular coordinate pair $(X(\xi), Y(\xi))$.

Then let a two-dimensional r.v. (X, Y) be defined on a probability space $(S, \mathcal{A}, P)$, with joint c.d.f. $F_{X,Y}$. (X, Y) maps each element $\xi \in S$ into a point $(x, y) \in \mathbb{R}^2$ in the usual manner—that is, according to the rectangular coordinate system. Another two-dimensional r.v. (R, Θ) is to be defined on $(S, \mathcal{A}, P)$, such that for each element $\xi \in S$, $(R, \Theta)(\xi)$ represents the polar coordinates of the point $(X, Y)(\xi) \in \mathbb{R}^2$. (R, Θ) is therefore *a function of* (X, Y): $(R, \Theta) = g(X, Y)$, where

$$\left.\begin{aligned} R &= g_1(X, Y) = \sqrt{X^2 + Y^2}, \\ \Theta &= g_2(X, Y) = \arctan\frac{Y}{X}. \end{aligned}\right\} \tag{48.3}$$

Before applying Rule 48.2 it is to be noted that R cannot be negative, and the values assigned by Θ should be confined to an interval of size 2π—usually taken to be $\overline{0, 2\pi}$ or $\overline{-\pi, \pi}$. Let $0 \le \Theta < 2\pi$. The range of (R, Θ) in the r,θ-plane is then inherently restricted to the semi-infinite strip shown shaded in Figure 48.3. If R and Θ were not meant to express polar coordinates, then the range of (R, Θ) could in principle be the entire plane. But with (R, Θ) restricted as shown in Figure 48.3, it can be seen at once that $F_{R,\Theta}(r, \theta) = 0$ for $r < 0$ or $\theta < 0$.

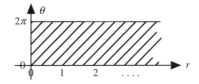

Figure 48.3   Range of (R, Θ) in the (r,θ)-plane

Now we proceed with Rule 48.2:

a) Pick some point $(r, \theta) \in \overline{0, \infty} \times \overline{0, 2\pi}$.

b) The set $U_{(r,\theta)} = \left\{ (x,y) : \sqrt{x^2 + y^2} \le r, \quad \arctan\left(\frac{y}{x}\right) \le \theta \right\}$ is shown in Figure 48.4.

c) Then $F_{R,\Theta}(r, \theta) = P_{X,Y}(U_{(r,\theta)})$.

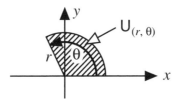

Figure 48.4   The set $U_{(r,\theta)}$

To complete the specification of $F_{R,\Theta}$ over the entire r,θ-plane (which is normally not of interest), we note that for $\theta \ge 2\pi$, Theorem 43.2 gives $F_{R,\Theta}(r, \theta) = F_{R,\Theta}(r, 2\pi) = F_R(r)$.

To illustrate this transformation for a specific two-dimensional r.v., let X and Y be s.i. standardized Gaussian r.v.'s with p.d.f.'s as given in Equation (40.2). Their joint p.d.f. is (from Theorem 46.1)

$$f_{X,Y}(x, y) = \frac{1}{2\pi} \exp\left(-\frac{x^2 + y^2}{2} \right).$$

Step (c), above, then becomes

$$F_{R,\Theta}(r, \theta) = \frac{1}{2\pi} \int_0^\theta \int_0^r \exp\left(-\frac{r^2}{2} \right) r\,dr\,d\theta$$

$$= \frac{1}{2\pi} \int_0^\theta \left[1 - \exp\left(-\frac{r^2}{2} \right) \right] d\theta$$

$$= \frac{\theta}{2\pi} \left[1 - \exp\left(-\frac{r^2}{2} \right) \right], \quad 0 \le \theta < 2\pi \text{ and } r \ge 0.$$

Theorem 45.2 shows that R and Θ are again s.i., with $F_\Theta(\theta) = \dfrac{\theta}{2\pi}$ for $\theta \in \overline{0,\ 2\pi}$, and $F_R(r) = 1 - \exp(-r^2/2)$ for $r \geq 0$. Thus, Θ is unif. $\overline{0,\ 2\pi}$, and since it is independent of R, the distribution induced by the two-dimensional r.v. (R, Θ) must have *circular symmetry*. The marginal p.d.f.'s are

$$f_\Theta(\theta) = \begin{cases} \dfrac{1}{2\pi}, & 0 \leq \theta < 2\pi \\ 0, & \text{otherwise} \end{cases} \quad \text{and} \quad f_R(r) = \begin{cases} r\exp\left(-\dfrac{r^2}{2}\right), & r \geq 0 \\ 0, & \text{otherwise.} \end{cases} \tag{48.4}$$

P_R is called a *Rayleigh* distribution.

Queries

48.1 Suppose the two-dimensional r.v. $Y = (Y_1, Y_2)$ is obtained from the two-dimensional r.v. $X = (X_1, X_2)$ by means of the transformation

$$Y_1 = 1 + 2X_2, \quad Y_2 = 2 - X_1.$$

In the (x_1, x_2)-plane, what is the set $\mathsf{U}_{(0,0)}$?   [181]

48.2 Same question as in the preceding Query, for the transformation

$$Y_1 = X_1 + X_2, \quad Y_2 = X_1 - X_2.$$ [27]

48.3 The two r.v.'s X, Y defined in Query **44.1** are transformed to two new r.v.'s Z, W:

$$(Z, W) = g(X, Y)$$

For each of the following three transformations, decide which of the statements given below apply (if any):

$$\text{i) } Z = X/2, \ W = 2Y$$
$$\text{ii) } Z = X + Y, \ W = X - Y$$
$$\text{iii) } Z = XY, \ W = X/Y.$$

a) $F_{Z,W}(0, 0) = F_{X,Y}(0, 0)$.
b) The line $y = x$ in the (x,y)-plane maps into the z-axis in the (z,w)-plane.
c) $F_Z(z) = (1/2)F_X(z)$.
d) g is a one-to-one function.
e) The first quadrant of the (x,y)-plane maps into the first quadrant of the (z,w)-plane.   [61]

The Sum of Two Discrete Random Variables

The approach used in the transformation rules of Section 48 does not require that both X and Y in Equation (48.1) be two-dimensional random variables. In fact, the transformations discussed in Section **47** can be viewed as a special case of Equation (48.1). It is also possible for either X or Y to be two-dimensional, while the other is scalar. The most commonly encountered transformation of this nature is the *addition* of two r.v.'s,

$$Y = X_1 + X_2. \tag{49.1}$$

The relation (49.1) signifies that a r.v. Y gets defined on the underlying probability space in such a manner that Y assigns to each element ξ the value $Y(\xi) = X_1(\xi) + X_2(\xi)$. In this Section we examine in detail the transformation (49.1) for the discrete case, and we also consider some other transformations from two dimensions to one.

Let $X = (X_1, X_2)$ be a *discrete* r.v. Thus, X assigns positive probability at various isolated points in the plane, such as indicated by dots in Figure 49.1a. To be determined is P_Y, where $Y = X_1 + X_2$. Interpreting Rule 48.1 for the case where Y is scalar, we set $P_Y(\{y\}) = P_X(\mathsf{K}_y)$ for each real number y in the range of Y. For any such choice of y, the set $\mathsf{K}_y = \{(x_1, x_2): x_1 + x_2 = y\}$ consists of a slanting line in the (x_1, x_2)-plane that passes through one or more points where positive probability is induced, as shown in Figure 49.1a.[2] $P_X(\mathsf{K}_y)$ is the sum of the probabilities assigned to these points. Then

$$P_Y(\{y\}) = P_X(\mathsf{K}_y) = \sum_{x_1} P_X(\{(x_1, y - x_1)\}) \tag{49.2}$$

the summation being over all those x_1-values for which $\{(x_1, y - x_1)\}$ has positive probability; i.e., over all positive probability points lying on the line K_y. The computation expressed in Equation (49.2) is easily visualized graphically. Beginning with a diagram such as that shown in Figure 49.1a, slide a ruler across the figure while maintaining a slope of -1. At any position at which

---

[2] To simplify the language, we call such points (x, y) 'positive probability points', although the probability is actually associated with the *set* $\{(x, y)\}$ and not with the *element* (x, y).

Probability Concepts and Theory for Engineers, First Edition. Harry Schwarzlander.
© 2011 John Wiley & Sons, Ltd. Published 2011 by John Wiley & Sons, Ltd.

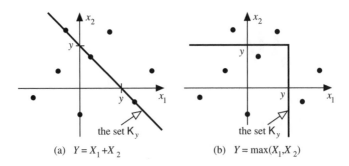

Figure 49.1   The set K_y for the sum of two r.v.'s and the larger of two r.v.'s

points of positive probability are intersected, sum these probabilities and assign them to the corresponding y-value (axis intercept).

If X_1, X_2 are *statistically independent* random variables, then the joint distribution P_X can be expressed as the product of the marginal distributions (Definition 45.1), so that Equation (49.2) becomes

$$P_Y(\{y\}) = \sum_{x_1} P_{X_1}(\{x_1\}) \, P_{X_2}(\{y - x_1\}). \tag{49.3}$$

Example 49.1

Let $X = (X_1, X_2)$ be a discrete r.v. that expresses the results obtained in the throw of two distinguishable dice, as in Example 41.2. A new r.v. Y is to express the sum of the dots obtained on the two dice, so that $Y = X_1 + X_2$. Since X_1 and X_2 are statistically independent r.v.'s, Equation (49.3) applies:

$$P_Y(\{y\}) = \sum_{x_1} \left(\frac{1}{6} \cdot \frac{1}{6} \right).$$

where the integers x_1 have to be chosen correctly for each integer y. In Equation (49.3), a value of $1/6$ is taken by *both* P_{X_1} and P_{X_2} only if $1 \le x_1 \le 6$ and also $1 \le y - x_1 \le 6$. Accordingly, we obtain:

$$P_Y(\{y\}) = \begin{cases} 0, & y \le 1 \text{ and } y \ge 13 \\ 1/36, & y = 2, \, 12 \\ 1/18, & y = 3, \, 11 \\ 1/12, & y = 4, \, 10 \\ 1/9, & y = 5, \, 9 \\ 5/36, & y = 6, \, 8 \\ 1/6, & y = 7. \end{cases}$$

This result is also easily obtained by proceeding in the manner of Figure 49.1a.

It is worth noting that if one of the r.v.'s X_1, X_2 in Equation (49.1) is replaced by a constant $b \in \mathsf{R}$, the effect is the same as if that r.v. were specified to be unary, so that Equation (49.3) applies. In other words, if $P_{X_1}(\{b\}) = 1$, then X_1, X_2 are s.i. and Equation (49.3) becomes

$$P_Y(\{y\}) = P_{X_1}(\{b\})\, P_{X_2}(\{y-b\}) = P_{X_2}(\{y-b\}).$$

This shows that the result is a shifted version of P_{X_2}, the shift being to the right by an amount b if $b > 0$, and to the left by an amount $|b|$ if $b < 0$.

The operation expressed in formula (49.3) is 'discrete convolution'. It can be illustrated graphically as in Figure 49.2, which shows a plot of a discrete distribution P_{X_1}. Under it is a plot of a discrete distribution P_{X_2} with the x_2-axis reversed, and the origin displaced by an amount y_0. The heights of every pair of bars that line up are multiplied, and all such products summed to give P_Y at the point $y = y_0$.

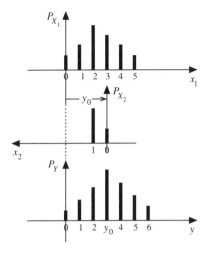

Figure 49.2   Discrete convolution

A similar approach can be used if Y is specified as some other function of X_1 and X_2. Suppose the r.v. Y is to express *the larger of* the two components of the discrete r.v. $X = (X_1, X_2)$,

$$Y = \max(X_1, X_2).$$

To determine P_Y, Rule 48.1 is again applied with Y a scalar r.v. We set $P_Y(\{y\}) = P_X(K_y)$ for each real number y in the range of Y, where $K_y = \{(x_1, x_2): \max(x_1, x_2) = y\}$. The set K_y in the (x_1, x_2)-plane consists of horizontal and vertical line segments that meet at the point (y, y), as shown in Figure 49.1b. Here, the general expression for P_Y, analogous to Equation (49.2), is

$$P_Y(\{y\}) = \sum_{x_1} P_X(\{(x_1, y)\}) + \sum_{x_2} P_X(\{(y, x_2)\}), \quad x_1, x_2 \le y \qquad (49.4)$$

the summations being over those x_1- and x_2-values for which positive probabilities are contributed.

Example 49.2

Let the r.v. X be defined as in Example 49.1, and let $Y = \max(X_1, X_2)$. This is equivalent to Y_2 in Example 48.1, but now the idea is to obtain $\max(X_1, X_2)$ directly as a scalar r.v. That is, we make a transformation from two dimensions to one, whereas in Example 48.1 $\max(X_1, X_2)$ was obtained as a component of the vector r.v. Y. Application of Equation (49.4) to the distribution depicted in Example 41.2 yields immediately

$$P_Y(\{y\}) = \begin{cases} \dfrac{2y-1}{36}, & y = 1, 2, \ldots, 6 \\ 0, & \text{otherwise.} \end{cases}$$

Of course, if two r.v.'s X_1, X_2 satisfy $X_1 > X_2$, that is, $X_1(\xi) > X_2(\xi)$ for every element ξ in the underlying probability space, then $\max(X_1, X_2)$ is simply X_1.

Queries

49.1 Two discrete r.v.'s X_1, X_2 induce a joint distribution that assigns positive probability only at the points (x_1, x_2) identified in Figure 49.3. Let $Y = X_1 + X_2$. For how many different numbers y is $P_Y(\{y\}) \neq 0$?   [200]

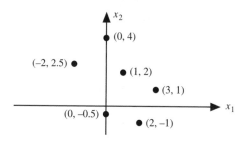

Figure 49.3   Points where the r.v. (X_1, X_2) in Query **49.1** assigns positive probability

49.2 A discrete r.v. X assigns positive probability to the sets $\{x\}$ where $x = 1, 2, \ldots, m$, and to no other one-element sets. A discrete r.v. Y assigns positive probability to the sets $\{y\}$ where $y = 1, 2, \ldots, n$, and to no other one-element sets. X and Y are statistically independent.
a) How many distinct one-element sets are there to which the r.v. $X + Y$ assigns positive probability?
b) Same question, if the distribution of Y is shifted to the right by $1/2$—i.e., Y is replaced by $Y + (1/2)$?   [162]

49.3 Consider a probability space $(S, \mathscr{A}, P)$ where $S = \{\xi_1, \xi_2, \xi_3, \ldots\}$ is a countably infinite set. Let the real-valued point functions X and Y be defined on S in the following way:

$$X(\xi_i) = i, \quad i = 1, 2, 3, \ldots.$$
$$Y(\xi_i) = 1/i, \quad i = 1, 2, 3, \ldots.$$

For each of the following statements, determine whether it is true, possibly true, or false:
a) X and Y are random variables.
b) If X and Y are indeed r.v.'s, they are statistically independent.
c) $Z = X \cdot Y$ is a random variable.   [235]

The Sum of Two Arbitrary Random Variables

Generalizing Section 49 to arbitrary random variables, we again consider the sum

$$Y = X_1 + X_2$$

but $X = (X_1, X_2)$ is no longer assumed to be discrete. The joint distribution function F_X is assumed given. Following Rule 48.2, we set $F_Y(y) = P_X(\mathsf{U}_y)$, where

$$\mathsf{U}_y = \{(x_1, x_2) : x_1 + x_2 \leq y\}. \tag{50.1}$$

The set U_y consists of all points on, and to the left of (or below), a slanting line in the (x_1, x_2)-plane, as shown in Figure 50.1.

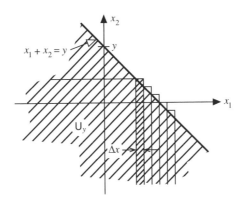

Figure 50.1 Approximating U_y with rectangular strips

In order to express $P_X(\mathsf{U}_y)$ in terms of F_X, we use the fact that the two-dimensional distribution function at a given point (x_1, x_2) expresses the probability associated with the set of all points to the left and below that point (including the boundary of that set).

Probability Concepts and Theory for Engineers, First Edition. Harry Schwarzlander.
© 2011 John Wiley & Sons, Ltd. Published 2011 by John Wiley & Sons, Ltd.

Referring to Figure 50.1 we see that $P_X(U_y)$ can be approximated as follows:

$$P_X(U_y) \approx \sum_{x_1} [F_X(x_1 + \Delta x_1, \, y - x_1) - F_X(x_1, \, y - x_1)] \tag{50.2}$$

where the summation is over all x_1-values spaced Δx_1 apart. Each term in the summation represents the probability of a semi-infinite strip of width Δx_1, such as the one shown shaded in Figure 50.1. Expression (50.2) becomes exact in the limit as $\Delta x_1 \rightarrow 0$:

$$F_Y(y) = \lim_{\Delta x_1 \rightarrow 0} \sum_{x_1} [F_X(x_1 + \Delta x_1, \, y - x_1) - F_X(x_1, y - x_1)]. \tag{50.3}$$

In Section **74** we will see that this expression suggests the general form of a *Stieltjes Integral*.

Rather than proceeding with a formulation in terms of the Stieltjes integral, we will at this point assume X to be *absolutely continuous*. In that case the joint density function for X_1, X_2 exists and can be integrated over the set U_y to yield

$$F_Y(y) = \int_{-\infty}^{\infty} \int_{-\infty}^{y - x_1} f_X(x_1, \, x_2) \, dx_2 \, dx_1. \tag{50.4}$$

Differentiation gives

$$\left. \begin{aligned} f_Y(y) = \frac{dF_Y(y)}{dy} &= \int_{-\infty}^{\infty} f_{X_1, X_2}(x_1, \, y - x_1) \, dx_1 \\ &= \int_{-\infty}^{\infty} f_{X_1, X_2}(y - x_2, \, x_2) \, dx_2. \end{aligned} \right\} \tag{50.5}$$

the second integral being obtained by a simple change of variable. The integrals in Equation (50.5) can be interpreted as integrations of $f_X(x_1, x_2)$ along the line $x_2 = y - x_1$ in the (x_1, x_2)-plane, which is illustrated in Figure 50.2; the integration being with respect to x_1 or x_2 (not with respect to the path direction).

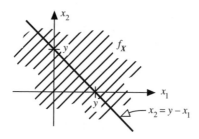

Figure 50.2   Integration along the line $x_2 = y - x_1$

Example 50.1

We consider a pair of r.v.s X_1, X_2 with a joint p.d.f. $f_{X_1, X_2}(x_1, \, x_2)$ that is constant in the square region identified in Figure 50.3, and zero elsewhere. Let $Y = X_1 + X_2$. Use of Equation (50.5) gives

$$f_Y(y) = \int_{-\infty}^{\infty} f_{X_1,X_2}(x_1, y - x_1)dx_1$$

$$= \int_{(y-1)/2}^{(y+1)/2} \frac{1}{2} dx_1 = \frac{1}{2}, \quad -1 \leq y \leq 1$$

and $f_Y = 0$ outside this range. The limits of integration identify the range of x_1-values for which the diagonal line of Figure 50.2 crosses the shaded region where $f_{X_1,X_2} = 1/2$.

Of course, we can also go back to the transformation rule (Rule 48.2), which in this example is simple to apply. The set U_y is made up of all the points that lie below the diagonal line $x_2 = y - x_1$ as shown in Figure 50.2. By noting how the set U_y changes as y increases in the range $-1 \leq y \leq 1$ it becomes immediately apparent that F_Y increases *linearly* over $\overline{-1, 1}$, so that Y must induce a uniform distribution over $\overline{-1, 1}$.

Since f_{X_1,X_2} is constant in the shaded region shown below, an even simpler approach is to visually apply the integration path of Figure 50.2 to Figure 50.3. As long as the integration path passes through the shaded region, the length of the path within the shaded region remains the same, implying a uniform distribution.

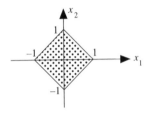

Figure 50.3   Region where f_{X_1,X_2} is positive

If X_1 and X_2 happen to be *statistically independent* r.v.'s, then their joint density function equals the product of their individual density functions and Equation (50.5) becomes

$$f_Y(y) = \int_{-\infty}^{\infty} f_{X_1}(x_1) f_{X_2}(y - x_1) \, dx_1$$

$$= \int_{-\infty}^{\infty} f_{X_1}(y - x_2) f_{X_2}(x_2) \, dx_2$$

(50.6)

We recognize the integrals (50.6) as convolution integrals, and thus have:

Theorem 50.1
If X_1 and X_2 are two statistically independent, absolutely continuous random variables, then the p.d.f. of their sum is the convolution of their individual density functions.

This theorem also expresses an interesting property of the convolution integral: Given two non-negative real functions both of which integrate to 1, then their convolution is also a non-negative real function that integrates to 1. There are many other instances where a particular line of reasoning required in Probability Theory brings into evidence a property that may be of broader interest.

It is well known that the convolution of two rectangular functions of equal width yields a triangular function that is symmetrical about its peak. From this it follows immediately that if X and Y are statistically independent and induce uniform distributions over intervals of equal measure, then $X + Y$ has a symmetrical triangular p.d.f. (See also Problem 50.1.)

Transform methods can simplify the convolution calculation (see Part VI). However, carrying out the integration (50.6) is sometimes not as demanding as specifying the correct limits of integration—when the density functions f_{X_1}, f_{X_2} are defined by different functional forms over different intervals.

Example 50.2

Let X_1, X_2 be identically distributed and statistically independent (i.d.s.i.) r.v.'s with density functions

$$f_{X_1}(x) = f_{X_2}(x) = \begin{cases} 2 - 2x, & 0 < x < 1 \\ 0, & \text{otherwise} \end{cases}$$

as sketched in Figure 50.4a. We wish to obtain $f_Y(y)$, where $Y = X_1 + X_2$.

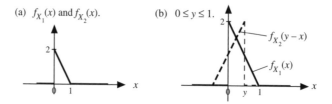

(a) $f_{X_1}(x)$ and $f_{X_2}(x)$.   (b) $0 \le y \le 1$.

Figure 50.4   Convolution of two triangular p.d.f.'s

A sketch of the functions $f_{X_1}(x)$ and $f_{X_2}(y - x)$ is helpful for finding the correct integration limits for the integral (50.6). For y in the range $0 < y < 1$, Figure 50.4b applies. The product $f_{X_1}(x) f_{X_2}(y - x)$ vanishes outside the interval $\overline{0, y}$; therefore integration proceeds from 0 to y.

For y in the range $1 \le y \le 2$, Figure 50.5a applies. In this case, $f_{X_1}(x) f_{X_2}(y - x)$ vanishes outside the interval $\overline{y - 1, 1}$; therefore integration is from $y-1$ to 1.

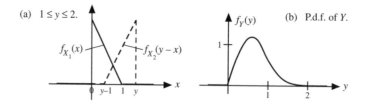

(a) $1 \le y \le 2$. $f_Y(y)$ (b) P.d.f. of Y.

Figure 50.5   Convolution of f_{X_1} and f_{X_2} concluded

For other values of y, the integrand vanishes altogether. Equation (50.6) therefore becomes

$$f_Y(y) = \begin{cases} \int_0^y (2-2x)(2-2y+2x)dx = 4y - 4y^2 + \dfrac{2}{3}y^3, & \text{for } y \in \overline{0,\,1} \\[3mm] \int_{y-1}^1 (2-2x)(2-2y+2x)dx = \dfrac{16}{3} - 8y + 4y^2 - \dfrac{2}{3}y^3, & \text{for } y \in \overline{1,\,2} \\[3mm] 0, & \text{for all other } y. \end{cases}$$

This result is sketched in Figure 50.5(b).

Given two random variables X_1, X_2, it is interesting to note that statistical independence is not a *necessary* condition for the density of $X_1 + X_2$ to be the convolution of f_{X_1} and f_{X_2}. This is illustrated in the following Example.

Example 50.3

In the (x,y)-plane, consider the set A shown shaded in Figure 50.6, that is,

$$A = \{(x, y): 0 \le y \le x \le 1\} \cup \{(x, y): 0 \le -y \le -x \le 1\} \cup \{(x, y): 0 \le -x \le y \le 1\}$$
$$\cup \{(x, y): 0 \le x \le -y \le 1\}.$$

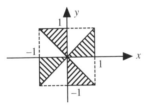

Figure 50.6   The set A

Let X, Y be two r.v.'s with joint p.d.f.

$$f_{X,Y}(x, y) = \begin{cases} \dfrac{1}{2}, & (x, y) \in A \\[2mm] 0, & \text{otherwise.} \end{cases}$$

The marginal densities are uniform (Figure 50.7a):

$$f_X(x) = f_Y(x) = \begin{cases} \dfrac{1}{2}, & -1 < x < 1 \\[2mm] 0, & \text{otherwise.} \end{cases}$$

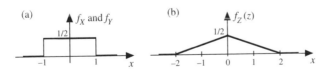

Figure 50.7   The two marginal densities and their convolution

Now let $Z = X + Y$. To find f_Z, Equation (50.6) cannot be used since X and Y are not statistically independent. Using Equation (50.5), it is easily verified that f_Z is triangular (Figure 50.7b); that is,

$$
f_Z(z) = \begin{cases}
\dfrac{z}{4} + \dfrac{1}{2}, & -2 < z < 0 \\[2mm]
-\dfrac{z}{4} + \dfrac{1}{2}, & 0 < z < 2 \\[2mm]
0, & \text{otherwise.}
\end{cases}
$$

But this is exactly what is obtained upon convolving f_X and f_Y.

Queries

50.1 An unbiased die is thrown twice in order to generate two randomly selected integers i, j in the range from 1 to 6. Define the random variables

$$X \leftrightarrow \text{`number of dots obtained on first throw'}$$
$$Y \leftrightarrow \text{`number of dots obtained on second throw'}$$
$$Z = X/Y.$$

Evaluate $F_Z(0.25)$.   [75]

50.2 Given the r.v.'s X, Y, let $Z = X + Y$. Indicate for each statement below whether it is true, possibly true, or false:
a) $F_Z = F_X + F_Y$
b) $f_Z = f_X * f_Y$ (where $*$ denotes convolution)
c) $Z = 2X$
d) $P_Z\left(\overline{0, \infty}\right) = P_{X,Y}\left(\overline{0, \infty} \times \overline{0, \infty}\right)$.   [243]

50.3 A random variable X induces a uniform distribution over $\overline{0, 1}$ and another r.v. Y induces a uniform distribution over $\overline{-1, 0}$. Let $Z = X + Y$. For each of the following statements, determine whether it is true, possibly true, or false:
a) Z also induces a uniform distribution
b) $X = Z - Y$
c) $Z \leq X - Y$
d) The maximum value assigned by Z is 0
e) Z is absolutely continuous.   [104]

n-Dimensional Random Variables

Up to this point we have treated only one- and two-dimensional random variables and two-dimensional vector r.v.'s, and their associated distributions. Clearly, situations arise where more than two numerical values are to be associated with each element of a probability space—so that more than two r.v.'s need to be defined on the space. We already have encountered examples of this when transforming a two-dimensional r.v. $X = (X_1, X_2)$ into another two-dimensional r.v. $Y = (Y_1, Y_2)$. In that case, four r.v.'s become defined on the underlying space. Now the definitions of earlier Sections will get extended to cover multidimensional and vector random variables with an arbitrary finite number of components.

Let $(S, \mathscr{A}, P)$ be a probability space on which several random variables, $X_1, X_2, \ldots, X_n$, are defined. These n random variables assign n real numbers $X_1(\xi), X_2(\xi), \ldots, X_n(\xi)$ to every element $\xi \in S$, and can be thought of as the components of an n–*dimensional* r.v. $X = (X_1, \ldots, X_n)$. Furthermore, it is sometimes convenient to regard such an n-dimensional r.v. as an n-dimensional

vector r.v. or 'random vector' $X = [X_1, \ldots, X_n]$ if a row vector, or $X = \begin{bmatrix} X_1 \\ \vdots \\ X_n \end{bmatrix}$ if a column vector. Here,

X maps every $\xi \in S$ into a point in n-dimensional Euclidian space, R^n.

A completely additive class of subsets of R^n needs to be specified which is of sufficient generality so that it can be used with any n-dimensional r.v. That is the class of n-dimensional Borel sets. This class can be defined by beginning with the class of all n-fold product sets of the one-dimensional Borel sets:

$$\underbrace{\mathscr{B} \times \mathscr{B} \times \ldots \times \mathscr{B}}_{n \text{ times}}$$

This is not an additive class; but it can be enlarged into a completely additive class by closing it with respect to countable set operations, giving *the completely additive class generated by* $\mathscr{B} \times \mathscr{B} \times \ldots \times \mathscr{B}$,

$$\mathscr{A}(\underbrace{\mathscr{B} \times \mathscr{B} \times \ldots \times \mathscr{B}}_{n \text{ times}}). \tag{51.1}$$

Probability Concepts and Theory for Engineers, First Edition. Harry Schwarzlander.
© 2011 John Wiley & Sons, Ltd. Published 2011 by John Wiley & Sons, Ltd.

(Definition) 51.1

For any specified integer $n \geq 2$, the class of subsets of R^n identified by Equation (51.1) is called *the class of n-dimensional Borel sets*, and denoted $\mathscr{B}^n$. Any set in $\mathscr{B}^n$ is called an *n-dimensional Borel set*.

Definition 42.1 can be seen to be just a special case of Definition 51.1.

In n-dimensional space the class $\mathscr{B}^n$ is again so comprehensive that it covers all subsets of R^n that we could possibly be interested in. $\mathscr{B}^n$ serves as the completely additive class on which an n-dimensional random variable must be able to induce a legitimate (n-dimensional) probability distribution. Thus, we have:

(Definition) 51.2

A point function X defined on a probability space $(S, \mathscr{A}, P)$ and assigning an n-tuple of real numbers to each element $\xi \in S$ is called an *n-dimensional random variable* if for every n-dimensional Borel set B, the set $\{\xi : X(\xi) \in B\}$ belongs to $\mathscr{A}$.

We see that the definition of an n-dimensional r.v. is completely analogous to the definition of an ordinary scalar (one-dimensional) r.v., and therefore applies also if $n = 1$. If an n-dimensional r.v. X is defined on a probability space $(S, \mathscr{A}, P)$, another probability space is automatically established. This is the (induced) probability space $(R^n, \mathscr{B}^n, P_X)$, which consists of the space R^n of n-tuples of real numbers, the completely additive class $\mathscr{B}^n$, and the probability function P_X induced on this class by the r.v. X according to the rule $P_X(B) = P(\{\xi : X(\xi) \in B\})$, for each $B \in \mathscr{B}^n$.

As in the one- and two-dimensional cases, the n-dimensional probability distribution P_X induced by an n-dimensional r.v. X on $\mathscr{B}^n$ is completely described by the n-dimensional distribution function for this r.v.

(Definition) 51.3

The *n-dimensional distribution function* $F_X(x_1, \ldots, x_n)$, or $F_{X_1,\ldots,X_n}(x_1, \ldots, x_n)$, which is associated with an n-dimensional random variable $X = (X_1, \ldots, X_n)$, is defined as

$$F_X(x_1, \ldots, x_n) \equiv P_X\left(\overline{-\infty, x_1} \times \overline{-\infty, x_2} \times \ldots \times \overline{-\infty, x_n}\right)$$

$$= P(\{\xi : X_1(\xi) \leq x_1, X_2(\xi) \leq x_2, \ldots, X_n(\xi) \leq x_n\}).$$

Clearly, for $n = 1$ this Definition reduces to the definition of a one-dimensional c.d.f. (Definition 38.1).

An argument analogous to that used in Section 43 will demonstrate why an n-dimensional c.d.f. completely describes an n-dimensional probability distribution. Consider the n-dimensional c.d.f. F_X that is associated with an n-dimensional r.v. X. According to Definition 51.3, F_X expresses the probabilities induced by X on the members of the class

$$\underbrace{\mathscr{V} \times \mathscr{V} \times \ldots \times \mathscr{V}}_{n \text{ times}} \qquad (51.2)$$

with $\mathscr{V}$ as defined in Section 37. Now, in each of the component spaces of R^n, the class $\mathscr{V}$ generates $\mathscr{B}$ (via countable set operations). Furthermore,

$$\underbrace{\mathscr{B} \times \mathscr{B} \times \ldots \times \mathscr{B}}_{n \text{ times}}$$

generates $\mathscr{B}^n$. Therefore, *any set* in $\mathscr{B}^n$ can be expressed in terms of members of the class (51.2), using countable set operations. Similarly, the *probability* of any set in $\mathscr{B}^n$ can be expressed in terms of a countable set of values of F_X—that is, either in terms of a finite number of such values or in terms of an infinite series of such values.

The *n*-dimensional distribution function $F_X(x_1, \ldots, x_n)$ associated with an *n*-dimensional r.v. $X = (X_1, \ldots, X_n)$ defined on a probability space (S, $\mathscr{A}$, P) completely expresses the behavior of the component random variables X_i ($i = 1, \ldots, n$). To see this, we note that the definition of $F_X(x_1, \ldots, x_n)$ given above can be rewritten as in Equation (43.1):

$$F_X(x_1, \ldots, x_n) = P\left[X_1^{-1}\left(\overline{-\infty, x_1}\right) \cap X_2^{-1}\left(\overline{-\infty, x_2}\right) \cap \ldots \cap X_n^{-1}\left(\overline{-\infty, x_n}\right)\right] \quad (51.3)$$

This shows that, at each point $(x_1, \ldots, x_n) \in \mathsf{R}^n$, F_X equals the probability of a joint event in (S, $\mathscr{A}$, P)—namely, the joint event which is the intersection of the events $X_1^{-1}\left(\overline{-\infty, x_1}\right)$, $X_2^{-1}\left(\overline{-\infty, x_2}\right)$ etc. Now suppose the c.d.f. of X_1 is to be determined, that is, $P\left[X_1^{-1}\left(\overline{-\infty, x_1}\right)\right]$ for all x_1. Recalling Equation (8.8b), we see that $x_2, x_3, \ldots, x_n$ in Equation (51.3) need to be specified in such a way that all the events appearing in Equation (51.3) equal S, except for $X_1^{-1}\left(\overline{-\infty, x_1}\right)$. This is achieved by setting $x_2 = x_3 = \ldots = x_n = +\infty$. Thus,

$$\left.\begin{aligned} F_X(x_1, +\infty, +\infty, \ldots, +\infty) &= P\left[X_1^{-1}\left(\overline{-\infty, x_1}\right) \cap \mathsf{S} \cap \mathsf{S} \cap \ldots \cap \mathsf{S}\right] \\ &= P\left[X_1^{-1}\left(\overline{-\infty, x_1}\right)\right] = F_{X_1}(x_1). \end{aligned}\right\} \quad (51.4)$$

Similarly, if any *one*, or any combination, of the arguments of F_X is set to $+\infty$, we obtain the joint c.d.f. of those component random variables whose corresponding arguments have not been so fixed. Of course, $F_X(+\infty, \ldots, +\infty) = 1$. On the other hand, $F_X(x_1, \ldots, x_n) = 0$ if any of the arguments are $-\infty$.

Example 51.1

Consider three r.v.'s X, Y, Z that jointly induce probability 1 uniformly throughout the unit cube, $\overline{0, 1} \times \overline{0, 1} \times \overline{0, 1}$. At any point within this cube, the joint c.d.f. is therefore

$$F_{X,Y,Z}(x, y, z) = xyz, \quad 0 \leq x, y, z \leq 1$$

This completely defines $F_{X,Y,Z}$ in all of R^3. For instance, we can obtain an expression for $F_{X,Y,Z}$ in the volume $\overline{0, 1} \times \overline{0, 1} \times \overline{1, \infty}$ by applying an appropriate modification of Equation (51.4). We note that if $z \geq 1$, then all the probability encountered when moving in the direction of increasing z has been accumulated, and $F_{X,Y,Z}$ does not change if z is increased further. Thus, for $0 \leq x, y \leq 1$ and $z > 1$,

$$F_{X,Y,Z}(x, y, z) = F_{X,Y,Z}(x, y, +\infty) = F_{X,Y,Z}(x, y, 1) = xy.$$

On the other hand, consider $F_{X,Y,Z}$ in the volume $\overline{0, 1} \times \overline{0, 1} \times \overline{-\infty, 0}$. If $z < 0$, then none of the probability that can be encountered when moving in the direction of increasing z has been reached. Thus, for $0 \le x,\, y \le 1$ and $z < 0$,

$$F_{X,Y,Z}(x,\, y,\, z) = F_{X,Y,Z}(x,\, y,\, -\infty) = 0.$$

The transformation rules of Section 48 are easily extended to n-dimensional r.v.'s. We will consider only extension of Rule 48.2. Let X be some n-dimensional r.v. defined on a probability space $(S, \mathscr{A}, P)$, and let Y be an m-dimensional r.v. that is defined in terms of a transformation of X:

$$Y = g(X) \tag{51.5}$$

This means that, for each element $\xi \in S$, $X(\xi)$ is some n-tuple of real numbers x and $Y(\xi)$ is some m-tuple of real numbers y, such that $y = g(x)$. In other words, the transformation g is a function defined on the points in ^{n} (the x-space) and takes on m-dimensional values (i.e., points in R^m, the y-space). Thus, for each x, $g(x)$ is an element of R^m. This can be made explicit by writing g in terms of its m components,

$$g(x) = (g_1(x),\, g_2(x), \ldots,\, g_m(x)).$$

As in the one-dimensional case (Section 47), g must also be a Borel function. In other words, the inverse image of every m-dimensional Borel set must be an n-dimensional Borel set; so that we require, for every $\mathsf{B} \in \mathscr{B}^m$, that

$$\{x : g(x) \in \mathsf{B}\} \in \mathscr{B}^n.$$

This requirement is satisfied by all functions normally encountered, and will always be assumed to hold.

The probability distributions induced by the r.v.'s X and Y appearing in Equation (51.5) are completely described by the n-dimensional c.d.f. F_X and the m-dimensional c.d.f. F_Y, respectively. F_X is assumed known, and F_Y is to be determined. Then we must ask, for each $y = (y_1, y_2, \ldots, y_m)$, what probability gets assigned to the semi-infinite product set $\overline{-\infty, y_1} \times \overline{-\infty, y_2} \times \ldots \times \overline{-\infty, y_m}$ (see Figure 48.2 for the case $m = 2$).

Rule 51.1

Given a r.v. $X = (X_1, \ldots, X_n)$ defined on a probability space $(S, \mathscr{A}, P)$, with c.d.f. F_X. To find the c.d.f. F_Y of the r.v. $Y = (Y_1, \ldots, Y_m)$, where $Y = g(X) = (g_1(X),\, g_2(X),\, \ldots,\, g_m(X))$:
a) Consider some point $y = (y_1, \ldots, y_m) \in \mathsf{R}^m$.
b) Determine the inverse image of the set $\overline{-\infty, y_1} \times \ldots \times \overline{-\infty, y_m}$, that is, the set

$$\mathsf{U}_y = \{x : g_1(x) \le y_1,\, g_2(x) \le y_2, \ldots,\, g_m(x) \le y_m\}.$$

c) Then $F_Y(y) = P_X(\mathsf{U}_y)$, for the particular point $y = (y_1, \ldots, y_m)$ being considered.
d) Do this for all points $y \in \mathsf{R}^m$.

Example 51.2

In R^3, consider the oblique regular hexagon H with vertices at $(1, 0, -1)$, $(1, -1, 0)$, $(0, -1, 1)$, $(-1, 0, 1)$, $(-1, 1, 0)$, and $(0, 1, -1)$. This hexagon can be visualized as inscribed in the cube

$\overline{-1, 1} \times \overline{-1, 1} \times \overline{-1, 1}$, as shown in Figure 51.1. The equation of the plane containing this hexagon is

$$x + y + z = 0.$$

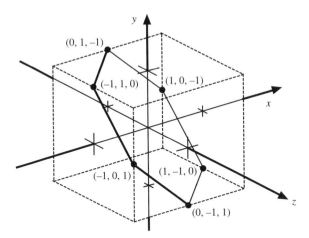

Figure 51.1   The hexagon H

Suppose that three r.v.'s X, Y, Z distribute all probability uniformly over this hexagon. Thus, $X + Y + Z = 0$, and also $-1 \leq X, Y, Z \leq 1$. What are the marginal distributions of X, Y, and Z?

Application of Equation (51.4) in this case would be messy, since $F_{X,Y,Z}$ would have to be found first. But it is easy to use the transformation rule. First, we note that, by symmetry, X, Y, Z must be identically distributed, so it suffices to determine P_X. Then, in order to be able to apply Rule 51.1, we express X as the result of a transformation, $X = g(X, Y, Z)$. (Here g is a scalar function of three variables.)

Consider $-1 \leq x \leq 0$. Projection of the x-coordinates onto the hexagon is illustrated in Figure 51.2a, where U_x is shown shaded. Using simple geometry, we have

$$P_X(U_x) = \frac{1}{3}(x+1) + \frac{1}{6}(x+1)^2$$

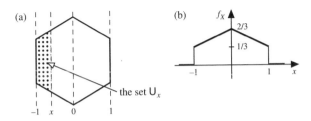

Figure 51.2   Determination of f_X

so that $F_X(x) = \dfrac{1}{6}x^2 + \dfrac{2}{3}x + \dfrac{1}{2}$. Then

$$f_X(x) = \frac{dF_X(x)}{dx} = \frac{x+2}{3}, \quad -1 \le x \le 0;$$

and from this it follows immediately that $f_X(x) = \dfrac{2-x}{3}$ for $0 \le x \le 1$. The graph of f_X is shown in Figure 51.2b.

Queries

51.1 Let $X = (X_1, \ldots, X_n)$ be an n-dimensional r.v. ($n \ge 2$) defined on some probability space. For each of the following statements, determine whether it is true, possibly true, or false:

a) $F_{X_1}(x_1) \le F_X(\infty, x_2, x_3, \ldots, x_n)$, where $x_1, \ldots, x_n$ are specified arguments (real numbers or $\pm\infty$).

b) $F_X(a, a, \ldots, a) < F_X(b, b, \ldots, b)$ if $a < b$.

c) If $X_1 = X_2 = \ldots = X_n$, then $F_{X_1}(a) = F_X(a, a, \ldots, a)$.

d) If $P_X(\overline{0, 1} \times \overline{0, 1} \times \ldots \times \overline{0, 1}) = 0$, then $F_X(0, 0, \ldots, 0) = F_X(1, 1, \ldots, 1)$.

e) If $P_X(\overline{0, 1} \times \overline{0, 1} \times \ldots \times \overline{0, 1}) = 1$, then $F_X(0, 0, \ldots, 0) = 0$.   [239]

51.2 Three r.v.'s X, Y, Z have the following joint c.d.f.:

$$F_{X,Y,Z}(x, y, z) = x^2 + \frac{z}{4}(1 + xy) - \frac{1}{2}$$

for $(x, y, z) \in \overline{0, 0.5} \times \overline{0, 0.5} \times \overline{2, 4}$. Determine:

a) $P_X(\overline{-\infty, 0.25})$ b) $P_Y(\overline{-\infty, 1})$

c) $P_Y(\overline{-\infty, 0.25})$ d) $P_{X,Z}(\overline{-\infty, 0.25} \times \overline{-\infty, 2.5})$.   [176]

Absolutely Continuous n-Dimensional R.V.'s

In R^2, 'Lebesgue measure' represents a generalization of the notion of 'area' (see Section **44**); in R^n ($n > 2$) it is a generalization of n-dimensional volume. An n-dimensional rectangle set $A = A_1 \times A_2 \times \ldots \times A_n \subset R^n$ (where $A_1, A_2, \ldots, A_n \subset R$) has measure $L(A) = L(A_1) \cdot L(A_2) \cdot \ldots \cdot L(A_n)$. For instance, the set of points

$$\overline{0, 1} \times \overline{0, 1} \times \overline{0, 1} \in R^3$$

(the unit cube) has measure 1. On the other hand, the same set has *zero* measure when considered as a subset of R^4 (more precisely, that would be a set such as $\overline{0, 1} \times \overline{0, 1} \times \overline{0, 1} \times \{0\}$). More generally, if $A^{(k)}$ and $A^{(n-k)}$ are arbitrary sets of k and $n - k$ dimensions, respectively, then $L(A^{(k)} \times A^{(n-k)}) = L(A^{(k)}) \cdot L(A^{(n-k)})$. In particular, if one of the sets $A^{(k)}$, $A^{(n-k)}$ has zero measure, then $A^{(k)} \times A^{(n-k)}$ also has zero measure.

With the meaning of Lebesgue measure thus extended to n-dimensional Euclidian space, Definition 38.2 applies also to n-dimensional random variables, and we have:

(Definition) 52.1

An n-dimensional random variable X ($n \geq 1$) is said to be *absolutely continuous* if it induces a probability distribution P_X that does not assign positive probability to any subset of R^n having zero measure. P_X is then likewise called absolutely continuous.

In order for an n-dimensional r.v. $X = (X_1, \ldots, X_n)$ to be absolutely continuous it is not sufficient that the component random variables $X_1, \ldots, X_n$ are absolutely continuous. On the other hand, if X is absolutely continuous, then so are $X_1, \ldots, X_n$ absolutely continuous (one-dimensional) random variables. $X_1, \ldots, X_n$ are then also called *jointly absolutely continuous*. Any r.v. of dimension lower than n obtained by deleting some components of X is then also absolutely continuous. (See Problem 52.4.)

Probability Concepts and Theory for Engineers, First Edition. Harry Schwarzlander.
© 2011 John Wiley & Sons, Ltd. Published 2011 by John Wiley & Sons, Ltd.

(Definition) 52.2

If the n-dimensional r.v. $X = (X_1, \ldots, X_n)$ is absolutely continuous, with n-dimensional distribution function $F_X(x_1, \ldots, x_n)$, then the n-dimensional probability density function $f_X(x_1, \ldots, x_n)$ is defined by

$$f_X(x_1, \ldots, x_n) = \frac{\partial^n}{\partial x_1 \partial x_2 \cdots \partial x_n} F_X(x_1, \ldots, x_n).$$

f_X is a non-negative function that integrates to 1 over R^n:

$$\int_{-\infty}^{\infty} \cdots \int_{-\infty}^{\infty} f_X(x_1, \ldots, x_n) dx_1 \ldots dx_n = \int_{-\infty}^{\infty} \cdots \int_{-\infty}^{\infty} \frac{\partial^n}{\partial x_1 \partial x_2 \cdots \partial x_n} F_X(x_1, \ldots, x_n) dx_1 \ldots dx_n$$

$$= F_X(+\infty, \ldots, +\infty) - F_X(-\infty, \ldots, -\infty)$$

$$= 1.$$

An absolutely continuous n-dimensional probability distribution P_X, induced by a r.v. X, is completely described by the corresponding n-dimensional p.d.f. $f_X(x_1, \ldots, x_n)$. The density function of any one of the component r.v.'s, or the joint density function of any combination of the component r.v.'s, is obtained from $f_X(x_1, \ldots, x_n)$ by 'integrating out' the unwanted dimensions. For instance,

$$\left. \begin{aligned} \int_{-\infty}^{\infty} \cdots \int_{-\infty}^{\infty} f_X(x_1, x_2, \ldots, x_n) dx_2 \ldots dx_n &= \frac{d}{dx_1} F_X(x_1, +\infty, \ldots, +\infty) \\ &= \frac{d}{dx_1} F_{X_1}(x_1) = f_{X_1}(x_1). \end{aligned} \right\} \tag{52.1}$$

Here, dimensions 2 through n are integrated out, leaving the one-dimensional or 'marginal' density function f_{X_1}.

Example 52.1

Consider three r.v.'s X, Y, Z whose joint p.d.f. $f_{X,Y,Z}(x, y, z)$ is constant in that subset of the unit cube for which $y \leq x$ and $z \leq y$, and is zero elsewhere. In other words,

$$f_{X,Y,Z}(x, y, z) = \begin{cases} c, & (x, y, z) \in A \\ 0, & \text{otherwise} \end{cases}$$

where A is the set

$$A = \{(x, y, z) : 0 \leq x \leq 1, 0 \leq y \leq x, \text{ and } 0 \leq z \leq y\}. \tag{52.2}$$

Suppose that $f_Z(z)$ is to be found. Equation (52.1) then becomes

$$f_Z(z) = \int_{-\infty}^{\infty} \int_{-\infty}^{\infty} f_{X,Y,Z}(x, y, z) dx dy.$$

In $f_{X,Y,Z}(x, y, z)$, the statistical dependence between X, Y, and Z shows up in the shape of the region A, which is a pyramid (see Figure 52.1) and not a rectangle set. Care must be taken to set up the limits of integration correctly, so that the integration completely removes all dependence on x and y. We expect to end up with a function of z that is defined over a certain range of z-values, and a different function of z (namely $z = 0$) outside this range. To proceed, the description of A needs

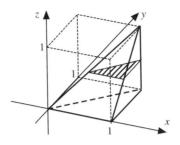

Figure 52.1   The region A defined in Equation (52.2)

to get modified so that the range of z no longer depends on x or y. Inspection of Equation (52.2) reveals that A can be written:

$$A = \{(x,y,z) : y \leq x \leq 1,\ z \leq y \leq 1,\ \text{and } 0 \leq z \leq 1\}.$$

Now it is possible to integrate, for each $z \in \overline{0,\,1}$, over a triangle such as shown shaded in Figure 52.1:

$$f_Z(z) = \int_z^1 \int_y^1 c\, dx\, dy = c \int_z^1 (1-y)\, dy = \frac{c}{2}(1-z)^2,\ 0 \leq z \leq 1.$$

For other values of $z, f_Z(z) = 0$. c can be found by setting $\int_0^1 f_Z(z)\, dz = 1$, giving $c = 6$. This can also be obtained from Equation (52.2) by noting that $L(A) = 1/6$.

In the next Example we have an application of the transformation rule, Rule 51.1, to an absolutely continuous r.v.

Example 52.2: The largest of three r.v.'s.
Given a three-dimensional r.v. $X = (X_1, X_2, X_3)$ which induces a uniform distribution over the unit cube, we wish to find $f_Y(y)$, where $Y = \max(X_1, X_2, X_3)$. In other words, Y assigns to every element ξ in the underlying space the largest of the three numbers $X_1(\xi)$, $X_2(\xi)$, $X_3(\xi)$. As was the case in Example 51.2, the transformation is here from a three-dimensional to a one-dimensional (or scalar) r.v.
We apply Rule 51.1 to first find F_Y, then differentiate to obtain f_Y. For every y, U_y is the set

$$U_y = \{x : x_1, x_2, x_3 \leq y\}.$$

For $y < 0, F_Y(y) = P_X(U_y) = 0$. For $y \in \overline{0,1}$, $P_X(U_y)$ is the probability of the cube shown shaded in Figure 52.2. Then $F_Y(y) = P_X(U_y) = y^3, 0 \leq y \leq 1$. Also, $F_Y(y) = 1$ for $y > 1$. Differentiation gives

$$f_Y(y) = \begin{cases} 3y^2, & 0 \leq y \leq 1 \\ 0, & \text{otherwise.} \end{cases}$$

If an n-dimensional r.v. assigns probability 1 to a zero-measure subset of R^n, it is not necessarily discrete—as was already seen for $n = 2$ in Section 44. Definition 44.4 is now extended to an arbitrary number of dimensions.

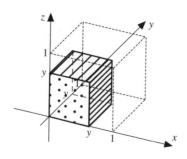

Figure 52.2   The set U_y

(Definition) 52.3
An n-dimensional r.v. ($n \geq 1$) that induces probability 1 on a set of Lebesgue measure zero in $\mathbb{R}^n$ but induces zero probability on every countable set is said to be *singular*. Its induced probability distribution is then also called *singular*.

Scalar r.v.'s that are singular (i.e., the case of $n = 1$ in Definition 52.3) will be considered in Section 56.

Queries

52.1 Consider these subsets of $\mathbb{R}^3$:

$$A = \lceil 0,1 \rceil \times \lceil 0.5,1 \rceil \times \lceil 0,0.5 \rceil, \quad \text{and} \quad B = \lceil 0,0.5 \rceil \times \lceil 0,1 \rceil \times \lceil 0.5,1 \rceil.$$

Determine $L(A)$, $L(B)$, $L(A \cap B)$, $L(A \cup B)$. [31]

52.2 The six-dimensional discrete r.v. $X = (X_1, X_2, \ldots X_6)$ assigns equal probability to the six points $(1, 0, 0, 0, 0, 0)$, $(0, 1, 0, 0, 0, 0)$, $(0, 0, 1, 0, 0, 0)$, ..., $(0, 0, 0, 0, 0, 1)$, and zero probability everywhere else. Determine the following:
a) $P_{X_1}(\{0\})$
b) $P_{X_1,X_2}(\{(1, 1)\})$
c) $P_{X_1,X_6}(\{(0, 1)\})$
d) $F_X(1, 0, 0, 1, 0, 0)$. [143]

52.3 Consider the three-dimensional vector r.v. X of Example 52.2. Determine the following:
a) $F_X(0.5, 0.5, 0.5)$
b) $F_X(0.2, 0.5, 0.7)$
c) $P_X(\{(x_1, x_2, x_3): x_1 = x_2\})$
d) $P_X(\{(x_1, x_2, x_3): x_1 < x_2\})$. [168]

Coordinate Transformations

An important special case arises if a transformation is applied to an absolutely continuous random variable and the transformation has the nature of a *coordinate transformation*. By this is meant a one-to-one transformation that does not change the number of dimension of the random variable, and such that the *Jacobian* of the transformation (see Definition 53.1 below) exists and is nonzero.

Thus, suppose $Y = g(X)$, where $X = (X_1, \ldots, X_n)$ and $Y = (Y_1, \ldots, Y_n)$, $n > 1$, with

$$
\left.
\begin{aligned}
Y_1 &= g_1(X_1, \ldots, X_n) \\
Y_2 &= g_2(X_1, \ldots, X_n) \\
&\cdot \ \cdot \ \cdot \ \cdot \ \cdot \ \cdot \ \cdot \ \cdot \\
Y_n &= g_n(X_1, \ldots, X_n).
\end{aligned}
\right\}
\tag{53.1}
$$

Suppose further that X is absolutely continuous, and that g is a one-to-one function from R^n to R^n with continuous first partial derivatives. Then Y is also an absolutely continuous random variable, and the transformation can be carried out directly in terms of density functions rather than c.d.f.'s.

(Definition) 53.1
Given a transformation $g = (g_1, \ldots, g_n)$, as specified in Equation (53.1). The *Jacobian* of g is the determinant

$$
J_g \equiv \frac{\partial(y_1, \ldots y_n)}{\partial(x_1, \ldots, x_n)} =
\begin{vmatrix}
\dfrac{\partial y_1}{\partial x_1} & \dfrac{\partial y_1}{\partial x_2} & \cdots & \dfrac{\partial y_1}{\partial x_n} \\
\vdots & \vdots & & \vdots \\
\vdots & \vdots & & \vdots \\
\dfrac{\partial y_n}{\partial x_1} & \dfrac{\partial y_n}{\partial x_2} & \cdots & \dfrac{\partial y_n}{\partial x_n}
\end{vmatrix}.
$$

Now, at some point $y = (y_1, \ldots, y_n) \in R^n$, consider a differential n-dimensional volume element (a subset of R^n) containing the point y. This volume element will be denoted Δy. Let $A_{\Delta y}$ be the

Probability Concepts and Theory for Engineers, First Edition. Harry Schwarzlander.
© 2011 John Wiley & Sons, Ltd. Published 2011 by John Wiley & Sons, Ltd.

inverse image of Δy, that is, the set $g^{-1}(\Delta y) \in R^n$ (see Figure 53.1). The ratio of the Lebesgue measures of Δy and $A_{\Delta y}$ (in the limit as all dimensions of Δy go to 0) is expressed by the absolute value of the Jacobian of the transformation, evaluated at y:

$$\lim_{L(\Delta y) \to 0} \frac{L(\Delta y)}{L(A_{\Delta y})} = |J_g(y)|.$$

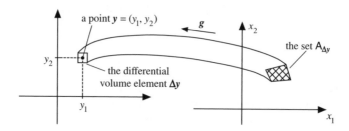

Figure 53.1   Coordinate transformation

The same probability is assigned to both Δy and $A_{\Delta y}$,

$$P_Y(\Delta y) = P_X(A_{\Delta y}). \tag{53.2}$$

Therefore, the absolute value of the Jacobian also expresses the change in probability *density* accompanying the transformation g, in going from any point x to the point $g(x)$. This can also be seen by rewriting Equation (53.2) in the following way:

$$\int_{\Delta y} f_Y(y)dy = \int_{A_{\Delta y}} f_X(x)dx, \quad \text{or}$$

$$(f_Y(y)|_{y \in \Delta y}) \cdot L(\Delta y) \approx (f_X(x)|_{x \in A_{\Delta y}}) \cdot L(A_{\Delta y})$$

so that

$$f_Y(y) \approx \frac{L(A_{\Delta y})}{L(\Delta y)} \cdot f_X(x)$$

which becomes an exact equality as $L(\Delta y) \to 0$. A general expression relating f_Y and f_X, at every point $y = g(x)$ where J_g exists and is nonzero, is therefore

$$\boxed{f_Y(y) = \frac{1}{|J_g|} f_X(x)} \tag{53.3}$$

Example 53.1

The transformation rule (53.3) is now applied to the conversion of a two-dimensional polar coordinate r.v. (R, Θ) to the two-dimensional rectangular coordinate r.v. (X, Y):

$$X = R \cos \Theta, \qquad Y = R \sin \Theta$$

where $R > 0$ and $0 \leq \Theta < 2\pi$. We assume R and Θ to be jointly absolutely continuous. The Jacobian of the transformation is

$$J = \begin{vmatrix} \dfrac{\partial x}{\partial r} & \dfrac{\partial x}{\partial \theta} \\[2mm] \dfrac{\partial y}{\partial r} & \dfrac{\partial y}{\partial \theta} \end{vmatrix} = r$$

so that, using Equation (48.3),

$$f_{X,Y}(x,y) = \frac{1}{r} f_{R,\Theta}(r,\theta) = \frac{1}{\sqrt{x^2 + y^2}} f_{R,\Theta}\left(\sqrt{x^2 + y^2}, \arctan \frac{y}{x} \right). \tag{53.4}$$

It should be noted that this method applies also in the one-dimensional case if the stated conditions hold, and reduces to the relation (47.6):

$$f_Y(y) = \frac{1}{\left| \frac{dy}{dx} \right|} f_X(x)$$

Or, writing both sides as functions of y, we have

$$\boxed{f_Y(y) = \frac{1}{\left| \frac{dy}{dx} \right|} f_X(g^{-1}(y))} \tag{53.5}$$

Applied to the transformation $Y = aX + b$, $a \neq 0$, this yields immediately

$$f_Y(y) = \frac{1}{|a|} f_X\left(\frac{y - b}{a} \right). \tag{53.6}$$

Actually, the conditions under which Equation (53.3) is applicable are not quite as limited as suggested at the beginning of this Section. Consider again a coordinate transformation applied to an n-dimensional absolutely continuous r.v. X. Let $B_0 \subset R^n$ be a set over which $f_X = 0$. Then f_Y is not at all affected by the way in which g is defined over the set B_0. In other words, in B_0 the transformation g need not have continuous first partial derivatives that are nonzero—it need not even be single-valued. Besides (see Problem 52.3), over a set of zero measure, a probability density function can be redefined arbitrarily without affecting the distribution. Thus, in the set B_0 can be included the boundary of a set in R^n over which f_X is positive (as well as any other zero-measure subset of R^n). The Theorem below incorporates these considerations.

Theorem 53.1

Let $X = (X_1, \ldots, X_n)$ be an absolutely continuous r.v. defined on a probability space $(S, \mathscr{A}, P)$, and let $B \in \mathscr{B}^n$ be the subset of R^n over which f_X does not vanish. Let g be a transformation from R^n to R^n satisfying the following conditions everywhere on B, except possibly on a set B_0 that is a zero-measure subset of B:

a) it is single-valued; and
b) its first partial derivatives are continuous, and the Jacobian J_g as given in Definition 53.1 exists and is nonzero.

Then, the r.v. $Y = g(X)$ has the density function f_Y that at points $y = g(x)$ is given by

$$f_Y(y) = \begin{cases} \dfrac{1}{|J_g|} f_X(x), & \text{for all } x \in (B - B_0) \\ 0, & x \notin B. \end{cases}$$

At times, a required transformation of an absolutely continuous random variable that does not fit the requirements of Theorem 53.1 can be adjusted in such a way that the Theorem can be applied after all.

Example 53.2

Suppose an absolutely continuous r.v. $X = (X_1, X_2)$ is given and its two-dimensional density function $f_X(x_1, x_2)$ is known. The p.d.f. of the r.v. $Z = X_1 X_2$ is to be found. As it stands, this is a transformation from two dimensions to one. However, by introducing an extraneous r.v. $W = g_2(X_1, X_2)$, the problem can be changed to a coordinate transformation. The function g_2 can be arbitrarily chosen, as long as the requirements of Theorem 53.1 are satisfied. For instance, let $W = X_1$. Then,

$$J_g = \begin{vmatrix} x_2 & x_1 \\ 1 & 0 \end{vmatrix} = -x_1$$

giving

$$f_{Z,W}(z, w) = \frac{1}{|x_1|} f_{X_1, X_2}(x_1, x_2) = \frac{1}{|w|} f_{X_1, X_2}\left(w, \frac{z}{w}\right).$$

Using Equation (44.2), the w–dimension can now be integrated out to yield

$$f_Z(z) = \int_{-\infty}^{\infty} f_{Z,W}(z, w)\, dw = \int_{-\infty}^{\infty} \frac{1}{|w|} f_{X_1, X_2}\left(w, \frac{z}{w}\right) dw.$$

If X_1, X_2 are s.i., this reduces to the expression that appears in Problem 50.6.

If the transformation function g is not one-to-one, it is still possible to apply Theorem 53.1, provided the domain of g can be partitioned into subsets on each of which g is one-to-one. This is most conveniently illustrated in the one-dimensional case.

Example 53.3

Given an absolutely continuous random variable X, let $Y = X^{2n}$, where n is any positive integer. For every n, the transformation $g(x) = x^{2n}$ consists of two 'branches', namely,

$$y = x^{2n}, \text{ for } x > 0; \text{ and}$$

$$y = x^{2n}, \text{ for } x < 0.$$

Within each branch domain, $g(x)$ is one-to-one. We omit the boundary between the two branch domains, namely, the zero-measure set $\{0\}$, since $J_g = 0$ at this point. Elsewhere, $|J_g| = |dy/dx| = 2nx^{2n-1} = 2ny^{1-1/(2n)}$. The range for both branches of g is the positive y-axis. Summing the contributions of the two branches yields

$$f_Y(y) = \begin{cases} \dfrac{1}{2ny^{1-1/(2n)}} \left[f_X\left(y^{1/(2n)}\right) + f_X\left(-y^{1/(2n)}\right) \right], & y > 0 \\[2ex] 0, & y < 0. \end{cases}$$

This result was previously obtained in Example 47.2 for the case $n = 1$.

The approach in Example 53.3 can also be understood in terms of 'conditional distributions', as will be seen in Section 57.

Queries

53.1 An absolutely continuous r.v. $X = (X_1, X_2, X_3)$ is transformed into the r.v. $Y = (X_1 + X_2, X_2 + X_3, X_3 + X_1)$. What is the Jacobian of this transformation?   [9]

53.2 The joint density function of the three r.v.'s X, Y, Z has value 1 in the unit cube ($0 \leq x, y, z \leq 1$), and is zero elsewhere. If the r.v.'s U, V, W are defined by

$$U = X + Y + Z$$
$$V = -X + 2Y - Z$$
$$W = X + Y - 2Z,$$

what is the joint density of U, V, W evaluated at $(u, v, w) = (1, 0, 0)$?   [74]

53.3 Given an n-dimensional absolutely continuous r.v. X, and an n-dimensional r.v. $Y = g(X)$ such that $f_Y(y) = f_X(y)$ for all y. For each of the following statements, decide whether it is true, possibly true, or false.
a) g is the identity transformation; i.e., $Y = X$
b) $J_g = -1$
c) $X = g(Y)$
d) g has n branches.   [125]

Rotations and the Bivariate Gaussian Distribution

A special kind of coordinate transformation of a two-dimensional r.v. is a *rotation about the origin*. Let g be a transformation that rotates the absolutely continuous r.v. $X = (X_1, X_2)$ counterclockwise about the origin by an angle φ,[3] resulting in the new r.v. $Y = (Y_1, Y_2)$:

$$Y = g(X).$$

Rotation of an arbitrary point (x_1, x_2) is illustrated in Figure 54.1. To express the result of this rotation, polar coordinates can be used, giving:

$$\left. \begin{aligned} y_1 &= r\cos(\theta + \varphi) = r(\cos\theta\,\cos\varphi - \sin\theta\,\sin\varphi) = x_1\cos\varphi - x_2\sin\varphi \\ y_2 &= r\sin(\theta + \varphi) = r(\sin\theta\,\cos\varphi + \cos\theta\,\sin\varphi) = x_1\sin\varphi + x_2\cos\varphi. \end{aligned} \right\} \tag{54.1}$$

For every differential element Δy (see Figure 53.1), the set $A_{\Delta y}$ that maps into Δy has the same size as Δy. Therefore, the Jacobian of the transformation must be $J_g = \pm 1$, or $|J_g| = 1$, so that Equation (53.3) reduces to

$$f_Y(y) = f_X(x) = f_X[g^{-1}(y)]. \tag{54.2}$$

The argument of f_X is obtained in terms of y_1 and y_2 by solving Equations (54.1) for x_1 and x_2:

$$\left. \begin{aligned} x_1 &= y_1\cos\varphi + y_2\sin\varphi \\ x_2 &= -y_1\sin\varphi + y_2\cos\varphi. \end{aligned} \right\} \tag{54.3}$$

Of course, the effect of this transformation is the same as rotating the *coordinate axes* by the same angle in the opposite direction.

---

[3] In other words, for every ξ in the underlying probability space, the point $X(\xi)$ gets rotated about the origin by the angle φ. Rotation of X, of course, also produces a corresponding rotation of P_X. However, rotation of P_X can also arise in other ways than through a rotation of X.

Probability Concepts and Theory for Engineers, First Edition. Harry Schwarzlander.
© 2011 John Wiley & Sons, Ltd. Published 2011 by John Wiley & Sons, Ltd.

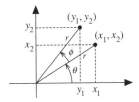

Figure 54.1   Rotation about the origin

Example 54.1: Uniform distribution.

Suppose the r.v. X induces a uniform distribution over the unit square; that is,

$$f_X(x) = \begin{cases} 1, & 0 < x_1 < 1, \ 0 < x_2 < 1 \\ 0, & \text{otherwise.} \end{cases}$$

A counterclockwise rotation about the origin through an angle of $30°$ results in a new r.v. Y whose p.d.f. is found from Equations (54.2) and (54.3) to be:

$$f_Y(y) = 1, \text{ for } 0 < 0.866y_1 + 0.5y_2 < 1, \text{ and } 0 < 0.866y_2 - 0.5y_1 < 1$$

and $f_Y(y) = 0$, otherwise. This density function is sketched in Figure 54.2.

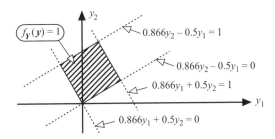

Figure 54.2   Result of a $30°$ rotation applied to the r.v. X

We now apply a rotation about the origin to the joint distribution of two statistically independent Gaussian r.v.'s X_1, X_2. Assuming both f_{X_1} and f_{X_2} to be centered at the origin, then Equation (40.1) with $a = 0$ gives

$$f_{X_1}(x_1) = \frac{1}{\sqrt{2\pi}\sigma_1} \exp\left(-\frac{x_1^2}{2\sigma_1^2}\right), \quad f_{X_2}(x_2) = \frac{1}{\sqrt{2\pi}\sigma_2} \exp\left(-\frac{x_2^2}{2\sigma_2^2}\right)$$

so that the joint p.d.f. is

$$f_X(x) = \frac{1}{2\pi\sigma_1\sigma_2} \exp\left[-\frac{1}{2}\left(\frac{x_1^2}{\sigma_1^2} + \frac{x_2^2}{\sigma_2^2}\right)\right]. \tag{54.4}$$

The density function (54.4) can be represented graphically by means of a set of contours in the (x_1, x_2)-plane. These turn out to be concentric ellipses (see Problem 46.2). A single one of these is adequate to suggest the shape of the density function. It can be the ellipse shown in Figure 54.3a, obtained by setting the negative of the exponent in Equation (54.4) equal to 1:

$$\frac{x_1^2}{2\sigma_1^2} + \frac{x_2^2}{2\sigma_2^2} = 1.$$

We denote this the 'standard ellipse' for the p.d.f. (54.4). It is the locus of points in the (x_1, x_2)-plane where the two-dimensional p.d.f. (54.4) has the value $1/(2\pi e \sigma_1 \sigma_2)$, or $1/e$ times its value at the origin.

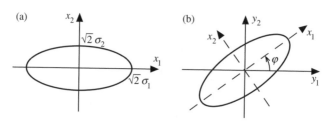

Figure 54.3   Standard ellipse for a two-dimensional Gaussian p.d.f., (a) based on Equation (54.4), and (b) after rotation through an angle φ

Now let X be rotated counterclockwise about the origin through an angle φ. Using Equations (54.2) and (54.3), the two-dimensional p.d.f. of the resulting r.v. Y is found to be

$$f_Y(y_1, y_2) = \frac{1}{2\pi\sigma_1\sigma_2} \exp\left\{ -\frac{1}{2}\left[\left(\frac{\cos^2\varphi}{\sigma_1^2} + \frac{\sin^2\varphi}{\sigma_2^2}\right) y_1^2 \right.\right.$$

$$\left.\left. + 2\cos\varphi \sin\varphi \left(\frac{1}{\sigma_1^2} - \frac{1}{\sigma_2^2}\right) y_1 y_2 + \left(\frac{\cos^2\varphi}{\sigma_2^2} + \frac{\sin^2\varphi}{\sigma_1^2}\right) y_2^2 \right]\right\}. \tag{54.5}$$

To express this more simply, let the coefficients within the square bracket in the exponent in Equation (54.5) be denoted α, 2β, and γ, respectively. Then, Equation (54.5) becomes

$$f_Y(y) = \frac{1}{2\pi\sigma_1\sigma_2} \exp\left[-\frac{1}{2}\left(\alpha y_1^2 + 2\beta y_1 y_2 + \gamma y_2^2\right)\right]. \tag{54.6}$$

The constraints on α, β, and γ are the following. Clearly, $\alpha > 0$ and $\gamma > 0$. Also, since f_Y is a rotated version of f_X (see Figure 54.3b), $\alpha y_1^2 + 2\beta y_1 y_2 + \gamma y_2^2 = 1$ must again be the equation for an ellipse. This requires that $\beta^2 < \alpha\gamma$.

We began with the joint density function (54.4) whose marginal densities are Gaussian. What are the marginal densities of the rotated distribution, f_{Y_1} and f_{Y_2}? Using Equation (44.2),

$$f_{Y_1}(y_1) = \int_{-\infty}^{\infty} f_Y(y_1, y_2)\, dy_2 = \frac{1}{2\pi\sigma_1\sigma_2} \int_{-\infty}^{\infty} \exp\left[-\frac{1}{2}\left(\alpha y_1^2 + 2\beta y_1 y_2 + \gamma y_2^2\right)\right] dy_2.$$

The argument of the exponential, which is a quadratic in y_2, can be converted into a complete square by writing

$$f_{Y_1}(y_1) = \frac{1}{2\pi\sigma_1\sigma_2} \int_{-\infty}^{\infty} \exp\left[-\frac{1}{2}\left(\alpha y_1^2 - \frac{\beta^2}{\gamma}y_1^2 + \frac{\beta^2}{\gamma}y_1^2 + 2\beta y_1 y_2 + \gamma y_2^2 \right) \right] dy_2$$

$$= \frac{1}{2\pi\sigma_1\sigma_2} \exp\left[-\frac{1}{2}\left(\alpha - \frac{\beta^2}{\gamma} \right)y_1^2 \right] \int_{-\infty}^{\infty} \exp\left[-\frac{1}{2}\left(\frac{\beta}{\sqrt{\gamma}}y_1 + \sqrt{\gamma}y_2 \right)^2 \right] dy_2.$$

Now the integrand has the form of Equation (40.1) with x replaced by y_2, σ^2 replaced by $1/\gamma$, and a replaced by $-\beta y_1/\gamma$. Since Equation (40.1) integrates to 1, we see that the above integral evaluates to $\sqrt{2\pi/\gamma}$, giving

$$f_{Y_1}(y_1) = \frac{1}{\sqrt{2\pi\gamma}\,\sigma_1\sigma_2} \exp\left[-\frac{1}{2}\left(\alpha - \frac{\beta^2}{\gamma} \right)y_1^2 \right].$$

Upon substituting for α, β, and γ, this becomes

$$f_{Y_1}(y_1) = \frac{1}{\sqrt{2\pi}\sqrt{\sigma_1^2 \cos^2\varphi + \sigma_2^2 \sin^2\varphi}} \exp\left[-\frac{1}{2}\frac{y_1^2}{\sigma_1^2 \cos^2\varphi + \sigma_2^2 \sin^2\varphi} \right]. \tag{54.7a}$$

Similarly,

$$f_{Y_2}(y_2) = \frac{1}{\sqrt{2\pi}\sqrt{\sigma_2^2 \cos^2\varphi + \sigma_1^2 \sin^2\varphi}} \exp\left[-\frac{1}{2}\frac{y_2^2}{\sigma_2^2 \cos^2\varphi + \sigma_1^2 \sin^2\varphi} \right]. \tag{54.7b}$$

This is an interesting result. It can be seen that Y_1 and Y_2 are again Gaussian random variables. Therefore, a *rotation* of the two-dimensional Gaussian distribution characterized by the density function (54.4) results again in the joint distribution of two Gaussian r.v.'s. This property does not hold for an arbitrary two-dimensional distribution. In Example 54.1, for instance, a two-dimensional *uniform* distribution was rotated by 30°. It is easily seen that the marginal distributions after the rotation are not uniform.

It should be noted, however, that whereas the two Gaussian r.v.'s X_1, X_2 in Equation (54.4) are statistically independent, Y_1 and Y_2 will in general not be statistically independent. This would only be the case if $\beta = 0$; that is, if φ is a multiple of $\pi/2$, or if $\sigma_1 = \sigma_2$. In other words, after the rotation the axes of the standard ellipse would again have to coincide with the coordinate axes, or the standard ellipse would have to be actually a circle.

Since Equation (54.6) includes Equation (54.4) as a special case, we see that the distribution characterized by Equation (54.6) is a more general form of the two-dimensional Gaussian distribution, also called a *bivariate Gaussian distribution*. The bivariate Gaussian distribution with density function as given in Equation (54.5) or Equation (54.6) has its 'center' (the point of maximum density) at the origin. The most general form of the bivariate Gaussian distribution allows for an arbitrary location of the center, i.e., arbitrary shifts in y_1 and y_2. A concise specification is the following.

‖ **(Definition) 54.1**
‖ Let X_1, X_2 be two statistically independent Gaussian r.v.'s with joint density function

$$f_X(x_1, x_2) = \frac{1}{2\pi} \exp\left[-\frac{1}{2}(x_1^2 + x_2^2) \right]. \tag{54.8}$$

If $Y_1 = a_{11}X_1 + a_{12}X_2 + b_1$ and $Y_2 = a_{21}X_1 + a_{22}X_2 + b_2$ where the a's and b's are constants and such that the Jacobian of the transformation is nonvanishing, then Y_1, Y_2 are called *jointly Gaussian r.v.'s*. The probability distribution induced in $\mathbb{R}^2$ by jointly Gaussian r.v.'s is called a *bivariate Gaussian*, or *bivariate normal*, distribution.

Theorem 54.1
The sum of two jointly Gaussian r.v.'s is a Gaussian r.v.

Proof of this important property is left as an exercise (Problem 54.2).

In many problems of practical interest that give rise to two Gaussian r.v.'s, these r.v.'s are also jointly Gaussian, so that their joint distribution is bivariate Gaussian. However, two arbitrary Gaussian r.v.'s are not necessarily jointly Gaussian.

Example 54.2
Consider the r.v.'s X, Y with joint p.d.f.

$$f_{X,Y}(x,y) = \begin{cases} \dfrac{1}{\pi}\exp\left[-\dfrac{1}{2}\left(x^2 + y^2\right)\right], & xy \geq 0 \\ 0, & xy < 0 \end{cases}$$

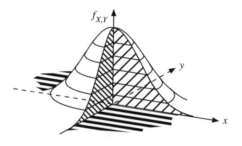

Figure 54.4   Two Gaussian r.v.'s that are not jointly Gaussian

as sketched in Figure 54.4. $f_{X,Y}$ has the form of Equation (54.8) in the first and third quadrant, and is zero in the second and fourth quadrant. X and Y are Gaussian r.v.'s, but they are not jointly Gaussian. Furthermore, it can be inferred from Figure 54.4 that the sum of the Gaussian r.v.'s X and Y is not a Gaussian r.v.

Queries

54.1 Let $Y = (Y_1, Y_2)$ be obtained by applying a rotational transformation to $X = (X_1, X_2)$, where X_1, X_2 are statistically independent Gaussian r.v.'s. For each of the following statements, determine whether it is true, possibly true, or false:

a) If $\sigma_1 = \sigma_2$ in Equation (54.4), then f_Y is the same function as f_X.
b) If Y_1, Y_2 are identically distributed, then the angle of rotation is $\varphi = \pi/4 + n\pi/2$, where $n = 0, 1, 2$ or 3.

c) If $\varphi = \pi/2$, then $Y_1 = -X_2$ and $Y_2 = X_1$.

d) The r.v. $\mathbf{Z} = (Y_1, -Y_2)$ can be obtained by applying a rotational transformation to X.   [126]

54.2 Consider the Gaussian r.v.'s X, Y of Example 54.2, and let $Z = X + Y$, and $W = X - Y$. For each of the following statements, determine whether it is true, possibly true, or false:

a) f_Z has its maximum at the origin.

b) f_W has its maximum at the origin.

c) $Z + W$ is a Gaussian r.v.

d) $Z \geq W$.   [247]

SECTION 55

Several Statistically Independent Random Variables

In Section 24 the concept of independence for several events was examined. There we saw that a number of events could be *pairwise independent* without being *mutually independent*. For several *random variables*, statistical independence may likewise hold on a mere pairwise basis, or the r.v.'s may be mutually statistically independent. That this must be the case follows from the fact that statistical independence of r.v.'s is defined in terms of independence of events, as shown in Section 45.

The ideas of Sections 24 and 45 can be combined to yield the following basic definition for mutually independent r.v.'s.

> **(Definition) 55.1**
> Let $X_1, \ldots, X_n$ be n r.v.'s defined on a probability space $(n > 2)$. $X_1, \ldots, X_n$ are *mutually statistically independent* (m.s.i.) r.v.'s if and only if
>
> $$P_{X_{i_1}, \ldots, X_{i_k}}(\mathsf{B}_1 \times \ldots \times \mathsf{B}_k) = \prod_{j=1}^{k} P_{X_{i_j}}(\mathsf{B}_j)$$
>
> for all choices of distinct subscripts $i_1, \ldots, i_k \in \{1, \ldots, n\}$, all choices of $\mathsf{B}_1, \ldots, \mathsf{B}_k \in \mathscr{B}$, and for each $k \in \{2, \ldots, n\}$.

In other words, in order for $X_1, \ldots, X_n$ to be mutually statistically independent, the k-dimensional joint distribution induced by any k of the r.v.'s must be the product of their marginal distributions, and this must hold for $k = 2, 3, \ldots, n$.

Example 55.1

Consider the three-dimensional discrete r.v. $X = (X_1, X_2, X_3)$ that induces the following probabilities (see Figure 55.1a):

$$P_X(\{(0, 0, 1)\}) = 1/4$$
$$P_X(\{(0, 1, 0)\}) = 1/4$$
$$P_X(\{(1, 0, 0)\}) = 1/4$$
$$P_X(\{(1, 1, 1)\}) = 1/4.$$

Probability Concepts and Theory for Engineers, First Edition. Harry Schwarzlander.
© 2011 John Wiley & Sons, Ltd. Published 2011 by John Wiley & Sons, Ltd.

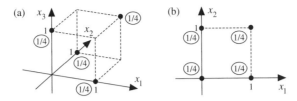

Figure 55.1 Probabilities induced by the three-dimensional r.v. X, and the joint distribution of two component r.v.'s

Here, the component r.v.'s are *pairwise s.i.* but not m.s.i. We have, for $i, j \in \{0, 1\}$ (see Figure 55.1b),

$$P_{X_1,X_2}(\{(i,j)\}) = \frac{1}{4} = P_{X_1}(\{i\})P_{X_2}(\{j\})$$

where $P_{X_1}(\{0\}) = P_{X_1}(\{1\}) = P_{X_2}(\{0\}) = P_{X_2}(\{1\}) = \frac{1}{2}$; likewise for P_{X_1,X_3} and P_{X_2,X_3}. On the other hand,

$$P_X(\{(0,0,1)\}) = \frac{1}{4} \neq P_{X_1}(\{0\})P_{X_2}(\{0\})P_{X_3}(\{1\}).$$

We recall that a unary r.v. is statistically independent of every r.v. defined on the same space S. n such r.v.'s jointly assign probability 1 to a single point in R^n. It follows from Definition 55.1 that they are *mutually* statistically independent (see Problem 55.2).

Theorem 45.2 and the Theorems of Section 46 can all be extended to the n-dimensional case. In place of Theorem 45.2, for instance, we have the following requirements in order for n r.v.'s $X_1, \ldots, X_n$ to be m.s.i.:

$$\left.\begin{array}{l} F_{X_i,X_j} = F_{X_i}F_{X_j}, \text{ for all distinct } i,j \in \{1,2,\ldots,n\} \\[4pt] F_{X_i,X_j,X_k} = F_{X_i}F_{X_j}F_{X_k}, \text{ for all distinct } i,j,k \in \{1,2,\ldots,n\} \\[4pt] \ldots\ldots\ldots\ldots \\[4pt] F_{X_1,\ldots,X_n} = F_{X_1}F_{X_2}\ldots F_{X_n}. \end{array}\right\} \quad (55.1)$$

These equations are analogous to Equation (24.1). However, in the case of random variables it turns out that if the last equation is satisfied then all of the other equations are automatically satisfied as well. For instance, the first of the above equations can be obtained from the last one by writing

$$\begin{aligned} F_{X_i,X_j}(x_i, x_j) &= F_X(\infty, \ldots, \infty, x_i, \infty, \ldots, \infty, x_j, \infty, \ldots, \infty). \\ &= F_{X_1}(\infty)\ldots F_{X_{i-1}}(\infty)F_{X_i}(x_i)F_{X_{i+1}}(\infty)\ldots F_{X_{j-1}}(\infty)F_{X_j}(x_j)F_{X_{j+1}}(\infty)\ldots F_{X_n}(\infty) \\ &= F_{X_i}(x_i)F_{X_j}(x_j). \end{aligned}$$

This demonstrates:

Theorem 55.1
The r.v.'s $X_1, \ldots, X_n$ are m.s.i. if and only if

$$F_{X_1,\ldots,X_n}(x_1, \ldots, x_n) = F_{X_1}(x_1)F_{X_2}(x_2)\ldots F_{X_n}(x_n). \quad (55.2)$$

If $X_1, \ldots, X_n$ are mutually statistically independent, all the equations (55.1) hold, so that the r.v.'s are also pairwise statistically independent. In other words, pairwise statistical independence is a weaker form of statistical independence. On the other hand, if only some but not all pairs of the r.v.s $X_1, \ldots, X_n$ are s.i., then $X_1, \ldots, X_n$ are *not* called pairwise s.i.

In the jointly absolutely continuous case, Equation (55.2) can be differentiated with respect to $x_1, \ldots, x_n$, which results in the following:

Corollary 55.1a

n jointly absolutely continuous r.v.'s $X_1, \ldots, X_n$ are m.s.i. if and only if, with probability one,[4]

$$f_{X_1,\ldots,X_n}(x_1,\ldots,x_n) = f_{X_1}(x_1)\, f_{X_2}(x_2)\ldots f_{X_n}(x_n). \tag{55.3}$$

The manner in which Theorem 55.1 and Corollary 55.1a are stated requires that the n–dimensional distribution as well as the n marginal distributions are known. However, it is possible to go one step further and say: If $F_X(x)$ factors into n functions of one variable, each satisfying the requirements for a c.d.f., then those factors must be the c.d.f.'s $F_{X_1}, F_{X_2}, \ldots, F_{X_n}$, and $X_1, \ldots, X_n$ are mutually s.i. For, suppose $F_X(x)$ factors into n c.d.f.'s: $F_X(x) = F_1(x_1)\, F_2(x_2)\, \ldots\, F_n(x_n)$. Then $F_1(x_1) = F_X(x_1, \infty, \ldots, \infty) = F_{X_1}(x_1)$, and similarly for $X_2, \ldots, X_n$. Thus, we have:

Corollary 55.1b

The r.v.'s $X_1, \ldots, X_n$ are m.s.i. if and only if $F_X(x)$ factors into n one-dimensional c.d.f.'s or, in the absolutely continuous case, if and only if $f_X(x)$ factors into n one-dimensional density functions.

Example 55.2

Let $X = (X_1, X_2, X_3)$, where X_1, X_2, X_3 are identically distributed r.v.'s that are unif. $\overline{0,1}$. Let $f_X(x) = 2$ for $x \in A \uplus B \uplus C \uplus D$, where

$$A = \overline{0,0.5} \times \overline{0,0.5} \times \overline{0,0.5}, \quad B = \overline{0.5,1} \times \overline{0,0.5} \times \overline{0.5,1},$$

$$C = \overline{0,0.5} \times \overline{0.5,1} \times \overline{0.5,1}, \quad \text{and} \quad D = \overline{0.5,1} \times \overline{0.5,1} \times \overline{0,0.5}.$$

This joint distribution is illustrated in Figure 55.2. X_1, X_2, X_3 are pairwise s.i. r.v.'s but are not mutually s.i.

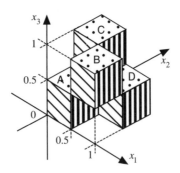

Figure 55.2   Region in $\mathbb{R}^3$ where the r.v. X induces positive probability density

---

[4] The latter specification allows for the fact that p.d.f.'s can be arbitrarily redefined on a set of zero-measure.

Next, we consider the effect of *transformations* on the property of statistical independence. First we note that two multidimensional r.v.'s $X = (X_1, \ldots, X_n)$ and $Y = (Y_1, \ldots, Y_m)$ are s.i. if each of the pairs of r.v.'s X_i, Y_j ($i = 1, \ldots, n; j = 1, \ldots, m$) are s.i., so that

$$F_{X,Y}(x_1, \ldots, x_n, y_1, \ldots, y_m) = F_X(x_1, \ldots, x_n) F_Y(y_1, \ldots, y_m).$$

Similarly, X, Y, and $Z = (Z_1, \ldots, Z_l)$ are m.s.i. if $F_{X,Y,Z} = F_X F_Y F_Z$. This extends in the manner of Theorem 55.1 to more than three multidimensional r.v.'s.

Now we have:

Theorem 55.2
If the two r.v.'s X, Y are s.i., and g_1, g_2 denote any two transformations (Borel functions), then $g_1(X)$ and $g_2(Y)$ are also s.i. r.v.'s.

Proof
It is sufficient to prove that the r.v.'s $Z = g_1(X)$ and Y are s.i. Assume X to be l-dimensional, Y to be m-dimensional, and Z to be n-dimensional. Now consider any $B_1 \in \mathscr{B}^n$, $B_2 \in \mathscr{B}^m$, and let $B_3 = \{x: g_1(x) \in B_1\}$. Then

$$P_{Z,Y}(B_1 \times B_2) = P_{X,Y}(B_3 \times B_2) = P_X(B_3) P_Y(B_2) = P_Z(B_1) P_Y(B_2).$$

The converse of this Theorem does not hold. It is possible for $g_1(X)$ and $g_2(Y)$ to be s.i. even if X and Y are not s.i. This is illustrated with two scalar r.v.'s in the following Example.

Example 55.3
Let X and Y have a joint p.d.f. that equals 1 in the region shown shaded in Figure 55.3, and is zero elsewhere. By Theorem 46.3, X and Y are not statistically independent. However, the r.v.'s $|X|$ and Y have a joint density function that is 1 in the unit square, zero elsewhere, and therefore they are s.i.

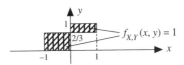

Figure 55.3 Joint p.d.f. of X, Y

Theorem 55.2 is easily extended to multiple r.v.'s, which for the case of n scalar r.v.'s results in the following:

Theorem 55.3
If the r.v.'s $X_1, \ldots, X_n$ are m.s.i., and $g_1, \ldots, g_n$ are n one-dimensional transformations, then $g_1(X_1), \ldots, g_n(X_n)$ are also m.s.i.

Queries

55.1 Each of the ten r.v.'s $X_1, \ldots, X_{10}$ induces a uniform distribution over $\overline{-1, 1}$. What is the probability of the event that all ten r.v.'s assign a positive value, if:

a) $X_1 = X_2 = \ldots = X_{10}$

b) The r.v.'s $X_1, \ldots, X_{10}$ are m.s.i.

c) The r.v.'s $X_1, \ldots, X_9$ are m.s.i., and $X_{10} = -X_9$.   [3]

55.2 In each of the following cases, determine whether it is true, possibly true, or false that the r.v.'s X, Y, Z are m.s.i.

a) X, Y, Z induce probability 1 uniformly over the region $\overline{-1, 0} \times \overline{5, 10} \times \overline{-2, 2}$.

b) X, Y, Z induce probability 1 uniformly over the interior of a sphere of radius 2 centered at the origin.

c) $P_{X,Y,Z}(\{(0, 0, 0)\}) = P_X(\{0\})\, P_Y(\{0\})\, P_Z(\{0\})$.

d) Each of the r.v.'s X, Y, Z is statistically independent of a fourth r.v. W.

e) $X + Y + Z = 0$.   [237]

Singular Distributions in One Dimension

One-dimensional singular r.v.'s are rather unusual and occur only in very special cases. Nevertheless, such cases can arise in conceptual models for some quite straightforward engineering problems.

Example 56.1: Random square-wave applied to a low-pass filter.

A first-order low-pass filter (for time-continuous voltage waveforms) is a device for which the input and output voltages, $v_i(t)$ and $v_o(t)$, are functions of time t that are related by the differential equation

$$\tau \frac{dv_o(t)}{dt} + v_o(t) = v_i(t).$$

Thus, the response to a step input of size a is an exponential that approaches the value a gradually, as indicated in Figure 56.1. Furthermore, if the input $v_i(t)$ is a sequence of step inputs occurring at different time instants, then the output $v_o(t)$ is the superposition of the corresponding decaying exponentials.

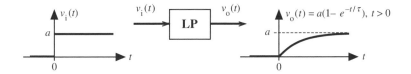

Figure 56.1   Response of a first-order low-pass filter to a step input

With the time scale divided into consecutive 1-second intervals, we consider a voltage waveform that takes on a constant value of $+1$ V or -1 V in each consecutive interval according to whether 'head' or 'tail' is obtained in consecutive tosses of a coin. We call this a *random square-wave*. Suppose such a random square-wave is applied to the first-order low-pass filter. Representative portions of this input waveform and the corresponding output waveform are shown in Figure 56.2.

Probability Concepts and Theory for Engineers, First Edition. Harry Schwarzlander.
© 2011 John Wiley & Sons, Ltd. Published 2011 by John Wiley & Sons, Ltd.

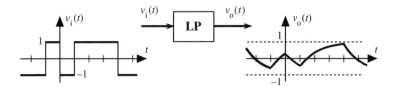

Figure 56.2   A random square-wave applied to the first-order low-pass filter

During periods of constant input voltage, the output approaches the input voltage value exponentially with time constant τ that, for simplicity, we take to be 1.

Of interest is the magnitude of the output waveform at some time instant t_0 that coincides with one of the 1-second marks. Therefore, we consider a p.e. in which a random square wave is applied to the input of a first-order low-pass filter with $\tau = 1$, and the output is observed at time $t = t_0$. Independent repetitions of this p.e., in principle, cannot be carried out in succession, since the entire past history of the input influences the output. Instead, 'repetitions' can be thought of as multiple simultaneous, but independent, implementations of the same p.e.

On the probability space characterizing this p.e., let a r.v. V_o be defined that expresses the output voltage at time t_0. What is the probability distribution induced by V_o? To begin to answer this question, we determine the *range* of the r.v. V_o.

The lowest possible output value is -1, which results from an input $v_i(t) = -1$ for all $t < t_0$. Similarly, the highest possible output value is $+1$, which results from an input $v_i(t) = +1$ for all $t < t_0$. These two extreme cases are illustrated in Figure 56.3, with the interval $\overline{-1, 1}$ the apparent range of V_o.

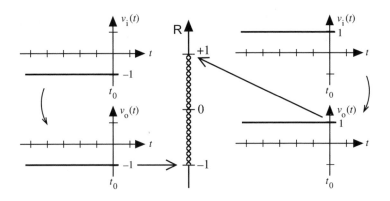

Figure 56.3   Extreme values for the filter output at time t_0

This is not the whole story, however. Let the 1-second intervals prior to t_0 be denoted as follows:

$$N_1 = \overline{t_0 - 1, t_0}; \quad N_2 = \overline{t_0 - 2, t_0 - 1}; \quad N_3 = \overline{t_0 - 3, t_0 - 2}; \quad \text{etc.}$$

Now suppose that $v_i(t) = -1$ for $t \in N_1$. Then the lowest possible value of V_o is -1, as noted above. This arises if $v_i(t) = -1$ also for all $t < t_0 - 1$. But what is the *highest* possible value V_o if $v_i(t) = -1$ for $t \in N_1$? This occurs if $v_i(t) = +1$ for all $t < t_0 - 1$, as illustrated on the left side of

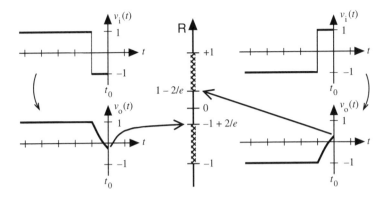

Figure 56.4   Range of filter output values at time t_0 if the input has been constant for $t < t_0 - 1$

Figure 56.4, and results in $V_0 = -1 + 2/e$. Similarly, if $v_i(t) = +1$ for $t \in N_1$, then the highest possible output value at time t_0 is $+1$, as pointed out before. This arises if $v_i(t) = +1$ also for all $t < t_0 - 1$. On the other hand, the *lowest* possible value of V_0, given that $v_i(t) = +1$ for $t \in N_1$, is $V_0 = 1 - 2/e$.

So we see that the interval

$$\overline{-1 + \frac{2}{e}, \ 1 - \frac{2}{e}}$$

is *not* included in the range of V_0 (see Figure 56.4).

Next, suppose $v_i(t) = +1$ for $t \in N_1 \cup N_2$. The lowest possible value of V_0 is then $1 - 2/e^2$, and arises when $v_i(t) = -1$ for $t < t_0 - 2$. On the other hand, if

$$v_i(t) = \begin{cases} +1, & \text{for} \quad t \in N_1 \\ -1, & \text{for} \quad t \in N_2 \end{cases}$$

then the highest possible output is $1 - 2/e - 2/e^2$. Thus, the interval

$$\overline{1 - \frac{2}{e} + \frac{2}{e^2}, \ 1 - \frac{2}{e^2}}$$

is not included in the range of V_0 (see Figure 56.5). An analogous argument shows that the interval

$$\overline{-1 + \frac{2}{e^2}, \ -1 + \frac{2}{e} - \frac{2}{e^2}}$$

is also not part of the range of V_0.

Proceeding in this fashion, we find that more and more subsets of $\overline{-1, 1}$ have to be deleted because they do not contain points that are part of the range of V_0. Now, the interval $\overline{-1, 1}$ has Lebesgue measure $L(\overline{-1, 1}) = 2$. For the first interval that was removed we find

$$L\left(\overline{-1 + \frac{2}{e}, \ 1 - \frac{2}{e}}\right) = 2 - \frac{4}{e}.$$

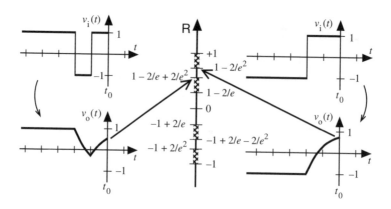

Figure 56.5   Range of filter output at t_o if the input is assumed constant for $t < t_o - 2$

Next, two intervals were removed whose combined measure is

$$L\left(\overline{-1 + \frac{2}{e^2}, -1 + \frac{2}{e} - \frac{2}{e^2}} \uplus \overline{1 - \frac{2}{e} - \frac{2}{e^2}, 1 - \frac{2}{e^2}}\right) = \frac{4}{e} - \frac{8}{e^2}.$$

By taking into account only the constraint that the input has *constant* value $+1$ or -1 during both N_1 and N_2, we have reduced the range of V_o to a set of measure $2 - (2 - 4/e) - (4/e - 8/e^2) = 8/e^2$. When taking into account the constraints on the input waveform during the intervals N_1, $N_2, N_3, \ldots, N_i$, then the range of V_o is found to have measure $2(2/e)^i$. In the limit as $i \to \infty$, the true range of V_o is obtained and can be seen to have *measure zero*. However, V_o is not a discrete r.v. From the preceding discussion it should be clear that no two distinct input waveforms can produce the same value of $v_o(t_0)$. The collection of all random square-waves (running from $t = -\infty$ to $t = t_0$) can be put into one-to-one correspondence with the set of all infinite-duration coin-toss sequences and therefore is uncountable, so that the range of V_o is uncountable. Therefore, V_o is a *singular* r.v.

The c.d.f. of a one-dimensional singular distribution cannot be described in a simple manner, and its graph can be drawn only in approximation. Loosely speaking we may say that such a c.d.f. consists of an uncountable set of steps, or jumps, all of which have zero height.[5] Rather than trying to describe this c.d.f. by specifying its steps, it is more appropriate to specify some of the intervals over which it is constant.

Consider the r.v. V_o of Example 56.1. Its c.d.f. $F_{V_o}(v_o)$ must be 0.5 over the interval $\overline{-1 + \frac{2}{e}, 1 - \frac{2}{e}}$. This follows because:

a) V_o cannot assign values in that interval, therefore F_{V_o} does not increase over this interval; and
b) $V_o \le -1 + 2/e$ if and only if $v_i(t) = -1$ for $t \in N_1$, which is an event of probability 1/2.

In a similar fashion we find, for instance,

$$F_{V_o}(v_o) = \frac{1}{4}, \quad v_o \in \overline{-1 + \frac{2}{e^2}, -1 + \frac{2}{e} - \frac{2}{e^2}};$$

$$F_{V_o}(v_o) = \frac{3}{4}, \quad v_o \in \overline{1 - \frac{2}{e} + \frac{2}{e^2}, 1 - \frac{2}{e^2}}.$$

---

[5] The set of points that constitute the range of V_o is called a 'Cantor set'. Cantor sets are usually considered to be a mere mathematical curiosity. We see in Example 56.1 how such a set can arise in a straightforward, though idealized, model of a simple engineering situation.

An approximate sketch of F_{V_o} appears in Figure 56.6. Only *some* of the constant segments have been drawn in, which might give the impression of discontinuity. Actually, this c.d.f. is continuous.[6]

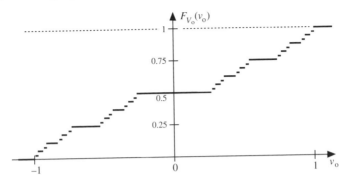

Figure 56.6   C.d.f. of a singular distribution

We have noted earlier that an absolutely continuous r.v. cannot arise directly out of real-world experience. The same is true for a singular r.v. Several idealizations entered into the conceptual model of Example 56.1. One idealization is postulating an infinite past, i.e., a random square wave that has persisted forever. Without this assumption V_o would be a discrete r.v. Another idealization is the assumption of a mathematically perfect input waveform, with transitions from positive to negative and vice versa occurring instantly and exactly at the specified instants, without timing jitter. A further assumption is the absence of noise—at the input as well as at the output of the filter. And finally, there is the assumption that the filter output can be observed at precisely the instant t_o, and with infinite precision. Thus, the singular r.v. in Example 56.1 is an idealized limiting case that in practice can only be approximated. It is useful, however, when the practical limits are not known or difficult to incorporate into the model.

Queries

56.1 Consider the r.v. V_o of Example 56.1. For each of the following, determine whether it is true, possibly true, false, or meaningless.

a) V_o^2 is a singular r.v.

b) $V_o + X$ is a singular r.v., where X is statistically independent of V_o.

c) Let V_1 be the r.v., which characterizes the output waveform at time $t_1 = t_0 + 1$. Then V_o and V_1 are s.i. r.v.'s.

d) Let $g(x) = \begin{cases} 1/4, & x > 1/4 \\ x, & -1/4 \le x \le 1/4 \\ -1/4, & x < -1/4 \end{cases}$. Then $g(V_o)$ is a discrete r.v.

e) A r.v. that distributes probability 1 equally over all rational numbers in $\overline{0,1}$ is singular.

                                                                              [97]

---

[6] This follows from the fact that the set of center points of the nonconstant segments of F_{V_o} can be put in one-to-one correspondence with the set of all infinite-duration coin-toss sequences. Thus it is an uncountable set. However, the set of discontinuities of a monotone function must be countable. This is because every discontinuity, being an interval in the vertical direction or ordinate, must contain at least one rational point; and the set of rational numbers is countable.

Conditional Induced Distribution, Given an Event

As we have seen, after a probability space $(S, \mathcal{A}, P)$ has been defined for a given problem, it is sometimes desirable to be able to introduce a revised probability distribution that assigns *certainty* to some subset of S. Conditional probability makes it possible to do this. Similarly, when a random variable is defined on the probability space, it can be useful to be able to express the effect of such a condition on the random variable. This is accomplished by a straightforward application of the concept of conditional probability to the c.d.f. of the random variable.

Let a r.v. X be defined on a probability space $(S, \mathcal{A}, P)$, and let an event $A \in \mathcal{A}$ have nonzero probability. Suppose the underlying experiment is to be limited by the condition that only those possible outcomes belonging to A are permitted to occur. Rather than revising the probability space, we know that we can accommodate this condition by introducing an additional probability distribution on the space, namely, the conditional probability distribution $P(\cdot|A)$. Similarly, rather than define a new random variable, it may be simpler to retain X but let it induce probabilities based on the condition 'A must occur'. In other words, X now maps the probabilities $P(\cdot|A)$ into the induced space, and thus induces the *conditional probability distribution given the event* A.

(Definition) 57.1[7]
Given a r.v. X on $(S, \mathcal{A}, P)$, and a set $A \in \mathcal{A}$ with $P(A) \neq 0$. The *conditional distribution induced by X, given the event* A, $P_X(\cdot|A)$, assigns to each $B \in \mathcal{B}$ the probability

$$P_X(B|A) = P(\{\xi : X(\xi) \in B\}|A) = \frac{P(\{\xi : X(\xi) \in B\} \cap A)}{P(A)}.$$

The conditional distribution $P_X(\cdot|A)$ is completely specified by the *conditional c.d.f. associated with the r.v. X, given* A, denoted $F_{X|A}$, which expresses for each $x \in R$,

$$F_{X|A}(x) = P(\{\xi : X(\xi) \leq x\}|A) = P_X(\overline{-\infty, x}|\{X(\xi): \xi \in A\}). \tag{57.1}$$

---

[7] The notation $P_X(\cdot|A)$ is somewhat inconsistent, since the conditioning event A does not belong to the induced space $(R, \mathcal{B}, P_X)$ on which P_X is defined, but to the basic space $(S, \mathcal{A}, P)$. However, since X is a r.v., there exists a unique Borel set A' such that $\{\xi: X(\xi) \in A'\} = A$, so that no ambiguity can arise. The same notation is used if, instead, a subset of R is specified as the conditioning event.

Probability Concepts and Theory for Engineers, First Edition. Harry Schwarzlander.
© 2011 John Wiley & Sons, Ltd. Published 2011 by John Wiley & Sons, Ltd.

Although the conditional distribution $P_X(\cdot|A)$ is not defined if $P(A)=0$, it will be seen in Section 59 that a different kind of conditional distribution can be defined which can accommodate zero-probability events as conditions. Clearly, a conditional distribution $P_X(\cdot|A)$ has all the properties of an ordinary, or unconditional, induced probability distribution. If X is discrete, then $P_X(\cdot|A)$ will be a discrete distribution.

(Definition) 57.2
Given a r.v. X on $(S, \mathscr{A}, P)$, and a set $A \in \mathscr{A}$ with $P(A) \neq 0$. If $F_{X|A}(x)$ can be differentiated, then

$$f_{X|A}(x) = \frac{d}{dx} F_{X|A}(x)$$

is the *conditional p.d.f.* associated with X, given the event A.

Example 57.1
Let X be an exponential r.v. with p.d.f.

$$f_X(x) = \begin{cases} \alpha e^{-\alpha x}, & x \geq 0 \\ 0, & x < 0. \end{cases}$$

In the underlying space $(S, \mathscr{A}, P)$, for some real number $b > 0$, let $A_b \in \mathscr{A}$ be the event $\{\xi: X(\xi) > b\}$. Then

$$P(A_b) = P_X(\overline{b, \infty}) = \int_b^\infty \alpha e^{-\alpha x} dx = e^{-\alpha b}.$$

From Definition 57.1,

$$F_{X|A_b}(x) = \frac{P_X(\overline{0, x} \cap \overline{b, \infty})}{P_X(\overline{b, \infty})} = \begin{cases} 1 - e^{-\alpha(x-b)}, & x \geq b \\ 0, & x < b. \end{cases}$$

Then the conditional p.d.f. is

$$f_{X|A_b}(x) = \begin{cases} \alpha e^{-\alpha(x-b)}, & x \geq b \\ 0, & x < b \end{cases}$$

which is sketched in Figure 57.1. Thus, $f_{X|A_b}(x) = f_X(x-b)$, which is a special property of the exponential distribution. (See also Problem 39.4, part (c).)

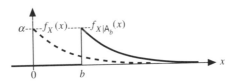

Figure 57.1 Conditional p.d.f. associated with X, given the event A_b

In the absolutely continuous case, the conditional p.d.f. can also be expressed directly in terms of the unconditional p.d.f. To show this it is helpful to let A' denote the set $\{X(\xi) : \xi \in A\}$ in Equation (57.1); in other words, $A \in \mathcal{A}$ gets mapped by X into the set $A' \in \mathcal{B}$. Then Equation (57.1) can be written

$$F_{X|A}(x) = \frac{P_X\left(\overline{-\infty, x]} \cap A'\right)}{P_X(A')}.$$

Since $P_X\left(\overline{-\infty, x]} \cap A'\right)$, considered as a function of x, is constant for those x that lie in an interval contained in A'^c, we have

$$\frac{d}{dx} P_X\left(\overline{-\infty, x]} \cap A'\right) = f_X(x) \cdot I_{A'}(x)$$

where $I_{A'}(x)$ is the *indicator function* of the set A', $I_{A'}(x) = \left\{ \begin{matrix} 1, & x \in A' \\ 0, & x \notin A' \end{matrix} \right\}$. Accordingly,

$$f_{X|A}(x) = \frac{f_X(x) I_{A'}(x)}{P_X(A')}. \tag{57.2}$$

Thus, $f_{X|A}(x)$ is proportional to $f_X(x)$ on the set A', and zero elsewhere.

Suppose a r.v. X is defined on a probability space $(S, \mathcal{A}, P)$, and some event $A \in \mathcal{A}$ gets specified for which X induces the conditional distribution $P_X(\cdot|A)$. It is then possible to define a new random variable Y on $(S, \mathcal{A}, P)$ in such a way that it induces precisely the probability distribution $P_X(\cdot|A)$ on the subsets of R; that is, for every Borel set B, $P_Y(B) = P_X(B|A)$, or $F_Y(x) = F_{X|A}(x)$ for every $x \in$ R. We may then regard Y as a 'conditional random variable', namely 'the r.v. X under the condition A'.

Conditional distributions often arise quite naturally in problems that involve r.v.'s. For instance, a p.e. may be such that certain conditional distributions are initially specified, or easily inferred, and the *unconditional* distribution of the r.v. of interest needs to be found.

Example 57.2

A manufacturer produces an equipment whose operating life depends on a critical component whose time till failure is characterized by an exponential r.v. X_1 with p.d.f.

$$f_{X_1}(x) = \begin{cases} \alpha e^{-\alpha x}, & x \geq 0 \\ 0, & x < 0. \end{cases}$$

It turns out that the manufacturer's stock of this critical component has mixed in an inferior quality version of the same component, whose time till failure is also characterized by an exponential r.v. X_2, but with density function

$$f_{X_2}(x) = \begin{cases} 2\alpha e^{-2\alpha x}, & x \geq 0 \\ 0, & x < 0. \end{cases}$$

If 20% of the supply of this critical component consists of the inferior kind, what is the r.v. that characterizes the operating life of an equipment coming off the assembly line?

An idealized p.e. can be postulated where an equipment is operated till failure and the operating time till failure is measured with infinite precision. On the probability space that has as its elements all possible operating times t till failure of an equipment (without regard as to whether its critical component is of good or inferior quality), i.e., $0 \leq t < \infty$, we can define a r.v. Y that expresses the time to failure of the equipment. But on that space we cannot specify X_1 and X_2, and therefore are unable to utilize the information that is given. Instead, a refined probability space is required whose elements are pairs (t, q), expressing the time till failure t, and also whether the quality q of the critical component is good or inferior. In principle, the underlying experiment therefore requires identification of the quality of the critical component as well as the life test of the equipment being observed.

On this refined probability space, the r.v.'s X_1, X_2 and Y can be defined. f_{X_1} and f_{X_2} are conditional density functions, which we now write $f_{Y|G}$ and $f_{Y|H}$, the conditioning events being $G \leftrightarrow$ 'good' and $H \leftrightarrow$ 'inferior', with $P(G) = 0.8$ and $P(H) = 0.2$. The rule of total probability now yields

$$f_Y(x) = f_{Y|G}(x)P(G) + f_{Y|H}(x)P(H) \tag{57.3}$$

$$= \begin{cases} 0.8\alpha e^{-\alpha x} + 0.4\alpha e^{-2\alpha x}, & x \geq 0 \\ 0, & x < 0. \end{cases}$$

We see that Y is not an exponential r.v. When a p.d.f. consists of a weighted sum of exponentials with differing coefficients in the exponent, the r.v. and the induced distribution are called 'hyperexponential'.

(Definition) 57.3
Let n distinct r.v.'s $X_1, \ldots, X_n$ ($n \geq 2$) be defined on a probability space $(S, \mathscr{A}, P)$. Another r.v., Y, defined on the same space, induces the distribution P_Y satisfying $P_Y = a_1 P_{X_1} + a_2 P_{X_2} + \cdots + a_n P_{X_n}$, where $a_1, \ldots, a_n$ are positive constants summing to 1. Then P_Y is referred to as a *mixture* of the distributions $P_{X_1}, \ldots, P_{X_n}$.

In Equation (57.3), for example, the distribution induced by Y is expressed as a mixture of two conditional distributions.

In the above discussion and Definitions 57.1 and 57.2, a one-dimensional r.v. X was indicated. Conditional distributions are used in the same way with multidimensional random variables. For $X = (X_1, \ldots, X_n)$, we have the conditional c.d.f. given the event A, $F_{X|A}(x_1, \ldots, x_n)$. If $P_{X|A}$ is absolutely continuous, then the derivative of $F_{X|A}(x_1, \ldots, x_n)$ with respect to all n variables is the conditional p.d.f. given A, $f_{X|A}(x_1, \ldots, x_n)$. Since these functions are of the same nature as ordinary n-dimensional c.d.f.'s and density functions, respectively, the same computational rules apply. For instance,

$$\int_{-\infty}^{\infty} f_{X|A}(x_1, \ldots, x_n) dx_n = f_{X'|A}(x_1, \ldots, x_{n-1})$$

where $X' = (X_1, \ldots, X_{n-1})$.

So far, the conditioning event has been taken to be a subset of the underlying probability space on which the random variable in question is defined. The conditioning event can just as easily arise as a subset of the induced space, i.e., as a Borel set. It is customary to use the same notation in both cases.

Example 57.3
Consider two statistically independent r.v.'s X, Y both of which induce the same exponential distribution with density function $f_X(x) = f_Y(x) = e^{-x}$, $x \geq 0$. We wish to find the density function for X under the condition that $Y < X$.

Here, the conditioning event A is specified in terms of a relationship between random variables. In the (x,y)-plane it is the half-plane below the line $y = x$ (see Figure 57.2). However, since all probability is induced in the first quadrant, the event A can be regarded as confined to the first quadrant also, as shown in the figure.

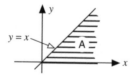

Figure 57.2   The conditioning event A

The joint density function in the first quadrant is $f_{X,Y}(x, y) = e^{-(x+y)}$, and it is zero elsewhere. The conditional c.d.f. is

$$F_{X|A}(x) = P_X(\overline{-\infty, x}|A) = P_{X,Y}(\overline{-\infty, x} \times R|A)$$

$$= \frac{P_{X,Y}(\overline{-\infty, x} \times R \cap A)}{P_{X,Y}(A)}.$$

From the symmetry of $f_{X,Y}$ it follows that the denominator must equal 1/2. The numerator yields

$$\int_0^x \int_0^x e^{-(x+y)} dy dx = \int_0^x e^{-x}(1 - e^{-x}) dx = \frac{1}{2}(1 - e^{-x})^2, \ x \geq 0.$$

Differentiation of $F_{X|A}(x)$ gives the conditional p.d.f.:

$$f_{X|A}(x) = \frac{d}{dx} \frac{\frac{1}{2}(1 - e^{-x})^2}{\frac{1}{2}} = 2e^{-x}(1 - e^{-x}), \ x \geq 0.$$

The conditioning event can also be the specification that a particular value is taken on by a discrete r.v. Thus, if X_1, X_2 are discrete r.v.'s defined on a probability space $(S, \mathscr{A}, P)$, then

$$P_{X_1}(\{x_1\}|\{\xi : X_2(\xi) = x_2\}) = \frac{P_{X_1,X_2}(\{(x_1, x_2)\})}{P_{X_2}(\{x_2\})} \tag{57.4}$$

where x_2 is a point in the range of X_2. An abbreviated notation is convenient for the left side of Equation (57.4), namely, $P_{X_1}(\{x_1\}|X_2 = x_2)$. This notation is now used to state the following *chain rule* for expressing joint probabilities of an n-dimensional discrete r.v. $X = (X_1, \ldots, X_n)$. This chain rule is a natural extension of Equation (57.4):

$$P_X(\{x_1, \ldots, x_n\}) = P_{X_1}(\{x_1\})P_{X_2}(\{x_2\}|X_1 = x_1) \ldots P_{X_n}(\{x_n\}|X_1 = x_1, \ldots, X_{n-1} = x_{n-1}) \tag{57.5}$$

where for $i = 1, \ldots, n$, x_i denotes a point in the range of X_i, such that $P_{X_i}(\{x_i\}) > 0$.

Through the use of conditional distributions, coordinate transformations involving several 'branches' can be expressed more concisely. This is best illustrated by referring to Example 53.3. There, the transformation $Y = X^{2n}$ involves two branches, namely $x > 0$ and $x < 0$. In each of these regions, the transformation is one-to-one. Then let $A = \{x: x > 0\}$, and $B = \{x: x < 0\}$, so that $f_Y(y) = f_{Y|A}(y)P(A) + f_{Y|B}(y)P(B)$, which leads to the same result as obtained before.

Queries

57.1 The r.v. X induces a uniform distribution over $\overline{-1,0}$. Evaluate the conditional p.d.f. $f_{X|A}(x)$ at $x = -0.1$, where A is the event $\{\xi: X(\xi) > -0.2\}$.   [217]

57.2 The following information is given about the r.v. X, where A is an event with $P(A) = 0.4$:

$$f_{X|A}(x) = \begin{cases} \text{constant,} & 0 < x < 2 \\ \text{zero,} & \text{elsewhere} \end{cases}$$

$$f_{X|A^c}(x) = \begin{cases} \text{constant,} & -1 < x < 1 \\ \text{zero,} & \text{elsewhere.} \end{cases}$$

Find $f_X(0.5)$; i.e., the *unconditional* probability density at $x = 0.5$.   [86]

57.3 A two-dimensional r.v. $X = (X_1, X_2)$ distributes all probability uniformly over the circumference of the unit circle (Figure 57.3). Thus, X is a singular r.v.

Figure 57.3   Unit circle

a) What is the value of $F_X(0, 0)$?

b) Let $A = \{\xi: X_1(\xi) \leq 0\}$. Evaluate $F_{X|A}(0, 0)$.

c) Let $B = \{\xi: X_1(\xi) > X_2(\xi)\}$. Evaluate $F_{X|B}(0, 0)$.   [142]

Resolving a Distribution into Components of Pure Type

We have become acquainted with three distinct 'pure' types of r.v.'s and induced distributions, namely discrete, absolutely continuous, and singular types. Beside these pure types, there are those r.v.'s that exhibit the characteristics of two (or all three) of the pure types in different portions of their range. We referred to such r.v.'s as being of 'mixed type'. Through the use of conditional distributions, the induced probability distribution of a r.v. of mixed type can be resolved into components that are of pure type.

> **Theorem 58.1**
> Given an arbitrary n-dimensional r.v. X, where $n \geq 1$. The c.d.f. of X can be written in the form
>
> $$F_X(x) = a\, F_{X_a}(x) + d\, F_{X_d}(x) + s\, F_{X_s}(x) \qquad (58.1)$$
>
> where $F_{X_a}(x), F_{X_d}(x), F_{X_s}(x)$ are the c.d.f.'s of an absolutely continuous, a discrete, and a singular r.v., respectively, and a, d, s are non-negative constants satisfying $a + d + s = 1$.

For any given r.v. X, the expression (58.1) is arrived at as follows. First let B_d denote that *countable* subset of R^n that satisfies:

a) B_d^c has no countable subset on which X induces positive probability; and
b) B_d has no nonempty subset on which X induces zero probability.

Therefore, the set of all points on which X induces positive probability must be B_d. If X does not induce positive probability on any countable set, then $\mathsf{B}_d = \phi$.

Next, let B_s denote a *zero-measure* subset of B_d^c that satisfies:

a) X induces positive probability on B_s; and
b) $\mathsf{B}_d^c - \mathsf{B}_s$ has no zero-measure subset on which X induces positive probability.

Therefore, if X induces positive probability on an uncountable set of measure zero but zero probability on its countable subsets, then this uncountable set is B_s or a subset of B_s. B_s is not

---

Probability Concepts and Theory for Engineers, First Edition. Harry Schwarzlander.

defined uniquely (except if X does not induce any such probability assignment, in which case $B_s = \phi$).[8]

Finally, let B_a denote the set $(B_d \cup B_s)^c$. The class $\{B_a, B_d, B_s\}$ is a partition (see Figure 58.1) that, for a given r.v. X, may degenerate to size 2 or even size 1—that is, if one or both of the sets B_d, B_s are empty. The rule of total probability can be used to write

$$F_X(x) = P_X(B_a)F_{X|B_a}(x) + P_X(B_d)F_{X|B_d}(x) + P_X(B_s)F_{X|B_s}(x)$$

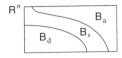

Figure 58.1   Venn diagram suggesting the partitioning of R^n by $\{B_a, B_d, B_s\}$

which has the desired form. The c.d.f.'s $F_{X_a}, F_{X_d}, F_{X_s}$ in Equation (58.1) thus are conditional c.d.f.'s of the r.v. X.[9]

If one or both of the sets B_d, B_s are empty, or if $P_X(B_a) = 0$, the corresponding conditional distribution is undefined. For instance, if $B_d = \phi$, $F_{X|B_d}(x)$ and $F_{X_d}(x)$ are undefined. But this causes no difficulty, since then $P(B_d) = 0$ so that the term $dF_{X_d}(x)$ in Equation (58.1) vanishes.

It should be kept in mind that a mixture of distributions does *not* correspond to a sum of r.v.'s. Summing a discrete and an absolutely continuous r.v., for instance, yields an absolutely continuous r.v., not one of mixed type.

In the case of a one-dimensional induced distribution that has no singular component, Equation (58.1) can of course be written in terms of density functions:

$$f_X(x) = P_X(B_a)f_{X|B_a}(x) + P_X(B_d)f_{X|B_d}(x) \tag{58.2}$$

where $f_{X|B_d}$ consists of a sum of weighted δ-functions.

Example 58.1

A r.v. X induces a distribution that is characterized by the p.d.f.

$$f_X(x) = 0.4e^{-|x|} + 0.2\,\delta(x).$$

This expression for f_X is already in the form of Equation (58.2). δ-functions appearing in a p.d.f. identify the set B_d. In the case at hand, $B_d = \{0\}$, so that $f_{X|B_d}(x) = \delta(x)$. Also, $B_a = \{0\}^c$ so that, strictly speaking,

$$f_{X|B_a}(x) = \begin{cases} 0.5e^{-|x|}, & x \neq 0 \\ 0, & x = 0. \end{cases}$$

Instead, however, following the convention stated in Section **39**, we write $f_{X|B_a}(x)$ as the continuous function $0.5e^{-|x|}$ for all x.

---

[8] For example, suppose a suitable set B_s has been found, then $B_s \cup (R_a - B_d)$ also satisfies the specifications for B_s, where R_a is the set of all rational numbers.

[9] Alternately, these c.d.f.'s may be thought of as describing the distributions induced by three 'conditional random variables,' which can be denoted $X|B_a$, $X|B_d$, and $X|B_s$. (See comment in Section **57** about conditional r.v.'s.)

Reflecting on this resolution of P_X into $P_{X|B_a}$ and $P_{X|B_d}$, we note: The underlying probability space $(S, \mathscr{A}, P)$ is partitioned by $\{A_a, A_d\}$, where $A_a = \{\xi: X(\xi) \in B_a\}$ and $A_d = \{\xi: X(\xi) \in B_d\}$, with $P(A_a) = P_X(B_a) = 0.8$ and $P(A_d) = P_X(B_d) = 0.2$. Under the condition A_a, X induces an absolutely continuous distribution with (conditional) p.d.f. $0.5e^{-|x|}$. Under the condition A_d, X maps all probability into the origin of the x-axis.

Queries

58.1 Let $F_X(x) = \begin{cases} 0, & x < -2 \\ 0.2 + 0.1x, & -2 \leq x < 0 \\ F_{V_o}(x), & x \geq 0 \end{cases}$

where F_{V_o} is the c.d.f. depicted in Figure 56.6. Determine the coefficients a, d, and s, if F_X is written in the form of Equation (58.1).   [169]

58.2 Consider the two-dimensional r.v. $X = (X_1, X_2)$, where X_1 and X_2 are statistically independent r.v.'s with identical c.d.f.'s:

$$F_{X_1}(x) = F_{X_2}(x) = \begin{cases} 0, & x < 0 \\ \dfrac{1+x}{2}, & 0 \leq x < 1 \\ 1, & x \geq 1. \end{cases}$$

Determine the coefficients a, d, and s, if F_X is written in the form of Equation (58.1).   [145]

Conditional Distribution Given the Value of a Random Variable

In Section 57 we saw how a conditioning event with positive probability can be applied to a r.v., resulting in a conditional distribution. Another kind of condition that can arise is that some absolutely continuous r.v. is required to take on a specified value, which is an event that has zero probability. This type of condition, and determination of the resulting conditional distribution, requires a different approach.

Suppose two r.v.'s, X and Y, are defined on a probability space $(\mathsf{S}, \mathscr{A}, P)$. For some specified $x \in \mathsf{R}$, consider the event $\mathsf{C}_x = \{\xi \colon X(\xi) = x\}$. If $P_X(\mathsf{C}_x) > 0$, then according to Definition 57.1, use of C_x as conditioning event for the r.v. Y results in the conditional distribution

$$P_Y(\,|\mathsf{C}_x). \tag{59.1a}$$

In (59.1a), x can also be regarded as a *parameter*, rather than as some fixed value that is initially specified. If this parameter is the value assigned by a r.v. X, then we speak of the *conditional distribution of Y given the value of X*, or simply 'the conditional distribution of Y given X', and use the notation

$$P_{Y|X}(\,|x). \tag{59.1b}$$

According to Definition 57.1, the expressions (59.1) are meaningful as functions of x only for those x-values (if any) satisfying $P_X(\{x\}) > 0$. Considering this situation in more detail, let X first of all be a *discrete r.v.* As we have noted, $P_{Y|X}(\,|x)$ exists and is defined in accordance with Definition 57.1 for all those real numbers x to which X assigns positive probability. For such x only and for every $\mathsf{B} \in \mathscr{B}$, since

$$P(\xi \colon X(\xi) = x, Y(\xi) \in \mathsf{B}) = P_{X,Y}(\{x\} \times \mathsf{B}).$$

it is possible to write

$$P_{Y|X}(\mathsf{B}|x) = P_Y(\mathsf{B}|\mathsf{C}_x) = \frac{P_{X,Y}(\{x\} \times \mathsf{B})}{P_X(\{x\})}. \tag{59.2}$$

Probability Concepts and Theory for Engineers, First Edition. Harry Schwarzlander.
© 2011 John Wiley & Sons, Ltd. Published 2011 by John Wiley & Sons, Ltd.

Now at first glance, Equation (59.2) may not seem like a very satisfactory 'function of x', since it remains undefined for all x other than some countable set. However, how likely is it that the underlying experiment produces an outcome for which $P_{Y|X}(B|x)$ is undefined? Since it is defined for all $\{x\}$ having positive probability, it is defined over a set (in S) of probability 1. A statement or specification that is valid 'with probability 1' is frequently adequate for purposes of probability theory. This is the case here; and we say that $P_{Y|X}(B|x)$ is defined 'with probability 1'.

Example 59.1

Consider the experiment described in Example 57.2 and let I_G be the indicator r.v. of the event $G \leftrightarrow$ 'good'. Then I_G assigns only the values 0 or 1. The conditional probability $P_{Y|I_G}(B|x)$, for $B \in \mathscr{B}$, is then only defined for $x = 0$ and $x = 1$. Since $P_{I_G}(\{0,1\}) = 1$, $P_{Y|I_G}(B|x)$ is defined 'with probability 1'.

Let X now be *absolutely continuous*. We again wish to define $P_{Y|X}(B|x)$ with probability 1 for every $B \in \mathscr{B}$. This can be achieved by expressing $P_{Y|X}(B|x)$ in a slightly different manner than in Equation (59.2). We write

$$P_{Y|X}(B|x) = \lim_{\Delta x \to 0} \frac{P_{X,Y}(\overline{x - \Delta x, x + \Delta x} \times B)}{P_X(\overline{x - \Delta x, x + \Delta x})} \tag{59.3}$$

provided that the intervals $\overline{x - \Delta x, x + \Delta x}$, for all $\Delta x > 0$, have positive probability assigned to them by X. In that case the limit (59.3) exists. Those x-values for which that condition does not hold comprise a set of zero probability. For those x-values, $P_{Y|X}(B|x)$ remains undefined. Since in Equation (59.3), $P_{Y|X}(B|x)$ is defined as the limit of a sequence of conditional probabilities, Equation (59.2) can be considered as a special case of Equation (59.3). The interpretation of $P_{Y|X}(B|x)$ given in Equation (59.3) can therefore be applied also if the r.v. X has a discrete and an absolutely continuous component.

If X is singular or has a singular component, the limit (59.3) in general does not exist. However, it can be shown that even in this case Equation (59.3) can be used, provided that the limiting operation is carried out in a specific way. It is therefore possible to consider Equation (59.3) as the defining expression for $P_{Y|X}(B|x)$, with a special interpretation of $\lim_{\Delta x \to 0}$ implied in those cases where X has a singular component. The details of this are not considered here.

We have now defined $P_{Y|X}(|x)$ with probability 1. Only if X assigns positive probability to every interval (i.e., if $F_X(x)$ is not constant over any interval) will $P_{Y|X}(|x)$ be defined for *all* x. For those x-values for which it is defined, $P_{Y|X}(|x)$ is a probability distribution and therefore is described by a c.d.f.—the conditional c.d.f. of Y, given a value of X:

(Definition) 59.1

The conditional c.d.f. of a r.v. Y, given the value of a r.v. X, is the function

$$F_{Y|X}(y|x) = P_{Y|X}(\overline{-\infty, y}|x) \tag{59.4}$$

which is a function of the two variables x, y, and is defined for all $y \in \mathbb{R}$ and for those x-values for which $P_{Y|X}(|x)$ is defined in accordance with Equation (59.2) or Equation (59.3), whichever is applicable. Thus, $F_{Y|X}(y|x)$ is defined with probability 1.

If, for given x, $F_{Y|X}(y|x)$ is defined and differentiable with respect to y, then we can introduce, for such x, the *conditional p.d.f.*

$$f_{Y|X}(y|x) = \frac{\partial}{\partial y} F_{Y|X}(y|x). \tag{59.5}$$

Furthermore, if X, Y are jointly absolutely continuous, then making use of Equations (59.4) and (59.3) and Definition 44.3 results in

$$f_{Y|X}(y|x) = \frac{\partial}{\partial y} \lim_{\Delta x \to 0} \frac{F_{X,Y}(x+\Delta x, y) - F_{X,Y}(x-\Delta x, y)}{F_X(x+\Delta x) - F_X(x-\Delta x)}$$

$$= \frac{\partial}{\partial y} \frac{\frac{\partial}{\partial x} F_{X,Y}(x,y)}{f_X(x)} = \frac{f_{X,Y}(x,y)}{f_X(x)}$$

for all x, y where the derivatives exist and $f_X(x) \neq 0$. The final expression for the conditional p.d.f. can be seen to be strongly reminiscent of the definition of conditional probability in Definition 21.1. We restate this result in the following:

Theorem 59.1
If X and Y are jointly absolutely continuous r.v.'s, then the conditional density function of Y, given X, is

$$\boxed{f_{Y|X}(y|x) = \frac{f_{X,Y}(x,y)}{f_X(x)}} \tag{59.6}$$

for all x, y where f_X and $f_{X,Y}$ exist and $f_X(x) > 0$.

In order to graphically interpret the conditional density function (59.6) we consider two jointly absolutely continuous r.v.'s X, Y and their joint density function $f_{X,Y}$. Then, for each x-coordinate for which it exists, the conditional density function $f_{Y|X}(y|x)$ is a slice through $f_{X,Y}$ parallel to the y-axis, at the x-coordinate, scaled by $1/f_X(x)$.

Example 59.2
We consider again the r.v.'s X, Y of Example 50.1. Their joint density equals 1/2 over the square region identified in Figure 59.1, and is zero elsewhere. This p.d.f. can be thought of as plotted in the third dimension, with the axis rising up from the page.

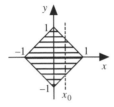

Figure 59.1 Region where the joint density equals 1/2

The conditional p.d.f. $f_{Y|X}(y|x)$ exists for $x \in -1, 1$. Pick a value $x = x_0$ in this interval. Then $f_{X,Y}(x_0, y)$ is a vertical slice through Figure 59.1 along the coordinate $x = x_0$, and its rectangular shape can easily be visualized. To obtain the conditional p.d.f. at $x = x_0$, this rectangular function needs to be scaled (divided by $f_X(x_0)$). It can be seen that as x_0 is increased (or decreased) from 0, $f_{Y|X}(y|x)$ becomes a successively narrower and taller rectangular function.

If a conditional distribution $P_{Y|X}(\cdot|x)$ is given for all x for which it is defined, then the rule for total probability can be applied to determine the unconditional distribution P_Y. If X is discrete, this results in

$$F_Y(y) = \sum_x F_{Y|X}(y|x) P_X(\{x\}) \tag{59.7}$$

where the summation is over all x for which $P_X(\{x\}) > 0$. If X is absolutely continuous, the corresponding expression is

$$F_Y(y) = \int_{-\infty}^{\infty} F_{Y|X}(y|x) f_X(x) \, dx. \tag{59.8}$$

Furthermore, if X and Y are jointly absolutely continuous, then the following relation follows from Theorem 59.1, or upon differentiating Equation (59.8) with respect to y:

$$\boxed{f_Y(y) = \int_{-\infty}^{\infty} f_{Y|X}(y|x) f_X(x) \, dx} \tag{59.9}$$

Equations (59.3) to (59.6) can be generalized to n-dimensional r.v.'s. Thus, let X, Y be jointly absolutely continuous (i.e., the components of X and the components of Y, altogether, are assumed jointly absolutely continuous). Equation (59.6) then becomes

$$f_{Y|X}(y|x) = \frac{f_{X,Y}(x, y)}{f_X(x)}. \tag{59.10}$$

From Equation (59.10), the following *chain rule* for conditional densities follows readily. Let $X_1, \ldots, X_n$ be jointly absolutely continuous r.v.'s, then

$$\left. \begin{array}{l} f_{X_1, \ldots, X_n}(x_1, \ldots, x_n) = f_{X_1|X_2, \ldots, X_n}(x_1|x_2, \ldots, x_n) f_{X_2|X_3, \ldots, X_n}(x_2|x_3, \ldots, x_n) \cdots \\ \qquad\qquad \cdots f_{X_{n-1}|X_n}(x_{n-1}|x_n) f_{X_n}(x_n). \end{array} \right\} \tag{59.11}$$

This can be seen to resemble the chain rule for conditional probabilities, Equation (22.6). Formula (59.11) can also be written more compactly as follows:

$$f_{X_1, \ldots, X_n}(x_1, \ldots, x_n) = \prod_{i=1}^{n} f_{X_i|X_{i+1}, \ldots, X_n}(x_i|x_{i+1}, \ldots, x_n)$$

where the factor corresponding to $i = n$ is simply the unconditional density $f_{X_n}(x_n)$.

Conditional densities can be useful in carrying out transformations. For example, suppose the p.d.f. of a r.v. Z is to be determined, where $Z = X/Y$ and X, Y are absolutely continuous s.i. r.v.'s whose densities f_X and f_Y are given. Y and Z are then jointly absolutely continuous. For each y such that

$f_Y(y) \neq 0$, $f_{Z|Y}(z|y)$ is simply the p.d.f. of X/y. From Equation (53.6) we obtain $f_{Z|Y}(z|y) = |y| f_X(yz)$. Now, Equation (59.9) yields

$$f_Z(z) = \int_{-\infty}^{\infty} f_{Z|Y}(z|y) f_Y(y) \, dy = \int_{-\infty}^{\infty} |y| f_X(yz) f_Y(y) \, dy.$$

Queries

59.1 Two statistically independent r.v.'s, X and Y, include a joint distribution that is uniform over the unit square $[0, 1] \times [0, 1]$. Determine which of the following density functions, if any, are rectangular over their region of definition:
a) The conditional density function of X, given Y
b) The conditional density function of $X + Y$, given Y
c) The conditional density function of X, given $X + Y$.   [71]

59.2 For the same r.v.'s as in the preceding Query, evaluate:
a) $f_{X|Y}(1.2 \mid 0.5)$ b) $f_{X|Y}(0.5 \mid 1.2)$
c) $f_{X+Y|Y}(1.2 \mid 0.5)$ d) $f_{X|X+Y}(0.5 \mid 1.2)$.   [114]

59.3 If, in the expression on the right side of Equation (59.11), the factor $f_{X_n}(x_n)$ is deleted, determine whether the result is
a) a conditional density function
b) an unconditional density function, but for $n - 1$ r.v.'s
c) valid only if X_n is a function of $X_1, \ldots, X_{n-1}$
d) meaningless.   [202]

Random Occurrences in Time

In this Section we examine in some detail a probabilistic experiment that serves as the prototype for a large class of experiments all of which are probabilistically equivalent (except for the value of a parameter). Consider real-world situations such as a stream of electrons arriving at some interface, the generation of α-particles in radioactive decay, the impinging of cosmic-rays on a radiation detector, the arrival of cars at a toll gate, or the initiation of telephone calls by the subscribers of a branch exchange. What all these situations have in common is a succession of 'happenings' (arrivals of electrons, initiations of telephone calls, etc.); and each such happening is triggered more or less independently of all the other happenings, by one member of a large collection of triggering entities (the electrons making up the electron stream, the subscribers of the branch exchange, etc.).

In any such situation, suppose an experiment is performed in order to determine the time instants at which the successive happenings occur. The essential features of such an experiment can usually be adequately modeled without making reference to the specific physical circumstances. Instead, a prototype p.e. can be used that is simply concerned with the observation of 'random occurrences in time'. A considerable variety of real-world circumstances can be modeled by such a p.e. We shall derive some of the r.v.'s that arise in the analysis of such situations.

The prototype experiment, then, has as its *purpose* the observation of 'random occurrences in time'. An 'occurrence' can mean the arrival of an electron at an interface, or the arrival of a car at a service station, or whatever kind of happening the actual physical situation is concerned with. More must be said, however, about the kind of 'randomness' that is assumed to exist: What sort of random mechanism is at work while the experiment is in progress? We assume this important feature: The timing of the occurrences is not affected by what has happened in the past. For instance, if observation starts at time $t = 0$, then the time until the next occurrence is not affected by whether or not there is an occurrence at $t = 0$, or how long before $t = 0$ the previous occurrence has taken place. In many of the physical contexts mentioned above it is reasonable to say that any single occurrence is 'independent' of all other occurrences, in the sense that the suppression of a particular occurrence would have no influence on the timing of the other occurrences. It is true that in those real-world contexts where the 'occurrences' involve the actions of human beings, such as arrivals at a check-out counter, this assumption may not be completely accurate. Nevertheless, a conceptual model based on this assumption may lead to results that agree adequately with experimental data.

A further specification is needed. We assume that there is a 'constant average rate' of occurrences, which we denote α (per second). This means that *the number of occurrences per unit time*

Probability Concepts and Theory for Engineers, First Edition. Harry Schwarzlander.
© 2011 John Wiley & Sons, Ltd. Published 2011 by John Wiley & Sons, Ltd.

approaches a fixed value as the observation interval is made large. This specification is closely related to statistical regularity, and may be thought of as 'statistical regularity in time'—in other words, the statistical description of the random mechanism does not change with time.

A random mechanism with this property is also called *stationary*. Again, in particular applications the assumption of stationarity may not be accurate. For instance, there are diurnal, weekly, seasonal and other fluctuations in the behavior and activities of groups of human beings. Some care must therefore be used in applying the prototype p.e. that is introduced here to situations involving human activities. More precisely, in an actual physical situation, it needs to be clear for how long a time period the stationarity assumption is justified. In our prototype p.e., however, stationarity will not be restricted and we assume that the experiment, once begun, continues 'for all time'. Now, this would seem to make it impossible to carry out repetitions of the p.e., as required to satisfy statistical regularity. But instead of thinking of 'repetitions' as getting carried out in time succession, these can also be regarded as *multiple executions* that are carried out in various places and perhaps beginning at different time instants—with all relevant circumstances being identical—as discussed also in Example 56.1.

The *procedure* for our prototype p.e. is very straightforward: Observation starts at time $t = 0$ and continues indefinitely. The values of t at which successive 'occurrences' take place are assumed to be precisely defined and measurable, and are recorded. If there is an occurrence at exactly $t = 0$, it will be ignored. Each possible outcome of the experiment is therefore an ascending sequence of positive real numbers. The collection of all possible outcomes is represented by the sample space S, which is uncountable. It will become clear that this model also takes care of finite observation times, and also of simpler experiments—for instance, an experiment for the purpose of merely counting the number of occurrences in a specified time interval.

We define now some random variables on the probability space for this prototype experiment. It will be possible to arrive at the probability distributions associated with these r.v.'s by considering the nature of the probabilistic experiment and our assumption about the kind of randomness encountered in this experiment.

Each of the possible outcomes of our p.e. represents a distinct sequence of *occurrence times*, or 'epochs',

$$\xi \leftrightarrow (t_1, t_2, t_3, \ldots).$$

This suggests defining r.v.'s T_1, T_2, T_3, etc., which associate with each element ξ the corresponding values t_1, t_2, t_3, etc., respectively. In addition, the elapsed time between successive occurrences, such as $t_3 - t_2$, might be of interest. Therefore, we also define the r.v.'s $W_k = T_k - T_{k-1}$ ($k = 2, 3, 4, \ldots$), representing the *waiting time* from the $(k-1)$th to the kth occurrence. For later convenience, we also define $W_1 = T_1$.

Another numerical value that can be associated with each possible outcome is the number of occurrences that have taken place up until some specified instant. This suggests a collection of r.v.'s M_t, where the subscript t is a parameter ($t > 0$) specifying the time instant up to which occurrences are to be counted. Other r.v.'s could be defined, but we will confine our attention to those mentioned. We will first determine the probability distribution associated with the r.v. M_t, for some specified $t > 0$.

Let the positive t-axis be divided into small increments of width Δt, as shown in Figure 60.1. Now consider a contraction of the prototype p.e. $\mathfrak{E}$ to another p.e. $\mathfrak{E}_{\Delta t}$ in which each occurrence time is identified only by the Δt-interval into which it falls. Since occurrences in one particular time interval are independent of occurrences elsewhere, we can consider the observation of what happens in the various small intervals as constituting independent, identical component experiments that together make up $\mathfrak{E}_{\Delta t}$:

$$\mathfrak{E}_{\Delta t} = \mathfrak{E}_1 \times \mathfrak{E}_2 \times \mathfrak{E}_3 \times \ldots$$

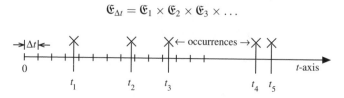

Figure 60.1   Random occurrences

The component experiment $\mathfrak{E}_i (i = 1, 2, 3, \ldots)$ thus calls for a count of the number of 'occurrences' during the time interval $\overline{(i-1)\Delta t, i\Delta t}$. Each of the component experiments therefore has the possible outcomes

> *no* occurrence,
>
> *one* occurrence,
>
> *two* occurrences, etc.

Now suppose that Δt is chosen small enough so that $\alpha \Delta t \ll 1$. Then in each of the experiments $\mathfrak{E}_i$ the event 'exactly one occurrence' has a small probability. Furthermore, the event 'more than one occurrence' has much smaller probability, which becomes negligibly small if Δt is made small enough. In that case, the situation can be approximated by changing the component experiments $\mathfrak{E}_i$ to binary trials, with the probability of 'success' (i.e., an occurrence) being $\alpha \Delta t$, and the probability of no occurrence equal to $1 - \alpha \Delta t$. We know from Section 35 that this approximation leads to a binomial distribution for the r.v.'s $M_{k\Delta t}$:

$$P_{M_{k\Delta t}}(\{m\}) = \binom{k}{m} p^m (1-p)^{k-m}, \quad m = 0, 1, 2, \ldots, k \tag{60.1}$$

with $p = \alpha \Delta t$. Next, consider successively finer subdivisions of the t-axis. This leads to successively better approximations of the experiments $\mathfrak{E}_i$ and of P_{M_t}. As $\Delta t \to 0$, the exact probability space associated with $\mathfrak{E}$ is approached, and (see Section 35) the approximate solution (60.1) becomes

$$P_{M_t}(\{m\}) = \frac{(kp)^m}{m!} e^{-kp} = \frac{(\alpha t)^m}{m!} e^{-\alpha t}. \tag{60.2}$$

Thus, the distribution induced by the r.v. M_t is the *Poisson* distribution with parameter αt.

Example 60.1: Radioactive decay

Consider a sample of radioactive material for which it is known that alpha particles are emitted within a specified solid angle (so as to impinge on a circular target) at the rate of 15 per minute. What is the probability that one or more alpha particles are emitted during a specific 1-second observation interval? The p.e. is one involving the observation of random occurrences in time. We consider the r.v. M_1 that expresses the number of occurrences during a 1-second interval. Its distribution is

$$P_{M_1}(\{m\}) = \frac{(\alpha t)^m}{m!} e^{-\alpha t}$$

with $\alpha t = (15/60) \cdot 1 = 1/4$, so that

$$P_{M_1}(\{m\}) = \frac{0.25^m}{m!} e^{-0.25}.$$

Defining the event $A \leftrightarrow$ 'one or more particles emitted during the 1-second interval', we have:
$P(A) = 1 - P(A^c) = 1 - P_{M_1}(\{0\}) = 1 - e^{-0.25} \doteq 0.221$.

In order to derive the distribution (60.2) it was necessary to go back to the conceptual model. This was because an adequate description of the probability space was not available. Now that we know P_{M_t}, we also know the probabilities associated with various events in the probability space, namely, events of the type

'm occurrences during the first t seconds'

for any positive t and $m = 0, 1, 2, \dots$. These can be used to find the distributions induced by the other r.v.'s that were identified earlier.

Thus, consider the waiting time from the first occurrence to the second, which is expressed by the r.v. W_2. For some given $t > t_1$, let $w = t - t_1$. Because of stationarity, the probability of *no* occurrence over $\overline{t_1, t}$ is

$$P_{M_w}(\{0\}) = P_{W_2}\overline{(w, \infty)} = 1 - F_{W_2}(w).$$

From Equation (60.2), $P_{M_w}(\{0\}) = e^{-\alpha w}$, so that $F_{W_2}(w) = 1 - e^{-\alpha w}$ for $w > 0$. The r.v. W_2 therefore induces an *exponential distribution* with p.d.f.

$$f_{W_2}(w) = \begin{cases} \alpha e^{-\alpha w}, & w > 0 \\ 0, & w < 0. \end{cases} \tag{60.3}$$

Because of stationarity and the independence of occurrences, the same development applies also to $W_3, W_4, \dots$. This leaves only f_{W_1} to be determined. Let t_0 be the time of the most recent occurrence prior to $t = 0$, and let the r.v. W_0 express the waiting time from t_0 to t_1, so that in accordance with the preceding development, $f_{W_0}(w) = f_{W_2}(w)$. Then $f_{W_1}(w) = f_{W_0|A_0}(w)$, where $A_0 \leftrightarrow$ 'no occurrence during $\overline{t_0, 0}$'. But in Problem 57.9b(i) we find that $f_{W_0|A_0}(w) = f_{W_0}(w)$. Therefore, all the waiting time r.v.'s $W_1, W_2, W_3, \dots$ are i.d.m.s.i. (identically distributed and mutually statistically independent) with p.d.f. (60.3).

From Equation (60.3) it is now easy to obtain the density functions of the r.v.'s $T_1, T_2, T_3, \dots$. First, since $T_1 = W_1$, its p.d.f. is given by Equation (60.3):

$$f_{T_1}(t) = \begin{cases} \alpha e^{-\alpha t}, & t > 0 \\ 0, & t < 0. \end{cases}$$

Furthermore, $T_m = \sum_{i=1}^{m} W_i$, $m = 2, 3, 4, \dots$. And since $W_1, W_2, W_3, \dots$ are identically distributed, then $f_{T_2} = f_{W_1} * f_{W_1}$; $f_{T_3} = f_{W_1} * f_{W_1} * f_{W_1} = f_{T_2} * f_{W_1}$; and in general, $f_{T_m} = f_{T_{m-1}} * f_{W_1}$. Carrying out the convolution, we have $f_{T_2}(t) = 0$ for $t < 0$, while for $t > 0$,

$$f_{T_2}(t) = \int_{-\infty}^{\infty} f_{W_1}(w) f_{W_1}(t - w) dw = \int_0^t \alpha e^{-\alpha w} \alpha e^{-\alpha(t-w)} dw$$

$$= \alpha^2 e^{-\alpha t} \int_0^t dw = \alpha^2 t e^{-\alpha t}. \tag{60.4}$$

Similarly, $f_{T_3}(t) = 0$ for $t < 0$, and for $t > 0$,

$$f_{T_3}(t) = \int_{-\infty}^{\infty} f_{T_2}(w) f_{W_1}(t-w) dw = \int_0^t \alpha^2 w e^{-\alpha w} \alpha e^{-\alpha(t-w)} dw$$

$$= \alpha^3 e^{-\alpha t} \int_0^t w\, dw = \alpha^3 \frac{t^2}{2} e^{-\alpha t}.$$

The following general formula, for $m = 1, 2, 3, \ldots$, is easily verified by induction:

$$f_{T_m}(t) = \begin{cases} \alpha^m \dfrac{t^{m-1}}{(m-1)!} e^{-\alpha t}, & t > 0 \\[2mm] 0, & t < 0. \end{cases} \tag{60.5}$$

The density function (60.5) describes the so-called *gamma distribution* of order m with parameter α.[10] Since the exponential distribution is obtained from Equation (60.5) with $m = 1$, it can be seen to be a special case of the gamma distribution.

Finally, we also obtain the joint density function of T_m and T_{m-1} ($m = 2, 3, 4, \ldots$). It is useful to first find the conditional density of T_m, given $T_{m-1}, f_{T_m|T_{m-1}}(t|\tau)$. If it is specified that the $(m-1)$th occurrence takes place at time τ ($\tau > 0$), then the mth occurrence is one waiting time later. Therefore, for fixed $\tau, f_{T_m|T_{m-1}}(t|\tau)$ must be an exponential density function with parameter α, but shifted to the right by an amount τ so that the exponential begins at $t = \tau$. Now, using Equations (59.6) and (60.5), the joint density of T_m and T_{m-1} is found to be

$$f_{T_m,T_{m-1}}(t, \tau) = f_{T_m|T_{m-1}}(t|\tau) f_{T_{m-1}}(\tau)$$

$$= \begin{cases} \alpha^m e^{-\alpha t} \dfrac{\tau^{m-2}}{(m-2)!}, & 0 < \tau < t \\[2mm] 0, & \text{otherwise}. \end{cases} \tag{60.6}$$

Example 60.2

In an experiment involving random occurrences in time, what is the probability that the second occurrence takes place no later than twice the time till the first occurrence; that is, what is the probability of the event $A \leftrightarrow T_2 < 2T_1$? Using Equation (60.6) we obtain

$$P(A) = \int_0^{\infty} \int_{\tau}^{2\tau} f_{T_2,T_1}(t, \tau) dt\, d\tau = -\alpha \int_0^{\infty} (e^{-2\alpha\tau} - e^{-\alpha\tau}) d\tau = 0.5.$$

Thus, $T_2 < 2T_1$ has the same probability as $T_2 > 2T_1$. This result can also be obtained in a different way. Let t denote the value taken on by T_2. Then, T_1 assigns a value in $\overline{0, t}$ in accordance with a *uniform* distribution:

$$f_{T_1|T_2}(\tau|t) = \frac{f_{T_2,T_1}(t, \tau)}{f_{T_2}(t)} = \frac{\alpha^2 e^{-\alpha t}}{\alpha^2 t e^{-\alpha t}} = \frac{1}{t}, \quad 0 < \tau < t.$$

Therefore, the conditions $T_1 < T_2/2$ and $T_1 > T_2/2$ arise with equal likelihood.

---

[10] The name of the Gamma distribution derives from the fact that the Gamma function $\Gamma(x)$ is defined by the integral

$$\Gamma(x) = \int_0^{\infty} t^{x-1} e^{-t} dt.$$

The model of 'random occurrences in time' that has been developed and analyzed in this Section imposes no constraints on occurrences except for stationarity and a given average rate. There are many other situations that can be said to involve 'random occurrences in time', but incorporate additional constraints. For example, occurrences might be constrained to a discrete time scale; or the time between occurrences might be constrained by some specified distribution. Problem 60.7 is an illustration of this.

Queries

60.1 An experiment is performed that involves the observation of random occurrences. In this experiment, the probability of one or more occurrences during a one-minute interval equals the probability of no occurrences in the same interval.

a) What is the probability that the time from the first until the second occurrence is at least one minute?

b) Given that an occurrence takes place during a particular one-minute interval, what is the probability that the next occurrence takes place during the next one-minute interval?   [24]

60.2 Consider the same experiment as in Query 1. What is the probability of no occurrences in a particular 90-second interval?   [156]

Part IV Summary

The basic understanding of random variables provided in Part III has now been broadened to include transformations, multiple r.v.'s and higher-dimensional vector r.v.'s, conditional induced distributions, and singular r.v.'s in one dimension.

We have seen in Sections 47, 48 and 51 how any transformation of random variables can be handled with the basic transformation rules that have been presented. However, a simpler approach is available in the absolutely continuous case if the transformation can be formulated as a coordinate transformation.

The property of mutual statistical independence was defined for multiple random variables, and distinguished from mere pairwise statistical independence.

Conditional induced distributions were found to be a straightforward extension of conditional probability if the conditioning event has positive probability. However, for situations where the condition is a particular value assigned by an absolutely continuous r.v., a new development was required in order to make it possible to specify a conditional distribution that is defined with probability 1.

Of all three pure types of random variables and induced distributions in one or more dimensions, a singular r.v. in one dimension is a somewhat special case. This was discussed in Section 56. Thereafter, in Section 58, it was possible to speak meaningfully of resolving any r.v. into pure types.

In many of the examples in Parts III and IV, a random variable was simply assumed as given, without justification. The discussion of random occurrences in time in Section 60, on the other hand, illustrated the development of random variables and their induced distributions from the nature of the p.e. under consideration. Previously, we have done this primarily for the die-throwing experiment, for the random real number selection and the spinning-pointer experiments in Examples 38.2 and 38.3, and also in Section 56. Yet, all commonly encountered induced distributions are associated with specific types of experiments. In Part VI it will be seen how the Gaussian distribution derives from elementary considerations.

Part V

Parameters for Describing Random Variables and Induced Distributions

Part V   Introduction

Upon completion of the basic study of random variables as presented in Parts III and IV, it is now possible to explore further aspects of random variables that are of great practical and also theoretical importance. In many probability problems requiring the use of random variables, a complete specification of the induced distributions either is not available or is not needed. Instead, certain descriptors may suffice to adequately characterize a r.v. and to lead to a solution. Such descriptors will now be introduced.

Of greatest interest, both in applications as well as in theoretical development, are those descriptors that constitute a 'statistical average' or 'expected value', as defined in Section 61. This definition, and the notion of the 'expectation of a function of a r.v.' discussed in Section 63—along with their extensions to multiple random variables—are the basis for all of Part V. As will be seen, descriptors of a r.v. apply equally well to the corresponding induced distribution.

Section 66 represents a slight digression from the main thread of Part V. It examines the relationship between a statistical average and the sample average in a sampling experiment. This idea is revisited from a different perspective in Section 73, which addresses basic principles of estimation.

Random variables with a singular component are excluded from consideration in most of Part V, due to reliance on the familiar Riemann integral. This limitation can be overcome either through the use of the Lebesgue integral or more simply—as is done in Section 74—by introducing the Stieltjes integral. Only the Stieltjes integral in one dimension is considered. Thus, Section 74 extends the earlier results to *all* scalar r.v.'s.

Some Properties of a Random Variable

In a probability problem it frequently happens that a random variable is defined but not known or specified in complete detail. Nevertheless, certain important features of the r.v. may be known and this knowledge may be sufficient to lead to the desired solution. In complicated problems, this can save time and effort. In this Section various features of a r.v. are considered, and the numerical descriptors of these features.

As we have seen, a random variable gets introduced into the mathematical model of a probability problem when numerical values are associated with the possible outcomes of an experiment. A useful descriptor of a r.v., therefore, ought to somehow characterize the numerical-valued outcomes of a p.e., and the way in which these outcomes arise in a sequence of independent, identical executions of the p.e.

Example 61.1

Consider once again the familiar experiment of throwing two dice, and let X be the r.v. that assigns to each element of the sample space the *absolute difference* between the number of dots showing on the two dice. Thus, X assigns the value

$x =$	0	1	2	3	4	5
with probability	3/18	5/18	4/18	3/18	2/18	1/18

Two obvious features of this r.v. are that the smallest number assigned by X is 0, and the largest number assigned by X is 5:

$$\min(X) = 0, \quad \max(X) = 5$$

The minimum possible value and maximum possible value of a function are examples of *deterministic parameters*. They are deterministic because they do not involve the probability distribution. They are based solely on the *range* of the r.v.; thus they express physical or operational

Probability Concepts and Theory for Engineers, First Edition. Harry Schwarzlander.
© 2011 John Wiley & Sons, Ltd. Published 2011 by John Wiley & Sons, Ltd.

limitations that are inherent in the underlying p.e. If X is an absolutely continuous scalar r.v., then $\min(X)$ and $\max(X)$ are taken as the greatest lower bound and the least upper bound, respectively, of the range of X, and of course these values 'occur' with probability zero.

Continuation of Example 61.1

Suppose the experiment is about to be performed. It is reasonable to ask: 'What x-value is most likely to arise?' It is that real number x for which $P_X(\{x\})$ is largest, that is, $x = 1$. We call this the *most probable value* of the r.v. X. When the experiment is executed many times, this is the value that X is likely to assign most frequently.

If a discrete r.v. does not induce greater probability at one point than at every other point in its range, then there is no 'most probable value'. If a most probable value does exist, it is also called the *mode* of the induced distribution. An absolutely continuous r.v., since it assigns all values in its range with zero probability, does not have a most probable value. However, if the p.d.f. of an absolutely continuous distribution has a unique maximum, then the abscissa value where that maximum occurs is the *mode* of the distribution.[1]

Further continuation of Example 61.1

Now imagine some board game in which a playing piece is advanced x units upon each throw of two dice, where x is the absolute difference between the two dice. By how much can a player expect to advance, 'on the average', upon each throw of the dice? For a *particular* sequence of n throws, this average is

$$\frac{1}{n}\sum_{k=1}^{n} x_k \quad \text{units} \tag{61.1}$$

with x_k denoting the absolute difference of the kth throw, where $k = 1, \ldots, n$. If, in a long sequence of throws, the various x-values were to occur in the exact proportion of their long-term relative frequencies, Equation (61.1) would evaluate to

$$\frac{3}{18} \cdot 0 + \frac{5}{18} \cdot 1 + \frac{4}{18} \cdot 2 + \frac{3}{18} \cdot 3 + \frac{2}{18} \cdot 4 + \frac{1}{18} \cdot 5 = \frac{35}{18}.$$

Normally, a sequence of n throws will not give rise to x-values in exactly these proportions. However, since probabilities are idealizations of relative frequencies it would seem that for *large* n the *sum* of the observed x-values should be close to $(35/18)n$, or the average (Equation (61.1)) close to 35/18.

We have just encountered two different *probabilistic parameters* that tell us something about a r.v. The first of these, the most probable value (or most likely value) assigned by a r.v., turns out to be of relatively little interest in Probability Theory. The other one is of great importance.

A sum of real numbers, where each term is weighted by the probability of its occurrence, is called a *statistical average*. (Actually, it is a 'probabilistic average'.) The statistical average of all the values assigned by a discrete scalar r.v. X is called the *expected value* or *expectation* of X, and is

---

[1] An absolutely continuous distribution is sometimes called 'bimodal' if its p.d.f. has two distinct maxima.

denoted **E**[*X*], or simply **E***X*. In the absolutely continuous case, this becomes a weighted integral. Specifically:

(Definition) 61.1

a) The *expected value*, or *expectation*, of a discrete scalar r.v. *X* is

$$\mathbf{E}X = \sum xP_X(\{x\}) \tag{61.2}$$

where the summation is over all real numbers *x* belonging to the range of *X*. If the range is infinite, then Equation (61.2) is an infinite series that must converge absolutely, otherwise **E***X* is undefined.

b) The *expected value*, or *expectation*, of a scalar r.v. *X* whose p.d.f. f_X exists is

$$\mathbf{E}X = \begin{cases} \int_{-\infty}^{\infty} x f_X(x)dx, & \text{if the integral converges absolutely} \\ \text{undefined}, & \text{otherwise.} \end{cases} \tag{61.3}$$

Case (b) of the Definition applies to any scalar r.v. *X* that is either discrete, absolutely continuous, or consists a discrete and an absolutely continuous component.[2] We continue to consider scalar r.v.'s unless otherwise noted.

The importance of the expected value of a r.v. will become more apparent in the Sections to follow. At this point, its connection to the underlying probability problem should be clearly understood. Figure 61.1 summarizes the relevant relationships between the conceptual and mathematical worlds.

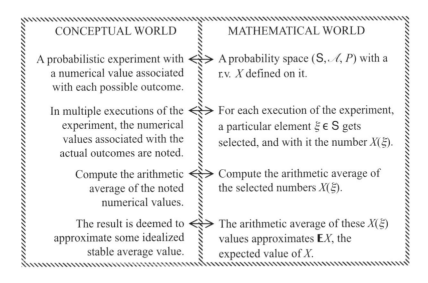

Figure 61.1   Averages in the conceptual and mathematical worlds

---

[2] Definition 61.1 does not apply to a r.v. that is singular or has a singular component. The expectation of a singular r.v. can only be expressed in terms of a Stieltjes integral or a Lebesgue integral (see Section 74).

As the continuation of Example 61.1 showed, the expected value of a r.v. can be a value that does not lie in the range of the r.v. It may or may not equal the most probable value. It can be interpreted as an idealization of the *average* of the actual values assigned by the r.v. in a long sequence of independent experiments. We will examine the nature of this idealization more closely in Part VI. But one simple property can be recognized at once: If a r.v. assigns all probability to a finite or semi-infinite interval, then its expectation cannot lie outside that interval.

For most r.v.'s of interest, the expectation exists. However, it is not difficult to find a r.v. that has no expectation.

Example 61.2

A discrete r.v. X induces the distribution

$$P_X(\{x\}) = x^{-1}, \text{ for } x = 2^1, 2^2, 2^3, \ldots$$

as illustrated by the bar graph in Figure 61.2.

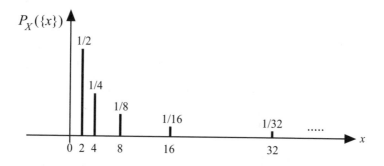

Figure 61.2   Bar graph of P_X in Example 61.2

Equation (61.2) gives

$$\sum x\, P_X(\{x\}) = 1 + 1 + 1 + \ldots = \lim_{n \to \infty} n$$

and this limit does not exist. Therefore, X has no expectation. This mathematical result has the following conceptual interpretation: The distribution P_X is scattered so widely that the *average* of the values that X assigns during the course of a long sequence of experiments will not tend toward any specific number.

The following two theorems give the expected value of a random variable in certain special cases.

Theorem 61.1

If a r.v. X induces a distribution that is symmetrical about some point $x_0 \in \mathbb{R}$, and if $\mathbf{E}X$ exists, then $\mathbf{E}X = x_0$.

Proof

> Suppose X is discrete. Then $\mathbf{E}X = \sum x P_X(\{x\}) = \sum (x + x_0) P_X(\{x + x_0\})$, the last expression being obtained through a change of variable. This can be written $\mathbf{E}X = x_0 + \sum x P_X(\{x + x_0\})$. Now, x is an odd function and $P_X(\{x + x_0\})$ is an even function, so that $x P_X(\{x + x_0\})$ is odd and $\sum x P_X(\{x + x_0\}) = 0$. Therefore, $\mathbf{E}X = x_0 + 0 = x_0$.
>
> The proof is analogous if the induced distribution is represented by a p.d.f.

A special case of Theorem **61**.1 arises if the distribution induced by X is concentrated solely at a point x_0, so that $X = x_0$. Thus, the expected value of a unary r.v. is the value to which it assigns all probability. We can also say that *the expected value of a constant* equals that constant. Since an expected value is a constant, the expectation of an expected value is again that constant:

$$\mathbf{E}[\mathbf{E}X] = \mathbf{E}X.$$

Theorem 61.2
The expected value of an indicator r.v. is the probability of the event that it indicates.

Proof

> Given a probability space $(\mathsf{S}, \mathscr{A}, P)$, let I_A be the indicator r.v. of an event $\mathsf{A} \in \mathscr{A}$. P_{I_A} is binary, with $P_{I_\mathsf{A}}(\{1\}) = P(\mathsf{A})$, and $P_{I_\mathsf{A}}(\{0\}) = 1 - P(\mathsf{A})$. Therefore, $\mathbf{E}[I_\mathsf{A}] = 0 \cdot [1 - P(\mathsf{A})] + 1 \cdot P(\mathsf{A}) = P(\mathsf{A})$.

Since an indicator r.v. takes on only the value 1 or 0, its expected value expresses the expected fraction of the time—in a large number of repetitions of the experiment—that the indicated event occurs. But this is the idealized value of the relative frequency of the event, that is, its probability.

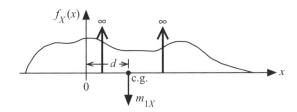

Figure 61.3   Density function of a distribution of mixed type

Descriptive parameters can also be defined for the *distribution* that a r.v. induces. Here, the analogy between a probability distribution and a one-dimensional mass distribution with a total mass of 1 (in whatever system of units is used) is helpful. Consider a probability distribution P_X consisting possibly of both a discrete and an absolutely continuous component, described by a p.d.f. such as shown in Figure 61.3. This can be viewed as a mass distribution that includes two 'point masses'. The *first moment* m_{1X} of this distribution (about the origin) is the product of total mass M and the distance d from the origin to the center of gravity. It is also obtained by summing (or integrating) the contributions of all individual particles. Since $M = 1$,

$$m_{1X} = d = \int_{-\infty}^{\infty} x f_X(x) dx \tag{61.4}$$

which is the *first moment of the probability distribution* P_X. A comparison with Equation (61.3) shows that for any given r.v. X, m_{1X} and $\mathbf{E}X$ are identical quantities. The same condition regarding existence applies. m_{1X} and $\mathbf{E}X$ can therefore be used interchangeably. Since $m_{1X} = d$, the first moment also identifies the location (on the x-axis) of the center of gravity, which in the case of a probability distribution is called the *mean* of the distribution. Thus, for any r.v. whose expected value exists, and for the distribution it induces,

$$\text{mean} = \text{first moment} = \text{expected value.}$$

We use the notation m_{1X} also for the mean of an induced distribution P_X.

If a r.v. X is such that $\mathbf{E}X = m_{1X} = 0$, then X and P_X are said to be *centralized*. The distribution shown in Figure 61.3 would become centralized if the origin is shifted to the center of gravity.

An induced distribution P_X that has no mean can still be regarded as having a 'center'—namely, the *median*, which always exists. It is the point x_m such that P_X induces the same probability on both sides of it. This can usually be expressed by the requirement that

$$F_X(x_m) = 0.5. \tag{61.5}$$

However, if Equation (61.5) has no solution as a result of $P_X(\{x_m\}) > 0$, x_m must instead satisfy $P_X(-\infty, x_m) \le 0.5 < F_X(x_m)$. If an entire interval of points satisfies Equation (61.5), then the median is taken to be the midpoint of this interval. If the mean does exist, it does not necessarily coincide with the median. Even though the concept of the median is simpler than that of the mean, the median turns out to have little analytic significance.

For an n-dimensional r.v. $X = (X_1, \ldots, X_n)$ whose component r.v.'s $X_1, \ldots, X_n$ all have expected values, the expectation is the n-tuple (or vector, if X is a vector r.v.) of those expected values:

$$\mathbf{E}X = (\mathbf{E}[X_1], \mathbf{E}[X_2], \ldots, \mathbf{E}[X_n]).$$

It is not so common to speak of the 'first moment' of an n-dimensional distribution.

The mean of an induced distribution P_X can also be expressed in terms of the c.d.f., F_X. Assume that P_X can be described by a p.d.f. f_X, and that its mean exists. Using integration by parts, we then have

$$m_{1X} = \int_{-\infty}^{\infty} x f_X(x)dx = \lim_{a \to \infty} \left[x F_X(x) \Big|_{-a}^{a} - \int_{-a}^{a} F_X(x)dx \right]$$

$$= \lim_{a \to \infty} \left\{ a \left[F_X(a) + F_X(-a) \right] - \left(\int_0^a + \int_{-a}^0 \right) F_X(x)dx \right\}.$$

Now, $\int_0^a F_X(x)dx$ can be written $a - \int_0^a [1 - F_X(x)]dx$, and $\lim_{a \to \infty} [F_X(a) + F_X(-a)] = 1$, so that

$$m_{1X} = \lim_{a \to \infty} \left\{ a - a + \int_0^a [1 - F_X(x)]dx - \int_{-a}^0 F_X(x)dx \right\}$$

$$= \int_0^{\infty} [1 - F_X(x)]dx - \int_{-\infty}^0 F_X(x)dx. \tag{61.6}$$

This shows that the mean is the area between $F_X(x)$ and 1 along the positive x-axis, minus the area between $F_X(x)$ and 0 along the negative x-axis (see Figure 61.4). When these two areas are equal,

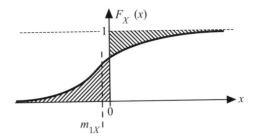

Figure 61.4 Relating the mean of a distribution to the c.d.f

then the distribution is centralized. Alternatively (see Problem 61.6), we may say that m_{1X} is that abscissa value for which

$$\int_{-\infty}^{m_{1X}} F_X(x)dx = \int_{m_{1X}}^{\infty} [1 - F_X(x)]dx. \tag{61.7}$$

Queries

61.1 Consider the experiment of throwing three dice. Let A denote the event that at least two sixes are obtained. Let I_A be the indicator r.v. for A. What are $\min(I_A)$, $\max(I_A)$, the most probable value of I_A, and the expected value of I_A? [187]

61.2 Which of the induced distributions described below have a mean?
a) $P_X(\{(-2)^n\}) = 2^{-n}$, $n = 1, 2, 3, \ldots$
b) $P_Y(\{2^{-n}\}) = 2^{-n}$, $n = 1, 2, 3, \ldots$
c) $f_Z(z) = \begin{cases} z^{-2}, & z \geq 1 \\ 0, & \text{otherwise} \end{cases}$
d) $f_V(v) = \dfrac{1}{\pi(1 + v^2)}$, all v. [48]

61.3 Given two binary r.v.'s X, Y with $\mathbf{E}X = 0.6$ and $\mathbf{E}Y = 0.4$. What is the smallest possible size of the class $\mathscr{A}$ of the probability space $(\mathsf{S}, \mathscr{A}, P)$ on which X, Y are defined? [82]

Higher Moments

In the preceding Section, the first moment of an induced distribution was defined by analogy to the first moment of a one-dimensional mass distribution. The same analogy applies to higher moments. The *second moment* m_2 (about the origin), or moment of inertia, of a (one-dimensional) mass distribution consisting of discrete point masses $M_1, M_2, M_3, \ldots$ at distances $d_1, d_2, d_3, \ldots$ from the origin, is given by the summation

$$m_2 = \sum M_i \, d_i^2 .$$

For a discrete probability distribution P_X, this becomes

$$m_{2X} = \sum x^2 P_X(\{x\}). \qquad (62.1a)$$

the summation being over those x-values to which P_X assigns positive probability. If P_X is described by a density function f_X, then the above summation gets replaced by an integral:

$$m_{2X} = \int_{-\infty}^{\infty} x^2 f_X(x) \, dx. \qquad (62.1b)$$

We call m_{2X} the *second moment of the probability distribution P_X*.

The moment of inertia of a mass distribution can also be expressed as the product of the total mass M and the square of the radius of gyration, r_g. Since a probability distribution has total 'probability mass' $M = 1$, the analogy with a mass distribution shows that m_{2X} corresponds to r_g^2.

Moments higher than the second moment are defined in a similar manner. The kth moment of P_X, $k = 3, 4, 5, \ldots$, is

$$m_{kX} = \int_{-\infty}^{\infty} x^k f_X(x) dx \qquad (62.2)$$

where, in the case of a discrete distribution, a summation takes the place of the integral. The same convergence requirements as in Definition 61.1 apply for all k.

Probability Concepts and Theory for Engineers, First Edition. Harry Schwarzlander.
© 2011 John Wiley & Sons, Ltd. Published 2011 by John Wiley & Sons, Ltd.

These concepts are summarized in the following.

(Definition) 61.1
Given an induced distribution P_X without singular component. If P_X is discrete, its kth moment $(k = 1, 2, 3, \ldots)$ is given by

$$m_{kX} = \begin{cases} \sum x^k P_X(\{x\}), & \text{the summation being over all } x \text{ in the range of } X, \text{ and absolute convergence holds;} \\ \text{undefined}, & \text{otherwise} \end{cases}$$

If P_X is described by a p.d.f. f_X, its kth moment $(k = 1, 2, 3, \ldots)$ is given by

$$m_{kX} = \begin{cases} \displaystyle\int_{-\infty}^{\infty} x^k f_X(x) dx, & \text{if the inegral converges absolutely;} \\ \text{undefined}, & \text{otherwise} \end{cases}$$

Example 61.1:
Which moments of the exponential distribution exist, and what are their values? Consider an exponential distribution with p.d.f.

$$f_X(x) = \begin{cases} \alpha e^{-\alpha x}, & x \geq 0, \alpha > 0 \\ 0, & x < 0. \end{cases}$$

We observe that $m_{kX} = \displaystyle\int_0^{\infty} x^k \alpha \, e^{-\alpha x} dx$ converges for all $k = 1, 2, 3, \ldots$. Thus, all moments exist, and integration by parts gives

$$\int_0^{\infty} x^k \alpha \, e^{-\alpha x} dx = -x^k e^{-\alpha x} \Big|_0^{\infty} + \int_0^{\infty} k x^{k-1} e^{-\alpha x} dx, \quad k = 1, 2, 3, \ldots$$

where the first term on the right vanishes. Then $m_{1X} = \displaystyle\int_0^{\infty} e^{-\alpha x} dx = \frac{1}{\alpha}$, and $m_{kX} = \dfrac{k}{\alpha} m_{(k-1)X}$, $k = 1, 2, 3, \ldots$. Thus, the general expression for the kth moment is

$$m_{kX} = k! \, \alpha^{-k}, \quad k = 1, 2, 3, \ldots$$

In addition to moments about the origin, 'central moments', i.e., moments about the mean or center of gravity, are of interest. In that case distance is measured from the mean of the distribution.

(Definition) 62.2
Given an induced distribution P_X without singular component, whose first moment m_{1X} exists.
a) If P_X is discrete, its kth *central moment* $(k = 1, 2, 3, \ldots)$ is given by

$$\mu_{kX} = \begin{cases} \sum (x - m_{1X})^k P_X(\{x\}), & \text{the summation being over all } x \text{ in the range of } X, \text{ and absolute convergence holds;} \\ \text{undefined}, & \text{otherwise} \end{cases}$$

b) If P_X is described by a p.d.f. f_X, its kth central moment $(k = 1, 2, 3, \ldots)$ is

$$\mu_{kX} = \begin{cases} \displaystyle\int_{-\infty}^{\infty} (x - m_{1X})^k f_X(x)\,dx, & \text{if the inegral converges absolutely;} \\ \text{undefined,} & \text{otherwise} \end{cases}$$

Clearly, if m_{1X} exists then μ_{1X} exists and equals 0. Also, if the mean of a distribution is zero, then each of its central moments equals the ordinary moment of the same order. From Definitions 62.1 and 62.2 it follows that m_{kX} and μ_{kX} are non-negative for k even:

$$m_{kX}, \ \mu_{kX} \geq 0, \quad k \text{ even.} \tag{62.5}$$

It should be noted that the dimensional units of the various moments differ. For instance, if the r.v. X expresses values of distance in centimetres, then the kth moments (m_{kX} and μ_{kX}, $k = 1, 2, 3, \ldots$) are in units of cm^k.

Of particular interest in Probability Theory is the second central moment, μ_{2X}, of a distribution P_X. It serves as an indicator of the 'spread' of the distribution about the mean. It is also called the *variance* of the distribution, and its square root is the *standard deviation*. The standard deviation of a probability distribution, therefore, can be seen to correspond to the radius of gyration about the center of gravity of a mass distribution.

The expression for the variance in Definition 62.2a can be rearranged as follows:

$$\begin{aligned} \mu_{2X} &= \sum (x - m_{1X})^2 P_X(\{x\}) = \sum (x^2 - 2x\,m_{1X} + m_{1X}^2)\, P_X(\{x\}) \\ &= m_{2X} - 2m_{1X}^2 + m_{1X}^2 = m_{2X} - m_{1X}^2. \end{aligned} \tag{62.6}$$

The same result is readily obtained with Definition 62.2b. We therefore have in Equation (62.6) a convenient alternative expression for the variance of an induced distribution.

Example 62.2: Uniform distribution.

The uniform distribution described by the density function (see Figure 62.1)

$$f_X(x) = \begin{cases} 1, & 0 \leq x \leq 1 \\ 0, & \text{otherwise} \end{cases}$$

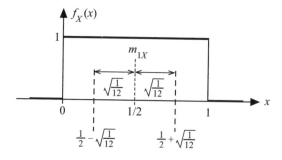

Figure 62.1   The p.d.f. of a uniform distribution

has mean $m_{1X} = 1/2$ (by symmetry), and variance

$$\mu_{2X} = \int_0^1 \left(x - \frac{1}{2} \right)^2 dx = \frac{1}{12}.$$

The variance cannot be indicated on the x-axis of this diagram because it has dimension of x^2. However, the standard deviation can be displayed as an increment to either side of m_{1X}, as shown in Figure 62.1. As can be seen, these increments are smaller than the *maximum* deviation from the mean that can arise (which would be $m_{1X} \pm 1/2$). On the other hand, they are slightly larger than the *average* of the possible deviations to either side of the mean (which would be $m_{1X} \pm 1/4$).

Suppose the r.v. Y induces a uniform distribution with zero mean and unit variance. How does f_Y differ from f_X? If $m_{1Y} = 0$, the distribution must be centered at the origin, as depicted in Figure 62.2. Furthermore, the width of the rectangle in f_X must be scaled by $\frac{1}{\sqrt{\mu_{2X}}} = \sqrt{12}$, which yields $a = \frac{\sqrt{12}}{2} = \sqrt{3}$.

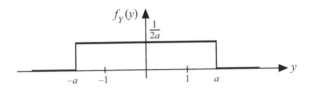

Figure 62.2   Uniform distribution with zero mean and unit variance

When discussing the properties of a probability distribution whose mean and variance exist, it is often convenient to shift the origin and change the scale of the real axis in such a way that the resulting distribution has *zero mean and unit variance*. The distribution is then said to be 'standardized'. Similarly, given any r.v. X that induces a distribution P_X whose mean and variance exist, then the r.v.

$$X_s = \frac{X - m_{1X}}{\sqrt{\mu_{2X}}} \tag{62.7}$$

is called the *standardized version* of the r.v. X. In Example 62.2, Y is the standardized version of X. If X in Equation (62.7) is absolutely continuous, then Equation (53.6) gives

$$f_{X_s}(x) = \frac{1}{\sqrt{\mu_{2X}}} \, f_X\left(\frac{x - m_{1X}}{\sqrt{\mu_{2X}}} \right).$$

Example 62.3: Gaussian distribution.

The p.d.f. associated with a standardized Gaussian r.v. X_s was given in Equation (40.2) as

$$f_{X_s}(x) = \frac{1}{\sqrt{2\pi}} \exp\left(-\frac{x^2}{2} \right).$$

This is consistent with the above meaning of 'standardized', since (see Problem 62.5) the first and second moments exist and are 0 and 1, respectively, so that the variance is 1. We further note that the parameters α and σ in the general expression for the Gaussian density function in Figure 39.1c are the *mean* and the *standard deviation* of the distribution, respectively. This follows since their use in Equation (62.7) converts the general Gaussian p.d.f. to the above standardized Gaussian p.d.f.

Taken individually, the shifting of the origin to the mean is called 'centralizing' (as noted also in Section 61), and changing the scale to achieve unit variance (or unit standard deviation) is called 'normalizing'. Therefore, a standardized r.v. or distribution is both centralized and normalized. A distribution for which the variance does not exist cannot be normalized and therefore cannot be standardized.

Theorem 62.1

If the kth moment of an induced distribution P_X exists, then all lower moments exist.

Proof (for the absolutely continuous case)

Let P_X be described by the p.d.f. f_X, and assume that m_{kX} exists. We write

$$\int_{-\infty}^{\infty} x^k f_X(x)dx = \int_{-\infty}^{-1} x^k f_X(x)dx + \int_{-1}^{1} x^k f_X(x)dx + \int_{1}^{\infty} x^k f_X(x)dx$$

where all integrals converge absolutely (Definition 62.1). Consider $i < k$. For $x \in \overline{-1, 1}$ we can say that $|x^i f_X(x)| = |x^i| f_X(x) \leq f_X(x)$. Therefore,

$$\int_{-1}^{1} |x^i f_X(x)|dx \leq \int_{-1}^{1} f_X(x)dx \leq 1$$

so that $\int_{-1}^{1} x^i f_X(x)dx$ converges absolutely. For $x \notin \overline{-1, 1}$ we have $|x^i f_X(x)| = |x^i| f_X(x) \leq |x^k| f_X(x)$. Therefore,

$$\int_{-\infty}^{-1} |x^i f_X(x)| \, dx \leq \int_{-\infty}^{-1} |x^k f_X(x)| \, dx$$

which exists, so that $\int_{-\infty}^{-1} x^i f_X(x)dx$ converges absolutely; and (by the same argument) so does $\int_{1}^{\infty} x^i f_X(x)dx$.

Queries

62.1 What are the first four moments of the symmetrical binary distribution $P_X(\{1\}) = P_X(\{-1\}) = 1/2$?

[113]

62.2 For each of the following statements determine whether it is true, possibly true, or false.

a) The mean of the distribution induced by a discrete r.v. is the most probable value of the r.v.

b) The second moment m_{2X} of an arbitrary distribution, if it exists, is non-negative.

c) Given a distribution, if the moment of order k exists, then the central moment of order k also exists.

d) If the mean is negative, then the standard deviation is also negative. [242]

62.3 If X_s is a standardized r.v. with p.d.f. $f_{X_s}(x)$, evaluate

$$\int_{-\infty}^{\infty} (x-1)^2 f_{X_s}(x) dx.$$ [219]

62.4 A r.v. X is defined on a probability space $(\mathsf{S}, \mathscr{A}, P)$ in such a way that it assigns only the value 0 or the value 1 to the various elements of the space:

for each $\xi \in \mathsf{S}$, $X(\xi) = 0$ or 1.

The standard deviation of X depends on the probability that X induces at the points $x = 0$ and $x = 1$ in R. Find the maximum and the minimum possible value of $\sqrt{\mu_{2X}}$. [7]

Expectation of a Function of a Random Variable

Since the expectation of a r.v. with a singular component has not been defined as yet, we continue to restrict our considerations in this and the following Sections to r.v.'s without singular component. Thus, let X be a scalar r.v. whose induced probability distribution is described by a p.d.f. $f_X(x)$ that possibly incorporates one or more δ-functions; and $\mathbf{E}X = \int_{-\infty}^{\infty} x f_X(x) dx$, provided the integral converges absolutely.

a) Scale change and shift of origin

Let $Y = aX + b$, $a \neq 0$, where X is a r.v. whose expectation exists. From Equation (53.6) it follows that

$$\mathbf{E}Y = \int_{-\infty}^{\infty} y f_Y(y) dy = \int_{-\infty}^{\infty} y \left[\frac{1}{|a|} f_X \left(\frac{y-b}{a} \right) \right] dy = \int_{-\infty}^{\infty} \frac{ax+b}{|a|} f_X(x) |a| dx$$

$$= a \int_{-\infty}^{\infty} x f_X(x) dx + b \int_{-\infty}^{\infty} f_X(x) dx.$$

Also, if $a = 0$, then $Y = b$ and $\mathbf{E}Y = b$. Thus:

Theorem 63.1
For any r.v. X whose expectation exists, and any real numbers a and b, $\mathbf{E}[aX + b] = a(\mathbf{E}X) + b$.

b) General formulation

Theorem 63.1 addresses a very simple instance of the *expectation of a function of a r.v.* We will now go to the general case of two *discrete* r.v.'s X and Y, which are related by some transformation g, $Y = g(X)$. P_X is assumed known. If $\mathbf{E}Y$ exists, then

$$\mathbf{E}Y = \sum_{\text{range of } Y} y P_Y(\{y\}).$$

Probability Concepts and Theory for Engineers, First Edition. Harry Schwarzlander.
© 2011 John Wiley & Sons, Ltd. Published 2011 by John Wiley & Sons, Ltd.

From step (c) of the transformation rule for a discrete r.v. (Rule 47.1) it follows that

$$\mathbf{E}Y = \sum_{\text{range of } Y} y\, P_X(\mathsf{K}_y)$$

where for any given y in the range of Y, K_y is the set of those x-values that get mapped into y by the function g. Therefore, it is possible to write

$$\mathbf{E}Y = \sum_{\text{range of } X} g(x) P_X(\{x\}). \tag{63.1}$$

This is an important formula, because it permits computation of $\mathbf{E}Y$ without first finding P_Y. Equation (63.1) generalizes as follows to r.v.'s that have p.d.f.'s:

Theorem 63.2
Given two r.v.'s X, Y with p.d.f.'s f_X, f_Y, and such that $Y = g(X)$ and $\mathbf{E}Y$ exists, then

$$\boxed{\mathbf{E}Y = \int_{-\infty}^{\infty} g(x) f_X(x)\, dx} \tag{63.2}$$

Proof
If X has a discrete and an absolutely continuous component, then $f_X(x)$ can be resolved into pure types in the manner of Equation (58.1), so that the right side of Equation (63.2) becomes the sum of two integrals. The integral involving the discrete component of X reduces to Equation (63.1). Therefore, only the absolutely continuous case needs to be considered. Thus, let X be an absolutely continuous r.v. The proof will be restricted to the case where g is a piecewise continuous and differentiable function.

Let $x_1, x_2, x_3, \ldots$ denote the endpoints of the intervals on the x-axis over which $g'(x) = dg(x)/dx$ is either positive, zero, or negative. Then

$$\int_{-\infty}^{\infty} g(x) f_X(x)\, dx = \sum_i \int_{x_{i-1}}^{x_i} g(x) f_X(x)\, dx$$

where $x_0 = -\infty$, and the upper limit of the last integral in the summation is $+\infty$.

(i) Suppose over the ith interval $g'(x) > 0$. Then, using Equation (47.6),

$$\int_{x_{i-1}}^{x_i} g(x) f_X(x)\, dx = \int_{y_{i-1}}^{y_i} y f_Y^{(i)}(y) \frac{dy}{dx}\, dx = \int_{-\infty}^{\infty} y f_Y^{(i)}(y)\, dy$$

where $y_{i-1} = g(x_{i-1})$, $y_i = g(x_i)$, and $f_Y^{(i)}(y)$ is the contribution to $f_Y(y)$ of $f_X(x)$ within x_{i-1}, x_i.

(ii) If, instead, over the ith interval $g'(x) < 0$, then similarly

$$\int_{x_{i-1}}^{x_i} g(x) f_X(x)\, dx = \int_{y_{i-1}}^{y_i} y f_Y^{(i)}(y) \left(-\frac{dy}{dx} \right) dx = \int_{-\infty}^{\infty} y f_Y^{(i)}(y)\, dy.$$

(iii) If over the ith interval, $g'(x) = 0$, then $f_Y^{(i)}(y)$ consists of a δ-function (possibly with zero weight) at $y = y_i$, and we can write

$$\int_{x_{i-1}}^{x_i} g(x) f_X(x) dx = \int_{x_{i-1}}^{x_i} y_i f_X(x) dx = \int_{-\infty}^{\infty} y f_Y^{(i)}(y) dy.$$

Since Y is a mixture of the contributions coming from the various x-intervals, summing over these intervals yields

$$\int_{-\infty}^{\infty} g(x) f_X(x) dx = \sum_i \int_{-\infty}^{\infty} y f_Y^{(i)}(y) dx = \int_{-\infty}^{\infty} y f_Y(y) dy.$$

If X is purely discrete, $f_X(x)$ consists of a sum of δ-functions, and therefore Equation (63.2) includes Equation (63.1) as a special case.

Theorem 63.2 still applies if Y is a function of several r.v.'s, $X_1, \ldots, X_n$. This can be seen by considering the case where $n = 2$: Let $Y = g(X)$, where $X = (X_1, X_2)$ is absolutely continuous, g has piecewise continuous partial derivatives with respect to x_1 and x_2, and g is such that X_2 and Y are jointly absolutely continuous. Then, the use of Equation (59.11) gives

$$\int_{-\infty}^{\infty} \int_{-\infty}^{\infty} g(x) f_X(x) \, dx_1 dx_2 = \int_{-\infty}^{\infty} \int_{-\infty}^{\infty} g(x_1, x_2) f_{X_1|X_2}(x_1|x_2) f_{X_2}(x_2) \, dx_1 dx_2.$$

Now, using steps similar to cases (i) and (ii) in the above proof, but conditioned by X_2, it can be shown that $\int y f_Y(y) dy$ equals the integral on the right (see also Problem 59.11).

Similarly, Equation (63.1) still applies if Y is a function of several r.v.'s, $X_1, \ldots, X_n$ that are *discrete*.

c) Sum of random variables

Theorem 63.2 can be used to find the expectation of the *sum* of two r.v.'s X_1 and X_2. However, X_1 and X_2 must be jointly absolutely continuous in order for the joint density function to be used. We have:

$$
\begin{aligned}
\mathbf{E}[X_1 + X_2] &= \int_{-\infty}^{\infty} \int_{-\infty}^{\infty} (x_1 + x_2) f_{X_1, X_2}(x_1, x_2) \, dx_1 dx_2 \\
&= \int_{-\infty}^{\infty} \int_{-\infty}^{\infty} x_1 f_{X_1, X_2}(x_1, x_2) \, dx_1 dx_2 + \int_{-\infty}^{\infty} \int_{-\infty}^{\infty} x_2 f_{X_1, X_2}(x_1, x_2) \, dx_1 dx_2 \\
&= \int_{-\infty}^{\infty} x_1 f_{X_1}(x_1) dx_1 + \int_{-\infty}^{\infty} x_2 f_{X_2}(x_2) dx_2 = \mathbf{E}[X_1] + \mathbf{E}[X_2].
\end{aligned}
$$

In the discrete case an analogous derivation applies:

$$
\begin{aligned}
\mathbf{E}[X_1 + X_2] &= \sum_{x_2} \sum_{x_1} (x_1 + x_2) P_{X_1, X_2}(\{(x_1, x_2)\}) \\
&= \sum_{x_2} \sum_{x_1} x_1 P_{X_1, X_2}(\{(x_1, x_2)\}) + \sum_{x_2} \sum_{x_1} x_2 P_{X_1, X_2}(\{(x_1, x_2)\}) \\
&= \sum_{x_1} x_1 P_{X_1}(\{x_1\}) + \sum_{x_2} x_2 P_{X_2}(\{x_2\}) = \mathbf{E}[X_1] + \mathbf{E}[X_2].
\end{aligned}
$$

The above summations are over those x_1 and x_2 values at which positive probability is assigned.

If there is both a discrete and an absolutely continuous component, these can be separated by the method of Section 58 and treated individually. We have:

Theorem 63.3
Given a two-dimensional r.v. $X = (X_1, X_2)$ that is absolutely continuous, discrete, or a mixture of these two types. If the expectations of X_1 and X_2 exist, then the expectation of $X_1 + X_2$ exists and is given by

$$\mathbf{E}[X_1 + X_2] = \mathbf{E}[X_1] + \mathbf{E}[X_2].$$

This result extends directly to the expectation of any finite sum of random variables satisfying the conditions of the Theorem. The restriction of Theorem 63.3 to r.v.'s X that do not have a singular component will be removed later (see Sections 70 and 74).

Example 63.1: Expected value of a binomial r.v.
Let B be a binomial r.v. that induces the distribution $P_B(\{m\})$ given in Equation (35.1):

$$P_B(\{m\}) = \binom{k}{m} p^m (1-p)^{k-m}, \quad m = 0, 1, 2, \dots, k.$$

Direct computation of $\mathbf{E}B$ from P_B, using Equation (61.2), is awkward. However, B can be expressed as the sum of statistically independent, identically distributed indicator r.v.'s,

$$B = I_1 + I_2 + \dots + I_k.$$

(In the development leading up to Equation (35.1), the I_i's are indicators of the events H_i $(i = 1, \dots, k)$. Thus, if B expresses the number of 'sixes' obtained in k throws of a die, then I_i indicates the event 'a six on the ith throw' $(i = 1, \dots, k)$, and $p = 1/6$.) An extension of Theorem 63.3 now yields:

$$\mathbf{E}B = \mathbf{E}[I_1] + \mathbf{E}[I_2] + \dots + \mathbf{E}[I_k] = \underbrace{p + p + \dots + p}_{k \text{ times}} = kp.$$

Theorems 63.1 and 63.3 can be combined by stating that the operation '$\mathbf{E}$' is a *linear operation*. That is, if $X = (X_1, \dots, X_n)$ satisfies the conditions of Theorem 63.3 (where one of the r.v.'s can also be taken as unary, assigning the constant 1), then

$$\mathbf{E}[a_1 X_1 + a_2 X_2 + \dots + a_n X_n] = a_1 \mathbf{E}[X_1] + a_2 \mathbf{E}[X_2] + \dots + a_n \mathbf{E}[X_n].$$

d) Powers of a random variable

Example 63.2
In Example 47.2, the r.v. X was taken to be uniform over $\overline{-2, 2}$, with $Y = X^2$. In order to compute $\mathbf{E}[X^2]$ we can apply Theorem 63.2 rather than using f_Y, and obtain

$$\mathbf{E}Y = \int_{-2}^{2} x^2 \frac{1}{4} dx = \frac{1}{12} x^3 \Big|_{-2}^{2} = \frac{4}{3}.$$

As an important consequence of Theorem 63.2, all the various moments of the probability distribution induced by a r.v. X are expressible as expectations of various powers and polynomials of X. In Section 61 we noted the equivalence of m_{1X} and $\mathbf{E}X$. Now, we have:

Theorem 63.4
Given a r.v. X. For $k = 1, 2, \ldots$, if $\mathbf{E}[X^k]$ exists, then $m_{kX} = \mathbf{E}[X^k]$, and

$$\mu_{kX} = \mathbf{E}[(X - m_{1X})^k] = \mathbf{E}[(X - \mathbf{E}X)^k].$$

In particular:

(Definition) 63.1
The expectation of the square of the centralized version of a r.v. X,

$$\mathbf{E}[(X - m_{1X})^2] \tag{63.3}$$

if it exists, is called the *variance of the r.v. X*, which we denote $\mathbf{V}[X]$, or simply $\mathbf{V}X$. Furthermore, $\sqrt{\mathbf{V}X}$ is then the *standard deviation* of the r.v. X.

$\mathbf{V}X$ equals μ_{2X}; in other words, no distinction needs to be made between the variance of a r.v. and the variance of the distribution induced by that r.v.—they are the same. It follows immediately, analogous to Equation (62.5), that if the variance of a r.v. exists, it is non-negative. A common notation for the standard deviation of a r.v. X is σ_X. The variance of X then becomes σ_X^2.

From Example 62.3 it follows immediately that if X is Gaussian with mean a and standard deviation σ, then

$$\mathbf{V}X = \frac{1}{\sqrt{2\pi}\sigma} \int_{-\infty}^{\infty} (x - a)^2 \exp\left[-\frac{(x - a)^2}{2\sigma^2} \right] dx = \sigma^2.$$

In Equation (62.7) we defined the standardized version X_s of a r.v. X whose variance exists. This can now also be written

$$\boxed{X_s = \frac{X - \mathbf{E}X}{\sqrt{\mathbf{V}X}}} \tag{63.4}$$

An alternative expression for the variance is obtained by expanding the square in Equation (63.3). This gives, using Theorems 63.1 and 63.3, a result equivalent to Equation (62.6):

$$\begin{aligned} \mathbf{V}X &= \mathbf{E}[X^2 - 2Xm_{1X} + (m_{1X})^2] = \mathbf{E}[X^2] - \mathbf{E}[2Xm_{1X}] + \mathbf{E}[(m_{1X})^2] \\ &= \mathbf{E}[X^2] - 2m_{1X}\mathbf{E}X + (m_{1X})^2 \\ &= \mathbf{E}[X^2] - (\mathbf{E}X)^2. \end{aligned} \tag{63.5}$$

$\mathbf{E}[X^2]$ is also called the *mean-square value* of X, or of the distribution P_X. Thus, the variance of a r.v. equals the mean-square value minus the square of the mean. Since $\mathbf{V}X$ cannot be negative, it follows that

$$(\mathbf{E}X)^2 \le \mathbf{E}[X^2]. \tag{63.6}$$

It can be seen from Equation (63.5) that the variance of a centralized r.v. equals the mean-square value.

Example 63.3

As an indicator of the 'spread' of an induced distribution about its mean, the standard deviation should not be confused with the expectation of the absolute deviation from the mean. For instance, let X be a symmetric binary r.v. with P_X as shown in Figure 63.1a. Then $\sqrt{\mathbf{V}X} = \sqrt{\mathbf{E}[X^2]} = a$, and also $\mathbf{E}[|X|] = a$.

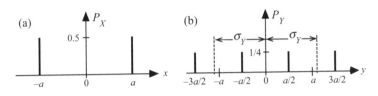

Figure 63.1 Bar graph of two different discrete distributions

Now let Y be a symmetric quaternary r.v. that assigns equal probability at four equally spaced points. The induced distribution is shown in Figure 63.1b. Here again, $\mathbf{E}[|Y|] = a$. But $\sqrt{\mathbf{V}Y} = \sqrt{0.5(a^2/4 + 9a^2/4)} = \sqrt{5}a/2 \doteq 1.118a$. Probability contributes more to the standard deviation the further out from the mean it is induced.

If the mean-square value of a r.v. X is given as zero, then this is equivalent to saying that $P_X(\{0\}) = 1$ (see Problem 63.15). Similarly, X has zero variance if and only if, for some $a \in \mathsf{R}$, $P_X(\{a\}) = 1$.

e) Product of random variables

By the *joint expectation of two r.v.'s* X, Y is meant the expected value of the *product* of the two r.v.'s, $\mathbf{E}[XY]$. If X and Y are jointly absolutely continuous and $\mathbf{E}[XY]$ exists, then Theorem 63.2 gives

$$\mathbf{E}[XY] = \int_{-\infty}^{\infty} \int_{-\infty}^{\infty} xy f_{X,Y}(x, y) dx dy. \tag{63.7}$$

The absolute value of the joint expectation of two r.v.'s is bounded as follows:

Theorem 63.5: Schwarz Inequality.

For two r.v.'s X, Y whose mean-square values exist,

$$|\mathbf{E}[XY]| \le \sqrt{\mathbf{E}[X^2]\,\mathbf{E}[Y^2]}.$$

Proof

Expansion of the non-negative quantity $\mathbf{E}[(aX + Y)^2]$, where a is a real-valued parameter, yields

$$0 \le \mathbf{E}[(aX + Y)^2] = a^2\mathbf{E}[X^2] + 2a\,\mathbf{E}[XY] + \mathbf{E}[Y^2]. \tag{63.8}$$

If $\mathbf{E}[X^2] > 0$, then the right side is a concave upward quadratic in a. Setting the derivative of this quadratic to zero gives the location of the minimum. Thus, $2a\mathbf{E}[X^2] + 2\mathbf{E}[XY] = 0$, or $a = -\mathbf{E}[XY]/\mathbf{E}[X^2]$. The right side of Equation (63.8) at this minimum, which must be ≥ 0, is then

$$\frac{\mathbf{E}[XY]^2}{\mathbf{E}[X^2]} - 2\frac{\mathbf{E}[XY]^2}{\mathbf{E}[X^2]} + \mathbf{E}[Y^2]$$

giving $\mathbf{E}[XY]^2 \leq \mathbf{E}[X^2]\,\mathbf{E}[Y^2]$, or $|\mathbf{E}[XY]| \leq \sqrt{\mathbf{E}[X^2]\,\mathbf{E}[Y^2]}$.

If $\mathbf{E}[X^2] = 0$, the right side of Equation (63.8) becomes $2a\mathbf{E}[XY] + \mathbf{E}[Y^2]$. Since a is arbitrary, this expression can be ≥ 0 if and only if $\mathbf{E}[XY] = 0$, and again the Schwarz Inequality holds.

According to Theorem 63.5, a sufficient condition for the existence of $\mathbf{E}[XY]$ is that $\mathbf{E}[X^2]$ and $\mathbf{E}[Y^2]$ exist.

$\mathbf{E}[XY]$ must not be confused with the expectation of the *pair* (X, Y), or of the vector r.v. $[X, Y]$. By the expected value of (X, Y), written $\mathbf{E}[(X, Y)]$ or $\mathbf{E}(X, Y)$, is meant the pair of expected values of the individual r.v.'s, as noted in Section 61:

$$\mathbf{E}[(X, Y)] = (\mathbf{E}X, \mathbf{E}Y).$$

This expected value identifies the 'center of gravity' of the *joint distribution* of X and Y. Similarly, the expectation of $[X, Y]$ is the (row) vector $\mathbf{E}[X, Y] = [\mathbf{E}X, \mathbf{E}Y]$. A column vector is treated in an analogous way.

If, in Equation (63.7), $f_{X,Y}(x, y)$ is replaced by a two-dimensional mass distribution, then the integration yields the 'product moment' of the distribution. Accordingly, $\mathbf{E}[XY]$ is also referred to as the *product moment* of the joint distribution. A 'higher-order product moment' is an expectation of the form $\mathbf{E}[X^n Y^m]$, where n, m are positive integers and $n + m > 2$.

Queries

63.1 An absolutely continuous r.v. X induces a distribution with density function

$$f_X(x) = \begin{cases} 2x & 0 \leq x \leq 1 \\ 0, & \text{otherwise.} \end{cases}$$

Determine: a) $\mathbf{E}X$ b) $\mathbf{E}[1/X]$.   [6]

63.2 Given two r.v.'s X, Y for which it is known that $\mathbf{E}X = 1/2$, $\mathbf{V}X = 1/3$, and $Y = X^2$. What is $\mathbf{E}Y$?   [76]

63.3 A discrete random variable X assigns the values $1/3$, $1/2$, 1, 2, 3 with equal probability of $1/5$ each. Furthermore, let $Y = X - \mathbf{E}X$, and $Z = 1/X$.

Determine: a) $\mathbf{E}[XYZ]$ b) $\mathbf{E}[X + Y + Z]$.   [148]

63.4 Given a two-dimensional r.v. $X = (X_1, X_2)$ that induces the joint distribution described in Figure 63.2.

Find: a) $\mathbf{E}[X_1 X_2]$ b) $\mathbf{E}[(X_1, X_2)]$ c) $\mathbf{E}[X_1 \mathbf{E}[X_2]]$.   [165]

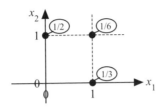

Figure 63.2   Probabilities induced by X in Query 63.4

The Variance of a Function of a Random Variable

In this Section we examine what happens to the *variance* as a result of some simple transformations. It is always assumed that the respective variances exist. First, let a scale factor a be applied to a r.v. X, giving the new r.v. $Y = aX$. Then,

$$\mathbf{V}Y = \mathbf{E}[(Y - \mathbf{E}[Y])^2] = \mathbf{E}[(aX - a\mathbf{E}[X])^2] = a^2\mathbf{E}[(X - \mathbf{E}[X])^2] = a^2\mathbf{V}X.$$

The significance of this result should be clear. Variance expresses the spread of the induced distribution. Multiplication by the scale factor, a, affects exactly the spread (and possibly axis reversal), nothing else. But since the dimensional units of the variance are those of x^2, the original variance $\mathbf{V}X$ is changed by the multiplier a^2. The standard deviation, on the other hand, is changed by the multiplier $|a|$:

$$\sqrt{\mathbf{V}[aX]} = |a|\sqrt{\mathbf{V}X}.$$

Next, consider a shift of X to give $Z = X + b$. In this case

$$\mathbf{V}Z = \mathbf{E}[(X + b - \mathbf{E}[X + b])^2] = \mathbf{E}[(X + b - \mathbf{E}[X] - b)^2] = \mathbf{V}X$$

as is to be expected, since a mere shift of the origin does not affect the spread (about the mean) of the induced distribution. Thus, we have:

> **Theorem 64.1**
> Given a r.v. X whose variance exists, and any real numbers a, b. Then
>
> $$\mathbf{V}[aX + b] = a^2\mathbf{V}X.$$

In particular, $\mathbf{V}[-X] = \mathbf{V}X$.

Probability Concepts and Theory for Engineers, First Edition. Harry Schwarzlander.
© 2011 John Wiley & Sons, Ltd. Published 2011 by John Wiley & Sons, Ltd.

The variance of the sum or difference of two r.v.'s X_1, X_2 can be expanded as follows.

$$
\begin{aligned}
\mathbf{V}[X_1 \pm X_2] &= \mathbf{E}[(X_1 \pm X_2 - \mathbf{E}[X_1 \pm X_2])^2] \\
&= \mathbf{E}[((X_1 - \mathbf{E}X_1) \pm (X_2 - \mathbf{E}X_2))^2] \\
&= \mathbf{E}[(X_1 - \mathbf{E}X_1)^2] \pm 2\mathbf{E}[(X_1 - \mathbf{E}X_1)(X_2 - \mathbf{E}X_2)] + \mathbf{E}[(X_2 - \mathbf{E}X_2)^2] \\
&= \mathbf{V}X_1 + \mathbf{V}X_2 \pm (2\mathbf{E}[X_1 X_2] - 2\mathbf{E}X_1 \mathbf{E}X_2).
\end{aligned}
$$

This proves:

Theorem 64.2

Given two r.v.'s X_1, X_2 whose variances exist, then

$$
\mathbf{V}[X_1 \pm X_2] = \mathbf{V}X_1 + \mathbf{V}X_2 \pm 2(\mathbf{E}[X_1 X_2] - \mathbf{E}X_1 \mathbf{E}X_2).
$$

If X_1, X_2 are *statistically independent* r.v.'s whose joint expectation exists and they are absolutely continuous, then their joint expectation can be expressed in the following way:

$$
\begin{aligned}
\mathbf{E}[X_1 X_2] &= \int_{-\infty}^{\infty} \int_{-\infty}^{\infty} x_1 x_2 f_{X_1, X_2}(x_1, x_2) dx_1 dx_2 \\
&= \int_{-\infty}^{\infty} x_1 f_{X_1}(x_1) dx_1 \int_{-\infty}^{\infty} x_2 f_{X_2}(x_2) dx_2 \\
&= \mathbf{E}[X_1]\, \mathbf{E}[X_2].
\end{aligned}
$$

The same result is obtained if one or both r.v.'s X_1, X_2 are discrete. Accordingly, we have:

Theorem 64.3

Given two statistically independent r.v.'s X_1, X_2.

a) If their expectations exist, then

$$
\boxed{\mathbf{E}[X_1 X_2] = \mathbf{E}X_1 \mathbf{E}X_2}
$$

b) If, furthermore, their variances exist then Theorem 64.2 reduces to

$$
\boxed{\mathbf{V}[X_1 \pm X_2] = \mathbf{V}X_1 + \mathbf{V}X_2}
$$

Example 64.1: Variance of a binomial r.v.
Consider again a binomial r.v. B as in Example 63.1, with

$$
P_B(\{m\}) = \binom{k}{m} p^m (1 - p)^{k - m}, \quad m = 0, 1, 2, \ldots, k
$$

and $B = I_1 + I_2 + \ldots + I_k$. The indicator r.v.'s I_i are associated with independent and identical trials so they are considered to be m.s.i., and we obtain

$$VB = V[I_1] + V[I_2] + \ldots + V[I_k].$$

Since $V[I_i] = E[I_i^2] - (E[I_i])^2 = p - p^2$, $i = 1, \ldots, k$, the variance of B is

$$VB = kp(1 - p).$$

If two r.v.'s X_1, X_2 are s.i., then it can be seen from Theorem 64.3b that $V[X_1 + X_2]$ is greater than each of the individual variances $V[X_1]$, $V[X_2]$ unless one of these variances is zero. This is not necessarily the case if X_1, X_2 are *not* s.i. The extreme counterexample would have $V[X_1]$, $V[X_2] > 0$ but $V[X_1 + X_2] = 0$. This will actually be the case if $X_2 = c - X_1$, for any constant c.

Part (b) of Theorem 64.3 can be extended as follows (see Problem 64.7):

Theorem 64.4
Let $X_1, \ldots, X_n$ be at least *pairwise* statistically independent r.v.'s whose variances exist. Then $V[X_1 + \ldots + Xn] = VX_1 + \ldots + VX_n$.

Going now to the general case of a r.v. X with p.d.f. $f_X(x)$, let $Y = g(X)$. If the variance of Y exists, a result analogous to Theorem 63.2 can be obtained by applying that Theorem:

$$\begin{aligned} VY &= E[Y^2] - (EY)^2 \\ &= E[g^2(X)] - (E[g(X)])^2 \\ &= \int_{-\infty}^{\infty} g^2(x) f_X(x) dx - \left[\int_{-\infty}^{\infty} g(x) f_X(x) dx \right]^2. \end{aligned} \quad (64.1)$$

This result, of course, reduces to just the first integral if Y is zero-mean.

Example 64.2
An exponential r.v. X induces a distribution described by the p.d.f.

$$f_X(x) = \begin{cases} \alpha e^{-\alpha x}, & x \geq 0, \ \alpha > 0 \\ 0, & x < 0. \end{cases}$$

Let $Y = X^2$. From Equation (64.1) it follows that $VY = m_{4X} - (m_{2X})^2$, and from Example 62.1 we obtain

$$VY = 4!\alpha^{-4} - 4\alpha^{-4} = 4\alpha^{-4}(3! - 1) = 20\alpha^{-4}.$$

Queries

64.1 Two standardized r.v.'s X, Y are statistically independent. Let $Z = a(X + Y) + b$. What are a and b if Z is also standardized?   [25]

64.2 Given the r.v.'s X, Y, Z where $Z = X + Y$ and all three have a standard deviation equal to 1. For each of the following statements, determine whether it is true, possibly true, or false.
a) The given information is self-contradictory

b) X and Y are statistically independent
c) All three r.v.'s have the same mean
d) If $W = X - Y$, then W has standard deviation $\sqrt{3}$.   [236]

64.3 Consider a three-dimensional discrete r.v. $X = (X_1, X_2, X_3)$ that can take on only the following triples of values, each with probability 1/4:

$$(1, 1, 1), (1, 0, 0), (0, 1, 0), (0, 0, 1)$$

If $Y = X_1 + X_2 + X_3$, what is the variance $\mathbf{V}Y$?   [90]

Bounds on the Induced Distribution

By the *shape* of a function is meant a complete description of the function except for possible shifts and scale factors. Thus, when a r.v. X induces a probability distribution whose variance exists, and the shape of the c.d.f. or the p.d.f. is known, then the only additional information needed to completely specify the distribution is the mean and the variance or standard deviation. Many families of distributions are characterized by their shape, such as the family of Gaussian distributions. In such families, the *standardized* distribution is often thought of as the prototype.

Now suppose that only the mean and variance of some r.v. are known. What can be said about the nature of the distribution? We can simplify our thinking by considering again the standardized case: mean $= 0$ and variance $= 1$. Since every distribution whose mean and variance exist can be standardized, it is clear that specification of mean and variance cannot convey any information about the *shape* of the c.d.f. or p.d.f., other than excluding those distributions for which at least the variance does not exist. However, various *bounds* on the exact form of the distribution can be derived, of which the Chebychev inequality is most broadly applicable.

If X_s is a standardized r.v., let $Y = X_s^2$, so that Y assigns zero probability to the negative axis: $P_Y(-\infty, 0) = 0$. We can write

$$\mathbf{V}X_s = \mathbf{E}Y = \int_0^\infty y f_Y(y)\, dy = \int_0^{a^2} y f_Y(y)\, dy + \int_{a^2}^\infty y f_Y(y)\, dy \tag{65.1}$$

where a is an arbitrary positive number. (If f_Y includes δ-functions and a is chosen such that $y = a^2$ coincides with a δ-function, the contribution of this δ-function is to be figured in the second integral on the right side of Equation (65.1).) Both integrals on the right side of Equation (65.1) have a lower bound:

$$\int_0^{a^2} y f_Y(y)\, dy \geq \int_0^{a^2} 0 \cdot f_Y(y)\, dy = 0$$

and

$$\int_{a^2}^\infty y f_Y(y)\, dy \geq \int_{a^2}^\infty a^2 f_Y(y)\, dy = a^2 \int_{a^2}^\infty f_Y(y)\, dy$$

$$= a^2 P_Y(\overline{a^2, \infty}) = a^2 P_{X_s}(\overline{-a, a^c}).$$

Probability Concepts and Theory for Engineers, First Edition. Harry Schwarzlander.
© 2011 John Wiley & Sons, Ltd. Published 2011 by John Wiley & Sons, Ltd.

Therefore, $VX_s \geq a^2 P_{X_s}(\overline{-a, a^c})$, which proves:

Theorem 65.1: Chebychev Inequality.
For any standardized r.v. X_s and any positive real number a,

$$\boxed{P_{X_s}(\overline{-a, a^c}) \leq \frac{1}{a^2}} \tag{65.2}$$

In words, the probability assigned by a standardized r.v. X_s to the set of real numbers having magnitude greater than or equal to some positive a is no greater than $1/a^2$.

Since a probability cannot exceed 1, this Theorem is only useful for $a > 1$. Expressed in terms of the c.d.f. F_{X_s} it says:

$$\left. \begin{array}{ll} \text{For } x < -1, & F_{X_s}(x) \leq \dfrac{1}{x^2} \\[2mm] \text{for } -1 \leq x \leq 1, & \text{no useful information} \\[2mm] \text{for } x > 1, & F_{X_s}(x) \geq 1 - \dfrac{1}{x^2}. \end{array} \right\} \tag{65.3}$$

These bounds are sketched in Figure 65.1.

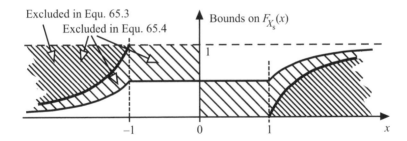

Figure 65.1 Bounds on a standardized c.d.f

In the special case where it is known that the distribution induced by the standardized r.v. X_s is *symmetrical*, the Chebychev Inequality provides a tighter bound. The probability assigned by X_s outside any interval $\overline{-a, a}$ must now be divided *equally* between $\overline{-\infty, -a}$ and $\overline{a, \infty}$. The bounds (65.3) then become:

$$\left. \begin{array}{ll} \text{For } x < -1, & F_{X_s}(x) \leq \dfrac{1}{2x^2} \\[2mm] \text{for } -1 \leq x < 0, & F_{X_s}(x) \leq \dfrac{1}{2} \\[2mm] \text{for } 0 < x \leq 1, & F_{X_s}(x) \geq \dfrac{1}{2} \\[2mm] \text{for } x > 1, & F_{X_s}(x) \geq 1 - \dfrac{1}{2x^2}. \end{array} \right\} \tag{65.4}$$

The second and third lines of Equation (65.4) simply express the symmetry condition. These bounds are also sketched in Figure 65.1. Furthermore, the following example demonstrates that the bound (65.4) for symmetrical r.v.'s cannot be tightened.

Example 65.1
Let X be a symmetrical ternary discrete r.v. that assigns the probabilities

$$P_X(\{-a\}) = P_X(\{a\}) = \frac{1}{2a^2}, \quad \text{and } P_X(\{0\}) = 1 - \frac{1}{a^2}, \quad a \geq 1.$$

We note that $\mathbf{E}X = 0$ by symmetry. Also, $\mathbf{V}X = \mathbf{E}[X^2] = \dfrac{a^2}{2a^2} + \dfrac{a^2}{2a^2} = 1$. Thus, X is standardized, $X = X_s$.

Now it can be seen that for each $a \geq 1$, the c.d.f. for the above distribution touches the boundary of the allowed region as specified in Equation (65.4) and sketched in Figure 65.1. For $a = 1$, X becomes a symmetrical binary r.v. with c.d.f. that coincides with the boundary over $\overline{-1, 1}$.

The Chebychev inequality is easily adjusted to apply to an arbitrary (nonstandardized) r.v. It is only necessary to scale the width of the interval that appears in Equation (65.2) in accordance with the standard deviation, and to shift the interval so that it is centered at the mean. This gives, for an arbitrary r.v. X whose mean and variance exist,

$$P_X\overline{(m_{1X} - a\sigma_X, \, m_{1X} + a\sigma_X}^c) \leq \frac{1}{a^2} \tag{65.5a}$$

or, replacing $a\,\sigma_X$ by a new constant b,

$$P_X\overline{(m_{1X} - b, \, m_{1X} + b}^c) \leq \frac{\mathbf{V}X}{b^2}. \tag{65.5b}$$

It is also possible to remove the complement and reverse the inequality, giving:

$$P_X\overline{(m_{1X} - b, \, m_{1X} + b)} \geq 1 - \frac{\mathbf{V}X}{b^2}. \tag{65.5c}$$

Tighter bounds on an induced distribution can be derived if the distribution satisfies additional requirements, beyond merely having a standardized version.

If a r.v. X is *non-negative*, then $\mathbf{E}X \geq 0$, i.e., it cannot be a centralized r.v. (unless $X = 0$ [P]). For such a r.v., the *mean* (rather than the variance) provides a bound on the c.d.f., as given by the following.

Theorem 65.2: Markoff Inequality.
Let X be a non-negative r.v. with expectation $\mathbf{E}X = m$. Then, for every $\gamma > 0$,

$$P_X(\overline{\gamma m, \, \infty)} \leq \frac{1}{\gamma}. \tag{65.6}$$

Proof

For $0\ \gamma \leq 1$, Equation (65.6) is trivially satisfied and provides no useful bound. For $\gamma > 1$, we prove Equation (65.6) for r.v.'s X whose p.d.f. exists (possibly including δ-functions). We have

$$\gamma m\, P_X(\overline{\gamma m, \infty}) = \int_{\gamma m}^{\infty} \gamma m f_X(x)\,dx$$

$$\leq \int_{\gamma m}^{\infty} x f_X(x)\,dx \leq \int_{0}^{\infty} x f_X(x)\,dx = m$$

where the contribution of a δ-function at $x = \gamma m$ gets included in the first two integrals. Division by γm yields Equation (65.6).

Queries

65.1 X_s is a standardized r.v. What is a bound on $P_{X_s}(\overline{2,\ \infty})$ if
a) nothing further is known about X_s?
b) P_{X_s} is symmetrical about the origin?   [151]

65.2 A r.v. X whose variance exists induces probabilities $P_X(\overline{-\infty,\ -1}) = P_X(\overline{1,\ \infty}) = 0.1$. Obtain a lower bound for $\mathbf{V}X$ if
a) it is also known that X is zero-mean
b) nothing is known about $\mathbf{E}X$.   [30]

Test Sampling

The need for collecting data arises in many areas of human endeavor, from social science research to industrial quality control, from economic analysis to public opinion polls. It might be of interest to determine the length of life of the light bulbs produced by a particular manufacturer (when operated under specified conditions), the quantity of milk produced by the cows at some dairy farm, the daily number of vehicles crossing a certain bridge, or the yearly income of the residents of a city. A common feature in these and similar examples is that there exists a *collection of objects or situations*, each having a numerical attribute that is to be measured.

If the total number of objects or situations that are under consideration is large, the question arises how the overall effort might be reduced. For instance, it may be sufficient to know the *average value* of the attribute to be measured, the average being over the whole collection of objects. The task at hand is then a simpler one, because it is possible to *estimate* this average value without measuring every single object in the collection. An experiment that can be performed to arrive at such an estimate involves (unordered) *sampling* without replacement, as described in Section 29. However, the situation is now turned around. The sampling considered in Part II dealt with known collections of objects, and the composition of the sample was of interest. Now we want to use a sample as a *test* for obtaining information about the collection from which it is drawn.

a) A simple random sample

The particular physical context is not important for this discussion. We will speak simply of 'objects', and refer to the numerical attribute of an object as 'its value'. The collection of all objects under consideration is called the *population*, and the number of objects in the population will be denoted by n. In the sampling experiment, k objects ($k < n$) are selected in such a way that prior to performing the experiment, each of the n objects has the same chance of being selected, and none is chosen more than once. Those members of the population that do get selected by such a sampling procedure constitute what is called a *simple random sample*—we will also refer to it as a 'test sample', or simply a 'sample'. The values of the objects in the simple random sample are to be used to estimate the average of the values of the objects making up the population, which is the '*population average*'.

The kind of experiment we are concerned with, then, includes in its procedure the selection of a test sample from the given population. In addition, the procedure requires the determination of the *value* of each member of the sample. Some calculation is then to be performed on the set of values

Probability Concepts and Theory for Engineers, First Edition. Harry Schwarzlander.
© 2011 John Wiley & Sons, Ltd. Published 2011 by John Wiley & Sons, Ltd.

thus obtained, which is to yield the desired estimate. The calculation that we shall specify is the calculation of the *sample average*, the arithmetic average of the values observed in the test sample. We say then that the sample average is to be used as the *estimator* of the population average.

Given some population of n objects, let the value of the ith object be denoted x_i ($i = 1, \ldots, n$). The population average is then

$$a_p = \frac{1}{n} \sum_{i=1}^{n} x_i.$$

Let a sampling experiment be performed, with specified sample size k. Each possible outcome of the experiment has associated with it a numerical value, namely, the test sample average a_T (which is to be an estimate of a_p). On the probability space for this experiment, therefore, we define the r.v. A_T that assigns to each element of the space the value a_T associated with the corresponding possible outcome.

Example 66.1

At a certain toll bridge for road traffic, the question has arisen how many people cross the bridge west to east on a particular day. We can consider each vehicle (car, truck, bus, motorcycle, etc.) an 'object' whose 'value' is the number of persons carried. The 'population' is all the trips made by vehicles going east across the bridge on the specified day. (We define the population in terms of 'trips' rather than vehicles, since some vehicles may cross over more than once during the day.)

The toll booths have a count of the number of vehicles crossing in a 24-hour period. This number has to be multiplied by the population average to give the total number of people crossing west-to-east. The population average is to be estimated from a simple random sample of specified size k. The procedure for obtaining the simple random sample might be to check, say, each 100th vehicle, up to a total of k; or to check the first vehicle arriving at a toll booth after each of k pre-specified time instants.

It is to be noted that for this experiment, statistical regularity cannot be assumed if repetitions are performed on different days. The population size and population average can be expected to vary with the day of the week and with season, and to be affected by holidays, school schedules, etc. The experiment, therefore, is concerned only with the population of trips *on the specified day*. Statistical regularity can be assumed to hold if a number of independent, identical experiments were to be carried out concurrently, by independent teams of observers, each using their own sampling schedule but with identical k. In this formulation, randomness does not reside in the population but in the sample. (The question of what conclusions might be drawn from the results thus obtained about people crossings on other days and over longer time periods is not addressed here.)

Now suppose that the *actual* number of people per vehicle trip is as follows:

10 000	vehicles carry	1	person each
2950	" "	2	persons each
800	" "	3	" "
150	" "	4	" "
50	" "	10	" "
30	" "	20	" "
20	" "	50	" "

Thus, $n = 14\,000$ and $a_p = 1.500$. What can be said about the test sample average, A_T, as an estimator of the quantity a_p (which is not known to the experimenter)? It is easy to see that the goodness of the estimate depends on k, since setting $k = n$ results in $A_T = a_p = 1.500$; whereas if $k = 1$, A_T has a range that extends from 1.0 to 50.0. Also, with $k = 1$, A_T cannot take on the value 1.5 at all. On the other hand, if $k = 2$ then A_T yields the value 1.500 with the moderate probability of 0.301, although the range of A_T still extends from 1.0 to 50.0.

b) Unbiased estimators

It is instructive to compute the expected value of the sample average. This can be done for an arbitrary experiment involving simple random sampling of some population of size n, with sample size k. As noted before, the value associated with the ith object is x_i. A typical test sample is then represented by the unordered k-tuple

$$(x_{j_1}, x_{j_2}, x_{j_3}, \ldots, x_{j_k}).$$

The subscripts $j_1, \ldots, j_k$ are distinct integers selected from $\{1, \ldots, n\}$, whereas not all the values $x_{j_1}, \ldots, x_{j_k}$ are necessarily distinct. The sample average for the above sample is

$$a_T = \frac{1}{k} \sum_{m=1}^{k} x_{j_m}.$$

There are $\binom{n}{k}$ distinct samples that can arise, and the sampling experiment is designed to make them equally likely. Thus,

$$\mathbf{E}\,A_T = \frac{1}{\binom{n}{k}} \sum_{j=1}^{\binom{n}{k}} \left(\frac{1}{k} \sum_{m=1}^{k} x_{j_m} \right), \tag{66.1}$$

where j indexes through all possible samples of size k (unordered k-tuples). Each x_i appears in exactly $\lambda = \binom{n-1}{k-1}$ different distinct samples, and therefore appears λ times in the overall sum in Equation (66.1). We also note that

$$k \binom{n}{k} = \frac{n!\,k}{(n-k)!\,k!} = \frac{n!}{(n-k)!\,(k-1)!}$$

$$= \frac{n\,(n-1)!}{[n-1-(k-1)]!\,(k-1)!} = n \binom{n-1}{k-1} = n\lambda. \tag{66.2}$$

Therefore,

$$\mathbf{E}\,A_T = \frac{1}{\lambda n} \sum_{i=1}^{n} \lambda x_i = \frac{1}{n} \sum_{i=1}^{n} x_i = a_p.$$

When the *expected value* of an estimator equals the true value that is being estimated, then the estimator is called *unbiased*, and its estimates are likewise called unbiased. So we have:

Theorem 66.1
Given the probability space for an experiment involving simple random sampling of some specified population. A random variable defined on this space, expressing the *sample average* of the simple random sample, is an *unbiased estimator* of the population average. The values this r.v. assigns are therefore unbiased estimates.

In Example 66.1, therefore, A_T is an unbiased estimator. However, since an experimenter who seeks to estimate a_p is not in a position to compute $\mathbf{E}A_T$, one may wonder what significance attaches to the fact that the estimator A_T is unbiased. Presumably, the experimenter begins in a state of complete ignorance about the values of the various members of the population, of the average a_p, or of any other parameter based on the population values. The knowledge that A_T is an unbiased estimator of a_p then does provide useful information, namely, that the probability distribution of the r.v. A_T is 'centered' at a_p. Actually the usefulness of this knowledge lies not so much in the fact that the distribution is centered *at* a_p, but in knowing *where* it is centered with respect to a_p.

c) Variance of the sample average

Further information about the quality of the estimate is obtained from the variance of the estimate, which we examine next. Consider first the mean-square value of A_T:

$$\mathbf{E}\left[A_T^2\right] = \frac{1}{\binom{n}{k}} \sum_{j=1}^{\binom{n}{k}} \left(\frac{1}{k}\sum_{m=1}^{k} x_{j_m}\right)^2 = \frac{1}{k^2\binom{n}{k}} \sum_{j=1}^{\binom{n}{k}} \left(\sum_{m=1}^{k}\sum_{\ell=1}^{k} x_{j_m} x_{j_\ell}\right)$$

For each i and $h \in \{1, \ldots, n\}$, the term $x_i x_h$ occurs in the overall sum a total of

$$\lambda = \binom{n-1}{k-1} \text{ times, if } h = i; \text{ and}$$

$$\omega = \binom{n-2}{k-2} \text{ times, if } h \neq i.$$

Therefore, making use of Equation (66.2), we can write

$$\mathbf{E}[A_T^2] = \frac{1}{k^2\binom{n}{k}} \left(\lambda\sum_{i=1}^{n} x_i^2 + \omega\sum_{i=1}^{n}\sum_{\substack{h=1 \\ h\neq i}}^{n} x_i x_h\right)$$

$$= \frac{1}{kn}\sum_{i=1}^{n} x_i^2 + \frac{1}{kn}\frac{k-1}{n-1}\sum_{i=1}^{n}\sum_{\substack{h=1 \\ h\neq i}}^{n} x_i x_h$$

$$= \frac{1}{kn}\left(1 - \frac{k-1}{n-1}\right)\sum_{i=1}^{n} x_i^2 + \frac{nk-1}{kn-1}a_p^2.$$

The variance of the estimate is then

$$\mathbf{VA_T} = \mathbf{E}\left[A_T^2\right] - \left(\mathbf{EA_T}\right)^2 = \frac{1}{kn}\left(1 - \frac{k-1}{n-1}\right)\sum_{i=1}^{n} x_i^2 - \left(1 - \frac{nk-1}{kn-1}\right)a_p^2$$

$$= \frac{1}{k}\frac{n-k}{n-1}\left(\frac{1}{n}\sum_{i=1}^{n} x_i^2 - a_p^2\right).$$

(66.3)

By analogy to the variance of a discrete probability distribution, we now introduce the following:

(Definition) 66.1

The variance of a collection of k numbers, $\{x_1, x_2, \ldots, x_k\}$, is the quantity

$$\text{var}(x_1, \ldots, x_k) = \frac{1}{k}\sum_{m=1}^{k}(x_m - a)^2 = \frac{1}{k}\sum_{m=1}^{k} x_m^2 - a^2$$

(66.4)

where a is the arithmetic average of the k numbers, $a = \dfrac{1}{k}\displaystyle\sum_{m=1}^{k} x_m$.

Letting v_p denote the population variance, Equation (66.3) then becomes

$$\mathbf{VA_T} = \frac{n-k}{n-1}\frac{v_p}{k}.$$

(66.5)

Thus, we see that the variance of A_T depends on the spread of the population values, the population size n, and the sample size k. For $k \ll n$, Equation (66.5) simplifies to

$$\boxed{\mathbf{VA_T} \approx \frac{v_p}{k},}$$

(66.6)

so that for a given population, the variance of A_T is approximately inversely proportional to the sample size k. The standard deviation of A_T is then inversely proportional to $\sqrt{k}$. Suppose, for example, that in a particular sampling experiment with $k \ll n$ the spread of P_{A_T} is to be cut in half, then the sample size has to be increased by a factor of 4.

The result stated in Equation (66.6), and the fact that the sample average is an unbiased estimator of the population average, represent a special case of a group of theorems called the Laws of Large Numbers (see Section 85). Furthermore, the distribution of A_T, for k reasonably large (say, $k > 20$) and $n \gg k$, is approximately normal, or Gaussian. This property derives from the so-called Central Limit Theorem, which will be discussed in Section 86.

Example 66.1 (continued)

From Definition 66.1 we find that the population variance (which is also the variance of A_T if $k = 1$) evaluates to $v_p \doteq 4.7786$.

Suppose the variance of the estimate is to be no more than 0.01 (or the standard deviation ≤ 0.1). Then $k \approx 478$ is required, based on Equation (66.6); or more precisely, using Equation (66.5), we find that $k = 463$ is sufficient. A computer simulation of the experiment, using the population data that has been specified, with $k = 463$, resulted in a test sample average of $a_T = \underline{1.4471}$.

If this were the result of an actual experiment, it would then become the estimate of the population average. It differs from the actual value of a_p by 0.0529, which happens to be considerably less than one standard deviation of the estimate. Based on the above value of a_T, the total number of people crossing the bridge that day (which is actually 21 000) thus gets estimated as 20 259.

d) Estimating the population variance

Of course, the experimenters in this Example do not know the population data and therefore cannot compute v_p. How then can they apply Equation (66.5) or Equation (66.6)? Some information can be obtained from the dispersion of the values in the test sample. Consider the *variance* of the set of numbers that constitute a simple random sample. Since this is a numerical value associated with each possible outcome of the sampling experiment, a r.v. V_T can be defined that expresses the variance of the test sample. Can V_T serve as an estimator for v_p?

We have

$$\mathbf{E}V_T = \frac{1}{\binom{n}{k}} \sum_{j=1}^{\binom{n}{k}} \left(\frac{1}{k} \sum_{i=1}^{k} x_{j_i}^2 \right) - \mathbf{E}[A_T^2].$$

The first term reduces in the same manner as Equation (66.1), giving

$$\mathbf{E}V_T = \frac{1}{n} \sum_{i=1}^{n} x_i^2 - [\mathbf{V}A_T + (\mathbf{E}A_T)^2] = v_p - \mathbf{V}A_T = \frac{(k-1)n}{k(n-1)} v_p.$$

For large n,

$$\boxed{\mathbf{E}V_T \approx \frac{k-1}{k} v_p} \qquad (66.7)$$

We see that V_T is a *biased* estimator of v_p. Instead, therefore, $\frac{k}{k-1} V_T$ should be used, which is an unbiased estimator of v_p if n is large. In other words, if a simple random sample of size k yields the k-tuple of values $(x_1, \ldots, x_k)$, then the computation

$$\frac{1}{k-1} \sum_{i=1}^{k} \left(x_i - \frac{1}{k} \sum_{j=1}^{k} x_j \right)^2$$

yields an unbiased estimate of the population variance.

It is now possible to specify the steps that can be followed to arrive at an estimate of a population average based on simple random sampling. Suppose a population of size n is given, where n is large compared to the number of objects whose value can reasonably be observed or measured. Also, an upper limit d_T on the variance of the test sample mean, $\mathbf{V}A_T$, is specified. Then:

1) Obtain an initial test sample of size k_1, which yields a k_1-tuple of values, $(x_1, x_2, \ldots, x_{k_1})$, where $k_1 \ll n$.

2) Compute $\dfrac{1}{k_1}\sum\limits_{i=1}^{k_1} x_i = a_{T_1}$, the sample average.

3) Compute $\dfrac{1}{k_1}\sum\limits_{i=1}^{k_1} (x_i - a_{T_1})^2 = v_{T_1}$, the sample variance (Definition 66.1).

4) The estimate of v_p is then $\dfrac{k_1}{k_1 - 1} v_{T_1}$. Use of this estimate of the population variance in Equation (66.6) yields $v_{T_1}/(k_1 - 1)$ as the estimated $\mathbf{V}A_T$ for sample size k_1. Therefore, if $v_{s_1}/(k_1 - 1) \le d_T$, then a_{T_1} is the desired estimate.

5) Otherwise, increase the test sample size to k_2, where (again using Equation (66.6))

$$k_2 \ge \frac{k_1}{(k_1 - 1)d_T} v_{T_1}.$$ Compute the sample average a_{T_2}.

Example 66.2

Suppose the life expectancy of a particular batch of light bulbs is to be determined under certain specified operating conditions. There are 10 000 bulbs. A test sample of ten light bulbs is subjected to a life test under the required conditions and the following lifetimes are observed, to the nearest ten hours:

$$280,\ 320,\ 350,\ 360,\ 420,\ 430,\ 510,\ 580,\ 600,\ 690$$

The sample average a_T is 454 hours. This is an unbiased estimate of the population average. How good an estimate is it?

To answer this question, we first compute the sample variance, which turns out to be 16 564 hours^2. An unbiased estimate of the population variance is then

$$(10/9)\ 16\,564 \doteq 18\,404.$$

Now, from Equation (66.6), the variance of A_T, the estimator of the population average, is $18\,404/10 = 1840.4$. Without knowing the distribution of A_T it is still possible to proceed by applying the Chebychev inequality. For instance, we might ask: What is the probability that the estimate of the population average lies within ± 100 hours of the true population average? Using Equation (65.5c),

$$P_{A_T}\overline{(a_p - 100,\ a_p + 100)} \ge 1 - \frac{\mathbf{V}A_T}{10,000} \doteq 1 - 0.184 = \underline{0.816}.$$

This is actually a rather loose bound on the probability in question. The Gaussian approximation to the distribution P_{A_T} (with the same mean and variance) gives a probability of 0.980, as obtained from the table of the Gaussian integral. However, this cannot be assumed to be a very good approximation for $k = 10$.

e) Sampling with replacement

As explained at the beginning of this Section, the procedure for obtaining a 'simple random sample' requires that no object is chosen more than once—i.e., sampling without replacement. Sampling *with* replacement does not satisfy this requirement because it allows an object to get included in the sample more than once. However, for n large (and $k \ll n$), the likelihood of this happening is very small, and goes to zero as $n \to \infty$. In other words, as the population size gets very large compared to the sample size, the difference between simple random sampling (sampling without replacement) and sampling *with* replacement becomes negligible.

In an experiment involving sampling with replacement, let X_i $(i = 1, \ldots, k$, where k can be greater than $n)$ represent the value of the ith object drawn. Then $\mathbf{E}X_i = a_{\mathrm{p}}$ (for $i = 1, \ldots, k$) and we have immediately

$$\mathbf{E}\,A_{\mathrm{T}} = \mathbf{E}\left[\frac{1}{k}\sum_{i=1}^{k}X_i\right] = \frac{1}{k}\sum_{i=1}^{k}\mathbf{E}X_i = a_{\mathrm{p}}. \tag{66.8}$$

We also note that in the limit as $n \to \infty$, formula (66.6) becomes exact. Furthermore, Equation (66.6) is an exact expression for the variance of the sample mean when the sampling procedure involves replacement—i.e., when a sample of size k can include some members of the population more than once—even if n is finite (see Problem 66.6).

Theorem 66.2
When sampling *with* replacement, using a test sample of size k, the variance of the sample mean A_{T} is $\mathbf{V}A_{\mathrm{T}} = v_{\mathrm{p}}/k$.

One way in which 'sampling with replacement' can be characterized is to say that the sampling procedure does not deplete or diminish any subset of the population. Thus, an experiment involving k tosses of a coin, for instance, can be thought of as sampling with replacement: If the 'population' is deemed to consist of just the two objects 'head' and 'tail', then the occurrence of a head in one of the trials does not purge the population of heads, nor is the likelihood of a head in later trials affected.

Queries

66.1 Which of the experiments suggested by the following descriptions involve simple random sampling?
  a) An ordinary die is thrown ten times and the number of dots obtained on each throw is recorded.
  b) A deck of cards is shuffled face down, and then four cards are removed and laid on the table face up.
  c) An urn contains balls of different colors. You blindly withdraw a ball, examine its color, return the ball and mix up the contents of the urn. You do this ten times.
  d) A large bin contains 1000 transistors of a certain type. 150 are removed for testing. This still being too large a number, ten transistors are picked from the lot of 150.
  e) In order to assess the reliability of different makes of automobiles, the makes of all the automobiles that are brought to a particular repair shop in a given month are recorded.
  f) From the final exam papers in a biology class, the papers with the five highest and the five lowest grades are removed. All the other papers are retained.   [65]

66.2 In a certain town a simple random sample of size 100 is used to estimate the average income tax paid by 10 000 households. In another town, the same size sample is used to estimate the average income tax paid by 5000 households. In the second town, is the standard deviation of the sample average

a) significantly smaller than,   b) essentially the same as,
c) significantly larger than, or   d) in no clear relationship to,

the standard deviation of the sample average in the first town?   [40]

66.3 A die is thrown ten times.
a) By how much can the sample average differ from the population average?
b) What is the largest possible sample variance?   [2]

Conditional Expectation with Respect to an Event

If a *conditional* distribution given an event with nonzero probability gets inserted into Definition 61.1, then a *conditional expectation* is obtained:

Definition 67.1
Let a r.v. X be defined on a probability space $(S, \mathscr{A}, P)$, and let A be an event in $\mathscr{A}$, with $P(A) \neq 0$.

a) If X is discrete, then the *conditional expectation* of X with respect to A is

$$\mathbf{E}[X|A] = \sum x P_X(\{x\}|A) \tag{67.1}$$

the summation being over all real numbers that belong to the range of X under the condition A. If this range is infinite, then Equation (67.1) is an infinite series that must converge absolutely, otherwise $\mathbf{E}[X|A]$ is undefined.

b) If the conditional p.d.f. $f_{X|A}$ is defined, then the *conditional expectation* of X with respect to A is

$$\mathbf{E}[X|A] = \begin{cases} \displaystyle\int_{-\infty}^{\infty} x f_{X|A}(x)dx, & \text{if the integral converges absolutely} \\ \text{undefined, otherwise.} \end{cases} \tag{67.2}$$

Formula (67.2) also applies if $f_{X|A}$ incorporates one or more δ-functions. When interpreted in this way, Equation (67.2) includes Equation (67.1) as a special case.

From Definition 67.1 it should be clear that a conditional expectation $\mathbf{E}[X|A]$ can be regarded as the mean of an induced probability distribution, namely, the mean of the conditional distribution of the r.v. X under the condition A, $P_X(\,|A)$. Therefore, all the properties of ordinary—or 'unconditional'—expectations also apply to conditional expectations.

Conditional moments, conditional standard deviation, and conditional expectation of a function of a r.v. are defined analogously.

Example 67.1

A salesman drives regularly on the New York Thruway between Buffalo and New York City. On each trip he watches for accidents. Let it be assumed that he has an acceptable notion of what constitutes a single accident, that accidents can be modeled as 'random occurrences', and that adequate statistical regularity exists (for instance, he does not drive in very bad weather), so that the model described in Section 60 applies. The number of accidents observed by the salesman during one trip can then be represented by a Poisson r.v., X. Suppose that, on the average, the salesman encounters *one* accident per trip. From Equation (60.2) we obtain, upon substituting $m_{1X} = \alpha t = 1$, the distribution sketched in Figure 67.1(a), namely

$$P_X(\{x\}) = \frac{1}{x!\,e}, \quad x = 0, 1, 2, \ldots \tag{67.3}$$

Now suppose that the salesman modifies his procedure in that, on each trip, he watches for accidents only upon encountering the first accident. Prior to actually seeing an accident he does not even think about accidents and about his experiment. This change in procedure does not affect the occurrence of accidents. Only the events of interest have been modified by the condition that at least one accident must occur. Therefore, the same model can be used and the probabilities expressed by the conditional distribution $P_X(\cdot|A)$, where $A = \{\xi \colon X(\xi) \geq 1\}$. This conditional distribution is easily computed. Since $P(A) = P_X(\{0\}^c) = 1 - e^{-1}$, we have

$$P_X(\{x\}|A) = \begin{cases} \dfrac{1}{x!\,e} \cdot \dfrac{1}{P_X(\{0\}^c)} = \dfrac{1}{x!\,e} \cdot \dfrac{1}{1 - e^{-1}} = \dfrac{1}{x!(e-1)}, & x = 1, 2, 3, \ldots \\ 0, & \text{otherwise.} \end{cases}$$

This is sketched in Figure 67.1(b).

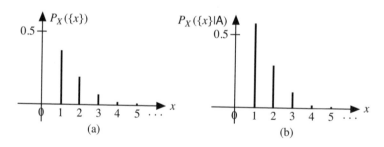

Figure 67.1   Discrete distribution without, and with, conditioning

How many accidents can the salesman expect to encounter per trip, if he disregards those trips on which he sees no accidents? In our mathematical model this is expressed by the conditional expectation

$$\mathbf{E}[X|A] = \sum_{x=1}^{\infty} x\, P_X(\{x\}|A) = \sum_{x=1}^{\infty} \frac{1}{(x-1)!(e-1)} = \frac{1}{e-1} \sum_{x=0}^{\infty} \frac{1}{x!}$$

$$= \frac{e}{e-1} \doteq \underline{1.582} \,. \tag{67.4}$$

Therefore, by restricting observations to those trips during which at least one accident is encountered, the average number of accidents increases by 0.582.

In some problems it may turn out that conditional expectations are more readily computed than an unconditional expectation. It is then possible, through a computation analogous to the Rule of Total Probability (Theorem 22.1) to compute the unconditional expectation from various conditional expectations.

Theorem 67.1
Let X be a r.v. defined on a probability space $(S, \mathcal{A}, P)$, and let $\mathcal{P} \subset \mathcal{A}$ be a (countable) partition consisting of sets with nonzero probability. If the conditional expectation $\mathbf{E}[X|A]$ exists for every $A \in \mathcal{P}$, then $\mathbf{E}X$ exists and is given by

(67.5)

$$\boxed{\mathbf{E}X = \sum_{A \in \mathcal{P}} \mathbf{E}[X|A]P(A)}$$

provided (if $\mathcal{P}$ is countably infinite) that the summation is an absolutely convergent infinite series.

Proof (for the case where X, and therefore all the conditional distributions $P_X(\ |A)$, are discrete): From the definition of conditional expectations and conditional distributions, under the assumption that Equation (67.5) is an absolutely convergent infinite series (which includes the possibility of a finite sum, and which also implies all the conditional expectations $\mathbf{E}[X|A]$ exist), we have:

$$\sum_{A \in \mathcal{P}} \mathbf{E}[X|A]P(A) = \sum_{A \in \mathcal{P}} \sum_{x \in Q} x\, P_X(\{x\}|A)P(A)$$

where Q denotes the range of X. Because of absolute convergence, the summations can be interchanged, giving

$$\sum_{A \in \mathcal{P}} \mathbf{E}[X|A]P(A) = \sum_{x \in Q} \sum_{A \in \mathcal{P}} x\, P_X(\{x\}|A)\, P(A) = \sum_{x \in Q} x P_X(\{x\}|A)$$

which is still absolutely convergent and defines $\mathbf{E}X$.

Definition 67.1 extends in an obvious way to the expectation of a *function* of a r.v. And since a variance is also an expectation, it is possible to introduce the *conditional variance* of a r.v. X. However, here it is important that the same conditioning event that conditions the outer expectation in Equation (63.3) is also used to condition m_{1X}. Thus,

$$\mathbf{V}[X|A] = \mathbf{E}[(X - \mathbf{E}[X|A])^2 |A] = \mathbf{E}[X^2|A] - (\mathbf{E}[X|A])^2. \tag{67.6}$$

This becomes clear when the event A is understood to be conditioning X rather than $\mathbf{V}X$.

Example 67.2: A Decision Problem.
We return to the situation described in Example **26**.4, and assume that it is known that a throw of the biased die results in a 'one', 'two' or 'three' with probability 0.3, and results in a 'four', 'five' or 'six' with probability 0.7. The technician again picks a die from the mixture of fair and biased

dice and throws the selected die a sufficient number of times so that she can decide, with a probability of correctness of 0.90, whether the selected die is fair or biased. How many throws of the die does this require?

We have here a conditional product experiment similar to the one in Example **26**.4, except that the number of die throws is now k, which is to be determined. The sample space for this experiment, then, is $S = S_0 \times S_1 \times S_2 \times \ldots \times S_k$, where S_0 models the selection of a die and S_1, ..., S_k represent the k throws. Since only two probabilities associated with the throw of the biased die are available, we specify in space S_i (for $i = 1, \ldots, k$) the event $H_i \leftrightarrow$ 'a high number of dots on the ith throw' (i.e., four, five, or six dots). Furthermore, in S_0, let $B \leftrightarrow$ 'a biased die is chosen', and $F = B^c \leftrightarrow$ 'a fair die is chosen'.

The decision as to whether the event B or F has occurred has to be made on the basis of how many of the die throws result in a high number of dots. Therefore, on the probability space for this experiment we define the indicator r.v.'s I_{H_i} $(i = 1, \ldots, k)$ for the events H_i. Now let

$$A = \frac{I_{H_1} + I_{H_2} + \ldots + I_{H_k}}{k}.$$

The r.v. A is analogous to the 'sample average' of Section 65(e) (sampling with replacement), except that the objects making up the 'population' ('more than three dots' and 'three or less dots') are not necessarily drawn with equal likelihood. We have $\mathbf{E}[A|B] = \mathbf{E}[I_{H_i}|B] = 0.7$, and $\mathbf{E}[A|F] = \mathbf{E}[I_{H_i}|F] = 0.5$. Also, the *conditional variances* are, from Theorem **64**.1 and Example **64**.1: $\mathbf{V}[A|B] = 0.7(0.3)/k = 0.21/k$, and $\mathbf{V}[A|F] = 0.5(0.5)/k = 0.25/k$.

We note that the conditional means are not affected by k, whereas the conditional standard deviations are proportional to $k^{-1/2}$. Thus, with increasing k the conditional distributions become more concentrated near their respective means. Some real number t must be specified as a *threshold* so that if A takes on a value less than the threshold, then the die that has been thrown is declared 'fair', and otherwise it is declared 'biased'. Clearly, the threshold needs to lie between $\mathbf{E}[A|F]$ and $\mathbf{E}[A|B]$.

What is the smallest k and the corresponding threshold t so that the die gets identified correctly with probability 0.9? Let $W \leftrightarrow$ 'wrong decision'. Then, we require

$$P(W) = P_A\overline{(0,\ t}|B)\,P(B) + P_A(\overline{t,\ \infty}|F)\,P(F) \leq 0.1$$

Refer to Figure 67.2.

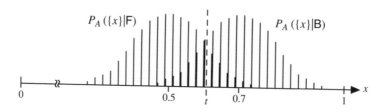

$P_A(\{x\}|F)$ $P_A(\{x\}|B)$

0   0.5 t 0.7   1 x

Figure 67.2   Conditional distributions

We will now use the Chebychev inequality to obtain an upper bound on the required k. $P_A(\ |F)$ is symmetrical, but $P_A(\ |B)$ is not. For large k, however, $P_A(\ |B)$ becomes very nearly symmetrical, so that Equation (65.4) will be used. Using the result of Problem **65**.2, and assuming that $\mathbf{E}[A|B] - t > \sqrt{\mathbf{V}[A|B]}$, the following is obtained:

$$P_A(\overline{0,\ t}|B) \le \frac{\mathbf{V}[A|B]}{2(t - \mathbf{E}[A|B])^2} = \frac{0.21}{2k\,(t - 0.7)^2}.$$

Similarly, with $t - \mathbf{E}[A|F] > \sqrt{\mathbf{V}[A|F]}$,

$$P_A(\overline{t,\ \infty}|F) \le \frac{\mathbf{V}[A|F]}{2(t - \mathbf{E}[A|F])^2} = \frac{0.25}{2k(t - 0.5)^2}.$$

From Example 26.4, $P(B) = 0.4$. Therefore,

$$P(W) = \frac{0.21 \times 0.4}{2k(t - 0.7)^2} + \frac{0.25 \times 0.6}{2k(t - 0.5)^2} = \frac{1}{k}\left[\frac{0.042}{(t - 0.7)^2} + \frac{0.075}{(t - 0.5)^2}\right].$$

As can be seen, t can be optimized independently of k. Setting $\frac{d}{dt}kP(W) = 0$ yields $t \doteq 0.6096$. The requirement $P(W) \le 0.1$ then results in $k = 114$ as the upper bound on the minimum k.

A better estimate for the required k can be obtained by approximating $F_{A|B}$ and $F_{A|F}$ by Gaussian c.d.f.'s having the same means and standard deviations (see Problem 67.7). This is justified for large k, as shown in Section 86.

Queries

67.1 A red die and a green die are thrown. The r.v.s R, G express the number of dots obtained with the red and green die, respectively. Let

> A $\leftrightarrow$ 'sum of the dots obtained is seven'
> B $\leftrightarrow$ 'sum of the dots obtained is eight'
> C $\leftrightarrow$ 'at least one die shows more than four dots'
> D $\leftrightarrow \{\xi : R(\xi) \ge G(\xi)\}$.

Determine:

a) $\mathbf{E}[R|A]$ b) $\mathbf{E}[R|B]$

c) $\mathbf{E}[R|C]$ d) $\mathbf{E}[R|D]$

e) $\mathbf{E}[G|D]$.   [164]

67.2 X, Y jointly induce a uniform distribution over $\overline{0,\ 1} \times \overline{0,\ 1}$. Let

$$A \leftrightarrow \{\xi : X(\xi) \ge Y(\xi)\}.$$

Determine:

a) $\mathbf{E}[X|A]$ b) $\mathbf{E}[X|A^c]$.   [92]

67.3 Let X be a r.v. that induces a uniform distribution over $\overline{0,\ 1}$. Specify an event A with $P(A) = 1/2$ such that $\mathbf{E}[X|A] = \mathbf{E}[X|A^c]$.   [189]

Covariance and Correlation Coefficient

The joint expectation of two r.v.'s X, Y, denoted $\mathbf{E}[XY]$, was introduced in Section 63. $\mathbf{E}[XY]$ is an example of a '*product moment*'—the expectation of a product involving two (or more) r.v.'s. The centralized version of $\mathbf{E}[XY]$ is of particular interest.

(Definition) 68.1

Given two r.v.'s X, Y whose variances exist, then the central product moment $\mathbf{E}[(X - m_{1X})(Y - m_{1Y})] = \mathbf{E}[XY] - m_{1X}m_{1Y}$ is called the *covariance* of X and Y. It will be denoted $\mathbf{C}[X, Y]$ or $c_{X,Y}$.

The following properties of the covariance follow directly from this Definition:

$$\mathbf{C}[Y, X] = \mathbf{C}[X, Y] \tag{68.1}$$

$$\mathbf{C}[aX + b,\ cY + d] = \mathbf{C}[aX, cY] = ac\,\mathbf{C}[X, Y]. \tag{68.2}$$

We also have:

Theorem 68.1

Given two r.v.'s X, Y whose variances exist, then

$$|\mathbf{C}[X, Y]| \leq \sqrt{\mathbf{V}X\,\mathbf{V}Y}. \tag{68.3}$$

Proof

Let X_c, Y_c be the centralized versions of X and Y, respectively. From Equation (68.2), the Schwarz inequality (Theorem 63.5), and Theorem 64.1, $\mathbf{C}[X, Y] = \mathbf{C}[X_c, Y_c] = |\mathbf{E}[X_c Y_c]| \leq \sqrt{\mathbf{E}[X_c^2]\,\mathbf{E}[Y_c^2]} = \sqrt{\mathbf{V}X\,\mathbf{V}Y}$.

Probability Concepts and Theory for Engineers, First Edition. Harry Schwarzlander.
© 2011 John Wiley & Sons, Ltd. Published 2011 by John Wiley & Sons, Ltd.

(Definition) 68.2
Given two r.v.'s X, Y whose variances exist and are nonzero, then the *correlation coefficient* of X and Y, denoted $\boldsymbol{\rho}[X, Y]$ or $\rho_{X,Y}$, is defined by

$$\boldsymbol{\rho}[X, Y] = \mathbf{E}[X_s Y_s]$$

where X_s and Y_s are the standardized versions of X and Y.

If one or both of the variances of X, Y are zero, $\boldsymbol{\rho}[X, Y]$ is undefined.[3] We note that

$$\rho_{X,Y} = \frac{\mathbf{C}[X, Y]}{\sqrt{\mathbf{VX}\,\mathbf{VY}}} = \rho_{Y,X}. \tag{68.4}$$

Example 68.1
Consider two binary r.v.'s X, Y. X assigns equal probability to $\{-1\}$ and $\{+1\}$, so that $\mathbf{VX} = 1$. Y assigns equal probability to $\{0\}$ and $\{+1\}$, giving $\mathbf{VY} = 1/4$. Let their joint distribution be as shown in Figure 68.1. Then

$$c_{X,Y} = \frac{3}{8}(-1)\left(-\frac{1}{2}\right) + \frac{1}{8}(-1)\frac{1}{2} + \frac{1}{8}(1)\left(-\frac{1}{2}\right) + \frac{3}{8}(1)\frac{1}{2} = \frac{1}{4}$$

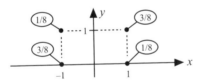

Figure 68.1 Joint distribution of X and Y in Example 68.1

and $\rho_{X,Y} = \dfrac{c_{X,Y}}{\sqrt{\mathbf{VX}\,\mathbf{VY}}} = 1/2$. On the other hand if X, Y are s.i., so that the four probabilities in the diagram are $1/4$ each, then $\rho_{X,Y} = c_{X,Y} = 0$.

Theorem 68.2
If the correlation coefficient of two r.v.'s X, Y exists, it satisfies:

a) $\boxed{-1 \le \boldsymbol{\rho}[X, Y] \le 1}$

b) $\boldsymbol{\rho}[X, Y] = 1$ if and only if $Y_s = X_s$ [P]

c) $\boldsymbol{\rho}[X, Y] = -1$ if and only if $Y_s = -X_s$ [P].

---

[3] If both X and Y are constants, $\boldsymbol{\rho}[X, Y]$ is sometimes defined to be 0.

Here, X_s, Y_s are the standardized versions of X, Y, respectively; and [P] indicates that the statement which precedes it holds *with probability 1*.

Proof

Part (a) follows immediately from Equations (68.3) and (68.4). That is, the correlation coefficient of two r.v.'s is simply the normalized version of the covariance.

To prove (b) and (c), suppose first that $\mathbf{E}[X_s Y_s] = \pm 1$, and consider:

$$\mathbf{E}[(X_s \mp Y_s)^2] = \mathbf{E}[X_s^2] + \mathbf{E}[Y_s^2] \mp 2\,\mathbf{E}[X_s Y_s] = 1 + 1 - 2 = 0.$$

From Problem 63.15 it then follows that $X_s \mp Y_s = 0$ [P], or $Y_s = \pm X_s$ [P]. Now suppose that $Y_s = \pm X_s$ [P]. Let $\mathsf{A} = \{\xi : Y_s(\xi) = \pm X_s(\xi)\}$. Then,

$$\mathbf{E}[X_s Y_s] = \mathbf{E}[X_s Y_s | \mathsf{A}]P(\mathsf{A}) + \mathbf{E}[X_s Y_s | \mathsf{A}^c]P(\mathsf{A}^c) = \mathbf{E}[\pm X_s^2] \cdot 1 + 0 = \pm 1.$$

The correlation coefficient is a measure of the 'correlation', or *degree of linear dependence*, of two r.v.'s. To get some insight into its significance, we note first that linear operations performed on either of two r.v.'s X, Y do not affect their correlation coefficient. This is because $\rho_{X,Y}$ is defined in terms of the standardized versions of X and Y.

Two r.v.'s X, Y with nonzero variance are said to be *linearly dependent* if

$$Y(\xi) = a\,X(\xi) + b \tag{68.5}$$

(a, b real and $a \neq 0$), for all elements ξ of the probability space on which X and Y are defined. This means that, jointly, they assign pairs of values only along the straight line $y = ax + b$ in the (x,y)-plane (i.e., within a one-dimensional subspace of R^2).[4] Figure 68.2a illustrates this for the case of two discrete r.v.'s, where the dots indicate points of positive probability. In many case of practical interest, two r.v.'s are not linearly dependent but their joint distribution exhibits a definite linear trend that may be clearly discernible by visual inspection of a suitable plot. Figure 68.2b illustrates such a situation where there is 'partial' linear dependence, as compared to the 'strict' linear dependence suggested in Figure 68.2a.

The correlation coefficient expresses how close two r.v.'s are to being linearly dependent. As seen in Theorem 68.2, strict linear dependence (with probability 1) is indicated by a correlation coefficient

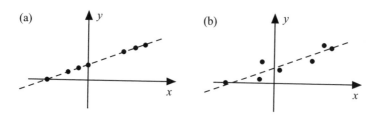

Figure 68.2   Strict and partial linear dependence

---

[4] The notion of linear dependence extends to more than two r.v.s. Let $X_1, \ldots, X_n$ ($n > 2$) have nonzero variance. Then $X_1, \ldots, X_n$ are *linearly dependent* if $\sum a_i X_i = b$ for suitable real numbers $a_1, \ldots, a_n, b$, where $a_1, \ldots, a_n \neq 0$.

of ± 1, where the sign is positive if the constant a in Equation (68.5) is positive, and is negative if $a < 0$. For instance, inspection of Figure 68.2a suggests immediately that $\rho_{X,Y} = +1$. Deviation from strict linear dependence results in a correlation coefficient whose magnitude is less than 1.

If X and Y are statistically independent r.v.'s whose variances exist, then it follows from Definition 68.1 that $c_{X,Y} = 0$ and (if $\mathbf{V}X$ and $\mathbf{V}Y$ are nonzero) $\rho_{X,Y} = 0$. On the other hand, if it is only known that $\rho_{X,Y} = 0$ or $c_{X,Y} = 0$, then it cannot be concluded that X and Y are s.i. Two r.v.'s X, Y for which $c_{X,Y} = 0$ are said to be *uncorrelated*. In other words, for two r.v.'s X, Y to be uncorrelated means that either $\rho_{X,Y} = 0$ or at least one of them has zero variance, i.e., at least one of them assigns a single real value with probability 1.

The variance of a sum of two r.v.'s (Theorem 64.2) can be rewritten in terms of the covariance:

$$\mathbf{V}[X + Y] = \mathbf{V}X + \mathbf{V}Y + 2\mathbf{C}[X, Y]. \tag{68.6}$$

From this and Definition 68.1 follows:

Theorem 68.3
For two r.v.'s X, Y whose variances exist the following are equivalent statements:

a) X and Y are uncorrelated;
b) $\mathbf{E}[X\,Y] = \mathbf{E}X\,\mathbf{E}Y$; and
c) $\mathbf{V}[X \pm Y] = \mathbf{V}X + \mathbf{V}Y$.

Now suppose two r.v.'s X, Y have been defined on a probability space. What is the relationship between X and Y if their correlation coefficient is neither 0 nor ± 1? Besides depending *in part* linearly on X as in Equation (68.5), Y then also incorporates another effect that is *not* linearly dependent on X. This other effect can be represented by an additional r.v., Z, which is uncorrelated with X. Thus, Y can be expressed as follows:

$$Y = aX + b + Z$$

where $\rho_{X,Z} = 0$. Z can be thought of as the remainder, after that part of Y has been subtracted that depends linearly on X (including any constant): $Z = Y - (aX + b)$. The coefficient a evaluates to $\rho_{X,Y}\sqrt{\mathbf{V}Y/\mathbf{V}X}$ (see Problem 68.1) and b is chosen to make $\mathbf{E}Z = 0$.

The relationship between X, Y and Z is expressed more clearly if the r.v.'s get replaced by their standardized versions:

$$Y_s = aX_s + cZ_s. \tag{68.7}$$

In this formulation, $a = \rho_{X,Y}$, and the constant c in Equation (68.7) is determined by the fact that Y_s and Z_s are standardized, which requires that $a^2 + c^2 = 1$ (see Problem 68.1). Accordingly, Equation (68.7) can be written

$$Y_s = \rho X_s + \sqrt{1 - \rho^2} Z_s, \tag{68.8}$$

where $\rho = \boldsymbol{\rho}[X, Y]$. Equation (68.8) exhibits clearly how $\boldsymbol{\rho}[X, Y]$ expresses the *extent to which* Y_s *depends linearly on* X_s. It also follows from Equation (68.8) that $\boldsymbol{\rho}[Y, Z] = \sqrt{1 - \rho^2} = \sqrt{1 - (\boldsymbol{\rho}[X, Y])^2}$.

Example 68.2

A radio signal is transmitted to a receiver. At the receiver's antenna appears a weak version of the transmitted signal, combined with signals from many other sources and various other undesirable effects. After amplification and filtering, there remains the desired signal plus a residual disturbance we will call 'noise'. Suppose the receiver output is observed at an arbitrarily selected time instant. At this instant, let the voltage due to the transmitted signal be described by the r.v. S, and the voltage due to the noise by N, making the total voltage $V = S + N$, where $\mathbf{E}S = \mathbf{E}N = \mathbf{E}V = 0$.

The noise can normally be assumed to be independent of the signal, so that

$$\mathbf{V}V = \mathbf{V}S + \mathbf{V}N.$$

Good reception requires a sufficiently large 'signal-to-noise ratio', $\mathbf{V}S/\mathbf{V}N$. We show that this is equivalent to requiring a high correlation between the transmitted signal and the received voltage. The correlation coefficient is given by

$$\rho[V, S] = \frac{\mathbf{E}[(V - \mathbf{E}V)(S - \mathbf{E}S)]}{\sqrt{\mathbf{V}V\,\mathbf{V}S}}$$

$$= \frac{\mathbf{E}[S^2 + SN] - \mathbf{E}(S + N)\mathbf{E}S}{\sqrt{(\mathbf{V}S)^2 + \mathbf{V}N\,\mathbf{V}S}} = \frac{1}{\sqrt{1 + (\mathbf{V}N/\mathbf{V}S)}}.$$

Now, solving for the signal-to-noise ratio, we obtain

$$\frac{\mathbf{V}S}{\mathbf{V}N} = \frac{\rho^2[V, S]}{1 - \rho^2[V, S]}.$$

Since the polarity of the signal is generally not of significance, it can be seen that an increase in $|\rho_{V,S}|$ corresponds to an increase in signal-to-noise ratio.

Theorem 68.4

If the joint distribution of two r.v.'s X, Y whose variances exist is symmetrical about a line parallel to one of the coordinate axes, then $\mathbf{C}[X, Y] = 0$ and (if it exists) $\rho[X, Y] = 0$.

Proof

Because of Equation (68.1), it is sufficient to suppose the joint distribution of X and Y is symmetrical about a line parallel to the y-axis. Then, $X = m_{1X} - (X - m_{1X})$, giving

$$\mathbf{C}[X, Y] = \mathbf{C}[(-X + 2m_{1X}), Y] = -\mathbf{C}[X, Y] \text{ (using Equation (68.2))}$$

so that $c_{X,Y} = 0$, and also $\rho_{X,Y} = 0$ if it exists.

Theorem 68.4 frequently makes it easy to recognize by inspection of the joint distribution that two r.v.'s are uncorrelated. However, Theorem 68.4 does not express a necessary condition; two r.v.'s may be uncorrelated even if their joint distribution exhibits no such symmetry.

Queries

68.1 Given two r.v.'s X, Y with $\rho[X, Y] = 1/2$. Determine:

a) $\rho[X, -Y]$ b) $\rho[X/2, Y]$

c) $\rho[1 + Y, 1 - Y]$ d) $\rho[X_s, Y_s]$.   [115]

68.2 X, Y and Z are standardized r.v.'s. Also, $\rho[X, Y] = 1$. If, furthermore, $Z = X - 2Y$, what are $\rho[X, Z]$ and $\rho[Y, Z]$?   [22]

68.3 Given two r.v.'s X, Y with $\mathbf{E}X = 1$ and $\mathbf{E}Y = \mathbf{E}[X\,Y] = -1$. For each of the following statements, determine whether it is true, possibly true, or false.

a) $\rho[X, Y] = -1$ b) $P_{X,Y}$ is singular

c) X and Y are s.i.   d) $X = -Y$

e) X and Y are uncorrelated.   [127]

The Correlation Coefficient as Parameter in a Joint Distribution

In this Section the role of the correlation coefficient in various joint distributions gets explored further, and the effect of linear transformations on the correlation coefficient. Only distributions for which the correlation coefficient exists are considered.

We have seen how one-dimensional induced distributions can be grouped into 'families', such as the family of uniform distributions, the family of exponential distributions, etc. In such a family there is usually a unique reference distribution (in many cases it is the standardized version), and the other members of the family are obtained from the reference distribution by shifts and/or scale changes. Two-dimensional distributions can also be grouped into families. In those families the correlation coefficient often serves as one of the parameters, in addition to scale factors and shift parameters.

Example 69.1: Two symmetric binary r.v.'s.
Similar to Example 68.1, we consider the joint distribution of two symmetric binary r.v.'s X, Y, but we take X and Y to be standardized. The joint distribution is then as shown in Figure 69.1, with only the parameter p remaining to be specified. If $p = 1/4$, then X and Y are statistically independent. Can the correlation coefficient $\boldsymbol{\rho}[X,Y] = \rho$ be used as parameter, in place of p?

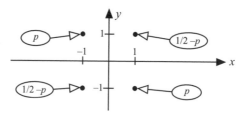

Figure 69.1   Joint distribution of two standardized binary r.v.'s

For $0 \leq p \leq 1/2$ we have

$$\rho = \mathbf{E}[XY] = 1 \cdot \left(\frac{1}{2} - p\right) + 1 \cdot \left(\frac{1}{2} - p\right) - 1 \cdot p - 1 \cdot p = 1 - 4p.$$

Probability Concepts and Theory for Engineers, First Edition. Harry Schwarzlander.
© 2011 John Wiley & Sons, Ltd. Published 2011 by John Wiley & Sons, Ltd.

Thus, the correlation coefficient depends linearly on the value of p, the probability assigned to the points in the second and fourth quadrants. The joint distribution thus can be expressed in terms of p, as shown in Figure 69.2. If ρ is increased toward $+1$, it can be visualized how most of the probability becomes concentrated along the line $y = x$. Similarly, if ρ is decreased toward -1, most of the probability becomes concentrated along the line $y = -x$.

Since shifts and scale changes do not affect the correlation coefficient, the distribution given in Figure 69.2 can be generalized as follows:

$$P_{X,Y}(\{(x_1, y_1)\}) = P_{X,Y}(\{(x_2, y_2)\}) = \frac{1 + \rho}{4}$$

$$P_{X,Y}(\{(x_1, y_2)\}) = P_{X,Y}(\{(x_2, y_1)\}) = \frac{1 - \rho}{4}$$

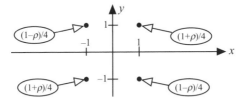

Figure 69.2   Joint distribution with ρ as parameter

for coordinate values x_1, x_2, y_1, y_2, such that $x_1 < x_2$ and $y_1 < y_2$. This is the general expression for the joint distribution of two symmetrical binary r.v.'s X, Y, with the correlation coefficient serving as one of the five parameters.

For two r.v.'s X, Y, Equation (68.2) expresses how their covariance is affected by shifts and scale changes performed on X and Y individually. Since these operations do not change $\mathbf{E}[X_0 Y_0]$ (assuming $a, c \neq 0$), it follows from Definition 68.2 that

$$\boldsymbol{\rho}[aX + b,\ cY + d] = \boldsymbol{\rho}[X, Y]. \tag{69.1}$$

We consider now a general linear transformation in two dimensions applied to a r.v. $\boldsymbol{X} = (X_1, X_2)$, yielding the new r.v. $\boldsymbol{Y} = (Y_1, Y_2)$:

$$\left. \begin{array}{l} Y_1 = aX_1 + bX_2 \\ Y_2 = cX_1 + dX_2 \end{array} \right\}. \tag{69.2}$$

There is no need to incorporate a shift of the origin, since this has no effect on the covariance nor on the correlation coefficient. Straightforward substitution yields the following expressions for the variances and covariance of Y_1, Y_2:

$$\left. \begin{array}{l} \mathbf{V}Y_1 = a^2\,\mathbf{V}X_1 + b^2\,\mathbf{V}X_2 + 2ab\,\mathbf{C}[X_1, X_2] \\ \mathbf{V}Y_2 = c^2\,\mathbf{V}X_1 + d^2\,\mathbf{V}X_2 + 2cd\,\mathbf{C}[X_1, X_2] \\ \mathbf{C}[Y_1, Y_2] = (ad + bc)\,\mathbf{C}[X_1, X_2] + ac\,\mathbf{V}X_1 + bd\,\mathbf{V}X_2 \end{array} \right\} \tag{69.3}$$

where the first two equations are generalizations of Equation (68.6).

If the linear transformation is a *rotation* about the origin (see also Section 54), then Equations (69.2) can be written

$$\left.\begin{array}{l} Y_1 = X_1 \cos\varphi + X_2 \sin\varphi \\ Y_2 = -X_1 \sin\varphi + X_2 \cos\varphi \end{array}\right\} \tag{69.4}$$

where the rotation is counterclockwise by an angle φ. Equations (69.3) then become

$$\left.\begin{array}{l} \mathbf{V}Y_1 = \cos^2\varphi\,\mathbf{V}X_1 + \sin^2\varphi\,\mathbf{V}X_2 - 2\sin\varphi\cos\varphi\,\mathbf{C}[X_1, X_2], \\ \mathbf{V}Y_2 = \sin^2\varphi\,\mathbf{V}X_1 + \cos^2\varphi\,\mathbf{V}X_2 + 2\sin\varphi\cos\varphi\,\mathbf{C}[X_1, X_2], \\ \mathbf{C}[Y_1, Y_2] = \cos2\varphi\,\mathbf{C}[X_1, X_2] + \dfrac{1}{2}\sin2\varphi\,(\mathbf{V}X_1 - \mathbf{V}X_2). \end{array}\right\} \tag{69.5}$$

Here, it can be seen that $\mathbf{V}Y_1 + \mathbf{V}Y_2 = \mathbf{V}X_1 + \mathbf{V}X_2$. Since shifts of the origin do not affect variances and the covariance, changing the point about which the rotation takes place does not influence this result. Thus, the following has been established:

Theorem 69.1
If a rotation is applied to a 2-dimensional random variable $X = (X_1, X_2)$ for which $\mathbf{V}X_1$ and $\mathbf{V}X_2$ exist, then the sum of the variances remains unchanged by the rotation.

Furthermore, this next result also follows directly from Equation (69.5).

Theorem 69.2
Given a random variable $X = (X_1, X_2)$ such that $\mathbf{V}X_1$ and $\mathbf{V}X_2$ exist, and $Y = (Y_1, Y_2) = gX$, where g is a *rotation* through any angle, then

$$\mathbf{V}Y_1\,\mathbf{V}Y_2 - (\mathbf{C}[Y_1, Y_2])^2 = \mathbf{V}X_1\,\mathbf{V}X_2 - (\mathbf{C}[X_1, X_2])^2$$

or equivalently, $\mathbf{V}Y_1\mathbf{V}Y_2\,(1 - \boldsymbol{\rho}^2[Y_1, Y_2]) = \mathbf{V}X_1\mathbf{V}X_2\,(1 - \boldsymbol{\rho}^2[X_1, X_2])$.

Thus, we can say: The product of the variances minus the square of the covariance is a quantity that remains invariant under shifts and rotations.

In the special case of a $45°$ counterclockwise rotation of the x_1 and x_2 axes about the origin, Equation (69.4) becomes

$$\left.\begin{array}{l} Y_1 = \dfrac{X_1 - X_2}{\sqrt{2}} \\[2mm] Y_2 = \dfrac{X_1 + X_2}{\sqrt{2}} \end{array}\right\} \tag{69.6}$$

so that Equations (69.3) evaluate to

$$\left.\begin{array}{l} \mathbf{V}Y_1 = \dfrac{\mathbf{V}X_1 + \mathbf{V}X_2}{2} - \mathbf{C}[X_1, X_2] \\[2mm] \mathbf{V}Y_2 = \dfrac{\mathbf{V}X_1 + \mathbf{V}X_2}{2} + \mathbf{C}[X_1, X_2] \\[2mm] \mathbf{C}[Y_1, Y_2] = \dfrac{\mathbf{V}X_1 - \mathbf{V}X_2}{2}. \end{array}\right\} \tag{69.7}$$

The last line of Equation (69.7) demonstrates:

Theorem 69.3
A 45° rotation applied to the joint distribution of two r.v.'s X_1, X_2 whose variances exist results in the joint distribution of two *uncorrelated* random variables if and only if $VX_1 = VX_2$.

If X_1, X_2 are two random variables whose variances exist and their joint distribution is *symmetric* about a ±45° line, then of course X_1 and X_2 must have equal variances. A 45° rotation then changes the line of symmetry to be in parallel with one of the coordinate axes and Theorem 68.4 also applies. The joint distribution at the beginning of Example 69.1 illustrates this situation.

Example 69.2: Bivariate Gaussian distribution.
Let X_1, X_2 be two statistically independent (and therefore uncorrelated) zero-mean Gaussian r.v.'s with variances $VX_1 = \sigma_1^2$ and $VX_2 = \sigma_2^2$. Writing $X = (X_1, X_2)$, their joint density function, as given in Equation (54.3), is

$$f_X(x) = \frac{1}{2\pi\sigma_1\sigma_2} \exp\left[-\frac{1}{2}\left(\frac{x_1^2}{\sigma_1^2} + \frac{x_2^2}{\sigma_2^2}\right)\right].$$

Let $Y = (Y_1, Y_2)$ be obtained by rotating this joint distribution counterclockwise through an angle φ about the origin. Equation (69.4) applies, and Equation (69.5) simplifies to

$$\left.\begin{aligned}
VY_1 &= \cos^2\varphi\, \sigma_1^2 + \sin^2\varphi\, \sigma_2^2, \\
VY_2 &= \sin^2\varphi\, \sigma_1^2 + \cos^2\varphi\, \sigma_2^2, \\
C[Y_1, Y_2] &= \sin\varphi \cos\varphi\, (\sigma_1^2 - \sigma_2^2).
\end{aligned}\right\} \tag{69.8}$$

Turning now to f_Y as given in Equation (54.5), we express the coefficients of y_1^2, y_2^2, and y_1y_2 in terms of VY_1, VY_2 and $C[Y_1, Y_2]$. Using Equations (69.8), Theorem 69.2 and the fact that $C[X_1, X_2] = 0$, the coefficient of y_1^2 can be written

$$\frac{\cos^2\varphi}{\sigma_1^2} + \frac{\sin^2\varphi}{\sigma_2^2} = \frac{\cos^2\varphi\, \sigma_2^2 + \sin^2\varphi\, \sigma_1^2}{\sigma_1^2\sigma_2^2} = \frac{VY_2}{VY_1\, VY_2 - C^2[Y_1, Y_2]}$$

$$= \frac{1}{VY_1}\frac{1}{1 - \rho^2[Y_1, Y_2]}.$$

In an analogous way the coefficient of y_2^2 becomes

$$\frac{\cos^2\varphi}{\sigma_2^2} + \frac{\sin^2\varphi}{\sigma_1^2} = \frac{1}{VY_2}\frac{1}{1 - \rho^2[Y_1, Y_2]}.$$

Finally, for the coefficient of y_1y_2 we obtain, using Theorem 69.2,

$$2\cos\varphi \sin\varphi \frac{\sigma_2^2 - \sigma_1^2}{\sigma_1^2\sigma_2^2} = \frac{-C[Y_1, Y_2]}{VY_1\, VVY_2 - C^2[Y_1, Y_2]}$$

$$= -2\frac{\rho[Y_1, Y_2]}{\sqrt{VY_1\, VY_2}}\frac{1}{1 - \rho^2[Y_1, Y_2]}.$$

Now, it is possible to restate Equation (54.5) in a form that uses only the parameters of P_Y. Writing $\mathbf{V}Y_1 = \sigma_{Y_1}^2$, $\mathbf{V}Y_2 = \sigma_{Y_2}^2$, and $\rho[Y_1,Y_2] = \rho_Y$, and using Theorem 69.2 we obtain

$$f_Y(y) = \frac{1}{2\pi\sigma_{Y_1}\sigma_{Y_2}\sqrt{1-\rho_Y^2}}\exp\left\{-\frac{1}{2(1-\rho_Y^2)}\left[\frac{y_1^2}{\sigma_{Y_1}^2} - 2\frac{\rho_Y y_1}{\sigma_{Y_1}\sigma_{Y_2}} + \frac{y_2^2}{\sigma_{Y_2}^2}\right]\right\}. \tag{69.9}$$

The most general form of the joint p.d.f. for two jointly Gaussian r.v.'s is obtained by allowing a shift of the center of the joint distribution to an arbitrary point (a_1, a_2), that is, by replacing y_1 and y_2 in Equation (69.9) by $y_1 - a_1$ and $y_2 - a_2$, respectively. This change will not affect the coefficients.

If $\varphi = 45°$, then the correlation coefficient ρ_Y in Equation (69.9) expresses in a simple way the 'thinness' of the elliptic cross section of the density function (see Problem 69.5). In the limiting case where $\rho_Y = \pm 1$, Equation (69.9) does not apply because the joint distribution is then collapsed into a singular distribution (and is not designated as jointly Gaussian or bivariate Gaussian). On the other hand, if $\rho_Y = 0$ then it can be seen that Equation (69.9) reduces to the joint density function of two s.i. Gaussian r.v.'s as given at the beginning of Example 69.2. Thus:

Theorem 69.4
If two jointly Gaussian r.v.'s are uncorrelated, they are also statistically independent.

The property expressed in this Theorem also applies to some other families of joint distributions but is, of course, not true in general. For instance, two symmetric binary r.v.'s also have this property, as can be seen from Example 69.1.

In Example 69.2, if $\sigma_1 = \sigma_2$, then f_X has circular symmetry. Therefore, any rotation leaves the joint distribution unchanged, so that Y_1, Y_2 are then also s.i. The sum and difference r.v.'s $X_1 - X_2$ and $X_1 + X_2$ are then s.i. as well (Problem 69.2).

From Example 69.2 we also see how a bivariate Gaussian distribution with nonzero correlation coefficient can be obtained simply by rotating the joint distribution of two statistically independent Gaussian r.v.'s. With most other distributions this is not possible. For instance, rotating the joint distribution induced by two statistically independent exponential distributions results in marginal distributions that in general are no longer exponential.

Queries

69.1 Two indicator r.v.'s I_A and I_B have correlation coefficient $\rho[I_A, I_B] = 0.6$. What is $P(A - B)$? [88]

69.2 In each of the distributions shown in Figure 69.3, probability 0.2 is assigned at each of the five points that are identified. Order the distributions according to increasing correlation coefficient, from -1 to $+1$. [63]

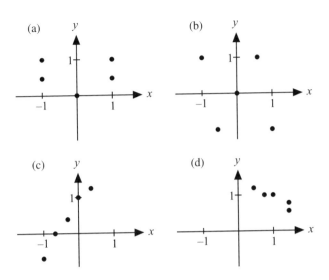

Figure 69.3   Four discrete distributions referred to in Query 69.2

69.3 Two zero-mean r.v.'s X_1, X_2 are jointly Gaussian with $\rho[X_1, X_2] = 0.5$. The joint distribution is rotated counterclockwise about the origin by an angle φ, resulting in a new distribution P_Y. For each of the following statements, determine whether it is true, possibly true, or false:
a) If $\varphi = 90°$, then $\rho[Y_1, Y_2]$ is negative
b) If $\varphi = 180°$, then $\rho[Y_1, Y_2] = -0.5$
c) If $\varphi = 45°$, then $\rho[Y_1, Y_2] = 0$
d) If $\rho[Y_1, Y_2] = \rho[Y_1, -Y_2]$, then $\rho[Y_1, Y_2] = 0$.   [123]

More General Kinds of Dependence Between Random Variables

We have seen that the correlation coefficient expresses the degree of *linear* dependence that exists between two r.v.s. If there is a different type of dependence between two r.v.'s, then their correlation coefficient will not reflect this.

Example 70.1

Let X be uniform over $-1, 1$. Let $Y = g(X) = \sqrt{1 - X^2}$ (see Figure 70.1). Then Y is uniquely determined by X, which is the strongest form of dependence. Yet, from Theorem 68.4 we see that $\rho[X, Y] = 0$.

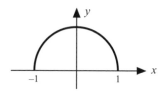

Figure 70.1   Joint distribution concentrated on a semi-circular arc

Is there a way to identify the kind of dependence that exists between two r.v.'s, if any: linear, or of some other functional form? It turns out this information can be obtained by examining the expectation of one of the two r.v.'s under the condition that the other r.v. takes on various values within its range—in other words, we need a *conditional expectation*. In Example 70.1, for instance, the condition that $X = x$ (for $|x| \le 1$) happens to imply a *unique* y-value, so that the conditional expectation of Y under the condition $X(\xi) = x$ must yield just this value, namely, $y = \sqrt{1 - x^2}$. If x is allowed to traverse the whole range of the r.v. X, that is, $-1 \le x \le 1$, then the 'conditional expectation of Y given X' traces out the graph shown in Figure 70.1. This is the graph of the transformation g in

Probability Concepts and Theory for Engineers, First Edition. Harry Schwarzlander.
© 2011 John Wiley & Sons, Ltd. Published 2011 by John Wiley & Sons, Ltd.

Example 70.1 and thus expresses exactly the relation that exists between Y and X. However, the conditioning event in this case has zero probability, so that Definition 67.1 is not applicable and the conditional expectation has to be defined in a different manner.

First, a *discrete* r.v. X will be considered, and some other r.v. Y. Let x be any real number in the range of X, with $P_X(\{x\}) > 0$. Then $\{\xi : X(\xi) = x\}$ is an event with positive probability in the underlying probability space. Therefore, the conditional expectation

$$\mathbf{E}[Y | \{\xi : X(\xi) = x\}] \tag{70.1}$$

which is also written $\mathbf{E}[Y | X = x]$, exists in accordance with Definition **67**.1a if the requirement stated there is satisfied.

It should be noted that if the expression (70.1) exists over the entire range of X, then Equation (70.1) defines a real-valued function over the range of X. In other words, it constitutes a transformation of the r.v. X, yielding a new r.v. that is denoted $\mathbf{E}[Y | X]$. This can be summarized as follows:

(Definition) 70.1
Let X be a discrete r.v., and Y an arbitrary r.v., both defined on a probability space $(\mathsf{S}, \mathscr{A}, P)$. The *conditional expectation* of Y given X, denoted $\mathbf{E}[Y|X]$, is a transformation of the r.v. X that assigns to each x in the range of X a value $\mathbf{E}[Y | X = x]$ (if that value is defined in accordance with Definition **67**.1).

Now it is desirable to circumvent the requirement that the conditioning event must have positive probability. This is possible by working with the *conditional distribution of Y given the value of X*, as defined in Section 59. Definition 70.1 can then be extended as follows.

(Definition) 70.2
Let X and Y be two r.v.'s defined on a probability space $(\mathsf{S}, \mathscr{A}, P)$. The *conditional expectation of Y given X*, denoted $\mathbf{E}[Y|X]$, is a transformation of the r.v. X that assigns to each $x \in \mathsf{R}$ the value

a) $\displaystyle\sum_y y \, P_{Y|X}(\{y\}|x)$, if $P_{Y|X}$ is discrete; or

b) $\displaystyle\int_{-\infty}^{\infty} y \, f_{Y|X}(y|x) dy$, if $f_{Y|X}$ exists

provided that the sum or integral is absolutely convergent.

We recall that in general, $P_{Y|X}$ and $f_{Y|X}$ are defined with probability 1. Therefore, Definition 70.2 may leave $\mathbf{E}[Y|X]$ undefined over some set $Q \subset \mathsf{R}$ for which $P_X(Q) = 0$.

Since $\mathbf{E}[Y|X]$ is some transformation r of the r.v. X, it is possible to write $\mathbf{E}[Y|X] = r(X)$. In the terminology of Statistics, this function r is called the *regression function of Y on X*. Example 70.1 illustrates the following property of the regression function:

Theorem 70.1
Let X and Y be two r.v.'s such that Y is some function of X, $Y = g(X)$, and $\mathbf{E}[Y|X]$ exists over the range of X. Then $\mathbf{E}[Y | X = x] = g(x)$ over this range.

Example 70.2
Let the r.v.'s X, Y have a joint p.d.f. that is uniform within the semi-circle shown in Figure 70.2, and zero elsewhere. We wish to find $\mathbf{E}[Y|X]$. The value of $f_{X,Y}$ in the shaded region is $2/\pi$. We have

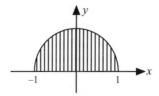

Figure 70.2   Uniform distribution over a semi-circle

$$f_X(x) = \begin{cases} \int_0^{\sqrt{1-x^2}} \dfrac{2}{\pi}\,dy = \dfrac{2}{\pi}\sqrt{1-x^2}, & -1 < x < 1 \\[2ex] 0, & \text{elsewhere.} \end{cases}$$

Then, for $-1 < x < 1$,

$$f_{Y|X}(y|x) = \frac{f_{X,Y}(x,y)}{f_X(x)} = \begin{cases} \dfrac{1}{\sqrt{1-x^2}}, & 0 < y < \sqrt{1-x^2} \\[2ex] 0, & \text{elsewhere.} \end{cases}$$

The conditional density function $f_{Y|X}(y|x)$ is defined only for $-1 < x < 1$; i.e., with probability 1. Now applying Definition 70.2,

$$E[Y|X = x] = \int_0^{\sqrt{1-x^2}} \frac{y}{\sqrt{1-x^2}}\,dy = \frac{1}{\sqrt{1-x^2}} \int_0^{\sqrt{1-x^2}} y\,dy = \frac{1}{2}\sqrt{1-x^2}$$

so that $E[Y|X] = \dfrac{1}{2}\sqrt{1-x^2}$. A plot of the values taken on by $E[Y|X = x]$ as a function of x, or the 'regression curve of Y on X', results in a semi-ellipse, as shown in Figure 70.3. In this simple example, the result is clear by inspection since it expresses for each x-value the mean of the conditional y-distribution at that x-value. Of course, a completely different plot is obtained for $E[X|Y=y]$; namely, a vertical line segment from $(0, 0)$ to $(0, 1)$. Thus, it turns out that $E[X|Y] = E[X]$; in other words, $E[X|Y]$ is a unary r.v.

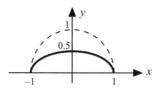

Figure 70.3 $E[Y|X]$ for the joint distribution in Figure 70.2

Since for two r.v.'s X, Y the conditional expectation $E[Y|X]$ is a r.v., its expected value can be found. This leads to a result similar to Theorem 67.1:

Theorem 70.2
Given two r.v.'s X, Y, then $E[E[Y|X]] = EY$ (provided the various expectations exist).

Proof (Absolutely continuous case)
Since $E[Y|X]$ is a function of the r.v. X, Theorem 63.2 can be used:

$$E[E[Y|X]] = \int_{-\infty}^{\infty} E[Y|X=x]f_X(x)dx = \int_{-\infty}^{\infty}\int_{-\infty}^{\infty} y f_{Y|X}(y|x)f_X(x)dy\,dx$$

$$= \int_{-\infty}^{\infty}\int_{-\infty}^{\infty} y f_{X,Y}(x,y)dy\,dx = \int_{-\infty}^{\infty} y f_Y(y)dy = EY.$$

In Theorem 70.1, if g is the identity function (over a set to which X assigns probability 1), that is, in the special case where $Y = X$ [P], the conditional expectation $E[Y|X] = E[X|X] = X$. If Y in Theorem 70.1 gets multiplied by another r.v. Z and we consider $E[YZ|X]$, the following generalization of Theorem **70**.1 is obtained:

Theorem 70.3
Let X, Y, Z be three r.v.'s such that $Y = g(X)$ [P]. Then if the expectations exist,

$$E[YZ|X] = Y\,E[Z|X]\ [P].$$

This result follows from the fact that for each x-value assigned by X, the r.v. $Y = g(X)$ is a constant multiplier that, by linearity of the expectation operation, can be brought outside the expectation operation.

Example 70.3
As an illustration of the use of Theorems 70.2 and 70.3, we show that

$$E[Y\,E[Y|X]] = E[(E[Y|X])^2] \tag{70.2}$$

where X, Y are any r.v.'s such that the expressions in Equation (70.2) exist with probability 1. Theorem 70.2 allows the left side of Equation (70.2) to be written $E\{E[YE[Y|X]|X]\}$, where $E[YE[Y|X]|X]$ is of a form to which Theorem 70.3 applies, since $E[Y|X]$ is a function of X. Thus, $E[YE[Y|X]|X] = E[Y|X]\,E[Y|X] = E[Y|X]^2$, and Equation (70.2) follows.

A conditional expectation can be useful in determining a joint expectation. Suppose $E[g(X, Y)]$ is to be found, where X, Y are jointly absolutely continuous r.v.'s and g is a function of two variables, such that $g(X, Y)$ is also absolutely continuous. Then Theorem 59.1 can be used to obtain

$$E[g(X,Y)] = \int\int g(x,y)f_{X,Y}(x,y)dx\,dy = \int\int g(x,y)f_{Y|X}(y|x)f_X(x)dy\,dx$$

$$= \int\int [g(x,y)f_{Y|X}(y|x)dy]f_X(x)dx = E[E[g(X,Y)|X]]. \tag{70.3}$$

This result can also be written $E_X[E_{Y|X}[g(X, Y)|X]]$, where the subscripts make it explicit that the inner expectation is a conditional expectation requiring averaging with respect to $f_{Y|X}$, and the outer expectation calls for averaging with respect to f_X.

Now, suppose that X, Y are statistically independent r.v.'s. Then $\mathbf{E}[Y|X]$ reduces to $\mathbf{E}[Y]$, since in this case specifying a value assigned by X does not affect the distribution induced by Y. The r.v. $\mathbf{E}[Y|X]$ then is unary and assigns the fixed value $\mathbf{E}[Y]$. Also, Equation (70.3) then becomes $\mathbf{E}[g(X, Y)] = \mathbf{E}_X \mathbf{E}_Y[g(X, Y)]$; i.e., the expected value can be computed first with respect to one of the r.v.'s, then with respect to the other. However, $\mathbf{E}[Y|X] = \mathbf{E}[Y]$ does not imply that X, Y are s.i. (as was seen in Example **70.2**).

What about the conditional expectation of a conditional expectation? This can arise in an expression of the form $\mathbf{E}[\mathbf{E}[Y|X]|X]$, where both expectations have the same conditioning r.v X. To understand the meaning of this, consider first the outer expectation. It specifies the expected value of the r.v. $\mathbf{E}[Y|X]$ under the condition that X assigns some value, $X = x$. But for every x in the range of X, if $X = x$, then $\mathbf{E}[Y|X]$ is simply the constant $\mathbf{E}[Y \mid X = x]$, so that

$$\mathbf{E}[\mathbf{E}[Y|X]|X] = \mathbf{E}[Y|X]. \tag{70.4}$$

Now let $Z = g(X)$, a function of X, and consider $\mathbf{E}[\mathbf{E}[Y|X]|Z]$. Here, the outer expectation specifies the expected value of the r.v. $\mathbf{E}[Y|X]$ under the condition that Z assigns some value, $Z = z$. For any z in the range of Z, $g^{-1}(z)$ is a set of one or more x-values. Thus, $\mathbf{E}[\mathbf{E}[Y|X]|Z = z]$ takes the expectation of the r.v. $\mathbf{E}[Y|X]$ over all of $g^{-1}(z)$. For instance, suppose that $g^{-1}(z) = \{x_1, \ldots, x_n\}$; then the expectation of the r.v. $\mathbf{E}[Y|X]$ is over $\mathbf{E}[Y|X = x_1], \ldots, \mathbf{E}[Y|X = x_n]$. Therefore, it is possible to write

$$\mathbf{E}[\mathbf{E}[Y|X]|Z] = \mathbf{E}[Y|Z]. \tag{70.5}$$

Clearly, Equation (70.4) is a special case of Equation (70.5).

Queries

70.1 A red and a white die are thrown. Let X express the number of dots obtained on the red die, and Y the sum of the dots on both dies. Determine:
 a) $\mathbf{E}[Y|X = 2]$
 b) $\mathbf{E}[X|Y = 2]$.   [171]

70.2 $\mathbf{E}[Y|X]$ is uniformly distributed over $\overline{-1, 1}$.
 a) What is m_{1Y} ?
 b) Is X necessarily absolutely continuous?
 c) Are X and Y necessarily jointly absolutely continuous?   [57]

70.3 Given two jointly absolutely continuous r.v.'s X, Y such that $\mathbf{E}[Y|X] = \mathbf{E}[X]$. For each of the following statements, determine whether it is true, possibly true, or false.
 a) X is uniformly distributed
 b) X and Y are linearly related
 c) X and Y are statistically independent
 d) X and Y have identical means
 e) X and Y are uncorrelated.   [137]

The Covariance Matrix

If X is an n-dimensional random variable ($n \geq 2$) and is treated as a column vector, $X = [X_1, X_2, \ldots, X_n]^{\mathrm{T}}$ (where the exponent ^{T} denotes the transpose), then by the expectation of X is meant the vector

$$\mathbf{E}X = \begin{bmatrix} \mathbf{E}X_1 \\ \mathbf{E}X_2 \\ \vdots \\ \mathbf{E}X_n \end{bmatrix}. \tag{71.1}$$

$\mathbf{E}X$ identifies a specific point in R^n whose coordinates are the expectations of all the components of X. It represents the 'center of gravity' of the n-dimensional distribution induced by X. In Equation (71.1) it is assumed that all the expectations on the right exist, otherwise $\mathbf{E}X$ is not defined.

The various second-order central moments of the component random variables of X make up the *covariance matrix*:

> **(Definition) 71.1**
> Given an n-dimensional r.v. X ($n \geq 2$), represented as a column vector as in Equation (71.1), with component r.v.'s whose variances exist. Then the *covariance matrix* of X, denoted $\mathbf{C}[X, X]$ or simply c_X, is the matrix
>
> $$\mathbf{C}[X, X] = c_X = \mathbf{E}[(X - \mathbf{E}X)(X - \mathbf{E}X)]^{\mathrm{T}}$$
> $$= \begin{bmatrix} \mathbf{V}X_1 & \mathbf{C}[X_1, X_2] & \mathbf{C}[X_1, X_3] & \cdots & \mathbf{C}[X_1, X_n] \\ \mathbf{C}[X_2, X_1] & \mathbf{V}X_2 & \mathbf{C}[X_2, X_3] & \cdots & \cdots \\ \cdots & \cdots & \cdots & \cdots & \cdots \\ \mathbf{C}[X_n, X_1] & \mathbf{C}[X_n, X_2] & \mathbf{C}[X_n, X_3] & \cdots & \mathbf{V}X_n \end{bmatrix}.$$

When only a single vector r.v. is under consideration, the notation for the covariances can be simplified to $\mathbf{C}[X_i, X_j] = c_{ij}$. Since the diagonal terms in c_X can then also be written c_{ii}, for $i = 1, \ldots, n$, the covariance matrix can as well be thought of as the matrix whose elements are c_{ij}, for $i, j = 1, \ldots, n$.

It can be seen from Definition 71.1 that c_X is a symmetric matrix whose diagonal elements are non-negative. If every pair of the component r.v.'s $X_1, \ldots, X_n$ is uncorrelated, then c_X is a diagonal matrix. Otherwise, the off-diagonal elements of the matrix have to satisfy Equation (68.3). However, the

---

Probability Concepts and Theory for Engineers, First Edition. Harry Schwarzlander.
© 2011 John Wiley & Sons, Ltd. Published 2011 by John Wiley & Sons, Ltd.

values that can appear in the off-diagonal positions of the matrix must satisfy a more stringent condition, and this is expressed by the property of *positive semi-definiteness*.

(Definition) 71.2

An $n \times n$ matrix c is positive semi-definite (or non-negative definite) if, for every column vector of real numbers $q = \begin{bmatrix} q_1 \\ \vdots \\ q_n \end{bmatrix}$, it is true that

$$q^{\mathrm{T}} c q \geq 0. \tag{71.2}$$

If the above-mentioned notation c_{ij} is used in Definition 71.2 to denote the element of c in the ith row and jth column, then the requirement (71.2) can also be written $\sum_{i,j} q_i q_j c_{ij} \geq 0$. We now have:

Theorem 71.1

The covariance matrix of a multidimensional or vector random variable is positive semi-definite.

Proof

Let $X = \begin{bmatrix} X_1 \\ \vdots \\ X_n \end{bmatrix}$ be a vector r.v. with covariance matrix c_X, and $q = \begin{bmatrix} q_1 \\ \vdots \\ q_n \end{bmatrix}$ a vector of n arbitrary real numbers. Then,

$$\begin{aligned} q^{\mathrm{T}} c_X q &= \mathbf{E}[q^{\mathrm{T}}(X - \mathbf{E}X)(X - \mathbf{E}X)^{\mathrm{T}} q] \\ &= \mathbf{E}\{[q_1(X_1 - \mathbf{E}X_1) + q_2(X_2 - \mathbf{E}X_2) + \ldots + q_n(X_n - \mathbf{E}X_n)]^2\}, \end{aligned}$$

which is non-negative, since it is the expectation of a non-negative r.v.

While Definition 71.2 does define positive semi-definiteness, it provides no convenient method for checking whether a given $n \times n$ matrix satisfies the requirement of positive semi-definiteness and therefore can be the covariance matrix of some vector r.v. If $n = 2$, i.e., $X = \begin{bmatrix} X_1 \\ X_2 \end{bmatrix}$, the situation is simple because in that case positive semi-definiteness of c_X is equivalent to the requirement expressed by Equation (68.3) (see Problem 71.6). Before considering the general case, we examine the effect of a linear transformation on the covariance matrix.

Theorem 71.2

Let $X = \begin{bmatrix} X_1 \\ \vdots \\ X_n \end{bmatrix}$ be a vector r.v. whose covariance matrix exists. Let a *linear transformation a* be

applied to X, resulting in the new vector r.v. $Y = \begin{bmatrix} Y_1 \\ \vdots \\ Y_n \end{bmatrix}$; i.e.,

$$Y = a X \qquad (71.3)$$

where a is the transformation matrix. Then,

$$c_Y = a c_X a^{\mathrm{T}}. \qquad (71.4)$$

Proof

$$
\begin{aligned}
c_Y &= \mathbf{E}[(Y - \mathbf{E}Y)(Y - \mathbf{E}Y)^{\mathrm{T}}] = \mathbf{E}[(a X - \mathbf{E}[a X])(a X - \mathbf{E}[a X])^{\mathrm{T}}] \\
&= \mathbf{E}[a(X - \mathbf{E}X)[a(X - \mathbf{E}X)]^{\mathrm{T}}] = \mathbf{E}[a(X - \mathbf{E}X)(X - \mathbf{E}X)^{\mathrm{T}} a^{\mathrm{T}}] \\
&= a\,\mathbf{E}[(X - \mathbf{E}X)(X - \mathbf{E}X)^{\mathrm{T}}] a^{\mathrm{T}} = a\, c_X\, a^{\mathrm{T}}.
\end{aligned}
$$

In Theorem 71.2, if c_X is singular or a is singular—i.e., of rank less than n—then c_Y will be singular as well.

In the two-dimensional case, if the transformation matrix is written

$$a = \begin{bmatrix} a & b \\ c & d \end{bmatrix}$$

then Equation (71.3) reduces to Equation (69.2). The elements of c_Y are then given by Equation (69.3).

Example 71.1: Density function of n jointly Gaussian r.v.'s.

We begin with a vector r.v. X whose components are n (mutually) s.i. zero-mean Gaussian r.v.'s $X_1, \ldots, X_n$ with positive variances $\sigma_1^2, \ldots, \sigma_n^2$, respectively, so that the covariance matrix of X is

$$
\mathbf{C}[X, X] = c_X = \begin{bmatrix}
\sigma_1^2 & 0 & \cdots & 0 \\
0 & \sigma_2^2 & \cdots & 0 \\
\vdots & \vdots & & \vdots \\
0 & 0 & \cdots & \sigma_n^2
\end{bmatrix}.
$$

From Equation (40.1), the joint p.d.f. is

$$
f_X(x) = \prod_{i=1}^{n} \frac{1}{\sqrt{2\pi}\sigma_i} \exp\left(-\frac{x_i^2}{2\sigma_i^2} \right)
$$

$$
= (2\pi)^{-n/2} \left(\prod_{i=1}^{n} \sigma_i \right)^{-1} \exp\left[-\frac{1}{2} \sum_{i=1}^{n} \left(\frac{x_i}{\sigma_i} \right)^2 \right]. \qquad (71.5a)
$$

Since c_X^{-1} is a diagonal matrix with elements $1/\sigma_1^2, \ldots, 1/\sigma_n^2$ along the major diagonal, this joint p.d.f. can be written more compactly as

$$
f_X(x) = (2\pi)^{-n/2} [\det(c_X)]^{-1/2} \exp\left[\left(-\frac{1}{2} \right) x^{\mathrm{T}} c_X^{-1} x \right] \qquad (71.5b)
$$

where $\det(c_X)$ denotes the determinant of c_X, which is positive since the σ_i's are positive.

Proceeding in a manner similar to Example 69.2, suppose that the n-dimensional distribution P_X is to get *rotated* in an arbitrary manner. That is, a new r.v. Y is specified by $Y = aX$, where a is an *orthogonal* matrix; i.e., $\det(a) = \pm 1$. An arbitrary orthogonal transformation, besides producing a rotation, may also introduce a reversal of one or more axes. Since $X_1, \ldots, X_n$ are zero-mean and therefore symmetrical about the origin, such a reversal is of no consequence as long as only the result of the transformation is of interest, and not the relationship between X and Y.

Since a is orthogonal, the Jacobian $J_a = \det(a) = \pm 1$, and $X = a^{-1}aX = a^{T}aX = a^{T}Y$, so that (53.3) gives

$$f_Y(y) = f_X(a^{T}y)$$
$$= (2\pi)^{-n/2}[\det(c_X)]^{-1/2}\exp\left[\left(-\frac{1}{2}\right)y^{T}a\,c_X^{-1}\,a^{T}y\right].$$

c_Y, the covariance matrix of Y, satisfies $\det(c_Y) = \det(c_X)$. Using Equation (71.4) and noting that $a\,c_X^{-1}\,a^{T} = c_Y^{-1}$, we see that f_Y can be written in terms of c_Y in the manner of Equation (71.5), the difference being that c_Y is not necessarily diagonal:

$$f_Y(y) = (2\pi)^{-n/2}[\det(c_Y)]^{-1/2}\exp\left[\left(-\frac{1}{2}\right)y^{T}\,c_Y^{-1}\,y\right]. \tag{71.6}$$

This is the general form of the density function for an n-dimensional zero-mean multivariate Gaussian distribution. Verification that the marginal densities are indeed Gaussian is left as an exercise (see Problem 71.7).

It is important to note in Equation (71.6) that c_Y has rank n, which follows from the stipulation that $\sigma_1^2, \ldots, \sigma_n^2 > 0$. If this is not satisfied, then X and Y are singular n-dimensional r.v.'s that induce so-called 'degenerate' multivariate Gaussian distributions in $\mathbb{R}^n$, and Equation (71.6) does not apply since $\det(c_Y) = 0$ in that case and c_Y^{-1} does not exist. In other words, an n-dimensional multivariate Gaussian distribution must have a covariance matrix that is positive *definite*, which means that Equation (71.2) must be a strict inequality.

We have seen in Example 71.1 that $\det(c_Y) \geq 0$ if c_Y is the covariance matrix for an n-dimensional Gaussian or degenerate Gaussian distribution. Actually, this is true for *any* covariance matrix since a positive semi-definite matrix has a non-negative determinant.[5]

At the beginning of Example 71.1, 'mutually' appears in parentheses. That is motivated by the following result, which follows from Problem 71.3:

Theorem 71.3
If n Gaussian r.v.'s are pairwise statistically independent, then they are mutually statistically independent.

Consider now an arbitrary distribution P_X in n dimensions, induced by a vector r.v. X, such that c_X exists and is not a diagonal matrix. Is it possible to find a linear transformation (71.3) that will cause the resulting distribution P_Y to be such that the components of Y will all be uncorrelated? The answer is yes. The transformation that accomplishes this must have a transformation matrix a that diagonalizes c_X.

Diagonalization of a real symmetric $n \times n$ matrix c is achieved by a *similarity transformation*, $a^{-1}ca$, where a is an $n \times n$ matrix that has the eigenvectors of c as columns. But the problem here is not merely to transform the covariance matrix c_X into a diagonal matrix; it is to transform the r.v. X so

---

[5] See Sections 7.1 and 7.2 in [HJ1].

that the resulting covariance matrix is diagonal. According to Theorem 71.2, if a transformation a^T is applied to X, this changes the covariance matrix to $c_Y = a^T c_X a$, whereas, as noted above, diagonalization of c_X calls for $a^{-1} c_X a$. This discrepancy can be eliminated by requiring $a^{-1} = a^T$; i.e., by requiring a to be an *orthogonal* matrix (such as the transformation a in Example 71.1). This is possible since the eigenvectors of c_X, which are to constitute the columns of a (or the rows of a^T) are specified only to within an arbitrary constant multiplier. The desired orthogonal matrix is obtained by scaling the eigenvectors that make up the columns of a so that they are unit vectors. Furthermore, c_Y will then have on its major diagonal the eigenvalues of c_X.

We return now to the earlier question: Given a real symmetric $n \times n$ matrix c with non-negative diagonal elements, then how can positive semi-definiteness of c be established? If c is assumed to be the covariance matrix of some n-dimensional r.v. X, then the transformation described above will yield a diagonal matrix c'. But this transformation need not actually be carried out since we know that c' has the eigenvalues of c on its principal diagonal. Since c is real and symmetric, its eigenvalues are real. Then, if those eigenvalues are non-negative, they can be the variances of the n uncorrelated components of a r.v. Y. In that case, according to Theorem 71.2, c is also a covariance matrix and therefore is positive semi-definite (by Theorem 71.1). This shows:

Theorem 71.4
A real symmetric $n \times n$ matrix c is positive semi-definite if and only if all the eigenvalues of c are non-negative.

The above principles are illustrated in Example 71.2.

Example 71.2: Diagonalizing a covariance matrix.

A r.v. $X = \begin{bmatrix} X_1 \\ X_2 \\ X_3 \\ X_4 \end{bmatrix}$ has the covariance matrix $c_X = \begin{bmatrix} 1/9 & -2/9 & 2/9 & 0 \\ -2/9 & 4/3 & 4/9 & 4/9 \\ 2/9 & 4/9 & 4/3 & 4/9 \\ 0 & 4/9 & 4/9 & 2/9 \end{bmatrix}$.

We begin by verifying that c_X is a valid covariance matrix. Visual inspection establishes that it is a symmetric matrix with non-negative elements on the principal diagonal. Now the eigenvalues of c_X are computed by solving the equation $\det|c_X - \lambda I| = 0$, where I denotes the unit matrix. This yields

$$\begin{vmatrix} 1/9 - \lambda & -2/9 & 2/9 & 0 \\ -2/9 & 4/3 - \lambda & 4/9 & 4/9 \\ 2/9 & 4/9 & 4/3 - \lambda & 4/9 \\ 0 & 4/9 & 4/9 & 2/9 - \lambda \end{vmatrix} = \lambda^2 (\lambda^2 - 3\lambda + 2) = 0$$

giving the eigenvalues $\lambda_1 = \lambda_2 = 0$, $\lambda_3 = 1$, and $\lambda_4 = 2$, where the subscripts have been assigned arbitrarily to arrange the eigenvalues in increasing order. Since the eigenvalues are non-negative, c_X as given is a valid covariance matrix.

Next, we ask what rotational transformation, when applied to X, will decorrelate all components of the r.v. In other words, an orthogonal transformation matrix a is to be found such that $Y = aX$ has a diagonal covariance matrix with diagonal elements 0, 0, 1, and 2 (in any order). This can be accomplished easily using a suitable computer routine, but here the computational steps are carried out in order to exhibit the principles involved.

To construct a, a set of four orthogonal eigenvectors e_i must be found, where e_i corresponds to λ_i ($i = 1, 2, 3, 4$). This is accomplished by solving, for $i = 1, \ldots, 4$, the equation

$$[c_X - \lambda_i I] e_i = 0 \tag{71.7}$$

where $\mathbf{0}$ denotes the zero column vector.

Since $\lambda_1 = \lambda_2$, i.e., $\lambda = 0$ is a double eigenvalue, the solution of Equation (71.7) for $i = 1, 2$ will not yield a pair of vectors e_1, e_2 whose directions in R^4 are completely specified. Orthogonality to e_3 and e_4 will be satisfied, but the orthogonality between e_1 and e_2 needs to be expressed as a supplementary condition. Writing

$$e_i = [e_{i_1}, e_{i_2}, e_{i_3}, e_{i_4}]^{\mathrm{T}}$$

then for $\lambda_1 = \lambda_2 = 0$, Equation (71.7) becomes

$$c_X\, e_i = \mathbf{0}. \tag{71.8}$$

Multiplying out Equation (71.8) gives these equations:

$$e_{i1} = 2(e_{i2} - e_{i3}), \quad \text{and} \quad e_{i4} = -2(e_{i2} + e_{i3})$$

for $i = 1, 2$. The remaining two equations which can be derived from Equation (71.8) are redundant. Normalization requires that $e_i^{\mathrm{T}} e_i = 1$, or

$$e_{12}{}^2 + e_{13}{}^2 = 1/9, \text{ and} \tag{71.9a}$$

$$e_{22}{}^2 + e_{23}{}^2 = 1/9. \tag{71.9b}$$

Introduction of the orthogonality requirement, $e_1^{\mathrm{T}} e_2 = 0$, now yields the condition

$$e_{12}\, e_{22} + e_{13}\, e_{23} = 0. \tag{71.9c}$$

One way of satisfying Equations (71.9) is to set $e_{12} = e_{23} = 0$, and $e_{13} = e_{22} = 1/3$. This results in

$$e_1 = [-2/3 \quad 0 \quad 1/3 \quad -2/3]^{\mathrm{T}}, \text{ and } e_2 = [2/3 \quad 1/3 \quad 0 \quad -2/3]^{\mathrm{T}}.$$

The remaining two eigenvectors of c_X are readily obtained to within an arbitrary multiplicative constant by solving Equation (71.7) for $i = 3$ and 4, respectively. Normalization then gives these results:

$$e_3 = \pm[-1/3 \quad 2/3 \quad -2/3 \quad 0]^{\mathrm{T}}, \text{ and } e_4 = \pm[0 \quad 2/3 \quad 2/3 \quad 1/3]^{\mathrm{T}}.$$

Arbitrarily choosing the $+$ sign for e_3 and e_4, a transformation matrix a of the desired form can now be specified by setting

$$a = \begin{bmatrix} e_1^{\mathrm{T}} \\ e_2^{\mathrm{T}} \\ e_3^{\mathrm{T}} \\ e_4^{\mathrm{T}} \end{bmatrix} = \begin{bmatrix} -2/3 & 0 & 1/3 & -2/3 \\ 2/3 & 1/3 & 0 & -2/3 \\ -1/3 & 2/3 & -2/3 & 0 \\ 0 & 2/3 & 2/3 & 1/3 \end{bmatrix}. \tag{71.10}$$

Clearly, the order in which the e_i's are arranged in Equation (71.10) is arbitrary, and determines the order in which the eigenvalues appear along the major diagonal of c_Y. It should be noted that Equation (71.10) may differ from a strict rotational transformation by one or two axis reversals. Another observation to be made is that, in the notation of Theorem 71.2, diagonalization of c_X would require that the transformation matrix a use the eigenvectors as *columns*. However, as discussed earlier, the matrix a in this Example really corresponds to a^T in Theorem 71.2, and therefore has to be constructed with the eigenvectors as *rows*.

Finally, it is now possible to express the relationship between the components of Y and the components of X, for a as given in Equation (71.10). Since $Y = aX$, we see that

$$Y_1 = \frac{1}{3}(-2X_1 + X_3 - 2X_4)$$

$$Y_2 = \frac{1}{3}(2X_1 + X_2 - 2X_4)$$

$$Y_3 = \frac{1}{3}(-X_1 + 2X_2 - 2X_3)$$

$$Y_4 = \frac{1}{3}(2X_2 + 2X_3 + X_4).$$

By referring back to c_X, it is easy to verify that $VY_i = \lambda_i$, $i = 1, \ldots, 4$, and that the components of Y are uncorrelated.

We note that the r.v. Y in Example 71.2 is singular, and therefore X is singular as well. More specifically, both P_X and P_Y distribute all probability in a two-dimensional subspace of R^4. This is more easily seen in the case of P_Y, which distributes all probability in the third and fourth dimensions, while in dimensions one and two P_Y assigns probability 1 to a single point. Of course, not every singular n-dimensional r.v. induces a distribution that within some subspace of R^n collapses to a point; so that not every singular n-dimensional r.v. has a singular covariance matrix. An example is a r.v. that induces probability 1 uniformly over the surface of an n-dimensional sphere.

Queries

71.1 Given a 6×6 covariance matrix $\mathbf{C}[X, X]$ with a 0 in the lower right-hand corner. This implies which of the following:
  a) The last row and last column must be all zeros
  b) All the entries along the major diagonal must be zero
  c) The given matrix cannot be a covariance matrix.   [206]

71.2 Consider the covariance matrix $\mathbf{C}[X, X]$ of the six-dimensional r.v. X defined in Query 52.2.
  a) What are the entries on the major diagonal of $\mathbf{C}[X, X]$?
  b) What are the values of the off-diagonal elements of $\mathbf{C}[X, X]$?   [149]

71.3 Two s.i. n-dimensional r.v.'s X, Y have covariance matrices with identical sets of eigenvalues. For each of the following, determine whether it is true, possibly true, or false:
  a) The components of X, Y have identical sets of variances
  b) $X = Y$
  c) Y can be obtained from X by applying an orthogonal transformation to X
  d) Let $X' = aX$ and $Y' = bY$, where a, b are orthogonal transformation matrices such that $c_{X'}$ and $c_{Y'}$ are diagonal. Then $c_{X'} = c_{Y'}$.   [120]

Random Variables as the Elements of a Vector Space

A vector space, or linear space, is a collection of elements on which certain operations are defined and these operations must satisfy a particular set of rules. It is a generalization of the idea of vectors as used, for instance, in problems of mechanics. The 'elements' of a vector space can therefore always be thought of as vectors. Specifically:

(Definition) 72.1

A collection V of elements $x, y, \ldots$ is called a *vector space* if for this collection is defined a binary operation denoted ' $+$ ', and another operation that allows associating with any element x a 'scalar' α (for our purposes, a real number), the result being denoted αx. Furthermore, the following rules must be satisfied:

1. If $x \in V$ and $y \in V$, then $x + y \in V$.
2. If $x \in V$, then $\alpha x \in V$.
3. To V belongs a 'zero element', denoted $\mathbf{0}$.
4. If $x \in V$, then an element '$-x$' also belongs to V, where $(-x) + x = \mathbf{0}$.

Various kinds of mathematical objects might be represented by the elements of a vector space. For instance, each element might represent a function of a real variable—in which case the vector space is a *function space*. It turns out that (scalar) random variables can be regarded as elements of a vector space. That can be of use in problems where linear combinations of r.v.'s arise. Since r.v.'s are functions, such a space is again a type of function space. We will not develop here the general properties of vector spaces but merely examine several pertinent features of a space whose elements are r.v.'s.

Underlying the discussion is (as always when we speak of r.v.'s) some probability space $(S, \mathscr{A}, P)$ on which all r.v.'s associated with a given problem or discussion are defined. Thus, let some probability space $(S, \mathscr{A}, P)$ be given, on which is defined a zero-mean r.v. X whose variance exists. We wish to regard X as an element or point of a vector space V. In order to do this, Definition 72.1 has to be satisfied.

Probability Concepts and Theory for Engineers, First Edition. Harry Schwarzlander.
© 2011 John Wiley & Sons, Ltd. Published 2011 by John Wiley & Sons, Ltd.

Condition (2) of Definition 72.1 is satisfied if αx is interpreted as ordinary multiplication of a r.v. by a real constant, since the definition of X immediately allows scaled versions of X to be defined on $(S, \mathscr{A}, P)$, and these are again zero-mean and therefore represented in V. Furthermore, if the ' $+$ ' operation is taken as ordinary addition, condition (1) is satisfied since for two zero-mean r.v.'s X, Y the sum $X + Y$ exists on $(S, \mathscr{A}, P)$ and is again a zero-mean r.v. Also, multiplication of X by 0 yields a unary r.v. that satisfies condition (3), and interpreting '$-x$' as $-X$ then satisfies condition (4).

It is necessary to introduce two other properties that are familiar from the application of vectors in mechanics: length and direction. In a more general formulation, these two properties are referred to as the 'norm' and 'inner product', respectively:

(Definition) 72.2
A number associated with every element x of a vector space V, satisfying the following requirements, is called the *norm* of x, and is denoted $||x||$:

1. $||x||$ is real and ≥ 0, with $||x|| = 0$ if and only if $x = 0$.
2. For any scalar α, $||\alpha x|| = |\alpha| \cdot ||x||$.
3. (Triangle inequality) For $x, y \in$ V, $||x + y|| \leq ||x|| + ||y||$.

A vector space on which a norm is defined is called a *normed vector space*.

(Definition) 72.3
A number associated with any pair of element x, y of a (real) normed vector space V, satisfying the following requirements, is called the *inner product* of x and y, and is denoted $\langle x, y \rangle$:

1. $\langle y, x \rangle = \langle x, y \rangle$.
2. $\langle x, x \rangle = ||x||^2$.
3. For any real constant α, $\langle \alpha x, y \rangle = \alpha \langle x, y \rangle$.
4. $\langle x + z, y \rangle = \langle x, y \rangle + \langle z, y \rangle$.

Consider again a probability space $(S, \mathscr{A}, P)$ and some zero-mean r.v. X defined on this space, which therefore is represented by a point in the associated vector space V (see Figure 72.1a). It is possible to make V a normed vector space by specifying the norm of X as $\sqrt{E[X^2]}$. This quantity satisfies Definition 72.2 (for the triangle inequality, see Problem 64.11). Since $EX = 0$, $\sqrt{E[X^2]} = \sqrt{VX}$, and we see that it is the standard deviation of X that expresses the distance of the point X from the point 0, or the length of the vector X. Furthermore, a r.v. aX, where a is any real number, is

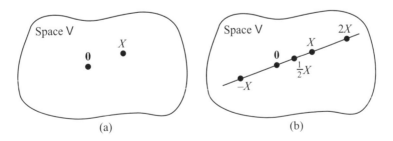

Figure 72.1   A random variable X as an element in the space V

represented by a point whose distance from the origin is proportional to a. This is illustrated in Figure 72.1b for $a = 2$, 1/2, and -1.

However, the familiar idea that all these elements are to lie on a single line through the origin, and that $-X$ lies in the opposite direction from X, does not follow from the norm. For this, an appropriate inner product is needed. V can be made into an inner product space by using $\mathbf{E}[XY]$ as the inner product of two zero-mean r.v.'s $X, Y \in$ V. It is easily verified that $\mathbf{E}[XY]$ satisfies the requirements of Definition 72.3.

How does the inner product express the angle θ between two vectors? Let $X, Y \in$ V have nonzero variance. Clearly, the *length* of the vectors X and Y should have no effect on the angle. So it appears that the inner product needs to be normalized before it can express an angle. This gives (keeping in mind that $\mathbf{E}X = \mathbf{E}Y = 0$)

$$\frac{\langle X, Y \rangle}{||X|| \, ||Y||} = \frac{\mathbf{E}[XY]}{\sqrt{\mathbf{V}X}\sqrt{\mathbf{V}X}} = \rho_{X,Y}. \tag{72.1}$$

Now, if $Y = aX$, where a is a nonzero real constant, then $\rho_{X,Y} = \pm 1$, the sign depending on whether a is positive or negative. The angle θ between the vectors X and aX should be 0 if $a > 0$, and π if $a < 0$. This result can be achieved by setting Equation (72.1) equal to $\cos \theta$.

Next, suppose that Y is statistically independent of X. With X as shown in Figure 72.1a, what point in V represents Y? We want the rules of vector addition to hold. The length of the vector Y must be $\sqrt{\mathbf{E}[Y^2]}$, whereas the length of vector $W = X + Y$ (lying in the plane defined by $\mathbf{0}$, X and Y) is

$$\sqrt{\mathbf{E}[W^2]} = \sqrt{\mathbf{E}[X^2] + \mathbf{E}[Y^2]}. \tag{72.2}$$

Thus, the length of the vector W will obey the rule for vector addition if Y is placed in a direction that is *orthogonal to the vector* X (see Figure 72.2). Actually, statistical independence is not necessary for Equation (72.2) to hold. If it is merely known that $\mathbf{E}[XY] = 0$, then the r.v.'s X and Y are called *orthogonal*. Thus, orthogonal r.v.'s are those that need to be represented as orthogonal vectors in the space V. Furthermore, two or more orthogonal r.v.'s can serve to define a set of coordinate axes in V.

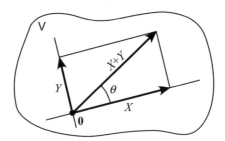

Figure 72.2   Addition of orthogonal random variables

Clearly, if at least one of two r.v.'s is zero mean and they are orthogonal, then they are also uncorrelated. Since we restrict V to zero-mean r.v.'s, the properties of orthogonality and uncorrelatedness are equivalent *in* V.

Finally, with X and Y still orthogonal, what is the angle θ between X and W (see Figure 72.2)? Here, we want basic vector geometry to be satisfied, that is, $\cos \theta = ||X||/||W||$. From Equation (72.2) and Example 68.2,

$$\frac{||X||}{||W||} = \frac{\sqrt{\mathbf{E}[X^2]}}{\sqrt{\mathbf{E}[X^2] + \mathbf{E}[Y^2]}} = \frac{1}{\sqrt{1 + \dfrac{\mathbf{V}Y}{\mathbf{V}X}}} = \rho_{W,X}.$$

In other words, $\rho_{W,X}$ is the 'direction cosine' of the vector W relative to X.[6] We conclude that for two arbitrary r.v.'s X, Y the relation $\cos\theta = \rho_{X,Y}$ correctly specifies the angle θ between X and Y in V.

It can be seen that the dimensionality of V is determined by how many mutually orthogonal r.v.'s can be defined on the given probability space. Furthermore, every point in V corresponds to a centralized r.v. whose variance exists, and that can be defined on the underlying probability space.

Example 72.1

Let I_1, I_2 be the indicator r.v.'s of two independent events A_1, $\mathsf{A}_2 \subset \mathsf{S}$, with $P(\mathsf{A}_1) = P(\mathsf{A}_2) = p$. I_1 and I_2 do not belong to the inner product space V for the underlying probability space $(\mathsf{S}, \mathscr{A}, P)$, because they are not zero-mean. Let I_{1c} and I_{2c} be the centralized versions of I_1 and I_2. Then, I_{1c} and I_{2c} are orthogonal vectors in V, with lengths $\sqrt{\mathbf{V}[I_{1c}]} = \sqrt{\mathbf{V}[I_{2c}]} = \sqrt{p(1-p)}$ (see Example 64.1). Their sum is a vector of length $\sqrt{2p(1-p)}$, representing the centralized version, B_c, of the binomial r.v. $B = I_1 + I_2$, as depicted in Figure 72.3.

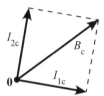

Figure 72.3   Vector addition of I_{1c} and I_{2c}

We see that both I_1 and I_2 are positively correlated with B, with the same correlation coefficient

$$\rho[I_1, B] = \rho[I_{1c}, B_c] = \cos\frac{\pi}{4} \doteq 0.707.$$

Returning to Equation (68.8), it is now possible to interpret that relation from the perspective of the vector space representation of r.v.'s. Equation (68.8) states a relationship between *standardized* r.v.'s, which are represented by vectors of unit length in V. X_s and Y_s are correlated r.v.'s, and therefore their vectors are separated by an angle $\theta = \arccos\rho \neq \pi/2$. Z_s is specified orthogonal to X_s, in the plane defined by X_s and Y_s, as illustrated in Figure 72.4. Y_s is then decomposed into a component collinear with X_s and a residual component collinear with Z_s.

The resolution of a given r.v. Y into orthogonal components can be extended to an arbitrary number of dimensions. Thus, let $X_{1s}, \ldots, X_{ns}$ be n orthogonal unit vectors, and for convenience let Y be standardized so that $Y = Y_s$. Then, Equation (68.7) becomes

$$Y_s = a_1 X_{1s} + a_2 X_{2s} + \ldots + a_n X_{ns} + c Z_s \tag{72.3}$$

---

[6] Given a vector X and n orthogonal coordinate vectors, then there are n direction cosines associated with X, which are the cosines of the angles between X and each of the coordinate vectors.

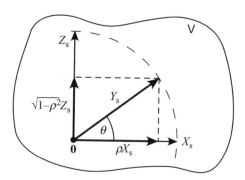

Figure 72.4   Representation of Y_s in terms of X_s and Z_s

where Z_s is orthogonal to the r.v.'s $X_{1s}, \ldots, X_{ns}$. The coefficients $a_1, \ldots, a_n$ are then the direction cosines of Y_s relative to $X_{1s}, \ldots, X_{ns}$. If these satisfy $a_1^2 + \ldots + a_n^2 = 1$, then Y_s lies entirely within the n dimensions spanned by $X_{1s}, \ldots, X_{ns}$, and $c = 0$. Otherwise, cZ_s represents the residual component of Y_s, which lies outside the subspace spanned by $X_{1s}, \ldots, X_{ns}$, and c is the direction cosine of Y_s relative to Z_s—i.e., relative to the residual component.

Queries

72.1 The class $\mathscr{A}$ of a certain probability space $(S, \mathscr{A}, P)$ consists of four sets. What is the number of dimensions of the inner product space V associated with $(S, \mathscr{A}, P)$?   [218]

72.2 Three zero-mean r.v.'s X, Y, Z have correlation coefficients $\rho_{X,Y} = \rho_{X,Z} = \rho_{Y,Z} = \rho$. For each of the following statements, determine whether it is true, possibly true, or false.
a) $\rho = -1$
b) V is a three-dimensional space
c) X, Y and Z are mutually statistically independent
d) $X + Y + Z = 0$.   [133]

72.3 On a probability space $(S, \mathscr{A}, P)$ are defined two r.v.'s X, Y with means $m_{1X} = 1$ and $m_{1Y} = -2$. Let X_c, Y_c be their centralized versions. In the inner product space V for $(S, \mathscr{A}, P)$, X_c is a unit vector and Y_c is a vector of length 3 orthogonal to X_c. Evaluate $\mathbf{E}[XY]$.   [154]

Estimation

In this Section, we encounter applications of the vector space representation of r.v.'s and also of conditional expectations. But first, we address the basic idea of the *estimation* of a r.v.

a) The concept of estimating a random variable

Consider a p.e. $\mathfrak{E}$ with associated probability space (S, $\mathscr{A}$, P). On this space a r.v. Y is defined. Each time $\mathfrak{E}$ is performed, the outcome is characterized by some element $\xi \in$ S, and gives rise to a numerical value $Y(\xi)$. The basic question in which we are interested here is the following: *Before the numerical value $Y(\xi)$ is known, what is the best possible 'guess', i.e. estimate, of this value?* This question can arise in two different ways:

  I) The value $Y(\xi)$ to be estimated is the one that will arise *the next time* $\mathfrak{E}$ is performed; or it is the value $Y(\xi)$ that arose in a previous execution of $\mathfrak{E}$ but the results of that execution are not available to the person doing the estimating.
 II) $\mathfrak{E}$ has been performed but only *some* of the results are known to the person doing the estimating, not the value $Y(\xi)$. The difference is that in this situation, more information should be available than in case I on which to base the estimate.

b) Optimum constant estimates

We begin with case I and consider the question stated above: What is the best possible estimate of the value $Y(\xi)$ that will arise the next time $\mathfrak{E}$ is performed? Since executions of $\mathfrak{E}$ are assumed independent, then as long as (S, $\mathscr{A}$, P) is available, values of $Y(\xi)$ that have arisen in earlier executions of $\mathfrak{E}$ are of no help. The only useful information on which the estimate can be based is P_Y. Since P_Y is the same each time $\mathfrak{E}$ is performed, the optimum estimate of $Y(\xi)$ cannot depend on ξ and therefore must be a constant, c.

Now, in order to actually answer the above question, it must be made more precise: The meaning of 'best' has to be made explicit; in other words, a 'criterion of optimality' needs to be specified. The value of c will depend on this optimization criterion.

Probability Concepts and Theory for Engineers, First Edition. Harry Schwarzlander.
© 2011 John Wiley & Sons, Ltd. Published 2011 by John Wiley & Sons, Ltd.

Example 73.1

You find yourself in a game in which a specially marked die is thrown. It is a normal six-sided fair die, but two of its faces are marked with a '1' and the remaining four faces with a '0'. Before the die is thrown you are to write down on a piece of paper a number (not necessarily an integer). The closer this guess is to the value that will show on the die after it gets thrown, the greater your winning. The specifics of how this pay-off is computed, and the circumstances under which you play the game, may affect your choice of criterion (see Problem 73.1).

A binary r.v. Y defined on the binary probability space for this experiment expresses the numerical result of the die throw and induces the distribution $P_Y(\{0\}) = 2/3$ and $P_Y(\{1\}) = 1/3$. What is wanted is the *optimum estimate*, c. Of interest will also be the *estimation error*, which is the r.v. $Q = Y - c$. We consider several different plausible optimization criteria.

a) Suppose the criterion is to *minimize the maximum error*. From P_Y as given above it follows immediately that $c = 0.5$, giving $\max(|Q|) = |Q| = 0.5$.

b) To minimize the *expected magnitude of the error*, we write

$$E[|Q|] = E[|Y - c|] = \frac{2}{3}|0 - c| + \frac{1}{3}|1 - c| = \frac{1}{3}(c + 1)$$

assuming $0 \le c \le 1$. This gives $c = 0$, resulting in $E[|Q|] = 1/3$.

c) Another possibility is to minimize the *magnitude of the expected error*, $|EQ|$. Here, $|EQ| = |E[Y - c]| = |E[Y] - c|$, which clearly is minimized by setting $c = EY = 1/3$ since this reduces $|EQ|$ to 0.

d) We can also minimize the *mean-square value of the error* or *m.s. error*, $E[Q^2]$. In this case we find that

$$E[Q^2] = E\left[(Y - c)^2\right] = \frac{2c^2 + (1 - c)^2}{3} = c^2 - \frac{2}{3}c + \frac{1}{3}.$$

This is minimized by setting $2c - 2/3 = 0$, which again yields $c = EY = 1/3$. Furthermore, $E[Q^2] = 2/9$.

The results obtained in Example 73.1 can be generalized to an arbitrary r.v. Y, although different optimization criteria may place different restrictions on P_Y. Thus, the *maximum error* can be minimized only if the range of Y is finite. Then, $c = [\min(Y) + \max(Y)]/2$.

To minimize the *expected magnitude of the error*, EY must exist, otherwise $E[|Q|]$ does not exist. Assuming that P_Y is described by the p.d.f. f_Y, then the expression to be minimized is

$$E[|Q|] = \int_c^\infty (y - c) f_Y(y)\, dy + \int_{-\infty}^c (c - y) f_Y(y)\, dy.$$

Application of Leibnitz's rule for differentiating an integral where the differentiation variable appears in the limits yields

$$\frac{dE[|Q|]}{dc} = 0 = \int_{-\infty}^c f_Y(y)\, dy - \int_c^\infty f_Y(y)\, dy$$

so that c is the *median* of P_Y (see Equation (61.5)), which always exists. This agrees with the result obtained in Example 73.1 for this optimization criterion.

To minimize the *magnitude of the expected error*, the existence of $\mathbf{E}Y$ is required also. Since $\mathbf{E}Q = \mathbf{E}[Y - c] = \mathbf{E}Y - c$, it is easily seen that $|\mathbf{E}Q|$ is minimized if $\mathbf{E}Q$ is zero, so that $c = \mathbf{E}Y$. Finally, to minimize the *mean-square error*, it is necessary that $\mathbf{E}[Y^2]$ exists. The m.s. error is

$$\mathbf{E}[Q^2] = \mathbf{E}[(Y - c)^2] = \mathbf{E}[Y^2] - 2c\mathbf{E}Y + c^2.$$

Now setting $d\,\mathbf{E}[Q^2]/dc = 0$ gives $-2\mathbf{E}Y + 2c = 0$, so that again $c = \mathbf{E}Y$. Substitution into the above expression for $\mathbf{E}[Q^2]$ then gives $\mathbf{E}[Q^2] = \mathbf{E}[Y^2] - (\mathbf{E}Y)^2 = \mathbf{V}Y$.

c) Mean-square estimation using random variables

Next, we consider the situation that was described at the beginning of this Section as case II: A p.e. $\mathfrak{E}$ has been performed, resulting in a particular outcome that we identify by the element ξ in the sample space. Furthermore, one or more numerical values have been observed, but not the numerical value $Y(\xi)$ that is wanted. $\mathfrak{E}$ therefore involves one or more numerical observations or measurements—we will say n measurements. On the probability space for this p.e., n r.v.'s $X_1, \ldots, X_n$ are defined that express the results of the n measurements, as well as the r.v. Y that expresses the values taken on by the quantity to be estimated. The actually observed values are sample values of the 'data' r.v.'s, $X_1(\xi), \ldots, X_n(\xi)$, and these are to be used to obtain an optimum estimate of $Y(\xi)$. Since this estimate will depend on $X_1(\xi), \ldots, X_n(\xi)$, and therefore on ξ, this estimate is also a r.v. that will be denoted $\tilde{Y}$.

For instance, $\mathfrak{E}$ might serve to measure some quantity that is described by the r.v. Y. However, due to practical limitations, only n 'noisy' (imperfect, approximate) measurements can actually be obtained. The specific physical context will not affect the discussion—we might be dealing with the measurement of physical phenomena, the interpolation of waveforms from sampled values, extrapolation of current weather data, etc.

The discussion will be restricted to minimization of the *mean-square error*,

$$\mathbf{E}[Q^2] = \mathbf{E}\{[Y - g(X)]^2\} \tag{73.1}$$

where $X = [X_1, X_2, \ldots, X_n]^\mathrm{T}$, and g is a function of n variables that is to be chosen so as to minimize Equation (73.1). Using Equation (70.3), the m.s. error can be written

$$\mathbf{E}\{[Y - g(X)]^2\} = \mathbf{E}_X[\mathbf{E}\{[Y - g(X)]^2 | X\}].$$

Since $\mathbf{E}\{[Y - g(X)]^2 | X\}$ is non-negative for all vector-values taken on by X, it is sufficient to minimize this conditional expectation. For each vector x in the range of X, we have

$$\mathbf{E}\{[Y - g(X)]^2 | X = x\} = \mathbf{E}[Y^2 | X = x] - 2g(x)\,\mathbf{E}[Y | X = x] + g^2(x).$$

Thus, for each fixed x, $g(x)$ is optimized by setting

$$\frac{d\,\mathbf{E}\{[Y - g(X)]^2 | X = x\}}{dg} = 0 = -2\,\mathbf{E}[Y | X = x] + 2g(x)$$

giving $g(x) = \mathbf{E}[Y | X = x]$. The optimum m.s. estimate of Y in terms of X is therefore $g(X) = \mathbf{E}[Y | X]$, the regression function of Y on X.

Example 73.2

Let $Z = [X_1, \ldots, X_n, Y]^\mathrm{T}$ be zero-mean multivariate Gaussian, with covariance matrix c_Z. The r.v. Y is to be estimated in terms of $X = [X_1, \ldots, X_n]^\mathrm{T}$. The optimum m.s. estimate, as derived above, is $g(x) = \mathbf{E}[Y|X = x]$. The p.d.f.'s $f_Z(z)$ and $f_X(x)$ have the form Equation (71.6), that is,

$$f_Z(z) = (2\pi)^{-(n+1)/2}[\det(c_Z)]^{-1/2}\exp\left[\left(-\frac{1}{2}\right)z^T c_Z^{-1} z\right]$$

and

$$f_X(x) = (2\pi)^{-n/2}[\det(c_X)]^{-1/2}\exp\left[\left(-\frac{1}{2}\right)x^T c_X^{-1} x\right].$$

Therefore,

$$f_{Y|X}(y|x) = \frac{f_Z(z)}{f_X(x)}$$

$$= \left(\frac{\det(c_X)}{2\pi \det(c_Z)}\right)^{1/2}\exp\left[-\frac{1}{2}(z^T c_Z^{-1} z - x^T c_X^{-1} x)\right].$$

Since x is assumed fixed, the only variable in this function is y, which appears in the homogeneous second-order polynomial in the exponent. Therefore, $f_{Y|X}(y|x)$ is Gaussian, and by completing the square in the exponent it can be put into the form (40.1). The exponent contains only second-order terms involving the components of z and x, from which it follows that the constant a in Equation (40.1) must be linear in $x_1, \ldots, x_n$. Since this constant is the mean of the conditional distribution (see Problem 61.3), we see that the optimum mean-square estimate of Y in terms of X is of the form

$$\tilde{Y} = g(X) = \mathbf{E}[Y|X] = a_1 X_1 + a_2 X_2 + \ldots + a_n X_n$$

where the coefficients $a_1, \ldots, a_n$ depend on c_Z. These coefficients can be obtained in a simple manner, as derived below.

d) Linear mean-square estimation

Sometimes, a mean-square estimate is desired that is restricted to a linear combination of the data r.v.'s. This is called a '*linear* mean-square (l.m.s.) estimate'. In general, a l.m.s. estimate will result in a larger m.s. error than the unrestricted mean-square estimate (which is also called a 'nonlinear' m.s. estimate). One motivation for pursuing a l.m.s. estimate is that the nonlinear m.s. estimate can be difficult to derive. However, Example 73.2 shows that in a given situation it is possible for a nonlinear m.s. estimate to be in fact linear.

Thus, suppose that a r.v. Y is to be estimated in terms of n r.v.'s $X_1, X_2, \ldots, X_n$ as follows:

$$Y \approx a_1 X_1 + a_2 X_2 + \ldots + a_n X_n = \tilde{Y}. \tag{73.2}$$

Here we assume that $X_1, X_2, \ldots, X_n$ and Y have been *centralized*, and that $X_1, X_2, \ldots, X_n$ are not linearly dependent. The error of approximation is the r.v.

$$Q = Y - \tilde{Y} = Y - (a_1 X_1 + a_2 X_2 + \ldots + a_n X_n).$$

To be found are the constants $a_1, \ldots, a_n$ that minimize $\mathbf{E}[Q^2]$.

L.m.s. estimation is easily visualized in terms of the space V defined in Section 72. Y and the estimate $\tilde{Y} = a_1X_1 + \ldots + a_nX_n$ are each represented by a vector in V. The error, Q, is the vector difference whose magnitude is to be minimized.

Suppose that by some method, the *direction* of the vector $\tilde{Y}$ has been optimized, leaving only the magnitude to be specified. If the optimum direction of $\tilde{Y}$ coincides with the direction of Y, then the mean-square error $\mathbf{E}[Q^2]$ can be made zero by setting the vector $\tilde{Y}$ equal to the vector Y, so that the r.v.'s $\tilde{Y}$ and Y are equal with probability 1. However, in a typical l.m.s. estimation problem this is not possible. Therefore, assume that optimization of the direction of Y still leaves a nonzero angle between the directions of $\tilde{Y}$ and Y. Since the shortest path from a point to a line is a perpendicular to the line (Figure 73.1), it can be seen that the length of Q is minimized by making Q *orthogonal* to $\tilde{Y}$, i.e., by adjusting the length of the vector $\tilde{Y}$ so as to make

$$\mathbf{E}[Q\tilde{Y}] = \mathbf{E}[Q(a_1X_1 + \ldots + a_nX_n)] = 0. \tag{73.3}$$

This result is referred to as the *'Orthogonality Principle'*.

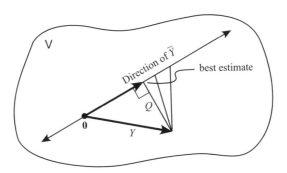

Figure 73.1   L.m.s. estimation of Y

We ignore the trivial solution of Equation (73.3), which is $a_1 = a_2 = \ldots = a_n = 0$. Therefore, and because the X_i's are not linearly dependent, the parenthesis in Equation (73.3) cannot vanish. Furthermore, since Q here plays the role of cZ_0 in Equation (72.3), according to the discussion of that equation Q must be orthogonal to *each* of the r.v.'s $X_1, X_2, \ldots, X_n$. Thus, we have the following requirement:

$$\mathbf{E}[X_iQ] = \mathbf{E}[X_iY - a_1X_iX_1 - a_2X_iX_2 - \ldots - a_nX_iX_n] = 0, \quad i = 1, \ldots, n.$$

This results in the following set of simultaneous equations for the coefficient values:

$$a_1\mathbf{E}[X_1^2] + a_2\mathbf{E}[X_1X_2] + \ldots + a_n\mathbf{E}[X_1X_n] = \mathbf{E}[X_1Y]$$

$$a_1\mathbf{E}[X_2X_1] + a_2\mathbf{E}[X_2^2] + \ldots + a_n\mathbf{E}[X_2X_n] = \mathbf{E}[X_1Y]$$

$$\cdots\cdots\cdots\cdots\cdots\cdots\cdots\cdots\cdots\cdots\cdots\cdots\cdots$$

$$\cdots\cdots\cdots\cdots\cdots\cdots\cdots\cdots\cdots\cdots\cdots\cdots$$

$$a_1\mathbf{E}[X_nX_1] + a_2\mathbf{E}[X_nX_2] + \ldots + a_n\mathbf{E}[X_n^2] = \mathbf{E}[X_nY].$$

These equations can be written more compactly in terms of the covariance matrix:

$$\mathbf{C}[X, X]a = \mathbf{E}[X\,Y] \tag{73.4}$$

where $a = [a_1, \ldots, a_n]^{\mathrm{T}}$, and $\mathbf{E}[XY] = [\mathbf{E}[X_1Y], \ldots, \mathbf{E}[X_nY]]^{\mathrm{T}}$. Making use of Equation (73.3), the minimum mean-square error then becomes

$$\mathbf{E}[Q^2] = \mathbf{E}[Q\,(Y - \tilde{Y})] = \mathbf{E}[Q\,Y] = VY - a^{\mathrm{T}}\,\mathbf{E}[X\,Y]. \tag{73.5}$$

Example 73.3

A radio signal is observed with a receiver at equally spaced instants of time $t_1, t_2, t_3, \ldots, t_n$. Without going into the specifics of the p.e., we assume that a probability space can be specified on which it is possible to define r.v.'s X_i that express the signal voltage at time t_i ($i = 1, \ldots, n$). Suppose that all X_i's have zero mean and identical distribution. The correlation coefficient relating various observations is given by $\rho[X_i, X_{i+k}] = \rho^k$, where $\rho > 0$. Suppose the observations at times t_1, t_2, t_3 are to be used to obtain an l.m.s. estimate of the signal at time t_4, that is, to 'predict' the signal at the next observation instant on the basis of the present and preceding two observations.

$$\text{Let } X = \begin{bmatrix} X_1 \\ X_2 \\ X_3 \end{bmatrix}, \text{ so that } \mathbf{C}[X, X] = \mathbf{E}[X_1^2] \begin{bmatrix} 1 & \rho & \rho^2 \\ \rho & 1 & \rho \\ \rho^2 & \rho & 1 \end{bmatrix} \text{ and } \mathbf{E}[X\,X_4] = \mathbf{E}[X_1^2] \begin{bmatrix} \rho^3 \\ \rho^2 \\ \rho \end{bmatrix}.$$

Substitution into Equation (73.4) gives

$$\begin{bmatrix} 1 & \rho & \rho^2 \\ \rho & 1 & \rho \\ \rho^2 & \rho & 1 \end{bmatrix} \begin{bmatrix} a_1 \\ a_2 \\ a_3 \end{bmatrix} = \begin{bmatrix} \rho^3 \\ \rho^2 \\ \rho \end{bmatrix}$$

yielding $a_1 = a_2 = 0$ and $a_3 = \rho$. Thus, for the particular relationship between correlation coefficients which was specified, the l.m.s. estimate of X_4 makes no use of X_1 and X_2. It is $\tilde{X}_4 = \rho X_3$.

The mean-square error of this estimate is, from Equation (73.5),

$$\mathbf{V}[X_4] - \rho\mathbf{E}[X_3X_4] = \mathbf{V}[X_4] - \rho^2\mathbf{V}[X_4] = (1 - \rho^2)\mathbf{V}[X_4].$$

If no measurements were available at all, then (as in Example 73.1) the l.m.s. estimate of X_4 reduces to a constant c and Equation (73.3) becomes

$$\mathbf{E}[(X_4 - c)c] = c\mathbf{E}[X_4] - c^2 = 0$$

so that the optimum estimate is $c = \mathbf{E}[X_4] = 0$. In that case, the m.s. error is simply $\mathbf{V}X_4$. Therefore, utilizing the measurement X_3 results in a reduction of the m.s. error by the factor $(1 - \rho^2)$. If $\rho = 1$, then the m.s. error is zero, or $\tilde{X}_4 = X_4[P]$.

Revisiting Example 73.2, it can now be seen that Equation (73.4) and Equation (73.5) also apply to the particular nonlinear estimation problem considered there. In other words, we have:

Theorem 73.1

If X and Y are jointly Gaussian r.v.'s, then the optimum ('nonlinear') m.s. estimate of Y in terms of X is the l.m.s. estimate of Y in terms of X.

Queries

73.1 Given two standardized i.d.s.i. binary r.v.'s X, Y. If aX is the l.m.s. estimate of Y, what is the value of a, and what is the mean-square error?   [225]

73.2 Let X_s, Y_s be two standardized s.i. r.v.'s. Furthermore, let $Z = X_s^2 + Y_s^2$. When the underlying experiment is performed, the value of X_s turns out to be 0.5. What is the optimum (nonlinear) m.s. estimate of Z in terms of X_s?   [155]

73.3 The l.m.s. estimate of a centralized r.v. Y in terms of the centralized r.v. X results in a m.s. error $\mathbf{E}[Q_Y^2] = 0.01$. The l.m.s. estimate of another centralized r.v. Z in terms of the r.v. Y results in a m.s. error $\mathbf{E}[Q_Z^2] = 0.04$. For each of the following statements, determine whether it is true, possibly true, false, or meaningless:

a) $\mathbf{V}Z > \mathbf{V}Y$

b) The l.m.s. estimate of $2Y$ in terms of X results in a m.s. error of 0.04

c) The l.m.s. estimate of Z in terms of $2Y$ results in a m.s. error of 0.16

d) X, Y and Z are mutually orthogonal

e) The l.m.s. estimate of Y in terms of Z results in a m.s. error of 0.04

f) The m.s. error of the l.m.s. estimate of Z in terms of X is at least 0.04.   [138]

The Stieltjes Integral

In accordance with Definition 61.1, it has so far been possible to express and compute the expected value of a r.v. only if that r.v. does not have a singular component. This limitation is not inherent in the notion of expected value, but rather is due to the type of integral we have been using, which is the Riemann integral. An arbitrary probability distribution induced by some random variable on the real line can always be described by a c.d.f., but not necessarily by a density function. This provides a clue as to how a more general definition of the expected value needs to be formulated: The integral needs to be expressed in terms of the c.d.f., not the density function. This is accomplished by means of the Stieltjes integral.

We recall that integration in the Riemann sense of a real function of a real variable, $g(x)$, over some interval $\overline{a,\ b}$, is defined by partitioning the interval into small increments of width Δx. As shown in Figure 74.1, the value of $g(x)$ in the center of each such increment gets multiplied by the width Δx of the increment. Adding up all these products yields an estimate of $\int_a^b g(x)dx$. The exact value of the integral is defined as the limit as $\Delta x \rightarrow 0$.

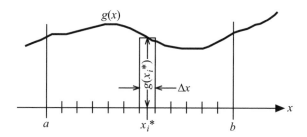

Figure 74.1 Riemann integration

The *Stieltjes integral* (or 'Riemann–Stieltjes integral') of $g(x)$ differs from the Riemann integral mainly in the fact that the value of $g(x)$ in any one of the small intervals does not get multiplied by the width of this interval but by the *increase*, over this interval, of a nondecreasing reference function that is initially specified. For our purposes, this other function will be a c.d.f. Another difference between the two integrals is that for the Stieltjes integral, the partitioning of the interval of integration does not require increments of equal size.

Probability Concepts and Theory for Engineers, First Edition. Harry Schwarzlander.
© 2011 John Wiley & Sons, Ltd. Published 2011 by John Wiley & Sons, Ltd.

Suppose then, that some c.d.f. $F(x)$ is given, and over a finite interval $\overline{a, b}$ of x-values there is defined a function $g(x)$. The Stieltjes integral of $g(x)$ with respect to $F_X(x)$, over $\overline{a, b}$, is obtained by first partitioning $\overline{a, b}$ into any n subintervals $\overline{x_{i-1}, x_i}$, where $i = 1, \ldots, n$ and $x_0 = a$, $x_n = b$. Next, the sum

$$\sum_{i=1}^{n} g(x_i^*) \, [F(x_i) - F(x_{i-1})] \tag{74.1}$$

is formed, where x_i^* is any number satisfying $x_{i-1} \leq x_i^* \leq x_i$. Figure 74.2 illustrates the significance of a representative term in this sum. If the subintervals $\overline{x_{i-1}, x_i}$ are chosen to have equal length Δx, x_i^* is the midpoint of the ith subinterval, and $F(x)$ is replaced by the identity function x, then Figure 74.2 becomes equivalent to Figure 74.1.

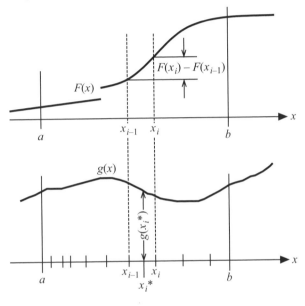

Figure 74.2   The Stieltjes integral

Now we consider the limit of the sum (74.1) as the partition of $\overline{a, b}$ is increased and as, simultaneously, the size of the largest subinterval goes to zero. If that limit exists we have:

(Definition) 74.1
The *Stieltjes integral* from a to b, of some function $g(x)$ with respect to a nondecreasing function $F(x)$ is defined as the limit

$$\int_a^b g(x) \, dF(x) = \lim_{\substack{n \to \infty \\ \mu \to 0}} \sum_{i=1}^{n} g(x_i^*) \, [F(x_i) - F(x_{i-1})],$$

where $x_0 = a$, $x_n = b$, $x_i^* \in \overline{x_{i-1}, x_i}$, and $\mu = \max_i (x_i - x_{i-1})$; provided that this limit exists when $g(x_i^*)$ is replaced by $|g(x_i^*)|$.

It is to be noted that the definition requires *absolute convergence* of the integral. Also, as mentioned above, we will always assume that the function $F(x)$ is a c.d.f. To simplify the notation, the requirement that $\mu \to 0$ will henceforth not be explicitly indicated.

Example 74.1
Given $g(x) = 1$, and $F(x) = \left\{ \begin{array}{l} 0, \; x < 0.5 \\ 1, \; x \ge 0.5 \end{array} \right\}$. The integral of $g(x)$ is to be determined over the interval $\overline{0, \, 1^|}$,

$$\int_0^1 dF(x) = \lim_{n \to \infty} \sum_{i=1}^n [F(x_i) - F(x_{i-1})].$$

As can be seen, only one of the subintervals $\overline{x_{i-1}, \, x_i^|}$ contributes to this sum, namely the one for which $x_{i-1} < 0.5$ and $x_i \ge 0.5$. For that particular interval, $F(x_{i-1}) = 0$, $F(x_i) = 1$, and therefore

$$\int_0^1 dF(x) = \lim_{n \to \infty} [1 - 0] = 1.$$

This Example shows that a Stieltjes integral over an interval that approaches measure zero as $n \to \infty$ can yield a nonzero result. More generally, we note that in fixing the limits of integration it is important to specify whether or not a step discontinuity in F_X at either endpoint of the interval of integration is to be included in the integration. Thus, let the notations b^+ and b^- indicate explicitly that

$$\int_a^{b^+} g(x)dF(x) = \lim_{\varepsilon \to 0} \int_a^{b+\varepsilon} g(x)dF(x), \; \text{and} \; \int_a^{b^-} g(x)dF(x) = \lim_{\varepsilon \to 0} \int_a^{b-\varepsilon} g(x)dF(x).$$

The same applies to the lower limit of integration. Now if, for instance, $F(x)$ has a step discontinuity at $x = b$ and $g(b) \ne 0$, then

$$\int_a^{b^+} g(x)dF(x) \ne \int_a^{b^-} g(x)dF(x).$$

We will assume in the following, unless indicated otherwise, that the lower limit of integration a is to be interpreted as a^+, and the upper limit b as b^+.

If the integration interval is infinite, the resulting integral is an *improper* Stieltjes integral. This is defined in a manner analogous to an improper Riemann integral. Thus,

$$\int_{-\infty}^\infty g(x) \, dF(x) = \lim_{\substack{a \to -\infty \\ b \to \infty}} \int_a^b g(x) \, dF(x).$$

The reader should be able to supply proofs of the following basic properties of a Stieltjes integral by using Definition 74.1 along with arguments similar to those employed in the theory of ordinary Riemann integrals (see Problems 74.1–74.3).

Theorem 74.1
The Stieltjes integral

$$\int_a^b g(x) \, dF(x)$$

is linear in both $g(x)$ and $F(x)$.

Theorem 74.2

Let $a = c_0 < c_1 < c_2 < \ldots < c_n = b$, then

$$\int_a^b g(x)\, dF(x) = \sum_{i=1}^{n} \int_{c_{i-1}}^{c_i} g(x)\, dF(x).$$

Theorem 74.3

If $g(x) \geq 0$ for $a \geq x \geq b$, where $b > a$, then

$$\int_a^x g(x)\, dF(x)$$

is a non-negative, nondecreasing function of x over $\overline{a,\ b}$.

Returning now to probability theory, we note first of all how the Stieltjes integral expresses very simply the probability assigned to some interval $\overline{a,\ b}$ by a random variable X:

$$P_X\overline{(a,\ b)} = F_X(b) - F_X(a) = \int_a^b dF_X(x). \tag{74.2}$$

Since a c.d.f. is defined for every r.v., Equation (74.2) applies irrespective of the type of the r.v. X. If the density function $f_X(x) = dF_X(x)/dx$ exists, then the mean-value theorem can be used to write

$$\int_a^b dF_X(x) = \lim_{n \to \infty} \sum_{i=1}^{n} [F_X(x_i) - F_X(x_{i-1})]$$

$$= \lim_{n \to \infty} \sum_{i=1}^{n} f_X(x_i^*)(x_i - x_{i-1}) = \int_a^b f_X(x)\, dx.$$

The right-hand integral is the Stieltjes integral of $f_X(x)$ with respect to the linear function x. However, if the x_i of Definition 74.1 are equally spaced, so that for each n,

$$x_i - x_{i-1} = \frac{b - a}{n} \quad \text{and}$$

$$x_i^* = \frac{x_i - x_{i-1}}{2} \quad \text{for} \quad i = 1, \ldots, n$$

then the right-hand integral can also be taken as an ordinary Riemann integral. This is always possible if X is absolutely continuous. In this way, Equation (74.2) reduces to Equation (39.1). More generally, if f_X exists and $g(x)$ is Riemann integrable, then

$$\int_a^b g(x)\, dF_X(x) = \int_a^b g(x) f_X(x)\, dx \tag{74.3}$$

i.e., the Stieltjes integral becomes a Riemann integral.

It is now possible to state a general definition of the expected value of a r.v. in terms of the Stieltjes integral that applies without restrictions to any (scalar) r.v.:

(Definition) 74.2
The expectation of a r.v. X with c.d.f. $F_X(x)$ is

$$\mathbf{E}X = \int_{-\infty}^{\infty} x dF_X(x)$$

provided this integral exists. Otherwise, X does not have an expectation.

This definition includes Definition 61.1 as a special case. In particular, using Equation (74.3), it is easy to see how Definition 74.2 reduces to Definition 61.1(b). Theorem 63.2 can similarly be restated by replacing Equation (63.2) with

$$\mathbf{E}Y = \int_{-\infty}^{\infty} g(x) dF_X(x). \tag{74.4}$$

Therefore, if X is a singular r.v. in one dimension, $\mathbf{E}[g(X)]$ is now defined—namely, as the Stieltjes integral (74.4)—provided the integral converges absolutely. This does not imply, of course, that a simple analytic solution exists. In some cases, however, it is possible to arrive at a closed form solution, as illustrated below in Example 74.2.

The method of solution in Example 74.2 makes use of the following property of Stieltjes integrals, which derives from the fact that in these integrals the integration depends on changes in $F(x)$, not in x.

Theorem 74.4
Into the Stieltjes integral $\int_a^b g(x) dF(x)$ can be incorporated shifts and scale factors so that

$$\int_a^b g(x) dF(x) = \frac{1}{\delta} \int_{(a-\beta)/\alpha}^{(b-\beta)/\alpha} g(\alpha x + \beta)\, d[\gamma + \delta F(\alpha x + \beta)]$$

where $\alpha, \delta > 0$, and β, γ are any real constants.

Proof
Consider first the parameters α, β: Substitution into Definition 74.1 changes the representative term in the summation to $g(\alpha x_i^* + \beta)\, [F(\alpha x_i + \beta) - F(\alpha x_{i-1} + \beta)]$, where $(\alpha x_i^* + \beta) \in \overline{\alpha x_{i-1} + \beta, \ \alpha x_i + \beta}$, and $x_0 = \dfrac{a - \beta}{\alpha}$, $x_n = \dfrac{b - \beta}{\alpha}$. Using y_i to denote $\alpha x_i + \beta$, this becomes $g(y_i^*)\, [F(y_i) - F(y_{i-1})]$, where $y_i^* \in \overline{y_i, \ y_{i-1}}$, and the integration limits then are $\alpha \dfrac{a - \beta}{\alpha} + \beta = a$, and $\alpha \dfrac{b - \beta}{\alpha} + \beta = b$.

Now consider the parameters γ, δ: It follows from Definition 74.1 that adding a constant γ to $F(x)$ has no effect on the integration. And according to Theorem 74.1, the multiplier δ can be taken outside the integral sign, where it gets canceled by the factor $1/\delta$.

Example 74.2: Variance of a singular r.v.
Let X be a singular r.v. whose c.d.f. is specified as follows:

$$F_X(x) = \begin{cases} 0, & x < 0 \\ 1, & x \geq 1 \\ 1/2, & x \in \overline{1/3,\ 2/3} \\ 1/4, & x \in \overline{1/9,\ 2/9} \\ 3/4, & x \in \overline{7/9,\ 8/9} \\ 1/8, & x \in \overline{1/27,\ 2/27} \\ 3/8, & x \in \overline{7/27,\ 8/27} \\ 5/8, & x \in \overline{19/27,\ 20/27} \\ 7/8, & x \in \overline{25/27,\ 26/27} \\ \cdots \\ \cdots \\ \cdots \end{cases} \qquad (74.5)$$

We wish to find $\mathbf{V}X$. For reference, a sketch of $F_X(x)$ reflecting only the first five lines of Equation (74.5) is shown in Figure 74.3. F_X is similar to the c.d.f. F_{V_o} sketched in Figure 56.6. The distribution induced by X has a point of symmetry at $x = 0.5$, so that $\mathbf{E}X = 0.5$. The mean-square value of X is given by Equation (74.4) with $g(x) = x^2$.

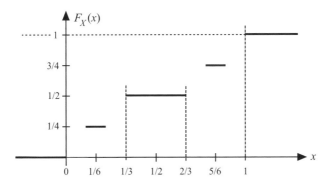

Figure 74.3   Partial representation of $F_X(x)$

Since there are no contributions to the integral outside the interval $\overline{0,\ 1}$ and over $\overline{1/3,\ 2/3}$, the m.s. value of X can be written

$$\mathbf{E}[X^2] = \int_{0^-}^{1} x^2 dF_X(x) = \int_{0^-}^{1/3} x^2 dF_X(x) + \int_{2/3^-}^{1} x^2 dF_X(x). \tag{74.6}$$

The integrals on the right can be expressed as follows (Theorem 74.4):

$$\int_{0^-}^{1/3} x^2 dF_X(x) = \int_{0^-}^{1} \left(\frac{x}{3}\right)^2 dF_X\left(\frac{x}{3}\right)$$

and

$$\int_{2/3^-}^{1} x^2 dF_X(x) = \int_{0^-}^{1} \left(\frac{x}{3} + \frac{2}{3}\right)^2 dF_X\left(\frac{x}{3} + \frac{2}{3}\right).$$

Now taking advantage of the recursive nature of the description of F_X, we note that for $x \in \overline{0, 1}$, $F_X(x/3) = (1/2)F_X(x)$, so that

$$\int_{0^-}^{1/3} x^2 dF_X(x) = \frac{1}{2}\int_{0^-}^{1} \left(\frac{x}{3}\right)^2 dF_X(x) = \frac{1}{18}\mathbf{E}[X^2].$$

Similarly, for $x \in \overline{0, 1}$, $F_X(x/3 + 2/3) = 1/2 + (1/2)F_X(x)$ so that, using Theorem 74.4 again,

$$\int_{2/3^-}^{1} x^2 dF_X(x) = \frac{1}{2}\int_{0^-}^{1} \left(\frac{x}{3} + \frac{2}{3}\right)^2 dF_X(x)$$

$$= \frac{1}{18} \mathbf{E}[X^2] + \frac{4}{18}\mathbf{E}X + \frac{4}{18} = \frac{1}{18} \mathbf{E}[X^2] + \frac{1}{3}.$$

Substitution into Equation (74.6) gives $\mathbf{E}[X^2] = \mathbf{E}[X^2]/9 + 1/3$, so that $\mathbf{E}[X^2] = 3/8$. The variance of X is therefore

$$\mathbf{V}X = \mathbf{E}[X^2] - (\mathbf{E}X)^2 = 0.375 - 0.25 = 0.125.$$

Queries

74.1 Let $g(x)$ be a continuous function of x and $F(x)$ a c.d.f., and assume that all the integrals given below exist. For each statement, determine whether it is true, possibly true, or false.

a) $\displaystyle\int_{-\infty}^{\infty} dF(x) = 1$
b) $\displaystyle\int_{-\infty}^{\infty} d[F(x) - 1] = 0$

c) $\displaystyle\int_{-\infty}^{\infty} g(x)dF(x) = \int_{-\infty}^{0^-} g(x)dF(x) + \int_{0^+}^{\infty} g(x)dF(x).$

d) If $f(x) = dF(x)/dx$ exists,

then $\displaystyle\int_{-\infty}^{\infty} g(x)dF(x) = \int_{-\infty}^{\infty} f(x)g(x)dx.$ [131]

74.2 If $F_X(x)$ is as given in Example 74.2, find the variance of the r.v. Y with c.d.f.

$$F_Y(y) = \begin{cases} 2F_X\left(\dfrac{y}{3}\right), & y < 1 \\ 1, & y \geq 1. \end{cases} \qquad [91]$$

Part V Summary

From among a variety of descriptors of scalar r.v.'s mentioned in Section 61, the *expected value* emerged as the one having the greatest significance. It is equivalent to the *mean* or 'center of gravity' of the induced distribution. Since the variance of a r.v. is simply the expectation of a function of that r.v., we saw that the variance and the standard deviation are also derived from expected values.

Mean and standard deviation often serve as the parameters for a particular family of one-dimensional distributions, such as the Gaussian distribution. Changing the mean of a given distribution to zero and the standard deviation to 1 results in the 'standardized version' of the distribution. Some bounds on standardized distributions were investigated in Section 65, which could then be extended to nonstandardized versions. In a family of two-dimensional distributions, the covariance or the correlation coefficient, introduced in Section 68, can serve as an additional parameter. This is of particular relevance in the case of bivariate Gaussian distributions, as illustrated in Section 69. Extension of this idea to more than two dimensions was carried out in Section 71, where the covariance matrix was introduced.

A brief excursion into Statistics was undertaken in Section 66, with an exploration of the use of a test sample to estimate the mean and variance of a population.

It was seen that there are two different kinds of conditional expectations, depending on whether the conditioning results from specifying an event or a r.v. These two distinct cases were treated in Sections 67 and 70, respectively.

To pave the way for the topic of Estimation as treated in Section 73, the mathematical notion of a vector space was introduced in Section 72. It was then shown how all the r.v.'s that can be defined on a given probability space can be characterized as elements of a vector space. This characterization is useful for the formulation of linear mean-square estimation. When mean-square estimation is not restricted to a linear function of the data, we found that the optimum estimate involves a conditional expectation.

Finally, in Section 74, the Stieltjes integral was introduced in order to 'clean up' a minor restriction carried along from the beginning of Part V, namely, excluding consideration of r.v.'s that have a singular component. By defining the expected value in terms of a Stieltjes integral, this restriction was removed in Section 74—explicitly for scalar r.v.'s, but in principle for all r.v.'s.

Part VI

Further Topics in Random Variables

Part VI Introduction

This final Part of the book provides the opportunity to explore some slightly more specialized aspects of random variables. In Section 75 complex r.v.'s are introduced. These serve as the basis for the transform methods that can be applied in various manipulations of r.v.'s, as described in Sections 76–79.

Section 80 is devoted to a broader treatment of Gaussian r.v.'s that brings together some properties presented earlier and also introduces some additional aspects. This leads naturally into an exploration of spherically symmetric vector r.v.'s in Section 81. A r.v of this type is obtained by applying a modification to a Gaussian vector r.v. with i.d.s.i. components. In Section 82 the notion of randomness or entropy, previously discussed in Section 32, gets applied to r.v.'s. Here, Gaussian r.v.'s also play a role—as a result of an interesting property of the Gaussian distribution.

Section 83 is devoted to a characterization of the relationship, or coupling, between two r.v.'s that is called their 'copula'. From the marginal c.d.f.'s of two r.v.'s and their copula, the joint c.d.f. can be found.

The remaining three Sections are devoted to infinite sequences of r.v.'s. Their basic features are discussed in Section 84 and illustrated in the context of simple gambling situations. This is followed by an examination of different types of convergence and the 'Laws of Large Numbers'. Section 86, the final Section, addresses convergence in distribution and the Central Limit Theorem. That theorem establishes the central role played by the Gaussian distribution in Probability Theory.

Complex Random Variables

In Part III we saw how the introduction of random variables into the mathematical probability model permits a more effective treatment of probabilistic experiments that have numerical-valued outcomes. This led to the exploration of a variety of mathematical topics and techniques that apply to r.v.'s. We noticed that r.v.'s could be useful even in those situations where the outcomes of the p.e. are not directly associated with numerical values—for example, as indicator r.v.'s. The many rules and operations applicable to r.v.'s can then be useful for solving a given problem, or perhaps merely to get more insight into it. In a similar vein it sometimes turns out to be useful to introduce r.v.'s that associate *complex numbers* with the elements of the given probability space. This is not because the corresponding real-world experiment produces complex-valued results, but because calculations with complex numbers and complex-valued functions can provide analytic insight or computational conveniences.

Let C denote the space of complex numbers. The elements or points of C are ordered pairs of real numbers, (x, y), so that the space C is quite similar to the space R^2 and can be represented geometrically by a plane (the complex plane) with two perpendicular coordinate axes. C differs from R^2 in the computational rules that apply to its elements. These computational rules are suggested by the customary notation for a complex number, $x + jy$, where x and y are real numbers and j is the imaginary unit, $j \equiv \sqrt{-1}$. Familiarity with these rules is assumed.

We describe a rectangle set in C in the same manner as in R^2. For instance,

$$\overline{-1, 1} \times \overline{0, 1} \subset \mathsf{C}$$

is the set of complex numbers $z = x + jy$ for which $x \in \overline{-1, 1}$ and $y \in \overline{0, 1}$.

For a complex number $z = x + jy$, $x = \mathrm{Re}z$ and $y = \mathrm{Im}z$ are called the real and imaginary *components* of z, respectively; and the complex number $z^* = x - jy$ is called the *complex conjugate* of z. Also, the *magnitude* of z is denoted $|z|$, where $|z|^2 = z z^*$.

> **(Definition) 75.1**
> A *complex r.v.* Z is a two-dimensional r.v. (X, Y) defined on a probability space $(S, \mathcal{A}, P)$. It assigns to each element $\xi \in S$ a pair of real numbers, $(X(\xi), Y(\xi))$, which represent the real and imaginary components of a complex number. This complex number is denoted $Z(\xi)$, so that $Z(\xi) = X(\xi) + jY(\xi)$; and the r.v. Z is expressed in terms of X and Y by writing $Z = X + jY$.

From this Definition it should be clear that most of the rules and properties of two-dimensional real r.v.'s apply to complex r.v.'s. Thus, a complex r.v. Z induces a probability distribution P_Z on the two-dimensional Borel sets of the complex plane. The c.d.f. of $Z = X + jY$ is completely analogous to the c.d.f. of the two-dimensional r v. (X, Y). At an arbitrary point $z_0 = x_0 + jy_0 \in C$ it expresses the probability

$$F_Z(z_0) = P(\{\xi : X(\xi) \le x_0, Y(\xi) \le y_0\}) = P_Z(\{(x, y): x \le x_0, y \le y_0\}). \tag{75.1}$$

The complex r.v. $Z = X + jY$ is absolutely continuous if and only if its real and imaginary components, X and Y, are jointly absolutely continuous. Two complex r.v.'s $Z_1 = X_1 + jY_1$ and $Z_2 = X_2 + jY_2$ are statistically independent if and only if the two-dimensional r.v.'s (X_1, Y_1) and (X_2, Y_2) are statistically independent.

Example 75.1:   Random observation of a sinusoidal waveform.
  Consider an experiment that involves observation of a sinusoidally oscillating quantity (voltage, deflection, etc.) at a random instant of time. Suppose it is a sinusoidally varying voltage, $v(t) = a \cos \omega t$ volts $(a > 0)$, so that the outcome of the experiment is a measured voltage value in the range from $-a$ to $+a$ volts. By 'random instant of time' we mean that the observation instant is not selected in a way that favors a particular portion of a cycle of the oscillation; that is, we may represent the observation instant by a r.v. T that is *uniformly distributed* over a time interval of duration $2\pi n/\omega$ (where n is a positive integer), that is, over one period of the oscillation or some multiple thereof. The observed voltage is then characterized by a r.v. $V = v(T) = a \cos \omega T$.
  Since $a \cos \omega t = \text{Re}(ae^{j\omega t})$, we can introduce the complex r.v. $Z = ae^{j\omega T} = V + jY$, so that V is expressed by $V = \text{Re}\, Z$. The probability distribution induced by Z is easily described. $|ae^{j\omega T}| = a$, so that Z can only assign values on the circle of radius a centered at the origin of the complex plane (Figure 75.1). Furthermore, since T is uniformly distributed over one or more entire periods, Z is *uniformly* distributed on this circle (and is a singular r.v.).

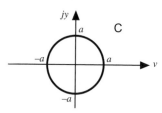

Figure 75.1   Range of the r.v. Z

(Definition) 75.2
The *expectation* of a complex r.v. $Z = X + jY$ is the complex number

$$\mathbf{E}Z = m_{1Z} = \mathbf{E}X + j\,\mathbf{E}Y$$

provided both expectations on the right side exist. Otherwise, $\mathbf{E}Z$ does not exist.

  It follows that $\mathbf{E}[Z^*] = (\mathbf{E}Z)^*$. A complex r.v. Z can also be expressed in exponential form as $Z = Re^{j\Theta}$, where $R = |Z|$. The expectation of Z can therefore also be written (see Theorem 70.3)

$$\mathbf{E}Z = \mathbf{E}[Re^{j\Theta}] = \mathbf{E}_\Theta[\mathbf{E}[Re^{j\Theta}|\Theta]] = \mathbf{E}_\Theta[e^{j\Theta}\,\mathbf{E}[R|\Theta]] \tag{75.2a}$$

$$= \mathbf{E}_R[\mathbf{E}[Re^{j\Theta}|R]] = \mathbf{E}_R[R\,\mathbf{E}[e^{j\Theta}|R]]. \tag{75.2b}$$

Example 75.1 (continued)

For the r.v. $Z = a\,e^{j\omega T}$ as defined in this Example, $\mathbf{E}Z = \mathbf{E}[a\,e^{j\Theta}] = a\mathbf{E}[e^{j\Theta}]$, where Θ is unif. $\overline{0, 2\pi}$. It follows from Theorem 61.1 that $\mathbf{E}[e^{j\Theta}] = 0$, so that $\mathbf{E}Z = 0$. Or, writing $Z = a\,(\cos\Theta + j\sin\Theta)$, then it follows immediately from symmetry that $\mathbf{E}Z = a\,(\mathbf{E}[\cos\Theta] + j\mathbf{E}[\sin\Theta]) = a\,(0 + j0) = 0$.

$Z = X + jY = Re^{j\Theta}$ is a *complex Gaussian r.v.* if X, Y are jointly Gaussian. If, in particular, X, Y are i.d.s.i. Gaussian, then R and Θ are s.i. with R a Rayleigh r.v. and Θ unif. $\overline{0, 2\pi}$ (Example 48.2).

Various properties established for expected values of real r.v.'s carry over to complex r.v.'s. For instance, it follows readily from Definition 75.2 that $\mathbf{E}$ is a linear operation also for complex random variables. As a generalization of Definition 75.2 we have, analogous to Theorem 63.2:

Theorem 75.1

Given an n-dimensional real r.v. $X = (X_1, \ldots, X_n)$. If g is a complex-valued (Borel-) function of n real variables, then the expectation of the complex r.v. $Z = g(X)$ is expressed by

$$\mathbf{E}Z = \int_{-\infty}^{\infty} g(x)\,dF_X(x) = \int_{-\infty}^{\infty} \operatorname{Re} g(x)\,dF_X(x) + j\int_{-\infty}^{\infty} \operatorname{Im} g(x)\,dF_X(x) \tag{75.3}$$

integration in each case being over the n dimensions of X.

In particular, if X reduces to an absolutely continuous one-dimensional r.v. X, we get

$$\mathbf{E}Z = \int_{-\infty}^{\infty} g(x)f_X(x)\,dx = \int_{-\infty}^{\infty} \operatorname{Re} g(x)f_X(x)\,dx + j\int_{-\infty}^{\infty} \operatorname{Im} g(x)f_X(x)\,dx \tag{75.4}$$

where as usual absolute convergence of the integrals is required.

An important distinction arises in second-order moments of complex r.v.'s, compared to real r.v.'s, which is evident from the following Definition:

(Definition) 75.3

The variance of a complex r.v. Z, and the covariance of two complex r.v.'s Z_1, Z_2, are defined by

$$\mathbf{V}Z = \mathbf{E}[|Z - m_{1Z}|^2] = \mathbf{E}[|Z|^2] - |\mathbf{E}Z|^2$$

$$\mathbf{C}[Z_1, Z_2] = \mathbf{E}[(Z_1 - m_{1Z_1})(Z_2 - m_{1Z_2})^*] = \mathbf{E}[Z_1 Z_2^*] - \mathbf{E}[Z_1]\mathbf{E}[Z_2^*]$$

provided the various expectations exist.

Thus, although the expected value of a complex r.v. is a complex number, the variance of a complex r.v. is real and non-negative. $\mathbf{V}Z$ can also be written $\mathbf{C}[Z, Z]$. The covariance of two complex r.v.'s, however, is generally a complex number, with $\mathbf{C}[Z_2, Z_1] = \mathbf{C}[Z_1, Z_2]^*$. But since a complex number with zero imaginary part equals its complex conjugate, Definition 75.3 is consistent with the earlier definition of the covariance for real r.v.'s (Definition 68.1).[1]

Expanding now the variance of $Z = X + jY$,

$$\begin{aligned} \mathbf{V}Z &= \mathbf{E}[(Z - m_{1Z})(Z - m_{1Z})^*] \\ &= \mathbf{E}\{[X - m_{1X} + j(Y - m_{1Y})][X - m_{1X} - j(Y - m_{1Y})]\} \end{aligned}$$

yields the interesting result

$$\mathbf{V}Z = \mathbf{V}X + \mathbf{V}Y. \tag{75.5}$$

Referring to Theorem 68.3, the above might at first glance suggest that X and Y are uncorrelated. However, that Theorem applies to *real* r.v.'s. If we consider the general case of the sum of two arbitrary complex r.v.'s, $Z = Z_1 + Z_2$, the variance can be written

$$\begin{aligned} \mathbf{V}Z &= \mathbf{V}[Z_1 + Z_2] = \mathbf{E}[(Z_1 + Z_2)(Z_1 + Z_2)^*] - (\mathbf{E}Z_1 + \mathbf{E}Z_2)(\mathbf{E}Z_1 + \mathbf{E}Z_2)^* \\ &= \mathbf{V}Z_1 + \mathbf{V}Z_2 + \mathbf{C}[Z_1, Z_2] + \mathbf{C}[Z_2, Z_1]. \end{aligned} \tag{75.6}$$

First, we note that $\mathbf{C}[Z_2, Z_1] = \mathbf{C}[Z_1, Z_2]^*$, so that $\mathbf{V}Z$ is indeed real, as required. Now, under what condition do the last two terms in Equation (75.6) vanish? Certainly if $\mathbf{C}[Z_1, Z_2] = \mathbf{C}[Z_2, Z_1] = 0$; i.e., if Z_1 and Z_2 are uncorrelated. On the other hand, merely setting $\mathbf{C}[Z_1, Z_2] + \mathbf{C}[Z_2, Z_1] = 0$ is equivalent to stipulating Re $\mathbf{C}[Z_1, Z_2] = 0$ (since the two covariances are complex conjugates). With $Z_i = X_i + jY_i$, $i = 1, 2$, and using subscript-c to indicate centralized versions, we come to the requirement

$$\text{Re } \mathbf{E}[Z_{1c} Z_{2c}^*] = \mathbf{E}[X_{1c} X_{2c}] + \mathbf{E}[Y_{1c} Y_{2c}] = 0.$$

This is satisfied in the steps leading to Equation (75.5) because there, $X_{2c} = 0$ and $Y_{1c} = 0$. Thus, one of the conditions under which Equation (75.6) reduces to $\mathbf{V}Z = \mathbf{V}Z_1 + \mathbf{V}Z_2$ is that the pair (Z_1, Z_2) consists of a purely real and a purely imaginary r.v.

If $\mathbf{Z}$ denotes a *complex vector r.v.*, $\mathbf{Z} = \begin{bmatrix} Z_1 \\ \vdots \\ Z_n \end{bmatrix}$, then the covariance matrix of $\mathbf{Z}$ is, analogous to Definition 71.1,

$$\begin{aligned} \mathbf{C}[\mathbf{Z}, \mathbf{Z}] = c_{\mathbf{Z}} &= \mathbf{E}[(\mathbf{Z} - \mathbf{E}\mathbf{Z})(\mathbf{Z} - \mathbf{E}\mathbf{Z})^{*T}] \\ &= \begin{bmatrix} \mathbf{V}[Z_1] & \mathbf{C}[Z_1, Z_2] & \mathbf{C}[Z_1, Z_3] & \cdots & \mathbf{C}[Z_1, Z_n] \\ \mathbf{C}[Z_2, Z_1] & \mathbf{V}[Z_2] & \mathbf{C}[Z_2, Z_3] & \cdots & \mathbf{C}[Z_2, Z_n] \\ \cdots & \cdots & \cdots & \cdots & \cdots \\ \cdots & \cdots & \cdots & \cdots & \cdots \\ \mathbf{C}[Z_n, Z_1] & \mathbf{C}[Z_n, Z_2] & \mathbf{C}[Z_n, Z_3] & \cdots & \mathbf{V}[Z_n] \end{bmatrix}. \end{aligned}$$

It can be seen that $c_{\mathbf{Z}}$ is a *Hermitian matrix*; in other words, $c_{\mathbf{Z}}^{*T} = c_{\mathbf{Z}}$.

---

[1] The correlation coefficient of two complex r.v.'s is not of much interest and will not be considered.

Queries

75.1 Given a complex r.v. $Z = X + jY$ such that $\mathbf{E}Z = 1$. For each of the following statements, determine whether it is true, possibly true, or false:

a) $Z = X - jY$

b) $\mathbf{E}[Z^*] = -1$

c) Z is purely real

d) If $W = jZ$, then $\mathbf{E}W = j$

[106]

e) $\mathbf{E}[W^*] = -j$.

75.2 The p.d.f. describing the distribution induced by a complex r.v. Z is

$$f_Z(z) = \begin{cases} 1/2, \ z \in \overline{1,2} \times \overline{0,2} \\ 0, \quad \text{otherwise.} \end{cases}$$

What are the mean and variance of the distribution? [172]

75.3 The real r.v. X induces a uniform distribution over $\overline{0, 2\pi}$. Let

$$Z = \cos X + j \sin X, \text{ and } W = -\sin X + j \cos X.$$

What is $\mathbf{C}[W, Z]$? [45]

The Characteristic Function

When manipulating functions of a real variable, the *Fourier transform* can provide some conveniences and insights. If $g(t)$ is a real or complex function of the real variable t, then the Fourier transform of $g(t)$ is usually defined as

$$G(\omega) = \int_{-\infty}^{\infty} g(t)e^{-j\omega t}dt \qquad (76.1)$$

provided the integral exists. $G(\omega)$ is a complex function of the real variable ω. Performing another Fourier transformation on $G(\omega)$, but this time with a scale factor of $1/2\pi$ and without the minus sign in the exponent, restores the original function (except possibly on a set of t-values that has measure 0):

$$g(t) = \frac{1}{2\pi} \int_{-\infty}^{\infty} G(\omega)e^{j\omega t}d\omega. \qquad (76.2)$$

When applying the Fourier transform to a probability density function it turns out to be more appropriate to use formula (76.2), i.e., the 'inverse Fourier transform', but without the scale factor of $1/2\pi$. Thus, if f_X is the p.d.f. of an absolutely continuous r.v. X, then we define a new function

$$\psi_X(\tau) = \int_{-\infty}^{\infty} e^{j\tau x}f_X(x)dx. \qquad (76.3)$$

The function ψ_X is called the *characteristic function* of the r.v. X. Strictly speaking, of course, it is not 'a function of' the r.v. X but a function 'associated with' the distribution P_X induced by that r.v. From Equations (76.1) and (76.2) it is clear that f_X can be recovered from ψ_X by the integral

$$f_X(x) = \frac{1}{2\pi} \int_{-\infty}^{\infty} \psi_X(\tau)e^{-j\tau x}d\tau. \qquad (76.4)$$

Comparison with Equation (75.4) shows that Equation (76.3) can be written differently, namely,

$$\psi_X(\tau) = \mathbf{E}[e^{j\tau X}] \qquad (76.5)$$

the expectation of $e^{j\tau X}$, a complex function of the r.v. X, where τ is a real-valued parameter. With the characteristic function expressed in this manner, there is no need to restrict it to absolutely continuous r.v.'s, and using a Stieltjes integral we have:

Probability Concepts and Theory for Engineers, First Edition. Harry Schwarzlander.
© 2011 John Wiley & Sons, Ltd. Published 2011 by John Wiley & Sons, Ltd.

(Definition) 76.1

The *characteristic function* of a (real) r.v. X is the function

$$\psi_X(\tau) = \int_{-\infty}^{\infty} e^{j\tau x} dF_X(x) = \mathbf{E}[e^{j\tau X}].$$

Having defined the characteristic function as an expectation, we must examine the possibility of this expectation failing to exist. Fortunately, although $\mathbf{E}X$ may not exist, $\mathbf{E}[e^{j\tau X}]$ always does (for all τ) because the range of $e^{j\tau X}$ is finite—in fact, for each τ it is some point on or within the unit circle in the complex plane.

Not only is it true that ψ_X exists for every r.v. X, but it describes the induced distribution. This is a basic result of Fourier–Stieltjes transform theory (cf. [Re2] Chapter VI, Section 4). Thus, there is a one-to-one correspondence between c.d.f.'s and characteristic functions.

Example 76.1

Let the r.v. X be unif.$\overline{0,1}$ (Figure 76.1a) and consider the complex r.v.'s $Z_\tau = e^{j\tau X}$ for different τ.

For $\tau = 0$, $Z_0 = e^0 = 1$; i.e., all of S (the space on which X is defined) gets mapped into the point $(1, 0)$ in $\mathbf{C}$. Thus,

$$\psi_X(0) = \mathbf{E}[Z_0] = 1.$$

Clearly, this result holds for any r.v. X !

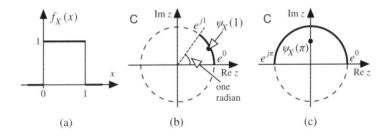

(a) (b) (c)

Figure 76.1 P.d.f. and characteristic function of the r.v. X

For $\tau = 1$, Z_1 is a r.v. that takes on values on the unit circle in the range from e^0 to e^{j1}, with probability distributed uniformly over this arc (Figure 76.1b). $\psi_X(1)$ locates the center of gravity of this distribution:

$$\psi_X(1) = \mathbf{E}Z_1 = \mathbf{E}[\cos X] + j\mathbf{E}[\sin X]$$

$$= \int_0^1 \cos x \, dx + j \int_0^1 \sin x \, dx = \sin 1 + j(1 - \cos 1).$$

For $\tau = \pi$, Z_π distributes probability uniformly over the upper half-circle, i.e., the arc from e^0 to $e^{j\pi}$ (see Figure 76.1c), giving

$$\psi_X(\pi) = \mathbf{E}Z_\pi = \mathbf{E}[\cos \pi X] + j\mathbf{E}[\sin \pi X]$$

$$= \int_0^1 \cos \pi x \, dx + j \int_0^1 \sin \pi x \, dx = 0 + j\frac{2}{\pi}.$$

By proceeding in this way, the behavior of any characteristic function can be understood. For instance, for the above r.v. X it can be seen at once that $\psi_X(\tau) = 0$ if $\tau = n2\pi$ ($n = \pm 1, \pm 2, \pm 3, \ldots$), because for those values of τ the r.v.'s Z_τ induce probability uniformly around the unit circle. $\psi_X(\tau)$ is easily found as a function of τ by using Equation (76.3):

$$\psi_X(\tau) = \int_{-\infty}^{\infty} e^{j\tau x} f_X(x)\, dx = \int_0^1 e^{j\tau x}\, dx = \begin{cases} \dfrac{1}{j\tau}(e^{j\tau} - 1), & \tau \neq 0 \\ 1, & \tau = 0 \end{cases}$$

which can also be written

$$\psi_X(\tau) = e^{j\tau/2}\,\frac{\sin(\tau/2)}{\tau/2}. \qquad (76.6)$$

From the discussion in the above Example it can be seen that for any r.v. X, the characteristic function $\psi_X(\tau)$ must satisfy the inequality

$$|\psi_X(\tau)| \le 1, \text{ for all } \tau. \qquad (76.7)$$

If X has a density function, this is verified by writing

$$|\psi_X(\tau)| = \left| \int_{-\infty}^{\infty} e^{j\tau x} f_X(x)\, dx \right| \le \int_{-\infty}^{\infty} |e^{j\tau x}|\, f_X(x)\, dx = \int_{-\infty}^{\infty} f_X(x)\, dx = 1.$$

If f_X does not exist, the above line can be rewritten in terms of Stieltjes integrals. Furthermore, it follows from Equation (76.3) that for $\tau = 0$, Equation (76.7) becomes $\psi_X(0) = 1$.

Example 76.2

As another example, we obtain the characteristic function of a standardized Gaussian r.v. X. With $f_X(x) = (2\pi)^{-1/2} \exp(-x^2/2)$, Equation (76.3) becomes

$$\psi_X(\tau) = \frac{1}{\sqrt{2\pi}} \int_{-\infty}^{\infty} e^{j\tau x} \exp\left(-\frac{x^2}{2}\right) dx$$

$$= \frac{1}{\sqrt{2\pi}} \int_{-\infty}^{\infty} (\cos \tau x + j\sin \tau x) \exp\left(-\frac{x^2}{2}\right) dx.$$

The imaginary part of this integral is

$$\frac{1}{\sqrt{2\pi}} \int_{-\infty}^{\infty} \sin \tau x \, \exp\left(-\frac{x^2}{2}\right) dx = 0$$

since the integrand is odd if $\tau \neq 0$, and vanishes if $\tau = 0$. Upon expanding the cosine in a power series and multiplying through by $\exp(-x^2/2)$, it is possible to integrate term by term (because after the multiplication the series in the integrand still converges uniformly for each τ in every finite x-interval). Making use of the moments of the standardized Gaussian distribution (see Problem 62.5), there results

$$\psi_X(\tau) = \frac{1}{\sqrt{2\pi}} \int_{-\infty}^{\infty} \left(1 - \frac{(\tau x)^2}{2!} + \frac{(\tau x)^4}{4!} - \frac{(\tau x)^6}{6!} + \cdots \right) \exp\left(-\frac{x^2}{2}\right) dx$$

$$= 1 - \frac{\tau^2}{2!} \cdot 1 + \frac{\tau^4}{4!} \cdot 1 \cdot 3 - \frac{\tau^6}{6!} \cdot 1 \cdot 3 \cdot 5 + \cdots$$

$$= 1 - \left(\frac{\tau}{\sqrt{2}}\right)^2 + \frac{1}{2!}\left(\frac{\tau}{\sqrt{2}}\right)^4 - \frac{1}{3!}\left(\frac{\tau}{\sqrt{2}}\right)^6 + \cdots = \exp\left(-\frac{\tau^2}{2}\right).$$

Thus, the characteristic function of the standardized Gaussian density function is the same function, scaled by the factor $\sqrt{2\pi}$ so that it equals 1 at the origin.

Another elementary property of characteristic functions follows from Equation (76.3):

$$\psi_X(-\tau) = [\psi_X(\tau)]^* \tag{76.8}$$

so that $\mathrm{Re}\,\psi_X$ is an even function, and $\mathrm{Im}\,\psi_X$ is an odd function.

One of the uses of characteristic functions arises in the calculation of moments.

Theorem 76.1: 'Moment generating property' of the characteristic function.
Let X be a r.v. for which $m_{1X}, \ldots, m_{nX}$ exist, $n \geq 1$. Then $\psi_X(\tau)$ is n times differentiable, and furthermore

$$\psi_X^{(k)}(0) = j^k m_{kX}, \quad \text{for } k = 1, \ldots, n \tag{76.9}$$

where $\psi_X^{(k)}(0)$ is the kth derivative of ψ_X, evaluated at $\tau = 0$.

Proof

This proof assumes the existence of $f_X(x)$—a restriction that can be removed by changing all Riemann integrals to Riemann–Stieltjes integrals. Formally, we can write

$$\psi_X'(\tau) = \frac{d}{d\tau}\int_{-\infty}^{\infty} e^{j\tau x} f_X(x) dx = \int_{-\infty}^{\infty} jx\, e^{j\tau x} f_X(x) dx \tag{76.10}$$

so that $\psi_X'(0) = \int_{-\infty}^{\infty} jx f_X(x) dx = j m_{1X}$. However, it remains to be verified that it is actually possible to carry out the differentiation indicated in Equation (76.10). Differentiation under the integral sign is valid provided the integral converges absolutely, and provided it does so uniformly over all values of the independent variable. This is the case for the first integral in Equation (76.10). Furthermore, if m_{1X} exists then $\int xf_X(x)dx$ is absolutely convergent by definition. Therefore, $\int_{-\infty}^{\infty} xe^{j\tau x} f_X(x) dx$ is also absolutely convergent, and it is so uniformly with τ. Repeated application of this argument gives Equation (76.9), for $k = 1, \ldots, n$.

An absolutely continuous two-dimensional r.v. $X = (X_1, X_2)$ has a *two-dimensional characteristic function* defined as the two-dimensional (inverse) Fourier transform of the joint density function $f_X(x_1, x_2)$:

$$\psi_X(\tau_1, \tau_2) = \int_{-\infty}^{\infty}\int_{-\infty}^{\infty} f_X(x_1, x_2) e^{j(\tau_1 x_1 + \tau_2 x_2)} dx_1 dx_2 = \mathbf{E}[e^{j(\tau_1 X_1 + \tau_2 X_2)}]. \tag{76.11}$$

In turn, f_X can be recovered from ψ_X in a manner analogous to Equation (76.5):

$$f_X(x_1, x_2) = \frac{1}{4\pi^2} \int_{-\infty}^{\infty} \int_{-\infty}^{\infty} \psi_X(\tau_1, \tau_2) e^{j(\tau_1 x_1 + \tau_2 x_2)} d\tau_1 d\tau_2. \tag{76.12}$$

The characteristic functions of the component r.v.'s X_1 and X_2 are easily obtained from ψ_X, as can be seen from Equation (76.11):

$$\psi_{X_1}(\tau_1) = \psi_X(t_1, 0); \quad \psi_{X_2}(\tau_2) = \psi_X(0, \tau_2).$$

Moments of X_1 and of X_2 can therefore also be obtained from ψ_X. In addition, however, joint moments can be computed. Noting that

$$\frac{\partial^{h+k}}{\partial \tau_1^h \, \partial \tau_2^k} \psi_X(0,0) = \frac{\partial^{h+k}}{\partial \tau_1^h \, \partial \tau_2^k} \mathbf{E}[\exp[j(\tau_1 X_1 + \tau_2 X_2)]] \Big|_{\tau_1 = \tau_2 = 0}$$

$$= \mathbf{E}[j^h X_1^h j^k X_2^k]$$

where interchanging the expectation and differentiation operations is justified in the same way as in the Proof of Theorem 76.1 if $\mathbf{E}[X_1^h X_2^k]$ exists, and we obtain

$$\mathbf{E}\left[X_1^h X_2^k\right] = j^{-(h+k)} \frac{\partial^{h+k}}{\partial \tau_1^h \, \partial \tau_2^k} \psi_X(0,0). \tag{76.13}$$

Clearly, Equation (76.13) applies also for moments of X_1 and moments of X_2.

We conclude this Section by noting first that for a *complex* $n \times n$ matrix, Definition 71.2 must be modified so that $\boldsymbol{q}$ is a column vector of *complex* numbers, and Equation (71.2) reads $\boldsymbol{q}^{\mathrm{T}} \boldsymbol{c} \boldsymbol{q}^* \geq 0$. Now, we state the following property:

Theorem 76.2
The characteristic function ψ_X of any r.v. X is a *positive semi-definite* function. That is, for every integer $n > 1$, if $\tau_1, \ldots, \tau_n$ are arbitrary real numbers and if $s_1, \ldots, s_n$ are arbitrary complex numbers, then

$$\sum_{k=1}^{n} \sum_{i=1}^{n} \psi_X(\tau_k - \tau_i) s_k s_i^* \geq 0. \tag{76.14}$$

Proof
For given n, let $\boldsymbol{\tau} = \begin{bmatrix} \tau_1 \\ \vdots \\ \tau_n \end{bmatrix}$, $\boldsymbol{s} = \begin{bmatrix} s_1 \\ \vdots \\ s_n \end{bmatrix}$, and $\boldsymbol{Z}(\tau) = \begin{bmatrix} Z_1 \\ \vdots \\ Z_n \end{bmatrix} = \begin{bmatrix} e^{j\tau_1 X} \\ \vdots \\ e^{j\tau_n X} \end{bmatrix}$. Now,

$$\psi_X(\tau_k - \tau_i) = \mathbf{E}[e^{j(\tau_k - \tau_i)X}] = \mathbf{E}[e^{j\tau_k X} e^{-j\tau_i X}] = \mathbf{E}[Z_k Z_i^*].$$

Therefore, the left side of Equation (76.14) can be written

$$\sum_{k=1}^{n}\sum_{i=1}^{n}\psi_X(\tau_k - \tau_i)s_k s_i^* = s^{\mathsf{T}}\mathbf{E}[Z(\tau)^{\mathsf{T}}Z(\tau)^*]s^*$$

$$= \mathbf{E}[|s_1 e^{j\tau_1 X} + s_2 e^{j\tau_2 X} + \ldots + s_n e^{j\tau_n X}|^2]$$

which is the expectation of a real, non-negative r.v. and therefore is real and non-negative.

Queries

76.1 Let $\psi_X(\tau)$, $\psi_Y(\tau)$ be the characteristic functions of two r.v.'s X, Y, respectively. For each of the following expressions it is claimed that it is also a characteristic function. Indicate in each case whether this is true, possibly true, or false.

a) $\psi_X(\tau) + \psi_Y(\tau)$ b) $\psi_X(\tau)/2$

c) $e^{-j\tau}\psi_X(\tau)$ d) $\psi_X(\tau)/\psi_Y(\tau)$.   [246]

76.2 Let $\psi_X(\tau)$ be the characteristic function of some r.v. X, and let some value of τ be given. For each of the following statements, indicate whether it is true, possibly true or false:

a) $\psi_X(\tau) + \psi_X(-\tau)$ is real

b) $\psi_X(\tau) + \psi_X(-\tau)$ is purely imaginary

c) $|\psi_X(\tau) - \psi_X(-\tau)| \leq 2$

d) $\mathrm{Re}[\psi_X(\tau) + \psi_X(-\tau)] \geq -2$.   [140]

Characteristic Function of a Transformed Random Variable

In this Section we examine how characteristic functions are influenced by the transformation of r.v.'s. First, here is how scale change and shift applied to a r.v. X affects its characteristic function. Let $Y = aX + b$ (a, b real, $a \neq 0$), then

$$\psi_Y(\tau) = \mathbf{E}[e^{j\tau Y}] = \mathbf{E}[e^{j\tau(aX+b)}] = e^{j\tau b}\,\mathbf{E}[e^{j\tau aX}] = e^{j\tau b}\psi_X(a\tau). \qquad (77.1)$$

Thus, once the characteristic function of one member of a family of distributions is known, it can easily be modified so as to apply to another member of that family. For instance, if Y is uniformly distributed over $\overline{-1,1}$, then its characteristic function can be obtained from Equation (76.6) by writing $Y = 2X - 1$ and then applying Equation (77.1) to give

$$\psi_Y(\tau) = e^{-j\tau}e^{j\tau}\left[\frac{\sin\tau}{\tau}\right] = \frac{\sin\tau}{\tau}. \qquad (77.2)$$

Of particular interest is the characteristic function of a *sum of s.i. r.v.'s*. Beginning with two arbitrary r.v.'s X, Y, let $V = X + Y$. This sum has the characteristic function

$$\psi_V(\tau) = \mathbf{E}[e^{j\tau(X+Y)}] = \mathbf{E}[e^{j\tau X}e^{j\tau Y}].$$

Now, if X, Y are s.i. then by Theorem 55.2 $e^{j\tau X}$ and $e^{j\tau Y}$ are also a pair of s.i. r.v.'s. Therefore, the product rule for expectations can be applied, giving

$$\psi_V(\tau) = \mathbf{E}[e^{j\tau X}]\,\mathbf{E}[e^{j\tau Y}] = \psi_X(\tau)\psi_Y(\tau). \qquad (77.3)$$

This important result is restated in the following theorem.

Theorem 77.1
The characteristic function of the sum of two statistically independent r.v.'s is the product of their characteristic functions.

This Theorem extends in an obvious way to products of more than two r.v.'s that are m.s.i.

Probability Concepts and Theory for Engineers, First Edition. Harry Schwarzlander.
© 2011 John Wiley & Sons, Ltd. Published 2011 by John Wiley & Sons, Ltd.

Example 77.1

Let X and Y be statistically independent r.v.'s that induce Cauchy distributions, with

$$f_X(x) = \frac{1}{\pi b_1}\left[1 + \left(\frac{x - a_1}{b_1}\right)^2\right]^{-1}, \quad f_Y(y) = \frac{1}{\pi b_2}\left[1 + \left(\frac{x - a_2}{b_2}\right)^2\right]^{-1}$$

where $b_1, b_2 > 0$. What is the characteristic function of $W = X + Y$? If $a_1 = 0$, we find

$$\psi_X(\tau) = \int_{-\infty}^{\infty} \frac{e^{j\tau x}}{\pi b_1\left[1 + (x/b_1)^2\right]}dx = \int_{-\infty}^{\infty} \frac{\cos \tau x}{\pi b_1\left[1 + (x/b_1)^2\right]}dx$$

$$= \frac{1}{\pi}\int_{-\infty}^{\infty} \frac{\cos \tau b_1 v}{1 + v^2}dv = \exp(-|b_1\tau|).$$

For arbitrary a_1, making use of Equation (77.1), the above becomes $\psi_X(\tau) = \exp(j\tau a_1 - |b_1\tau|)$. Similarly, $\psi_Y(\tau) = \exp(j\tau a_2 - |b_2\tau|)$. Then

$$\psi_W(\tau) = \psi_X(\tau) \cdot \psi_Y(\tau) = \exp[j\tau(a_1 + a_2) - |(b_1 + b_2)\tau|].$$

It can be seen that W is again a Cauchy r.v., and we conclude: The sum of n m.s.i. Cauchy r.v.'s with parameters $(a_1, b_1), (a_2, b_2), \ldots, (a_n, b_n)$ is again a Cauchy random variable with parameters $(a_1 + a_2 + \ldots + a_n), (b_1 + b_2 + \ldots + b_n)$.

Since a characteristic function is a function of the distribution induced by a r.v., it can also be defined for a *conditional distribution given an event*. Thus, replacing $f_X(x)$ in Equation (76.3) by a conditional density $f_{X|A}(x)$ as in Definition 57.2 results in

$$\int_{-\infty}^{\infty} e^{j\tau x} f_{X|A}(x)dx = \psi_{X|A}(\tau)$$

which is a *conditional characteristic function*. In order to write this in the form of an expectation as in Equation (76.5), a 'conditional r.v.' Y can be introduced, such that $f_Y(x) = f_{X|A}(x)$. Then,

$$\psi_{X|A}(\tau) = \psi_Y(\tau) = \mathbf{E}[e^{j\tau Y}].$$

However, Definition 76.1 also allows us to write

$$\psi_{X|A}(\tau) = \int_{-\infty}^{\infty} e^{j\tau x}\, dF_{X|A}(x) = \mathbf{E}[e^{j\tau X}|A].$$

In other words, $\psi_{X|A}(\tau)$ is a conditional expectation, so that the rules for conditional expectations apply. For instance, given $\psi_{X|A}(\tau)$ and $\psi_{X|A^c}(\tau)$, then

$$\psi_X(\tau) = P(A)\psi_{X|A}(\tau) + P(A^c)\psi_{X|A^c}(\tau).$$

More generally, since the Fourier transform is a linear transformation, it follows that a mixture of induced distributions has a characteristic function that can be expressed as a corresponding mixture of characteristic functions.

Example 77.2

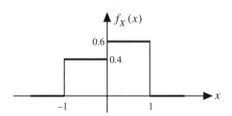

Figure 77.1   P.d.f. for the r.v. X

Let X be a r.v. that induces a distribution P_X characterized by the p.d.f. whose graph is shown in Figure 77.1. P_X can be expressed as a mixture of two uniform distributions—for instance, in terms of one distribution that is unif. $\overline{-1, 1}$, and the other unif. $\overline{0, 1}$. Denoting the corresponding density functions f_1 and f_2, we have

$$f_X(x) = 0.8 f_1(x) + 0.2 f_2(x).$$

From Equation (77.2) and Equation (76.6) it then follows that

$$\psi_X(\tau) = 0.8 \frac{\sin \tau}{\tau} + 0.2\, e^{j\tau/2} \frac{\sin (\tau/2)}{\tau/2}.$$

The above discussion of conditional characteristic functions leads directly to the determination of the characteristic function of a *product* of two statistically independent r.v.'s. Let X, Y be two s.i. r.v.'s, where at least Y is discrete, and let $W = XY$. If y_1, y_2, $y_3, \ldots$ are the values that Y can take on, let $A_k = \{\xi: Y(\xi) = y_k\}$, for every y_k in the range of Y. Then it is possible to express ψ_W in the following way:

$$\psi_W(\tau) = \sum_k \psi_{W|A_k}(\tau)\, P(A_k)$$

where the summation is over all those k for which a set A_k is defined. But $\psi_{W|A_k}(\tau) = \mathbf{E}[e^{j\tau XY}|A_k]$, and since X, Y are s.i. this can be written $\psi_{W|A_k}(\tau) = \mathbf{E}[e^{j\tau X y_k}] = \psi_X(y_k\tau)$, giving

$$\psi_W(\tau) = \sum_k \psi_X(y_k\tau)\, P_Y(\{y_k\}). \tag{77.4}$$

The characteristic function of the product r.v. W is therefore a weighted sum of characteristic functions.

The result (Equation (77.4)) can be generalized to a product of two independent r.v.'s of arbitrary types by using the conditional expectation of $e^{j\tau XY}$ given Y. For instance, if Y is absolutely continuous, then

$$\psi_W(\tau) = \mathbf{E}[e^{j\tau XY}] = \int_{-\infty}^{\infty} \mathbf{E}[e^{j\tau XY}|Y = y]\, f_Y(y)dy$$
$$= \int_{-\infty}^{\infty} \mathbf{E}[e^{j\tau Xy}] f_Y(y)dy = \int_{-\infty}^{\infty} \psi_X(y\tau)\, f_Y(y)dy. \tag{77.5}$$

In order to avoid the need to restrict X or Y to be of a particular type, Equation (77.5) can be rewritten as a Stieltjes integral. Assuming only that X and Y are s.i., then

$$\psi_W(\tau) = \int_{-\infty}^{\infty} \psi_X(y\tau)\, dF_Y(y). \tag{77.6}$$

If W is any other function $g(X, Y)$ of two s.i. r.v.'s X, Y, a similar approach can be used to obtain an expression for the characteristic function of W. However, if X, Y are not statistically independent, then in Equation (77.5) the conditional expectation $\mathbf{E}[e^{j\tau g(X,Y)}|Y = y] = \mathbf{E}[e^{j\tau g(X,y)}|Y = y]$ must be evaluated in terms of $P_{X|Y}$.

Queries

77.1 Let $\psi_X(\tau)$ be the characteristic function of the discrete ternary r.v. X that induces probability 1/3 on each of the sets $\{-1\}$, $\{0\}$, and $\{+1\}$. Determine:
a) $\psi_X(\pi)$
b) $\psi_Y(\pi)$, where $Y = 2X$
c) $\psi_Z(\pi)$, where $Z = X + Y$
d) $\psi_W(\pi)$, where $W = XY$.   [178]

77.2 X and Y are two i.d.s.i. r.v.'s. Let $V = X + Y$. If $\psi_V(1) = 0.01j$, what is $\psi_X(1)$?   [111]

77.3 X and Y are any two i.d.s.i. r.v.'s. Which of the following statements are correct:
a) $|\psi_X(\tau)|^2$ is the characteristic function of $X - Y$
b) $|\psi_X(\tau)|^2$ is the characteristic function of X^2 and of Y^2
c) $|\psi_X(\tau)|^2$ is the characteristic function of XY
d) $|\psi_X(\tau)|^2$ is not a characteristic function.   [203]

Characteristic Function of a Multidimensional Random Variable

In Section 76, the characteristic function of an absolutely continuous two-dimensional r.v. was introduced. Here, we extend this idea to an arbitrary number of dimensions and to r.v.'s of arbitrary type. Some of the significant applications of characteristic functions arise in the multidimensional case.

> **(Definition) 78.1**
> Let $X = (X_1, \ldots, X_n)$ be an n-dimensional r.v. The characteristic function of X is a complex function of n real variables $\tau_1, \ldots, \tau_n$, defined by[2]
>
> $$\psi_X(\tau_1, \ldots, \tau_n) = \mathbf{E}[\exp[j(\tau_1 X_1 + \tau_2 X_2 + \ldots + \tau_n X_n)]]. \qquad (78.1)$$

Thus, if X is n-dimensional, then the characteristic function ψ_X is n-dimensional, and its value at any point $(\tau_1, \ldots, \tau_n)$ in R^n is the expectation of the random variable

$$\exp[j(\tau_1 X_1 + \tau_2 X_2 + \ldots + \tau_n X_n)].$$

For all values of $\tau_1, \ldots, \tau_n$ this r.v. distributes probability only onto the unit circle in the complex plane, so that $\psi_X(\tau_1, \ldots, \tau_n)$ always exists and $|\psi_X(\tau_1, \ldots, \tau_n)| \leq 1$ for all values of the arguments. Furthermore, it follows from Equation (78.1) that $\psi_X(0, 0, \ldots, 0) = 1$. It should be noted, however, that if X is not absolutely continuous, then the n-dimensional Fourier transform implied by Equation (78.1) must be expressed as a Stieltjes integral.

From Definition 78.1 it is readily apparent how to obtain the characteristic function of any component r.v. of X, or of a lower-dimensional r.v. made up of components of X: As we saw in Section **76** for the two-dimensional case, it is merely necessary to replace by zero those arguments of ψ_X that correspond to unwanted components of X.

---

[2] The significance of the notation $\tau_1, \ldots, \tau_n$ here must not be confused with its use in Theorem 76.2, where it represents n values taken on by a single variable τ.

Probability Concepts and Theory for Engineers, First Edition. Harry Schwarzlander.
© 2011 John Wiley & Sons, Ltd. Published 2011 by John Wiley & Sons, Ltd.

Example 78.1

Suppose that $X = (X_1, X_2, X_3, X_4, X_5)$, and $Y = (X_1, X_3)$. How is ψ_Y related to ψ_X? For convenience, denote the two variables of ψ_Y by τ_1 and τ_3. Then,

$$\psi_Y(\tau_1, \tau_3) = \mathbf{E}[\exp j(\tau_1 X_1 + \tau_3 X_3)] = \psi_X(\tau_1, 0, \tau_3, 0, 0).$$

If the components of X are m.s.i., then it follows from Equation (78.1) and Theorem 64.3 that ψ_X is the *product* of the characteristic functions of these components (but each with a distinct independent variable):

$$\begin{aligned}
\psi_X(\tau_1, \ldots, \tau_n) &= \mathbf{E}[\exp j(\tau_1 X_1)] \, \mathbf{E}[\exp j(\tau_2 X_2)] \ldots \mathbf{E}[\exp j(\tau_n X_n)] \\
&= \psi_{X_1}(\tau_1) \psi_{X_2}(\tau_2) \ldots \psi_{X_n}(\tau_n).
\end{aligned} \tag{78.2}$$

The converse of this result also holds:

Theorem 78.1

Let $\psi_X(\tau_1, \ldots, \tau_n)$ be the characteristic function of an n-dimensional r.v. $X = (X_1, X_2, \ldots, X_n)$. If ψ_X factors into n functions of one variable, such each of these factors is a characteristic function, then these factors are $\psi_{X_1}(\tau_1), \psi_{X_2}(\tau_2), \ldots, \psi_{X_n}(\tau_n)$, and $X_1, X_2, \ldots, X_n$ are mutually statistically independent.

Proof (for the jointly absolutely continuous case)

Suppose ψ_X factors into n characteristic functions that we denote $\psi_{Y_1}(\tau_1), \psi_{Y_2}(\tau_2), \ldots, \psi_{Y_n}(\tau_n)$. Using the n-dimensional Fourier transform,

$$\begin{aligned}
f_X(x_1, \ldots, x_n) &= \\
&= \frac{1}{(2\pi)^n} \int_{-\infty}^{\infty} \cdots \int_{-\infty}^{\infty} \psi_X(\tau_1, \ldots, \tau_n) \exp[-j(\tau_1 x_1 + \ldots + \tau_n x_n)] \, d\tau_1 \ldots d\tau_n \\
&= \left[\frac{1}{2\pi} \int_{-\infty}^{\infty} \psi_{Y_1}(\tau_1) \exp(-j\tau_1 x_1) \, d\tau_1 \right] \cdots \left[\frac{1}{2\pi} \int_{-\infty}^{\infty} \psi_{Y_n}(\tau_n) \exp(-j\tau_n x_n) \, d\tau_n \right] \\
&= f_{Y_1}(x_1) \cdots f_{Y_n}(x_n).
\end{aligned}$$

Then, by Corollary 55.1b, $f_{Y_k}(x_k) = f_{X_k}(x_k)$ so that $\psi_{Y_k} = \psi_{X_k}$, for $k = 1, \ldots, n$, and (again by Corollary 55.1b) the r.v.'s $X_1, \ldots, X_n$ are m.s.i.

Some basic properties of multidimensional characteristic functions are stated in the following Theorems, proofs of which are left as an exercise (see Problem 78.1).

Theorem 78.2: Scale change and shift of origin.

Let $X = (X_1, \ldots, X_n)$ be an n-dimensional r.v. with characteristic function ψ_X. A scale change and shift is applied to each of the components of X, giving $Y = (a_1 X_1 + b_1, a_2 X_2 + b_2, \ldots, a_n X_n + b_n)$. Then,

$$\psi_Y(\tau_1, \ldots, \tau_n) = \exp\left(j \sum_{k=1}^{n} b_k \tau_k \right) \psi_X(a_1 \tau_1, \ldots, a_n \tau_n).$$

Theorem 78.3

Let $X = (X_1, \ldots, X_n)$ be an n-dimensional r.v. whose characteristic function is $\psi_X(\tau_1, \ldots, \tau_n)$, and let $Y = X_1 + X_2 + \ldots + X_n$. Then, $\psi_Y(\tau) = \psi_X(\tau, \tau, \ldots, \tau)$.

Of course, if $X_1, \ldots, X_n$ in Theorem 78.3 are m.s.i. r.v.'s then, according to Equation (78.2), $\psi_X(\tau, \tau, \ldots, \tau)$ can be expressed as the product $\psi_{X_1}(\tau)\psi_{X_2}(\tau) \ldots \psi_{X_n}(\tau)$.

Theorem 78.4

Let $X = (X_1, \ldots, X_n)$ and $Y = (Y_1, \ldots, Y_n)$ be two statistically independent n-dimensional r.v.'s with characteristic functions $\psi_X(\tau_1, \ldots, \tau_n)$ and $\psi_Y(\tau_1, \ldots, \tau_n)$, respectively. If $V = X + Y$, that is, $V = (X_1 + Y_1, X_2 + Y_2, \ldots, X_n + Y_n)$, then

$$\psi_V(\tau_1, \ldots, \tau_n) = \psi_X(\tau_1, \ldots, \tau_n)\psi_Y(\tau_1, \ldots, \tau_n).$$

Equation (78.1) can also be expressed in matrix notation. If the n real variables $\tau_1, \ldots, \tau_n$ are arranged into a column vector $\tau = [\tau_1, \ldots, \tau_n]^T$, and X is expressed as a column vector $[X_1, \ldots, X_n]^T$, then

$$\psi_X(\tau) = \mathbf{E}[\exp(j\tau^T X)]. \tag{78.3}$$

Suppose now that a linear coordinate transformation and shift of origin is applied to X, resulting in a new n-dimensional vector r.v. $Y = aX + b$, where a is a nonsingular transformation matrix and b is a column vector of real constants. Then,

$$\psi_Y(\tau) = \mathbf{E}[\exp(j\tau^T\, Y)] = \mathbf{E}[\exp[j\tau^T(aX + b)]] = \exp(j\tau^T\, b)\psi_X(a^T\, \tau). \tag{78.4}$$

If $X_1, \ldots, X_n$ are m.s.i. and $b = 0$, this can be written out to yield

$$\psi_Y(\tau) = \mathbf{E}\left[\exp j\left(\sum_{k=1}^{n}\sum_{i=1}^{n}\tau_k a_{ki} X_i\right)\right] = \prod_{i=1}^{n}\mathbf{E}\left[\exp j\left(\sum_{k=1}^{n}\tau_k a_{ki}\right)X_i\right]$$

$$= \prod_{i=1}^{n}\psi_{X_i}\left(\sum_{k=1}^{n}\tau_k a_{ki}\right). \tag{78.5}$$

Example 78.2

A r.v. $X = \begin{bmatrix} X_1 \\ X_2 \end{bmatrix}$ distributes probability uniformly over the square $\overline{-1, 1} \times \overline{-1, 1}$. From Equation (77.2) and statistical independence of X_1, X_2 it follows that

$$\psi_X(\tau_1, \tau_2) = \frac{(\sin \tau_1)(\sin \tau_2)}{\tau_1 \tau_2}.$$

Suppose we wish to find the characteristic function of a r.v. Y that induces a distribution P_Y obtained by rotating P_X counterclockwise about the origin through an angle φ. Then, it is possible to specify $Y = aX$, where a is the rotational transformation. (Y need not actually be related to X in this way; it could be statistically independent of X, for instance.) Here,

$$a = \begin{bmatrix} \cos\varphi & -\sin\varphi \\ \sin\varphi & \cos\varphi \end{bmatrix}$$

and making use of Equation (78.5) we obtain

$$\psi_Y(\tau_1, \tau_2) = \psi_{X_1}(\tau_1 \cos \varphi + \tau_2 \sin \varphi) \cdot \psi_{X_2}(-\tau_1 \sin \varphi + \tau_2 \cos \varphi)$$

$$= \frac{\sin(\tau_1 \cos \varphi + \tau_2 \sin \varphi) \, \sin(-\tau_1 \sin \varphi + \tau_2 \cos \varphi)}{\tau_1 \tau_2 \cos 2\varphi + 0.5(\tau_2^2 - \tau_1^2) \sin 2\varphi}.$$

Direct derivation of ψ_Y from Equation (76.11) would have been quite a bit messier.

The determination of a product moment from an n-dimensional characteristic function is made in accordance with a straightforward extension of Equation (76.13). For instance, suppose that $\psi_X(\tau_1, \tau_2, \tau_3, \tau_4)$ is the characteristic function of a r.v. $X = (X_1, X_2, X_3, X_4)$. Then for any positive integers q, r, s,

$$E[X_1^q X_2^r X_3^s] = j^{-(q+r+s)} \frac{\partial^{q+r+s}}{\partial \tau_1^q \partial \tau_2^r \partial \tau_3^s} \psi_X(\tau_1, \tau_2, \tau_3, 0) \Bigg|_{\tau_1 = \tau_2 = \tau_3 = 0}.$$

Finally, in the following Example we obtain the characteristic function for a multidimensional r.v. that is singular.

Example 78.3

Consider the r.v. $X = (X_1, X_2)$ where $X_2 = -X_1$. It is still possible to write $\psi_X(\tau_1, \tau_2) = E[\exp j(\tau_1 X_1 + \tau_2 X_2)]$, where the expected value is now understood to be a Stieltjes integral. However, properties of expected values discussed in Part V still apply. Then,

$$E[\exp j(\tau_1 X_1 + \tau_2 X_2)] = E[\exp j(\tau_1 - \tau_2) X_1] = \psi_{X_1}(\tau_1 - \tau_2).$$

It can be seen that the linear dependence between X_1 and X_2 has the effect of making $\psi_X(\tau_1, \tau_2)$ actually a function of a single variable, $\tau_1 - \tau_2 = \tau$. Now, suppose X_1 is unif. $\overline{-1, 1}$. Making use of Equation (77.2), we then obtain

$$\psi_X(\tau_1, \tau_2) = \frac{\sin(\tau_1 - \tau_2)}{\tau_1 - \tau_2}.$$

Queries

78.1 The characteristic function $\psi_X(\tau_1, \tau_2, \tau_3)$ of a three-dimensional absolutely continuous r.v. X is such that every permutation of the three arguments τ_1, τ_2, τ_3 leaves ψ_X unchanged. For each of the following statements, determine whether it is true, possibly true, false, or meaningless:

a) $\psi_X(\tau_1, \tau_2, \tau_3) = [\psi_X(\tau_1)]^3$
b) The components r.v.'s of X are identically distributed
c) A rotational transformation of X leaves $\psi_X(\tau_1, \tau_2, \tau_3)$ unchanged
d) $\psi_{X_1}(\tau) = \psi_{X_2}(\tau)$
e) $[\psi_X(\tau_1, \tau_2, \tau_3)]^2$ is also unchanged by a permutation of its three arguments.   [109]

78.2 Two two-dimensional r.v.'s $X = (X_1, X_2)$ and $Y = (Y_1, Y_2)$ have characteristic functions $\psi_X(\tau_1, \tau_2)$ and $\psi_Y(\tau_1, \tau_2)$, respectively. X and Y are statistically independent. Let $W = (X, Y) = (X_1, X_2, Y_1, Y_2)$. Suppose $\psi_X(1, 0.5) = 0.2 - j0.2$, and $\psi_Y(1, 0.5) = 0.5 + j0$. Evaluate $\psi_W(1, 0.5, 1, 0.5)$.   [180]

The Generating Function

In this Section we consider a slightly different transformation of a probability distribution, which results in the so-called 'generating function'. It is applicable only to distributions induced by integer-valued discrete r.v.'s. As we have seen, there are many Probability problems that are concerned with counting in some way and therefore can give rise to integer-valued discrete r.v.'s.

(Definition) 79.1

Let X be a discrete r.v. whose range is a subset of the set of integers. The *generating function* of the r.v. X (also called *probability generating function* or *moment generating function*) is denoted G_X and is defined by

$$G_X(z) = \sum_{k=-\infty}^{\infty} P_X(\{k\})z^k \tag{79.1}$$

for all complex numbers z for which the series converges.

Since $\sum_{k=-\infty}^{\infty} P_X(\{k\}) = G_X(1) = 1$, the series (79.1) always converges at least on the unit circle in the complex plane, that is, for $|z| = 1$. Equation (79.1) brings to mind the z-transform; as a matter of fact, $G_X(1/z)$ can be recognized as expressing the (two-sided) z-transform of the discrete distribution $P_X(\{k\})$, or equivalently, $G_X(z)$ is the z-transform of $P_X(\{-k\})$. Therefore, P_X can be recovered from $G_X(z)$ by inverting the z-transform.[3]

The definition of the generating function is often restricted to non-negative integer-valued r.v.'s, because its application is more straightforward in that case. For instance, as can be seen from Equation (79.1), if X is a non-negative integer-valued r.v. then $P_X(\{0\}) = G_X(0)$, and for $k = 1$, 2, 3, ..., $P_X(\{k\}) = G_X^{(k)}(0)$, where $G_X^{(k)}$ denotes the kth derivative of G_X. This is not usually a convenient way to do the inversion, however (see Problem 79.1).

For values of z as specified in Definition 79.1, Equation (79.1) can also be written

$$G_X(z) = \mathbf{E}[z^X]. \tag{79.2}$$

From this it follows immediately that $G_X(1/z)$ is the generating function of $-X$.

---

[3] For a detailed discussion of the z-transform, see Chapter 4 of [OS1].

Probability Concepts and Theory for Engineers, First Edition. Harry Schwarzlander.
© 2011 John Wiley & Sons, Ltd. Published 2011 by John Wiley & Sons, Ltd.

There is a close connection between the generating function and the characteristic function, as is evident from Equation (79.2). Writing $e^{j\tau}$ in place of z in Equation (79.1) or Equation (79.2) yields the characteristic function of the r.v. X. And for $|z| = 1$, that is, at any point on the unit circle in the z-plane, we do have $z = e^{j\tau}$ (τ real). So it can be seen that, along the unit circle in the z-plane, the generating function of X equals the characteristic function of X.

Example 79.1:

Consider the indicator r.v. I_A of an event A. Its generating function is

$$G_{I_A}(Z) = P(A^c)z^0 + P(A)z^1 = 1 - P(A) + P(A)z. \tag{79.3}$$

Figure 79.1 shows a sketch of $|G_{I_A}(z)|$ over the unit circle, for a representative value of $P(A)$.

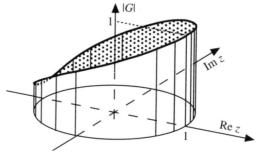

Figure 79.1   Generating function for the indicator r.v. I_A

The characteristic function of I_A is

$$\psi_{I_A}(\tau) = P(A^c)e^{j\tau 0} + P(A)e^{j\tau 1} = 1 - P(A) + P(A)e^{j\tau}. \tag{79.4}$$

A comparison of the sketch of $|\psi_{I_A}(\tau)|$ in Figure 79.2 with Figure 79.1, and of Equation (79.4) with Equation (79.3), shows how the characteristic function can be thought of as an 'unwrapping' of $|G_{I_A}|$ along the unit circle in the z-plane.

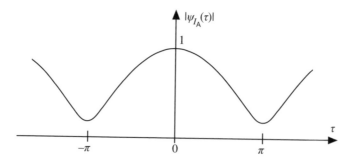

Figure 79.2   Magnitude of the characteristic function of I_A

From the discussion in Example 79.1 it follows that a discrete r.v. of the type specified in Definition 79.1 has a characteristic function that is periodic with period 2π. This can be verified by writing

$$\psi_X(\tau+n2\pi) = \sum_{k=-\infty}^{\infty} P_X(\{k\})e^{j(\tau+n2\pi)k} = \sum_{k=-\infty}^{\infty} P_X(\{k\})e^{j\tau k} = \psi_X(\tau).$$

Everything that can be accomplished with generating functions can also be done with characteristic functions. Besides, the characteristic function exists for every r.v. However, manipulation of the generating function is often simpler.

From Equation (79.2) follows a theorem analogous to Theorem 77.1:

Theorem 79.1
Let X and Y be s.i. discrete r.v.'s whose ranges consist of only integer values. If $W = X + Y$, then $G_W(z) = G_X(z)G_Y(z)$.

Example 79.2: A binomial r.v.
To find the generating function of a binomial r.v. B that induces the distribution

$$P_B(\{b\}) = \binom{n}{b}p^b(1-p)^{n-b}, \; b = 0, 1, 2, \ldots, n$$

we recall that B can be expressed as the sum of n statistically independent indicator r.v.'s I_1, $I_2, \ldots, I_n$, each of which takes the value 1 with probability p. From Equation (79.3), $G_{I_k}(z) = 1 - p + pz$, so that Theorem 79.1 gives

$$G_B(z) = [G_{I_k}(z)]^n = (1-p+pz)^n.$$

From this, the characteristic function of B is also obtained immediately as the function

$$\psi_B(\tau) = [1 + (e^{j\tau} - 1)p]^n.$$

Different moments can be computed from $G_X(z)$. Thus,

$$G_X'(1) = \sum_{k=-\infty}^{\infty} kP_X(\{k\}) = m_{1X}.$$

Also, since

$$G_X''(1) = \sum_{k=-\infty}^{\infty} k(k-1)P_X(\{k\}) = m_{2X} - m_{1X}.$$

it follows that $m_{2X} = \mathbf{E}X^2 = G_X''(1) + G_X'(1)$. Higher moments can also be expressed in terms of derivatives of the generating function, but the expressions become more complicated. This can be seen by noting that

$$G_X^{(n)}(z) = \mathbf{E}[X(X-1)(X-2)\ldots(X-n+1)z^X] \tag{79.5}$$

so that $G_X^{(n)}(1) = \mathbf{E}[X(X-1)\ldots(X-n+1)]$.

Now let $Y = aX$, where X is an integer-valued r.v. and a an integer. Then,

$$G_Y(z) = \mathbf{E}[z^Y] = \mathbf{E}[z^{aX}]$$

so that $G_Y(z) = G_X(z^a)$. Also, if $Y = X + b$, we have $G_Y(z) = \mathbf{E}[z^Y] = \mathbf{E}[z^{X+b}] = z^b G_X(z)$. Thus, integer-valued shifts and scalings are easily accommodated. Actually, Definition 79.1 can be extended so as to allow shifts and scaling that are not integer-valued. However, such modified forms of the generating function are not so convenient to work with. Instead, in such cases it is usually simpler to first transform the shifted or scaled r.v. into an integer-valued r.v., and then apply Definition 79.1.

Analogous to the conditional characteristic function (Section 77), a *conditional generating function* can be defined. Thus, let X be a r.v. of the kind specified in Definition 79.1 and defined on a probability space $(\mathsf{S}, \mathscr{A}, P)$, and let $\mathsf{A} \in \mathscr{A}$ with $P(\mathsf{A}) > 0$ (or let A denote a set with positive probability in the induced space). Then, A can be used as a conditioning event in Equation (79.2), which results in the conditional generating function of X given A,

$$G_{X|\mathsf{A}}(z) = \mathbf{E}[z^X | \mathsf{A}]. \tag{79.6}$$

Example 79.3:

Let X be a geometric r.v. that induces the distribution

$$P_X(\{x\}) = (1-p)^{x-1} p, \ x = 1, 2, 3, \ldots$$

(see also Figure 34.5). I_A is the indicator r.v. of an event A with $P(\mathsf{A}) = 1 - p$, where X, I_A are s.i. What is the generating function of $Y = X I_\mathsf{A}$?

To obtain $G_Y(z)$ we can write

$$G_Y(z) = G_{X|\mathsf{A}^c}(z) P(\mathsf{A}^c) + G_{X|\mathsf{A}}(z) P(\mathsf{A}).$$

Under the condition A^c, $Y = 0$ so that $G_{X|\mathsf{A}^c}(z) = 1$. And since

$$G_X(z) = \sum_{x=1}^{\infty} (1-p)^{x-1} p z^x = \frac{pz}{1-(1-p)z}$$

we obtain

$$G_Y(z) = p + (1-p) \frac{pz}{1-(1-p)z} = \frac{p}{1-(1-p)z} = \frac{1}{z} G_X(z).$$

Thus, it can be seen that $P_Y(\{y\}) = P_X(\{y+1\})$ for all integers y; i.e., the bar graph of the distribution P_Y is obtained by shifting the graph of P_X to the left by one unit. Of course, it is not correct to say that $Y = X - 1$ in this case, since each x-value can give rise to two different y-values.

As noted earlier, the generating function of a *non-negative* integer-valued r.v. is manipulated more easily. Also, the generating function of a *nonpositive* integer-valued r.v. is easily converted to the former type through replacement of z by $1/z$. If X is an integer-valued r.v. whose range includes both positive and negative integers, then it can be helpful to resolve G_X into *two* generating functions— one representing a non-negative distribution and the other a nonpositive distribution. This can be accomplished either by using conditional generating functions, or by expressing P_X as a mixture (see Problem 79.7).

Queries

79.1 Which of the following functions of z are generating functions:

    a) $g(z) = z + (1/z)$
    b) $g(z) = 1$
    c) $g(z) = 1 - z/2 + (z/2)^2 - (z/2)^3 + (z/2)^4 - \ldots$
    d) $g(z) = [g_1(z)]^2$, where $g_1(z)$ is a generating function
    e) $g(z) = g_1(z) + g_2(z)$, where $g_1(z)$, $g_2(z)$ are generating functions.
    f) $g(z) = g_1(z/2)$, where $g_1(z)$ is a generating function.   [66]

79.2 Let X be an integer-valued r.v. with generating function $G_X(z)$. If $\mathbf{V}X = 2$, what is $\mathbf{V}Y$, where Y is an integer-valued r.v. with generating function $zG_X(z)$?   [194]

SECTION **80**

Several Jointly Gaussian Random Variables

Gaussian r.v.'s and distributions have been discussed several times so far. The one-dimensional case, for instance, was considered in Sections 39 and 40, bivariate Gaussian distributions in Section 54, and a multidimensional Gaussian distribution in Example 71.1. Here, some of the earlier results are revisited and enlarged upon, around the common context of an n-dimensional Gaussian r.v.

Let the induced space be R^n, for a given $n \geq 2$. Let $X_s = [X_{1s}, \ldots, X_{ns}]^\mathsf{T}$ be an n-dimensional vector r.v. whose component r.v.'s $X_{1s}, \ldots, X_{ns}$ are m.s.i. *standardized* Gaussian r.v.s.[4] We call X_s a *standardized n-dimensional Gaussian r.v.*, which induces the *standardized n-dimensional Gaussian distribution*. The p.d.f. for this induced distribution is

$$f_{X_s}(x) = \prod_{k=1}^{n} \frac{1}{\sqrt{2\pi}} \exp\left(\frac{-x_k^2}{2}\right) = (2\pi)^{-n/2} \exp\left[-\frac{1}{2} \sum_{k=1}^{n} x_k^2 \right] \tag{80.1a}$$

where $x \in \mathsf{R}^n$. In the exponential on the right side of Equation (80.1a), setting $\sum x_k^2 = r^2$ yields the equation for an n-dimensional sphere with center at the origin. For every $r > 0$ such a sphere is a contour of constant probability density. It follows that f_{X_s} exhibits *spherical symmetry*. In other words, any rotational transformation about the origin leaves the induced distribution unchanged. If x denotes an n-dimensional column vector, $x = [x_1, \ldots, x_n]^\mathsf{T}$, then Equation (80.1a) can also be written

$$f_{X_s}(x) = (2\pi)^{-n/2} \exp\left(-\frac{1}{2} x^\mathsf{T} x\right). \tag{80.1b}$$

The covariance matrix of X_s is the identity matrix: $\mathbf{C}[X_s, X_s] = \mathbf{E}[X_s X_s^\mathsf{T}] = \mathbf{I}$.

From Example 76.2 and Equation (78.2) it follows that the characteristic function of X_s is

$$\psi_{X_s}(\tau) = \prod_{k=1}^{n} \exp\left(\frac{-\tau_k^2}{2}\right) = \exp\left[-\frac{1}{2} \sum_{k=1}^{n} \tau_k^2 \right] \tag{80.2}$$

where $\tau \in \mathsf{R}^n$. We note that the functions (80.1) and (80.2) differ only by a scale factor.

---

[4] We recall from Section 71 that for multivariate Gaussian r.v.s there is no distinction between pairwise and mutual statistical independence.

Probability Concepts and Theory for Engineers, First Edition. Harry Schwarzlander.
© 2011 John Wiley & Sons, Ltd. Published 2011 by John Wiley & Sons, Ltd.

In order to extend Definition 54.1 to n dimensions, we consider now an n-dimensional vector r.v. Y obtainable from X_s by means of a linear transformation whose transformation matrix has a nonvanishing Jacobian. Any such Y is called an *n-dimensional (jointly) Gaussian r.v.* Such a r.v. Y induces an *n-dimensional Gaussian distribution* (or multivariate Gaussian distribution) on $\mathbb{R}^n$. Thus, let $Y = [Y_1, \ldots, Y_n]^T$, then Y is an n-dimensional Gaussian (vector) r.v. if

$$Y = aX_s + b \tag{80.3}$$

where a is $n \times n$ with nonvanishing determinant and b is any real n-element column vector:

$$a = \begin{bmatrix} a_{11} & a_{12} & \cdots & a_{1n} \\ a_{21} & a_{22} & \cdots & a_{2n} \\ \cdots & \cdots & \cdots & \cdots \\ a_{n1} & a_{n2} & \cdots & a_{nn} \end{bmatrix}, \quad b = \begin{bmatrix} b_1 \\ b_2 \\ \cdots \\ b_n \end{bmatrix}.$$

Applying Equation (53.3) directly to Equation (80.1b) and noting that the Jacobian is $\det(a)$, the determinant of a, we obtain

$$f_Y(y) = (2\pi)^{-n/2} |\det(a)|^{-1} \exp\left(-\frac{1}{2} x^T x\right).$$

In order to express the right side as a function of y, the substitution $x = a^{-1}(y - b)$, needs to get applied to the quadratic form in the exponent, giving $(y - b)^T (a^{-1})^T a^{-1} (y - b)$. Now making use of Equation (71.4), we note that $(a^{-1})^T a^{-1} = (a a^T)^{-1} = (a I a^T)^{-1} = c_Y^{-1}$, the inverse of the covariance matrix of Y. Also, $\det(c_Y) = \det(aa^T) = \det(a) \cdot \det(a^T) = [\det(a)]^2$. The n-dimensional density function for Y is therefore

$$f_Y(y) = \frac{1}{(2\pi)^{n/2} |\det(c_Y)|^{1/2}} \exp\left[-\frac{1}{2} (y - b)^T c_Y^{-1} (y - b)\right]. \tag{80.4}$$

This is the most general form for the joint p.d.f. of an n-dimensional Gaussian distribution.

To obtain the characteristic function of Y, Equation (78.4) can be used. From Equation (80.2), $\psi_{X_s}(\tau) = \exp[-\tau^T \tau/2]$. Therefore,

$$\psi_Y(\tau) = \exp(j \tau^T b) \exp\left(-\frac{1}{2} \tau^T a a^T \tau\right) = \exp(j \tau^T b) \exp\left(-\frac{1}{2} \tau^T c_Y \tau\right). \tag{80.5}$$

‖ **Theorem 80.1**
‖ The sum of n jointly Gaussian r.v.'s is a Gaussian r.v.

Proof

With Y as above, let $W = Y_1 + \ldots + Y_n$. From Theorem (78.2), $\psi_W(\tau) = \psi_Y(\tau, \tau, \ldots, \tau)$. In Equation (80.5), replacing τ by $(\tau, \tau, \ldots, \tau)$ changes $\tau^T b$ into $(\sum b_i)\tau$, and $\tau^T c_Y \tau$ into $(\sum\sum c_{ik})\tau^2$, where c_{ik} are the elements of c_Y ($1 \leq i, k \leq n$). Therefore, $\psi_W(\tau)$ has the form $e^{ib\tau} \exp(-a\tau^2/2)$ that, from Equation (77.1) and Example (76.2), is the characteristic function of a scaled and shifted version of a standardized Gaussian r.v., i.e., the characteristic function of a Gaussian r.v.

In Section 54 it was observed that the marginal distributions of an arbitrary bivariate Gaussian distribution are Gaussian. Is this also the case for an n-dimensional Gaussian distribution? The answer is yes, since each component of the r.v. Y in Equation (80.3) is defined as a sum of n jointly

Gaussian r.v.'s plus a constant,

$$Y_i = a_{i1}X_1 + \ldots + a_{in}X_n + b_i, \quad i = 1, \ldots n.$$

Then, from Theorem 80.1, Y_i is a Gaussian r.v. plus a constant, which is still a Gaussian r.v.

We consider next the conditional distribution that results if the conditioning and conditioned r.v.'s are jointly Gaussian. The simplest situation is the case $n = 2$. Then, $Y = [Y_1, Y_2]^T$, Equation (80.4) is a two-dimensional p.d.f., and of interest is the distribution induced by Y_2 conditioned on values assigned by Y_1. From Theorem 59.1, for all y_1 and y_2,

$$f_{Y_2|Y_1}(y_2|y_1) = \frac{f_{Y_1,Y_2}(y_1, y_2)}{f_{Y_1}(y_1)}. \tag{80.6}$$

Expanding Equation (80.4) with $c_Y = \begin{bmatrix} \mathbf{V}Y_1 & \mathbf{C}[Y_1, Y_2] \\ \mathbf{C}[Y_1, Y_2] & \mathbf{V}Y_2 \end{bmatrix}$ and

$$c_Y^{-1} = \frac{1}{\mathbf{V}Y_1 \, \mathbf{V}Y_2 - \mathbf{C}[Y_1, Y_2]^2} \begin{bmatrix} \mathbf{V}Y_2 & -\mathbf{C}[Y_1, Y_2] \\ -\mathbf{C}[Y_1, Y_2] & \mathbf{V}Y_1 \end{bmatrix}$$

and letting $\boldsymbol{\rho}[Y_1, Y_2] = \rho$, results in $f_{Y_1, Y_2}(y_1, y_2) =$

$$\frac{1}{2\pi \, \sigma_{Y_1}\sigma_{Y_2} \sqrt{1-\rho^2}} \exp\left\{ -\frac{1}{2(1-\rho^2)} \left[\frac{(y_1 - b_1)^2}{\mathbf{V}Y_1} - \rho\frac{(y_1 - b_1)(y_2 - b_2)}{\sigma_{Y_1}\sigma_{Y_2}} + \frac{(y_2 - b_2)^2}{\mathbf{V}Y_2} \right] \right\}.$$

This is a bivariate Gaussian density function as in Equation (69.9), but with nonzero means. From Equation (80.3) we know that Y_1 is Gaussian with mean b_1 and variance $\mathbf{V}Y_1$, so that Equation (40.1) becomes

$$f_{Y_1}(y_1) = \frac{1}{\sqrt{2\pi}\,\sigma_{Y_1}} \exp\left[-\frac{1}{2}\frac{(y_1 - b_1)^2}{\mathbf{V}Y_1} \right].$$

Now, substitution into Equation (80.6) results in $f_{Y_2|Y_1}(y_2|y_1) =$

$$\frac{1}{\sqrt{2\pi}\,\sigma_{Y_2}\sqrt{1-\rho^2}} \exp\left\{ -\frac{1}{2(1-\rho^2)} \left[\frac{\rho^2(y_1 - b_1)^2}{\mathbf{V}Y_1} - \rho\frac{(y_1 - b_1)(y_2 - b_2)}{\sigma_{Y_1}\sigma_{Y_2}} + \frac{(y_2 - b_2)^2}{\mathbf{V}Y_2} \right] \right\}. \tag{80.7}$$

An examination of this result shows that, for every fixed y_1, $f_{Y_2|Y_1}(y_2|y_1)$ is a Gaussian density function in y_2 (see also Problem 69.8). Since the roles of y_1 and y_2 in the above derivation can be interchanged, the following result has been obtained:

Theorem 80.2
If two r.v.'s are jointly Gaussian, then the conditional distribution induced by one of the r.v.'s, conditioned on the value taken by the other r.v., is Gaussian.

From Theorem 80.2 it follows that every 'slice' through a bivariate Gaussian density function, parallel to one of the axes, results in a scaled version of a (scalar) Gaussian density function. What about a slice along an oblique straight line? Since the above analysis applies equally to any rotated version of a given bivariate Gaussian density function, the rotation angle can be chosen so as to make the oblique line parallel to one of the axes. We conclude:

Theorem 80.3
A slice through a bivariate Gaussian density function along any straight line in R^2 yields a function that is a scaled version of a scalar Gaussian density function.

Theorem 80.2 can be extended to multiple jointly Gaussian r.v.'s. In other words, suppose $Y = (Y_1, \ldots, Y_n)$ is an n-dimensional jointly Gaussian r.v. Then, the joint density of $(Y_{k+1}, \ldots, Y_n)$, conditioned on $(Y_1, \ldots, Y_k)$, is $(n{-}k)$-dimensional Gaussian. This follows from the fact that for each vector value $(y_1, \ldots, y_k)$, $f_{Y_{k+1}, \ldots, Y_n | Y_1, \ldots, Y_k}$ is simply a scaled version of the $(n{-}k)$-dimensional function obtained from f_Y by fixing the values $y_1, \ldots, y_k$. The exponent in that function consists then of a quadratic form in the variables $y_{k+1}, \ldots, y_n$, thus assuring that the function is an $(n{-}k)$-dimensional Gaussian p.d.f.

As shown in Problem 53.10, if X_s is a standardized Gaussian r.v., then $W_1 = X_s^2$ is first-order chi-square. Furthermore (Problem 77.8), if $X_{1s}, \ldots, X_{ns}$ $(n \geq 2)$ are statistically independent standardized Gaussian r.v.'s, then $W_n = X_{1s}^2 + \ldots + X_{ns}^2$ is *chi-square* of order n. We consider now the r.v.'s $R_k = \sqrt{W_k}$, for $k = 1, 2, 3, \ldots$. When the underlying p.e. is performed, then the value taken on by R_k expresses the distance from the origin of the point $(x_1, \ldots, x_k) \in R^k$ assigned by the k r.v.'s $X_{1s}, \ldots, X_{ks}$.

The case $k = 1$ is simple, since $R_1 = |X_{1s}|$, so that

$$f_{R_1}(r) = \begin{cases} \sqrt{\dfrac{2}{\pi}} \exp\left(-\dfrac{r^2}{2}\right), & r \geq 0 \\ 0, & r < 0. \end{cases}$$

If $k = 2$, a transformation to polar coordinates applied to (X_{1s}, X_{2s}) yields (R_2, Θ) (see Example 48.2). From Equation (48.4) we then have

$$f_{R_2}(r) = r \exp\left(-\frac{r^2}{2}\right), \quad r \geq 0. \tag{80.8}$$

Thus, R_2 induces a *Rayleigh* distribution (see Problem 47.2).

For $k = 3$ we note that $\sqrt{X_{1s}^2 + X_{2s}^2 + X_{3s}^2} = \sqrt{R_2^2 + X_{3s}^2}$, so that the same approach as above can be used. In other words, a transformation to polar coordinates can be applied to (R_2, X_{3s}), giving (R_3, Θ), from which the p.d.f. of R_3 can be obtained. The joint density of R_2 and X_{3s} is

$$f_{R_2, X_{3s}}(v, x) = f_{R_2}(v) f_{X_{3s}}(x) = \frac{v}{\sqrt{2\pi}} \exp\left[-\frac{1}{2}(v^2 + x^2)\right], \quad v \geq 0.$$

Theorem 53.1 can be applied over the set $B = \overline{0, \infty} \times \overline{-\infty, \infty}$. The transformation rule is given in Equation (48.3), and within B, the Jacobian is readily found to be $(v^2 + x^2)^{-1/2}$. Therefore,

$$f_{R_3, \Theta}(r, \theta) = \sqrt{v^2 + x^2}\, f_{R_2, X_{3s}}(v, x) = r^2 \frac{\cos\theta}{\sqrt{2\pi}} \exp\left(-\frac{r^2}{2}\right), \quad -\frac{\pi}{2} \leq \theta \leq \frac{\pi}{2}, r \geq 0.$$

Here, the integration with respect to θ is only over $\overline{-\pi/2, \pi/2}$ (since $v \geq 0$), and we have

$$f_{R_3}(r) = \int_{-\pi/2}^{\pi/2} r^2 \frac{\cos\theta}{\sqrt{2\pi}} \exp\left(-\frac{r^2}{2}\right) d\theta = \sqrt{\frac{2}{\pi}} r^2 \exp\left(-\frac{r^2}{2}\right), \quad r \geq 0. \tag{80.9}$$

The corresponding result for larger k can be obtained by induction (see Problem 80.3).

A somewhat different situation is sometimes of interest, where the center of the standardized Gaussian distribution is shifted away from the origin. We return to the case where $k = 2$, but replace

X_{1s} by $X_1 = X_{1s} + b$, where $b > 0$. Then X_1 and X_{2s} are still normalized statistically independent Gaussian r.v.'s, but only X_{2s} is centralized. The joint distribution, which originally was centered at the origin, is now centered to the right of the origin, at $(b, 0)$, and has the density function

$$f_{X_1,X_{2s}}(x_1, x_2) = \frac{1}{2\pi} \exp\left\{-\frac{1}{2}\left[(x_1 - b)^2 + x_2^2\right]\right\}.$$

Let $R_2^{(b)} = \sqrt{X_1^2 + X_{2s}^2}$. To find the p.d.f. of $R_2^{(b)}$ we can proceed as above by applying Example **53**.1 to obtain

$$f_{R_2^{(b)}}(r) = \frac{r}{2\pi} \int_0^{2\pi} \exp\left\{-\frac{1}{2}\left[(r\cos\theta - b)^2 + r^2\sin^2\theta\right]\right\} d\theta$$

$$= \frac{r}{2\pi} \int_0^{2\pi} \exp\left[-\frac{1}{2}(r^2 - 2br\cos\theta + b^2)\right] d\theta$$

$$= r\exp\left[-\frac{1}{2}(r^2 + b^2)\right] \frac{1}{2\pi} \int_0^{2\pi} e^{br\cos\theta} d\theta$$

$$= r\exp\left[-\frac{1}{2}(r^2 + b^2)\right] I_0(br), \quad r \geq 0. \tag{80.10}$$

Here, I_0 is the modified Bessel function of the first kind of order 0 (cf. [AS1], Chapter 9).[5] If b is set to 0 in Equation (80.10), since $I_0(0) = 1$, it can be seen that $f_{R_2^{(b)}}$ becomes $f_{R_2^{(0)}} = f_{R_2}$ as given in Equation (80.8).

This derivation of $f_{R_2^{(b)}}$ assumed that the joint distribution of two s.i. standardized Gaussian r.v.'s, $P_{X_{1s}, X_{2s}}$, has been shifted *to the right* of the origin (i.e., along the x_1-axis) by a distance b. As b is made larger ($b \gg 1$), most of the probability of that joint distribution becomes concentrated in a small angular range $-\varepsilon \leq \theta \leq \varepsilon$. Then, the above integration with respect to θ can be approximated by an integration in the vertical (or x_2) direction, as illustrated in Figure 80.1. But integrating out X_{2s} leaves the shifted version of $P_{X_{1s}}$. This suggests that as b is increased, $P_{R_2^{(b)}}$ approaches a normalized Gaussian distribution with mean b. We note also that $R_2^{(b)}$ induces a *radial* distribution with respect to the origin; the directional r.v. Θ has been integrated out. Therefore, $f_{R_2^{(b)}}$ will be as given in

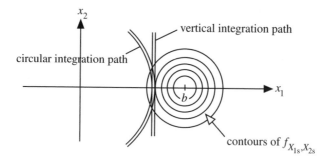

Figure 80.1   Integration of two-dimensional Gaussian distribution

---

[5] The integral of $f_{R_2^{(b)}}$ over $\overline{a, \infty}$, or $1 - F_{R_2^{(b)}}(a)$, has been called the 'Q-function'. [Ma1]

Equation (80.10), irrespective of the direction of that shift from the origin by a distance b. This is restated here as follows:

Theorem 80.4
Let X_1 and X_2 be s.i. normalized Gaussian r.v.'s with means b_1 and b_2, respectively, and let $b = \sqrt{b_1^2 + b_2^2}$. Then, the p.d.f. of $R_2^{(b)} = \sqrt{X_1^2 + X_2^2}$ is

$$
f_{R_2^{(b)}}(r) = \begin{cases} r \exp\left[-\frac{1}{2}(r^2 + b^2)\right] I_0(br), & r \geq 0 \\ 0, & r < 0. \end{cases}
$$

Queries

80.1 Let X_1 and X_2 be jointly Gaussian r.v.'s with zero means and with variances σ_1^2 and σ_2^2, respectively. For each of the following properties, indicate whether it is true, possibly true, or false: The conditional p.d.f. $f_{X_2|X_1}(x_2|x_1)$ describes a distribution that
  a) is Gaussian
  b) is zero-mean
  c) has zero variance
  d) has variance σ_2^2 if X_1 and X_2 are uncorrelated.   [234]

80.2 X_{1s} and X_{2s} are standardized statistically independent Gaussian r.v.'s. What is the radius of the circle, centered at the origin, to which their joint distribution assigns probability
  a) 0.9?   b) 0.99?   [146]

80.3 Let $n = 3$ in Equation (80.3), and let $a = \begin{bmatrix} 1 & 0.5 & 0.5 \\ 1 & 2 & 0.5 \\ 2 & 1 & 1 \end{bmatrix}$. Determine whether Y is

  three-dimensional jointly Gaussian.   [28]

SECTION 81

Spherically Symmetric Vector Random Variables

In this Section we broaden our discussion to a much larger class of r.v.'s that may loosely be described as 'Gaussian-like' vector r.v.'s. Specifically, we consider vector r.v.'s that, like the standardized n-dimensional Gaussian r.v. with p.d.f. (80.1), induce an absolutely continuous distribution that is not restricted to a finite region and exhibits spherical symmetry. Such r.v.'s are of interest in radar clutter modeling and other applications.

On a given probability space $(S, \mathscr{A}, P)$ let there be defined the s.i. standardized Gaussian r.v.'s X_{1s}, X_{2s}, X_{3s}, ... For each integer $n \geq 1$, the r.v. $X_s^{[n]} = [X_{1s}, \ldots, X_{ns}]^T$ is then a standardized n-dimensional Gaussian vector r.v. with p.d.f. as given in Equation (80.1b). $X_s^{[1]}$ is actually a scalar r.v. $X_s = X_{1s}$ with p.d.f. (40.2), but it can be thought of as a one-element vector. We recall from Section 80 that, for each n, the induced distribution exhibits spherical symmetry. (For $n = 1$, this simply means symmetry about $\{0\} \in \mathsf{R}$; and for $n = 2$, it is circular symmetry.) The collection of r.v.'s $X_s^{[1]}$, $X_s^{[2]}$, $X_s^{[3]}$, ..., will now be used as the 'seed' for generating other r.v.'s that also exhibit spherical symmetry.

> **(Definition) 81.1**
>
> Let $X_s^{[1]}$, $X_s^{[2]}$, $X_s^{[3]}$, ..., be a collection of standardized Gaussian vector r.v.'s with m.s.i. components, defined on a probability space $(S, \mathscr{A}, P)$ in such a way that components bearing the same subscript are identical. On the same probability space a new collection of r.v.'s $Y^{[1]}$, $Y^{[2]}$, $Y^{[3]}$, ..., can be defined by specifying for each integer $n \geq 1$,
>
> $$Y^{[n]} = SX_s^{[n]} \qquad (81.1)$$
>
> where S is a positive scalar r.v. and furthermore, S and the components of $X_s^{[n]}$ are m.s.i. Once S is specified to within one or more parameters, then the resulting r.v.'s $Y^{[n]}$ constitute what will be called a *family of spherically symmetric vector random variables* (s.s.v.r.v.'s).

Since each X_{is} gets multiplied by S, the components Y_1, ..., Y_n of $Y^{[n]}$ all induce the same distribution, which is described by the p.d.f. $f_{SX_{1s}}$. Clearly, these components have zero mean. Their variance is $VY_1 = V[SX_{1s}] = E[S^2]E[X_{1s}^2] = E[S^2]$.

Probability Concepts and Theory for Engineers, First Edition. Harry Schwarzlander.
© 2011 John Wiley & Sons, Ltd. Published 2011 by John Wiley & Sons, Ltd.

The components Y_1, Y_2, Y_3, ..., in general are not statistically independent, since they all depend on S. But we note that, for $i \neq j$,

$$\mathbf{E}[Y_i Y_j] = \mathbf{E}[S^2 X_{is} X_{js}] = \mathbf{E}[S^2]\mathbf{E}[X_{is} X_{js}] = \mathbf{E}[S^2]\mathbf{E}X_{is}\mathbf{E}X_{js} = 0$$

and also $\mathbf{E}Y_i\,\mathbf{E}Y_j = \mathbf{E}[SX_{is}]\,\mathbf{E}[SX_{js}] = (\mathbf{E}S)^2\,\mathbf{E}[X_{is}]\,\mathbf{E}[X_{js}] = 0$. Therefore (Theorem 68.3), the Y_i's are uncorrelated. Mere uncorrelatedness, however, does not lead to a simple relationship between $f_{Y^{[n]}}$ and the marginal densities, as would be the case if the component r.v.'s were m.s.i.

We also note that for any given S and all n, $P_{Y^{[n]}|S}(\,|s)$ for each s is a scaled version of $P_{X_s^{[n]}}$; that is, an n-dimensional centralized Gaussian distribution with covariance matrix $s^2\mathbf{I}$ (or, for $n=1$, a zero-mean Gaussian distribution with variance s^2). Therefore, using Equation (53.6),

$$f_{Y^{[n]}}(\mathbf{y}) = \int_0^\infty f_{Y^{[n]}|S}(\mathbf{y}|s)f_S(s)ds$$

$$= \int_0^\infty (2\pi)^{-n/2}s^{-n}\exp\left[-\frac{1}{2s^2}\left(y_1^2 + \ldots + y_n^2\right)\right]f_S(s)ds. \tag{81.2}$$

The integrand in Equation (81.2) is a function of the sum $y_1^2 + \ldots y_n^2$, which we denote by γ. Since γ is not involved in the integration, $f_{Y^{[n]}}(\mathbf{y})$ can be considered as a function of just γ. Therefore, $f_{Y^{[n]}}(\mathbf{y})$ is *spherically symmetric*—thus justifying the terminology in Definition 81.1. More specifically, we have the following.

Theorem 81.1
The p.d.f. of $Y^{[n]}$ has the form

$$f_{Y^{[n]}}(\mathbf{y}) = (2\pi)^{-n/2}h_n(\gamma) \tag{81.3}$$

where for each n, h_n is a non-negative function defined for non-negative argument.

For any positive integer k, h_k should not be confused with the p.d.f. of the radial r.v. R_k discussed in Section 80, which would be obtained if $Y^{[k]}$ were transformed to k-dimensional spherical coordinates.

For a particular p.d.f. f_S, Equation (81.2) may be difficult to evaluate. One approach would be to first obtain the characteristic function of $Y^{[n]}$. We note from Equation (80.5) that

$$\psi_{Y^{[n]}|S}(\tau_1, \ldots, \tau_n|s) = \exp\left[-s^2(\tau_1^2, \ldots, \tau_n^2)/2\right].$$

Then

$$\psi_{Y^{[n]}}(\tau_1, \ldots, \tau_n) = \int_0^\infty \psi_{Y^{[n]}|S}(\tau_1, \ldots, \tau_n|s)f_S(s)ds. \tag{81.4}$$

It can be seen that $\psi_{Y^{[n]}}$ is of the form

$$\psi_{Y^{[n]}}(\tau_1, \ldots, \tau_n) = g_n(\tau_1^2 + \ldots + \tau_n^2). \tag{81.5}$$

In other words, regarded as a real function (since its imaginary component is zero), $\psi_{Y^{[n]}}$ also exhibits spherical symmetry.

Another possibility for evaluating $f_{Y^{[n]}}$ derives from an examination of the function h_n in Equation (81.3). Comparison of Equation (81.2) and Equation (81.3) leads to

$$h_n(\gamma) = \int_0^\infty s^{-n} \exp\left[-\frac{\gamma}{2s^2}\right] f_S(s)\, ds.$$

It can be seen that h_n is differentiable, even if f_S includes δ-functions. Differentiation with respect to γ gives

$$\frac{dh_n(\gamma)}{d\gamma} = -\frac{1}{2}\int_0^\infty s^{-(n+2)}\exp\left[-\frac{\gamma}{2s^2}\right]f_S(s)\,ds = -\frac{1}{2}h_{n+2}(\gamma).$$

This shows that there is a recurrence relationship [CL1],

$$h_{n+2}(\gamma) = -2\frac{dh_n(\gamma)}{d\gamma} \tag{81.6}$$

which for given f_S allows h_n to be obtained for all $n > 2$ through differentiation, once h_1 and h_2 are known.[6]

From Equation (81.6) it follows that $dh_n(\gamma)/d\gamma \le 0$ is necessary for all n; i.e., $h_n(\gamma)$ is a nonincreasing function for $\gamma > 0$. This shows that the families of s.s.v.r.v.'s generated in accordance with Definition 81.1 do not encompass *all* n-dimensional r.v.'s that induce a spherically symmetric distribution. For instance, a r.v. that distributes all probability uniformly over the surface of an n-dimensional sphere cannot belong to such a family.

Example 81.1

In Definition 81.1, let S be the unary r.v. $S = \sigma$, where σ is any positive real constant. Then, $Y^{[n]}$ is a vector r.v. having n zero-mean s.i. Gaussian components with identical variances σ^2 ($n \ge 2$) or, if $n = 1$, it is a scalar zero-mean Gaussian r.v. with variance σ^2. That is, $Y^{[n]} = \sigma X_s^{[n]}$. Thus, we have identified a one-parameter family of s.s.v.r.v.'s with σ the parameter. $X_s^{[n]}$ itself belongs to this family, corresponding to $\sigma = 1$.

In this case, $h_1(\gamma) = \sigma^{-1}\exp[-\gamma/2\sigma^2]$. Similarly, from Equation (54.3), $h_2(\gamma) = \sigma^{-2}\exp[-\gamma/2\sigma^2]$. Application of Equation (81.6) then gives:

$$h_3(\gamma) = -\frac{2}{\sigma}\left(-\frac{1}{2\sigma^2}\right)\exp\left[-\frac{\gamma}{2\sigma^2}\right] = \frac{1}{\sigma^3}\exp\left[-\frac{\gamma}{2\sigma^2}\right]$$

$$h_4(\gamma) = -\frac{2}{\sigma^2}\left(-\frac{1}{2\sigma^2}\right)\exp\left[-\frac{\gamma}{2\sigma^2}\right] = \frac{1}{\sigma^4}\exp\left[-\frac{\gamma}{2\sigma^2}\right], \text{ etc.}$$

which, when multiplied by the appropriate normalizing coefficient, yield the joint p.d.f.'s of 3, 4, ... i.d.s.i. centralized Gaussian r.v.'s with variance σ^2.

We have already seen that the components of $Y^{[n]}$ are not necessarily m.s.i., but now Example 81.1 has demonstrated that if $Y^{[n]}$ belongs to the *Gaussian* family of s.s.v.r.v.'s, then the components of $Y^{[n]}$ are m.s.i. Are there other families for which $Y^{[n]}$ has m.s.i. components?

---

[6] We suppressed the dimensionality n in the parameter γ. On both sides of Equation (81.6), γ represents $y_1^2 + \ldots + y_n^2$, whereas h_{n+2} requires $y_1^2 + \ldots + y_{n+2}^2$ as its argument. This apparent incongruity is no problem, however, since the recurrence relation (81.6) simply yields the *form* of h_{n+2}. Naturally, we use h_{n+2} with its correct argument.

Theorem 81.2
The only family of s.s.v.r.v.'s with mutually statistically independent component r.v.'s is the Gaussian family.

Proof
A necessary condition for mutual statistical independence is pairwise statistical independence. Thus, consider Definition 81.1 and let Y_i and Y_j be any pair of components of $Y^{[n]}$ ($n \geq 2$), and suppose Y_i and Y_j are s.i. Then, Y_i^2 and Y_j^2 are s.i. as well (Theorem 55.2), and therefore are uncorrelated, so that (Theorem 68.3) $\mathbf{E}[Y_i^2 Y_j^2] = \mathbf{E}[Y_i^2]\,\mathbf{E}[Y_j^2]$. Now,

$$\mathbf{E}[Y_i^2 Y_j^2] = \mathbf{E}[X_i^2 S^2 X_j^2 S^2] = \mathbf{E}[X_i^2]\,\mathbf{E}[X_j^2]\,\mathbf{E}[S^4]$$

and

$$\mathbf{E}[Y_i^2]\,\mathbf{E}[Y_j^2] = \mathbf{E}[X_i^2 S^2]\,\mathbf{E}[X_j^2 S^2] = \mathbf{E}[X_i^2]\,\mathbf{E}[X_j^2]\,(\mathbf{E}[S^2])^2.$$

Therefore, a necessary condition for Y_i and Y_j to be s.i. is that $\mathbf{E}[S^4] = (\mathbf{E}[S^2])^2$; i.e., that S^2 has zero variance. This requires S to be unary. But that is exactly the case considered in Example 81.1, so that the condition is also sufficient and we have mutual statistical independence.

In the following Example another family of s.s.v.r.v.'s is considered.

Example 81.2
In Definition 81.1, let S be an absolutely continuous r.v. with p.d.f. of the form

$$f_S(s) = \begin{cases} \dfrac{1}{(\beta - \alpha)s}, & e^\alpha < s < e^\beta, \text{ where } 0 < \alpha < \beta \\ 0, & \text{otherwise.} \end{cases}$$

Since there are two unspecified parameters, α and β, this specification of f_S should yield a two-parameter family of s.s.v.r.v.'s. However, to simplify the presentation, $\alpha = 0$ and $\beta = 1$ will be used in what follows. Then, from Equation (81.2),

$$f_{Y_1}(y_1) = \int_1^e \frac{1}{\sqrt{2\pi s^2}} \exp\left[-\frac{y_1^2}{2s^2}\right] ds.$$

Using the substitution $v = y_1/s$, and making use of Equation (40.4), this becomes

$$f_{Y_1}(y_1) = \int_{y_1/e}^{y_1} \frac{1}{\sqrt{2\pi} y_1} \exp\left[-\frac{v^2}{2}\right] dv = \frac{1}{y_1}\left[\Phi(y_1) - \Phi\left(\frac{y_1}{e}\right)\right]$$

where $f_{Y_1}(0)$ can be set to $(e-1)/e\sqrt{2\pi}$ in order to make f_{Y_1} continuous at the origin. This p.d.f. describes the distribution induced by each of the component r.v.'s of $Y^{[n]}$, and it is the p.d.f. of $Y^{[1]} = Y_1$. Furthermore, $\mathbf{V}Y_1 = \mathbf{V}[SX_{1s}] = \mathbf{E}[S^2] = (e^2 - 1)/2$, so that the standardized version of Y_1 is $Y_{1s} = \sqrt{2}(e^2 - 1)^{-1/2} Y_1$.

A graph of $f_{Y_{1s}}$ is shown in Figure 81.1. Superimposed on the same plot, for comparison, is the standardized Gaussian p.d.f., $f_{X_{1s}}$. As can be seen, the body or central peak of $f_{Y_{1s}}$ is slightly narrower than that of $f_{X_{1s}}$, which makes up for the higher peak and somewhat different tails of $f_{Y_{1s}}$.

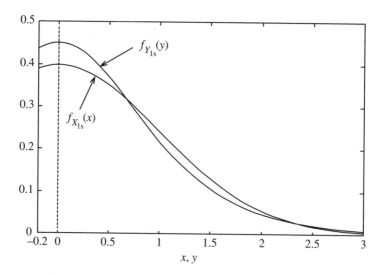

Figure 81.1   Comparison of $f_{Y_{1s}}$ with a standardized Gaussian p.d.f

Since the component r.v.'s Y_1, Y_2, ... are not mutually s.i., the joint p.d.f. associated with $Y^{[2]}$ must again get computed from Equation (81.2):

$$f_{Y^{[2]}}(y_1, y_2) = \int_1^e \frac{1}{2\pi s^3} \exp\left[-\frac{y_1^2 + y_2^2}{2s^2}\right] ds$$

Using the substitution $v = \sqrt{y_1^2 + y_2^2}/s$ and denoting $y_1^2 + y_2^2$ by γ results in

$$f_{Y^{[2]}}(y_1, y_2) = \int_{\sqrt{\gamma}/e}^{\sqrt{\gamma}} \frac{v}{2\pi \gamma} \exp\left[-\frac{v^2}{2}\right] dv = \frac{1}{2\pi \gamma}\left[\exp\left(-\frac{\gamma}{2e^2}\right) - \exp\left(-\frac{\gamma}{2}\right)\right].$$

The appropriate joint p.d.f.'s for three or more of the r.v.'s Y_1, Y_2, ... can now be found using Equation (81.6). For instance, since

$$h_1(\gamma) = \frac{\sqrt{2\pi}}{\sqrt{\gamma}}\left[\Phi(\sqrt{\gamma}) - \Phi(\sqrt{\gamma}/e)\right]$$

it follows from Equation (81.6) that

$$h_3(\gamma) = -2\frac{dh_1(\gamma)}{d\gamma}$$

$$= \frac{\sqrt{2\pi}}{\gamma^2}\left[\Phi(\sqrt{\gamma}) - \Phi(\sqrt{\gamma}/e)\right] - \frac{1}{\gamma}\left[\exp(-\gamma/2) - \exp(-1-\gamma/2e^2)\right].$$

Therefore,

$$f_{Y^{[3]}}(y_1, y_2, y_3) = (2\pi)^{-3/2} h_3(\gamma)$$

$$= \frac{1}{2\pi\,\gamma^{3/2}}[\Phi(\sqrt{\gamma}) - \Phi(\sqrt{\gamma}/e)] - \frac{1}{(2\pi)^{3/2}\gamma}[\exp(-\gamma/2) - \exp(-1-\gamma/2e^2)]$$

where now $\gamma = y_1^2 + y_2^2 + y_3^2$. This p.d.f. has spherical symmetry since it is a function of $y_1^2 + y_2^2 + y_3^2$, as is also true for the-higher dimensional p.d.f.'s obtained through further application of Equation (81.6).

It should be kept in mind that the set of r.v.'s $Y^{[1]}$, $Y^{[2]}$, $Y^{[3]}$, ... described here represents just one member of the two-parameter family of s.s.v.r.v.'s defined at the beginning of this Example.

At the beginning of this Section the notion of 'Gaussian-like' distributions was mentioned. This implies an absolutely continuous distribution with a body consisting of a single peak or hump, and tails (or 'skirt' in two or more dimensions) that fall off very rapidly (more rapidly than a simple exponential) without vanishing at any finite point. Figure 81.1 illustrates the Gaussian-like nature (in one dimension) of a member of the family of s.s.v.r.v.'s considered there.

It must be noted, however, that such a characterization does not fit all families of s.s.v.r.v.'s, and in some cases not all the members of such a family. The distribution induced by the product $SX_s^{[1]} = SX_{1s}$ in Definition 81.1 can be influenced as much by S as by X_s. If the range of S is restricted to a finite interval $\overline{0,b}$, then the tails of Y will be Gaussian-like for $y > b$, approximately. (Of course, in the extreme case of Example 81.1, Y is actually Gaussian.) However, the body of Y still can deviate from a Gaussian-like nature if, for instance, the distribution of S consists of quite disparate discrete components that may produce the appearance of a noticeable 'blip' at the center of f_Y (Problem 81.5). On the other hand, if $f_S(s)$ falls off rather slowly with increasing s, the distribution induced by SX_s will also have rather slowly declining tails. Thus, in Problem 81.7 we have an example of an s.s.v.r.v. that is 'Cauchy-like' rather than Gaussian-like. This is interesting, since a vector r.v. whose components are i.d.m.s.i. Cauchy r.v.'s does *not* exhibit spherical symmetry.

Another question that can be considered is whether a *mixture* of several s.s.v.r.v.'s is again an s.s.v.r.v. If the mixture involves only members of the Gaussian family of s.s.v.r.v.'s (Example 81.1), then the answer clearly is yes, since Definition 81.1 leads to such a result if S is specified as a discrete r.v. whose range is a set of points $\{s_1, ..., s_k\}$, where $0 < s_1 < ... < s_k$, with $P_S(\{s_i\}) = p_i > 0$, $i = 1, ..., k$, and $p_1 + ... + p_k = 1$. This suggests the following property (Problem 81.1).

Theorem 81.3
A mixture of s.s.v.r.v.'s (derived from the same seed sequence $X_s^{[1]}$, $X_s^{[2]}$, ...) is again an s.s.v.r.v.

If the linear coordinate transformation (80.3) gets applied to $Y^{[n]}$ as defined in Equation (81.1), then the resulting vector r.v.

$$W^{[n]} = a\,Y^{[n]} + b \tag{81.7}$$

has contours of constant probability density that are *concentric ellipsoids* centered at $w = b$, and $W^{[n]}$ is called a *spherically invariant vector r.v.* For example, the general n-dimensional Gaussian r.v. with p.d.f. (80.4) is a spherically invariant vector r.v. Furthermore, since $W^{[n]} = aSX_s^{[n]} + b$, we see from Equation (80.4) that

$$f_{W^{[n]}|S}(w|s) = \frac{1}{(2\pi)^{n/2} |\det c_{Y^{[n]}}|^{1/2} s^n} \exp\left[-\frac{1}{2} \frac{(w-b)^{\mathrm{T}} c_Y^{-1}(w-b)}{s^2} \right]$$

so that, for given n, $f_{W^{[n]}}$ is completely specified by $c_{Y^{[n]}}$, b, and f_S.

Queries

81.1 A collection of standardized Gaussian vector r.v.'s $X_s^{[1]}, X_s^{[2]}, X_s^{[3]}, \ldots$, is defined on a probability space $(S, \mathscr{A}, P)$ in the manner of Definition 81.1. On the same space are defined two r.v.'s S_1, S_2 satisfying the requirements for S in Definition 81.1. Let $Y_1^{[n]} = S_1 X_s^{[n]}$, and $Y_2^{[n]} = S_2 X_s^{[n]}$. For each of the following statements, with arbitrary $n \geq 1$, determine whether it is true, possibly true, or false.

a) $Y_1^{[n]} + Y_2^{[n]}$ is a s.s.v.r.v.

b) $Y_1^{[n]} - Y_2^{[n]}$ is a s.s.v.r.v.

c) $Y_3^{[n]} = S_2 Y_1^{[n]}$ is a s.s.v.r.v.

d) $Y_4^{[n]} = Y_1^{[n]}/S_2$ is a s.s.v.r.v.   [121]

81.2 Let $f_S(s) = 0.5[\delta(s-1) + \delta(s-5)]$.

a) What is the variance of $Y^{[1]}$?

b) How many maxima does the p.d.f. $f_{Y^{[1]}}$ have?   [228]

Entropy Associated with Random Variables

In Definition 32.1 we saw that the amount of randomness or uncertainty inherent in a countable probability space $(S, \mathscr{A}, P)$, also called the *entropy* of the space, can be expressed as a function of the probabilities $p_1, p_2, p_3, \ldots$ of the sets making up the partition that generates $\mathscr{A}$. In this Section we examine how this notion translates to induced probability spaces.

Clearly, Definition 32.1 cannot simply get applied to an induced space $(R, \mathscr{B}, P_X)$ since we have set things up in such a way that an induced space is never countable, and there is no countable partition that generates $\mathscr{B}$. However, it should be noted that the important feature in Definition 32.1 is not the partition, but the probabilities $p_1, p_2, p_3, \ldots$. These probabilities define the probability distribution P, and it is P that determines the entropy. The manner in which we have defined induced spaces requires that the space (R) and the completely additive class $(\mathscr{B})$ are always the same (in the one-dimensional case); only the induced distribution differs. Therefore, the entropy must now be expressed in terms of the induced distribution. And since the induced distribution results from a r.v., the entropy of the induced space can also be thought of as the entropy that is conveyed by, or associated with, the inducing r.v. Accordingly, we proceed to define the entropy associated with a r.v.

a) Discrete random variables

In the case of a discrete r.v. X, the induced distribution P_X is completely defined once all the positive probabilities are specified that get assigned to various one-element sets. If these probabilities are denoted $p_1, p_2, p_3, \ldots$, then the entropy associated with P_X is readily expressed in a manner that follows directly from Definition 32.1:

> **(Definition) 82.1**
> The entropy H_X associated with a discrete r.v. X is
>
> $$H_X = H(p_1, p_2, p_3, \ldots) = -\sum_i p_i \log_2 p_i = \mathbf{E}[-\log_2 P_X(\{X\})] \text{ bits,} \qquad (82.1)$$
>
> $p_1, p_2, p_3, \ldots$ denoting the positive probabilities induced onto one-element sets by X.

Probability Concepts and Theory for Engineers, First Edition. Harry Schwarzlander.
© 2011 John Wiley & Sons, Ltd. Published 2011 by John Wiley & Sons, Ltd.

The note in Definition 32.1 concerning the case where $p_i = 0$ for some i need not be included here since the p_i's are assumed positive. Definition 82.1 makes it clear that H_X depends only on the numerical values of the induced probabilities, not on *where* X induces these probabilities. This is because the induced distribution is assumed known and has no uncertainty associated with it. The uncertainty, prior to performing the underlying p.e., comes from not knowing *which* of the points in the range of the r.v. will get selected when the experiment is performed.

We see from Equation (82.1) that H_X cannot be negative since $P_X \leq 1$. Furthermore, it follows from Theorem 32.1 that if X assigns positive probabilities $p_1, p_2, \ldots, p_k$ to only a finite number of points $a_1, a_2, \ldots, a_k$ in R, then H_X is maximized if $p_1 = p_2 = \ldots = p_k = 1/k$, and that maximum entropy is $\log_2 k$ bits.

Definition 82.1 extends in an obvious way to discrete multidimensional or vector r.v.'s:

(Definition) 82.2

Let X be a discrete n-dimensional r.v. that assigns positive probabilities $p_1, p_2, p_3, \ldots$ (where $\sum_i p_i = 1$) to a countable class of one-element sets in R^n. The entropy H_X associated with X is then

$$H_X = H(p_1, p_2, p_3, \ldots) = -\sum_i p_i \log_2 p_i = \mathbf{E}[-\log_2 P_X(\{X\})] \text{ bits.} \tag{82.2}$$

Here, the question arises how Equation (82.2) is related to the entropies of the component r.v.'s of X. We have the following result:

Theorem 82.1

If the components of $X = (X_1, X_2, \ldots, X_n)$ are m.s.i., then for $k = 1, \ldots, n$,

$$H_X = \sum_k H_{X_k}.$$

Proof

Since the components of X are m.s.i.,

$$\begin{aligned}
H_X &= \mathbf{E}[-\log_2 P_X(\{X\})] = \mathbf{E}[-\log_2[P_{X_1}(\{X_1\})P_{X_2}(\{X_2\}) \ldots P_{X_n}(\{X_n\})]] \\
&= \mathbf{E}[-\log_2 P_{X_1}(\{X_1\}) - \log_2 P_{X_2}(\{X_2\}) - \ldots - \log_2 P_{X_n}(\{X_n\})] \\
&= H_{X_1} + H_{X_2} + \ldots + H_{X_n}.
\end{aligned}$$

So we see that in this case each component r.v. contributes a certain amount of entropy and those contributions *add* to give the entropy of X.

If the components of $X = (X_1, X_2, \ldots, X_n)$ are not m.s.i., conditional distributions can be used in order to express the entropy of X. In Section 57 we noted the chain rule,

$$\begin{aligned}
P_X(\{(x_1, \ldots, x_n)\}) &= \\
&= P_{X_1}(\{x_1\})P_{X_2|X_1}(\{x_2\}|\{x_1\}) \cdots P_{X_n|X_1, \ldots, X_{n-1}}(\{x_n\}|\{(x_1, \ldots, x_{n-1})\}).
\end{aligned} \tag{82.3}$$

This can be rephrased in terms of entropies upon first defining a *conditional entropy* as follows.

(Definition) 82.3

Let X, Y be any two discrete r.v.'s defined on a probability space. The *conditional entropy of Y given X*, denoted $H_{Y|X}$, is

$$H_{Y|X} = -\sum P_{X,Y}(\{(x,y)\}) \log_2 P_{Y|X}(\{y\}|\{x\})$$
$$= -\mathbf{E}[\log_2 P_{Y|X}(\{Y\}|\{X\})] \text{ bits}$$

where the summation is over all points $(x, y) \in \mathbf{R}^2$ for which $P_{X,Y}(\{(x, y)\}) > 0$.

$H_{Y|X}$ expresses the entropy or average uncertainty associated with Y that remains after X is known or observed. If X and Y are s.i., then $H_{Y|X} = H_Y$. The difference $H_Y - H_{Y|X}$, which is non-negative, expresses how much entropy is shared by X and Y. It follows that $H_X - H_{X|Y}$ must equal $H_Y - H_{Y|X}$ (see Problem 82.6).[7] Either of these differences, therefore, serves as another indicator of the *degree of relatedness* of two r.v.'s.

Definition 82.3 extends in an obvious way to multidimensional and vector r.v.'s. Upon introducing Equation (82.3) into Definition 82.2, and making use of Definition 82.3, H_X (in bits) can now be expressed in terms of the following chain rule, the summations being over all points $(x_1, \ldots, x_n) \in \mathbf{R}^n$ for which the joint probability $P_X(\{(x_1, \ldots, x_n)\}) > 0$:

$$H_X = -\sum P_X(\{(x_1, \ldots, x_n)\}) \log_2 P_X(\{(x_1, \ldots, x_n)\})$$
$$= -\sum P_X(\{(x_1, \ldots, x_n)\})$$
$$\log_2 \left[P_{X_1}(\{X_1\}) P_{X_2|X_1}(\{x_2\}|\{x_1\}) \ldots P_{X_n|X_1}(\{x_n\}|\{(x_1, \ldots, x_{n-1})\}) \right]$$
$$= H_{X_1} + H_{X_2|X_1} + H_{X_3|X_1,X_2} + \ldots + H_{X_n|X_1,\ldots,X_{n-1}}. \tag{82.4}$$

Each successive term on the right expresses the additional entropy associated with a particular component r.v. given that the components with lower index are known or specified.

Example 82.1

Consider the three-dimensional discrete r.v. $X = (X_1, X_2, X_3)$ defined in Example 55.1. X assigns probability 1/4 to each of four points (one-element sets) in $\mathbf{R}^3$, so that from Definition 82.1,

$$H_X = -\frac{1}{4}\log_2\frac{1}{4} - \frac{1}{4}\log_2\frac{1}{4} - \frac{1}{4}\log_2\frac{1}{4} - \frac{1}{4}\log_2\frac{1}{4} = -\log_2\frac{1}{4} = 2 \text{ bits}.$$

To obtain this result from H_{X_1}, H_{X_2}, and H_{X_3}, we note first that since X_1, X_2, X_3 are not m.s.i., Equation (82.4) must be used in place of Theorem 82.1. However, as noted in Example 55.1, X_1 and X_2 are s.i., so that $H_{X_2|X_1} = H_{X_2}$ and

$$H_X = H_{X_1} + H_{X_2} + H_{X_3|X_1,X_2}.$$

X_1 and X_2 both assign probabilities 1/2 to $\{0\}$ and $\{1\}$. Thus,

$$H_{X_1} = H_{X_2} = -\frac{1}{2}\log_2\frac{1}{2} - \frac{1}{2}\log_2\frac{1}{2} = -\log_2\frac{1}{2} = 1 \text{ bit}.$$

On the other hand, once the values taken on by X_1 and X_2 are known, then X_3 is determined and no uncertainty remains about X_3 (as is easily seen from Figure 55.1a). In other words, $H_{X_3|X_1,X_2} = 0$, and we have again

$$H_X = H_{X_1} + H_{X_2} = 2 \text{ bits}.$$

---

[7] In Information Theory this quantity is called the 'average mutual information'.

As discussed previously in Section 32, an entropy of *1 bit* is associated with a probability space consisting of two equally likely elementary events. Such a space therefore can be thought of as the 'reference space', conveying unit entropy. Similarly, when considering discrete r.v.'s, a binary r.v. B that induces the pair of probabilities $(1/2, 1/2)$ can be thought of as defining unit entropy, and thus can be regarded as the basic reference:

$$H_B = H\left(\frac{1}{2}, \frac{1}{2}\right) = -\frac{1}{2} \log_2 \frac{1}{2} - \frac{1}{2} \log_2 \frac{1}{2} = 1 \text{ bit.} \tag{82.5}$$

Thus, the result $H_X = 2$ bits in Example 82.1 indicates that the three r.v.'s X_1, X_2, X_3 can be 'encoded' (represented without ambiguity) in terms of *two* s.i. equally likely binary r.v.'s B_1, B_2, since $H_{(B_1, B_2)} = H_{B_1} + H_{B_2} = 2$ bits. In fact, in Example 82.1, X_1 and X_2 can serve that function. In Problem 82.1, this idea is developed further.

b) Absolutely continuous random variables

If X is an absolutely continuous r.v., the first inclination might be to try to define H_X by introducing a discrete approximation to P_X and then using Definition 82.1. Thus, let x_i be real numbers spaced Δx apart and let $P_{X_d}(\{x_i\}) = f_X(x_i)\, \Delta x$, with associated entropy H_{X_d}. However, if this formulation gets inserted into Definition 82.1, the limit of H_{X_d} as $\Delta x \to 0$ does not exist.

Instead, let X_r denote an absolutely continuous 'reference r.v.' with p.d.f. $f_{X_r}(x)$, whose entropy we approximate in the manner just described:

$$H_{X_{rd}} = -\sum_i P_{X_{rd}}(\{x_i\}) \log_2 P_{X_{rd}}(\{x_i\}) = -\sum_i f_{X_r}(x_i)\Delta x \log_2 [f_{X_r}(x_i)\Delta x]$$

$$= -\Delta x \sum_i f_{X_r}(x_i)\log_2 f_{X_r}(x_i) - (\Delta x \log_2 \Delta x) \sum_i f_{X_r}(x_i).$$

Now consider the discrete approximation to H_X *relative to* the discrete approximation to H_{X_r}:

$$H_{X_d} - H_{X_{rd}} = -\Delta x \sum_i f_X(x_i)\log_2 f_X(x_i) + \Delta x \sum_i f_{X_r}(x_i)\log_2 f_{X_r}(x_i)$$

$$- (\Delta x \log_2 \Delta x) \sum_i f_X(x_i) + (\Delta x \log_2 \Delta x) \sum_i f_{X_r}(x_i).$$

In the limit as $\Delta x \to 0$, the last two terms cancel, and therefore it is possible to define a '*relative entropy*' or '*differential entropy*' of X (relative to X_r) as

$$H_X = -\int_{-\infty}^{\infty} f_X(x) \log_2 f_X(x)\, dx + \int_{-\infty}^{\infty} f_{X_r}(x) \log_2 f_{X_r}(x)\, dx.$$

For convenience, X_r can be defined as unif. $\overline{0, 1}$. In the second integral, the integration over $\overline{0, 1}$ is then zero, and outside that interval we replace the indeterminate integrand $0 \log_2 0$ by $\lim_{\varepsilon \to 0} \varepsilon \log_2 \varepsilon = 0$ (for positive ε). Thus, the second integral vanishes, giving:

> **(Definition) 82.4**
> The *relative entropy* or *differential entropy* associated with an absolutely continuous r.v. X, in bits, is given by
>
> $$H_X = -\int_{-\infty}^{\infty} f_X(x) \log_2 f_X(x)\, dx = \mathbf{E}[-\log_2 f_X(X)]. \tag{82.6}$$

The word 'relative' or 'differential' is usually suppressed when it is clear from the context that the entropy under consideration is associated with an absolutely continuous r.v.

Example 82.2

a) Let X be unif. $\overline{a,b}$, so that $-f_X(x)\log_2 f_X(x) = (b-a)^{-1}\log_2(b-a)$, and integrating over $\overline{a,b}$ yields $\log_2(b-a)$. Over the remainder of R, the integration contributes 0 (as noted above). Therefore, $H_X = \underline{\log_2(b-a)}$ bits.

It should be noted that, depending on the values of a and b, H_X can be positive as well as zero or negative, neither of which indicates the absence of randomness. In particular, $H_X = 0$ if $b - a = 1$, which is the case for the distribution assigned to X_r above. For $b - a < 1$, the differential entropy is negative—in other words, $H_X < H_{X_r}$ in that case.

b) Now let Y be triangular over the interval $\overline{a,b}$, with

$$f_Y(y) = \begin{cases} \dfrac{2(y-a)}{(b-a)^2}, & a < y < b \\ 0, & \text{otherwise.} \end{cases}$$

Then,

$$H_Y = -\int_a^b f_Y(y)\log_2 f_Y(y)\,dy$$

$$= -\int_a^b \frac{2(y-a)}{(b-a)^2}\left[\log_2\frac{2}{(b-a)^2} + \log_2(y-a)\right]dy$$

$$= \log_2\frac{(b-a)^2}{2} - \frac{2}{(b-a)^2\ln 2}\int_a^b (y-a)\ln(y-a)\,dy$$

and after a change of variable the integral becomes

$$\int_0^{b-a} v\ln v\,dv = (b-a)^2\left[\frac{\ln(b-a)}{2} - \frac{1}{4}\right].$$

Therefore,

$$H_Y = \log_2\frac{(b-a)^2}{2} - \log_2(b-a) + \frac{1}{2\ln 2}$$

$$= 2\log_2(b-a) - \log_2 2 - \log_2(b-a) + \frac{1}{2}\log_2 e$$

$$= \log_2(b-a) + \frac{1}{2}\log_2\frac{e}{4}\ \text{bits.}$$

In the limit as $b - a \to 0$, X and Y become unary r.v.'s that convey no randomness at all, and both H_X and $H_Y \to -\infty$. Keeping in mind that H_X and H_Y are *relative* entropies, this can be interpreted as an indication that the reference r.v., which is unif. $\overline{0,1}$, induces 'infinite' entropy (and in fact so does every absolutely continuous r.v.)—i.e., infinitely more entropy than a unary (or any discrete) r.v. This reflects the fact (mentioned in Example 38.3) that measurement on a continuous scale implies an infinite sequence of readings of successive decimal positions.

An absolutely continuous r.v. is a mathematical idealization, as was discussed in Section 38. In an actual real-world situation, observations or measurements can be made only up to a certain precision, so that strictly speaking the entropy associated with a physical phenomenon is always finite. But since the (idealized) modeling in terms of continuous scales is so convenient and effective, we have now seen how the entropy concept can be extended to such mathematical representations.

Both H_X and H_Y in Example 82.2 depend only on the difference $b - a$. This suggests that the entropy associated with an absolutely continuous r.v. is not influenced by the mean of the induced distribution, but does depend on the standard deviation. Indeed, let X_s be a standardized absolutely continuous r.v., with relative entropy H_{X_s}, and let $X = \sigma X_s + m$. Using Equation (53.6) with Definition 82.4 yields

$$H_X = -\int_{-\infty}^{\infty} f_X(x) \log_2 f_X(x)\, dx = -\int_{-\infty}^{\infty} \frac{1}{\sigma} f_{X_s}\left(\frac{x-m}{\sigma}\right) \log_2 \left[\frac{1}{\sigma} f_{X_s}\left(\frac{x-m}{\sigma}\right)\right] dx$$

$$= \log_2 \sigma - \int_{-\infty}^{\infty} \frac{1}{\sigma} f_{X_s}(\nu) \log_2 f_{X_s}(\nu)\sigma\, d\nu = \log_2 \sigma + H_{X_s}.$$

Thus:

Theorem 82.2
If an absolutely continuous r.v. X has mean m and standard deviation σ, and X_s is its standardized version, then $H_X = H_{X_s} + \log_2 \sigma$ (in bits).

Example 82.3
Theorem 82.2 shows that the entropy associated with an absolutely continuous r.v. X can be made arbitrarily large by increasing the standard deviation. But if X is standardized, what is the maximum possible entropy?

This is a variational problem,[8] with constraints $\mathbf{V}X = 1$ and $\int f(x)dx = 1$. According to Theorem 82.2, $\mathbf{E}X$ has no effect on H_X and therefore can be specified as 0. The constraint on the variance can then be rewritten as $\mathbf{E}[X^2] = 1$. The two constraints get incorporated into the integral to be maximized with Lagrange multipliers (cf. [Hi1], Sections 2–6), giving

$$I = \int_{-\infty}^{\infty} F(f)\, dx = \int_{-\infty}^{\infty} [-f(x)\log_2 f(x) + \lambda_1 x^2 f(x) + \lambda_2 f(x)]\, dx.$$

Setting $\partial F/\partial f = 0$ yields $\log_2 f(x) = \lambda_1 x^2 + \lambda_2 - \log_2 e$, or changing to natural logarithms, $\ln f(x) = (\lambda_1 x^2 + \lambda_2)\ln 2 - 1$, so that

$$f(x) = \exp(\lambda_1 x^2 \ln 2) \exp(\lambda_2 \ln 2 - 1).$$

---

[8] Let $f(x)$ be a function of a real variable, and let $F(f)$ be an expression involving $f(x)$ such that the integral $I = \int_{x_1}^{x_2} F(f)dx$ exists. Then I can be maximized with respect to $f(x)$ by supposing that $f(x)$ is the actual maximizing function and writing instead $f(x) + \varepsilon\, \eta(x)$, where ε is a real number. The term $\varepsilon\, \eta(x)$ is called the *variation* of $f(x)$. Then the integral $I(\varepsilon) = \int_{x_1}^{x_2} F(f + \varepsilon \eta)dx$ is maximized, for any given η, when $\varepsilon = 0$. Thus, we want $dI(\varepsilon)/d\varepsilon = 0$ when $\varepsilon = 0$. Carrying out that differentiation gives

$$\frac{dI(\varepsilon)}{d\varepsilon} = \int_{x_1}^{x_2} \frac{dF(f + \varepsilon\eta)}{d\varepsilon}\, dx = \int_{x_1}^{x_2} \frac{\partial F(f + \varepsilon\eta)}{\partial(f + \varepsilon\eta)} \frac{dF(f + \varepsilon\eta)}{d\varepsilon}\, dx = \int_{x_1}^{x_2} \frac{\partial F(f)}{\partial(f)} \eta\, dx.$$

Since this must equal zero for any η, the requirement for an extremum is $\partial F/\partial f = 0$.

The values of λ_1 and λ_2 are obtained by substitution into the two constraint equations

$$\int_{-\infty}^{\infty} x^2 f(x)\, dx = 1 \quad \text{and} \quad \int_{-\infty}^{\infty} f(x)\, dx = 1$$

and solving simultaneously. This gives $\lambda_1 \ln 2 = -1/2$, and $\exp(\lambda_2 \ln 2 - 1) = 1/\sqrt{2\pi}$. Thus, $f(x) = \frac{1}{\sqrt{2\pi}} \exp\left(-\frac{x^2}{2}\right)$. Combined with Theorem 82.2, this shows that among all absolutely continuous r.v.'s with the same standard deviation, the largest entropy is associated with a *Gaussian* r.v.

A similar approach can be used to determine the maximum entropy if the *range* of an absolutely continuous r.v. is constrained to a specified interval $\overline{a, b}$. That maximum is reached if the r.v. is unif. $\overline{a, b}$ (see Problem 82.3), with an entropy of $\log_2(b - a)$ as shown in Example 82.2. In the expression for H_Y in Example 82.2, the second term expresses the reduction from maximum entropy that results when going from a uniformly distributed r.v. to a r.v. that is triangularly distributed over the same interval.

Now, we briefly consider multidimensional absolutely continuous r.v.'s.

(Definition) 82.5
If $X = (X_1, \ldots, X_n)$ is an n-dimensional absolutely continuous r.v., then the relative entropy associated with X, in bits, is

$$H_X = -\int_{-\infty}^{\infty} \cdots \int_{-\infty}^{\infty} f_X(x_1, \ldots, x_n) \log_2 f_X(x_1, \ldots, x_n) dx_1 \ldots dx_n$$
$$= -\mathbf{E}[\log_2 f_X(X)].$$

Theorem 82.3
If $X = (X_1, \ldots, X_n)$ is an n-dimensional absolutely continuous r.v. with m.s.i. components, then for $k = 1, \ldots, n$,

$$H_X = \sum_k H_{X_k}.$$

The proof is analogous to the proof of Theorem 82.1. And in a manner analogous to Definition 82.3, a conditional entropy can be defined as follows:

(Definition) 82.6
Let X, Y be any two jointly absolutely continuous r.v.'s defined on a probability space. The *conditional (relative) entropy of Y given X*, denoted $H_{Y|X}$, is

$$H_{Y|X} = -\int\int f_{X,Y}(x, y) \log_2 f_{Y|X}(y|x)\, dxdy = -\mathbf{E}[\log_2 f_{Y|X}(Y|X)] \text{ bits}$$

where the integration is over all regions in $\mathbf{R}^2$ over which $f_{Y|X}$ exists.

The significance of $H_{Y|X}$ in the absolutely continuous case is the same as in the discrete case.

It is now possible to make use of the chain rule (59.11) to express the entropy of an arbitrary absolutely continuous n-dimensional r.v. $X = (X_1, \ldots, X_n)$ as a sum of conditional entropies, showing that Equation (82.4) holds also for jointly absolutely continuous r.v.'s, except that all the entropies in that case are relative entropies (see Problem 82.4).

Queries

82.1 A discrete three-dimensional r.v. X assigns equal probability to the eight points that are the corners of the unit cube.
a) What is H_X in bits?
b) How many bits of entropy is associated with each of the component r.v.'s?   [152]

82.2 An absolutely continuous three-dimensional r.v. X distributes all probability uniformly over the volume of the unit cube. Answer the same questions as in Query 82.1.   [16]

82.3 If X and Y are any two absolutely continuous s.i. r.v.'s, then $H_{Y|X}$ equals
a) 0   b) $-\infty$ c) H_X d) H_Y
e) none of the above.   [211]

SECTION *83*

Copulas

Several ways of characterizing the relationship that exists between two r.v.'s X, Y have been discussed so far: the correlation coefficient, conditional expectations, and the difference between the entropy of Y and the conditional entropy of Y given X. In this Section we examine yet another characterization, called the *copula*, which expresses how one r.v. is 'coupled' to another one. A copula provides a more comprehensive characterization of the relationship between two r.v.'s because, as will be seen, their marginal distributions and their copula together completely define their joint distribution.

The discussion in this Section will be restricted to two-dimensional r.v.'s with marginal distributions that are absolutely continuous, but whose joint distribution can be singular or can have a singular component.

We begin by considering two r.v.'s U, V both of which are unif. $\overline{0,\ 1}$. Together, they can give rise to any one of an infinite variety of joint distributions. Since their joint distribution can have a singular component, we will represent it by the two-dimensional c.d.f., $F_{U,V}(u,\ v)$. We have previously encountered instances of such two-dimensional distribution functions. For instance, if we set $V = 1 - U$, then $F_{U,V}(u,\ v)$ has the form described in Example 44.2. If, on the other hand, $V = U$, then $F_{U,V}(u,\ v)$ is the function described in Problem 44.7. These two different c.d.f.'s, it turns out, have a special significance, as expressed in the following Theorem.

Theorem 83.1
Given two r.v.'s U, V both of which are unif. $\overline{0,\ 1}$, with joint c.d.f $F_{U,V}(u,\ v)$. Let

$$F_m(u,\ v) = F_{U,V}(u,\ v) \text{ for the case where } V = U, \text{ and}$$
$$F_w(u,\ v) = F_{U,V}(u,\ v) \text{ for the case where } V = -U.$$

Then an arbitrary joint c.d.f. for U and V satisfies

$$F_w(u,\ v) \le F_{U,V}(u,\ v) \le F_m(u,\ v).$$

Proof
Outside the unit square $\overline{0,\ 1} \times \overline{0,\ 1}$, all joint c.d.f.'s $F_{U,V}$ are identical, so that it is only necessary to prove the inequality on the unit square. There we have, from the basic properties of c.d.f.'s,

---

Probability Concepts and Theory for Engineers, First Edition. Harry Schwarzlander.
© 2011 John Wiley & Sons, Ltd. Published 2011 by John Wiley & Sons, Ltd.

$F_{U,V}(u, v) \leq F_{U,V}(u, 1) = F_{U,V}(u, \infty) = F_U(u) = u$; and similarly, $F_{U,V}(u, v) \leq F_{U,V}(1, v) = v$. Then $F_{U,V}(u, v) \leq \min(u, v) = F_m(u, v)$.

For any point (u, v) in the unit square we must have $P_{U,V}(\overline{u, 1} \times \overline{v, 1}) \geq 0$. From Example 43.2, $P_{U,V}(\overline{u, 1} \times \overline{v, 1}) = 1 - F_{U,V}(u, 1) - F_{U,V}(1, v) + F_{U,V}(u, v)$, giving $F_{U,V}(u, v) \geq u + v - 1$. Since $F_{U,V}(u, v)$ must also be non-negative, we obtain $F_{U,V}(u, v) \geq \max(0, u + v - 1) = F_w(u, v)$.

Thus, when visualized in a three-dimensional representation as in Figure 44.1b, we see that the infinite variety of possible c.d.f.'s $F_{U,V}(u, v)$ differ only on the unit square, and there they are constrained to a rather thin sliver of the unit cube, consisting of the points lying on and between the boundaries defined by F_m and F_w. This sliver therefore represents the portion of the unit cube within which every possible relationship between two r.v.'s that are unif. $\overline{0, 1}$ can be uniquely represented.

Example 83.1

Consider two s.i. r.v.'s U, V both of which are unif. $\overline{0, 1}$. From Equation (45.6) it follows that on the unit square,

$$F_{U,V}(u, v) = F_U(u) F_V(v) = uv. \tag{83.1}$$

It is easily verified that this result agrees with Theorem 83.1 (see Problem 83.1). In this case, of course, the joint p.d.f. exists. It has the constant value 1 over the unit square, which is easier to visualize than the joint c.d.f.

In this Section we will continue to use the names U, V to denote two r.v.'s that are unif. $\overline{0, 1}$. Still restricting our attention to two such r.v.'s, we now introduce the following:

(Definition) 83.1

Given a pair of r.v.'s U, V both of which are unif. $\overline{0, 1}$. Their *copula*, denoted $C_{U,V}(u, v)$, is the portion of their joint c.d.f. that is defined on the closed unit square, $\overline{0, 1} \times \overline{0, 1}$:

$$C_{U,V}(u, v) = F_{U,V}(u, v),\ 0 \leq u, v \leq 1.$$

$C_{U,V}(u, v)$ is not defined for other values of u, v.

It should be noted that no information has been lost in going from $F_{U,V}(u, v)$ to $C_{U,V}(u, v)$, since outside the unit square, the function $F_{U,V}(u, v)$ is determined solely by the fact that U, V are both unif. $\overline{0, 1}$. Thus,

$$F_{U,V}(u, v) = \begin{cases} 1, & \text{for } u, v \geq 1 \\ v, & \text{for } u \geq 1 \text{ and } 0 \leq v \leq 1 \\ u, & \text{for } v \geq 1 \text{ and } 0 \leq u \leq 1 \\ 0, & \text{for } u < 0 \text{ or } v < 0. \end{cases}$$

Now, we examine how any given pair of r.v.'s X, Y of the kind specified at the beginning of this Section can be transformed into a pair of r.v.'s U, V, both of which are unif. $\overline{0, 1}$. If that transformation is one-to-one, then upon applying the inverse transformation, the joint distribution of X, Y can be recovered from the joint distribution of U, V. In that case, then, the nature of the *relationship* between X and Y gets preserved in the joint distribution of U and V, even though individually, U and V are unif. $\overline{0,\ 1}$; and we can regard the copula of U, V as the copula of X, Y as well.

Again, let X, Y be some pair of r.v.'s that individually are absolutely continuous but whose joint distribution can be singular or have a singular component. The task is then to shift, stretch and/or compress their two-dimensional c.d.f. in both the x and the y direction as needed so as to reshape both *marginal* distributions into uniform distributions over $\overline{0,\ 1}$. Considering just the x-dimension, this transformation can be visualized as illustrated in Figure 83.1. For any given $x=x_0$, we want $u=u_0 \in \overline{0,\ 1}$ such that $F_U(u_0)=F_X(x_0)$. Since $F_U(u)=u$ for $u \in \overline{0,\ 1}$, we obtain $u=F_X(x)$, or $x=F_X^{-1}(u)$. Thus, the desired F_U is obtained by setting $F_U(u)=F_X[F_X^{-1}(u)]$, for $u \in \overline{0,\ 1}$.

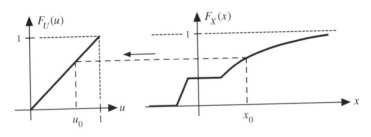

Figure 83.1 Transforming a r.v. to make it uniform over the unit interval

If the function F_X is constant over one or more intervals, as suggested in Figure 83.1, this transformation is not one-to-one. However, it should be noted that since no probability is induced on an interval over which F_X is constant, such an interval has no effect on the transformation and can be ignored (as long as that interval gets reinserted if the transformation is to be inverted).

In a similar way, the transformation required in the y-dimension is seen to be $F_V(v)=F_Y[F_Y^{-1}(v)]$, for $v \in \overline{0,\ 1}$. Combining these two transformations, we have:

(Definition) 83.2
Let X, Y be a pair of absolutely continuous r.v.'s whose joint distribution may have a singular component. Let U, V be a pair of r.v.'s that are unif. $\overline{0,\ 1}$ and whose joint c.d.f. is given by

$$F_{U,V}(u,\ v)=F_{X,Y}[F_X^{-1}(u),\ F_Y^{-1}(v)]. \qquad (83.2)$$

Then, over $\overline{0,\ 1} \times \overline{0,\ 1}$ the function $F_{U,V}(u,\ v)$ is the *copula* of X, Y and is denoted $C_{X,Y}(u,\ v)$.

First, we examine how this works for a pair of r.v.'s X, Y that are s.i. In that case, Equation (83.2) yields

$$F_{U,V}(u,\ v)=F_{X,Y}[F_X^{-1}(u),\ F_Y^{-1}(v)]=F_X[F_X^{-1}(u)]\ F_Y[F_Y^{-1}(v)]=uv.$$

We have thus shown the following.

Theorem 83.2
The copula $C_{X,Y}(u, v)$ for any pair of absolutely continuous r.v.'s X, Y that are s.i. is the function uv over the unit square,

$$C_{X,Y}(u, v) = uv. \tag{83.3}$$

Of course, the collection of pairs of absolutely continuous s.i. r.v.'s X, Y includes the pair U, V considered in Example 83.1, and we see that the result expressed in Equation (83.1) is in agreement with Theorem 83.2. In other words, on the closed unit square, $C_{X,Y}(u, v) = F_{U,V}(u, v)$ as found in Example 83.1. A contour diagram of this copula is made up of portions of rectangular hyperbolas, as sketched in Figure 83.2.

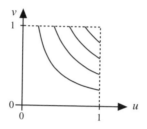

Figure 83.2   Contour diagram for the copula of a pair of statistically independent random variables

The inverse of the transformation (83.2) is given by

$$F_{X,Y}(x, y) = C_{X,Y}[F_X(x), F_Y(y)]. \tag{83.4}$$

That is, in the expression for $C_{X,Y}$, $F_X(x)$ gets substituted for u and $F_Y(y)$ gets substituted for v. Although the domain of this inverse transformation is the closed unit square, its range can be taken as $\mathbb{R}^2$. Thus, given the copula (83.3) and the marginal c.d.f.'s of the two r.v.'s X and Y, F_X and F_Y, then the joint c.d.f. of two s.i. r.v.'s,

$$F_{X,Y}(x, y) = F_X(x) F_Y(y)$$

is obtained.

Among all the different relationships that can exist between two r.v.'s X, Y, the opposite extreme from statistical independence is linear dependence. Suppose $Y = aX + b$, with $a > 0$. Then all probability is distributed along the line $y = ax + b$, which has positive slope. In that case,

$$F_{X,Y}(x, y) = \begin{cases} F_Y(y), & y \le ax + b \\ F_X(x), & y \ge ax + b \end{cases}$$

where equality is included with both inequalities since $F_{X,Y}$ is continuous. Then application of Definition 83.2 gives

$$C_{X,Y}(u, v) = \begin{cases} v, & y \le ax + b \\ u, & y \ge ax + b. \end{cases}$$

To express the conditions on the right in terms of u and v, the fact that F_Y is a monotone nondecreasing function can be utilized to change $y \le ax + b$ into

$$F_Y(y) \le F_Y(ax + b) = F_X(x)$$

and similarly for $y \ge ax + b$. Definition 83.2 then yields $v \le u$ and $v \ge u$, respectively, so that

$$C_{X,Y}(u, v) = \begin{cases} v, & v \le u \\ u, & v \ge u. \end{cases} \tag{83.5}$$

Over the unit square, this result can be seen to equal the function $F_m(u, v)$ in Theorem 83.1. This, then, is the copula for any pair of absolutely continuous r.v.'s X, Y with correlation coefficient $+1$.

On the other hand, if $Y = aX + b$ but with $a < 0$, so that the correlation coefficient is -1, then (see Problem 83.2) the copula is $C_{X,Y}(u, v) = F_w(u, v)$ over the unit square. Just as correlation coefficients range between the extremes of $+1$ and -1, with zero correlation (and therefore also statistical independence) at the midpoint, so do copulas range between the extremes F_m and F_w that correspond to correlation $+1$ and -1, respectively, and the copula (83.3) for s.i. r.v.'s represents a surface in the unit cube lying between these two extremes.

In the following Example two r.v.'s are considered that are not s.i.

Example 83.2

Let two *exponential* r.v.'s have the following joint density:

$$f_{X,Y}(x, y) = \begin{cases} e^{-(x+y)}[1 + \beta(1 - 2e^{-x})(1 - 2e^{-y})], & x, y \ge 0 \\ 0, & \text{otherwise} \end{cases} \tag{83.6}$$

where $-1 \le \beta \le 1$, so that we are actually considering a one-parameter *family* of two-dimensional distributions that includes the s.i. case (when $\beta = 0$). Upon integrating $f_{X,Y}$ the joint c.d.f. is found to be

$$F_{X,Y}(x, y) = (1 + \beta e^{-(x+y)})(1 - e^{-x} - e^{-y} + e^{-(x+y)}), \quad x, y \ge 0 \tag{83.7}$$

and zero otherwise. It should be noted that $F_X(x) = F_{X,Y}(x, \infty) = 1 - e^{-x}$ for $x \ge 0$, and $F_Y(y) = F_{X,Y}(\infty, y) = 1 - e^{-y}$ for $y \ge 0$, which represent exponential distributions. Furthermore, to verify that Equation (83.6) is a legitimate two-dimensional p.d.f. we also note that

1) $F_{X,Y}(\infty, \infty) = 1$; and
2) $|\beta(1 - 2e^{-x})(1 - 2e^{-y})| \le |\beta| |1 - 2e^{-x}| |1 - 2e^{-y}| \le 1$ so that $f_{X,Y} \ge 0$ over the specified range of β.

Now it is possible to apply Definition 83.2 and write

$$C_{X,Y}(u, v) = F_{X,Y}[F_X^{-1}(u), F_Y^{-1}(v)]$$

where $F_X^{-1}(u) = -\ln(1 - u)$, and $F_Y^{-1}(v) = -\ln(1 - v)$, so that substitution into Equation (83.7) gives

$$\begin{aligned} C_{X,Y}(u, v) &= [1 + \beta(1 - u)(1 - v)][1 - (1 - u) - (1 - v) + (1 - u)(1 - v)] \\ &= uv + \beta uv(1 - u)(1 - v). \end{aligned} \tag{83.8}$$

To get a better sense of the distributions that this family of copulas describes on the unit square, $C_{X,Y}$ can be extended over $\mathbf{R}^2$ as described earlier, giving the joint c.d.f. of two r.v.'s U and V. Then their joint p.d.f. can be found:

$$f_{U,V}(u,\ v) = \frac{\partial^2}{\partial u\, \partial v}\, C_{X,Y}(u,\ v) = 1 + \beta\,(1-2u)\,(1-2v).$$

This shows at once that $\beta = 0$ corresponds to U, V being s.i. ($f_{U,V} = 1$ over the unit square). The term $\beta(1-2u)(1-2v)$ tilts the p.d.f. along the edges of the unit square, but maintains $f_{U,V} = 1$ along the lines $u = 1/2$ and $v = 1/2$, giving a contour diagram for $f_{U,V}$ consisting of hyperbolic arcs with the point of symmetry at $(0.5, 0.5)$ as shown in Figure 83.3. For $\beta > 0$, U and V are positively correlated, for $\beta < 0$ negatively correlated.

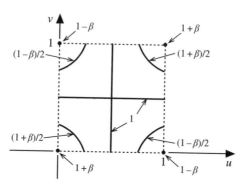

Figure 83.3   Contours of the joint p.d.f. of U and V as functions of β

Given a particular copula, any number of pairs of r.v.'s can be imbued with the specific coupling that is expressed by that copula. This is illustrated in the next Example, where the family of copulas found in Example 83.2 gets applied to a pair of *Gaussian* r.v. In this way, a family of two-dimensional distributions is found that are not jointly Gaussian (except for the case $\beta = 0$) but with marginal distributions that are Gaussian.

Example 83.3

Let W, Z be two standardized Gaussian r.v.'s, so that $F_W(w) = F_Z(w) = \Phi(w)$. Then, Equations (83.4) and (83.8) yield

$$\begin{aligned}
F_{W,Z}(w,\ z) &= C_{X,Y}[F_W(w),\ F_Z(z)] \\
&= \Phi(w)\Phi(z) + \beta\,\Phi(w)\Phi(z)\,[1 - \Phi(w)]\,[1 - \Phi(z)].
\end{aligned}$$

To compare this result with a bivariate Gaussian distribution as in Equation (69.9), it is useful to obtain the joint p.d.f. of W and Z. This is readily found to be

$$\begin{aligned}
f_{W,Z}(x,\ z) &= \frac{\partial^2}{\partial w\, \partial z} F_{W,Z}(w,\ z) \\
&= \frac{1}{2\pi}\exp\left(-\frac{w^2 + z^2}{2}\right)\{1 + \beta\,[1 - 2\Phi(w)]\,[1 - 2\Phi(z)]\}.
\end{aligned}$$

If $\beta = 0$, W and Z are s.i. standardized Gaussian r.v.'s, so that the joint distribution has circular symmetry. Departure from circular symmetry is governed by the factor in braces. Figure 83.4 shows a plot of several contours of $f_{W,Z}(w, z)$ for the case $\beta = +1$. Although these contours are ovals, they differ noticeably from ellipses.

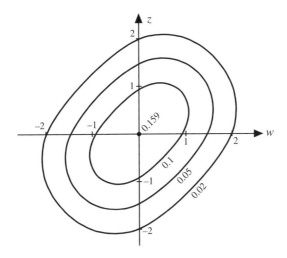

Figure 83.4   Contours of $f_{W,Z}$ where W and Z are Gaussian but not jointly Gaussian

A simple joint distribution of two Gaussian r.v.'s that are not jointly Gaussian was described in Example 54.2. In general, however, it is not easy to define a joint distribution for two given marginal distributions unless statistical independence or strict linear dependence is assumed. Copulas make it possible to overcome this difficulty. Any copula can be mapped into R^2 using Gaussian or any other marginal distributions. The only difficulty is that the analytics may get complicated, and a closed-form expression in terms of simple functions may not exist.

In Example 83.4 the copula is found for a pair of r.v.'s whose joint distribution is singular.

Example 83.4

Let X, Y be i.d. r.v.'s that induce probability uniformly over the two line segments identified in Figure 83.5a, with probability 0.5 associated with each line segment. Then

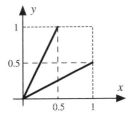

(a) Line segments on which
     probability is induced uniformly.

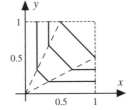

(b) Contours of the joint c.d.f.
     of X and Y.

Figure 83.5   Two absolutely continuous r.v.'s X, Y with singular joint distribution

$$f_X(x) = f_Y(y) = \begin{cases} 1.5, & 0 < x < 0.5 \\ 0.5, & 0.5 < x < 1 \\ 0, & \text{otherwise.} \end{cases}$$

The joint c.d.f over the unit square is easily found, giving

$$F_{X,Y}(x, y) = \begin{cases} \dfrac{3x}{2}, & \text{for } x \le \dfrac{y}{2} \\ \dfrac{3y}{2}, & \text{for } y \le \dfrac{x}{2} \\ \dfrac{x+y}{2}, & \text{for } x \ge \dfrac{y}{2} \text{ and } y \ge \dfrac{x}{2}. \end{cases} \tag{83.9}$$

Figure 83.5b shows a contour plot of this c.d.f. with contour lines at $F_{X,Y} = 1/4$, 1/2, and 3/4. The marginal c.d.f.'s are:

$$F_X(x) = F_{X,Y}(x, \infty) = F_{X,Y}(x, 1) = \begin{cases} \dfrac{3x}{2}, & x \in \overline{0, 0.5} \\ \dfrac{x+1}{2}, & y \in \overline{0.5, 1}. \end{cases}$$

and $F_Y(y) = F_X(y)$.

To obtain $C_{X,Y}$, the marginal distributions have to made uniform. Some reflection will make it clear that this cannot be accomplished by merely redistributing the probability on the two line segments. The transformation (83.2) must be applied, with careful attention given to the region over which each particular expression applies. We have

$$F_X^{-1}(x) = \begin{cases} \dfrac{2x}{3}, & x \in \overline{0, 0.75} \\ 2x - 1, & x \in \overline{0.75, 1} \end{cases} \qquad F_Y^{-1}(y) = \begin{cases} \dfrac{2y}{3}, & y \in \overline{0, 0.75} \\ 2y - 1, & y \in \overline{0.75, 1}. \end{cases}$$

Replacing x and y in the first line of Equation (83.9) by $F_X^{-1}(u)$ and $F_Y^{-1}(v)$, respectively, yields:

a) $C_{X,Y}(u, v) = u$, for $u \le \dfrac{v}{2}$ if $v \in \overline{0, 0.75}$; and for $u \le \dfrac{3v}{2} - 0.75$ if $v \in \overline{0.75, 1}$.

For the second line of Equation (83.9) the result is similar:

b) $C_{X,Y}(u, v) = v$, for $v \le \dfrac{u}{2}$ if $u \in \overline{0, 0.75}$; and for $v \le \dfrac{3u}{2} - 0.75$ if $u \in \overline{0.75, 1}$.

To transform the third line of Equation (83.9) it is necessary to distinguish four different regions:

c) For $u \ge \dfrac{v}{2}$ and $v \ge \dfrac{u}{2}$ and $u, v \in \overline{0, 0.75}$, $C_{X,Y}(u, v) = \dfrac{u+v}{3}$.

d) For $\dfrac{3v}{2} - 0.75 \le u \le 0.75$ and $v \in \overline{0.75, 1}$, $C_{X,Y}(u, v) = \dfrac{u}{3} + v - 0.5$.

e) For $\dfrac{3u}{2} - 0.75 \le v \le 0.75$ and $u \in \overline{0.75, 1}$, $C_{X,Y}(u, v) = u + \dfrac{v}{3} - 0.5$.

f) For $u, v \in \overline{0.75, 1}$, $C_{X,Y}(u, v) = u + v - 1$.

A contour plot of the copula $C_{X,Y}(u, v)$ is shown in Figure 83.6a, with the regions identified by (a) through (f) according to which of the above expressions applies. The plot in Figure 83.6b conveys a better sense of the joint distribution of (U, V) represented by this copula. U and V distribute all probability over *four* straight-line segments, with probability 1/4 distributed uniformly over each of the line segments.

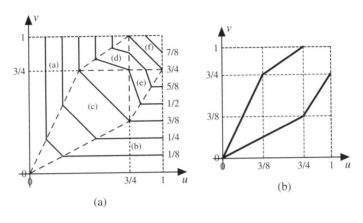

(a)

(b)

Figure 83.6 Copula for the r.v.'s X, Y in Example 83.4

It is left as an exercise (Problem 83.4) to carry out the inverse transformation and recover the joint distribution of the r.v.'s X, Y from their copula.

The fact that the copula in this Example represents a distribution that assigns equal probability to *four* line segments, whereas X, Y distribute probability equally over only two line segments, suggests that the distribution induced by X, Y is a special case among all the distributions sharing that copula: We can think of the distribution of X, Y as being concentrated also on two *pairs* of line segments, but with the lines of each pair having identical slope and thus becoming a single straight line. In Problem 83.3 the reader is asked to verify that the junction point for each pair is indeed the midpoint of the entire straight line segment, so that the probability assigned to each of the *four* subsegments is 1/4, just as in the copula.

We examine now what sort of joint distribution is characterized by the copula of Example 83.4 if that distribution is not restricted to a square or rectangle in R^2.

Example 83.5

Let X_e, Y_e be exponential r.v.'s with marginal c.d.f.'s belonging to this two-parameter family of pairs of marginal c.d.f.'s:

$$F_{X_e}(x) = \begin{cases} 1 - e^{-\alpha x}, & x \geq 0 \\ 0, & x < 0, \end{cases} \quad \text{and} \quad F_{Y_e}(y) = \begin{cases} 1 - e^{-\beta y}, & y \geq 0 \\ 0, & y < 0 \end{cases}$$

where α, $\beta > 0$. What is the joint distribution, if the copula is $C_{X,Y}(u, v)$ as obtained in Example 83.4?

Using Equation (83.4), we set $F_{X_e, Y_e}(x, y) = C_{X,Y}[F_{X_e}(x), F_{Y_e}(y)]$; that is, in $C_{X,Y}(u, v)$, u and v get replaced by $1 - e^{-\alpha x}$ and $1 - e^{-\beta y}$, respectively. Writing $\ln 4 = c$, this yields the following expressions for $F_{X_e, Y_e}(x, y)$:

a) $1 - e^{-\alpha x}$,   for $\beta y \geq -\ln(2 e^{-\alpha x} - 1)$ if $\beta y < c$, and for $\beta y \geq -\ln[(2/3) e^{-\alpha x} - 1/6]$ if $\beta y \geq c$;

b) $1 - e^{-\beta y}$,   for $\alpha x \geq -\ln(2 e^{-\beta y} - 1)$ if $\alpha x < c$, and for $\alpha x \geq -\ln[(2/3) e^{-\beta y} - 1/6]$ if $\alpha x \geq c$;

c) $(2 - e^{-\alpha x} - e^{-\beta y})/3$,   for $\beta y < \min[-\ln(2 e^{-\alpha x} - 1), c]$ and $\alpha x < \min[-\ln(2 e^{-\beta y} - 1), c]$;

d) $5/6 - (e^{-\alpha x})/3 - e^{-\beta y}$,   for $c \leq \beta y < -\ln[(2/3) e^{-\alpha x} - 1/6]$ and $\alpha x < c$;

e) $5/6 - e^{-\alpha x} - (e^{-\beta y})/3$,   for $c \leq \alpha x < -\ln[(2/3) e^{-\beta y} - 1/6]$ and $\beta y < c$;

f) $1 - e^{-\alpha x} - e^{-\beta y}$,   for $\alpha x \geq c$ and $\beta y \geq c$.

In the order in which these expressions are listed above, they apply to regions (a), (b), (c), (d), (e), and (f), respectively, in Figure 83.7. As can be seen, similar to U, V in Example 83.4, X_e, Y_e distribute all probability over four line segments, but these are now curved line segments. (They are drawn with solid lines in Figure 83.7.) The curved boundary of region (d) approaches $\alpha x = \ln 4$ as $y \to \infty$; similarly, the curved boundary of region (e) approaches $\beta y = \ln 4$ as $x \to \infty$. Thus, the probability induced on the curved boundary of region (e) is

$$F_{X_e}(\infty) - F_{X_e}\left(\frac{1}{\alpha} \ln 4\right) = 1 - (1 - e^{-\ln 4}) = e^{-\ln 4} = \frac{1}{4}.$$

From symmetry it follows that X_e and Y_e induce probability 1/4 on each of the four curved line segments. It can also be verified that the other two arcs, when extended, have as asymptotes $\alpha x = \ln 2$ and $\beta y = \ln 2$, respectively.

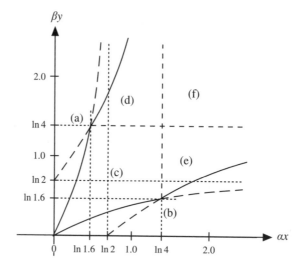

Figure 83.7   Boundaries separating the regions (a) through (f) in which $F_{X_e, Y_e}(x, y)$ has distinct functional forms

Example 83.5 shows that given a particular copula, even a rather simple one, it may not be easy to discern the essential features shared by all the joint distributions that are characterized by that copula. Nor is it clear what features of a joint distribution arise from the particular coupling embodied in that joint distribution. Thus, a prominent feature of the joint distribution of two i.d.s.i Gaussian r.v.'s is circular symmetry, but this feature is obviously not present in their copula, which is $C(u, v) = uv$.

Again, at first glance the copula in Example 83.4 may suggest that it represents singular joint distributions that assign all probability to four straight line segments. That they are singular distributions appears correct (at least, as long as the marginal distributions are absolutely continuous, which we assume here). But as we see in Example 83.5, probability will not necessarily be concentrated on *straight* line segments—which can also be surmised by noting the distortion that can arise, in both the x and y directions, when reversing the transformation illustrated in Figure 83.1. On the other hand, equal probability assigned to the four line segments appears to be preserved. And what about the junction point at the origin? Problem 83.6 addresses this question. There, the copula of Example 83.4 is to be applied to two *Gaussian* r.v.'s.

One application of copulas is the generation of random number pairs. It is relatively simple to generate pairs of uniformly distributed random numbers, even if the joint distribution is not uniform. Therefore, if random numbers get generated in accordance with the *copula* of the desired joint distribution, then it is only necessary to weight those numbers in accordance with Equation (83.4) in order to obtain the desired random number pairs.

In this introduction to copulas, attention has been restricted to two-dimensional r.v.'s with marginal distributions that are absolutely continuous. This limitation can be overcome, however [Ne1]. Copulas can also be defined for joint distributions of more than two r.v.'s.

Queries

83.1 What are the minimum and maximum of the range of values a copula can take on at the center of the unit square, i.e., at the point (0.5, 0.5)?   [190]

83.2 Let $A = \overline{0, 1} \times \overline{0, 1}$, the closed unit square in R^2, and let $C_{X,Y}$ be the copula for some pair of absolutely continuous r.v.'s X, Y. For each of the following statements, determine whether it is true, possibly true, or false.

a) $C_{X,Y}(u, v) = C_{X,Y}(v, u)$.
b) There is a point $(u_1, v_1) \in A$ where $C_{X,Y} = F_w$, and also a point $(u_2, v_2) \in A$, where $C_{X,Y} = F_m$.
c) In the family of pairs of r.v.'s that have $C_{X,Y}$ as their copula, there is a pair X, Y that induces zero probability on A.
d) On A, the function $1 - C_{X,Y}(1 - u, 1 - v)$ is also a copula.
e) The mixture of two distinct copulas is again a copula; i.e., if C_1 and C_2 are distinct copulas then for $0 < \alpha < 1$, $\alpha C_1 + (1 - \alpha)C_2$ is again a copula.   [119]

Sequences of Random Variables

a) Preliminaries

In some probability problems it is useful to define an *infinite* collection of r.v.'s on a probability space. This was done in Section 60 where random occurrences in time were considered. There, the r.v.'s W_k express the elapsed time between the $(k-1)$st and the kth 'occurrences', for $k = 1, 2, 3, \ldots$. These W_k's therefore comprise a *countably infinite* collection. The same is true for the r.v.'s T_k that express the total time until the kth occurrence. On the other hand, r.v.'s M_t were defined for all real $t > 0$, thus forming an *uncountable* collection.[9]

We concern ourselves now in more detail with countably infinite collections of r.v.'s. When the r.v.'s in such a collection are *ordered*, as indicated by distinct integer subscripts, then such a collection is also called a *sequence* of random variables. The abbreviated notation $\{X_k\}$ stands for a sequence $X_1, X_2, X_3, \ldots, X_k, \ldots$. Analogous to n-dimensional r.v.'s, the kth r.v. in the sequence, X_k, is then called the kth *component* of $\{X_k\}$. Upon a single execution of the underlying p.e., each of the r.v.'s X_k $(k = 1, 2, 3, \ldots)$ assigns a particular sample value x_k. The sequence of sample values is called a *sample sequence* of $\{X_k\}$ and is denoted $\{x_k\}$.

Suppose the underlying conceptual model is a product experiment consisting of an unending sequence of *independent*, *identical* component experiments or trials. The mathematical model for this is an infinite-dimensional product space $S = S_1 \times S_2 \times S_3 \times \ldots$. If each trial generates a numerical value (as would be the case when a die is thrown repeatedly, for instance), then the value generated in the kth trial can be expressed by a r.v. X_k $(k = 1, 2, 3, \ldots)$. In other words, on the kth component space S_k, the r.v. X_k is defined that induces on R a distribution P_{X_k}, and for all k these distributions are identical. Thus, the r.v.'s $X_1, X_2, X_3, \ldots, X_k, \ldots$ constitute a sequence $\{X_k\}$ of *i.d.m.s.i. r.v.'s*. Analogous to Definition 27.2, the distributions P_{X_k} can be brought together to form the infinite product probability distribution $P_{\{X\}} = P_{X_1} \bullet P_{X_2} \bullet P_{X_3} \bullet \ldots$ on the infinite-dimensional product space R^∞.

---

[9] Underlying these infinite collections of r.v.'s is an idealized conceptual model that allows for an infinite time scale over which statistical regularity is maintained. This in itself, of course, is physically meaningless. But, it allows the mathematical examination of features that pertain also to a truncated time scale, without specifying the truncation point.

Probability Concepts and Theory for Engineers, First Edition. Harry Schwarzlander.
© 2011 John Wiley & Sons, Ltd. Published 2011 by John Wiley & Sons, Ltd.

The following simple result for sequences of i.d.m.s.i. r.v.'s may at first glance seem surprising.

Theorem 84.1
Given a sequence $\{X_k\}$ of i.d.m.s.i. r.v.'s, and a set $A \subset R$ with $P_{X_k}(A) > 0$. Let $K_A R^\infty$ be the set of all those sample sequences $\{x_k\}$ that have A occurring at least once at some position in the sequence. Then $P_{\{X\}}(K_A) = 1$.

Proof
K_A^c is the set of sequences in which A does not occur at all. Then

$$P_{\{X\}}(K_A^c) = P_{X_1}(A^c)\, P_{X_2}(A^c)\, P_{X_3}(A^c) \cdots = \lim_{k \to \infty} [P_{X_1}(A^c)]^k = 0.$$

Corollary 84.1
Given a sequence $\{X_k\}$ of i.d.m.s.i. r.v.'s. Suppose that for each $k = 1, 2, 3, \ldots$ there is defined a set $A_k \subset R$, where $P_{X_k}(A_k) < 1$ for at least all those k's belonging to an infinite subset of I_+, the set of all positive integers. Then the probability is 0 that a sample sequence $\{x_k\}$ will satisfy $x_k \in A_k$ for all $k = 1, 2, 3, \ldots$.

For example, in the (idealized) product experiment consisting of an infinite sequence of die throws, let X_k indicate the number of dots showing on the kth throw, $k = 1, 2, 3, \ldots$. Then, the probability is 0 that a sample sequence will arise that satisfies: 'less than four dots on every odd-numbered throw and more than three dots on every even-numbered throw'. It should be noted that Corollary 84.1 applies also when a constraint is specified for, say, only the even values of k, or for only every tenth k. For all the remaining k, A_k is then taken to be R.

b) Simple gambling schemes

An early application of sequences of r.v.'s was in the analysis of gambling. We consider a simple gambling situation involving two players, A and B, who carry out a series of trials (such as tossing a coin). It is assumed that the trials are independent and identical, and involve activating a random mechanism that exhibits statistical regularity. Each trial leads to two possible results: 'A receives an amount a from B', or 'B receives an amount b from A'. The mathematical model for a single trial is a binary probability space with $S = \{\xi_1, \xi_2\}$, where we let $\xi_1 \leftrightarrow$ 'A wins' and $\xi_2 \leftrightarrow$ 'B wins', and we denote $P(\{\xi_1\}) = p$, and $P(\{\xi_2\}) = q = 1 - p$.

If X is the r.v. expressing the amount won by player A in one trial, then $\mathbf{E}X = ap - bq$. A bet or game is 'unbiased' or *fair* if $\mathbf{E}X = 0$, so that $a = bq/p$ (or $a = b$ if the two possible outcomes are equally likely). Fairness implies that on the average neither player has an advantage, and winnings are only due to 'luck'—i.e., chance—rather than some structural feature of the game.

The notion of fairness can be carried over to sequences. If the two players A and B carry out repeated trials and for $k = 1, 2, 3, \ldots$, and we let the r.v. X_k express the amount gained by A in the kth trial (which will be negative if there is a loss), then the sequence $\{X_k\}$ can be thought of as characterizing the game. Then we have:

(Definition) 84.1
A sequence $\{X_k\}$ is called *absolutely fair* if $\mathbf{E}X_1 = 0$ and for $k = 1, 2, 3, \ldots$,

$$\mathbf{E}[X_{k+1}|X_1, \ldots, X_k] = 0.$$

For instance, if $\{X_k\}$ is a sequence of i.d.m.s.i. r.v.s, and $\mathbf{E}X_1 = 0$, then $\{X_k\}$ is absolutely fair. It is important to note, however, that a gambler whose trial-by-trial pay-off is characterized by an absolutely fair sequence should not expect that his or her accumulated losses at a certain point in the game will necessarily be cancelled later by corresponding winnings (or vice versa), as seen in the following Example.

Example 84.1

Consider an unbiased coin toss, where player A receives \$1 from player B if the result of the toss is 'heads' and pays \$1 to player B if the result is 'tails'. Here, $p = q = 1/2$, and $a = b = 1$, making it a fair game. Multiple executions of such a Bernoulli trial can be gathered together into a product experiment, with a corresponding product probability space serving as the mathematical model. On the kth component space we define the r.v. X_k that expresses the amount won by A in the kth of n component experiment ($k = 1, 2, 3, \ldots, n$); i.e., $X_k = \pm 1$ with equal probability. If, in place of a fixed n, we let $n \to \infty$, then the amount won by A in each successive trial is expressed by the sequence $X_1, X_2, X_3, \ldots = \{X_k\}$, which satisfies Definition 84.1.

Of interest to each of the players, however, is not so much the result of each individual coin toss, but the total amount won or lost after some number of tosses. To express this, let $W_k = X_1 + X_2 + \ldots + X_k$ ($k = 1, 2, 3, \ldots$). Then $\{W_k\}$ is called the sequence of *partial sums* of $\{X_k\}$. We know from Example 35.1 that the W_k are *binomial* r.v.'s; but here they are *centralized* ($\mathbf{E}W_k = 0$ for all k), and W_k takes on only odd integer values if k is odd, and only even integer values if k is even. More important is the fact that the variance of W_k is proportional to k (Theorem 64.4):

$$\mathbf{V}W_k = k\mathbf{V}W_1 = k$$

or equivalently, the standard deviation equals $\sqrt{k}$. This shows that if the players plan to stop the game after n tosses, then the larger n, the more player A's winnings (or losses) are likely to deviate from zero at the end of the game.

The next Example illustrates how gambling becomes more precarious if the probability p is very small.

Example 84.2

At a gambling casino you are invited to play the following game at \$1 a try, in which you have a chance to win \$1 000 000. The game is based on a reliable chance mechanism with two possible outcomes, with successive trials identical and independent. One possible outcome pays you \$1 000 000 and occurs with probability $p = 10^{-6}$. The other possible outcome pays nothing. Are you willing to play this game? A quick calculation tells you that the expected pay-off per trial is \$1, so that a payment of \$1 per trial makes this a fair game. (This is a 'hypothetical' casino, of course, since an actual casino is out for earnings and no game will be totally fair.) But there is a hitch: You are likely to run out of money (and out of time as well)[10] long before one of those \$1 000 000 pay-offs occurs.

---

[10] The nature of the time constraint depends on how the game is physically implemented. For instance, the game could be carried out by computer in the following way: The computer generates six-digit random numbers (in the range from 000 000 to 999 999). You pick a number you want to bet on and pay \$1000 for the first 1000 trials. The computer generates 1000 numbers. You initiate a search for your chosen number. If it is present and occurs (for the first time) in position n, then you receive \$1 000 000 plus a refund of \$(1000–n), assuming you planned to terminate playing upon winning.

With the \$1 payment per trial the product probability space is similar to the one in Example 84.1, but we now have $p = 10^{-6}$, $q = 1 - 10^{-6}$, $a = 10^6 - 1$, $b = 1$. Let the sequence $\{X_k\}$ express the amount you gain in successive trials (which for most k will be -1). As in Example 84.1, let the sequence of partial sums of $\{X_k\}$ be denoted by $\{W_k\}$. In Example 84.1 a typical sample sequence of $\{W_k\}$ will be a sequence of integers that begins with $+1$ or -1 and meanders up and down. But here, a typical sample sequence will show a regular decline in unit steps that is very likely to continue over hundreds of thousands of trials.

Now let c denote your available capital when you start the game. If $c < 10^6$ and you decide to play the game until you either win once or lose all your capital, what is the probability of the event $L \leftrightarrow$ 'you lose your entire capital'? We have

$$P(L) = q^c = \left(1 - 10^{-6}\right)^c$$

If, for instance, you enter the game with \$1000, then $P(L) \doteq 0.9990$. Equivalently, the chance of a win in those first 1000 trials (or any 1000 trials) is one in a thousand. We might ask how much capital you need to reduce $P(L)$ to 0.5; that is, you want winning a million and losing all your capital to be equally likely. Setting $P(L) = 0.5$ in the above equation yields $c = \$735\,320$. You will have to be prepared to spend that much money just to guarantee yourself an even chance of winning or losing.

Having seen the effect of limited starting capital in Example 84.2, we examine also the impact of limited starting capital in the coin tossing game of Example 84.1.

Example 84.1 (continued)

Suppose both players start out with c dollars, and they agree to continue to play for a certain number of tosses that we denote k^*, or until one of them runs out of money. The same product space can be used as before, and the same probability space, on which the sequence $\{X_k\}$ is defined. Of course, in a given execution of the experiment, the model ceases to be relevant once the game ends.

It follows from the symmetry of this form of the game that neither player has an advantage. Thus, if L_A, L_B denote the events 'A loses all' and 'B loses all', respectively, then $P(L_A) = P(L_B)$ must hold. But what are those probabilities for any given k^*?

If $k^* < c$ then neither player can go broke and $P(L_A) = P(L_B) = 0$. To examine what happens for larger k^*, we consider the simplest case of interest: $c = 2$. (If $c = 1$, the game consists of only one toss.) Let the r.v.'s A_k express how much money player A has after the kth toss ($k = 1, 2, 3, \ldots$); and let $A_0 = 2$, the starting capital. Then, for $k = 1$ and 2, $A_k = W_k + 2$; but for larger k, this relationship no longer holds. To see what happens it is helpful to turn to a *state-transition diagram* as shown in Figure 84.1, where white nodes represent transition states and black nodes the terminal states, with the state probabilities indicated. All state transitions proceed left to right.

For *even* k^*, the probability of the event $L_A^c\, L_B^c$, i.e., that the game ends with neither player going broke (and thus both ending up with their starting capital of $c = 2$) is the probability marked on level 2 of the diagram ($A_k = 2$). Thus, for $k^* = 6$, this probability is 1/8 so that $P(L_A) = P(L_B) = 7/16$. For any even k^* it is easily seen that

$$P(L_A) = P(L_B) = \left(1 - 2^{-k^*/2}\right)/2$$

As $k^* \to \infty$, $P(L_A), P(L_B) \to 0.5$ and $P(L_A \cup L_B) \to 1$. Furthermore, convergence is fairly rapid. This can be seen by considering the r.v. K that expresses the number of trials till one of the players

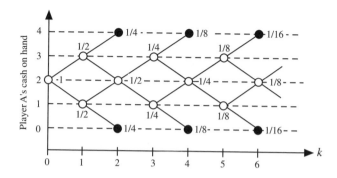

Figure 84.1   State-transition diagram

loses all. What is $\mathbf{E}K$? We note that $P_K(\{2\}) = 1/2$, $P_K(\{4\}) = 1/4$, etc., so that $\mathbf{E}K = 2(1/2)$ $+ 4(1/4) + 6(1/8) + \ldots = 4.0$. In other words, if the duration of the game is not delimited, then on the average, the game will end after 4 coin tosses.[11]

It turns out that exact calculation of the game's progress in this manner quickly gets much more complicated if the starting capital is increased or unequal (see also Problem 84.2). However, if starting capital is unequal and the game's duration is not limited ($k^* = \infty$), then the probabilities of L_A and L_B can be determined in the following way, using symmetry considerations.

Suppose players A and B start out with capital c_A and $c_B = 3c_A$, respectively. If at some point in the game A reaches $2c_A$, then from that point on the game proceeds as a symmetrical game so that L_A and L_B have conditional probability $1/2$. What is the probability that A attains $2c_A$? It is the same as A losing all, since A has an equal chance of winning or losing an amount c_A. Thus, let $W \leftrightarrow$ 'the capital A has on hand increases to $2c_A$', where $P(W) = 1/2$. Then $P(L_B) = P(L_B|W)P(W) = (1/2)(1/2) = 1/4$, and $P(L_A) = 3/4$.

This approach can be used to show that in the general case, with players A and B starting out with capital c_A and c_B, respectively,

$$P(L_A) = \frac{c_B}{c_A + c_B}, \quad \text{and} \quad P(L_B) = \frac{c_A}{c_A + c_B}. \tag{84.1}$$

c) Operations on sequences

Many types of sequences of r.v.'s can arise. The sequence $\{W_k\}$ of exponential r.v.'s mentioned at the beginning of this Section, and the sequence $\{X_k\}$ of Example 84.1, are both sequences of *i.d.m.s.i.* r.v.'s. Such sequences are often the starting point in the derivation or definition of other kinds of sequences, such as the sequence $\{W_k\}$ of partial sums in Example 84.1.

Unless otherwise specified, the index variable k in the notation $\{X_k\}$ is always assumed to range over all positive integers. Therefore, given a sequence $\{X_k\}$, then the notation $\{X_{k+1}\}$ refers to the sequence $X_2, X_3, X_4, \ldots$, which is obtained from $\{X_k\}$ by deleting the first component. On the other hand, the notation $\{X_{k-1}\}$ is only meaningful if a component X_0 is defined. Similarly, $\{X_{2k}\} = X_2, X_4, X_6, \ldots$, which is the sequence of only the even-numbered components of $\{X_k\}$. When an operation

---

[11] The kind of problem considered here is also called a 'random walk with absorbing boundaries.' In this case the random walk is characterized by the A_ks and the boundaries are c and $-c$. A boundary can also be a 'reflecting' boundary. This is illustrated in Problem 84.3 by the manner in which the B_ks are constrained. (See also Chapter 15 of [Fe1].)

involving a constant and a sequence is specified, the operation is meant to be applied to every element of the sequence. Thus, $a\{X_k\} = aX_1, aX_2, aX_3, \ldots$. An operation involving two sequences of r.v.'s (defined on the same probability space) is meant to be applied to pairs of corresponding components. For instance, if two sequences $\{U_k\}$ and $\{V_k\}$ are given, then $\{U_k\} + \{V_k\}$ is the sequence

$$U_1 + V_1, U_2 + V_2, U_3 + V_3, \ldots$$

Also, $\{X_{k+1}\} - \{X_k\} = X_2 - X_1, X_3 - X_2, X_4 - X_3, \ldots$, the sequence of increments or differences between successive components of $\{X_k\}$. Two sequences $\{X_k\}$, $\{Y_k\}$ are *identical* if and only if $X_k = Y_k$ for all k; that is, if corresponding components in the two sequences are identical.

$E\{V_k\}$ is the sequence of numbers $EV_1, EV_2, EV_3, \ldots$, assuming these expectations exist. Similarly, $V\{V_k\}$ is the sequence of numbers $VV_1, VV_2, VV_3, \ldots$, assuming these variances exist.

Given some sequence $\{X_k\}$, we have already seen in Example 84.1 how a sequence $\{W_k\}$ of *partial sums* of $\{X_k\}$ can be defined. In a similar way, a sequence $\{Y_k\}$ of *partial averages* of $\{X_k\}$ is obtained by setting, for $k = 1, 2, 3, \ldots$,

$$Y_k = \frac{X_1 + X_2 + \ldots + X_k}{k} = \frac{W_k}{k}.$$

If $\{X_k\}$ is a sequence of i.d.m.s.i. r.v.'s, then of course the partial sums W_k are not identically distributed (unless $X_k = 0$ for all k) nor are they m.s.i. (unless all X_k's are unary). Similarly, the partial averages Y_k are not i.d., nor are they m.s.i., unless all X_k's are unary. If $\{X_k\}$ is a sequence of i.d.m.s.i. r.v.'s and VX_1 exists, then

$$EW_k = kEX_1 \text{ and } VW_k = kVX_1;$$

$$EY_k = EX_1 \text{ and } VY_k = (VX_1)/k.$$

Now consider some sequence $\{X_k\}$, and suppose for each component the same event $A \subset R$ is of interest. For instance, for each k, A might be the event that X_k takes on a value greater than 2. Once the underlying experiment has been performed, the resulting sample sequence $\{x_k\}$ is examined for the occurrence of A. Suppose those components of $\{x_k\}$ that do not satisfy A get deleted. Thus, if x_{k_1} is the first component in the sequence $\{x_k\}$ that satisfies A, x_{k_2} is the second component in $\{x_k\}$ that satisfies A, and so on, we form the sequence $x_{k_1}, x_{k_2}, x_{k_3}, \ldots$ and forget about those components of $\{x_k\}$ that do not satisfy A. Of course, when the experiment is performed again, the subscripts $k_1, k_2, k_3, \ldots$ are likely to be different from the first time around, and another execution of the experiment will again result in different numbers $k_1, k_2, k_3, \ldots$; and so on. In other words, the numbers $k_1, k_2, k_3, \ldots, k_m, \ldots$ obtained in any one execution of the experiment are really sample values of a sequence of r.v.'s, $\{K_m\}$. These r.v.'s are called the *order statistics of the event* A on $\{X_k\}$. They express where, in $\{X_k\}$, A is satisfied for the first time, the second time, and so on.

Example 84.3

A fair die is thrown repeatedly, with successive throws considered independent. Of interest is the occurrence of a 'six'. This can be modeled as an infinite product experiment. On the infinite product probability space the sequence $\{X_k\}$ can then be defined, which expresses the die-throw results. For each of the component r.v.'s, let $A = \{6\}$, so that it is now possible to define the sequence of order statistics $\{K_m\}$ of the event $\{6\}$ on $\{X_k\}$.

The probability distribution induced by K_1 has been obtained previously, in Example 27.3. It is the geometric distribution

$$P_{K_1}(\{k\}) = \frac{5^{k-1}}{6^k}, \quad k = 1, 2, 3, \ldots \tag{84.2}$$

What about the distribution of K_2, which expresses the number of die throws till the second occurrence of A? This can be thought of as the number of throws till the first occurrence of A, plus the number of additional throws till the second occurrence. Let the latter be expressed by a r.v. K_1', so that $K_2 = K_1 + K_1'$. Since the die throws are independent and identical component experiments, K_1 and K_1' are i.d.s.i. The generating function for K_2 is then (Theorem 79.1):

$$G_{K_2}(z) = G_{K_1}(z) \cdot G_{K_1'}(z) = [G_{K_1}(z)]^2$$

From the solutions to Problems 79.5 and 79.6 it follows that K_2 induces a negative binomial distribution:

$$P_{K_2}(\{k\}) = (k-1)\frac{5^{k-2}}{6^k}$$

Applying a similar argument to $K_3, K_4, \ldots$, we find that K_m ($m = 2, 3, 4, \ldots$) induces the negative binomial distribution with parameter m (see also Problem 27.14):

$$P_{K_m}(\{k\}) = \binom{k-1}{m-1}\frac{5^{k-m}}{6^k}, \quad k = m, m+1, m+2, \ldots$$

Queries

84.1 Let $\{B_k\}$ be a sequence of i.d.m.s.i. binary r.v.'s that take on the values ±1 with equal probability. For each of the following statements, determine whether it is true, possibly true, or false:

a) $|\{B_k\}|$ is an absolutely fair sequence.

b) If $\{W_k\}$ is the sequence of partial sums of $\{B_k\}$, then $\{W_k\}$ is an absolutely fair sequence.

c) Let $X_k = \begin{cases} B_k, & \text{for } k = 1,2 \\ (1 + |B_{k-1} + B_{k-2}|)B_k, & \text{for } k \geq 3. \end{cases}$
Then $\{X_k\}$ is an absolutely fair sequence.

d) Let $D_k = \begin{cases} 1, & \text{for } k = 1 \\ \text{the number of consecutive components of } \{B_k\} \text{ just prior to } B_k, \\ \quad \text{that agree in value, for } k > 1. \end{cases}$
Thus, if a sample sequence of $\{B_k\}$ is $-1, 1, 1, -1, -1, -1, .-1, 1, \ldots$, then D_k takes on the values $1, 1, 2, 1, 2, 3, 4, 1, \ldots$. Now let $\{X_k\} = \{B_k\} \cdot \{D_k\}$. Then $\{X_k\}$ is an absolutely fair sequence. [118]

84.2 A p.e. involves repeatedly tossing, without bias, a nickel and a dime together. Let $\{K_m\}$ be the sequence of order statistics for the event 'heads on the nickel', and let $\{L_m\}$ be the sequence of order statistics for the event 'heads on the dime'. For each of the following statements, determine whether it is true, possibly true, or false:

a) $\mathbf{E}\{K_m\} = \mathbf{E}\{L_m\}$

b) $\{K_m\} = \{L_m\}$

c) $\mathbf{V}K_{m+1} > \mathbf{V}K_m$, $m = 1, 2, 3, \ldots$.

d) $K_1(\xi) = L_2(\xi)$, where ξ is an element of the underlying probability space.

e) If an execution of the p.e. results in 'tails' on both coins for the first three tosses, then the sequences of order statistics for that execution become $\{K_m\} + 3$ and $\{L_m\} + 3$. [232]

Convergent Sequences and Laws of Large Numbers

a) Convergence of Sequences

What can it mean to speak of convergence of a sequence of r.v.'s? Does it mean that their c.d.f.'s form a sequence of functions that converge to some limiting function? Does it mean that the sequence of numerical values assigned by the r.v.'s to an element of S is a convergent sequence of numbers? It turns out that both of these kinds of convergence can be of interest—and some other kinds as well.

Example 85.1

Consider a probability space $(S, \mathcal{A}, P)$ on which is defined a sequence of r.v.'s $\{X_k\}$. Suppose that for each $k = 1, 2, 3, \ldots$, the r.v. X_k has as its *range* the interval $[0, 1/k]$. Then, even without knowing anything else about the r.v.'s, we can say that the sequence $\{X_k\}$ assigns to each element $\xi \in S$ a sample sequence $\{x_k\}$ that converges to 0. For, taking any $\varepsilon > 0$ we can specify an n (namely, any $n > 1/\varepsilon$) such that the numbers $x_n, x_{n+1}, x_{n+2}, \ldots$ all differ from 0 by less than ε.

The kind of convergence exhibited by $\{X_k\}$ in Example 85.1 can be expressed by writing $\lim_{k \to \infty} X_k(\xi) = 0$, or simply,

$$X_k(\xi) \to 0. \tag{85.1}$$

In that example, Equation (85.1) holds for every $\xi \in S$, and we say that $\{X_k\}$ *converges to a constant.* Some other sequence $\{Y_k\}$ might *converge to a random variable.* This means that every sample sequence of $\{Y_k\}$ converges to a limiting value, but not every sample sequence has *the same* limiting value. The various limits, one for each ξ, are therefore represented by a new r.v. Y. One way in which this could arise is the following. Let

$$\{Y_k\} = \{X_k\} + Y$$

where $\{X_k\}$ is the sequence defined in Example 85.1, and Y is some other r.v. Then the sequence $\{Y_k\}$ assigns to each element $\xi \in S$ a sample sequence $\{y_k\}$ that can also be written $\{x_k\} + y$. Since $x_k \to 0$, it follows that $y_k \to y$, a sample value of the r.v. Y. Thus, $\{Y_k\}$ converges to a r.v. and we write:

Probability Concepts and Theory for Engineers, First Edition. Harry Schwarzlander.
© 2011 John Wiley & Sons, Ltd. Published 2011 by John Wiley & Sons, Ltd.

$$Y_k(\xi) \to Y(\xi), \text{ for all } \xi \in \mathsf{S}. \tag{85.2}$$

Even in Equation (85.1), the limit can be regarded as a r.v., namely the unary r.v. that takes on the value 0 with certainty.

The type of convergence described in Equations (85.1) and (85.2) is the 'strongest' kind of convergence for r.v.'s; it is called *pointwise convergence*, because there is convergence at *every point* in the underlying probability space. In other words, $\{X_k\}$ in Example 85.1 assigns a convergent sequence of numbers to every element $\xi \in \mathsf{S}$. In many situations of interest, a sequence of r.v.'s does not exhibit pointwise convergence, or pointwise convergence cannot be proven for the sequence. However, a slightly weaker form of convergence may hold. One possibility is that pointwise convergence is not satisfied on all of S, but only on a set of probability 1.

(Definition) 85.1
A sequence of r.v.'s $\{X_k\}$ defined on a probability space (S, $\mathscr{A}$, P) is said to *converge with probability* 1 to a r.v. X if the set $A = \{\xi : X_k(\xi) \to X(\xi)\}$ has probability $P(A) = 1$.

This type of convergence is also referred to as *almost certain convergence*, or *almost sure convergence*. It includes pointwise convergence. Definition 85.1 can be seen to utilize two basic concepts:

i) for a given ξ, the ordinary convergence of a sequence of real numbers, $X_k(\xi) \to X(\xi)$; and
ii) the probability associated with an event, namely, the event made up of those elements ξ for which it is true that $X_k(\xi) \to X(\xi)$.

It turns out that, in many cases, this type of convergence is not easy to prove and may not hold. For that reason, still other modes of convergence of r.v.'s have been defined and are found useful in probability theory. One of these is the following:

(Definition) 85.2
A sequence of r.v.'s $\{X_k\}$ is said to *converge in probability* to the r.v. X if for each $\varepsilon > 0$,

$$\lim_{k \to \infty} P(\{\xi : |X(\xi) - X_k(\xi)| \geq \varepsilon\}) = 0.$$

We note that this Definition makes use of the following two concepts.

i) For each $\varepsilon > 0$, a sequence of *events* $\mathsf{E}_1(\varepsilon)$, $\mathsf{E}_2(\varepsilon)$, $\mathsf{E}_3(\varepsilon)$, ... $\in \mathsf{S}$ is considered, where

$$\mathsf{E}_k(\varepsilon) = \{\xi : |X(\xi) - X_k(\xi)| \geq \varepsilon\}.$$

Thus, $\mathsf{E}_k(\varepsilon)$ is the set of those elements ξ to which X_k assigns a value that differs by more than ε from the value assigned by X.
ii) There is the ordinary convergence of a sequence of real numbers, namely, convergence to 0 of the sequence of *probabilities* of the sets $\mathsf{E}_k(\varepsilon)$; and this for each $\varepsilon > 0$.

Definitions 85.1 and 85.2 are related as follows:

Theorem 85.1
Convergence with probability 1 implies convergence in probability.

Proof

Consider a sequence of r.v.'s $\{X_k\}$ that converges to X with probability 1. In the notation of Definitions 85.1 and 85.2, the set A contains those elements ξ for which the following condition is satisfied: Given any $\varepsilon > 0$, there is an integer k_ε such that $|X_k(\xi) - X(\xi)| < \varepsilon$ for all $k > k_\varepsilon$. This means that for $k > k_\varepsilon$, the sets $\mathsf{E}_k(\varepsilon) \subset \mathsf{A}^c$. Thus, $P(\mathsf{E}_k(\varepsilon)) \leq P(\mathsf{A}^c) = 0$ and $\lim_{k \to \infty} P(\mathsf{E}_k(\varepsilon)) = 0$. This holds for each $\varepsilon > 0$, so that $\{X_k\}$ converges in probability to X.

Convergence in probability does *not* imply convergence with probability 1, however, as is demonstrated by the following counterexample.

Example 85.2 [Pf1]

Let X be a r.v. that is unif. $\overline{0, 1}$. Now define the r.v.'s $Y_1, Y_2, Y_3, \ldots$ as indicator r.v.'s of events $A_1, A_2, A_3, \ldots$, respectively, where:

$$A_1 = \{\xi: 0 < X(\xi) \leq 1/2\} \qquad A_2 = \{\xi: 1/2 < X(\xi) \leq 1\}$$
$$A_3 = \{\xi: 0 < X(\xi) \leq 1/4\} \qquad A_4 = \{\xi: 1/4 < X(\xi) \leq 1/2\}$$
$$A_5 = \{\xi: 1/2 < X(\xi) \leq 3/4\} \qquad A_6 = \{\xi: 3/4 < X(\xi) \leq 1\}$$
$$A_7 = \{\xi: 0 < X(\xi) \leq 1/8\} \qquad A_8 = \{\xi: 1/8 < X(\xi) \leq 1/4\}$$

$$\cdots\cdots\cdots\cdots$$

$$A_{13} = \{\xi: 3/4 < X(\xi) \leq 7/8\} \quad A_{14} = \{\xi: 7/8 < X(\xi) \leq 1\}$$

$$\cdots\cdots\cdots\cdots$$

etc.

The resulting sequence $\{Y_k\}$ has a distinct sample sequence $\{y_k\}$ for every distinct sample value of X. Furthermore, each sample sequence has constraints among its components, so that the Y_k's are not m.s.i. nor even pairwise s.i. This can be seen by dividing a sample sequence $\{y_k\}$ into blocks having lengths $2^1, 2^2, 2^3, 2^4, \ldots$:

$$\{y_1, y_2\}, \{y_3, y_4, y_5, y_6\}, \{y_7, y_8, \ldots, y_{14}\}, \text{etc.}$$

Then the sample value 1 must occur exactly once in each block.

The distributions induced by the r.v.'s Y_k are shown in Figure 85.1. We want to show that the sequence $\{Y_k\}$ converges in probability but not with probability 1. First, in order to verify that Definition 85.2 applies, it is necessary to decide what the sequence $\{Y_k\}$ might converge to. From the distributions of the r.v.'s Y_k it can be seen that with increasing k the probability of $\{1\}$ gradually vanishes, while the probability of $\{0\}$ increases toward 1. Therefore, if the sequence $\{Y_k\}$ converges in any sense, then it must converge to 0. Now we apply Definition 85.2: Is it true that for

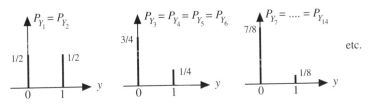

Figure 85.1 Distributions induced by the first 14 r.v.'s in the sequence $\{Y_k\}$

every $\varepsilon > 0$,

$$\lim_{k \to \infty} P(\{\xi: |Y_k - 0| \geq \varepsilon\}) = 0?$$

We have

$$P(\{\xi: |Y_k - 0| \geq \varepsilon\}) = P(\{\xi: |Y_k| \geq \varepsilon\}) = P(\{\xi: Y_k \geq 0\}). \tag{85.3}$$

The last equality is correct as long as $0 < \varepsilon < 1$, because the Y_k's can only take on the value 0 or 1, so that it makes no difference how small an ε we pick. Thus, the right-hand side of Equation (85.3) is simply $P(A_k)$, and by construction, $P(A_k) \to 0$ as $k \to \infty$. Therefore, $\{Y_k\}$ converges in probability.

Next, we test $\{Y_k\}$ with Definition 85.1 by examining the set $\{\xi: Y_k(\xi) \to 0\}$. By construction, for every ξ, $\{Y_k(\xi)\}$ is a sequence of 0's and 1's, with the 1's becoming very rare as k gets large. Nevertheless, given any k there will always be another 1, for some larger k. Therefore, $\{\xi: Y_k(\xi) \to 0\} = \phi$, so $\{Y_k\}$ does not converge with probability 1.

b) Laws of Large Numbers

Given a sequence of r.v.'s $\{X_k\}$, we consider the sequence $\{Y_n\}$ of *partial averages* of the X_k's:

$$Y_n = \frac{X_1 + X_2 + \ldots + X_n}{n}, \, n = 1, 2, 3, \ldots \tag{85.4}$$

If the sequence $\{Y_n\}$ converges to a constant, then the sequence $\{X_k\}$ is said to *obey the law of large numbers*. If it happens to be the case that the components of $\{X_k\}$ are *centralized r.v.'s* and $\{Y_n\}$ converges to a constant, then that constant is 0.

The type of convergence of the sequence $\{Y_n\}$ is important. If $\{Y_n\}$ converges with probability 1, then $\{X_k\}$ obeys the *Strong Law of large numbers*. If the convergence of $\{Y_n\}$ is in probability, which we have noted to be a weaker form of convergence, then $\{X_k\}$ obeys the *Weak Law of large numbers*. A variety of conditions are known under which a sequence $\{X_k\}$ obeys the strong law of large numbers. Some of these are contained in the following Theorem.

Theorem 85.2: Strong Law of large numbers.
Let $\{X_k\}$ be a sequence of *m.s.i.* r.v.'s whose expectations exist and are zero, and let

$$Y_n = \frac{1}{n} \sum_{k=1}^{n} X_k, \quad \text{for } n = 1, 2, 3, \ldots$$

The sequence $\{Y_n\}$ converges to zero with probability 1 if *any one* of the following conditions hold:

a) The X_k's are identically distributed.
b) $\mathbf{V}[X_k] < c$ for all k, where c is some constant.
c) $\displaystyle\sum_{k=1}^{\infty} \frac{\mathbf{V}X_k}{k^2} < \infty.$

For instance, we see at once from condition (a) that the sequence $\{X_k\}$ defined in Example 84.1 obeys the strong law of large numbers.

It follows from Theorem 85.1 that the *Weak Law* of large numbers must hold under the conditions stated in Theorem 85.2. But it also holds under more relaxed conditions. We have the following necessary and sufficient condition for the Weak Law, for sequences $\{X_k\}$ whose components are centralized—i.e., their expectations exist and are zero.

Theorem 85.3: Weak Law of large numbers. [Gn1]
Let $\{X_k\}$ be a sequence of centralized r.v.'s, and let

$$Y_n = \frac{1}{n}\sum_{k=1}^{n} X_k, \quad \text{for } n = 1, 2, 3, \ldots$$

The sequence $\{Y_n\}$ converges to zero in probability if and only if

$$\lim_{n\to\infty} \mathbf{E}\left[\frac{Y_n^2}{1+Y_n^2}\right] = 0. \tag{85.5}$$

Examining Equation (85.5) it can be seen that, since $Y_n^2 \geq 0$, we have

$$0 \leq \frac{Y_n^2}{1+Y_n^2} < Y_n^2.$$

Therefore,

$$\mathbf{E}[Y_n^2] = \mathbf{V}Y_n \to 0 \tag{85.6}$$

implies Equation (85.5). In other words, Equation (85.6) is a sufficient (but not necessary) condition for the Weak Law, but in many cases it is easier to apply than Equation (85.5).

It should be noted that Theorem 85.3 does not call for the r.v.'s X_k to be m.s.i. However, mere *pairwise* statistical independence of the r.v.'s X_k is sufficient for the Weak Law of large numbers to hold with condition (b) of Theorem 85.2. That is, $\mathbf{V}X_k < c$, $k = 1, 2, \ldots$, where c is some constant. Equation (85.6) is then satisfied, which is seen as follows. Assume the X_k's to be centralized and pairwise s.i., with variances $\mathbf{V}X_k < c$. Then

$$\begin{aligned}
\mathbf{V}Y_n &= \frac{1}{n^2}\,\mathbf{E}\left[\sum_{k=1}^{n}\sum_{\ell=1}^{n} X_k X_\ell\right] \\
&= \frac{1}{n^2}\sum_{k=1}^{n}\mathbf{V}X_k \quad (\text{since for } k \neq \ell,\ \mathbf{E}[X_k X_\ell] = \mathbf{E}[X_k]\mathbf{E}[X_\ell] = 0) \\
&\leq \frac{c}{n}
\end{aligned}$$

so that $\mathbf{V}Y_n \to 0$ as $n \to \infty$. Therefore, $\{X_k\}$ obeys the Weak Law of large numbers.

Example 85.3
It has already been noted that the sequence $\{X_k\}$ defined in Example 84.1 obeys the Strong Law of large numbers. Consider now a slight modification of the game described in that Example: Let the pay-off increase by \$1 with each coin toss, and assume that the players have unlimited capital. Then $X_k = \pm k$, for $k = 1, 2, 3, \ldots$. The X_k's are zero mean, and $\{X_k\}$ is an absolutely fair sequence. However, none of the conditions stated in Theorem 85.2 are satisfied.

Now let Y_n be defined as in Equation (85.4), giving

$$\mathbf{V}Y_n = \frac{1}{n^2}\sum_{k=1}^{n}\mathbf{V}X_k = \frac{1}{n^2}\sum_{k=1}^{n}k^2 > 1, \quad \text{for all } n > 1.$$

Thus, Equation (85.6) is not satisfied. This is not sufficient, however, to prove that the Weak Law does not hold for the sequence $\{X_k\}$.

To test for convergence in probability of $\{Y_n\}$ we note that for any n,

$$Y_{n+1} = (nY_n + X_{n+1})\frac{1}{n+1} = \frac{n}{n+1}Y_n \pm 1.$$

Therefore, for $0 < \varepsilon < 1/2$ and any n,

$$P_{Y_{n+1}}\big(\overline{-\varepsilon,\ \varepsilon}\big) =$$

$$= \frac{1}{2}P_{Y_n}\left(\frac{n+1}{n}(1-\varepsilon),\ \frac{n+1}{n}(1+\varepsilon)\right) + \frac{1}{2}P_{Y_n}\left(-\frac{n+1}{n}(1+\varepsilon),\ -\frac{n+1}{n}(1-\varepsilon)\right)$$

$$= P_{Y_n}\left(\frac{n+1}{n}(1-\varepsilon),\ \frac{n+1}{n}(1+\varepsilon)\right).$$

Since the interval on the left and the interval on the right are disjoint, we can write more simply:

$$P_{Y_{n+1}}\big(\overline{-\varepsilon,\ \varepsilon}\big) \le 1 - P_{Y_n}\big(\overline{-\varepsilon,\ \varepsilon}\big).$$

In other words, the larger the probability that Y_n assigns a value close to zero, the smaller the probability that Y_{n+1} assigns a value close to zero. Thus, there can be no convergence, and the weak law does not hold.

Conceptually, this result can be interpreted as an indication that for this game, a player's average cumulative gains/losses do not approach or favor any predictable value as the number of trials increases.

We have noted that Equation (85.6) is not a necessary condition for the Weak Law to hold. What might be a sequence $\{X_k\}$ that obeys the Weak Law even though it does not satisfy Equation (85.6)? The following Example is devoted to the design of such a sequence.

Example 85.4

In order to construct a sequence $\{X_k\}$ that satisfies Equation (85.5) but not Equation (85.6), we begin by designing a suitable sequence of partial averages $\{Y_n\}$, and from this derive $\{X_k\}$. Thus, suppose that a representative sample sequence $\{y_n\}$ of $\{Y_n\}$ begins with the sample values 1, 2, 3, 4, $\ldots$, m, after which come only zeros. The parameter m, however, is to be the sample value of a r.v. M, for which $P_M(\{m\})$ remains to be specified.

In that case,

$$0.5 \le \frac{y_n^2}{1+y_n^2} < 1 \text{ for } n \le m$$

and $\dfrac{y_n^2}{1+y_n^2} = 0$ for $n > m$. In order to compute $\mathbf{E}\left[\dfrac{Y_n^2}{1+Y_n^2}\right]$ for given n, we define the conditioning event $\mathsf{A}_n = \{\xi\colon M(\xi) > n\}$. Then,

$$\mathbf{E}\left[\frac{Y_n^2}{1+Y_n^2}\right] = \mathbf{E}\left[\frac{Y_n^2}{1+Y_n^2}\bigg|A_n\right]P(A_n) + \mathbf{E}\left[\frac{Y_n^2}{1+Y_n^2}\bigg|A_n^c\right]P(A_n^c)$$

$$= \frac{n^2}{1+n^2}P(A_n). \tag{85.7}$$

Furthermore, let the nonzero portion of a sample sequence be either positive or negative with equal probability, and independently of the events $A_1, A_2, A_3, \ldots$. In that way, $\mathbf{E}Y_n = 0$ for all n, so that $\mathbf{V}Y_n = \mathbf{E}[Y_n^2]$. Now, it is necessary to find a distribution P_M that yields

$$\mathbf{V}Y_n = \mathbf{E}[Y_n^2] \nrightarrow 0, \text{ as } n \to \infty \tag{85.8a}$$

$$\text{and } \mathbf{E}\left[\frac{Y_n^2}{1+Y_n^2}\right] \to 0, \text{ as } n \to \infty. \tag{85.8b}$$

Since $\mathbf{E}[Y_n^2] = n^2 P(A_n)$, then setting $P(A_n) = \alpha/n^2$ satisfies Equation (85.8a), where $\alpha = 6/\pi^2$ in order to make $\sum P(A_n) = 1$. And this also satisfies Equation (85.8b), since from Equation (85.7), $\mathbf{E}[Y_n^2/(1+Y_n^2)] = \alpha/(1+n^2) \to 0$. This concludes the specification of $\{Y_n\}$.

The question remains whether a sequence $\{X_k\}$ can be found for which $\{Y_n\}$ is the sequence of partial averages. It is easily seen that this is possible by setting

$$x_k = \pm(2k - 1), \text{ for } k \le m$$

where the leading minus sign applies if the nonzero samples in $\{y_n\}$ are negative. Furthermore, in order to make $y_n = 0$ for $n > m$, x_{m+1} must cancel $\sum_{1 \le k \le m} x_k$, with zeros thereafter. The complete specification of the component r.v.'s of the desired sequence $\{X_k\}$ is then

$$X_k = \begin{cases} B(2k-1), & k \le M, \\ -B(k-1)^2, & k = M+1, \\ 0, & k > M+1 \end{cases}$$

where B is a binary r.v. that assigns equal probability to $\{+1\}$ and $\{-1\}$ and is statistically independent of M, and

$$P_M(\{m\}) = 6/\pi^2 n^2, \, m = 1, 2, 3, 4, \ldots$$

A representative sample sequence is shown in Figure 85.2.

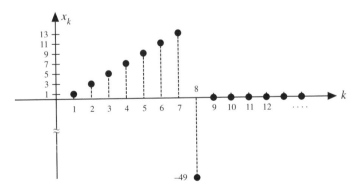

Figure 85.2 The sample sequence $\{y_k\}$ for $B = +1$ and $M = 7$

Thus, we have found a sequence $\{X_k\}$ that obeys the Weak Law but whose sequence of partial averages does not satisfy Equation (85.6). In fact, it also obeys the Strong Law (Problem 85.5).

Conceptually there is no obvious distinction between situations in which only the weak law holds and those in which the strong law holds. The important distinction is mathematical. The type of convergence exhibited by a particular sequence $\{X_k\}$ determines what other properties or results it is possible to establish for $\{X_k\}$.

c) Connection with Statistical Regularity

It is now possible to go back to the notion of 'statistical regularity' as introduced in Section 15 and examine it from the vantage point of the theory that has been developed. Statistical regularity, we recall, must be exhibited by events in multiple executions of a probabilistic experiment, if these events are to have probabilities assigned to them in the corresponding mathematical model. Thus, suppose a particular p.e. is specified, and in this p.e. an event A is of interest. This p.e. is to get executed n times. If, over these executions, statistical regularity holds relative to the event A, then its relative frequency rf(A) is deemed to tend toward a stable numerical value as n increases. In Section 16, that hypothetical stable value was taken to be $P(A)$, the probability of the set A representing A in the mathematical model—the probability space. We now ask in what manner rf(A) 'tends toward a stable value'.

Let the n (independent and identical) executions be denoted $\mathfrak{E}_1, \ldots, \mathfrak{E}_n$. These can be thought of as the component experiments or trials of a single product experiment $\mathfrak{E}$:

$$\mathfrak{E} = \mathfrak{E}_1 \times \mathfrak{E}_2 \times \ldots \times \mathfrak{E}_n.$$

In the corresponding product probability space, the set A is defined in each component space. We assume that the event A represented by A exhibits statistical regularity over the n trials. That is, the occurrence of A is governed by a random mechanism that is stable over all the trials, and is not influenced by carrying out those trials.

Now, let X_k ($k = 1, \ldots, n$) be the indicator r.v.'s of A in the n component spaces. Then, the actual occurrence or nonoccurrence of A observed in the n trials is expressed by a particular set of sample values of the X_k's. The average of these sample values is a sample value of the r.v.

$$Y_n = \frac{1}{n} \sum_{k=1}^{n} X_k. \tag{85.9}$$

If n in this expression is allowed to be any positive integer, then Y_n can be thought of as a component r.v. in the infinite sequence of partial averages $\{Y_n\}$. Although this sequence in its entirety has no real-world counterpart, it can always be truncated. Therefore, it incorporates in a single representation whatever number of trials might be deemed physically meaningful.

We have assumed that statistical regularity holds with respect to the event A and that therefore rf(A) tends toward some stable value, so that $P(A)$ is defined. Since each X_k is an indicator r.v. of A in one of the component spaces, we have $\mathbf{E}X_k = P(A)$ for each k. In order to put the X_k's into a form that is compatible with the manner in which Theorem 85.2 is stated, we centralize these r.v.'s, giving

$$X_{kc} = X_k - P(A), \quad k = 1, 2, 3, \ldots$$

Based on any of the three conditions in Theorem 85.2, we see that $\{X_{kc}\}$ obeys the Strong Law of large numbers. Furthermore, $\{Y_{nc}\} = \{Y_n\} - P(\mathsf{A})$ converges to zero, or

$$\{Y_n\} \to P(\mathsf{A}).$$

Thus, we find that the mathematical model is consistent with the original assumption that rf(A) tends toward a definite stable value. Furthermore, in the mathematical model the Strong Law makes precise what is implied by statistical regularity in the conceptual model.

Queries

85.1 Given a sequence of m.s.i. r.v.'s $\{X_k\}$. Determine whether this sequence obeys the Strong Law of large numbers if, for $k = 1, 2, 3, \ldots$:
a) $X_k = \pm 2^k$ with probability 1/2 each
b) $X_k = \pm 2^k$ with probability $2^{-(2k+1)}$ each, and $X_k = 0$ with probability $1 - 2^{-2k}$
c) $X_k = \pm k$ with probability $(k^{-1/2})/2$ each, and $X_k = 0$ with probability $1 - k^{-1/2}$.
[53]

85.2 A sequence of r.v.'s $\{V_k\}$ converges to 0 with probability 1. Another sequence $\{W_k\}$ converges to 1 with probability 1. In each of the following cases, does $\{X_k\}$ converge with probability 1, and if so, to what limit?
a) For each k, let X_k be chosen with equal likelihood from V_k and W_k, independently of all other selections and of the values of V_k and W_k.
b) For each k, let $F_{X_k} = aF_{V_k} + (1-a)F_{W_k}$, where a is some fixed number satisfying $0 < a < 1$.
c) $\{X_k\} = \{V_k\} - \{W_k\}$.
d) $X_k = \begin{cases} V_k, & k \text{ odd} \\ W_k, & k \text{ even.} \end{cases}$ [185]

Convergence of Probability Distributions and the Central Limit Theorem

Every r.v. X_k in a sequence $\{X_k\}$ induces a distribution P_{X_k}. With $\{X_k\}$ is therefore associated the *sequence of probability distributions* $\{P_{X_k}\}$. In Figure 85.1, for instance, the first few members of a sequence of discrete probability distributions are described. It is of considerable interest whether or not such a sequence of distributions converges. Here, it is necessary to be clear as to what exactly is meant by the convergence of a sequence of probability distributions. Since an induced distribution is uniquely described by its c.d.f., what we are really addressing is the convergence of the *sequence of c.d.f.'s*. Since a c.d.f. is simply a real-valued function of a real variable, this means the ordinary (or 'pointwise') convergence of a sequence of such functions to a function that is again a c.d.f.

(Definition) 86.1

A sequence ot r.v.'s $\{X_k\}$ is said to *converge in distribution* to the r.v. X if the *sequence of c.d.f.'s* $F_{X_1}(x), F_{X_2}(x), F_{X_3}(x), \ldots$ converges to the c.d.f. $F_X(x)$ for every real number x.

Convergence in distribution is a very weak form of convergence for a sequence of r.v.'s, but a very important one, nevertheless. Although it is commonly referred to as a form of convergence for a sequence *of r.v.'s*, we should keep in mind that it is really only convergence of a sequence of *c.d.f.'s*. For instance, a sequence of i.d.s.i. r.v.'s (which does not converge in any other sense) always converges in distribution, since all the c.d.f.'s are identical.

When we speak of convergence in distribution, it is not necessary to restrict ourselves to sequences of r.v.'s in which all members of the sequence are defined on a single probability space. Each r.v. in the sequence could possibly be defined on a different probability space, and would thus be associated with a different experiment. In that case, we would really have in mind a *sequence of experiments*, $\mathfrak{E}_1, \mathfrak{E}_2, \mathfrak{E}_3, \ldots$. In fact, we utilized such a construction in Section 60 in order to arrive at the Poisson distribution for the r.v. M_t. There, successive experiments $\mathfrak{E}_1, \mathfrak{E}_2$, etc., differed in that the time scale was divided into successively smaller intervals. For each of these experiments, a r.v. M_t was defined in an equivalent way, each inducing a binomial distribution. These binomial distributions were found to converge to the Poisson distribution.

Probability Concepts and Theory for Engineers, First Edition. Harry Schwarzlander.
© 2011 John Wiley & Sons, Ltd. Published 2011 by John Wiley & Sons, Ltd.

Example 86.1

A simple illustration of convergence in distribution is provided by a sequence of discrete r.v.'s $\{X_k\}$ whose marginal distributions are

$$P_{X_k}\left(\left\{\frac{m}{k}\right\}\right) = \frac{1}{k}, \; m = 1, 2, \ldots, k, \text{ where } k = 1, 2, 3, \ldots$$

Any dependence that might exist between members of the sequence $\{X_k\}$ has no bearing on convergence in distribution and therefore need not be specified. Bar graphs of the first four components of the sequence $\{P_{X_k}\}$ are shown in Figure 86.1. In Figure 86.2, the c.d.f.'s of those same four distributions are depicted. Here, we have the interesting result that a sequence of discrete r.v.'s converges (in distribution) to an absolutely continuous r.v., namely, a r.v. X that is uniformly distributed over $\overline{0, 1}$ (see Problem 86.2). The c.d.f. F_X of the limiting distribution is shown dashed in Figure 86.2.

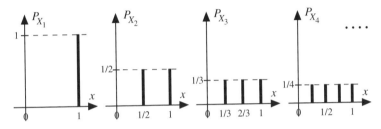

Figure 86.1   The discrete distributions $P_{X_1}, P_{X_2}, P_{X_3},$ and P_{X_4}

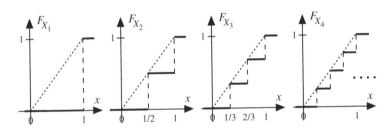

Figure 86.2   The c.d.f.s $F_{X_1}, F_{X_2}, F_{X_3},$ and F_{X_4}

We consider now a sequence of arbitrary r.v.'s $\{X_k\}$ whose means and variances exist, and let $\{Y_{ns}\}$ be the sequence of *standardized* partial averages of $\{X_k\}$. In other words, if

$$Y_n = \frac{1}{n}\sum_{k=1}^{n} X_k$$

is the nth partial average of the sequence $\{X_k\}$ (for $n = 1, 2, 3, \ldots$), then the standardized version of Y_n is

$$Y_{ns} = \frac{Y_n - \mathbf{E}[Y_n]}{\sqrt{\mathbf{V}[Y_n]}}$$

$$= \frac{\frac{1}{n}(X_1 + X_2 + \ldots + X_n) - \mathbf{E}\left[\frac{1}{n}(X_1 + X_2 + \ldots + X_n)\right]}{\sqrt{\mathbf{V}\left[\frac{1}{n}(X_1 + X_2 + \ldots + X_n)\right]}}$$

$$= \frac{(X_1 + X_2 + \ldots + X_n) - \mathbf{E}X_1 - \mathbf{E}X_2 - \ldots - \mathbf{E}X_n}{\sqrt{\mathbf{V}[X_1 + X_2 + \ldots + X_n]}}. \tag{86.1}$$

As is evident from Equation (86.1), $\{Y_{ns}\}$ can also be thought of as the sequence of standardized partial *sums* of $\{X_k\}$. Either way, the important feature of interest here is that the components of $\{Y_{ns}\}$ are *standardized*.

It turns out that under suitable conditions that are fairly general, a sequence $\{X_k\}$ exhibits this rather amazing property: Its associated sequence $\{Y_{ns}\}$ converges in distribution, and the limiting distribution is not affected by the detailed nature of the X_k's but is always the same. This limiting distribution is the standardized *Gaussian* or *normal* distribution. In other words,

$$F_{Y_{ns}}(y) \to \frac{1}{\sqrt{2\pi}} \int_{-\infty}^{y} \exp\left(-\frac{x^2}{2}\right) dx, \text{ for } n \to \infty. \tag{86.2}$$

For sequences $\{X_k\}$ exhibiting this property, we say that 'the Central Limit Theorem holds.'

Actually, various theorems exist that state known conditions under which Equation (86.2) holds. All such theorems are referred to as Central Limit Theorems. Here, we consider only the simplest case, where $\{X_k\}$ is a sequence of i.d.m.s.i. r.v.'s whose means and variances exist. In that case, Equation (86.1) becomes

$$Y_{ns} = \frac{(X_1 + X_2 + \ldots + X_n) - n\mathbf{E}X_1}{\sqrt{n\mathbf{V}X_1}} = \frac{1}{\sqrt{n}} \sum_{k=1}^{n} X_{ks} \tag{86.3}$$

X_{ks} being the standardized version of X_k, for $k = 1, 2, 3, \ldots$.

Theorem 86.1: Central Limit Theorem.
Given a sequence $\{X_k\}$ of i.d.m.s.i. r.v.'s with mean $\mathbf{E}X_1$ and variance $\mathbf{V}X_1$. Let $\{Y_{ns}\}$ be the sequence of standardized partial sums of $\{X_k\}$, as given by Equation (86.3). Then $\{Y_{ns}\}$ converges in distribution to a standardized Gaussian r.v. Y.

Proof

Let X_{ks}, $k = 1, 2, 3, \ldots$ denote the standardized versions of the components of the sequence $\{X_k\}$. The r.v.'s X_{ks} are identically distributed, so they all have the same characteristic function $\psi_{X_s}(\tau)$, and

$$\psi_{Y_{ns}}(\tau) = \psi_{X_s}{}^n(\tau/\sqrt{n}). \tag{86.4}$$

We are interested in determining the characteristic function

$$\psi_Y(\tau) = \lim_{n \to \infty} \psi_{Y_{ns}}(\tau).$$

According to Theorem 76.1, the first three terms of the Maclaurin series for ψ_{X_s} exist, so that

$$\psi_{X_s}(\tau) = \psi_{X_s}(0) + \psi'_{X_s}(0)\frac{\tau}{1!} + \psi''_{X_s}(0)\frac{\tau^2}{2!} + \cdots$$

$$= 1 + jm_{1X_s}\tau - m_{2X_s}\frac{\tau^2}{2!} + \cdots$$

Since $m_{1X_s} = 0$ and $m_{2X_s} = \mathbf{V}[X_s] = 1$, we have

$$\psi_{X_s}(\tau/\sqrt{n}) = 1 - \frac{\tau^2}{2n} + \cdots \tag{86.5}$$

The terms not shown in Equation (86.5) are of order higher than $1/n$. Let the notation $o(1/n)$ denote this remainder. Then

$$\psi_Y(\tau) = \lim_{n \to \infty}\left[\left(1 - \frac{\tau^2}{2n}\right) + o\left(\frac{1}{n}\right)\right]^n$$

$$= \lim_{n \to \infty}\left[\left(1 - \frac{\tau^2}{2n}\right)^n + n\left(1 - \frac{\tau^2}{2n}\right)^{n-1}o\left(\frac{1}{n}\right) + \cdots\right]$$

$$= \lim_{n \to \infty}\left(1 - \frac{\tau^2}{2n}\right)^n$$

$$= 1 - \frac{\tau^2}{2} + \frac{1}{2!}\left(\frac{\tau^2}{2}\right)^2 - \frac{1}{3!}\left(\frac{\tau^2}{2}\right)^3 + \cdots = \exp\left(-\frac{\tau^2}{2}\right). \tag{86.6}$$

The sequence $\{Y_{ns}\}$ thus converges in distribution to a r.v. Y whose characteristic function is given by Equation (86.6) that, as shown in Example 76.2, is the characteristic function of a standardized Gaussian r.v.

It is helpful to consider a specific example in order to examine the behavior that is expressed by the Central Limit Theorem.

Example 86.2

Let $\{X_k\}$ be a sequence of i.d.m.s.i. r.v.'s all of which are unif. $\overline{0,1}$, so that for all k,

$$f_{X_k}(x) = \begin{cases} 1, & 0 < x < 1 \\ 0, & \text{otherwise} \end{cases}$$

Since the partial sums of the sequence $\{X_k\}$ are all absolutely continuous, we can look at the *density functions* of some of those partial sums, rather than the c.d.f.'s. The density functions of the first three standardized partial sums of $\{X_k\}$ are sketched in Figure 86.3, giving a visual impression of how the p.d.f. of the limiting distribution (the standardized Gaussian p.d.f., shown dashed) is approached quite rapidly.

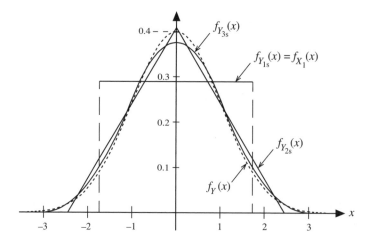

Figure 86.3   Convergence toward a Gaussian distribution when the X_k's are uniform

If the i.d.m.s.i. r.v.'s comprising a sequence $\{X_k\}$ are not absolutely continuous, then the convergence of the partial sums must be examined in terms of their c.d.f.'s, in accordance with Definition 86.1. However, in the discrete case the *envelopes* of the bar graphs for $P_{Y_{ns}}(\{y\})$ will exhibit a trend toward the shape of the Gaussian p.d.f.

Example 86.3

Consider the simple case where $\{X_k\}$ is a sequence of i.d.m.s.i. symmetrical *binary* r.v.'s, assigning probability 0.5 to $\{-1\}$ and $\{1\}$. The X_k's are standardized, so that $X_k = X_{ks}$ for all k. Then, from Equation (86.3),

$$Y_{ns} = \frac{1}{\sqrt{n}} \sum_{k=1}^{n} X_k.$$

Each Y_{ns} is a binomial r.v. (see Example 35.1), but with zero mean and unit variance. Similar to Example 86.2, we examine the convergence of $\{Y_{ns}\}$ to a Gaussian distribution. Since the Y_{ns}'s are discrete, their induced distributions will be described by c.d.f.'s. In this case, convergence cannot be as rapid as in Figure 86.3 because the functions $F_{Y_{ns}}$ consist only of steps. This is illustrated in Figure 86.4 where, superimposed on the standardized Gaussian c.d.f., is shown $F_{Y_{1s}} = F_{X_1}$, as well as $F_{Y_{5s}}$. For any given n, the number of steps in $F_{Y_{ns}}$ is $n + 1$. The step sizes therefore decrease as n increases. Figure 86.4 suggests how the step sizes become more nearly proportional to the slope of F_Y at the location of each step—i.e., proportional to f_Y. In this way, with increasing n, the discrete probabilities $P_{Y_{ns}}(\{y\})$ approach proportionality to $f_Y(y)$ at corresponding y-values.

Another factor in the rapidity of convergence in distribution of the sequence $\{Y_{ns}\}$ is how close the distribution of the X_k's is to being symmetrical. In both Examples 86.2 and 86.3, the X_k's induce a symmetrical distribution. A different case is considered in Problem 86.6, where the X_k's are absolutely continuous but induce an unsymmetrical triangular distribution.

An important application of the Central Limit Theorem is in determining the number of independent trials required in order to estimate the probability of some event within a specified tolerance. The bound given by the Chebychev Inequality (Theorem 65.1) is always applicable in

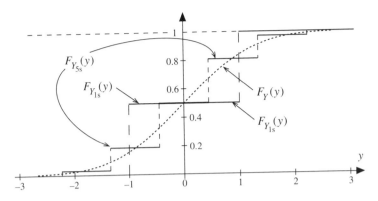

Figure 86.4 Convergence toward the Gaussian c.d.f. when the X_k's are binary r.v.'s

such a situation, since it is not restricted to particular probability distributions. However, we have now seen that the measurement of the relative frequency of occurrence of an event, as expressed by Equation (85.9), approaches a Gaussian distribution—and in many cases fairly rapidly. The Gaussian distribution provides a much tighter estimate, so that the number of trials required for a specified measurement tolerance is greatly reduced when the Gaussian estimate can be applied.

Example 86.4
Suppose a p.e. has been defined in which the relative frequency of an event A is to be estimated. Equivalently, in the associated probability space, the probability of the set or event A corresponding to A is to be estimated—i.e., to be determined experimentally. The probability is to be no greater than 0.05 that the estimate deviates from the true value of $P(A)$ by more than 0.01. We assume that the true value of $P(A)$ is neither extremely small, nor extremely close to 1. Thus, assume $0.05 \leq P(A) \leq 0.95$.

Repeated executions of the underlying p.e. can be combined into a single product experiment. In each component experiment we define the indicator r.v. I_A for the event A. In the kth component experiment ($k = 1, 2, 3, \ldots, n$) let this r.v. be denoted simply I_k. We seek the average

$$Y_n = \frac{1}{n}\sum_{k=1}^{n} I_k.$$

What value of n yields the desired estimate with the specified accuracy?

The Chebychev inequality can be applied by substituting into Equation (65.5b), giving

$$P_{Y_n}\overline{(P(A) - 0.01, P(A) + 0.01^c)} \leq 10^4 \mathbf{V}Y_n.$$

Now, the value of n must be chosen so as to make $\mathbf{V}Y_n \leq 5 \times 10^{-6}$, where $\mathbf{V}Y_n = (\mathbf{V}I_A)/n$. $\mathbf{V}I_A$ depends on $P(A)$. The worst case is when $P(A) = 0.5$, for which $\mathbf{V}I_A = 0.25$. Then $n = 50\,000$. This has to be assumed if it is only known that $P(A) \in 0.05, 0.95$. On the other hand, if $P(A) \approx 0.05$ can be assumed, then $\mathbf{V}I_A$ reduces to 0.0475, giving $n = 9\,500$. Nevertheless, this can still require a formidable effort, depending on the nature of the experiment.

But here we clearly have a situation where the Central Limit Theorem is applicable—that is, F_{Y_n} will be closely approximated by a Gaussian distribution. Proceeding as in Example 40.1, it can be seen from the table of the Gaussian integral that Y_{ns} assigns probability 0.95 to the interval

$-1.960, 1.960$. Then, Y_n needs to have a standard deviation of $0.01/1.960 \doteq 0.00510$, or a variance of 2.6×10^{-5}. The worst case, with $P(A) = 0.5$, now requires $n = 9615$. And if $P(A) = 0.05$, then $n = 1827$ is required.

So we see that the knowledge that Y_n will be very nearly Gaussian significantly reduces the experimental effort that must be specified in order to obtain the desired result, compared to application of the Chebychev inequality. In this Example it is about an 80% reduction.

In Example 86.4 we have excluded the possibility of $P(A)$ being close to zero or close to 1, or equivalently, of P_{I_A} being *extremely* lopsided. The reason is that in such a case an additional consideration enters into the picture. The situation described in Example 84.2 can serve to illustrate this. If

$$A \leftrightarrow \text{'the chance mechanism proclaims a win'}$$

and it is desired to determine experimentally whether indeed $P(A)$ has a value close to 10^{-6}, the acceptable range cannot be specified as ± 0.01, that is, $10^{-6} \pm 0.01$ (as in Example 86.4). After all, we know from the outset that $P(A) \geq 0$, so that ± 0.01 here really means the interval $\overline{0, 0.010001}$. Therefore, this is equivalent to specifying 0.0050005 ± 0.0050005. On the other hand, when estimating very small probabilities, much greater precision is usually desired. Thus, in Example 84.2, even $\pm 5 \times 10^{-7}$ would allow for an unacceptably large departure from the nominal value of $P(A) = 10^{-6}$, because of the large amount of money involved (see also Problem 86.5).[12]

Queries

86.1 In the sequence $\{X_k\}$ defined in Example 86.1, what is the smallest k for which $|F_{X_k} - F_X| \leq 0.01$ for all x?   [223]

86.2 Which of the following sequences of r.v.'s converge in distribution:

a) $\{X_k\}$, where $F_{X_k}(x) = \frac{1}{2}\left[u(x) + u\left(x - (-1)^k\right)\right]$, $u(x)$ denoting the unit step at the origin

b) $\{X_k\}$, where $X_{k+1} = X_k/2$, $k = 1, 2, 3, \ldots$

c) $\{X_k\}$ is a sequence of i.d.s.i. random variables

d) $\{Y_k\}$, where $Y_k = \cos kX$, and where X is uniformly distributed over $\overline{0, 2\pi}$

e) $\{X_k\}$, where $X_{k+1} = X_k + 1$, $k = 1, 2, 3, \ldots$.   [69]

86.3 Consider two sequences, $\{X_k\}$ and $\{Y_k\}$, both of which converge in distribution to a Gaussian r.v. For each of the following, determine whether it is true, possibly true, or false.

a) Let $Z_k = X_k - Y_k$, $k = 1, 2, 3, \ldots$. Then $\{Z_k\}$ converges in distribution to a Gaussian r.v.

b) The sequence $X_1, Y_2, X_3, Y_4, X_5, Y_6, \ldots$ converges in distribution to a Gaussian r.v.

c) The sequence $\{X_k\} + \{Y_k\}$ converges in distribution to a Gaussian r.v.

d) If $W_n = \frac{1}{\sqrt{n}}\sum_{k=1}^{n} X_k$, then $\{W_n\}$ converges in distribution to a Gaussian random variable.   [110]

---

[12] When working on the tails of the Gaussian distribution, a logarithmic plot or table of the Gaussian p.d.f. can be helpful.

Part VI Summary

In Part VI a variety of topics were treated that build on the earlier study of r.v.'s in Parts III–V. Complex random variables were introduced in Section 75, which served as a basis for the discussion of the characteristic function of a r.v., and of the closely related generating function. It was seen how these constructs, which are standard tools in Probability Theory, can be useful in some manipulations of r.v.'s.

Following up on earlier discussions of Gaussian r.v.'s, a more thorough coverage of these important functions was provided in Section 80. Section 81 opened the door to the design of entire families of n-dimensional r.v.'s that are 'Gaussian-like' in that they induce spherically symmetric distributions. The marginal distributions of such r.v.'s are not initially specified.

In Section 82, the concept of 'entropy' as a measure of randomness, first introduced in Section 32, was extended to discrete as well as absolutely continuous r.v.'s. This material is basic to the development of Information Theory, but it also provides insights into the nature of r.v.'s.

Copulas were explored in Section 83. It was seen how a copula expresses in full detail the relationship between any two r.v.'s. Copulas represent an interesting tool for designing a two-dimensional r.v. that not only has specified marginal distributions, but whose joint distribution incorporates a precisely specified dependence.

Finally, the last three Sections, concerned with infinite sequences of random variables, brought us back to topics of central importance to our study. First, gambling examples served to illustrate basic features of such sequences. Different kinds of convergence of sequences of r.v.'s, and the important Laws of Large Numbers, were discussed next. The Laws of Large Numbers were seen to give a mathematical stamp of approval, so to speak, to the conceptual model developed in Part I. Finally, convergence 'in distribution' and the Central Limit Theorem were treated in Section 86. The Central Limit Theorem makes it clear why Gaussian r.v.'s play such a fundamental role in Probability Theory.

Appendices

Answers to Queries

2. 2.5, 6.25
3. 0.5, 2^{-10}, 0
4. 0.402, 0.137
5. 0.7, 1.0
6. 2/3, 2
7. 1/2, 0
8. 4
9. 2
10. 3
11. 1.7
12. ± 1
13. 1/2
14. 7/8
15. 7
16. 0, 0
17. 1/8
18. 1/4
19. 2, 3
20. 27, 6
21. 1/120
22. $-1, -1$
23. 0, 6
24. 0.5, 0.5
25. $1/\sqrt{2}$, 0
26. 4 or 8 or 16
27. $\{(x_1, x_2): x_1 \leq -|x_2|\}$
28. no
29. 4, 16, 10, yes
30. 0.2, 0.2

31. 1/4, 1/4, 0, 1/2
32. $\{7\}$, $\{9\}$, $\{4, 6\}$, $\{1, \ldots, 10\}$
35. yyyyn; b, c, d
36. a, d
38. a, c, d
40. d
41. a, b
42. YNYYYN
43. NYYY
44. YNNN
45. j
46. e
47. a, c
48. b
49. a, f
50. d, e
52. NYYNY
53. NYN
55. YYYY
56. NYNN
57. 0, YN
59. a, b, c
61. ade, abd, e
62. c, d
63. d, b, a, c
64. c, e
65. a, b, d
66. b, d, f
67. c, d, e

Probability Concepts and Theory for Engineers, First Edition. Harry Schwarzlander.
© 2011 John Wiley & Sons, Ltd. Published 2011 by John Wiley & Sons, Ltd.

68.	a, c	124.	FTFT
69.	b, c, d	125.	PPPP
70.	b	126.	TPTF
71.	a, b, c	127.	FPPPT
73.	1/3	128.	TFTP
74.	1/9	129.	PPFT
75.	1/12	130.	FTFP
76.	7/12	131.	TFPT
77.	5/54	132.	TPTF
78.	B^c	133.	FPPP
81.	$B - \{\xi_3\}$	134.	PPFPT
82.	4	135.	FFTTF
84.	0.36	136.	TPTFT
86.	0.5	137.	PFPTT
87.	2.0	138.	PTFPPP
88.	0.1	139.	FTFFT
89.	0.6, 0.7	140.	TFTT
90.	0.75	141.	4951; 19405
91.	0.125	142.	1/4, 1/2, 1/4
92.	2/3, 1/3	143.	5/6, 0, 1/6, 1/3
94.	A_1, A_2, F	145.	1/4, 1/4, 1/2
95.	neither	146.	2.146, 3.035
96.	TTMTF	147.	0.234, 0.377
97.	TPFTM	148.	0, 41/15
98.	TMPMFTP	149.	5/36, $-1/36$
99.	PFPTPMMT	150.	13/36, 3/13
101.	PFPTTP	151.	0.25, 0.125
102.	PPMFPT	152.	3, 1
103.	MFTMT	153.	2^{-9}, 2^{-9}, 2^{-4}
104.	PTTPP	154.	-2
105.	TMPPP	155.	1.25
106.	PFPTT	156.	$1/\sqrt{8}$
107.	FFTMTP	158.	$x/2$, $x + 2$, $\sqrt{|x/2|}$
108.	FPFTTPT	161.	$1/n(n-1)$
109.	MTPTT	162.	$m + n - 1$, $m + n - 1$
110.	PPPP	164.	3.5, 4, 4.3, 13/3, 8/3
111.	$\pm(1+j)\, 0.1/\sqrt{2}$	165.	1/6, (0.5, 2/3), 1/3
113.	0, 1, 0, 1	166.	1/120, 0
114.	0, undef., 1, 1.25	167.	5, 1, $-$, $-1/3$, $-$
115.	-0.5, 0.5, -1, 0.5	168.	0.125, 0.07, 0, 0.5
116.	2, 1&3, $-$, 1&2	169.	0.2, 0.3, 0.5
117.	1, 2, n, b, b	170.	0.5, 1, 0
118.	FFTT	171.	5.5, 1
119.	PTTFT	172.	$1.5 + j$, 5/12
120.	PFFP	173.	1/4, 0, >0, -6
121.	TPTT	175.	4, ∞; 0, ∞
122.	n, s, n, e or s, s, s	176.	0.688, 1, 0, 0.266
123.	TFPT	178.	$-1/3$, 1, $-1/3$, 1

179. 01, 1234, 0, 34
180. $0.1(1 - j)$
181. $\overline{2, \infty} \times \overline{-\infty, -1/2}$
183. 8, n, n, n, n
184. 2; 123; 1; c; 2; c
185. n, n, −1, n
186. 0, 1, 0.5, 0.5, 0.1
187. 0, 1, 0, 2/27
188. 3, 1, 0, 2
189. $\overline{0.25, 0.75}$
190. 0, 0.5
191. 4
192. 4
193. 7
194. 2
195. 3
196. 0
197. 1
198. 1/2
199. 0
200. 5
201. 0.4
202. a
203. a
204. c
206. a
207. d
208. b
211. d
212. a
213. c
214. 6/21
215. 1/15
216. 1/4

217. 5
218. 1
219. 2
220. 14
221. 3
222. 128
223. 100
224. 27
225. 0, 1
226. 7, 1
228. 13, 1
229. 8, 4
230. 4, 5
231. YNNYY
232. TFTPT
233. N, Y
234. TPFT
235. PFT
236. FFPT
237. TFPPP
238. PTFTT
239. PPTTT
240. PTTPP
241. PPPTF
242. PTTF
243. FPPP
244. TPT
245. PTTP
246. FFTP
247. FTTF
248. FFF
249. TFT
250. FTPT

Table of the Gaussian Integral, $\Phi(x) = \int_{-\infty}^{x} \frac{1}{\sqrt{2\pi}} \exp\left(-\frac{x^2}{2}\right) dx.$

x	$\Phi(x)$	x	$\Phi(x)$	x	$\Phi(x)$	x	$\Phi(x)$
0.00	0.5000	0.74	0.7703	1.48	0.9306	2.22	0.9868
0.02	0.5080	0.76	0.7764	1.50	0.9332	2.24	0.9875
0.04	0.5160	0.78	0.7823	1.52	0.9357	2.26	0.9881
0.06	0.5239	0.80	0.7881	1.54	0.9382	2.28	0.9887
0.08	0.5319	0.82	0.7939	1.56	0.9406	2.30	0.9893
0.10	0.5398	0.84	0.7995	1.58	0.9429	2.32	0.9898
0.12	0.5478	0.86	0.8051	1.60	0.9452	2.34	0.9904
0.14	0.5557	0.88	0.8106	1.62	0.9474	2.36	0.9909
0.16	0.5636	0.90	0.8159	1.64	0.9495	2.38	0.9913
0.18	0.5714	0.92	0.8212	1.66	0.9515	2.40	0.9918
0.20	0.5793	0.94	0.8264	1.68	0.9535	2.42	0.9922
0.22	0.5871	0.96	0.8315	1.70	0.9554	2.44	0.9927
0.24	0.5948	0.98	0.8365	1.72	0.9573	2.46	0.9931
0.26	0.6026	1.00	0.8413	1.74	0.9591	2.48	0.9934
0.28	0.6103	1.02	0.8461	1.76	0.9608	2.50	0.9938
0.30	0.6179	1.04	0.8508	1.78	0.9625	2.52	0.9941
0.32	0.6255	1.06	0.8554	1.80	0.9641	2.54	0.9945
0.34	0.6331	1.08	0.8599	1.82	0.9656	2.56	0.9948
0.36	0.6406	1.10	0.8643	1.84	0.9671	2.58	0.9951
0.38	0.6480	1.12	0.8686	1.86	0.9686	2.60	0.9953
0.40	0.6554	1.14	0.8729	1.88	0.9699	2.62	0.9956
0.42	0.6628	1.16	0.8770	1.90	0.9713	2.64	0.9959
0.44	0.6700	1.18	0.8810	1.92	0.9726	2.66	0.9961
0.46	0.6772	1.20	0.8849	1.94	0.9738	2.68	0.9963
0.48	0.6844	1.22	0.8888	1.96	0.9750	2.70	0.9965
0.50	0.6915	1.24	0.8925	1.98	0.9761	2.72	0.9967
0.52	0.6985	1.26	0.8962	2.00	0.9772	2.74	0.9969
0.54	0.7054	1.28	0.8997	2.02	0.9783	2.76	0.9971
0.56	0.7123	1.30	0.9032	2.04	0.9793	2.78	0.9973
0.58	0.7190	1.32	0.9066	2.06	0.9803	2.80	0.9974
0.60	0.7257	1.34	0.9099	2.08	0.9812	2.82	0.9976
0.62	0.7324	1.36	0.9131	2.10	0.9821	2.84	0.9977
0.64	0.7389	1.38	0.9162	2.12	0.9830	2.86	0.9979
0.66	0.7454	1.40	0.9192	2.14	0.9838	2.88	0.9980
0.68	0.7517	1.42	0.9222	2.16	0.9846	2.90	0.9981
0.70	0.7580	1.44	0.9251	2.18	0.9854	2.92	0.9982
0.72	0.7642	1.46	0.9279	2.20	0.9861	2.94	0.9984

x	$\Phi(x)$	x	$\Phi(x)$	x	$\Phi(x)$
2.96	0.99846	3.40	0.999663	4.00	0.9999683
2.98	0.99856	3.50	0.999767	4.20	0.9999867
3.00	0.99865	3.60	0.999841	4.40	0.9999946
3.10	0.99903	3.70	0.999892	4.60	0.9999979
3.20	0.99931	3.80	0.999928	4.80	0.9999992
3.30	0.99952	3.90	0.999952	5.00	0.9999997

Part I Problems

Section 1

1.1. In Example 1.1 there is mention of conceptual, mathematical and procedural errors. Consider a situation where a waitress in a restaurant comes to the bartender with an order for five martinis. The bartender is very busy preparing a large number of drinks. Give an example of a conceptual error, a mathematical error, and a procedural error that might be made by the bartender in filling that order.

1.2. At a small hospital, on a certain day, there were 17 patients occupying beds all day (i.e., midnight to midnight). In addition, three patients were discharged during the morning, and two new patients were admitted in the afternoon. Also, one patient who had received care for several days died in the evening. Furthermore, in the morning a young woman was admitted and gave birth to twins around noon. She and the twins did not remain overnight.

　　As the hospital administrator, you have to fill out a form in which one of the questions asks for the number of patients on that day. Describe at least two different ways of conceptualizing the wording 'patients on that day' under the circumstances as described, leading to different answers to the question.

1.3. a) Produce a sketch of an object that you would expect about half the students in a class to interpret as a chair, the other half as something else.
　　b) Suppose you want to test your sketch on a class, or some other group of people. How might you go about doing this so that your method will not bias the subjects?
　　c) Actually carry out the test.

Section 2

2.1. You see a man come out of his house, pause, look up and about him, then quickly disappear back into his house only to reappear again shortly with an umbrella, and depart. You suspect that you have seen a probabilistic experiment in progress.

　　a) Describe what this probabilistic experiment may have been.
　　b) How can you confirm whether this was the actual experiment?

2.2. You are playing bridge. (It could also be some other card game that involves taking tricks.) With two cards left in your hand (call them A and B) it is your play. Describe the probabilistic experiment that you will conduct at this point. (Think this over carefully!)

2.3. Model parts (b) and (c) of Problem 1.3 as a probabilistic experiment.

Section 3

3.1. Consider the following experiment: Two dice, one green and one red, are thrown in order to come up with a rational number. The procedure is to throw the dice and then take the ratio of the number of dots showing on the red die to the number of dots showing on the green die. Each possible outcome is therefore a different rational number. How many different possible outcomes are there?

3.2. An experiment is performed for the purpose of obtaining a number 'at random'. The procedure is to throw n dice simultaneously and add the dots facing up on all n dice.

a) Fill in the complete description of the experimental plan.
  (Note: If n appears in your statement of the experimental plan, explain why this is not inconsistent with the requirement of Definition 2.1 that calls for an unambiguous specification of all the circumstances.)
b) How many different possible outcomes are there?

3.3. A secretary wants to find out whether a certain document is contained in her four-drawer file cabinet.

a) Knowing from experience that in a single search of the cabinet there is a 10 per cent chance of overlooking the desired item, she starts the search with the intention of repeating the search once if she does not come across the desired document the first time. Describe the experimental plan and identify the possible outcomes.
b) Rather than first searching the whole cabinet once, she decides to search the first drawer twice, then the second twice, etc. Again describe the experimental plan and identify the possible outcomes.
c) Is it possible that execution of the procedure you specified under (a) or (b) can result in an unsatisfactorily performed experiment, and if so, how?

3.4. Describe the experimental plan of two p.e.'s that illustrate the two extremes considered in Query 3.2.

Section 4

4.1. Consider the experimental plan for the die-throwing experiment in Section 2.

a) Which of the following describe possible outcomes of this experiment:

  'Less than six dots face up'
  'six dots face up'
  'More than six dots face up'
  'At least one dot faces up'.

b) Which of the above statements define events for this experiment?
c) Which of the statements listed under (a) are
  i) impossible events?
  ii) elementary events?
  iii) certain events?
d) Of the following events, what is the largest number and the smallest number that can occur simultaneously when the experiment is performed. Justify your answer.

  'At least one dot faces up'
  'An even number of dots faces up'
  'Five dots face up'
  'Seven dots face up'.

4.2. A game is played in which n black dice and m white dice are thrown simultaneously (where n and m are suitable numbers that have been specified). The total number of dots showing on the black dice and also the total number of dots showing on the white dice is observed.

a) How many different outcomes are possible upon carrying out this procedure?
b) If the playing of this game actually depends on the amount by which the two totals differ (and also which is the larger), how many *events* are of interest in the throw of the $n + m$ dice?

4.3. In a bowling alley, there are 100 pairs of bowling shoes for rent. At some time during an evening while bowling is in progress, the attendant decides to count the number of pairs left on his shelves (i.e., not rented out).

a) Model the attendant's activity as a probabilistic experiment.
b) How many possible outcomes are there?
c) Suppose he performs his count and finds that there are 21 1/2 pairs of shoes left on his shelves; that is 21 pairs and one single left shoe without a matching right shoe. Is this result one of your possible outcomes, is it an event, or how does it fit into your model?
d) Suppose that while the attendant is counting, someone returns a pair of shoes. How does your model take account of this—for instance, should the returned pair be included in the count or not?

Section 5

In these Problems, specify an appropriate conceptual model (not necessarily a p.e.) for the activity that is described, and the way in which that model is brought into correspondence with the mathematical world. Where appropriate, point out simplifications or 'idealizations' that take place in going from the real-world situation to the conceptual model. How do these play a role in the interpretation of computed results?

5.1. Counting the chairs in a particular room.
5.2. Determining the floor area of a rectangular living room, for the purpose of arriving at the cost of wall-to-wall carpeting.
5.3. Determining the rate of change of the liquid level in a circular conical tank as a function of the rate of liquid flow into and out of the tank.
5.4. Expressing in some suitable way the total cost of one's car in relation to its utility.

Section 6

6.1. Describe the following sets in words:

a) $\{\xi_1, \xi_2, \xi_3\}$
c) $\{\xi: \xi = \xi_1\}$

b) $\{\xi: \xi \neq \xi_1, \xi_2\}$
d) $\{\xi: \xi \in A \text{ and also } \xi \notin B\}$.

6.2. Write in symbolic form (using braces, as in Problem 6.1):

a) The set of elements that do not belong to the set A.
b) The set consisting of all elements except the element ξ_1.
c) The set of those elements in A that also belong to B.
d) The set of those elements of A that do not belong to A.

6.3. Simplify the following expressions:

a) $\{\xi: \xi \in \{\xi_1\}\}$
c) $\{\xi: \{\xi\} \subset \{\xi\}\}$.

b) $\{\xi: \{\xi\} \subset A\}$

6.4. Let A, B, C be subsets of a space S. Prove the following:

a) $A \subset B$ and $B \subset C$ implies $A \subset C$.
b) $A = B$ and $B = C$ implies $A = C$.
c) $A \subset B$ if and only if, for all elements ξ, $\xi \notin B$ implies $\xi \notin A$.

6.5. Prove that two sets are the same set if and only if they are composed of the same elements. More precisely, prove that the following two statements are equivalent (each implies the other):

a) $A = B$
b) $\xi \in A$ if and only if $\xi \in B$, for every element $\xi \in S$.

6.6. In each of the following three cases, develop a set-theoretic model that describes the specified situation. Explain the correspondences you are setting up, that is, what do the elements represent, what do various sets represent, what does the space represent.

a) A desk with six drawers. (Suggestion: You might consider a space with six subsets, or a space with 6 elements. Does your model distinguish between the collection of all six drawers, and the desk with all its drawers? Should it?) Arrange your model in such a way that you can incorporate in it the information 'the lower left drawer and the upper right drawer are open'. (Suggestion: Define a set $O \leftrightarrow$ 'the set of open drawers'.)
b) A specific English sentence; the words making up the sentence; and the letters of the alphabet.
c) Graphs of the following equations, for $0 \leq x \leq \pi$, $0 \leq y \leq 1$:

$$y = \sin x \qquad y = x/\pi$$
$$y = |\cos x| \qquad y = 0.5$$

The intercepts between any pair of these graphs is also to be expressed in the model.

Section 7

7.1. Give examples of hierarchies of real-world objects or concepts that can be made to correspond to elements, sets and classes. (For instance, the elements of a space might represent the apples from a particular orchard, sets represent crates of apples, and a class represents a truckload of crates full of apples.)

7.2. [Ha1] Given a space S, let $\mathscr{C}$ be the collection of all subsets of S (including the empty set ϕ and S itself). Furthermore, let ξ be some point in S, A some subset of S, and $\mathscr{E}$ some class of subsets of S. Thus, A is a member of $\mathscr{C}$, and $\mathscr{E}$ is a subclass of $\mathscr{C}$. Now suppose α and β can each stand for any of the five symbols ξ, A, S, $\mathscr{E}$, $\mathscr{C}$. Then, there are fifty different expressions that can be implied by

$$\alpha \in \beta \text{ or } \alpha \subset \beta.$$

Some of these expressions are necessarily true, some are possibly true, some are necessarily false, and some are meaningless. For instance, $\alpha \in \beta$ is meaningless unless the right term is a subset of a space in which the left term is a point, and $\alpha \subset \beta$ is meaningless unless both α and β are subsets of the same space.

For each expression, indicate which is the case by filling in the letter T (necessarily true), P (possibly true), F (necessarily false), or M (meaningless) in the appropriate box in Figure A.1.

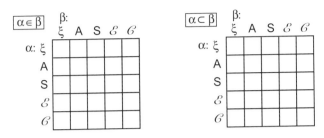

Figure A.1 Charts for Problem 7.2

7.3. *Partial ordering.* A relation r among the sets of a class $\mathscr{C}$ is a 'partial ordering', or $\mathscr{C}$ is a 'partially ordered class' if, for any sets A, B, C $\in \mathscr{C}$, the following three requirements are satisfied:

a) A r A
b) If A $\neq$ B and A r B, then it is not true that B r A
c) If A r B and B r C, then A r C.

Verify that the inclusion relation $\subset$ among the sets making up a class of sets $\mathscr{C}$ constitutes a partial ordering on $\mathscr{C}$.

7.4. Given a space S and a nonempty set A, let $\mathscr{P}_1$ and $\mathscr{P}_2$ be two partitions of the set A. Verify the following:

a) $\mathscr{P}_1 \subset \mathscr{P}_2$ if and only if $\mathscr{P}_1 = \mathscr{P}_2$.
b) The class $\mathscr{C} = \{C \colon C \in \mathscr{P}_1 \text{ but } C \notin \mathscr{P}_2, \text{ or } C \in \mathscr{P}_2 \text{ but } C \notin \mathscr{P}_1\}$ is a disjoint class if and only if $\mathscr{P}_1 = \mathscr{P}_2$.
c) The class $\mathscr{C}$ of part (b) cannot consist of exactly one set.

Section 8

Note: Venn diagrams can be used to guide your thinking but not as a method of proof.

8.1. Let S be the space that models a particular p.e., and let $\xi_o \in S$ be the element that represents the actual outcome of the p.e. If A, B, C denote any three events, then $\xi_o \in A$ means that the event A occurred, $\xi_o \in B$ means that the event B occurred, and so on. Express the following in symbolic form and illustrate each case with a Venn diagram:
Of the three events A, B, C,

a) all three occurred
b) at least two occurred
c) at least one occurred
d) exactly one occurred
e) the event A did not occur
f) none of the events occurred
g) either both A and B occurred or C occurred, but not all three of the events A, B, C
h) no more than two occurred.

8.2. Consider the checker board shown in Figure A.2, with checkers placed in the starting position for a game of checkers. Let the space S represent the 64 squares of the checker board. Consider

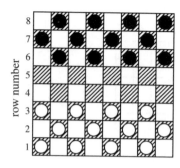

Figure A.2   Checker board for Problem 8.2

the following sets:

$$B \leftrightarrow \text{'the dark squares'}$$
$$R_i \leftrightarrow \text{'the } i\text{th row of squares' } (i = 1, \ldots, 8).$$

Express the following in terms of $B, R_1, R_2, \ldots, R_8, S,$ and ϕ:

a) The white squares
b) The squares occupied by a white checker
c) The squares occupied by no checkers
d) All those squares that are directly in front of a white checker (on the diagram, *above* a white checker)
e) The white squares occupied by a checker.

8.3. Let the space S be the collection of all positive integers, and consider the following subsets:

$$A = \{\xi : \xi \text{ is an even integer}\}$$
$$B = \{\xi : \xi \text{ is a multiple of 3}\}$$
$$C = \{\xi : \xi > 9\}$$
$$D = \{\xi : \xi < 7\}.$$

Express the following sets in terms of A, B, C, D:

a) $\{\xi : \xi \text{ is a multiple of 6}\}$ b) $\{1, 2, 3, 4, 5\}$
c) $\{1, 5, 7\}$ d) The set of odd integers greater than 10
e) $\{7\}$.

8.4. Given sets $A, B, C, D \subset S$, such that $A \subset B$ and $C \subset D$. Using the definition of a set and the meanings of $\cup$ and $\cap$ as given in Definition 8.1, show that:

a) $A \cup B = B$ b) $A \cap B = A$
c) $AC \subset BD$ d) $A \cup C \subset B \cup D$.

8.5. Verify Equations (8.1) through (8.10) (in any order) using the definition of a set, the meanings of $\cup$, $\cap$, and ^{c} as given in Definition 8.1, and the results of Problem 8.4. (Once a particular equation has been verified, it may be used in verifying others.)

8.6. Given sets $A, B, C \subset S$. Explain the significance of the relations given below. In other words, in each case try to arrive at a simpler statement than the one given, and having the same meaning.

a) $ABC = A$
b) $A \cup B \cup C = A$
c) $A^c \cup B^c = A \cup B$.

8.7. Given three sets A, B, C $\subset$ S. For each of the following statements, determine under what condition it is true. Illustrate each part with a Venn diagram.

  a) $A \cup B = A^c B$ (Answer: True if and only if $A = \phi$.)
  b) $ABC^c \subset A \cup B$
  c) $(A \cup B)^c C \subset A \cup B \cup C^c$
  d) $(AB) \cap (B^c C) = \phi$
  e) $ABC^c \cup A^c BC = AB \cup BC$.

8.8. a) Given the subsets A, B of a space S, where $A \subset B$. Show that $B^c \subset A^c$.
  b) Given three sets A, B, C $\subset$ S, where A, B are disjoint sets and also A, C are disjoint sets. Show that $A \cup B = A \cup C$ if and only if $B = C$.

8.9. Given subsets A, B of a space S. Show that the following four statements are equivalent:

  a) $B^c \subset A^c$ b) $A = AB$
  c) $B = A \cup (A^c B)$ d) $A \subset B$.

8.10. The 'Scheffer stroke' operation applied to two sets A, B $\subset$ S is defined as follows:

$$A|B \equiv \{\xi : \xi \notin A \cap B\} = (AB)^c$$

Use only the stroke operation to express:

  a) A^c b) $A \cap B$ c) $A \cup B$.

Section 9

9.1. Let S be the set of integers from 1 to 10, and consider the subsets of S

$$A_i = \{n : n \le i\}, \text{ for } i = 1, 2, \ldots, 10$$

so that $A_{10} = S$. Express the following in terms of two or more of the A_i's, using only directed difference and symmetric difference:

  a) $\{7\}$ c) $\{1, 3\}$
  b) $\{8, 9, 10\}$ d) $\{4, 6, 8\}$.

9.2. Given non-empty sets $A_1, A_2, A_3, \ldots, A_n \subset S$. Let $B_n = A_1 \triangle A_2 \triangle A_3 \triangle \ldots \triangle A_n$. For arbitrary $n \ge 2$, show that if B_n is non-empty, it is the set of all those elements that belong to an odd number of the sets $A_1, \ldots, A_n$.

9.3. Given three sets A, B, C $\subset$ S. Show that $(A - B) - C \subset A - (B - C)$. What relationship must hold among the sets A, B, C so that $(A - B) - C = A - (B - C)$?

9.4. Show that the following is an equivalent definition of an additive class:
  A nonempty class $\mathscr{A}$ of subsets of S is an additive class of sets if and only if

  a) $S \in \mathscr{A}$; and
  b) for every pair of sets A, B in $\mathscr{A}$, the set $A - B$ is also in $\mathscr{A}$.

9.5. *Lattice.* Let a class $\mathscr{C}$ with relation r among its sets be a partially ordered class (see Problem 7.3). Then $\mathscr{C}$ is a *lattice* if every two sets of $\mathscr{C}$ have a unique 'least upper bound' and a unique 'greatest lower bound'. For A, B $\in \mathscr{C}$, some set C $\in \mathscr{C}$ is an upper bound if A r C and B r C. It is a *least upper bound* of A, B if there is no other upper bound D of A, B such that D r C. For A, B $\in \mathscr{C}$, some set E $\in \mathscr{C}$ is a lower bound if E r A and E r B. It is a *greatest lower bound* of A, B if there is no other lower bound F of A, B such that E r F.

Verify that an additive class $\mathscr{A}$, with the inclusion relation $\subset$ among its members, is a lattice.

9.6. *Distributive lattice.* A class $\mathscr{C}$ is a 'distributive lattice' if it is a lattice (see Problem 9.5), and two binary operations x, y are defined on its elements such that x distributes over y, and y distributes over x. That is, if A, B, C $\in \mathscr{C}$, then

$$(A \times B) \, y \, C = (A \, y \, C) \times (B \, y \, C), \text{ and}$$

$$(A \, y \, B) \times C = (A \times C) \, y \, (B \times C).$$

Verify that an additive class $\mathscr{A}$, with the inclusion relation $\subset$ among its members, and with the operations $\cup$ and $\cap$ defined on $\mathscr{A}$, is a distributive lattice.

9.7. *Boolean algebra.* Consider a class $\mathscr{C}$ that is a distributive lattice, with relation r and binary operations x, y (see Problem 9.6). $\mathscr{C}$ is a *Boolean algebra* if there is a complementation operation such that for A, B $\in \mathscr{C}$, if B is the *complement* of A, then A x B equals the 'universal upper bound' and A y B equals the 'universal lower bound'. Some C $\in \mathscr{C}$ is a *universal upper bound* if for every set K $\in \mathscr{C}$, K r C. Some D $\in \mathscr{C}$ is a *universal lower bound* if for every K $\in \mathscr{C}$, D r K.

Verify that an additive class $\mathscr{A}$ with the inclusion relation $\subset$ among its members and the operations $\cup$, $\cap$ and ^{c} constitutes a Boolean algebra.

9.8. Verify Equations (9.1) through (9.8), by using Equations (9.9) and (9.10) and the properties given in Section 8. Once a particular equation has been established, it can be used in verifying the remaining ones.

9.9. Given sets A, B, C $\subset$ S. Explain the significance of each of the following statements:

a) $A \cup B = A - B^{\text{c}}$
b) $A \, \Delta \, B \, \Delta \, C = A$
c) $(A - B) \cup C = A.$

9.10. Given A, B $\subset$ S. Show that the following three statements are equivalent:

a) $A \, \Delta \, B \neq \phi$
b) At least one of the sets $A - B$, $B - A$ is nonempty
c) $A \neq B.$

9.11. Given a space S and sets A, B, C $\subset$ S. Prove that the following set equations are identities, and illustrate each by a Venn diagram.

a) $A \, \Delta \, B = (A \cup B) - AB = AB^{\text{c}} \, \Delta \, BA^{\text{c}} = A^{\text{c}} \, \Delta \, B^{\text{c}}$
b) $A \cup B = (A \, \Delta \, B) \, \Delta \, AB$
c) $A - B = (AB) \, \Delta \, A = (A \, \Delta \, B)A$
d) $ABC = A - [A - [B - (B - C)]].$

9.12. Given a space S and sets A, B, C, D $\subset$ S. For each of the following assertions, determine the conditions under which it is true.

a) $(A \cup B) - B = A$
b) $A \triangle B \triangle C = AB^c C^c \cup A^c BC^c \cup A^c B^c C$
c) $A \triangle B = C \triangle D$ implies $A \triangle C = B \triangle D$.

(Suggestion: For (c) make use of Problem 9.10.)

Section 10

10.1. Given the space $S = \{0, 1, 2, 3, 4, 5, 7, 10\}$. On S, let the real-valued point functions G_1 and G_2 be defined that assign values as follows:

$$G_1(x) = 1 + x/2, \text{ for } x \in S$$
$$G_2(x) = |x - 5|, \text{ for } x \in S.$$

a) Draw a graph of G_1. What is the range of G_1?
b) Draw a graph of G_2. What is the range of G_2?
c) Is it meaningful to speak of $G_1 G_2$? Of $G_2 G_1$?

10.2. A grocery shelf is stocked early in the morning with cans of different soups: 100 cans of one type, marked 'five for $1.76', and 100 cans of another kind marked 'three for $1.48'. (This means that if five cans of the first type or three cans of the other type are not bought together, they cost 36¢ and 50¢ each, respectively.) At the end of the day, the total amount of money taken in for soup is determined.

If the initial stock of cans of soup is represented by a space of elements, explain how a function might be defined so that it can be used to express the amount of money taken in for soup sold in one day. (Should it be a point function or a set function?)

10.3. Given a space S and a set $A \subset S$. Let I_A be a point function that assigns the value 1 to every element ξ for which it is true that '$\xi \in A$', and otherwise assigns the value 0. The point function I_A is called the *indicator function* of the set A.

Consider the indicator functions I_A and I_B of two arbitrary sets A, B $\subset$ S. Perform the various set operations of Section 8 and 9 on the sets A and B and in each case, express the indicator function of the resulting set in terms of I_A and I_B.

10.4. Set operations can be interpreted as functions. Given a space S and a set $A \subset S$. On the class of all subsets of S can be defined a set function G_1 that assigns to every subset B of S the set $A \cap B$. In other words, G_1 is a 'set-valued set function'. Similarly, let G_2 be the function that assigns to every subset B of S the set $A \cup B$.

a) Specify the *range* of each of the set functions G_1 and G_2.
b) What is the range of $G_1 G_2$? Of $G_2 G_1$?

Section 11

11.1. A space S has ten elements. Let $\mathscr{C}$ be the class of all the distinct subsets of S that are of size 5.

a) Determine $|\mathscr{C}|$.

b) Consider three distinct sets A, B, C $\in \mathscr{C}$. What is the smallest possible size of A$\cup$B$\cup$C? The largest possible size?

c) Consider again three distinct sets A, B, C $\in \mathscr{C}$. What is the smallest possible size of A$\cap$B$\cap$C? The largest possible size?

11.2. Specify an experiment involving the tossing of a coin, for which the sample space is countably infinite.

11.3. Prove the following:

a) A subset of a countable set is countable.

b) Let S be an uncountable space, and A a countable subset of S. Then A^c is uncountable.

11.4. Let n be the size of a finite space S.

a) For what values of n is the following true: There exists a partition $\mathscr{P}$ whose members have sizes 1, 2,..., k.

b) Demonstrate the following. If S is such that there exists a partition $\mathscr{P} = \{A_1,..., A_k\}$ of the type described in (a), then there exists also another partition $\mathscr{Q} = \{B_1,..., B_k\}$ such that every element $\xi \in$ S is completely identified by these two numbers: the size of the set $A_i \in \mathscr{P}$ to which ξ belongs, and the size of the set $B_j \in \mathscr{Q}$ to which ξ belongs.

Note: A generalization of this problem, and its application to the testing of telephone cables, is discussed in [Gr1]. Specifically, Graham proves that for every finite space S of size other than 2, 5 or 9, there exist two partitions, $\mathscr{P}$ and $\mathscr{Q}$, such that every element ξ of the space is uniquely identified by the sizes of the set in $\mathscr{P}$ and the set in $\mathscr{Q}$ to which that element belongs.

11.5. Explain how an infinite table of distinct infinite-length coin-toss sequences can be specified such that the sequence T H T H T H T H... cannot possibly be present in this table.

11.6. An infinite-length coin-toss sequence is *periodic* with period n ($n < \infty$) if n is the smallest positive integer such that the first n positions of the sequence form a pattern of T's and H's that repeats every n positions. For example, in the sequence

$$\text{T T H T T H T T H T T H...}$$

the pattern TTH keeps repeating, so that this sequence is periodic with period 3.

a) For $n = 1, 2, 3$, how many distinct periodic sequences of period n are there?

b) Verify that the collection of *all* periodic sequences is countably infinite.

c) Try out the diagonal method of Example 11.1 on the collection of all periodic sequences and show that it fails to demonstrate that this collection is uncountable.

Section 12

12.1. Consider the space I_+ of positive integers, and the sets $A_j \subset I_+$, where $A_j = \{1, 2, 3,..., j\}$, for $j = 1, 2, 3,...$. Express the following sets in terms of countable set operations performed on sets from the class $\{A_j: j = 1, 2, 3,...\}$.

a) B $= \{2\}$

b) C $= \{3, 4, 5\}$

c) D $= \{10, 11, 12, 13,...\}$

d) E $= \{1, 3, 5, 7, 9,...\}$.

12.2. Prove that a finite union of finite sets is finite.

12.3. Use Definition 12.1 to prove Theorem 12.1.

12.4. Consider a countable class $\mathscr{C}$ of subsets of a space S, and let B be another subset of S.

    a) If every set in $\mathscr{C}$ is a subset of B, show that

$$\bigcup_{A \in C} A \subset B.$$

    b) If B is a subset of every set in $\mathscr{C}$, show that

$$B \subset \bigcap_{A \in C} A.$$

12.5. Let $\mathscr{P} = \{B_1, B_2, B_3, \ldots\}$ be a partition. Show that

$$\bigcup_{i=1}^{\infty} A B_i = A.$$

12.6. Consider the class $\mathscr{C} = \{A_1, A_2, \ldots, A_n\}$ consisting of n subsets of a space S, as well as the classes $\mathscr{C}_1, \ldots, \mathscr{C}_n$, where $\mathscr{C}_k$ (for $k = 1, \ldots, n$) is the class of all k-fold intersections of sets in $\mathscr{C}$. In other words, for $k = 1, \ldots, n$,

$$C_k = \left\{ B : B = \bigcap_{j=1}^{k} A_{m_j}, m_j \in \mathsf{N} \text{ and } m_j \neq m_\ell \text{ for } j \neq \ell \right\}$$

where N denotes the set of integers $\{1, \ldots, n\}$. In particular, $\mathscr{C}_1 = \mathscr{C}$, and $\mathscr{C}_n$ consists of the one set $A_1 \cap A_2 \cap \ldots \cap A_n$ (which may be empty).

    a) Let C_k denote the set of all those elements which belong to *at least k of the sets* in $\mathscr{C}$ (for $1 \leq k \leq n$). Write a general expression for C_k in terms of $\mathscr{C}_1, \ldots, \mathscr{C}_n$. Then, for the cases $k = 1$ and $k = n$, simplify this general expression to obtain formulas involving only A_1, $A_2, \ldots, A_n$.

    b) Let E_k denote the set of all those elements that belong to *exactly k of the sets* in $\mathscr{C}$ (for $0 \leq k \leq n$). Write a general expression for E_k in terms of the sets $C_1, \ldots, C_n$.

    c) Let D_k denote the set of all those elements that do *not* belong to *at most $k - 1$ of the sets* in $\mathscr{C}$ (for $1 \leq k \leq n$). Express $D_k{}^c$ in terms of $E_0, E_1, \ldots, E_n$.

    d) Using set manipulations, show that $C_k = D_k$ (for $1 \leq k \leq n$).

Section 13

13.1. Given a space S, let A, B be two distinct nonempty proper subsets of S, with $B \neq A^c$. Now consider the partitions $\mathscr{P}_1 = \{A, A^c\}$, $\mathscr{P}_2 = \{B, B^c\}$.

    a) Show that the class $\mathscr{C} = \mathscr{A}(\mathscr{P}_1) \cup \mathscr{A}(\mathscr{P}_2)$ is not an additive class.

    b) What is the additive class $\mathscr{A}$ generated by $\mathscr{C}$, and the largest partition in $\mathscr{A}$?

13.2. Prove that a completely additive class $\mathscr{A}$ cannot be countably infinite. (Suggestion: Consider two cases: In one case, the largest partition in $\mathscr{A}$ is finite. In the other case, $\mathscr{A}$ has a disjoint subclass that is infinite.)

13.3. Consider two distinct finite subsets A_1, A_2 of an infinite space S, such that $A_1 \cap A_2 \neq \phi, A_1, A_2$. Let $\mathscr{A}$ be the additive class generated by $\{A_1, A_2\}$.

a) How many of the sets in $\mathscr{A}$ are finite, and how many are infinite?
b) Show that if a set in $\mathscr{A}$ is infinite, then its complement is finite.

Section 14

14.1. Let $S = \{\xi_1, \xi_2, \xi_3\}$ and let $\mathscr{C}$ be the class of all subsets of S. Assign a *different* integer value to each set in $\mathscr{C}$ in such a way that the resulting assignment constitutes an additive set function.

14.2. Given a space S and three sets $A, B, C \subset S$. Under what conditions is it true that

$$(A \uplus C) \cup (B \uplus C) = A \uplus B \uplus C$$

(without parentheses)?

14.3. Suppose that in a particular city, three newspapers are published—denote them A, B, and C. In a suburb of this city, 1500 households were surveyed by the publisher of paper A regarding their newspaper subscriptions. The results of the survey are:

Paper A goes to 810 households,
Paper B goes to 583 households,
Paper C goes to 707 households.

It was further determined that

194 households subscribe to both A and B but not C,
237 households subscribe to both A and C but not B,
49 households subscribe to all three papers, and
44 households subscribe to none of the papers.

It was not determined how many households subscribe to B and C but not to A.
Check whether these results satisfy the requirement of additivity, and if so, determine how many households subscribe to B and C only.

14.4. Let A, B, C denote three subsets of a space S, and let H be an additive set function defined on the additive class generated by $\{A, B, C\}$. Verify the following:

a) $H(A - B) = H(A) - H(AB)$
b) $H(A \triangle B) = H(A) + H(B) - 2H(AB)$
c) $H(A \cup B \cup C) = H(A) + H(B) + H(C) - H(AB) - H(AC) - H(BC) + H(ABC)$
d) If H is non-negative, then $H(AB) \leq H(A) \leq H(A \cup B) \leq H(A) + H(B)$.

14.5. Given an additive class of sets $\mathscr{C} = \{A_1, A_2, \dots, A_n\}$, with H a non-negative additive set function defined on $\mathscr{C}$. Verify 'Boole's inequality',

$$H\left(\bigcup_{i=1}^{n} A_i\right) \leq \sum_{i=1}^{n} H(A_i).$$

(Suggestion: Use induction on part (d) of the previous problem.)

14.6. Suppose a set function H is defined on a class $\mathcal{C}$ of subsets of a space $S = \{\xi_1, \xi_2, \xi_3, \xi_4\}$. Instead of *additivity*, suppose H is to satisfy the relation

$$H(\mathsf{AB}) = H(\mathsf{A}) \cdot H(\mathsf{B}) \tag{*}$$

for every pair of sets A, B such that $\mathsf{A}, \mathsf{B}, \mathsf{AB} \in \mathcal{C}$. This can be done in a trivial way by either (a) specifying $H = 0$ or 1; or (b) specifying $\mathcal{C}$ in such a way that it is a disjoint class.

However, try to give an example in which you choose $\mathcal{C}$ and H in such a way that (*) is satisfied in a nontrivial way. Can you state any conditions under which (*) can be satisfied, or cannot be satisfied (for instance, $\mathsf{A} \subset \mathsf{B}$, disjointness, etc.)? Is it possible for H to be additive as well?

Section 15

15.1. A coin is tossed repeatedly, and on each throw it is observed whether 'head' or 'tail' occurred. What values of relative frequency of a 'head' is it possible to observe

a) in a single throw,
b) in two throws,
c) in three throws?
d) In a very large number of throws, what would you *expect* the relative frequency of a 'head' to be?

15.2. Suppose an experiment is performed a large number of times in order to observe the relative frequencies of events A, B, and 'both A and B'. Suppose that in the first 100 trials, the relative frequency of A is observed to be $\mathrm{rf}(A) = 0.56$. In the next 100 trials, $\mathrm{rf}(B) = 0.31$ is observed. In another 100 trials, $\mathrm{rf}(A\ \text{and}\ B) = 0.35$ is observed. Can this data be combined in some way to result in an improved value of the relative frequency of one or more of the events A, B, and 'A and B'?

15.3. Suppose we have a p.e. in which there is statistical regularity with respect to an event A that is not the certain event. Show that:

a) Statistical regularity also holds with respect to the event 'not A'.
b) There exists at least one nondegenerate partition of the sample space of this p.e., such that statistical regularity holds with respect to every member of the partition.
c) If statistical regularity holds with respect to all the members of some partition $\mathscr{P}$, then statistical regularity also holds with respect to all the events in $\mathscr{A}\ (\mathscr{P})$.

15.4. Suppose we have a p.e. in which there is statistical regularity with respect to two events A, B.

a) Verify that there is not necessarily statistical regularity with respect to the event 'A and B both occur'.
b) Consider now the sample space for this p.e. Show that it is possible that statistical regularity exists with respect to the members of two different additive classes, $\mathscr{A}_1$ and $\mathscr{A}_2$ (see Problem 15.3), but not with respect to all the members of the additive class generated by $\mathscr{A}_1 \cup \mathscr{A}_2$.

15.5. In each of the following, determine which part of the basic model is violated. Then, modify the statement in such a way that it represents a legitimate probability problem.

a) In the throw of several dice, a total of six dots is observed. What is the probability that the number of dice thrown was three?

b) You have filed a document in a four-drawer file cabinet, but don't remember in which drawer. What is the probability that it is in the top drawer?

c) You have lent someone a certain amount of money. What is the probability that you will get this money back?

Section 16

16.1. Consider an experiment in which a coin is tossed repeatedly, without end (see Example 11.1). The possible outcomes of this experiment are infinite coin-toss sequences. All the different sequences are equally likely, so that one experiment can be thought of as picking 'at random' one sequence from among all possible coin-toss sequences. On the subsets of the sample space S for this experiment we therefore define a set function P such that $P(\{\xi\}) = 0$, for all $\xi \in S$, and $P(S) = 1$. Is this set function P a probability function? Justify your answer.

16.2. *Weather forecasting.*

a) When a weather forecaster says, 'the probability of rain tomorrow is 30 per cent' (i.e., is 0.3), this should refer to some experiment whose repeated executions result in a relative frequency of approximately 0.3 for a particular event. Describe two different experimental plans, each one of which the weather forecaster might have in mind. Compare the significance of the forecast for the two models.

b) Is anything wrong with the following method of weather forecasting: The forecaster collects data about the rainy days in his locality over the past ten years. He finds that over this period exactly 30 per cent of the days had rain. Now his forecast is the same every day: 'The probability of rain tomorrow is 30 per cent.'

16.3. *Random integers* (Project). Obtain relative frequencies of various events for the experiment in which 'a positive integer is picked at random'. Consider the experiment in which you, as the experimenter, ask another person to call out an integer (or whole number). Do this experiment with a large number of persons, in such a way that they will not be likely to influence each other, and without giving hints or explanations. Determine the relative frequencies for events such as

$$A \leftrightarrow \text{an integer from 1 to 10 is picked}$$
$$B \leftrightarrow \text{an integer from 11 to 100 is picked}$$
$$C \leftrightarrow \text{an integer from 101 to 1000 is picked.}$$

Section 17

Note: In Problems 17.1 to 17.3, refer to Example 17.1 and sketch the various events on a Venn diagram arranged in the manner shown in Figure 17.1.

17.1. In the experiment of throwing a red and a white die, what is the probability of each of the following events:

a) The number of dots showing on the red die is less than the number of dots showing on the white die.

b) The number of dots showing on one of the dice is less than the number of dots showing on the other die.

c) The white die shows no more than three dots and the red die shows more than twice as many dots as the white die.

d) The number of dots showing on the white die is raised to the rth power, where r is the number of dots showing on the red die. The result is less than 7.

e) The largest number thrown is a '3'.

f) The smallest number thrown is a '3'.

17.2. In the experiment of throwing two dice, let the set A represent the event that the sum of the dots showing on both dice is less than 7. Also, let B represent the event that the difference in the number of dots showing on the two dice is greater than 2. Describe the events characterized by the following expressions and determine their probabilities:

$$AB, A \cup B, B - A, A \cup B^c, A \Delta B$$

17.3. What is the probability that a throw of two dice results in the sum of the numbers of dots on the two dice

a) being greater than 1?

b) being the square of an integer?

c) being greater than the difference between the numbers of dots?

d) being less than twice the number of dots showing on the die with the larger number of dots?

17.4. In each of the following cases, fill in the details of the p.e. and completely describe a suitable probability space. State any assumptions.

a) Tossing two coins, a penny and a nickel, to observe for each whether it results in 'head' or 'tail'.

b) Throwing a dart at a target. The target is marked off into ten different regions in such a way that each region is hit with roughly equal likelihood.

c) Tossing a coin an unending number of times.

17.5. For each of the following probability spaces, describe two different p.e.'s that can be modeled by that space.

a) $(S, \mathcal{A}, P)$, where $S = \{\xi_a, \xi_b\}$

$$A = \{\phi, S, \{\xi_a\}, \{\xi_b\}\}$$
$$P(\{\xi_a\}) = 1/4, \text{ which completely specifies } P.$$

b) $(S, \mathcal{A}, P)$, where $|S| = 12$,

$$A = \text{the class of all subsets of } S$$
$$P(\{\xi\}) = 1/12, \text{ for all } \xi \in S.$$

c) $(S, \mathcal{A}, P)$, where $S = \{\xi_1, \xi_2, \xi_3, \ldots\}$ is countably infinite,

$$A = \text{the class of all subsets of } S$$
$$P(\{\xi_i\}) = 2^{-i}, \text{ for } i = 1, 2, 3, \ldots$$

17.6. Given the binary space $S = \{\xi_1, \xi_2\}$, let $\mathscr{C}_0 = \{\ \}$, $\mathscr{C}_1 = \{\{\xi_1\}\}$, $\mathscr{C}_2 = \{\{\xi_2\}, \phi, S\}$, and $\mathscr{C}_3 =$ the class of all subsets of S. Also, let the function $N(\mathscr{C}_i) = |\mathscr{C}_i|$, the size of the class $\mathscr{C}_i$ (for $i = 0, 1, 2, 3$). Verify that the following is a probability space:

$$(\mathscr{C}_3, \{\mathscr{C}_0, \mathscr{C}_1, \mathscr{C}_2, \mathscr{C}_3\}, N/4).$$

Section 18

18.1. Given a probability space $(S, \mathscr{A}, P)$ and sets $A, B \in \mathscr{A}$.

   a) Show that the probability that exactly one of the sets A, B occurs is

$$P(A) + P(B) - 2P(AB).$$

   b) Show that $P(AB) \le P(A) \le P(A \cup B) \le P(A) + P(B)$.

18.2. *Loaded die.* A die is loaded in such a way that, when it is thrown, the probability of a given number of dots coming to face up is proportional to that number. What is the probability of obtaining a 'six' with this die?

18.3. Referring to Example 18.2, obtain the range of possible values for $P(D)$ using the result (18.7).

18.4. Given a probability, space $(S, \mathscr{A}, P)$, with $A, B \in \mathscr{A}$. Express $P(A \triangle B)$ in terms of

   a) $P(A)$, $P(B)$, and $P(A \cup B)$
   b) $P(A)$, $P(A - B)$, and $P(A \cup B)$
   c) $P(A)$, $P(B)$, and $P(A - B)$.

18.5. Prove the following inequalities:

   a) $P(A \cup B \cup C) \ge P(A - B) + P(B - C) + P(C - A)$
   b) $P(A \cup C) \le P(A \cup B) + P(B \cup C)$
   c) $P(A \triangle C) \le P(A \triangle B) + P(B \triangle C)$.

18.6. Refer to Problem 8.1. Let a probability space be defined in such a way that

   the event described under   (a) is assigned probability 0.15
   ”   ”   ”   ”   (b)   ”   ”   ”   0.28
   ”   ”   ”   ”   (c)   ”   ”   ”   0.79
   ”   ”   ”   ”   (e)   ”   ”   ”   0.30
   ”   ”   ”   ”   (g)   ”   ”   ”   0.22.

   Check whether this is a legitimate assignment of probabilities, and if it is not, modify it to make it legitimate. Then find the probabilities of the events (d), (f), and (h).

18.7. Prove Equation (18.5). Then, by induction, extend it to the union of an arbitrary finite number of sets, $A_1 \cup A_2 \cup \ldots \cup A_n$.

18.8. Given a probability space $(S, \mathscr{A}, P)$ in which all elementary sets have positive probability. Consider a class $\mathscr{C} = \{A_1, A_2, \ldots, A_n\} \subset \mathscr{A}$, such that for $k = 1, 2, 3, \ldots, n$, the probability of the intersection of any k of the sets in $\mathscr{C}$ depends only on k. That is,

$$P(A_{i1}) = p_1 \text{ and } P(A_{i_1} \cap A_{i_2} \cap \ldots \cap A_{i_k}) = p_k$$

where $i_1, i_2, \ldots, i_k$ are any k distinct subscripts selected from the integers $1, \ldots, n$.
What can be said about the p_k's in each of the following cases:

a) No other information is given. (Answer: $0 \le p_n \le p_{n-1} \le \ldots \le p_2 \le p_1 \le 1$.)
b) $A_1 \cap A_2 = \phi$
c) $A_1 \subset A_2$
d) $A_1 \cup A_2 = A_3$
e) $P(A_1 A_2 \cup A_3 A_4) = P(A_5 A_6)$.

18.9. *Distance between sets.* Given a probability space $(S, \mathcal{A}, P)$, a 'distance' $d(A, B)$ between any two sets $A, B \in \mathcal{A}$ can be defined as follows:

$$d(A, B) = \begin{cases} \dfrac{P(A \triangle B)}{P(A \cup B)}, & \text{provided } P(A \cup B) \neq 0; \\ 0, & \text{otherwise} \end{cases}$$

Verify that this definition satisfies the requirements of a *distance measure*:

a) $d(A, B) \ge 0$
b) $d(A, A) = 0$
c) $d(A, B) = d(B, A)$
d) (Triangle inequality) $d(A, B) + d(B, C) \ge d(A, C)$.

Suggestion for part (d): For $P(A \cup B)$ and $P(B \cup C) \neq 0$, show that

$$d(A, B) + d(B, C) \ge \frac{P(A \triangle B) + P(B \triangle C)}{P(A \cup B \cup C)} \ge d(A, C).$$

To verify the second inequality, write each set as a disjoint union of sets of the type LMN, where $L = A$ or A^c, $M = B$ or B^c, $N = C$ or C^c. Thus,

$$A \triangle B = AB^c C^c \cup AB^c C \cup A^c BC^c \cup A^c BC.$$

You might simplify your notation by writing $P(A \cup B \cup C) = \Sigma$, and denoting

$P(AB^c C^c) = p_1$ $P(A^c BC^c) = p_2$ $P(A^c B^c C) = p_3$
$P(ABC^c) = p_4$ $P(A^c BC) = p_5$ $P(AB^c C) = p_6$
$P(ABC) = p_7$.

Then $P(A \triangle B) = p_1 + p_2 + p_5 + p_6$, and $P(A \cup C) = \Sigma - p_2$; etc.

Part II Problems

Section 19

19.1. A small chest with three drawers has had a ball placed in each drawer. Two balls have the same color, the third has a different color. A contestant is asked to open one drawer, in which she will find a ball of a certain color. Then she is to open another drawer to see if she finds a ball of the same color. If so, she wins a prize. What is the probability that she will win the prize?

19.2. Refer to Example 19.3. Suppose the contestant has some 'inside knowledge' that biases his/her initial choice of door toward the one that does have the prize behind it. Thus, suppose that the probability of the prize being behind the door that the contestant chooses initially is $\alpha > 1/3$. How large must α be so that the contestant's chance of winning is better *without* switching the choice of door?

19.3. You have a chance to win $100. You are given two fair dice that you are to throw. Before throwing them you are to call out one of the following predictions:

'both greater than three'
or 'sum is six or eight'

If your prediction is correct, you win $100. That is, if you say 'both greater than three', both dice have to show more than three dots. In the other case, the total number of dots showing has to be either six or eight. What is the probability that you win the $100?

19.4. In each of the following, determine whether it is a legitimate probability problem or not, and explain why.

a) A computer random number generator is used to generate 100 integers in the range from 1 to 6, i.e., to simulate 100 die throws. What is the probability that each time a 6 appears it is followed by a 1?

b) Suppose that in the sequence of 100 numbers generated in part (a), a 6 appears ten times, and each of these 6's is followed by a 1. What is the probability that this random number generator always produces a 1 after a 6?

c) Let the sequence of numbers be as described in (b). What is the probability that there are no other 1's in the sequence except for those which follow the sixes?

d) A die is thrown until the first 6 appears. If it is a fair die, what is the probability that it has to be thrown an unending (infinite) number of times? (Equivalently: What is the probability that a 6 never appears?)

e) A certain die is thrown 100 times and not a single 6 is observed. What is the probability that this is a biased die?

f) An urn contains 100 dice, of which 99 are fair and one is biased so that it will not generate a 6. A die is taken blindly from the urn and thrown. What is the probability that the chosen die is the biased one?

Section 20

20.1. You have brought home seven video films from the library but find that you will have time to watch only one of them. You want to pick one at random and without bias. You have a single coin in your pocket that you can toss. How many times must you toss the coin, and how will you relate films to coin toss results, so that each of the seven films has the same probability of being chosen?

20.2. Consider the real-world activity of 'selecting a card at random from a standard deck of 52 playing cards'.

a) Give a sufficiently detailed description of this activity so that it can reasonably be modeled by a p.e. with equally likely outcomes.

b) Now give a different description of this activity, so that it cannot be reasonably so modeled.

20.3. An experimenter selects a card blindly from a well-shuffled standard deck of 52 playing cards, and then examines the card. Determine $P(A)$ if A denotes the event:

a) 'A face card (king, queen, or jack) or any card from the suit of spades is drawn'

b) 'A spade but not a face card is drawn'

c) 'A face card but not a spade is drawn'

Arrive at the answers in two different ways:

(1) Use Equation (20.1) directly on A.

(2) Express A in terms of the events

$$B \leftrightarrow \text{'a face card is drawn'}$$
$$C \leftrightarrow \text{'a spade is drawn'}$$

and then express $P(A)$ in terms of $P(B)$, $P(C)$ and $P(BC)$.

20.4. A large bin contains 1000 ping-pong balls. n of the balls have a red mark, the others are completely white. The balls are thoroughly mixed and indistinguishable to the touch. Ten balls are withdrawn blindly and of these, four are found to have a red mark.

a) Upon again thoroughly mixing the remaining balls, another ball is to be picked blindly. What is the probability (as a function of n) that it will have a red mark?

b) Demonstrate that the following question cannot be answered: 'What is the probability that there are no more balls with a red mark in the bin?' (Cf. Example 19.1)

20.5. Two indistinguishable dice are thrown. The possible outcomes, therefore, are the number pairs (1, 1), (1, 2), (1, 3),..., (1, 6), (2, 2), (2, 3),..., (2, 6), (3, 3),..., (3, 6), (4, 4),..., (4, 6), (5, 5), (5, 6), (6, 6). Explain why these 21 possible outcomes are not equally likely.

20.6. A fair coin is tossed three times. Of interest is the number of heads that is obtained.

a) Suppose the p.e. is set up in such a way that the possible outcomes are taken to be the number of heads: 0, 1, 2, 3.
Explain why these four possible outcomes are not equally likely.

b) Instead, how can you set up the p.e. for this situation so that the possible outcomes can be considered equally likely?

c) What is the probability of obtaining an odd number of heads? Less than three heads?

20.7. Two persons want to play a game that requires throwing a die, but the die has been lost. They improvise in the following way. Instead of throwing the die they toss two coins, a penny and a nickel, and interpret the result as follows:

$$H_P \leftrightarrow \text{head on the penny} \leftrightarrow \text{a number larger than three}$$
$$T_P \leftrightarrow \text{tail on the penny} \leftrightarrow \text{a number no larger than three}$$

$H_N \leftrightarrow$ head on the nickel $\leftrightarrow$ an even number
$T_N \leftrightarrow$ tail on the nickel $\leftrightarrow$ an odd number

Various combinations of coin toss results then give:

$H_P H_N \leftrightarrow$ four or six $H_P T_N \leftrightarrow$ five

$T_P H_N \leftrightarrow$ two $T_P T_N \leftrightarrow$ one or three

In order to resolve the ambiguity that remains between four or six, and one or three, they decide to toss the penny once more when they obtain $H_P H_N$ or $T_P T_N$, and choose the smaller number if the penny shows tail, and the larger number if it shows head.

a) What are the probabilities of getting 1, 2, 3, 4, 5, 6?
b) Propose a different improvization using coin tosses, that will be probabilistically equivalent to the throw of a fair die.

Section 21

21.1. Two persons play a board game that involves the throw of two dice. At one point in the game, one player has available only three possible moves that require, respectively, a sum of eight, nine, or ten on the two dice. If she obtains a different sum, she must repeat the throw of the dice until a sum of eight, nine, or ten is obtained. What is the probability that she will in fact make the move that requires a total of ten?

21.2. A red and a green die are thrown. If it is known that the sum of the dots on both dice is greater than six, what is the probability that

a) two dots show on one of the two dice?
b) the sum of the number of dots is 11?

Before answering, restate the question in a manner consistent with the model developed in this Section.

21.3. A die has the usual markings but has been constructed in such a way that the probability of k dots facing up is proportional to k ($k = 1, 2, 3, 4, 5, 6$). What is the probability of obtaining six dots upon throwing this die, under the condition that a throw resulting in an odd number of dots will be ignored?

21.4. Given a probability space $(S, \mathscr{A}, P)$. Let $\{A_1, A_2, \ldots, A_n\} \subset \mathscr{A}$ be a partition of the set $B \in \mathscr{A}$. Prove that $P(A_1|B) + P(A_2|B) + \ldots + P(A_n|B) = 1$.

21.5. Given a probability space $(S, \mathscr{A}, P)$ and sets A, B, C $\in \mathscr{A}$ with $P(ABC) \neq 0$. Show that the specification $P(A|BC) = P(B|AC)$ is equivalent to $P(AB^cC) = P(A^cBC)$. Use a Venn diagram to illustrate this result.

21.6. Consider an experiment in which a pair of fair dice are thrown repeatedly and the sum of the dots showing is recorded each time. What is the probability that a sum of 9 dots will appear before a sum of 7 dots appears for the first time?

21.7. Suppose you are given two well-mixed standard decks of playing cards (52 cards each) that we will call the left-hand deck and the right-hand deck, respectively. The backs of the cards in the two decks are indistinguishable. Consider the following procedure:

(1) A coin is tossed to select one of the decks; then

(2) a card is picked blindly from the deck that has been selected.

a) What is the probability of picking an ace of hearts?

b) Suppose that the ace of hearts has been removed from the left-hand deck and added to the right-hand deck, prior to shuffling. What is the probability of picking an ace of hearts?

21.8. In a store is a bin containing 100-watt light bulbs. In this bin, four defective bulbs have inadvertently become intermixed with eight good bulbs. A clerk takes two bulbs from the bin in succession.

a) What is the probability that, of the two bulbs that have been taken, only the first one will be good?

b) What is the probability that exactly one of the two bulbs drawn will be good?

c) Suppose the clerk follows this procedure: He checks the first bulb and if it tests good he sells both bulbs to a customer. If it tests bad, he gets a replacement and tests both, and gets additional replacements as needed until he is assured he has two good bulbs. What is the probability that the customer will get one bad bulb?

d) With the procedure as in (c), consider only the case where at least one of the bulbs drawn is bad. Under this condition, what is the probability that the customer gets exactly one bad bulb?

21.9. Consider a p.e. that involves observing the life span of a randomly selected new-born of a certain animal species living in a given environment. In the probability space for this experiment, A_x ($x = 0, 1, 2, \ldots, 8$) is the event that the animal lives x years (i.e., still lives x years after birth but no longer lives $x + 1$ years after birth). Suppose that the following probabilities have been determined for the events A_x:

$x =$	0	1	2	3	4	5	6	7	8
$P(A_x) =$	0.2	0.05	0.05	0.1	0.15	0.2	0.2	0.04	0.01

a) Make a mortality table, that is, a table that lists the probabilities that an animal lives x years, given that it has already lived y years ($y \leq x$).

b) If an animal has lived x years, what is the probability that the animal will not be alive one year hence, and the probability that it will live at least another year?

Section 22

22.1. Given an experiment with n equally likely possible outcomes, let A be an event consisting of m ($0 < m < n$) of these possible outcomes. Show that, under the condition A, one-element events contained in A are again equally likely, with conditional probabilities $1/m$.

Since conditional probabilities are equivalent to the unconditional probabilities of a 'reduced' experiment (with fewer possible outcomes), we conclude: A p.e. $\mathfrak{E}$ with n equally likely outcomes can always be changed into a p.e. $\mathfrak{E}'$ with m ($m < n$) equally likely outcomes by mere 'reduction', i.e., by deleting $n - m$ of the possible outcomes of $\mathfrak{E}$ and considering them as representing unsatisfactorily performed experiments.

22.2. A machine of a certain type is known to fail with probability 0.6 within the first 1000 hours of operation. However, if a particular modification has been made by the manufacturer (we will call it modification A), then the machine will fail with probability 0.2 within the first 1000 hours of operation. A different modification (modification B), furthermore, results in a probability of 0.1 of failure within the first 1000 hours. Suppose that 4/10 of all machines produced contain modification A; and an additional 1/10 of all machines produced contain modification B. (No machine contains both modifications.)

   a) What is the probability that a machine that has been randomly selected from all machines that have been produced will fail within the first 1000 hours of operation?
   b) If such a randomly selected machine does fail within the first 1000 hours, what is the probability that it contains neither modification A nor modification B?
   c) Suppose that one of the machines fails within the first 1000 hours of operation. Nothing is known about the manner in which it was selected from the total of all machines produced. What is the probability that the machine contains neither modification A nor modification B?

22.3. Two boxes of transistors are received by a laboratory. Box no. 1 is supposed to contain 150 transistors of type A, and box no. 2 is supposed to contain 200 transistors of type B. Due to carelessness by the manufacturer, the boxes actually contain:

   Box no. 1: a mixture of 130 type A and 20 type B transistors;
   Box no. 2: a mixture of 185 type B and 15 type A transistors.

   A technician at the laboratory is not aware that there is any difference between the two boxes, opens one of them and removes a transistor. It happens to be type A. What is the probability that it came from box no. 1?
   (Note: Here it is reasonable to assume that the technician selected a box without bias, that is, with each box having the same chance of being selected.)

22.4. Refer to Problem 21.7. With the two decks modified as in (b), the procedure is changed as follows: A card is taken at random from the right-hand deck and added to the left-hand deck, so that both decks again have 52 cards. Both decks are shuffled again. The remainder of the procedure is as described in Problem 21.7. What is the probability of picking the ace of hearts?
   Is there a simple explanation for the result that is obtained here?

22.5. Consider a communication system that serves to convey a 'message' from a certain location (the source) to another location (the destination). Two messages, M_1 and M_2, are available for transmission over the communication system. Message M_1 is sent with probability 0.6, and message M_2 is sent with probability 0.4. Due to errors in transmission, when M_1 is transmitted it will be received as M_2 with probability 0.05. Similarly, when M_2 is transmitted it will be received as M_1 with probability 0.2.

   a) When a message is received at the destination, what is the probability that it is M_1?
   b) If M_2 is received at the destination, what is the probability that M_2 was actually transmitted?
   c) What is the probability that an error will be made in transmission (i.e., transmitted M_1 received as M_2, or vice versa)?

22.6. Consider the situation faced by the contestant in Example 19.3, but with the following modification. There is still only one prize, but instead of three doors, the number of doors is now n (where $n \geq 3$). Furthermore, after the contestant has made his or her choice, the quizmaster opens m doors $(1 \leq m < n - 1)$, all of which have goats behind them. The contestant has the choice of trading the door initially chosen against one of the unopened doors. Is it still of advantage for the contestant to do so, and what is the probability of winning the prize?

22.7. In a quiz show, a contestant is shown two urns containing green and white balls that are indistinguishable to the touch. The contestant is asked to toss a coin in order to select one of the urns. She then is to draw a ball blindly from the urn she selected. Suppose there are a total of 13 balls, five white and eight green. If you are the quizmaster, how would you distribute the balls among the two urns, leaving neither urn empty, in order that the probability of drawing a white ball is

a) maximized?
b) minimized?

22.8. (After [Pa1].) Given a probability space $(S, \mathscr{A}, P)$, let A, B $\in \mathscr{A}$. For each of the following statements, give conditions that are necessary and sufficient to make the statement meaningful and true.

a) $P(A|B) + P(A^c|B) = 1$
b) $P(A|B) + P(A|B^c) = 1$
c) $P(A|B) + P(A^c|B^c) = 1$
d) $P(A|B) + P(B^c|A) = 1$.

22.9. You are shown three barrels that have been randomly arranged in a row and you are told that:

one of the barrels is filled with red balls;
another one is filled with yellow balls;
and the remaining one contains an equal mixture of red and yellow balls.

You will win a reward if you correctly identify the barrel containing the mixture.
You cannot look into the barrels. However, an attendant is standing by to help you. The attendant is able to blindly remove a total of two balls from the barrels and show them to you. You must instruct the attendant from which barrel or barrels to remove the two balls.

a) Suppose you decide to tell the attendant to remove one ball from the left barrel and one ball from the center barrel. What is the probability that you will win the reward?
b) Would you do better if you tell the attendant to pick two balls from the center barrel? What is the probability of winning the reward in that case?

22.10. Given a probability space $(S, \mathscr{A}, P)$ and two arbitrary events A, B $\in \mathscr{A}$ with $P(A), P(B) \neq 0, 1$. State all the constraints that must be satisfied jointly by the four numbers

$$a = P(A|B), \; b = P(B|A), \; c = P(A|B^c), \; d = P(B|A^c).$$

(Suggestion: Simplify your notation by writing $P(AB) = p_1$, $P(AB^c) = p_2$, $P(A^cB) = p_3$, and $P(A^cB^c) = p_4$. Then express a, b, c, d in terms of the p_i's and make use of the fact that $p_1 + p_2 + p_3 + p_4 = 1$.)
Suppose $a = 0.2$ and $b + c = 1$. Can d be determined?

Section 23

23.1. Consider the following experiment in ESP (extrasensory perception). An experimenter blindly picks a ball from an urn containing a mixture of 15 red and five black balls. In an adjoining room, a subject tries to guess the color of the ball that has been drawn. Suppose it is found that the subject guesses a red ball correctly with probability 2/3, and a black ball correctly with probability 1/3. Verify that, in that case, the subject's guess is independent of the color of the ball that is drawn.

23.2. Given a probability space $(S, \mathscr{A}, P)$ with A, B $\in \mathscr{A}$. Prove the following:

  a) If $P(A) = 0$ or 1, then A and B are independent.
  b) If A and B are independent and also A $\subset$ B, then $P(A) = 0$ or $P(B) = 1$.
  c) If $A \cup B$ and $AB \cup A^c B^c$ are independent, then $P(A \cup B) = 1$ or $P(A \Delta B) = 0$.

23.3. *Conditional independence.* It is possible that two events that are not independent become independent as a result of conditioning by some other event. Consider the single die-throw experiment. In the probability space for this experiment, let

$$A \leftrightarrow \text{'a one, three, or six is thrown'}$$
$$B \leftrightarrow \text{'a one, three, or five is thrown'}.$$

  a) Specify an event C such that $P(AB \mid C) = P(A \mid C) P(B \mid C)$; i.e., so that A and B become conditionally independent events when conditioned on C.
  b) Explain the significance of this result conceptually.

23.4. Given a probability space $(S, \mathscr{A}, P)$, where $S = \{\xi_1, \xi_2, \xi_3\}$ and $\mathscr{A}$ is the class of all subsets of S. Also, $P(\{\xi_i\}) > 0$ for $i = 1, 2, 3$. Show that it is not possible for a pair of independent sets to belong to $\mathscr{A} - \{\phi, S\}$.

23.5. Given a probability space $(S, \mathscr{A}, P)$ with A, B $\in \mathscr{A}$, where A and B are independent and equiprobable. If $P(A \Delta B) = 0.18$, determine $P[A^c B^c \mid (AB)^c]$.

23.6. Given a probability space $(S, \mathscr{A}, P)$ with A, B $\in \mathscr{A}$, where A and B are independent and $P(AB) \neq 0$. Consider a third event C.

  a) Show that AC and BC are again a pair of independent events if and only if

$$P(C|A) P(C|B) = P(C|AB).$$

  b) Which of the sets generated by {A, B} could be the set C in the above equation? (Assume $P(A)$, $P(B) \neq 1$.)

23.7. Consider an experiment in which one card is drawn from each of two shuffled standard decks of playing cards—call them the left-hand and the right-hand decks. In each of the following cases, determine whether the two events are independent and find the probability of the joint event.

  a) A face card (jack, queen or king) is drawn from the left-hand deck, and a face card is drawn from the right-hand deck.
  b) At least one of the cards drawn is an ace, and at least one of the cards is a heart.

c) Both cards are number cards (ace ($=1$), 2, 3,..., or 10) with a sum > 5, and both cards are number cards with a sum < 10.

23.8. Given a probability space $(S, \mathscr{A}, P)$. $\mathscr{P}(\mathscr{A}) = \{A, B, C, D\}$ is the partition that generates $\mathscr{A}$, where $P(A) = 0.1$, $P(B) = 0.2$, $P(C) = 0.3$, $P(D) = 0.4$. Verify that there are no two sets belonging to $\mathscr{A} - \{\phi, S\}$ that are independent.

23.9. Given a probability space $(S, \mathscr{A}, P)$ and sets $A, B_1, B_2 \in \mathscr{A}$ such that A is independent of B_1 and of B_2. Denote $\{B_1, B_2\}$ by $\mathscr{C}$, and let $\mathscr{C}*$ be the closure of $\mathscr{C}$ with respect to set operations. In other words, $\mathscr{C}*$ is the additive class generated by the members of $\mathscr{C}$.

a) Let B_1, B_2 be nonempty and disjoint, and $B_1 \neq B_2{}^c$. Show that A is independent of every set in $\mathscr{C}*$.

b) Let $B_1 \subset B_2$, with $B_1 \neq B_2$, ϕ; and $B_2 \neq S$. Again show that A is independent of every set in $\mathscr{C}*$.

c) Now let B_1, B_2, their intersection and their complements be nonempty. Also, let neither set be a subset of the other. Use the experiment of throwing a single die to construct a counterexample, showing that in this case it cannot be concluded that A is independent of every set in $\mathscr{C}*$.

23.10. Given a probability space $(S, \mathscr{A}, P)$ and a class of nonempty sets $\mathscr{C} = \{B_1, B_2, ..., B_n\} \subset \mathscr{A}$. Another set, $A \in \mathscr{A}$, is independent of each set in $\mathscr{C}$. Let $\mathscr{C}*$ be the closure of $\mathscr{C}$ with respect to (finite) set operations.

a) If $\mathscr{C}$ is a disjoint class, use the results of Problem 23.9 and induction to show that A is independent of every set in $\mathscr{C}*$.

b) If $\mathscr{C}$ is a class of 'nested sets', i.e., $B_1 \supset B_2 \supset ... \supset B_n$, with $B_i \neq B_{i+1}$ and $B_1 \neq S$, again use the results of Problem 23.9 to show that A is independent of every set in $\mathscr{C}*$.

Section 24

24.1. The marquee of a movie theatre has 100 light bulbs. Independent failure of the bulbs can be assumed. If the probability of a bulb burning out in less than 1000 hours of operation is 0.01, what is the probability that all bulbs are still burning after 1000 hours of operation? Repeat for the case where a single bulb fails with probability 0.1 in the first 1000 hours. (Obtain numerical answers.)

24.2. Verify that $\{A, B, C\}$ in Example 24.2 is an independent class.

24.3. In a factory, a certain kind of device gets assembled in large numbers. Three different kinds of errors can occur in assembly, which will be denoted errors of type A, B, and C. The errors occur in three different assembly steps and therefore can be considered mutually independent. The device functions acceptably if only one of the three errors occurs during assembly, but is faulty if any two or all three errors occur. Assume statistical regularity holds. Suppose the probability of an error of type A is 0.1, of type B is 0.2, and of type C is 0.14.

a) What is the probability that a completed device is faulty?

b) What is the probability that a device is assembled without any error?

24.4. a) Given an independent class $\mathscr{C}$ of nonempty sets. Show that $\mathscr{C} \cup \{\phi\}$ is again an independent class.

b) Given an independent class $\mathscr{C}$ with $S \notin \mathscr{C}$. Show that $\mathscr{C} \cup \{S\}$ is again an independent class.

24.5. Consider the following experiment. A pointer is fastened to the center of a circle in a horizontal plane so that it is free to spin. The circle has been divided into 25 sectors of equal size, marked with the letters A through Y. In addition, a die is available. The purpose of the experiment is to obtain an impartial selection of a letter from A through Y, and an integer from 1 to 6. The pointer is spun and the die is thrown and the results noted.

On the sample space S, define the following events:

A $\leftrightarrow$ 'the pointer comes to rest in the field marked with letter A'
$Z_i \leftrightarrow$ 'the die comes to rest with i dots facing up' ($i = 1, \ldots, 6$)

Verify that the following events are pairwise independent but not mutually independent:

$$A \cup Z_1, A \cup Z_2, A \cup Z_3, A \cup Z_4, A \cup Z_5, A \cup Z_6.$$

24.6. Given a probability space (S, $\mathscr{A}$, P) with A, B, C $\in \mathscr{A}$, and $P(A) = P(B) = P(C) = 1/2$. Is it possible for A and B to be independent and C to be independent of AB and also of A$\cup$B, and yet for C *not* to be independent of A, nor of B? Justify your answer. (After [Re2].)

24.7. $\mathscr{C}$ and $\mathscr{D}$ are two independent classes. Furthermore, each set in $\mathscr{C}$ is independent of each set in $\mathscr{D}$. Show that $\mathscr{C} \cup \mathscr{D}$ may or may not be an independent class. That is, give an example where $\mathscr{C} \cup \mathscr{D}$ is an independent class, and another example where $\mathscr{C} \cup \mathscr{D}$ is not an independent class.

24.8. Suppose $\mathscr{C} = \{A_1, \ldots, A_n\}$ is an independent class. Let $\mathscr{C}^*$ be obtained from $\mathscr{C}$ by complementing some (or all, or only one) of the sets belonging to $\mathscr{C}$.

a) Prove that $\mathscr{C}^*$ is again an independent class.
b) Show that, therefore,

$$P\left(\bigcup_{i=1}^{n} A_i\right) = 1 - \prod_{i=1}^{n} P(A_i^c).$$

24.9. Show that the following is an alternative definition of an independent class:
Given a probability space (S, $\mathscr{A}$, P) and a class $\mathscr{C} \subset \mathscr{A}$, where $|\mathscr{C}| \geq 2$. $\mathscr{C}$ is an independent class if and only if for every pair of nonempty subclasses $\mathscr{C}_1$, $\mathscr{C}_2$ of $\mathscr{C}$ such that $\mathscr{C}_1 \cap \mathscr{C}_2 = \{\}$ it is true that

$$P(BC) = P(B) P(C)$$

where B is the intersection of all sets in $\mathscr{C}_1$, and C is the intersection of all sets in $\mathscr{C}_2$.

24.10. Given a probability space (S, $\mathscr{A}$, P), with $\mathscr{C} \subset \mathscr{A}$ an independent class. Prove the following propositions:

a) With $\mathscr{C}_1$ and $\mathscr{C}_2$ any two nonempty subclasses of $\mathscr{C}$ such that $\mathscr{C}_1 \cap \mathscr{C}_2 = \{\}$, let $\mathscr{P}_1 = \mathscr{P}(\mathscr{C}_1)$, the largest partition generated by $\mathscr{C}_1$. Then, every set in $\mathscr{P}_1$ is independent of every set in $\mathscr{C}_2$.

b) With $\mathscr{P}_1$ and $\mathscr{C}_2$ as in (a), Let $\mathscr{A}_1 = \mathscr{A}(\mathscr{P}_1)$, the additive class generated by $\mathscr{P}_1$. Then, every set in $\mathscr{A}_1$ is independent of every set in $\mathscr{C}_2$.

24.11. Make use of Problem 24.8 and 24.10 to prove Theorem 24.1.

Section 25

Note: In Problems 25.1–25.4 describe the k-tuples you use to determine the size of the sample space and of particular events.

25.1. A certain model of automobile is available in any one of 12 different standard exterior colors, five different interior colors, with or without tinted windows, with or without overhead window. It is also available with or without rear-passenger TV, and if it has rear-passenger TV, with or without rear-door window shades. Besides, three different engine types are available. How many cars of this model can be sold in a particular city if no two cars sold in that city are to be exactly alike?

25.2. A bowl contains marbles of four different colors, with the same number of marbles for each color. The bowl is shaken up, a marble picked blindly, its color noted and the marble is returned to the bowl. This is done four times. What is the probability that

a) all four draws resulted in the same color?
b) the first and third draw resulted in the same color?
c) all four draws resulted in different colors?
d) at least three different colors were obtained?

25.3. A triple representing a date, such as (Wednesday, August, 17) is randomly generated by picking a day of the week, a month of the year, and an integer from 1 to 31 to represent the day. Each selection is made with equal likelihood. What is the probability that the result corresponds to a valid date in the current year?

25.4. Three coins and two dice are tossed simultaneously.

a) What is the probability that the three coins all show the same face and the two dice the same number of dots?
b) What is the probability that the three coins all show the same face and the number of dots on each die is less than 4?
c) What is the probability that the number of dots showing on each die is the same and is equal to the number of heads showing on the three coins?

25.5. Consider an experiment whose possible outcomes are characterized by a set of k-tuples, k being fixed. The experiment is such that, for each position in the k-tuple, the possible entries that can appear in that position arise with equal likelihood and independently of the entries in other positions. Prove that the possible outcomes of the experiment are equally likely.

25.6. You join a group of n people, consisting of a random collection of persons (such as a waiting line at the post office). What is the smallest n so that the probability of at least one person in the group having the same birthday as you is greater than 1/2? (Neglect Feb. 29 for the purpose of this calculation.)

25.7. Each of k urns contains n balls. Each ball is marked with a number from 1 to n, such that all n numbers are represented in each urn. One ball is drawn blindly from each urn. What is the probability that m (where $1 \le m \le n$) is the largest number that appears on the balls that have been drawn? Describe the p.e. and probability space before computing the desired probabilities.

25.8. Show that for any positive integer n, $\sum_{i=0}^{n} \binom{n}{i} = 2^n$.

Section 26

26.1. From a standard deck of 52 playing cards, all jacks, queens, kings and aces are removed, leaving the number cards 'two' through 'ten' in each of the four suits—36 cards in all. Consider an experiment where the deck is shuffled, a card is drawn and the number noted (not the suit), and the card is returned to the deck. This is done four times. This situation can therefore be modeled as a product experiment with possible outcomes that are four-tuples of integers from 2 to 10.

a) What is the size of the space S that models this experiment?
b) Define the event $A \leftrightarrow$ 'a 5 is drawn first, not a five on the second draw, and a 5 is drawn again on the third draw'. Also, for each of the individual component experiments, let $B \leftrightarrow$ 'a 5 is drawn'. Express A in terms of B.
c) What is the size of A?

26.2. Consider the product experiments $\mathfrak{E} = \mathfrak{E}_1 \times \mathfrak{E}_2 \times \mathfrak{E}_3$ and $\mathfrak{F} = \mathfrak{F}_1 \times \mathfrak{F}_2$. Each of the experiments $\mathfrak{E}1, \mathfrak{E}_2, \mathfrak{E}_3, \mathfrak{F}_1, \mathfrak{F}_2$ has two possible outcomes.

a) How many component experiments (of the types $\mathfrak{E}_1, \mathfrak{E}_2, \mathfrak{E}_3, \mathfrak{F}_1, \mathfrak{F}_2$) make up the experiment $\mathfrak{E} \times \mathfrak{F} \times \mathfrak{F}$?
b) What is the number of possible outcomes of $\mathfrak{E} \times \mathfrak{F} \times \mathfrak{F}$?
c) In the product space $S = S_1 \times S_2$ corresponding to $\mathfrak{F}$, consider the product set $B = A_1 \times A_2$, where A_1, A_2 are elementary sets. Express B^c as a disjoint union of:
  i) three nonempty product sets;
  ii) two product sets.
d) How many distinct product sets are subsets of the space S defined in part (c)?

26.3. a) Show that if S_1, S_2 are two countable spaces, then $S_1 \times S_2$ is also countable.
b) Given a finite number of countable spaces $S_1, \ldots, S_n$, show that $S_1 \times S_2 \times \ldots \times S_n$ is also countable.

26.4. Show that the Cartesian product is distributive over the various binary set operations. That is, given two spaces S, T, let A, $B \subset S$ and $C \subset T$. Also, let $*$ denote any one of the operations $\cup, \cap, -, \Delta$. Then:

i) in $S \times T$ we have $(A*B) \times C = (A \times C) * (B \times C)$; and
ii) in $T \times S$ we have $C \times (A*B) = (C \times A) * (C \times B)$.

26.5. Given a product space $S = S_1 \times S_2 \times \ldots \times S_n$, and product sets A, $B \subset S$, where $A = A_1 \times \ldots \times A_n$, $B = B_1 \times \ldots \times B_n$.

a) Show that $A \cap B = (A_1 \cap B_1) \times \ldots \times (A_n \cap B_n)$, i.e., $A \cap B$ is again a product set.
b) Obtain a general expression for B^c as a disjoint union of product sets.

c) Theorem 8.1 states that A can be written $A = AB \cup AB^c$. Instead, suppose that A is to be expressed as a disjoint union of *product* sets. According to (a), AB is a product set; but AB^c is in general not a product set. Therefore, expand AB^c (which is also $A - B$) into a disjoint union of product sets.

Section 27

27.1. Given $n - 1$ spaces $S_2, S_3, \ldots, S_n$, where $n > 2$ and $|S_i| = i$ $(i = 2, 3, \ldots, n)$.

a) What is the size of $S = S_2 \times S_3 \times \ldots \times S_n$?
b) If in each of the spaces S_i the one-element sets are equally probable, what is the probability of a one-element set in S?
c) Can you answer (b) under the condition that S_3 does not arise; i.e., under the condition S_3^c? Explain your answer.

27.2. At a university, user access codes to a campus-wide computer network are to be assigned using randomly generated strings of n letters from the English alphabet. Up to $10\,000$ valid codes are expected to be assigned. If an unauthorized user is to have a probability no greater than 0.001 of hitting a valid access code when typing an arbitrary sequence of n letters, how large should n be?

27.3. What is the smallest number of fair dice that must be thrown in order that the probability of obtaining at least one 'six' is ≥ 0.9?

27.4. A product experiment $\mathfrak{E}$ with $|S| = 63$ and with equally likely possible outcomes consists of three component experiments.

a) What are the sizes of the sample spaces of the component experiments?
b) The sets $A, B, C \subset S$, none of which are empty or identical to S, are mutually independent. Specify possible values for $P(A), P(B), P(C)$ that are consistent with the information given.

27.5. Consider an experiment in which two fair dice are thrown repeatedly. Let n denote a positive integer that is specified. Find the probability that the nth throw is the first throw in which the sum of the dots on the two dice is greater than nine.

27.6. Let $(S_1, \mathscr{A}_1, P_1)$ be the probability space associated with the experiment of tossing a single (perfect) die.

a) What is the size of $\mathscr{A}_1$?
b) What is the size of $\mathscr{A}_1 \times \mathscr{A}_1$?
c) What is the size of $\mathscr{A}(\mathscr{A}_1 \times \mathscr{A}_1)$, the additive class generated by $\mathscr{A}_1 \times \mathscr{A}_1$?

27.7. In a product space $S = S_1 \times S_2 \times \ldots \times S_n$, for $i = 1, \ldots, n$ let $\mathscr{P}_i$ be a partition of S_i. Show that $\mathscr{P}_1 \times \ldots \times \mathscr{P}_n$ is a partition of S.

27.8. Set up the probability space $(S, \mathscr{A}, P)$ for ten coin tosses (Example 25.3) as a product probability space. Which of the following events are product events:

a) 'Heads and tails occurred in alternation'
b) 'No heads occurred'
c) 'The first and the last toss resulted in a head'
d) 'Exactly three heads occurred'.

Express each product event as a product set, and find the probability of each event.

27.9. An ordinary die is thrown five times.

On the first throw, it is of interest which of the six faces comes to rest on top.
On the second throw, only faces 1 through 5 are of interest.
On the third throw, only faces 1 through 4 are of interest.
On the fourth throw, only faces 1 through 3 are of interest.
On the fifth throw, only faces 1 and 2 are of interest.

a) What is the size of the sample space?
b) What is the probability that in the five die-throw results no number greater than 2 is obtained?
c) If the five-tuple representing the five die-throw results contains a single 'two', what is the probability that this 'two' appears in the fifth position?

27.10. A die is thrown four times.

a) What is the probability of the event A ↔ "at least one 'six' is obtained, and no 'one' is obtained prior to the first 'six' "?
b) What is the conditional probability, given A, of the event B ↔ "at least one 'three' is obtained"?

27.11. *Playing a fair game with an unfair die.* (See also [JJ1].) Normally, a die is used to make a totally unbiased random selection from the set of integers $\{1, 2, 3, 4, 5, 6\}$. However, an actual die may not be perfectly symmetrical in shape, so that one suspects that the different faces do not come up with equal likelihood—the die is 'biased' or 'unfair'.

Assume that you have a biased die for which the following probabilities are not known and must be assumed to be all different:

$$P(\{i\}) = p_i \neq 0, \ i = 1, 2, 3, 4, 5, 6; \ \text{where} \ \sum p_i = 1.$$

Is it still possible to use such a die to 'play a fair game', i.e., make an unbiased random selection from a set of six alternatives?

Show that, if the biased die is thrown three times, and successive throws can be assumed independent, then it is possible to define six mutually exclusive events that are equally likely. (This is not possible, in general, with only two independent throws of the die.)

27.12. A Bridge player plays a series of ten games of Bridge. What is the probability that in this series of ten games there are three successive games in which she is dealt no ace, while in the remaining seven games she is dealt at least one ace?

27.13. In Problem 22.5, suppose that successive uses of the system can be modeled as independent experiments. In ten successive uses of the system, what is the probability that

i) no error is made?
ii) exactly two errors are made?

27.14. In Example 27.3, suppose the question is changed to: "How many times must a fair die be thrown till the kth 'six' appears ($k > 1$)?" Let $A_i^{(k)}$ ↔ 'kth six occurs on ith throw' ($i = k$, $k+1, \ldots$). Show that

$$P(A_i^{(k)}) = \binom{i-1}{k-1} P_1(D_6^c)^{i-k} P_1(D_6)^k$$

which for any given k is a 'negative binomial distribution' over the partition $\{A_k^{(k)}, A_{k+1}^{(k)}, \ldots\}$.

27.15. Given some experiment $\mathfrak{E}$ with associated probability space $(S, \mathscr{A}, P)$ where P is not known, and let A be some event, $A \in \mathscr{A}$. Suppose independent repetitions of the p.e. $\mathfrak{E}$ are performed until the event A occurs for the first time. For any specified $i \geq 2$, what is the largest possible value of the probability that A occurs for the first time in the ith trial?

27.16. Given a probability space $(S, \mathscr{A}, P)$ that represents an experiment $\mathfrak{E}$, and an independent class $\mathscr{C} \subset \mathscr{A}$. Now consider the product probability space corresponding to the product experiment $\mathfrak{E} \times \mathfrak{E}$. Use the alternative definition of an independent class in Problem 24.9 to show that the class $\mathscr{C} \times \mathscr{C} \subset \mathscr{A} \times \mathscr{A}$ is again an independent class.

27.17. Given a space S, and two classes $\mathscr{C} = \{C_1, \ldots, C_n, S\}$, $\mathscr{D} = \{D_1, \ldots, D_m, S\}$ of subsets of S. Let $e_1(\mathscr{C})$ be an expression involving set operations performed on the members of $\mathscr{C}$, and let $e_2(\mathscr{D})$ be an expression involving set operations performed on the members of $\mathscr{D}$. Making use of the results of Problem 26.4, show that the following is true:

$$e_1(\mathscr{C}) \times e_2(\mathscr{D}) = e_3(\mathscr{C} \times \mathscr{D})$$

where the right side is an expression involving set operations performed on product sets $C_i \times D_j$, where $C_i \in \mathscr{C}$ and $D_j \in \mathscr{D}$. In other words, if $\mathscr{A}(\mathscr{C})$ and $\mathscr{A}(\mathscr{D})$ are the additive classes generated by $\mathscr{C}$ and $\mathscr{D}$, respectively, and $C \in \mathscr{A}(\mathscr{C})$ and $D \in \mathscr{A}(\mathscr{D})$, then $C \times D \in \mathscr{A}(\mathscr{C} \times \mathscr{D})$, the additive class generated by $\mathscr{C} \times \mathscr{D}$.

Section 28

28.1. An urn contains an equal number n ($n > 2$) of red balls and black balls. Four balls are blindly removed in succession from the urn. What is the probability that the successive balls drawn alternate in color?

28.2. The host at a dinner party uses an unbiased random mechanism to assign seats to the eight diners around a round table. What is the probability that the diners get seated in alphabetical order, either clockwise or counterclockwise?

28.3. n standard decks of 52 playing cards (Bridge cards) are individually shuffled. One card is picked from each deck. What is the smallest n so that the probability that at least one card face appears twice among the selected cards is greater than 1/2?

28.4. Consider an arbitrarily chosen group of n persons. What is the smallest n so that the probability of at least two of the persons having the same birthday is greater than 1/2? Suggestion: Neglect Feb. 29 for the purpose of this calculation.

(Compare the result with that obtained in Problem 25.6.)

28.5. A random mechanism generates four-tuples of binary digits ($\mathbf{0}$'s and $\mathbf{1}$'s). In any given position of a four-tuple, the probability of a $\mathbf{0}$ is 0.5^{1+z}, where z is the number of zeros in preceding positions of the same four-tuple. What is the probability that a four-tuple gets generated that contains three zeros?

28.6. A ballroom has six chandeliers in a row, with ten light bulbs in each chandelier. The light bulbs are all of the same type, have been installed at the same time, and burn out independently of each other. Therefore, each light bulb has the same chance of being the first to burn out, and once one has burned out, each remaining bulb has the same chance of being the next to burn out, etc. In this ballroom, which of the following is more likely:

a) The first six light bulbs to burn out are all from one chandelier.
b) The first six light bulbs to burn out are one light bulb from each chandelier, starting at one end of the ballroom and proceeding in order to the other end.

28.7. At an assembly line position is located a bin of n identical parts, one of which is to be installed into each assembly that comes down the line. Mixed in among the contents of the bin are n_d defective parts $(n_d \leq n)$, which are indistinguishable to the assembler. Show that the probability that the first defective part gets installed into the kth assembly is n_d/n if $k = 1$, and for $k = 2, 3, \ldots, n - n_d + 1$ is

$$\frac{n_d}{n - (k-1)} \left(\prod_{i=0}^{k-2} \frac{n - n_d - i}{n - i} \right).$$

28.8. An urn contains a mixture of three white balls, five red balls, and g green balls. They are indistinguishable to the touch. Three balls are removed blindly.

a) If $g = 8$, what is the probability that the three balls are all of the same color?
b) Explain why the probability that the three balls are all of the same color is less if $g = 1$ than if $g = 0$.
c) Explain why the probability that the three balls are all of the same color increases as g increases, for g sufficiently large.
d) Find the value of g for which the probability that the three balls are all of the same color will be smallest.

28.9. A large barrel contains an equal number q $(q > 1)$ of red, green, blue and yellow ping-pong balls that are thoroughly mixed. n balls are withdrawn blindly. What is the smallest value of n so that the probability is at least 0.5 that two of the balls withdrawn have the same color?

28.10. In transmission of data over a telephone circuit, error bursts can occur. Suppose that when a (binary) digit is received correctly, the probability of the next transmitted digit being received in error is 10^{-3}; but when a digit is in error, the probability of the next digit being also in error is 0.1.

a) What is the probability that an arbitrarily chosen digit is received incorrectly?
b) What is the probability that, within a long string of digits, a particular data word consisting of ten digits will contain at least one error?
c) What is the probability that such a data word will contain at least two errors?

Section 29

29.1. You wish to select six distinct integers randomly from the set of integers from 1 to 49, in such a way that all unordered six-tuples are equally likely.

a) How many possible outcomes are there?
b) If the only random mechanism you have available is a fair die, how will you arrange the procedure of the experiment to assure equally likely outcomes?

29.2. You are getting dealt 13 cards from a well-shuffled deck of 52 Bridge cards. What is the probability that you will get dealt all four aces?

29.3. Twelve checker pieces are placed arbitrarily and without bias into the squares of an ordinary checker board of 32 white and 32 black squares (Figure A.2), but not more than one checker per square. What is the probability that all 12 pieces end up on white squares?

29.4. Verify Equation (29.2).

29.5. A small-town lottery has a total of 1000 tickets for sale. If it is known that there are ten winning tickets distributed randomly among all the tickets, how many tickets must a person buy to bring her chance of receiving one or more winning tickets to at least 1/2?

29.6. In a class of 18 students, six homework problems are assigned each week. By throwing dice, the instructor independently and with equal likelihood selects six students every week, each of whom is to discuss one of the homework problems. Neglect the possibility of having to call on additional students (because of absences, assigned problems not done by everyone, etc.) in determining the following:

   a) What is the probability that after four weeks, i.e., after discussion of four assignments, a particular student has not been called upon at all?
   b) If after four weeks, exactly two students have not been called upon, what is the probability that after five weeks, only one student remains who has not yet been called upon?

29.7. In a certain country there are 10-cent coins of two different types but similar appearance in circulation. There are 1 000 000 coins of one type (call it type A) in circulation, and 2 000 000 coins of the other type (type B).

   a) What is the probability that of three arbitrarily selected coins, one is type A and the other two type B?
   b) Suppose the three coins are arbitrarily taken from a cash register that is known to contain a mixture of ten type A coins and 20 type B. What is the probability that one coin will be type A, the other two type B?
   c) Repeat (b) if the cash register has two type A and four type B coins.

29.8. An urn contains a mixture of red, green, and yellow balls. Three balls are removed blindly.

   a) If there is the same number n of red, green, and yellow balls, what is the probability that the sample drawn contains a ball of each color?
   b) If there are r red balls, g green balls, and y yellow balls, what is the probability that the sample drawn contains a ball from each color?

Section 30

30.1. In the p.e. $\mathfrak{E}$ of Example 30.1, let $A \leftrightarrow$ 'k_g green, k_r red, and k_y yellow balls are drawn', and let $B \leftrightarrow$ 'k_b blue balls are dawn'. Determine the conditional probabilities $P(A|B)$ and $P(B|A)$.

30.2. A fair die is thrown repeatedly. For any non-negative integer k, show that the probability of getting exactly k sixes goes to zero as the number of throws becomes arbitrarily large.

30.3. A low-probability event A $(0 < P(A) \ll 1)$ is associated with an experiment $\mathfrak{E}$.

   a) In n independent executions of $\mathfrak{E}$, what is the probability that A will arise at least once? Plot this probability as a function of n, for $P(A) = 0.1, 0.01, 0.001$. Use a log scale for n.
   b) Note that the curves have nearly identical shapes, except for a horizontal displacement. Why is this so?
   c) Repeat part (a), this time using a first-order approximation to the binomial expansion of the probability expression. Superimpose the plots on the graph of part (a).
   d) Show that, in the limit as $n \to \infty$, the probability of A arising at least once equals 1, irrespective of how small $P(A)$ is.

30.4. Re-derive Equation (30.2) by making use of the product rule (22.6) and applying the binomial distribution (30.4), with appropriate choice of parameters, to each of the factors in the resulting product.

30.5. In the probability space $(S_1, \mathscr{A}_1, P_1)$ of the p.e. $\mathfrak{E}$ in Example 30.1, let N_b, N_g, N_r, N_y denote the events 'no blue balls drawn', 'no green balls drawn', 'no red balls drawn', 'no yellow balls drawn', respectively. Use these events and their probabilities to obtain the following results:

a) $P_1(A)$, where $A \leftrightarrow$ 'at least one blue ball is drawn'.
b) $P_1(B)$, where $B \leftrightarrow$ 'at least one blue ball and at least one green ball is drawn'.
c) $P_1(C)$, where $C \leftrightarrow$ 'at least one blue ball, at least one green ball, and at least one yellow ball is drawn'.
d) Show that, if $D \leftrightarrow$ 'at least one ball of each color is drawn', then

$$
\begin{aligned}
P_1(D) = {} & 1 - (1-p_b)^{10} - (1-p_g)^{10} - (1-p_r)^{10} - (1-p_y)^{10} \\
& + (p_b+p_g)^{10} + (p_b+p_r)^{10} + (p_b+p_y)^{10} \\
& + (p_g+p_r)^{10} + (p_g+p_y)^{10} + (p_r+p_y)^{10} \\
& - p_b^{10} - p_g^{10} - p_r^{10} - p_y^{10}.
\end{aligned}
$$

Section 31

31.1. Consider a board game in which players take turns throwing a die, then advancing their marker (or 'pawn') along a path marked on the board by the number of steps equalling the number of dots obtained in the die throw. Suppose one of the players is in the following situation. If she manages, in one or more moves, to advance her marker to a position exactly four steps ahead of the current position of the marker, she wins. If she does not get to place her marker into that position, she loses. Thus, if she throws a 'six', she loses; if she throws a 'one' in each of four successive throws, she wins; etc. What is the probability that she wins?

31.2. In Example 31.1, determine the probability that exactly one man gets to dance with the same woman he danced with previously.

31.3. A coin is tossed ten times (as in Problem 27.8 or Example 25.3). What is the probability of the event $A \leftrightarrow$ 'no two consecutive heads occurred'?
   Suggestion: Use a tree diagram to determine $|A|$ for the case of five coin tosses. From this, determine $|A|$ for ten tosses by grafting the appropriate tree onto each terminal node (lowest level node) of the tree for five tosses.

31.4. A deck of six playing cards is made up of the two of spades, the three of spades, the four of spades, the two of hearts, the three of hearts, and the four of hearts. The deck is thoroughly mixed. Show that the probability is 2/3 that there is at least one pair of consecutive cards having the same face value.
   Suggestion: When constructing a tree, label each node either a, b, or c, where 'a' denotes the face value of the first card in the deck, 'b' the next (different) face value encountered, and 'c' the remaining face value.

31.5. *Random walk*. Suppose a robot has been designed so that it moves a unit step at a time, in one of four directions: forward, left, right, or backwards. If the movement of the robot is guided by a random signal generator so that with each step all four directions are equally likely irrespective of the directions of earlier steps, what is the probability that the robot returns for the first time to its initial position

a) after two steps?
b) after four steps?
c) after six steps?

31.6. A k-tuple of die throw results can also be represented as a graph, or path, connecting appropriate nodes in an array such as shown in Figure A.3. Successive columns represent successive die throws, and the rows indicate the number of points obtained.

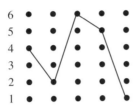

Figure A.3 Die throw results represented as a graph

Shown is the graph for the five-tuple $(4, 2, 6, 5, 1)$. This representation is sometimes useful for treating constraints among the components of a k-tuple.

a) Suppose that the path for a five-tuple is not allowed to have any upward sloping segment. Show that for five-tuples beginning with i and ending with j ($j \leq i$), the number of distinct paths satisfying this constraint is as follows:

$i - j$:	0	1	2	3	4	5
Number of distinct paths:	1	4	10	20	35	56

Note: Two paths are distinct if they differ in at least one segment.

Suggestion: For each unit increase in $i - j$, observe how many additional paths get added to the total.

b) The entries of a five-tuple are generated by means of separate throws of a fair die with the constraint that in each throw of the die, only those numbers are allowed that do not exceed numbers obtained on any of the preceding throws. Thus, if the first throw results in a 4, then a 5 on the second throw doesn't count and the die throw is repeated until a number not exceeding 4 is obtained, and so on. And if the first throw results in a 1, further throws of the die are unnecessary since all the remaining entries of the 5-tuple have to be 1's. Use the result of part (a) to show that the probability is 1/2 that a five-tuple obtained in this manner has a 1 in the last position.

Suggestion: Use the product probability space for five independent die throws.

31.7. Consider a situation just as described in Example 31.2, except for the following modification. Suppose that the job completion time is fixed and equals two time units. What is the probability that a job arrives during an interval in which the queue remains full, so that the job cannot be accepted into the queue?

31.8. *Clusters.* Let a low-probability event A ($0 < P(A) \ll 1$) be associated with an experiment $\mathfrak{E}$. During n independent executions of $\mathfrak{E}$, there may be isolated occurrences of A, but there may also be clusters. Many kinds of clusters can be defined. We will consider the simplest kind—a run. Thus, if A arises in k consecutive trials among the total of n trials ($k \leq n$), we have a run of length k. We wish to investigate how likely it is for such clusters to occur purely 'by chance', that is, without any hidden dependence between successive trials.

Let $\mathfrak{E}_n$ be the product experiment consisting of n independent consecutive executions of $\mathfrak{E}$. The possible outcomes of $\mathfrak{E}_n$ can be represented as n-tuples with each component either A or A^c. Let A_k be the event that $\mathfrak{E}_n$ results in an n-tuple that includes a string of k or more consecutive A's. Of interest is the probability $P_n(A_k)$ of that event. It turns out that there is no simple closed-form expression for this probability, as a function of k, n, and $P(A)$.

a) Why is the following argument and result incorrect: Among the n-tuples making up the event A_k there are those that begin with k A's. If all the remaining components of those n-tuples are arbitrary, the probability of this set of n-tuples is $P(A)^k \cdot 1 \cdot 1 \cdot \ldots \cdot 1 = P(A)^k$. Then there are sets of n-tuples all of which have k consecutive A's beginning at position 2, and at position 3, etc. Since each of these sets of n-tuples have probability $P(A)^k$, we obtain $P_n(A_k) = (n - k) P(A)^k$.

b) Draw the state diagram for the possible outcomes of $\mathfrak{E}_n$. (Note: A dynamic state diagram is required that makes a transition as each trial is performed. The different states represent different numbers of consecutive A's that have just occurred.)

c) Consider $k = 2$. Since A_2 encompasses those n-tuples that exhibit *at least one* instance of two consecutive A's, we work with the complementary event, A_2^c. In the state diagram of part (b), identify the portion of the diagram that generates A_2^c. From this portion of the state diagram, derive recursion relationships for the state probabilities and for $P_n(A_2^c)$, for $n = 1$, 2, 3,... Use a computer to determine from these recursion relationships the smallest n such that $P_n(A_2) > 0.5$, for $P(A) = 0.1$ and 0.01.

d) Repeat for $k = 3$, with $P(A) = 0.1$ only.

31.9. Refer to the situation described in Problem 29.6. What is the probability that after four weeks there is at least one student who has not been called upon at all?

Section 32

32.1. Suppose a biased coin is such that when it is thrown twice, this generates exactly the same amount of uncertainty as a single throw of a fair coin. What probabilities are associated with the two sides of this biased coin?

32.2. Prove Equation (32.7).

32.3. In Example 32.2, compute $H(\mathscr{P}_w | \mathscr{P}_1)$ and thus verify that $H(\mathscr{P}) = H(\mathscr{P}_1) + H(\mathscr{P}_w | \mathscr{P}_1)$ is satisfied.

32.4. A red and a white die are thrown.

a) By how much is the entropy reduced if only the absolute difference of the number of dots showing on each die is observed, as compared to the number of dots obtained on each of the dice individually?

b) Based on entropy computations, determine whether all uncertainty can be eliminated by observing, in addition, the number of dots showing on the white die. Interpret your result using a Venn diagram.

32.5. Consider an experiment $\mathfrak{E}$ in which a coin is tossed until a head appears for the first time. Of interest is the number of tosses required.

a) What is the entropy associated with this experiment?

b) Let $\mathfrak{E}_n$ ($n = 1, 2, 3, \ldots$) denote the experiment of tossing a coin n times and noting the result of each toss. For what value of n is the entropy of $\mathfrak{E}_n$ closest to the entropy of $\mathfrak{E}$?

c) As an alternative approach to finding the entropy associated with $\mathfrak{E}$, suppose that initially only $\mathfrak{E}_1$ gets performed; i.e., a partial observation is made. If that toss yields a 'tail', then the remainder of the experiment is again $\mathfrak{E}$. Set up Equation (32.7) appropriately and solve for $H_{\mathfrak{E}}$.

32.6. In a board game, when it is a player's turn, that player moves ahead as many steps as the number of dots obtained upon throwing a die. However, if a six is thrown the player gets to throw again and move ahead an additional number of steps equal to the result of the second throw. If the second throw is again a six, the player gets to throw a third time, and so on.

a) If the maximum number of throws allowed is four, define partial and conditional entropies and apply Equation (32.7) to obtain the entropy associated with a player's turn in the game.

b) If there is no limit on the number of throws, what is the entropy of a player's turn?

32.7. Prove the inequality $\ln x \leq x - 1$ ($x > 0$), which is used in the proof of Theorem 32.1. (Suggestion: Show that $y = \ln x$ is a convex upward function of x, and that $y = x - 1$ is a tangent to $y = \ln x$.)

Section 33

33.1. In a certain supermarket, equal shelf space has been allocated to the 15 different breakfast cereals that are stocked. The manager tells a clerk to allocate double the shelf space to the popular ones. How might the clerk arrive at a fuzzy set description of 'the popular cereals'?

33.2. Fuzziness can arise also as a result of different interpretations of an object (or situation or action) by different people. In Problem 1.3, fuzziness of the concept 'chair' is explored by considering a drawing that some viewers will accept as depicting a chair, others not.

a) Produce a sketch of a chair. Then vary one of the dimensions of the chair you have depicted, and draw four additional sketches in which the selected dimension is increased (decreased) somewhat, and increased (decreased) greatly.

b) Let S be the space representing a continuum of drawings of the kind you have produced in (a), with the variable dimension x ranging from 0 to ∞. On S define a fuzzy set C ('chair'), whose membership function $\mu_{\mathsf{C}}(x)$ is meant to express, for each x, what fraction of some population of viewers will interpret the picture as the drawing of a chair. Sketch the graph of what you think would be a suitable membership function.

33.3. Take an arbitrary passage of two to three sentences of text from a newspaper or magazine and identify some of the fuzzy concepts you can find therein. For each of these, explain whether, and how, it can be modeled by a suitable fuzzy set.

33.4. Let S be a set of elements ξ representing the students at a particular college. Suppose that 'the tall students' are characterized by the fuzzy set $\mathsf{T} \subset \mathsf{S}$, with

$$\mu_T(\xi) = \begin{cases} 0, & x \le 5 \\ 0.5\,(x-5)^2, & 5 < x \le 6 \\ 1 - 0.5\,(x-7)^2, & 6 < x \le 7 \\ 1, & x > 7, \end{cases}$$

where $x = x(\xi)$ is the person's height in feet. The president of the Student Council, George Garcia, is $5'6''$ tall. The students who are 'much taller than George (Garcia)' are characterized by the fuzzy set $\mathsf{G} \subset \mathsf{S}$ defined by the membership function

$$\mu_G(\xi) = \begin{cases} 0, & x \le 5.5 \\ 1 - \dfrac{(x-7)^2}{2.25}, & 5.5 < x \le 7 \\ 1, & x > 7. \end{cases}$$

a) Plot the graphs of the two membership functions.
b) With 'the tall students' and 'the students much taller than George' defined in this way, are the students much taller than George a subset of the tall students?
c) What is the set of tall students who are much taller than George?
d) What is the set of students who are either tall or much taller than George?
e) What is the set of students who are not tall but are much taller than George?

33.5. *Fuzzy Arithmetic.*
In the space of real numbers, let $\mathsf{A} = \{x\colon x \gg 1\}$, and $\mathsf{B} = \{y\colon y \ll 10\}$.

a) Specify suitable membership functions $\mu_A(x)$ and $\mu_B(x)$, and sketch their graphs.
b) Obtain $\mu_{A \cap B}(x)$.
c) Now let $z = x - y$; i.e., we consider the real numbers z obtained by taking the difference $x - y$, where $x \in \mathsf{A}$ and $y \in \mathsf{B}$. The numbers z, therefore, are members of a fuzzy set $\mathsf{C} = \{x - y\colon x \gg 1, y \ll 10\}$. To obtain $\mu_C(z)$, take first a particular z. This results if from a given x gets subtracted the y-value $y = x - z$; i.e., with $x \in \mathsf{A}$ must be paired the y-value $(x - z) \in \mathsf{B}$. Then, for this particular z and given x we have $\mu_C(z) = \mu_{A \cap B}(x) = \min[\mu_A(x), \mu_B(x - z)]$. But this expression holds for any x, i.e., we can take the union over all $\{x\}$, so that

$$\mu_C(z) = \max_x \min\,[\mu_A(x),\ \mu_B(x - z)].$$

And since this is true for every z, the above expression holds for all z. Evaluate $\mu_C(z)$ and sketch its graph.
d) To what degree is it true that $z \le 1$, or, what degree of membership does z have in the set of numbers that are ≤ 1?

Part III Problems

Section 34

34.1. Consider the experiment in which two fair (unbiased) dice are thrown and the number of dots facing up on each is noted. On the probability space for this experiment, define a discrete random variable M that expresses the maximum of the number of dots showing on the two dice. Sketch the bar graph that describes P_M, the probability distribution induced by M.

34.2. Consider the experiment of tossing a coin an unending number of times in order to record a 'coin-toss sequence', such as HTTHHHHTHTTT... On the probability space for this experiment a discrete random variable X is defined that expresses the length of the run (of either H's or T's) that begins with the result of the first coin toss. (For instance, in the outcome HTTHHHHTHTTT..., this run has length 1 since the first H is immediately followed by a T.)
   Find the probabilities induced by X on the one-element sets $\{n\}$, $n = 1, 2, 3, \ldots$, and sketch the bar graph that describes P_X.

34.3. You get dealt 13 cards from a well-shuffled standard bridge deck of 52 cards. You are interested in how many aces you receive. On the probability space for this experiment, define the discrete random variable A that expresses the number of aces you are dealt. Sketch the bar graph that describes P_A, the probability distribution induced by A.

34.4. Let the discrete random variable X express the number of cards that need to be drawn from a well-shuffled standard deck of 52 cards till the first ace is obtained.

   a) Obtain an expression for $P_X(\{n\})$, $1 \leq n \leq 49$.
   b) For which n is $P_X(\{n\})$ largest, and what is this largest value of $P_X(\{n\})$?

34.5. A die is thrown n times. Let X be a random variable that expresses the *smallest* number of dots obtained in the n throws.

   a) With $n = 2$, what is the distribution induced by X?
   b) Repeat for $n = 6$.

34.6. a) Find the probabilities of the following, for $n = 2, 3, 4, 5, 6, 7$: 'In the throw of n dice, a different number of dots shows on each die.'
   b) How can you define a single random variable that will induce these probabilities in a meaningful way into the real axis R?

Section 35

35.1. Suppose that a loaded (i.e., biased) die is thrown. In eight independent throws, the probability that an even number will appear four times is twice the probability that an even number will appear three times. What is the probability that an even number will not appear at all in eight independent throws of the die? (After [Pa1].)

35.2. A particular diner bakes and serves its own blueberry muffins. A large number n of muffins get baked at one time. How many blueberries need to get mixed into the dough if the probability that a muffin ends up with no blueberries at all is to be at most 0.01?
   (Suggestion: Use the Poisson approximation.)

35.3. Consider the experiment in which two persons each toss a coin ten times. What is the probability that they obtain the same number of 'heads'?

35.4. *Truncated binomial distribution.* A coin is to be tossed until six heads have been obtained or until ten tosses have been completed, whichever comes first. Let C be the random variable that expresses the number of heads obtained, and write an expression analogous to Equation (35.1) for the induced probabilities. C induces a 'truncated binomial distribution'.

35.5. You will win a prize if in a series of fair coin tosses, heads and tails come up an equal number of times. You have to decide in advance how many times the coin is to be tossed, which may be anywhere from two to twenty times.

 a) How many times should the coin be tossed to maximize your chance of winning?
 b) Sketch a graph showing the probability of winning as a function of the number of coin tosses.

35.6. Referring to Example 35.2, find the Poisson approximation to the probabilities of obtaining two, four, six, eight, and ten aces of hearts. Then compare these results with the actual values obtained by direct computation. Does the approximation become better or worse with increasing m?

Section 36

36.1. a) Express in set notation:
  i) The set of rational numbers greater than π and less than 2π
  ii) The set of irrational numbers greater than π and less than 2π
  iii) The set of all those irrational numbers that differ from an integer by less than 1/10.
 b) What is the Lebesgue measure of each of the sets in (a)?

36.2. When considering the positive real axis or the negative real axis, one often refers to these as 'the right half' and 'the left half' of the real axis, respectively. Show that these notions of 'right half' and 'left half' are arbitrary by demonstrating that for *any* real x, the intervals $\overline{-\infty, x}$ and $\overline{x, \infty}$ can be regarded as the left half and the right half of the real line.

36.3. Given sets A, B, C $\subset$ R. It is known that A is an interval, and that $L(A) = L(B) = L(C) = 1$. Also, $L(AB) = L(AC) = L(BC) = 1/2$. In each of the following cases, specify three sets A, B, C that satisfy these requirements and the stated additional one:

 a) $L(ABC) = 0$ b) $L(ABC) = 1/2$ c) $L(ABC) = 1/4$.

36.4. Determine the Lebesgue measure of the following sets in R:

 a) $\bigcup_{i=1}^{\infty} A_i$, where $A_i = \overline{i, i+2^{-i}}$

 b) $\bigcup_{i=1}^{10} A_i$, where $A_i = \overline{i, i+2}$

 c) $\bigcap_{i=1}^{\infty} A_i$, where $A_i = \overline{-i, i}$

 d) $\bigcup_{i=1}^{\infty} A_i$, where $A_i = \overline{i, i+\frac{1}{i}}$

 e) $\bigcap_{i=1}^{\infty} A_i$, where $A_i = \overline{1, 1+\frac{1}{i}}$

 f) $\bigcap_{i=1}^{\infty} A_i$, where $A_i = \overline{1, 2-\frac{1}{i}}$

 g) $\bigcup_{i=1}^{\infty} A_i$, where $A_i = \overline{1, 2-\frac{1}{i}}$.

36.5. Let $A = \overline{1, 2}$, and $B = \{x^2 : x \in A\}$. Is it correct to say that $B = \overline{1, 4}$?

36.6. Let a, b, c, d be real numbers, and let $A = \overline{a, b} - \overline{c, d}$, $B = \overline{a, b} \Delta \overline{c, d}$. If $A = B$, what must be the relationships between a, b, c and d? (Note that the specification of A and B already requires that $a < b$ and $c < d$.)

Section 37

37.1. On a probability space a random variable N is defined that induces the geometric distribution described in the graph of Figure 34.5. What probabilities does N assign to the sets in $\mathcal{V}$?

37.2. Given a probability space $(S, \mathcal{A}, P)$ and a partition $\mathcal{P} \subset \mathcal{A}$. A r.v. X is specified as follows: For each ξ, $X(\xi) = P(A_\xi)$, where A_ξ is that set in $\mathcal{P}$ that has ξ as one of its elements. Verify that X is a r.v. Then sketch a typical probability distribution induced by such a r.v. X over the sets $\{P(A_\xi)\}$, for the case where all the sets in $\mathcal{P}$ have different probabilities.

37.3. Consider the set R_a of all rational numbers.

a) Demonstrate that $R_a \subset \mathcal{B}$.
b) Show that R_i, the set of all irrational numbers, also belongs to $\mathcal{B}$.

37.4. Consider an induced probability space $(R, \mathcal{B}, P_X)$ where P_X distributes probability over the real line symmetrically with respect to the origin and assigns $P_X(\{0\}) = 0$. Show that for every real number $a > 0$, the pair of sets $A_a = \overline{-a, a}$ and $B = \overline{0, \infty}$ are independent.

37.5. Given a random variable X defined on some probability space $(S, \mathcal{A}, P)$. Referring to the class $\mathcal{V}$ defined in Equation (37.1), express each of the following sets in terms of sets belonging to $\mathcal{V}$ only. (After [Pf1].)

a) $\{\xi : X(\xi) \leq 0\}$ Answer: $X^{-1}(\overline{-\infty, 0})$. b) $\{\xi : X(\xi) < a\}$
c) $\{\xi : X(\xi) \geq a\}$ d) $\{\xi : X(\xi) \in \overline{a, b}\}$
e) $\{\xi : X(\xi) \in \overline{a, b}^c\}$.

37.6. Let X be a r.v. defined on a probability space $(S, \mathcal{A}, P)$, and B_1, B_2 any two Borel sets.

a) If $B_1 \subset B_2$, show that $X^{-1}(B_1) \subset X^{-1}(B_2)$.
b) If B_1, B_2 are disjoint, show that $X^{-1}(B_1)$, $X^{-1}(B_2)$ are also disjoint.

37.7. The class $\mathcal{V}$ defined in Equation (37.1) is not the only class that can be used to generate $\mathcal{B}$. For each of the following classes, determine whether it can be used to generate $\mathcal{B}$:

a) The class $\{\overline{-\infty, a} : a \in R\}$

b) The class $\{\overline{a, \infty} : a \in R\}$

c) The class $\{\overline{-a, a} : a \in R, a > 0\} \cup \{\{0\}\}$

d) The class $\{\{a\} : a \in R\}$.

Section 38

38.1. Consider the r.v. X whose c.d.f. is shown in Figure 38.1. Express each of the following probabilities in terms of F_X, and evaluate:

a) $P_X(\overline{-\infty, 2'})$

b) $P_X(\overline{1, 3'})$

c) $P_X(\overline{4, \infty})$

d) $P_X(\{2, 2.5, 7\})$.

38.2. A discrete r.v. X is defined on a probability space in such a way that it induces probability 1 on the set I_+ of positive integers, and induces the following probabilities on the one-element subsets of I_+:

$$P_X(\{k\}) = a^{-2k}, \; k = 1, 2, 3, \ldots$$

a) Determine the constant a.

b) Sketch the cumulative distribution function, $F_X(x)$.

c) Express the probability of each of the following events in terms of $F_X(x)$ and evaluate:

$$\overline{2, 3'}; \overline{2, 4}; \overline{4, \infty}; \overline{1, 3'}$$

38.3. An absolutely continuous r.v. X induces the following probabilities:

i) $P_X(\overline{-\infty, 0'}) = 0$.

ii) For any two positive numbers a, b, $P_X(\overline{0, b'}) = P_X(\overline{a, a+b} \,|\, \overline{a, \infty})$.

For instance, let X express the time until failure, in hours, of some device under test. If the device has operated without failure for a hours, then the probability that failure occurs in the next b hours equals the probability that failure occurs in the first b hours of operation. In other words, the time until failure is independent of how long the device has been in operation.

a) Show that the c.d.f. $F_X(x)$ that characterizes the induced distribution satisfies the following relation:

$$\frac{1 - F_X(a+b)}{1 - F_X(a)} = 1 - F_X(b)$$

b) Denoting $1 - F_X(x)$ by $g(x)$, show that $g(x)$ must be an exponential function.

c) Obtain the most general expression for $F_X(x)$.

38.4. Consider an Internet access provider who has introduced a rate structure where users are charged by connect time per session. Ten minutes after connecting, and every ten minutes thereafter (as long as the session continues), the customer receives a signal indicating that the connect charge is about to increase. As a result, a certain fraction of users terminate their session when they receive this prompt. Suppose then that the duration, in minutes, of a session is a random phenomenon with sufficient statistical regularity and sufficient independence between users so that a probability space can be defined that models the duration of a session. On this space, let the random variable X express the duration of a user's session, with P_X

characterized by the following c.d.f.:

$$F_X(x) = \begin{cases} 1 - \dfrac{2}{3}\exp\left(-\dfrac{x}{20}\right) - \dfrac{1}{3}\exp\left(-\dfrac{1}{2}\left\lfloor\dfrac{x}{10}\right\rfloor\right), & \text{for } x \geq 0 \\ 0, & \text{for } x < 0 \end{cases}$$

where the notation $\lfloor x/10 \rfloor$ stands for the largest integer less than or equal to $x/10$.

a) What is the probability that the duration in minutes of a session is:

    less than 20 minutes;
    more than 15 minutes;
    equal to ten minutes?

b) What is the conditional probability that a session is:

    less than 30 minutes, given that it is more than ten minutes?
    more than 15 minutes, given that it is less than 30 minutes?

(After [Pa1].)

38.5. On a certain probability space a r.v. X is defined in such a way that it induces these probabilities:

$$P_X\left(\overline{x, 2x}\right) = x, \quad \text{for } x\text{-values in some interval } \overline{a, b}.$$

For what intervals $\overline{a, b}$ does this specification uniquely define the c.d.f. $F_X(x)$, and what is that c.d.f.?

Section 39

39.1. A random variable X has a p.d.f. $f_X(x)$ that is zero outside the interval $\overline{1, 5}$; it rises linearly from 0 to a value a over $\overline{1, 2}$; and it decreases linearly from a to zero over $\overline{2, 5}$.

a) Sketch the c.d.f., $F_X(x)$.
b) Determine the probability $P_X(\overline{1.5, \infty})$.

39.2. An absolutely continuous random variable X is defined on a probability space $(S, \mathscr{A}, P)$. The probability distribution P_X that X induces on $\mathscr{B}$ is described by a p.d.f. that is triangular over $\overline{0, 1}$, with peak at 0.5. Over $\overline{0, 1}^c$, the p.d.f. is 0. Consider the following events belonging to $\mathscr{A}$:

    $A \leftrightarrow$ 'X assigns a value less than or equal to 0.5'
    $B \leftrightarrow$ 'X assigns a value greater than or equal to 0.5'
    $C \leftrightarrow$ 'X assigns a value greater than 0.25 but less than 0.75'.

a) Determine $f_X(0.5)$.
b) Find $P(A)$, $P(B)$, $P(C)$, and $P(A|B)$.

c) Which pairs of the events A, B, C, if any, are independent events? Justify your answer. (After [Pf1].)

39.3. Let X be a random variable of mixed type, which induces a distribution that is described by the probability density function

$$f_X(x) = \alpha f_1(x) + 0.4\,\delta\,(x - 0.5)$$

where $f_1(x)$ is the p.d.f. specified in Problem 39.2. Now do (a), (b) and (c) as given in Problem 39.2.

39.4. During a period of fairly consistent meteor activity, a group of students timed the occurrences of successive meteors visible during part of one night and found the time between occurrences, in minutes, to be described quite well by an *exponential* probability distribution with p.d.f.

$$f_X(x) = \begin{cases} ce^{-x/4}, & \text{for } x > 0 \\ 0, & \text{otherwise.} \end{cases}$$

The underlying p.e. in this case consists of measuring the time from a meteor sighting until the next sighting.

a) Determine c and sketch the density function.
b) Based on this p.d.f., find the probability that the time between successive meteors is:

more than eight minutes;
less than four minutes;
between four and eight minutes.

c) Let A_a denote the event that the time between two successive meteors is more than a minutes. What is $P(A_a)$? For some $b > 0$, show that irrespective of the value of b, $P(A_{a+b}|A_b) = P(A_a)$. Therefore, no matter at what instant of time the observation is begun, the likelihood that the next meteor will come more than a minutes later is the same.

39.5. The time until a radioactive atom decays is governed by an exponential distribution. The *half-life h* of the atom is the time interval $0, h$ over which the exponential distribution induces probability 1/2.

a) Express the parameter α in Figure 39.1b in terms of the half-life, h. Then write the p.d.f. using h as the parameter.
b) If the half-life of a certain atom is exactly 12 months, what is α? (Indicate the correct units.)

39.6. An experiment results in a numerical observation that is characterized by a r.v. X with p.d.f.

$$f_X(x) = \begin{cases} 0, & x < 0 \\ ax, & \le x < 1 \\ ax^{-2}, & x \ge 1. \end{cases}$$

The experiment is repeated ten times.

a) What is the probability that at least one of the ten independent observations results in a value greater than 5?
b) What is the probability that exactly five of the observations yield a value greater than 1?
c) What is the value of b that maximizes $P_X(b, b+1)$? In other words, what is the unit interval into which the numerical observation is most likely to fall?

39.7. A marquee uses 1000 light bulbs. The time to failure of any light bulb is described by an exponential r.v. X, with

$$f_X(x) = \begin{cases} \alpha e^{-\alpha x}, & x \geq 0 \\ 0, & x < 0 \end{cases}$$

where x is in hours. The marquee is to be operated for 200 hours without servicing (i.e. without replacement of burned out light bulbs). If the probability is 0.5 that a bulb burns out during the first 1000 hours, what is the probability that in those first 200 hours of operation of the marquee, no more than 100 light bulbs will fail?

39.8. A straight line is drawn from the origin of the x,y-plane at an angle θ with respect to the horizontal axis. Suppose that θ is chosen by a random mechanism from the range $0 \leq \theta \leq \pi/2$ without bias, so that it can be represented by a random variable Θ that is *uniformly* distributed over $\overline{0, 2\pi}$. For an arbitrary positive number y_o, what is the probability that the line drawn at angle θ will intersect a vertical line segment extending from $(1, 0)$ to $(1, y_o)$?

39.9. The random variable X induces a distribution characterized by the p.d.f.

$$f_X(x) = \frac{c}{d^2 + x^2}.$$

a) Show that X is a Cauchy r.v. by converting f_X into the form given in Figure 39.1d. Does f_X as given above characterize a two-parameter family of distributions?
b) Obtain the c.d.f. $F_X(x)$, and plot both $f_X(x)$ and $F_X(x)$ for the case where $P_X(\overline{-1, 1}) = 1/2$.

39.10. Show that a p.d.f. cannot have a tail that decays as $|x|^{-1}$. Specifically, consider

$$f(x) = \begin{cases} \text{arbitrary}, & x < t \\ \dfrac{c}{x}, & x \geq t \end{cases}$$

where t, c are positive constants.

Section 40

40.1. What probability does a standardized Gaussian r.v. X_0 assign to the interval bounded by the two points of inflection of f_{X_0}? Generalize the result to an arbitrary Gaussian r.v. X, showing that the answer is the same.

40.2. An electronic noise generator produces an output voltage such that, when the output-level control is kept at a constant setting, the instantaneous voltage at an arbitrary instant of time is characterized by a Gaussian r.v. with a distribution that is symmetrical about 0 V. Furthermore, with a probability of 0.9, the output voltage does not have an absolute value greater than 1.0 V. What is the probability that a particular measurement of the output voltage results in a positive voltage value greater than

a) $+2.0\,\text{V}$;
b) $+5.0\,\text{V}$?

40.3. Let X_0 be a standardized Gaussian r.v. Let α be the positive real number that results in

$$P_{X_0}\left(-\infty,\ -\alpha \cup \overline{\alpha,\ \infty}\right) = 10^{-5}.$$

Use the bound b_1 to estimate the value of the multiplier q that will make

$$P_{X_0}\left(-\infty,\ -q\alpha \cup \overline{q\alpha,\infty}\right) = 10^{-10}.$$

40.4. Let X be a r.v. that induces a Gaussian distribution described by the p.d.f. (40.1), where a is an arbitrary fixed constant and σ is an arbitrary fixed positive constant.

a) Express $P_X\overline{(1,\ 2)}$ in terms of Φ.
b) What is the largest value of σ for which it is possible to have $P_X\,\overline{(1,\ 2)}=0.5$?

40.5. Let X be a Gaussian r.v. Also, let A be the event that X assigns a value greater than 0, and let B be the event that X assigns a value greater than 1.

a) If the parameter a in Equation (40.1) is zero, find σ such that the conditional probability of B, given A, is 0.5.
b) If the parameter σ in Equation (40.1) equals 1, find a such that the conditional probability of B, given A, is 0.5.

40.6. This problem is intended to provide a better appreciation for the rapid decay of the tails of a Gaussian distribution.

a) Let $f_X(x)$ be the standardized Gaussian p.d.f. On the same set of axes, superimpose the following plots:
  i) $f_X(3 + x)/f_X(3)$ for $0 \le x \le 1$
  ii) $f_X(4 + x)/f_X(4)$ for $0 \le x \le 1$.
b) Suppose that additional plots are made of $f_X(n + x)/f_X(n)$ for $n = 5, 6, \ldots$ What will be the limiting value of the slope at $x = 0$?
c) Repeat parts (a) and (b) for a Cauchy distribution, using the Cauchy density function given in Figure 39.1d with $a = 0$ and $d = 1$.

Section 41

41.1. The two-dimensional probability distribution induced by two discrete r.v.'s X, Y is given by

$$P_{X,Y}(\{(x,\ y)\}) = \begin{cases} \dfrac{ax}{y}, & \text{for } x \in \{1,2,3\}, \quad y \in \{1,2,3\} \\ 0, & \text{otherwise.} \end{cases}$$

Find:
a) $P(\{\xi\colon X(\xi) + Y(\xi) = 4\})$
b) $P(\{\xi\colon X(\xi) > 2,\ Y(\xi) \ge 2\})$.

41.2. Let X and Y be two statistically independent discrete r.v.'s. It is known that X assigns probabilities 1/8, 1/2 and 3/8 to the point sets $\{0\}$, $\{1\}$, and $\{4\}$, respectively; and that Y assigns probabilities 2/3 and 1/3 to the sets $\{0\}$ and $\{2\}$, respectively. Determine $P_{X,Y}(\{(x,y)\})$ and sketch the bar graph describing this joint distribution.

41.3. Two discrete r.v.'s X, Y induce the two-dimensional distribution

$$P_{X,Y}(\{(x,y)\}) = c \min(x,y), \text{ where } x, y = 0.5, 1, 2$$

   a) Determine the value of c.
   b) Determine $P_X(\{x\})$ and $P_Y(\{y\})$.
   c) Are X and Y statistically independent r.v.'s?

41.4. Doreen tosses a coin ten times and Jack tosses a coin ten times.

   a) Let the r.v. X express the number of 'heads' obtained by Doreen, and Y the number of 'heads' obtained by Jack. Obtain an expression for $P_{X,Y}(\{(x,y)\})$.
   b) Now let the r.v. X express by how many 'heads' Jack's and Doreen's results differ, and Y by how many 'tails' Jack's and Doreen's results differ. Obtain an expression for $P_{X,Y}(\{(x,y)\})$, and plot.

41.5. Let G and H be two discrete r.v.'s defined on a probability space $(S, \mathscr{A}, P)$. G induces the probability space $(R, \mathscr{A}_G, P_G)$; H induces the probability space $(R, \mathscr{A}_H, P_H)$; and G and H jointly induce the probability space $(R^2, \mathscr{A}_{G,H}, P_{G,H})$. Show that if $\{(g,h)\} \in \mathscr{A}_{G,H}$, then $\{g\} \in \mathscr{A}_G$ and $\{h\} \in \mathscr{A}_H$.
   Note: $\{g\} \in \mathscr{A}_G$ if $G^{-1}(\{g\}) \in \mathscr{A}$.

41.6. A lady named Audrey is slightly gifted in extrasensory perception, so that she is able to correctly 'call' the result of a coin toss with probability 0.6. An experiment is performed in which Audrey is pitted against an impostor, Harvey, who has no detectable psychic ability. They are both, independently, asked to predict the outcome of ten successive coin tosses. Assume statistical regularity to hold (which is usually not the case in experiments involving extrasensory perception). What is the probability that Audrey's score (number of correct predictions) will be larger than Harvey's,

   a) for just the first coin toss;
   b) for all ten tosses?

Section 42

42.1. Express each of the sets described in Figure A.4 in terms of rectangle sets, and express each rectangle set as a Cartesian product. In each case, the set in question consists of the shaded area and the boundary points lying on solid line segments.
   Note: The set described in (e) includes the points $(2, 0)$ and $(1, 2)$.
   (Answer to case (a): $\overline{-2, 2} \times \overline{-2, 2} - \overline{-1, 1} \times \overline{-1, 1}$.)

42.2. A r.v. X is defined on a probability space $(S, \mathscr{A}, P)$. The range of X is $\overline{-1, 1} \cup \overline{2, 3}$. Another r.v., Y, is also defined on the same probability space. Y is a discrete r.v. whose range is $\{1/n: n = 1, 2, 3, \ldots\}$. The range of either r.v. is not influenced by the value assigned by the other. Describe the rectangle set consisting of all the pairs of numbers (x, y) that can be assigned by X, Y jointly. What is the size of the class of disconnected subsets of R^2 making up this rectangle set?

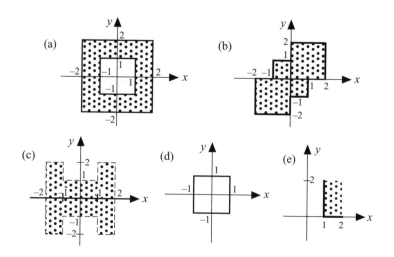

Figure A.4 Sets to be expressed in terms of rectangle sets

42.3. Refer to Example 42.2. Prove that

$$A = \bigcap_{n=2}^{\infty} A_n.$$

Suggestion: Clearly, $A \subset A_n$ for $n = 2, 3, 4, \ldots$ Therefore, for any point $(x, y) \notin A$, show that there is a set A_n such that $(x, y) \notin A_n$.

42.4. A two-dimensional r.v. X has as its range a subset of R^2 that is the line segment

$$\{(x, y): \ y = 2x, \ 0 \le x \le 1, \ 0 \le y \le 2\}.$$

Verify that this set is a two-dimensional Borel set.

42.5. Prove that the intersection of two rectangle sets, if it is nonempty, is a rectangle set.

Section 43

43.1. Obtain and sketch the two-dimensional c.d.f. $F_{X,Y}(x, y)$ that describes the distribution induced by the r.v.'s X, Y defined in Problem 41.2.

43.2. Consider the two random variables X, Y defined in Problem 41.1. Specify the set of points (x, y) in R^2 at which $F_{X,Y}(x, y) = 1/6$. Express this set in the form $A \times B$, i.e., as a product set.

43.3. Two r.v.'s X, Y have the following joint distribution function:

$$F_{X,Y}(x, y) = \begin{cases} 0, & x < -1, \ y < -1 \\ [1 + (x+1)(y+1)]/8, & -1 \le x, \ y < 1 \\ (y+2)/4, & -1 \le y < 1 \text{ and } 1 \le x \\ (x+2)/4, & -1 \le x < 1 \text{ and } 1 \le y \\ 1, & x, \ y \ge 1. \end{cases}$$

Identify the discrete component of the joint distribution induced by (X, Y).

43.4. Show that, in order for a function $F(x, y)$ to be a two-dimensional c.d.f., it is necessary that $F(x, -\infty) = 0$ for all x, and $F(-\infty, y) = 0$ for all y.

43.5. Evaluate the expression for $P_X(\mathsf{B})$ obtained in Example 43.3 if

$$
F_X(x_1, x_2) = \begin{cases}
(1+x_1)/2, & -1 \le x_1 < 1, \ x_2 \ge 0 \\
(1+x_1)(1+x_2)/2, & -1 \le x_1 < 1, \ -1 \le x_2 < 0 \\
1 + x_2, & x_1 \ge 1, \ -1 \le x_2 < 0 \\
1, & x_1 \ge 1, \ x_2 \ge 0 \\
0, & \text{otherwise.}
\end{cases}
$$

43.6. Two r.v.'s X, Y distribute probability 1/2 uniformly over the square $\overline{-1, 1} \times \overline{-1, 1}$. The remaining probability 1/2 is distributed uniformly over the line segment $y = -x$, for $x \in \overline{-1, 1}$.

a) Determine $F_{X,Y}(x, y)$.

b) With B as defined in Example 43.3, compare the result obtained for $P_{X,Y}(\mathsf{B})$ using the two formulas in that Example.

43.7. A pair of r.v.'s X, Y jointly induce probability uniformly over the circumference of the unit circle, $\{(x, y): x^2 + y^2 = 1\}$. Obtain $F_{X,Y}(x, y)$ and $f_X(x)$.

Section 44

44.1. Consider the following subsets of R^2:

$$
\mathsf{A} = \overline{0, 1} \times \overline{0, 1}; \quad \mathsf{B} = \overline{0.5, 1.5} \times \overline{0, 1}; \quad \mathsf{C} = \overline{0.25, 1.25} \times \overline{-1/3, 2/3}.
$$

Determine: $L(\mathsf{A})$, $L(\mathsf{B})$, $L(\mathsf{C})$, $L(\mathsf{AB})$, $L(\mathsf{AC})$, $L(\mathsf{BC})$, $L(\mathsf{ABC})$.

44.2. Two random variables X, Y jointly induce probability uniformly over the square $\overline{0, 2} \times \overline{0, 2}$. Write an expression for each of the following probabilities and evaluate:

a) The probability that the sum of the values assigned by the two random variables is greater than 1.

b) The probability that twice the value assigned by Y is greater than the value assigned by X.

c) The probability that both random variables assign the same value.

d) The probability that X and Y differ by less than 1.

44.3. Determine the Lebesgue measure of the following sets. (The sets in (a), (b), (c) are subsets of R^2, the set in (d) is a subset of R^3.)

a) $\displaystyle\bigcup_{i=1}^{\infty} (\mathsf{A}_i \times \mathsf{A}_i)$, where $\mathsf{A}_i = \overline{i, i + 2^{-i}}$

b) $\displaystyle\bigcap_{i=1}^{\infty} \left(\overline{1, 1 + \frac{1}{i}} \times \overline{1, i+1}\right)$

c) $\displaystyle\bigcup_{i=2}^{\infty} \left(\overline{1, 3 - \frac{1}{i!}} \times \overline{2, 2 + \frac{1}{i-1}}\right)$ (Ans. : $e - 1$)

d) $\displaystyle\bigcup_{i=1}^{\infty} \overline{(0, 2 - \frac{2}{i^2} \times i, i + \frac{1}{i}} \times \overline{1, 1 + \frac{1}{i})}.$

Note: $\displaystyle\sum_{i=1}^{\infty} \frac{1}{i^2} = \frac{\pi^2}{6}; \quad \sum_{i=1}^{\infty} \frac{1}{i^4} = \frac{\pi^4}{90}.$

44.4. Two r.v.'s X, Y have the following joint p.d.f.:

$$f_{X,Y}(x, y) = \begin{cases} 0.5, & (x,y) \in \overline{0, 0.8} \times \overline{0, 2.5} \\ 0, & \text{otherwise} \end{cases}$$

a) In the (x, y)-plane identify the region corresponding to the event $\{\xi: X(\xi) < Y(\xi)\}$, and determine its probability.

b) If X and Y were identically distributed, then by symmetry, $P(\{\xi: X(\xi) < Y(\xi)\}) = 1/2$. Use this fact to obtain the actual value of $P(\{\xi: X(\xi) < Y(\xi)\})$ in a different way, namely, using conditional probabilities and the rule of total probability (Theorem 22.2), with the conditioning events $A = \{\xi: Y(\xi) \leq 0.8\}$ and $A^c = \{\xi: Y(\xi) > 0.8\}$.

44.5. Suppose that the two-dimensional r.v. (X, Y) has the joint p.d.f.

$$f_{X,Y}(x, y) = \begin{cases} \alpha (2 - x), & 0 < x < 2, \ -x < y < x \\ 0, & \text{otherwise.} \end{cases}$$

a) Determine the value of α.

b) Find the marginal (i.e., one-dimensional) density functions $f_X(x)$ and $f_Y(y)$.

44.6. Two random variables X, Y have the following joint probability density function:

$$f_{X,Y}(x, y) = \begin{cases} 6e^{-c(1.5x + y)}, & \text{for } x > 0, \ y < 0 \\ 0, & \text{elsewhere} \end{cases}$$

a) Determine the value of c.

b) Find $F_{X,Y}(x, y)$.

c) What is the probability that both X and Y assign a value greater than 1? Obtain the answer in two different ways—from $f_{X,Y}(x, y)$ and from $F_{X,Y}(x, y)$.

d) What probability does each of the two r.v.'s individually induce on $\overline{1, \infty}$?

44.7. A r.v. $X = (X_1, X_2)$ induces the two-dimensional distribution described by the c.d.f. that is specified in Figure A.5. For each of the random variables X, X_1, X_2, determine whether it is

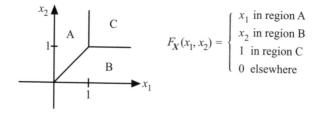

$$F_X(x_1, x_2) = \begin{cases} x_1 & \text{in region A} \\ x_2 & \text{in region B} \\ 1 & \text{in region C} \\ 0 & \text{elsewhere} \end{cases}$$

Figure A.5 A two-dimensional c.d.f

discrete, absolutely continuous, singular, or of mixed type, and describe the nature of the
probability distribution induced by each of these random variables.

Section 45

45.1. The c.d.f. describing the probability distribution induced by a r.v. X is

$$F_X(x) = \begin{cases} 0, & x \in \overline{-\infty, -1} \\ 0.4, & x \in \overline{-1, 0} \\ 0.8, & x \in \overline{0, 2} \\ 1, & x \in \overline{2, \infty}. \end{cases}$$

Defined on the same probability space is another r.v., Y, with c.d.f.

$$F_Y(y) = \begin{cases} 0, & y \in \overline{-\infty, 0} \\ 0.5y, & y \in \overline{0, 1} \\ 1, & y \in \overline{1, \infty}. \end{cases}$$

If X and Y are s.i. r.v.'s, determine and sketch the two-dimensional c.d.f. that describes the joint
distribution of X and Y.

45.2. The joint probability distribution $P_{X,Y}$ induced by two discrete random variables X and Y is
specified as follows:

$$P_{X,Y}(\{(x, y)\}) = \begin{cases} \dfrac{c^2}{xy}, & \text{for } (x, y) = (2^i, 2^j), \quad i, j = 0, 1, 2, 3, \ldots \\ 0, & \text{otherwise} \end{cases}$$

a) Are X and Y statistically independent r.v.'s?
b) Determine $F_{X,Y}(2, 3)$.

45.3. Given the indicator r.v.'s I_A, I_B of two events $A, B \subset S$, with $P(A) > 0$ and $P(B) > 0$. Prove the
following:

a) If A, B are independent events, then I_A, I_B are s.i. r.v.'s.
b) If $P_{I_A, I_B}(\{(1, 1)\}) = P(A)P(B)$, then I_A, I_B are s.i. r.v.'s.

45.4. Consider two r.v.'s X, Y defined on a probability space $(S, \mathscr{A}, P)$, where X is unary and Y is
arbitrary. Use Definition 45.1 to show that X and Y are s.i.
(Suggestion: Make use of Problem 23.2a.)

45.5. a) Let X, Y be discrete r.v.'s with $P_{X,Y}(\{(x, y)\}) = P_X(\{x\}) \cdot P_Y(\{y\})$ for every $x \in A_x$, the range
of X, and for every $y \in A_y$, the range of Y (where A_x, A_y are countable sets). Show that X, Y
are s.i.
b) More generally, let the joint distribution induced by X, Y satisfy the following relation, for
all $x \in A_x$ and all $y \in A_y$:

$$P_{X,Y}(\{(x, y)\}) = h_1(x)\, h_2(y)$$

where h_1 and h_2 are non-negative discrete functions with domain A_x and A_y, respectively. Show that X, Y are s.i.

45.6. Consider a r.v. X defined on a probability space $(S, \mathcal{A}, P)$ and let $B_1, B_2 \in \mathcal{B}$.

    a) Show that

$$X^{-1}(B_1) = X^{-1}(B_1 - B_2 \uplus B_1 B_2) = X^{-1}(B_1 - B_2) \uplus X^{-1}(B_1 B_2)$$

    i.e., $X^{-1}(B_1)$ can be expressed as the disjoint union of $X^{-1}(B_1 - B_2)$ and $X^{-1}(B_1 B_2)$.
    b) Justify Equation (45.1); that is, verify that $X^{-1}(B_1 B_2) = X^{-1}(B_1) \cap X^{-1}(B_2)$.
    c) Justify Equation (45.4).

Section 46

46.1. In each of the following cases, verify that the given function is a two-dimensional probability density function and check whether X and Y are s.i. r.v.'s:

    a) $f_{X,Y}(x, y) = \begin{cases} \dfrac{1}{3}(xy + x + y + 1), & \text{for } 0 \le x \le 1,\ |y| \le 1 \\ 0, & \text{otherwise} \end{cases}$

    b) $f_{X,Y}(x, y) = \begin{cases} \dfrac{\pi}{2}\cos\left(\dfrac{\pi y}{2x}\right), & \text{for } |y| \le x \le 1 \\ 0, & \text{otherwise} \end{cases}$

46.2. Given two s.i. Gaussian r.v.'s with p.d.f.'s

$$f_X(x) = \frac{1}{\sqrt{2\pi}\sigma_1}\exp\left[-\frac{x^2}{2\sigma_1^2}\right], \quad f_Y(y) = \frac{1}{\sqrt{2\pi}\sigma_2}\exp\left[-\frac{y^2}{2\sigma_2^2}\right].$$

    a) Show that the *contours* of the joint density function, or curves of constant $f_{X,Y}(x, y)$, are a family of ellipses.
    b) Show that the contours are circles if and only if $\sigma_1 = \sigma_2$.

46.3. Given two s.i. *Laplace* r.v.'s X, Y with density functions

$$f_X(x) = \frac{\alpha}{2}\, e^{-\alpha|x|}, \text{ for all } x; \qquad f_Y(y) = \frac{\beta}{2}\, e^{-\beta|y|}, \text{ for all } y$$

    where α, β are positive constants. Show that the constant magnitude contours of the joint p.d.f. $f_{X,Y}$ are a family of parallelograms.

46.4. Use Theorem 46.1 to prove Theorem 46.2.

46.5. Along a particular north–south road there is a one-lane bridge. Suppose that observations have demonstrated that there is sufficient statistical regularity, and have shown that the time until a car approaches the bridge from the south is adequately modeled by an exponential random variable X, with density function

$$f_X(t) = \begin{cases} 0.1\,e^{-0.1t}, & t \geq 0 \\ 0, & t < 0 \end{cases}$$

where t is in minutes. The time until a car approaches the bridge from the north is modeled by a random variable Y with the same distribution. Traffic going north can be assumed to be independent of traffic going south. It takes a car ten seconds to traverse the bridge in either direction.

What is the probability that the next car to arrive from the north and the next car to arrive from the south will get to the bridge within ten seconds of each other, so that one of the cars will have to wait?

46.6. X and Y are s.i. exponential r.v.'s (see Figure 39.1b) with parameters α and β, respectively. Determine:

a) the probability that X will assign a smaller value than Y when the underlying experiment is performed;
b) the probability that Y will assign a smaller value than X; and
c) the probability that X and Y will assign the same value.
d) How must α and β be related if the answer to (a) evaluates to 0.9?

46.7. Two computers each begin processing a job at time $t = 0$. The execution times for the two machines (machine #1 and machine #2) are characterized by s.i. r.v.'s X and Y, respectively, with the following density functions:

$$f_X(x) = \begin{cases} \alpha^2 x\,e^{-\alpha x}, & x \geq 0 \\ 0, & x < 0; \end{cases} \qquad f_Y(y) = \begin{cases} \beta^2 y\,e^{-\beta y}, & y \geq 0 \\ 0, & y < 0 \end{cases}$$

where $\alpha, \beta > 0$.

a) What is the probability that machine #1 finishes before machine #2?
b) Evaluate your result for the case where $\alpha = 2\beta$ (machine #1 is twice as fast as machine #2), and $\alpha = 10\beta$ (machine #1 is ten times as fast).

46.8. Consider two r.v.'s X, Y which are *jointly Gaussian*, so that their two-dimensional p.d.f. is given by

$$f_{X,Y}(x,y) = k\exp\left\{-\frac{1}{2(1-r^2)}\left[\frac{(x-a)^2}{\sigma_1^2} - 2r\frac{(x-a)(y-b)}{\sigma_1\sigma_2} + \frac{(y-b)^2}{\sigma_2^2}\right]\right\}$$

where a, b are any constants, $\sigma_1, \sigma_2 > 0$, and $|r| < 1$. Show that X, Y are s.i. if and only if $r = 0$.
Suggestion: Obtain f_X and f_Y by completing the square in the exponent and integrating $f_{X,Y}$.

46.9. Let X_1, X_2 be jointly absolutely continuous s.i. r.v.'s. In R^2, consider any rectangle $A = \overline{a, b} \times \overline{c, d}$, where a, b, c, d are real numbers (or $\pm\infty$) such that $a < b$, $c < d$, and $P_{X_1,X_2}(A) > 0$. If

$$f_{Y_1,Y_2}(y_1, y_2) = \begin{cases} \dfrac{f_{X_1,X_2}(y_1, y_2)}{P_{X_1,X_2}(A)}, & (y_1, y_2) \in A \\ 0, & \text{otherwise} \end{cases}$$

show that Y_1, Y_2 are also s.i. r.v.'s.

Part IV Problems

Section 47

47.1. As an alternative approach to obtaining the result given in Equation (47.5), consider again a binary r.v. X that assigns the values ± 2 with equal probability, and let $Y = X^2$. Obtain $f_Y(y)$ by beginning with (47.3).

47.2. The r.v. X induces a *Rayleigh* distribution, with p.d.f.

$$f_X(x) = \begin{cases} \dfrac{x}{h} \exp\left(-\dfrac{x^2}{2h}\right), & x \geq 0 \\ 0, & x < 0 \end{cases}$$

where $h > 0$. Let $Y = X^2$. Determine f_Y.

47.3. X induces a uniform distribution over $\overline{0, 1}$. Determine f_Y if $Y = -\ln X$.

47.4. X is an exponential r.v. with p.d.f.

$$f_X(x) = \begin{cases} \alpha e^{-\alpha x}, & x \geq 0 \\ 0, & x < 0 \end{cases}$$

where $\alpha > 0$.

a) Let $Y = e^{-2\alpha X}$. Using Rule 47.2 determine f_Y and sketch this density function.

b) Let $Z = e^{\alpha X} - 1$. Using (47.6), determine f_Z and sketch this density function.

c) Let $W = \begin{cases} X^2, & 0 < X < 1 \\ 0, & \text{otherwise} \end{cases}$. Use Equation (47.4) to find f_W and sketch this p.d.f.

47.5. An absolutely continuous r.v. X has c.d.f. $F_X(x)$. Let $Y = F_X(X)$. Show that if $f_X(x) \neq 0$ for all x, then Y is uniformly distributed over $\overline{0, 1}$.

47.6. A voltage that varies sinusoidally with time is observed at a random instant. The sinusoidal voltage has zero average value and a peak value of 1 V. What will be the value of the voltage at the moment of observation?

In order to answer this question, consider a single period of the cosine wave

$$\cos t, \ 0 \leq t < 2\pi$$

where the argument t is selected randomly and without bias, and is therefore described by a random variable T that induces a uniform distribution over $\overline{0, 2\pi}$. The observed value of the cosine function is expressed by another r.v., X, where $X = \cos T$. Use Rule 47.2 to find the p.d.f. $f_X(x)$.

47.7. The r.v. X distributes probability 3/4 uniformly over the interval $\overline{-1/2, 1}$, and assigns the remaining probability 1/4 to the one-element set $\{1/2\}$. Let $Y = X^2$. Determine F_Y and f_Y.

47.8. In a randomly drawn isosceles triangle, two sides have length 1. The angle between them is described by a r.v. that is unif. $\overline{0, \pi}$. Let Z be the random variable expressing the length of the third side of the triangle. Find f_Z.

Section 48

48.1. Consider the experiment of throwing a red and a green die. Let U be a two-dimensional r.v., $U = (X, Y)$, where the r.v. X expresses the number of dots obtained on the red die, and the r.v. Y expresses the number of dots obtained on the green die.

a) Identify the range of the random variable $V = (W, Z)$, where $W = X + Y$, and $Z = X - Y$.
b) Use Rule 48.1 to find the distribution induced by V.
c) Sketch $P_W(\{(w)\})$ and $P_Z(\{(z)\})$.
d) Are W and Z statistically independent r.v.'s?

48.2. Let two r.v.'s X, Y induce a joint distribution characterized by the p.d.f.

$$f_{X,Y}(x,y) = \begin{cases} \dfrac{3}{\pi}\left(1 - \sqrt{x^2 + y^2}\right), & x^2 + y^2 \leq 1 \\ 0, & \text{otherwise.} \end{cases}$$

In other words, $f_{X,Y}$ is a circular cone with base radius 1 centered at the origin. Convert X, Y to polar r.v.'s R, Θ and find $f_{R,\Theta}$, f_R and f_Θ. Are R and Θ s.i.? Are X, Y s.i.?

48.3. Given an absolutely continuous two-dimensional r.v. (X, Y) that induces an absolutely continuous distribution with circular symmetry about the origin; i.e., $f_{X,Y}$ is a function of $x^2 + y^2$:

$$f_{X,Y}(x,y) = h(x_2 + y_2)$$

If (X, Y) is transformed to the two-dimensional polar coordinate r.v. (R, Θ) in accordance with Equation (48.3), show that $f_R(r) = 2\pi r\, h(r^2)$, $r > 0$.

48.4. The voltage across two terminals varies sinusoidally as a function of time, according to the expression $v(t) = a \sin t$, where a is a positive constant. Suppose an instant t_0 is selected randomly without bias (i.e., according to a uniform distribution) in the interval $0, 2\pi$. Let the two r.v.'s V_1 and V_2 express the voltage at the instants t_0 and $t_0 + \pi/2$, respectively.

a) What is the range of $V = (V_1, V_2)$?
b) Describe P_V.

48.5. The two r.v.'s X, Y jointly distribute all probability over the circumference of the unit circle. It is also known that X is uniformly distributed over $-1, 1$, and that $P_{X,Y}$ is symmetrical about both the x-axis and the y-axis. Let (R, Θ) be the representation of (X, Y) in polar coordinates.

a) Determine $f_\Theta(\theta)$.
b) Determine $f_Y(y)$.

48.6. A pair of random variables (X, Y) induces the following joint distribution on $\mathscr{B}^2$: Probability $1/2$ is distributed uniformly over the unit square $0, 1 \times 0, 1$, and the remaining probability $1/2$ is distributed uniformly along the line segment $y = 1 - x$, for $0 \leq x \leq 1$. Now define the new r.v.'s $(Z, W) = g(X, Y)$, where $Z = Y^2$, and $W = 2X$.

a) Determine the joint c.d.f. $F_{Z,W}$.
b) Determine the marginal c.d.f.'s F_Z and F_W.

48.7. Two r.v.'s X, Y jointly induce probability uniformly over the two line segments identified in Figure A.6. Let $W = X^2 + Y^2$, and $Z = X + 1$. Determine the probability distribution induced jointly by W and Z.

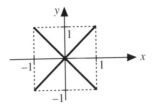

Figure A.6   Two line segments

48.8. An absolutely continuous joint distribution in $\mathbb{R}^2$ is induced by two r.v.'s X, Y defined on a probability space $(S, \mathscr{A}, P)$. Suppose $P_{X,Y}$ is known only in terms of the polar coordinate r.v.'s R, Θ defined on $(S, \mathscr{A}, P)$. In other words, $P_{X,Y}$ is described by the joint p.d.f. $f_{R,\Theta}(r, \theta)$.

a) By first finding $F_X(x)$ and then differentiating the integral, show that

$$f_X(x) = \int_0^{2\pi} f_{R,\Theta}(x \sec \theta, \theta) \, |\sec \theta| \, d\theta.$$

b) Apply the result in (a) to the joint p.d.f. $f_{R,\Theta}(r, \theta)$ obtained in Example 44.4.

Section 49

49.1. Consider the two discrete r.v.'s X and Y defined in Problem 41.2. Define a new r.v. $Z = X + Y$. Determine the probability distribution for Z and sketch.

49.2. Use Rule 48.1 to determine the probability distribution that characterizes by how much two independently selected random digits differ; i.e., integers from 0 to 9 selected independently and with equal likelihood. (For instance, the digits might be obtained by reading two successive positions at an arbitrary place in a random number table.)

49.3. X, Y are two discrete r.v.'s with finite range, and $Z = X + Y$. Let Q_X be the closed interval extending from the lowest value to the largest value taken on by X with positive probability, $Q_X = \overline{\min(X), \max(X)}$; except if X is unary, then $Q_X = \{X\}$. Q_Y and Q_Z are defined in the same way for Y and Z, respectively.

a) If X and Y are s.i., show that $L(Q_Z) \geq L(Q_X)$, $L(Q_Y)$. Under what condition is $L(Q_Z) = L(Q_X)$?

b) If X and Y are not s.i. and $L(Q_X)$, $L(Q_Y) \neq 0$, is it possible for $L(Q_Z)$ to equal 0?

49.4. Let X_1, X_2 be two arbitrary discrete r.v.'s defined on a probability space.

a) Obtain an expression analogous to (49.2) if $Y = X_1 X_2$.

b) Repeat if $Y = X_1/X_2$.

49.5. Let X_1, X_2 be positive discrete r.v.'s, and let $Y = \max(1/X_1, 1/X_2)$. Express $P_Y(\{y\})$ in terms of P_X where $X = (X_1, X_2)$.

49.6. Suppose X is a binomial r.v. assigning positive probability to the integers $m = 0, 1, \ldots, k$, so that P_X is given as in Equation (35.1) by

$$P_X(\{m\}) = \binom{k}{m} p^m (1-p)^{k-m}, \quad m = 0, 1, 2, \ldots, k.$$

Recall from Section 35 that such a r.v. arises in connection with a sequence of k independent, identical, binary trials. If the two possible results of each trial are denoted H and T, and the probability of any given trial resulting in H is p, then X expresses the number of trials (out of the k trials making up the experiment) resulting in H.

Now suppose one more trial gets appended onto the experimental procedure, yielding an experiment of $k + 1$ independent identical trials. On the probability space of the new experiment, let X be defined as before; i.e., X expresses the number of trials (out of the first k trials making up the experiment) that result in H. Also, let I be the indicator r.v. of the event 'the result of the $(k + 1)$th trial is H'. Then $P_I(\{1\}) = p$, and $P_I(\{0\}) = 1 - p$. Now let

$$Y = X + I$$

so that Y expresses the number of trials (out of all $k + 1$ trials) resulting in H. Clearly, Y must again be a binomial r.v. Verify this by using (49.3) to obtain P_Y from P_X and P_I.

49.7. Let X be a geometric r.v. whose range is the positive integers (as in Example 34.3), with parameter $P_X(\{1\}) = \alpha$. Another r.v. I, defined on the same probability space, is the indicator r.v. of an event A.

a) Let $P(A) = p$. If X, I are s.i. and $Z = X + I$, show that

$$P_Z(\{z\}) = \begin{cases} (1-p)P_X(\{1\}), & z = 1 \\ (1-\alpha+\alpha p)P_X(\{z-1\}), & z = 2, 3, 4, \ldots \end{cases}$$

b) Now, let $A = \{\xi: X(\xi) \text{ is even}\}$. If again $Z = X + I$, show that

$$P_Z(\{z\}) = (2-\alpha) P_X(\{z-1\}), \quad z = 2, 4, 6, \ldots$$

Section 50

50.1. The r.v.'s X, Y jointly distribute probability 1 uniformly over the rectangle $\overline{a, b} \times \overline{c, d}$, where $a < b$, $c < d$, and $b - a \geq d - c$. Let $Z = X + Y$. Determine and sketch $f_Z(z)$.

50.2. Let X and Y be identically distributed, statistically independent exponential r.v.'s with

$$f_X(x) = f_Y(x) = \begin{cases} \alpha e^{-\alpha x}, & x \geq 0 \\ 0, & \text{otherwise.} \end{cases}$$

a) Find $f_Z(z)$, where $Z = X - Y$. Note: $X - Y = X + (-Y)$.
b) Find $f_W(w)$, where $W = |X - Y|$.

50.3. X and Y are two s.i. exponential r.v.'s, with

$$f_X(x) = \begin{cases} \alpha\, e^{-\alpha x}, & x \geq 0 \\ 0, & \text{otherwise;} \end{cases} \qquad f_Y(y) = \begin{cases} \beta\, e^{-\beta y}, & y \geq 0 \\ 0, & \text{otherwise} \end{cases}$$

where $\alpha, \beta > 0$. If $\alpha \neq \beta$, show that the r.v. $H = X + Y$ induces a *hypoexponential distribution*, with a density function of the form

$$f_H(h) = \begin{cases} \dfrac{\mu_1 \mu_2}{\mu_2 - \mu_1}\, (e^{-\mu_1 h} - e^{-\mu_2 h}), & h \geq 0 \\ 0, & h < 0. \end{cases}$$

50.4. The random variables Θ_1 and Θ_2 both induce a uniform distribution over $\overline{0,\, 2\pi}$, and are statistically independent. Two new r.v.'s, X and Y, are defined as follows:

$$X = \cos{(\Theta_1 + \Theta_2)}, \text{ and } Y = \sin{(\Theta_1 + \Theta_2)}.$$

Determine the joint distribution induced by X and Y.

50.5. Two absolutely continuous r.v.'s X, Y are s.i. A new r.v. Z is defined as the ratio $Z = X/Y$. Show that

$$f_Z(z) = \int_{-\infty}^{\infty} |y| f_Y(y) f_X(zy)\, dy.$$

50.6. Two absolutely continuous r.v.'s X, Y are s.i. Z is another r.v., which is specified as the *product* of X and Y, i.e., $Z = XY$.

Show that $f_Z(z) = \displaystyle\int_{-\infty}^{\infty} \frac{1}{|x|} f_X(x) f_Y\left(\frac{z}{x}\right) dx$.

50.7. Two r.v.'s X, Y induce distributions that are described by the following p.d.f.'s:

$$f_X(x) = \begin{cases} \dfrac{1}{\pi\sqrt{1 - x^2}}, & |x| \leq 1 \\ 0, & \text{otherwise;} \end{cases} \qquad f_Y(y) = \begin{cases} y\, e^{-y^2/2}, & y \geq 0 \\ 0, & \text{otherwise.} \end{cases}$$

Use the result of the preceding problem to show that if X and Y are s.i., then their product is a Gaussian random variable.

50.8. Consider the r.v.'s X, Y specified in Problem 50.3 (without the requirement that $\alpha \neq \beta$).

a) Obtain the c.d.f. and the density function for the r.v. $Z = \min(X, Y)$; that is, the r.v. that assigns to every element ξ of the underlying probability space the smaller of the two values $X(\xi)$, $Y(\xi)$. Note that Z is also an exponential r.v.

b) Obtain the c.d.f. and the density function for the r.v. $W = \max(X, Y)$, and show that $f_W(w) = f_X(w) + f_Y(w) - f_Z(w)$.

c) By transformation of (X, Y), show that the result $f_W(w) = f_X(w) + f_Y(w) - f_Z(w)$ still holds if X and Y are two arbitrary absolutely continuous s.i. r.v.'s.

d) Explain why this should be so; i.e., justify the result in (c) in terms of more elementary considerations. (Suggestion: Work with the probabilities obtained by integrating the equation given in (b).)

50.9. Let X, Y be two arbitrary r.v.'s, and let $Z = \min(X, Y)$.

    a) Express $F_Z(z)$ in terms of F_X, F_Y, and $F_{X,Y}$.
    b) Now suppose X, Y are jointly absolutely continuous. Show that

$$f_Z(z) = \int_z^\infty f_{X,Y}(x, z)dx + \int_z^\infty f_{X,Y}(z, y)dy.$$

Suggestion: Express $F_{X,Y}(x, y)$ as an integral and apply Leibnitz's rule (see Note 1 in Section 47).

50.10. *Sum of two triangular r.v.'s.* Let X, Y be two r.v.'s that induce identical triangular distributions, with density functions

$$f_X(x) = f_Y(x) = \begin{cases} 2x, & 0 \le x \le 1 \\ 0, & \text{otherwise} \end{cases}$$

Let $Z = X + Y$.

    a) Suppose that X, Y are s.i. Find $F_Z(z)$ and $f_Z(z)$.
    b) Now suppose instead that X, Y have a joint distribution that is uniform over the triangle with vertices $(1, 0), (1, 1), (0, 1)$. Verify that the marginal densities are as stated above, and again find $F_Z(z)$ and $f_Z(z)$.

50.11. *Sum of two statistically independent Gaussian r.v.'s.* Let $Y = X_1 + X_2$ where X_1, X_2 are s.i. Gaussian r.v.'s with

$$f_{X_i}(x) = \frac{1}{\sqrt{2\pi}\sigma_i} \exp\left[-\frac{x^2}{2\sigma_i^2}\right], \quad i = 1, 2$$

Obtain f_Y and show that Y is again Gaussian.

50.12. Let $X_1 + X_2 = Y$, where $(X_1, X_2) = X$. If F_X is continuous (as is the case if X is absolutely continuous), then the summation and limiting operations in Equation (50.3) can be interchanged. In that case, show how Equation (50.5) can be derived from Equation (50.3). (Suggestion: Begin by multiplying the terms of the summation by $\Delta x_1 / \Delta x_1$.)

Section 51

51.1. Given a three-dimensional r.v. $X = (X_1, X_2, X_3)$ that distributes probability uniformly throughout the unit cube $C = \overline{0, 1} \times \overline{0, 1} \times \overline{0, 1}$. Specify $F_X(x_1, x_2, x_3)$ over all of $\mathbb{R}^3$.

51.2. An urn contains a mixture of five blue and 15 white poker chips. One chip is removed blindly from the urn and its color noted, and then returned and mixed in among the other chips. This is done three times. For $i = 1, 2, 3$, let the random variable X_i express the number of blue chips drawn (0 or 1) on the ith draw.

    a) Describe the probability distribution induced by (X_1, X_2, X_3).
    b) Consider three new r.v.'s Y_1, Y_2, Y_3, where $Y_1 = X_1 + X_2 + X_3$, $Y_2 = X_1 X_2 X_3$, and $Y_3 = \max(X_1, X_2, X_3)$. Describe the probability distribution induced by (Y_1, Y_2, Y_3).
    c) How will the answers to parts (a) and (b) differ if the urn contains one blue and three white chips?

51.3. In a probability space $(S, \mathscr{A}, P)$ are defined n distinct sets $A_1 \supset A_2 \supset A_3 \supset \ldots \supset A_n$, where $A_1 = S$, and $P(A_i) = p_i$, $i = 1, \ldots, n$. For $i = 1, \ldots, n$, let I_i be the indicator r.v. for A_i, and let $X = \sum I_i$. What probability distribution gets induced by X onto the integers from 0 to n? Verify that your result is a probability distribution.

51.4. On a probability space $(S_1, \mathscr{A}_1, P_1)$ are defined n r.v.'s $X_1, \ldots, X_n$ that induce distributions $P_{X_1}, \ldots, P_{X_n}$, respectively. By a *mixture* of the r.v.'s $X_1, \ldots, X_n$ is meant a r.v. Y that induces a *weighted sum* of the distributions $P_{X_1}, \ldots, P_{X_n}$; that is, $P_Y = a_1 P_{X_1} + a_2 P_{X_2} + \ldots + a_n P_{X_n}$, where $\sum_{1 \leq i \leq n} a_i = 1$. P_Y is then also called a mixture of the distributions $P_{X_1}, \ldots, P_{X_n}$. The a_i can be regarded as probabilities defined on the one-element sets of a supplementary space S_0. That is, let $S = S_0 \times S_1$, where S_0 is a space of n elements $\zeta_1, \ldots, \zeta_n$ with $P(\{\zeta_i\}) = a_i$, $i = 1, \ldots, n$.

Suppose S_1 is the binary space $\{\xi_T, \xi_H\}$ associated with the toss of a coin. Let X_1 be a r.v. that assigns $X_1(\xi_T) = 1$, $X_1(\xi_H) = 2$. Also, let $X_2 = X_1 + 2$, $X_3 = X_1 + 4$. Show that there is a mixture P_y of $P_{X_1}, P_{X_2}, P_{X_3}$ that is identical to the distribution P_G in Example 34.1. Can the r.v. inducing that mixture be defined on the same space S_1?

51.5. (Refer to Problem 51.4.) Suppose the r.v.'s $X_1, X_2, X_3, \ldots$ are binomial r.v.'s, where

$$P_{X_n}(\{k\}) = \binom{n}{k} p^k (1-p)^{n-k}, \quad k = 0, 1, \ldots, n; \quad n = 1, 2, 3, \ldots$$

Show that the mixture of the X_n's with weights $a_n = \lambda^n e^{-\lambda} (n!)^{-1}$ results in a Poisson r.v. [Re2]

Section 52

52.1. Consider a vector r.v. $X = [X_1, \ldots, X_n]$ that induces on $\mathscr{B}^n$ the absolutely continuous distribution described by the following density function:

$$f_X(x) = f_X(x_1, \ldots, x_n) = \begin{cases} k_n \prod_{i=1}^{n} x_i, & \text{for } 0 \leq x_i \leq 1, \, i = 1, \ldots, n \\ 0, & \text{otherwise.} \end{cases}$$

a) Determine the constant k_n.
b) Assuming $n > 3$, determine the three-dimensional density function $f_Y(y)$, where Y is the vector r.v. whose components are the first three components of X, that is, $Y = [X_1, X_2, X_3]$.

52.2. Express the following sets in R^3 in terms of three-dimensional Cartesian product sets:

a) The set of vertices of the unit cube.
b) The set of points making up the surface of the unit cube.
c) The set of points making up the edges of the unit cube.
d) The set of points making up the three faces of the unit cube that lie in the coordinate planes.

52.3. Let X be an absolutely continuous n-dimensional r.v. $(n \geq 1)$, with n-dimensional p.d.f. $f_X(x)$. Let Z be any zero-measure subset of R^n. Suppose that $f_X(x)$ is redefined as follows, where c is an arbitrary non-negative value:

$$f_X(x) = c, \text{ for } x \in Z; \text{ and } f_X(x) \text{ remains unchanged for } x \notin Z.$$

Show that the result is still a valid p.d.f., and that P_X is not affected by this change.

52.4. a) By means of an example, show that if $X_1, \ldots, X_n$ $(n \geq 2)$ are absolutely continuous r.v.'s, then $X = (X_1, \ldots, X_n)$ is not necessarily absolutely continuous.

b) Let $X = (X_1, \ldots, X_n)$ $(n \geq 2)$ be absolutely continuous. Let $X^{(i)}$ denote the $(n-1)$-dimensional r.v. obtained from X by deleting the ith component $(i = 1, \ldots, n)$. Show that $X^{(i)}$ is absolutely continuous.

52.5. Three r.v.'s A, B, and C jointly induce probability one uniformly throughout the unit cube $\overline{0, 1} \times \overline{0, 1} \times \overline{0, 1}$. When the underlying p.e. is performed, what is the probability that the equation $Ax^2 + Bx + C = 0$ will have real roots? (After [Re3].)

52.6. A pointer is spun n times in such a way that the angular position in which it comes to rest after any one spin is independent of the other spins and is governed by a uniform distribution over $\overline{0, 2\pi}$. Let X be the r.v. that expresses the *smallest* angular position obtained in the n trials.

a) With $n = 2$, determine f_X.

b) Repeat for $n = 6$. Here, note that 'the smallest angle is less than or equal to x' is the same event as 'not all six angles are greater than x'.

c) Based on the results of (a) and (b), deduce the density function of X for arbitrary n.

d) Again consider $n = 6$. Let $X^{(k)}$ (for $k = 1, \ldots, 6$) be the r.v. that expresses the kth smallest angle obtained. Clearly, $X^{(1)} = X$. Determine the densities of $X^{(2)}, \ldots, X^{(6)}$.

Note: Random variables such as $X^{(1)}, X^{(2)}, \ldots, X^{(6)}$, obtained by rearranging in ascending order the n-tuples assigned by an n-dimensional r.v. representing the results of independent trials, are called 'order statistics' (see also Section 84).

Section 53

53.1. An absolutely continuous r.v. $X = (X_1, X_2)$ is transformed into the r.v. $Y = g(X) = (Y_1, Y_2)$. In each of the following cases determine $|J_g|$.

a) $Y_1 = X_2$
$Y_2 = -X_1$;

b) $Y_1 = 10 - 0.1X_1$
$Y_2 = 10X_2 - 0.1$;

c) $Y_1 = (X_1 + X_2)/2$
$Y_2 = (X_1 - X_2)/2$;

d) $Y_1 = X_1/X_2$
$Y_2 = X_2/X_1$.

53.2. Let X be an absolutely continuous r.v., and let $Y = aX + b$, $a \neq 0$. The density functions f_X and f_Y are related as stated in Equation (53.6). Express the c.d.f. $F_Y(y)$ in terms of $F_X(x)$,

i) if $a > 0$; and
ii) if $a < 0$.

53.3. Given $f_X(x) = \frac{\alpha}{2} e^{-\alpha|x|}$, $-\infty < x < \infty$. Let $Y = |X|$. Determine f_Y.

53.4. Let the n r.v.'s $Y_1,\ldots, Y_n$ be related to n r.v.'s $X_1,\ldots, X_n$ by a set of nonhomogeneous linear equations:

$$Y_1 = a_{11}X_1 + a_{12}X_2 + \ldots + a_{1n}X_n + b_1$$
$$\cdot \quad \cdot \quad \cdot \quad \cdot \quad \cdot \quad \cdot \quad \cdot \quad \cdot \quad \cdot$$
$$Y_n = a_{n1}X_1 + a_{n2}X_2 + \ldots + a_{nn}X_n + b_n.$$

Show that the condition that the Jacobian of this transformation does not vanish is equivalent to the condition that the determinant of $\mathbf{A}$, $\det(\mathbf{A}) \neq 0$, where

$$\mathbf{A} = \begin{bmatrix} a_{11} & a_{12} & \cdots & a_{1n} \\ \cdots & \cdots & \cdots & \cdots \\ a_{n1} & a_{n2} & \cdots & a_{nn} \end{bmatrix}.$$

53.5. A two-dimensional distribution centered at the origin of the (x,y)-plane has circular symmetry around the origin. The radial distribution is exponential with parameter α, and the angular distribution is uniform over $0, 2\pi$. Express the joint p.d.f. in rectangular coordinates.

53.6. X is an exponential r.v. with density function

$$f_X(x) = \begin{cases} \alpha e^{-\alpha x}, & x \geq 0 \\ 0, & x < 0 \end{cases}$$

where $\alpha > 0$. Let $Y = X - \lfloor X \rfloor$, where $\lfloor z \rfloor$ denotes the largest integer $\leq z$. Determine $f_Y(y)$.

53.7. Let X be an exponential r.v. with $f_X(x)$ as given in Problem 53.6.

a) Find the p.d.f. $f_Y(y)$ if $Y = 2 - 3X$.
b) Find the p.d.f. $f_Z(z)$ if $Z = X^3$.

53.8. Two random variables X, Y that induce a joint probability distribution that is uniform over the triangular region having vertices at points $(0, 0)$, $(0, 1)$, and $(1, 1)$. Let $Z = X + Y$, and $W = X - Y$. Determine the density functions f_X, f_Y, f_Z, and f_W, and the joint density $f_{Z,W}$.

53.9. Let $Y = e^X$, where X is Gaussian with p.d.f. as given in Equation (40.1). Show that Y is *log-normal*, with density function

$$f_Y(y) = \begin{cases} \dfrac{1}{\sqrt{2\pi}\sigma\, y} \exp\left[-\dfrac{[\ln(y/\alpha)]^2}{2\sigma^2} \right], & y > 0 \\ 0, & y < 0 \end{cases}$$

where $\ln \alpha = a$.

53.10. Given a Gaussian r.v. X with density function

$$f_X(x) = \frac{1}{\sqrt{2\pi}\,\sigma} \exp\left(-\frac{x^2}{2\sigma^2} \right), \qquad -\infty < x < \infty.$$

Determine and sketch the density function of the r.v. $Y = aX^2$. What value of a will cancel the parameter σ in f_Y? (Note: The distribution induced by Y with $\sigma = 1$ is called a *chi-square*, or χ^2, distribution of order 1.)

53.11. The r.v. $X = (X_1, X_2)$ induces a uniform distribution over the unit square $\overline{0, 1} \times \overline{0, 1}$. A r.v. $Y = (Y_1, Y_2)$ is defined by the transformation

$$Y_1 = (-2\ln X_1)^{1/2}\cos 2\pi X_2$$
$$Y_2 = (-2\ln X_1)^{1/2}\sin 2\pi X_2.$$

Show that Y_1 and Y_2 are s.i. Gaussian r.v.'s (see Problem 46.8) with parameters $a = b = 0$ and $\sigma_1 = \sigma_2 = 1$. [Pa1]

53.12. X and Y are two i.d.s.i. exponential r.v.'s, the p.d.f. of X being

$$f_X(x) = \begin{cases} e^{-x}, & x > 0 \\ 0, & \text{otherwise.} \end{cases}$$

a) Obtain the joint p.d.f. of the pair of r.v.'s (Z, W), where $Z = X + Y$ and $W = X/Y$. Are Z and W s.i.?

b) Convert the pair (X, Y) to polar coordinates and express the joint density function in terms of the polar coordinate variables r, θ. Are R and Θ s.i.?

53.13. Consider a test station on a production line for electrical equipment. At this station, a particular current is measured on each equipment coming off the line. Suppose that this measurement results in a reading of I Amperes, where I is uniformly distributed between 0.3 and 0.5 Amperes. This current passes through a resistor whose resistance value is R Ohms, where R is uniformly distributed between 5 and 6 Ohms. Use coordinate transformation to find the p.d.f. of the power delivered to the resistor, $W = I^2 R$, assuming I and R are s.i.

(Suggestion: Use R as the second component of the transformed r.v.)

Section 54

54.1. What probability does a bivariate Gaussian r.v. $X = (X_1, X_2)$ assign to the interior of the standard ellipse? (Suggestion: First, consider X_1 and X_2 identically distributed and statistically independent.)

54.2. Let (X_1, X_2) be a bivariate Gaussian r.v. Show that $X_1 + X_2$ is a Gaussian r.v.

(Suggestion: Let $Y_1 = X_1 + X_2$, and $Y_2 = X_2 - X_1$, which amounts to a 45° rotation and scaling of the coordinate system. Then, apply the principles developed in this Section.)

54.3. Given two r.v.'s X_1, X_2 such that $X_1 + X_2$ and $X_1 - X_2$ are jointly Gaussian. Show that X_1, X_2 are jointly Gaussian.

54.4. Given the bivariate Gaussian p.d.f.

$$f_Y(y) = c\exp\left[-\frac{1}{2}(\alpha y_1^2 + 2\beta y_1 y_2 + \gamma y_2^2)\right]$$

where $\alpha, \gamma > 0$, and $\beta^2 < \alpha\gamma$. Determine the angle φ ($0 < \varphi < \pi/2$) through which P_Y must be rotated about the origin in a clockwise direction, so that the resulting distribution corresponds to two s.i. Gaussian r.v.'s.

54.5. Let $X = (X_1, X_2)$ induce a uniform distribution over the unit square, as defined in Example 54.1. Let $Z = (Z_1, Z_2)$ induce the distribution that results from rotating X through an arbitrary angle φ in the range $0 \leq \varphi < \pi/4$. Obtain expressions for the p.d.f.'s of Z_1 and Z_2 and sketch a representative case.

54.6. Consider the joint p.d.f. $f_Y(y)$ in Equation (54.6). Under what condition is the function $g(y_1) = f_Y(y_1, 0)$ (i.e., a slice through f_Y along the horizontal axis, with no additional scale factor) a Gaussian p.d.f.?

54.7. Obtain expressions for f_Z and f_W in Query 54.2, and from these determine

$$[f_Z(z) + f_W(z)]/2.$$

Section 55

55.1. In Example 54.1, X_1 and X_2 are s.i. r.v.'s. The r.v.'s Y_1 and Y_2 obtained after the rotational transformation are no longer s.i. since the joint distribution does not satisfy Theorem 46.3. Does this contradict Theorem 55.2? Explain.

55.2. $n-1$ unary r.v.'s $X_1 = a_1, X_2 = a_2, X_3 = a_3, \ldots, X_{n-1} = a_{n-1}$, and an arbitrary r.v. X_n are defined on a probability space. Use Definition 55.1 to show that they are m.s.i.

55.3. $X_1, \ldots, X_n$ are n m.s.i. exponential r.v.'s with parameters $\alpha_1, \ldots, \alpha_n$. What is the probability that X_1 assigns a value that is the smallest of the values assigned by the n r.v.'s?

  (Suggestion: The results of Problems 46.6(a) and 50.8(a) can be applied here.)

55.4. Consider two r.v.'s $X = (X_1, \ldots, X_n)$ and $Y = (Y_1, \ldots, Y_m)$.

  a) Suppose $X_1, \ldots, X_n$ are m.s.i., and $Y_1, \ldots, Y_m$ are also m.s.i. If X and Y are s.i., are $X_1, \ldots, X_n$, $Y_1, \ldots, Y_m$ m.s.i.?
  b) Suppose $X_1, \ldots, X_n$ are pairwise s.i., and $Y_1, \ldots, Y_m$ are also pairwise s.i. If X and Y are s.i., are $X_1, \ldots, X_n, Y_1, \ldots, Y_m$ pairwise s.i.?

55.5. Given three r.v.'s X, Y, Z that assign probability uniformly over two sides of the unit cube, as shown in Figure A.7.

  a) Determine which pairs of these r.v.'s are s.i. Justify your answer.
  b) Show that X, Y, Z are not m.s.i.
  c) For each of the three r.v.'s, determine whether it is statistically independent of the pair of the remaining two.

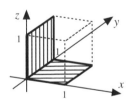

Figure A.7   A singular distribution in three dimensions

55.6. The three r.v.'s X_1, X_2, X_3 each induce a uniform distribution on $[0, 1]$. They are *not* pairwise statistically independent. Describe a possible joint distribution of X_1, X_2, X_3 for each of the following cases.

a) $X = (X_1, X_2, X_3)$ is singular.

b) $X = (X_1, X_2, X_3)$ is absolutely continuous. Note: An example in two dimensions is the joint p.d.f.

$$f_{X,Y}(x, y) = \begin{cases} 2(1 - x + y), & 0 < y < x < 1 \\ 2(y - x), & 0 < x < y < 1 \\ 0, & \text{otherwise.} \end{cases}$$

55.7. Three r.v.'s X, Y, Z are m.s.i. Jointly, they induce a uniform distribution over the unit cube, so that for the r.v. X, $f_X(x) = 1$ for $x \in \overline{0, 1}$, and similarly for Y and Z.

a) Sketch the density functions of the r.v.'s $X + Y$, $Y + Z$, and $X + Z$, and show that these r.v.'s are not pairwise s.i.

b) Determine the p.d.f. for the r.v. $X + Y + Z$.

c) Determine the probability of the event $A = \{\xi: X(\xi) < Y(\xi) < Z(\xi)\}$.

55.8. Generalize Problem 49.6 to the sum of two s.i. binomial r.v.'s X, Y with identical parameter p, where

$$P_X(\{m\}) = \binom{k}{m} p^m (1 - p)^{k - m}, \quad m = 0, 1, 2, \ldots, k, \text{ and}$$

$$P_Y(\{n\}) = \binom{\ell}{n} p^n (1 - p)^{\ell - n}, \quad n = 0, 1, 2, \ldots, \ell.$$

55.9. Use the result of Problem 55.8, and the discrete convolution expression (49.3), to prove the summation formula

$$\sum_{i=0}^{j} \binom{k}{i} \binom{\ell}{j - i} = \binom{k + \ell}{j}.$$

55.10. Example 51.1 suggests that for Theorem 55.1 and Corollaries 55.1a and 55.1b to hold, it may be sufficient if the requirement stated in each of these is satisfied within a closed n-dimensional rectangle set on which probability 1 gets induced. Show that this is the case.

Section 56

56.1. Consider the r.v. V_o defined in Example 56.1, with c.d.f. as sketched in Figure 56.6. Determine:

a) The probability that V_o takes on a value in the interval $\overline{-2/3, 2/3}$.

b) The probability that V_o takes on a value larger than 0.85.

56.2. Let V_o be the r.v. discussed in Example 56.1. Consider another r.v., V_1, which expresses the output voltage from the low-pass filter one second later. Describe the joint distribution of V_o and V_1. Is it also singular?

56.3. An alternative approach to specifying the r.v. V_o in Example 56.1 is to write it as an infinite
       sum of m.s.i. random variables:

$$V_o = X_1 + X_2 + X_3 + \cdots$$

Here, X_j is the contribution to the output $v_o(t_o)$ of only that portion of $v_i(t)$ that exists during
the time interval N_j, $j = 1, 2, 3, \ldots$. Show that this does indeed result in the same r.v. V_o.

56.4. In Example 56.1, suppose the time constant τ of the lowpass filter is specified as $(\ln 2)^{-1}$ and
       nothing else is changed. Describe F_{V_o}. Is V_o singular in this case?
       See also Problem 57.10.

Section 57

57.1. An integer n is selected with equal likelihood from the set $\{1, 2, \ldots, 10\}$. Then, an integer m is
       selected with equal likelihood from the set $\{n, n + 1, \ldots, 10\}$. (Note that if $n = 10$, then
       $m = 10$ with certainty.) What is the probability distribution induced by the r.v. M that expresses
       the selection of m?

57.2. Consider a game in which a red and a white die are thrown and then a coin is tossed. Depending
       on whether the coin toss results in 'head' or 'tail', the number of dots showing on the red die is
       either added to, or subtracted from, the number of dots showing on the white die. The number
       obtained in this way determines by how many spaces a playing piece gets moved forward. Use
       conditional distributions to determine the (unconditional) distribution that characterizes the
       number of spaces the playing piece gets moved ahead.

57.3. a) The r.v. X induces a uniform distribution over $\overline{0, 1}$. The r.v. Y is with equal probability
          either equal to X or equal to $1 - X$. Describe the joint distribution of X and Y.
       b) An indicator r.v. I assigns the values 0 and 1 with probabilities 0.8 and 0.2, respectively. A
          two-dimensional r.v. $X = (X_1, X_2)$ is distributed uniformly over $\overline{0.25, 0.75} \times \overline{-1, 1}$ if $I = 0$,
          and is distributed uniformly over $\overline{0, 1} \times \overline{-0.5, +0.5}$ if $I = 1$. Describe the (unconditional)
          distribution of X.

57.4. Two r.v.'s X and Y jointly induce the following probabilities on the one-element sets making
       up the product class $\{\{0\}, \{1\}, \{3\}\} \times \{\{-1\}, \{0\}, \{1\}\}$:

	$y = -1$:	$y = 0$:	$y = 1$:
$x = 0$:	0	0.2	0.02
$x = 1$:	0.3	0.08	0.1
$x = 3$:	0.15	0	0.15

Determine $P_X(\{x\}|\{\xi: Y(\xi) = y\})$ and $P_Y(\{y\}|\{\xi: X(\xi) = x\})$.

57.5. *Random sum.* Consider the r.v. $Y = X_1 + X_2 + X_3 + \ldots + X_N$, where the X_i ($i = 1, 2, \ldots$) are
       i.d.m.s.i. (identically distributed and mutually statistically independent) absolutely continu-
       ous r.v.'s, and N is a r.v. that is statistically independent of the X_i. Also, let $Y_i = X_1 + \ldots + X_i$,
       for $i = 1, 2, 3, \ldots$.

       a) Show that P_Y is a mixture with p.d.f. $f_Y(y) = \sum_i P_N(\{i\}) f_{Y_i}(y)$.
       b) Sketch $f_Y(y)$ for the situation where the X_i are unif. $\overline{0, 1}$ and N takes on the values 1 and 2
           with probabilities

      i) 0.5 each;

      ii) 0.9 and 0.1, respectively.

57.6. Consider the r.v. X and the event C as defined in Problem 39.2. First, determine $F_X(t)$, and sketch. Then obtain and sketch $F_{X|C}(t)$ and $f_{X|C}(t)$.

57.7. As in Problem 55.3, consider n m.s.i. exponential r.v.'s $X_1,\ldots,X_n$ with parameters $\alpha_1,\ldots,\alpha_n$. What is the conditional density of X_1, given that X_1 assigns a value that is the smallest of the values assigned by the n r.v.'s? Show that the result is the same if any X_i ($i=2,\ldots,n$) is substituted for X_1.

57.8. Consider some random variable X and a fixed number a. Let $Y=X-a$, and let A be the event $A=\{\xi: X(\xi)>a\}$. Assuming the conditioning events to have positive probability, express the following in terms of $F_X(x)$ and as a function of y:

a) $P(\{\xi: Y(\xi)>y\}|A)$;

b) $P(A|\{\xi: Y(\xi)>y\})$.

Sketch the results if X is unif. $\overline{-1,1}$ and $a=0.5$.

57.9. A manufacturer produces a particular type of device in large numbers. An experiment is conducted to measure the operating life of such a device. For this purpose, one of these devices, fresh off the production line, is operated continuously, under specified conditions, until it fails. The time at the start of the test is designated $t=0$. The time until failure is observed.

a) On the probability space for this experiment, let the r.v. X express the time till failure. If the test has proceeded over a time duration τ and the device has not yet failed, then the conditional distribution of X given that no failure occurs during $\overline{0,\tau}$ is of interest. Let this conditioning event be denoted A_τ. Express $f_{X|A_\tau}(t)$ in terms of f_X and F_X.

b) Sketch $f_{X|A_\tau}(t)$ for several values of τ, if:

    i) X is an exponential r.v., with p.d.f. as given in Figure 39.1b.

    ii) $X=|Y|$, where Y is standardized Gaussian with p.d.f. as given in Equation (40.2).

    iii) $X=|Y|$, where Y is Cauchy with f_Y as in Figure 39.1d, with $a=0$, $b=1$.

c) The probability that failure occurs in a short interval $\overline{\tau,\tau+\Delta\tau}$ (under the condition that no failure has occurred up till time τ) is approximately $(\Delta\tau)f_{X|A_\tau}(\tau)$, or equivalently, $(\Delta t)f_{X|A_t}(t)$. The function $f_{X|A_t}(t)$ is called the *hazard rate*. It expresses, as a function of t, the hazard of immediate failure given that the device in question has operated without failure for a time duration t. Determine and sketch the hazard rate for the three different distributions in (b).

57.10. Refer to Example 56.1.

a) Determine $F_{V_0|A}(v_0)$ and $F_{V_0|A^c}(v_0)$, where

$$A \leftrightarrow v_i(t) = +1 \text{ for } t_0-1<t<t_0, \text{ and}$$
$$A^c \leftrightarrow v_i(t) = -1 \text{ for } t_0-1<t<t_0.$$

b) Show that $F_{V_0|A}$ and $F_{V_0|A^c}$ combine to give the unconditional c.d.f. described in Figure 56.6.

c) Obtain the distribution of the filter output voltage at time $t_0 - \Delta$, where $\Delta = 0.2$ seconds, expressed in terms of $F_{V_0}(v_0)$ as given in Figure 56.6.

Section 58

58.1. A vector r.v. $X = (X_1, X_2, X_3, X_4)$ has a c.d.f. specified as follows:

$$F_X(x) = \begin{cases} 0, & x_1 < 0 \text{ or } x_2 < 0 \text{ or } x_3 < 0 \text{ or } x_4 < 0 \\ \min(x_1, 1) \cdot \min(x_2, 1) \cdot \min(x_3, 1), & x_1, x_2, x_3, x_4 \geq 0. \end{cases}$$

a) Decompose X into its components of pure type.
b) Do the same for (X_1, X_2) and for (X_3, X_4).

58.2. Express the joint c.d.f. of the two-dimensional r.v. (X, Y) defined in Problem 48.6 in the form of Equation (58.1).

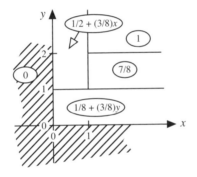

Figure A.8   A joint c.d.f

58.3. Decompose the vector r.v. $\begin{bmatrix} X \\ Y \end{bmatrix}$ whose joint c.d.f. is described in Figure A.8 into its discrete, absolutely continuous and singular components. Display the c.d.f. of each of the components present in this decomposition in the manner of Figure A.8.

58.4. Two statistically independent r.v.'s X, Y have c.d.f.s expressed in the form of Equation (58.1), with coefficients a_X, d_X, s_X, and a_Y, d_Y, s_Y, respectively. Determine the coefficients a_Z, d_Z, s_Z, where $Z = (X, Y)$.

58.5. a) A r.v. X induces a distribution consisting of a mixture of two pure types, as follows:
  i) a singular component of weight 0.5 whose c.d.f. equals F_{V_0} as shown in Figure 56.6; and
  ii) a discrete component that assigns probability 0.5 at the origin.
  Draw an approximate sketch of $F_X(x)$.

b) Repeat, if the discrete component is replaced by an absolutely continuous component of weight 0.5, which is unif. $-1, 1$.

58.6. Given a probability space $(S, \mathscr{A}, P)$ and a partition $\{A, B, C\} \subset \mathscr{A}$, with $P(A) = P(B) = 0.4$, $P(C) = 0.2$. A r.v. X is defined in such a way that

$$F_{X|A}(x) = \begin{cases} 0, & x < 0 \\ 0.3x + 0.3, & 0 \le x < 1 \\ 1, & x \ge 1; \end{cases} \quad F_{X|B}(x) = \begin{cases} 0, & x < -1 \\ 0.2x + 0.3, & -1 \le x < 1 \\ 1, & x \ge 1 \end{cases}$$

and

$$F_{X|C}(x) = \begin{cases} 0, & x < -1 \\ 0.1, & -1 \le x < -0.5 \\ 0.2, & -0.5 \le x < 0.5 \\ (x+2)/3, & 0.5 \le x < 1 \\ 1, & x \ge 1. \end{cases}$$

Resolve X into components of pure type.

58.7. The problem of resolving a distribution into components also arises in other contexts. For instance, an absolutely continuous distribution may have to be resolved into one component that is of a specified nature, and a residual component. Thus, suppose $f_X(x) = 0.5\, e^{-|x|}$ (for all x) is given and is to be resolved into a Gaussian and a residual component. The task, therefore, is to express $f_X(x)$ in the form (58.2) such that one of the conditional densities is Gaussian and the coefficient of the residual component is minimized. Note that because f_X is even, the parameter a in Equation (40.1) can be set to 0, leaving σ to be optimized. Show that the desired result is

$$f_X(x) = \sqrt{\frac{\pi}{2e}} f_G(x) + \left(1 - \sqrt{\frac{\pi}{2e}}\right) f_H(x)$$

where f_G is the Gaussian p.d.f. with $\sigma = 1$, and f_H is the p.d.f. of the residual component.

Note: It is sufficient to consider the positive x-axis. Considered graphically, the problem is then to inscribe a weighted Gaussian p.d.f. under the graph of $f_X(x)$ and to find the value of σ that maximizes the weight. At the point $x = x_0$ where the Gaussian p.d.f., weighted by g, touches $f_X(x)$, their magnitudes and their slopes must be equal. From these two conditions an expression for the weight g can be found and then maximized with respect to σ.

Section 59

59.1. Let X and Y be absolutely continuous s.i. r.v.'s, and let $Z = X + Y$. Show that $f_{Z|X}(z|x) = f_Y(z - x)$. Making use of this result, rederive Equation (50.6).

59.2. Derive the density function for the product of two absolutely continuous s.i. r.v.'s (Problem 50.6) by making use of the conditional p.d.f. $f_{Z|X}(z|x)$.

59.3. A r.v. X has an exponential distribution for specific values of another r.v. Y that is strictly positive. Specifically, let

$$f_{X|Y}(x|y) = \begin{cases} \dfrac{1}{y} e^{-(x/y)}, & x, y > 0 \\ 0, & x \leq 0. \end{cases}$$

Suppose that Y is a die-throw r.v., so that $P_Y(\{y\}) = 1/6$ for $y = 1, 2, \ldots, 6$.
a) Determine $f_X(x)$. Is X an exponential r.v.?
b) Of what type is the joint distribution of X and Y?

59.4. A real number x is obtained according to a uniform distribution over $\overline{0,1}$. Then, a real number y is obtained according to a uniform distribution over $\overline{x,1}$. Obtain the density function for the r.v. Y governing y.

59.5. The r.v. X induces a uniform distribution over $\overline{-1,1}$. Determine the conditional c.d.f. $F_{X|X^2}(x|y)$. Of what type is the conditional distribution described by this c.d.f. (i.e., absolutely continuous, discrete, singular, or mixed)?

59.6. A 1×1 square is placed, randomly and without bias, within a 3×3 square. The 1×1 square is oriented with sides parallel to the sides of the 3×3 square. ('Random placement without bias' means that the lower left corner of the 1×1 square, for instance, is just as likely to fall into one elemental region as any other of the same size, as long as the 1×1 square will be wholly within the 3×3 square.) A second 1×1 square, oriented in the same way, is also placed randomly within the 3×3 square, independently of the placement of the first 1×1 square.

What is the probability that the two 1×1 squares will not overlap?

59.7. A machine executes two jobs in succession. The first job has a duration that is exponentially distributed with parameter α (s^{-1}). The second job commences immediately when the first job is completed. Its duration is independent of the duration of the first job and is also exponentially distributed, with parameter β (s^{-1}). The durations of the two jobs are therefore characterized by two r.v.'s X, Y as specified in Problem 50.3. Let $X + Y = Z$.

a) What is the probability that the two jobs are completed within z seconds, given that the duration of the first job is x seconds?
b) What is the joint density function of X and Z?
c) Are X and Z statistically independent?
d) What is the probability that the first job is completed within x seconds, given that the two jobs together require exactly z seconds?

59.8. Three r.v.'s X, Y, Z are m.s.i., and all three induce a uniform distribution over $\overline{0,1}$. Let A be the event characterized by the relation $X < Y < Z$; that is, $A = \{\xi: X(\xi) < Y(\xi) < Z(\xi)\}$. Find:

a) $f_{X|A}(x)$;   b) $P(A|\{\xi: X(\xi) \leq 0.5\})$;   c) $P(A|\{\xi: X(\xi) = 0.5\})$.

59.9. The two-dimensional density function $f_{X,Y}(x, y)$ of two random variables X, Y is given by a semi-sphere centered at the origin and is zero elsewhere in the (x,y)-plane.

a) Find $f_X(x)$.
b) Find $f_{X|Y}(x|x)$; i.e., the conditional density function of X given that $Y = X$. Sketch this function.

59.10. If X and Y are s.i. Poisson r.v.'s, show that the conditional distribution induced by X, given $X + Y$, is binomial.

Suggestion: With $Z = X + Y$, express $P_{X|Z}(\{x\}|\{z\})$ as in Equation (59.2).

59.11. a) Rewrite Equation (47.6) and the associated discussion so that it applies to a transformation of *two* (jointly absolutely continuous but not necessarily s.i.) r.v.'s X_1, X_2 into a r.v. Y by means of a transformation $y = g(x_1, x_2)$, where g has piecewise continuous partial derivatives with respect to x_1 and x_2, and is such that X_2 and Y are also jointly absolutely continuous. Begin by expressing Equation (47.6) in terms of *conditional* densities, given X_2.

b) Apply the results of (a) to the transformation $y = g(x_1, x_2) = x_1 x_2$. Compare your result with Problem 50.6.

Section 60

60.1. Arrival of cars at a gas station is to be modeled in terms of random occurrences, with the time between arrivals being exponentially distributed, with probability density function f_W as given in Equation (60.3).

a) Suppose that there are two gas pumps, and cars get alternately routed to the first pump and the second pump. What is the density function governing the time between arrivals at the first pump?

b) Identify various factors that may cause the actual situation considered here to deviate from the random occurrences model.

60.2. An electronic pulse detector is used to observe and register pulses that are generated by some process. Suppose pulses arrive at the detector at random, with the number of pulses per second characterized by a Poisson distribution with parameter λt. Suppose the detector sometimes fails to record the arrival of a pulse. Let p denote the probability that any given arriving pulse gets detected. Under the assumption that the detection of a pulse is independent of the detection of any other pulse, show that the number of pulses per second that are detected is characterized by a Poisson distribution with parameter $\lambda p t$.

60.3. Verify that Equation (60.5) is the density function for T_m.

60.4. Show that the distribution induced by the r.v. T_2 is the limiting case of a hypoexponential distribution (see Problem 50.3) as $\mu_2 - \mu_1 \to 0$.

60.5. How does the density function (60.5) for the Gamma distribution have to be modified so that it applies for *all* real $m > 0$? Is your result consistent with Equation (60.5) for integer values of m?

60.6. In an experiment involving the observation of random occurrences, suppose that it is known that there are six occurrences in $\overline{0, 10}$. What is the probability that in $\overline{5, 10}$ there are

a) exactly three occurrences?

b) at least four occurrences?

60.7. Jobs arrive at a computer for processing, are stored in a queue, and get processed sequentially in the order in which they arrived in the queue. There are two types of jobs: Transfer from the queue and processing requires exactly ten seconds if it is a Type I job, and exactly one second if it is a Type II job. The two types arrive with equal probability, independently of previous arrivals. Assume the queue is never empty and processing starts at time $t = 0$. Thus, the initiation of a job is a random occurrence in time. Although all jobs start an integer number of seconds from $t = 0$, the time between these occurrences can be one second or ten seconds.

a) What is the expected completion time for the first 100 jobs?
b) What is the expected number of jobs that get completed in the first 20 seconds? (Suggestion: Use a state diagram as in Example 31.2.)
c) At an arbitrary instant of time $t_x \gg 10$ seconds that is unrelated to the processing being done by the computer, the computer is queried. What is the probability that at the instant t_x one of the Type I jobs is getting processed?
d) At the instant t_x, what is the expected time till completion of the currently running job?

Part V Problems

Section 61

61.1. For given n, an integer is randomly picked from among the integers 1, 2, 3,..., n, using a procedure that assures unbiased selection.

    a) If the r.v. X represents the integer that is picked, what is EX? What is the probability that EX equals the value actually picked?

    b) If the r.v. Y represents one-half the value that is picked, what is EY?

61.2. A discrete r.v. X induces the following probabilities on the integers:

$$P_X(\{n\}) = 6(\pi n)^{-2}, \text{ for } n = 1, -2, 3, -4, 5, \ldots$$

and $P_X(\{n\}) = 0$ for all other integers n. (Note: $1 + (1/2)^2 + (1/3)^2 + \ldots = \pi^2/6$.) Show that the first moment of P_X does not exist.

61.3. If X is a Gaussian r.v. with p.d.f. (40.1), show that $EX = a$.

61.4. For each of the following r.v.'s whose expectation exists, determine the constant c so that the expectation equals 1.

    a) X is absolutely continuous, with p.d.f. $f_X(x) = 0.5\, e^{-|x-c|}, \quad -\infty < x < \infty$

    b) X is of mixed type, with p.d.f. $f_X(x) = 0.5\, \delta(x-c) + \begin{cases} 0.05, & -5 < x < 5 \\ 0, & \text{otherwise} \end{cases}$

    c) X is of mixed type, with p.d.f. $f_X(x) = 0.5\, \delta(x-c) + \begin{cases} 0.5x^{-2}, & |x| \geq 2 \\ 0, & -2 < x < 2 \end{cases}$

    d) X is discrete, with $P_X(\{x\}) = 0.01$ for $x = c+1,\ c+2,\ c+3,\ldots,\ c+100$; and $P_X(\{x\}) = 0$ for all other x.

    e) X is discrete, with $P_X(\{x\}) = \begin{cases} 0.3, & x = 0, 0.5, 1 \\ 0.02, & x = -0.5, 3, 5, 8, c \\ 0, & \text{otherwise.} \end{cases}$

61.5. a) Obtain the mean of the geometric distribution with parameter p $(0 < p < 1)$; that is,

$$P_X(\{k\}) = \begin{cases} p(1-p)^{k-1}, & k = 1, 2, 3, \ldots \\ 0, & \text{otherwise.} \end{cases}$$

    b) Apply your result to Example 34.3: How many times should one expect to throw a die till the first 'six' appears? Compare this with the most likely number of throws till the first 'six' appears, i.e., the mode of P_X.

61.6. Prove Equation (61.7). (Suggestion: Write Equation (61.6) for the r.v. $Y = X - m_{1X}$, then apply a change of variable in the integrals.)

61.7. a) Determine the expected value of the exponential r.v. X with density function

$$f_X(x) = \begin{cases} \alpha e^{-\alpha x}, x \geq 0 \\ 0, & \text{otherwise.} \end{cases}$$

b) Consider another exponential r.v. Y with parameter β, statistically independent of X. Based on the results of Problem 50.8, determine the expected values of $Z = \min(X, Y)$, and of $W = \max(X, Y)$.

61.8. A Gamma distribution of order m with parameter α ($\alpha > 0$) is described by the p.d.f. (60.5). Determine the mode of this distribution, and plot as function of m.

61.9. A computer system has two identical processors, so that a job to be processed can be handled by either processor, when idle. When both processors are busy, newly arriving jobs are held in a first-in-first-out ('FIFO') queue. Suppose it has been determined that for this system, the processing time required by jobs can be characterized by i.d.s.i. exponential r.v.'s. Consider the situation where at a particular instant (time $t = 0$) the processors are busy with jobs J_1 and J_2, respectively; the queue is empty; and a new job J_3 arrives.

a) Find the probability that J_3 will be finished before one of the jobs running at $t = 0$.
b) What is the probability distribution governing the time t_3 of completion of J_3?
c) What is the expected time of completion of J_3? How does this compare to the expected time till completion for J_1 and J_2?

61.10. *Truncated exponential distribution.* The distribution induced by a r.v. X is described by the following p.d.f., for some $b > 0$:

$$f_X(x) = \begin{cases} ce^{-\alpha x}, & 0 \le x \le b \\ 0, & \text{otherwise.} \end{cases}$$

a) Show that $\mathbf{E}X = \dfrac{1}{\alpha} - \dfrac{b}{e^{\alpha b} - 1}$.
b) Verify that the result in (a) satisfies $0 < \mathbf{E}X < b$, as required.

Section 62

62.1. For each of the induced distributions described below, compute the mean and variance.

a) $f_X(x) = \begin{cases} x, & 0 < x \le 1 \\ 2 - x, & 1 < x < 2 \\ 0, & \text{otherwise} \end{cases}$
b) $f_Y(y) = \begin{cases} 0.5y^{-1/2}, & 0 < y < 1 \\ 0, & \text{otherwise} \end{cases}$

c) $F_Z(z) = \begin{cases} 0, & z < 0 \\ z^2, & 0 \le z < 1 \\ 1, & z \ge 1 \end{cases}$
d) $F_V(v) = \begin{cases} 0.5 e^v, & v \le 0 \\ 1 - 0.5 e^{-2v}, & v > 0. \end{cases}$

62.2. Refer to Problem 61.1. What is the standard deviation of the distribution P_X? Make a plot of P_X similar to the plot of f_X in Example 62.2, showing increments of one standard deviation to either side of the mean.

Note: $\displaystyle\sum_{k=1}^{n} k^2 = \frac{n^3}{3} + \frac{n^2}{2} + \frac{n}{6}$.

62.3. Consider the experiment of observing the operation of a device or equipment until failure occurs (at which time a new device or equipment is substituted). Assume that failures are 'random occurrences' in the sense of Section 60. The *reliability* of a device at time t_0, denoted $R(t_0)$, is the probability of no failure occurring in the period from the beginning of operation (time $t = 0$) up to time t_0. The *mean time between failures* (MTBF) is the mean of the exponential probability distribution expressing the time to failure.

a) Suppose a particular radar system has a MTBF of 200 hours. What is the 24-hour reliability of the system?
b) If the 24-hour reliability is to be at least 0.95, what is the minimum acceptable MTBF?

62.4. Consider an exponential distribution P_X with p.d.f. as given in Figure 39.1b.

a) Determine the standard deviation of P_X as function of m_{1X}. On a sketch of the p.d.f., enter the tangent to the exponential curve of the p.d.f. at the origin, and the intercept of that tangent with the x-axis. Then, similar to Figure 62.1, locate the points $x = m_{1X}$, and $x = m_{1X} \pm \sqrt{\mu_{2X}}$.
b) Obtain an expression for the standard deviation of X^n.

62.5. For the standardized Gaussian distribution with p.d.f. (40.2), obtain a recursion formula for the kth moment in terms of the $(k-2)$th moment, and in this way establish that all moments exist and have the values

$$m_k = \begin{cases} 0, & k \text{ odd} \\ 1 \cdot 3 \cdot 5 \cdot \ldots \cdot (k-1), & k \text{ even.} \end{cases}$$

62.6. Show that the variance of a Poisson distribution equals its first moment.
62.7. Given a distribution with the following density function, where k is an integer ≥ 2:

$$f_X(x) = \begin{cases} \alpha, & -1 \leq x \leq 1 \\ \alpha|x|^{-k}, & \text{otherwise} \end{cases}$$

For this distribution, what is the highest-order moment that exists?
62.8. The density function of an absolutely continuous distribution consists of two rectangles of width 0.5 and height 1, centered at $\pm c$ ($c > 0.25$).

a) For what c is the variance equal to 1?
b) Same question if the rectangles are replaced by δ-functions of weight 1/2 each.

62.9. Suppose the p.d.f. f_Y in Figure 62.2 is modified by a rectangular notch at the center. Thus, consider the r.v. Z with density function as shown in Figure A.9. P_Z is uniformly distributed over the two intervals $\overline{-a,-c}$ and $\overline{c,a}$. If this is again to be a standardized distribution, determine and plot the relationship between c and a.

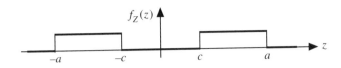

Figure A.9   P.d.f as considered in Problem 62.9

Section 63

63.1. With X inducing a uniform distribution over $\overline{-\pi/2, \pi/2}$, determine $\mathbf{E}Y$ if (a) $Y = \sin X$; and
(b) $Y = \cos X$.

63.2. Let N be a negative binomial r.v. that expresses the number of times a die has to be thrown till the kth 'six' appears (see Problem 27.14).

a) Determine $\mathbf{E}N$. (Suggestion: Let N_1 be the r.v. that expresses the number of times a die has to be thrown till the first six appears. Then N is the sum of k statistically independent r.v.'s inducing the same distribution as N_1.)

b) Generalize this result to obtain the expected value of a negative binomial r.v. N expressing the number of independent, identical trials required for a particular property to be observed k times, if in any one trial the property is observed with probability p.

63.3. Let X_k be the r.v. that expresses the number of red balls among k balls removed blindly from an urn containing r red balls and b black balls ($1 \leq k \leq r + b$). X_k induces a hypergeometric distribution.

a) Determine $\mathbf{E}X_k$.
(Suggestion: Let I_j be the indicator r.v. for the event that the jth ball drawn is red.)

b) Referring to Example 63.1, show that a binomial r.v. and a hypergeometric r.v. with corresponding parameters have the same expected value. Therefore, the expected value in (a) is unchanged if sampling is done *with* replacement.

63.4. Use Theorem 63.3 to obtain the expected value of the r.v. T_m with p.d.f. as given in Equation (60.5). (T_m induces a Gamma distribution of order m with parameter $\alpha > 0$.)

63.5. Consider the triangular r.v.'s X, Y specified in Problem 50.10, parts (a) and (b). For each of the two cases, find $\mathbf{E}[X/Y]$.

63.6. Given an induced distribution P_X for which moments exist up to order $2n + 2$ (for some positive integer n). Making use of the Schwarz inequality, show that

$$\left(m_{(2n+1)X}\right)^2 \leq m_{(2n)X}m_{(2n+2)X}.$$

63.7. The joint distribution induced by the two r.v.'s X, Y is uniform over the triangle with vertices at $(0, 0)$, $(2, 1)$, and $(0, 2)$. Furthermore, let $Z = X + Y$. Obtain the means and variances of X, Y and Z.

63.8. If X is a r.v. whose variance exists, show that an equivalent definition for the variance of X is $\mathbf{V}X = \min_a \mathbf{E}[(X - a)^2]$. This is the so-called 'minimum property of the center'.

63.9. Let X be any r.v. for which $\mathbf{E}[X^2]$ exists.

a) Show that $|\mathbf{E}X| \leq \mathbf{E}[|X|] \leq \sqrt{\mathbf{E}[X^2]}$.

b) Show that $\sqrt{\mathbf{V}X} \geq \mathbf{E}[|X - \mathbf{E}X|]$; that is, the expected value of the absolute deviation from the mean is bounded by the standard deviation.

c) Find an induced distribution P_X for which the inequalities in (a) become equalities.

63.10. Let X be a r.v. such that $P_X(\overline{-\infty, 0}) = 0$. Show that $\mathbf{E}[1/X] \geq 1/\mathbf{E}X$.

63.11. A bin in a department store contains a large number of knives, forks and spoons in equal proportion and thoroughly mixed. If implements are removed blindly, one at a time, how many can one expect to remove until a complete setting (knife, fork and spoon) has been obtained? (For the purpose of this problem, assume the supply of knives, forks and spoons to be inexhaustible.)

63.12. A delicate assembly task is performed by a skilled factory worker. The task requires ten parts to be assembled into a subassembly. Each part requires one minute to handle and assemble. Only after the tenth part has been inserted is the subassembly stable. If the assembly process gets interrupted prior to completion, it must be begun again from the beginning. (Assume that statistical regularity holds, and that the reassembly process starts at the end of the minute during which the interruption occurred.)

a) If the probability of an interruption while any one part is being assembled is 0.1, what is the expected assembly time required to complete a subassembly?

b) What probability of an interruption causes the expected time for completion of a subassembly to be two hours?

63.13. Given two r.v.'s X, Y whose second moments exist. Show the following:

a) $\mathbf{E}[X\,Y]$ cannot exceed the arithmetic average of the mean-square values:

$$|\mathbf{E}[X\,Y]| \leq (\mathbf{E}[X^2] + \mathbf{E}[Y^2])/2.$$

Suggestion: Consider $\mathbf{E}[(X \pm Y)^2]$.

b) The Schwarz inequality (Theorem 63.5) provides at least as tight a bound as the inequality in (a); i.e.:

$$\sqrt{\mathbf{E}[X^2]\mathbf{E}[Y^2]} \leq (\mathbf{E}[X^2] + \mathbf{E}[Y^2])/2$$

Suggestion: For $\mathbf{E}[X^2] > \mathbf{E}[Y^2]$, write $\mathbf{E}[X^2] = \mathbf{E}[Y^2] + b$.

63.14. Let P_X be the distribution induced by some r.v. X whose variance exists. Let a new r.v. Y be defined by $Y = (X - a)/b$, where a, b are real numbers and $b \neq 0$. If Y turns out to be standardized, is $Y = X_s$?

63.15. Prove the following two assertions that appear in this Section:

a) For a r.v. X, $m_{2X} = 0$ if and only if $P_X(\{0\}) = 1$.

b) For a r.v. X, $\mathbf{V}X = 0$ if and only if $P_X(\{a\}) = 1$ for some $a \in \mathsf{R}$.

63.16. Consider two absolutely continuous random variables X, Y such that, for some real number a,

$$f_X(x) \geq f_Y(x), \text{ for } x < a$$

$$f_X(x) \leq f_Y(x), \text{ for } x > a.$$

a) Prove that $\mathbf{E}X \le \mathbf{E}Y$.
b) If, in addition, $f_X(x) = f_Y(x) = 0$ for $x < 0$, and $a > 0$, show that $\mathbf{E}[X^k] \le \mathbf{E}[Y^k]$, for $k = 2, 3, 4, \ldots$.
    (Suggestion: Use the transformation $X' = X - a$, $Y' = Y - a$.)

Section 64

64.1.   The r.v. T_m induces a Gamma distribution of order m, with density function (60.5). For $m = 1$, 2, 3,..., obtain $\mathbf{V}T_m$ as functions of α $(\alpha > 0)$.

64.2. X induces a uniform distribution over $\overline{-1, 1}$. Let $Y = X^n$ $(n = 1, 2, 3,\ldots)$. What is the variance of Y?

64.3.   Let $X_1,\ldots, X_n$ be i.d.s.i. zero-mean Gaussian r.v.'s with standard deviation σ.

a) Determine the mean and variance of the r.v. $X_1{}^2$.
b) Determine the mean and variance of the χ^2-distributed r.v. $X_1^2 + X_2^2 + \ldots + X_n^2$.

64.4.   Let $X_1,\ldots, X_n$ be pairwise statistically independent r.v.'s, and $a_1,\ldots, a_n$ any real constants. Show that

$$\mathbf{V}\left[\sum_{i=1}^{n} a_i X_i\right] = \sum_{i=1}^{n} a_i{}^2 \mathbf{V}X_i.$$

64.5.   The r.v. X induces a uniform distribution over $\overline{0, b}$. Let $Y = X^2$. For what value of b is $\mathbf{V}Y = \mathbf{V}X$?

64.6.   Let Θ be a r.v. that is unif. $\overline{0, \pi}$. Let A be the area of an isosceles triangle whose vertices are the origin and two points on the circumference of the unit circle, separated by the angle Θ.

a) Determine $\mathbf{E}A$.
b) Determine $\sqrt{\mathbf{V}A}$.

64.7.   Use Theorem 64.3a to prove Theorem 64.4.

64.8.   The joint distribution induced by the r.v.'s X, Y is uniform over the interior of a circle of radius 1, centered at the origin.

a) Determine $\mathbf{V}X, \mathbf{V}Y$.
b) From the result in (a), determine $\mathbf{V}X$ and $\mathbf{V}Y$ if the joint distribution of X, Y is uniform over the interior of the ellipse $(x/a)^2 + (y/b)^2 = 1$.

64.9.   For the distributions induced by the r.v.'s X and Z in Problem 50.10 (in both parts (a) and (b)), find the mean and the variance.

64.10. *Random Walk.* A robot frog chooses a random angle X_1 (with uniform distribution over $\overline{0, 2\pi}$), and jumps a unit distance in that direction. Then, it independently chooses a second angle, X_2, and jumps a unit distance in that direction. After n such jumps, what is its expected squared distance (its mean-square distance) from the starting position?

64.11. Given any two r.v.'s X, Y for which the standard deviation exists.

a) Show that their standard deviations satisfy the triangle inequality,

$$\sqrt{\mathbf{V}[X + Y]} \le \sqrt{\mathbf{V}X} + \sqrt{\mathbf{V}Y}$$

Suggestion: Make use of Theorem 1; i.e., since the variance does not depend on the mean, assume centralized r.v.'s.

b) Show that $\left|\sqrt{VX} - \sqrt{VY}\right| \le \sqrt{V[X-Y]}$.

c) Do the above relationships still hold if $\sqrt{V}$ is replaced by V?

Section 65

65.1. Suppose that the number of vehicles arriving at a certain toll gate on weekdays between 12 noon and 12:15 p.m. is adequately modeled by a Poisson distribution with mean 100. Use the Chebychev inequality to determine a lower bound on the probability that, on a specified weekday, the number of vehicles arriving at this toll gate during this 15-minute period will be between 80 and 120. (Suggestion: See Problem 62.6.)

65.2. Rewrite Equation (65.4) so that it applies to nonstandardized symmetrical r.v.'s.

65.3. Give an example of a r.v. X for which the inequality (65.6) is satisfied with equality, for some $\gamma > 1$, thus showing that the bound (65.6) cannot be tightened. (Suggestion: Refer to the Proof of Theorem 65.2.)

65.4. The random variable X assigns zero probability outside the interval $\overline{-a, a}$. Show that $VX \le a^2$.

65.5. Verify that the c.d.f. of a standardized r.v. X_s cannot coincide with the bound (65.4) for all x.

65.6. It is known that the r.v. X induces a truncated exponential distribution (see Problem 61.10), but the truncation point is not known. Obtain an upper bound for $P_X(\overline{\gamma m, \infty})$ that is tighter than the bound (65.6).

65.7. Given a Cauchy distribution that is symmetrical about the origin, with p.d.f. $f_X(x) = \{\pi b [1 + (x/b)^2]\}^{-1}$, $b > 0$. Use Equation (65.4) to show that this distribution cannot be standardized.

Suggestion: Truncate the series expansion of $F_X(x)$ so as to obtain an *upper* bound on $F_X(x)$ for $x > 1$. Then show that for no b does this upper bound satisfy the lower bound in line 4 of (65.4) for all $x > 1$.

Section 66

66.1. A fair die is thrown 52 times. After each throw, the number of dots obtained is marked on one of the cards in a standard deck of 52 playing cards. Every card is thus marked with one of the integers 1, 2, 3, 4, 5 or 6. Suppose that the frequency with which each of these integers occurs in the deck is 12, 7, 10, 9, 5 and 9, respectively.

a) What is the population average?

b) A simple random sample of ten cards is taken from the deck. What is the expected value of the sample mean?

c) Can the random sample in (b) be regarded a sample of die throw results, with the 'population' considered to be the (infinite) population consisting of an equal mixture of ones, twos,..., sixes?

66.2. Given a population of size n. Suppose a random sample of size $k_1 < n$ is drawn, which yields the test sample average a_{T_1}. This sample is not returned to the population before another random sample of size $k_2 < n - k_1$ is drawn, with sample average a_{T_2}. What is the best estimate of the population mean, based on a_{T_1} and a_{T_2}?

66.3. The average age of the customers of the Avanti clothing store was estimated by a consultant to be 24.2, based on a random sample of size 100. The same consultant estimated the average customer age of Berman's Clothing to be 37.5, based on a sample size of 150. Berman's had twice as large a clientele as Avanti. The two companies have now merged and want to know the average age of their combined clientele.

Since the amount of overlap between the two clienteles before the merger is not known, obtain an estimate for the two extreme cases; i.e., estimate

a) the average age assuming no overlap;
b) the average age assuming the Avanti customers to have been a subset of the Berman customers.
c) Under the assumption in (b), what is the average age of those Berman customers who were not also Avanti customers?

66.4. Consider an experiment in which a single die is rolled a large number of times. Whenever a 'six' is thrown, this throw is ignored; otherwise, each value thrown is recorded. When 200 numbers have been recorded in this manner, the arithmetic average is computed. Let A be the r.v. that expresses this average.

a) What is the variance of A ?
b) Use the Chebychev inequality to estimate the range of values, centered at the mean, that A will take on with probability 0.9.
c) How is this range changed if 400 numbers are recorded?

66.5. Consider the experiment of picking a digit (from 0 to 9) at random—say, from a random number table.

a) By how much is a value thus picked expected to deviate from the population average; i.e., what is $E[|X - m_{1X}|]$, where X expresses the value of the digit picked?
b) Compare the result of (a) with the standard deviation of X.
c) What is the expected value of the sample average A_T if a sample of 12 digits is picked from the random number table, and how does it compare to the population average?
d) What is the standard deviation of A_T?

66.6. Show that if sampling with replacement is used, then the variance of the sample average for a sample of size k is v_p/k.

Suggestion: In this case it is simpler to represent the sample by an ordered k-tuple $(X_1, \ldots, X_k)$, where X_m ($m = 1, \ldots, k$) is a r.v. that assigns the value of the mth object selected, and $X_1, \ldots, X_k$ are m.s.i. r.v.'s.

66.7. Ten slabs of granite are delivered to a construction company. For $i = 1, \ldots, 10$, slab no. i has weight x_i (in thousands of pounds) as follows:

$$x_1 = 1.3 \quad x_2 = 1.2 \quad x_3 = 1.4 \quad x_4 = 1.4 \quad x_5 = 1.4$$
$$x_6 = 1.4 \quad x_7 = 1.6 \quad x_8 = 1.3 \quad x_9 = 1.0 \quad x_{10} = 1.5$$

A worker, initially unacquainted with the values listed above, performs an experiment to estimate a_p, the *average weight* of the slabs. His procedure is to select a simple random sample of size 2 and determine their values (i.e., weigh them), average the two values thus observed, and consider the result to be an *estimate* of a_p.

a) Plot a bar graph of the probability distribution induced by A_T, the sample average.
b) Determine $\mathbf{E}A_T$ and $\sqrt{VA_T}$.
c) What is the probability that the estimate of a_p will exactly equal a_p?
d) Plot a bar graph of the probability distribution induced by V_T, the sample variance.
e) Write the expression for an unbiased estimate of v_p.
f) What is the probability that the estimate computed in accordance with (e) exactly equals v_p?

Section 67

67.1. X and Y are s.i. r.v.'s. P_X is unif.$\overline{0,2}$. P_Y is discrete, with $P_Y(\{0\}) = P_Y(\{1\}) = P_Y(\{2\}) = 1/3$. Use conditional expectations to find $\mathbf{E}[X^Y]$.

67.2. In Example 67.1, apply Theorem 67.1 as a check on Equation (67.4).

67.3. *Random Sums.* Let

$$S = \sum_{k=1}^{N} X_k$$

where N, S, and the X_k are r.v.'s. Assuming N to take on only positive integer values and to be independent of the X_k's, use conditional expectations to verify that

$$\mathbf{E}[S] = \sum_{k=1}^{\infty} \mathbf{E}[X_k] P_N(\{k, k+1, k+2, \ldots\}).$$

67.4. In Problem 61.9, suppose three jobs arrive simultaneously at a time when both processors are idle, so that two of the jobs get picked randomly for immediate execution, while the third one has to wait till a processor becomes available. What is the expected time till completion for any one of these three jobs at arrival?

67.5. A r.v X is defined on a probability space $(S, \mathscr{A}, P)$, and $\{A_1, \ldots, A_n\}$ is a partition with $P(A_i) = p_i, i = 1, \ldots, n$, so that $\sum p_i = 1$. Suppose that the only information given about X is the conditional standard deviations $\sqrt{V[X|A_i]}$ and the conditional means $\mathbf{E}[X|A_i]$. Express $\sqrt{VX}$ in terms of these parameters.

67.6. A random phenomenon is described by a r.v. X with conditional p.d.f.

$$f_{X|A_i}(x) = \begin{cases} a_i \exp[-a_i x], & x \geq 0 \\ 0, & \text{otherwise.} \end{cases}$$

Here, a discrete r.v. A (statistically independent of X) assigns probabilities $p_1, \ldots, p_n$ to the positive numbers $a_1, \ldots, a_n$, and A_i is the event that $A = a_i$ ($i = 1, \ldots, n$).

a) Obtain an expression for $\mathbf{E}X$.
b) Apply the result of part (a) to the following situation. The marquee of a movie theatre uses 150 white light bulbs and 100 red bulbs. The life of a white bulb is exponentially distributed with a mean of 2000 hours, and the life of a red bulb is exponentially distributed with a mean of 1500 hours. What is the expected life of a bulb (irrespective of color) used in the marquee?
c) For the marquee in (b), what is the expected time until the first bulb burns out?

67.7. In Example 67.2, use a Gaussian approximation to estimate k, the required number of die throws.

Section 68

68.1. Given two r.v.'s X, Y with nonzero variances. Verify the following:

a) If Y is expressed in the form $Y = aX + b + Z$, where $\rho_{X,Z} = 0$, then the coefficient $a = \rho_{X,Y}\sqrt{VY/VX}$.

b) If Y is expressed in standardized form as in Equation (68.7), then $a = \rho_{X,Y}$, and $c^2 = 1 - a^2$.

68.2. The joint distribution of two r.v.'s X, Y is uniform over the two line segments shown in Figure A.10. Use conditional expectations to determine $\rho[X,Y]$.

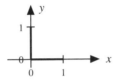

Figure A.10   Singular distribution considered in Problem 68.2

68.3. Obtain a general expression for the correlation coefficient for the sum and the difference of two s.i. r.v.'s with positive variance. That is, if X_1, X_2 are s.i., and $Y_1 = X_1 + X_2$, $Y_2 = X_1 - X_2$, determine $\rho[Y_1,Y_2]$. Under what condition is $\rho[Y_1,Y_2] = 0$?

68.4. Consider an experiment in which two noisy measurements of the same quantity are made. The quantity to be measured is described by the r.v. X. The measurements are described by the r.v.'s

$$Y = X + N_1, \text{ and}$$

$$Z = X + N_2$$

where N_1, N_2 are r.v.'s representing the superimposed measurement noise. Assume that N_1, N_2 have zero means and equal variances, and are statistically independent of each other and of X. Show that $\rho[Y,Z] = VX/VY$.

68.5. Given three standardized r.v.'s X, Y and Z. If $\rho[X,Y] = \rho[X,Z] = 0.5$, determine the range of values within which $\rho[Y,Z]$ must lie.

Suggestion: Write out (68.8) for Y and also for Z, multiply the two equations and take the expected value.

68.6. Given two arbitrary r.v.'s X_1, X_2, such that VX_1 and VX_2 exist. What conditions, if any, must be satisfied by X_1, X_2 so that it is possible to transform them into *uncorrelated* r.v.'s Y_1, Y_2 by means of a *linear* transformation, i.e., $Y_i = a_iX_1 + b_iX_2$, where $i = 1$, 2.

68.7. Two discrete r.v.'s X, Y jointly induce probability 1/4 at each of the two points $(-1, 0)$ and $(1, 0)$, and probability 1/2 at some other point (x, y). Determine the locus of points (x, y) such that $\rho[X,Y] = 1/2$.

68.8. Two discrete r.v.'s X, Y induces probability 1/4 at each of the two points $(0, 0)$ and $(1, 1)$. At a third point, probability 1/2 is induced. Determine the locus of possible positions of the third point if $\rho[X,Y] = 0$.

68.9. Prove that if two binary r.v.'s X, Y are uncorrelated, they are statistically independent.

68.10. Consider the r.v.'s X, Y that induce the joint distribution described in Figure 68.1.

   a) Express the relation between the standardized versions of X, Y in the manner of Equation (68.8).
   b) Convert your expression to nonstandardized form, i.e., in terms of X and Y.
   c) Determine the distribution of Z in this case, and sketch the joint distribution of X and Z, and of Y and Z.

68.11. A sinusoidally varying voltage of unit amplitude, $v(t) = \sin\omega t$ (where ω is a positive constant), is observed at two arbitrary instants of time t_1 and t_2. On the probability space for this experiment, let the two observed voltages be characterized by two r.v.'s X and Y.

   a) Determine $\rho_{X,Y}$ as a function of $\tau = t_2 - t_1$. (Suggestion: Express X and Y in terms of T and τ, where T is the random variable that characterizes the observation instant t_1. Assume T to be uniformly distributed over $0, 2\pi/\omega$.)
   b) Relate the result obtained in (a) to the joint distribution of X, Y for the cases where $\tau = \pi/4\omega$, $\pi/2\omega$, and π/ω. (Note: These joint distributions are singular and are concentrated on so-called Lissajous figures.)

Section 69

69.1. Verify the steps in the derivation of Equation (69.9) from the joint distribution of two statistically independent r.v.'s given at the beginning of Example 69.2.

69.2. Let X, Y be s.i. Gaussian r.v.'s, with identical variances σ^2. Show that $X + Y$ and $X - Y$ are also s.i. r.v.'s.

69.3. Using the results of Problem 64.8, determine VX, VY, and $\rho[X,Y]$ if the joint distribution induced by X and Y is uniform over the interior of an ellipse whose axes are at $45°$ angles with respect to the coordinate axes.

69.4. Generalize Example 69.1 to *unsymmetrical* binary r.v.'s X, Y that jointly induce probability at the same points in R^2 as do the r.v.'s X, Y in Example 69.1.

69.5. In Example 69.2, show that if $\varphi = 45°$ then

$$\rho_Y = \frac{\sigma_1^2 - \sigma_2^2}{\sigma_1^2 + \sigma_2^2}.$$

Sketch a graph of ρ_Y as function of σ_2, for fixed σ_1.

69.6. Consider the bivariate Gaussian distribution of Example 69.2. If VY_1, VY_2 and $C[Y_1,Y_2]$ are given, show that

$$\varphi = \frac{1}{2} \arctan \frac{2\mathbf{C}[Y_1, Y_2]}{\mathbf{V}Y_1 - \mathbf{V}Y_2}.$$

Does this result apply for all values of $\mathbf{V}Y_1$, $\mathbf{V}Y_2$ and $\mathbf{C}[Y_1,Y_2]$?

69.7. Two two-dimensional r.v.'s $X = (X_1, X_2)$, $Y = (Y_1, Y_2)$ are s.i. Let $Z = (Z_1, Z_2) = X + Y$.

a) Express $\mathbf{C}[Z_1,Z_2]$ in terms of $\mathbf{V}X_1$, $\mathbf{V}X_2$, $\mathbf{V}Y_1$, $\mathbf{V}Y_2$, $\mathbf{C}[X_1,X_2]$ and $\mathbf{C}[Y_1,Y_2]$.

b) If, furthermore, $\mathbf{V}X_1 = \mathbf{V}Y_1$ and $\mathbf{V}X_2 = \mathbf{V}Y_2$, show that

$$\boldsymbol{\rho}[Z_1, Z_2] = (\boldsymbol{\rho}[X_1, X_2] + \boldsymbol{\rho}[Y_1, Y_2])/2.$$

69.8. Y_1 and Y_2 are zero-mean bivariate Gaussian r.v.'s with joint p.d.f. as given by Equation (69.9). Making use of Equation (54.7b), obtain $f_{Y_2}(y_2)$. Then show that $f_{Y_1|Y_2}(y_1|y_2)$ is Gaussian. Express the mean and variance of this conditional density in terms of $\sigma_{Y_1}^2$, $\sigma_{Y_2}^2$, ρ_Y and y_2, where $Y = (Y_1, Y_2)$.

69.9. X_1, X_2 are uncorrelated zero-mean bivariate Gaussian r.v.'s, with variances σ_1^2 and σ_2^2, respectively. Z_1, Z_2 are uncorrelated r.v.'s that induce a distribution that is uniform over the interior of an ellipse L centered at the origin. If, furthermore, we set $\mathbf{V}Z_1 = \sigma_1{}^2$ and $\mathbf{V}Z_2 = \sigma_2^2$, then L becomes L_c, the *ellipse of concentration* for the distribution induced by X_1, X_2.

a) For what value of c does the equation $f_{X_1,X_2}(x_1, x_2) = c$ yield the ellipse L_c?

b) If the same rotation is applied to (X_1, X_2) and also to (Z_1, Z_2), will the covariance in both cases be the same?

69.10. Let X_1, X_2 be Gaussian r.v.'s with a standard ellipse that is a circle centered at the origin. Let Z be a zero-mean Gaussian r.v. that is statistically independent of both X_1 and X_2. Let $Y_1 = X_1 + Z$, and $Y_2 = X_2 + Z$.

a) Determine f_Y.

b) Is it possible to go backwards and determine f_Z from f_Y, if it is known that:

   i) $Y_1 = X_1 + Z$, and $Y_2 = X_2 + Z$;

   ii) X_1, X_2 are i.d.s.i. Gaussian r.v.'s; and

   iii) Z is zero-mean Gaussian, and is statistically independent of both X_1 and X_2?

Section 70

70.1. Suppose X and Θ are statistically independent r.v.'s, with P_Θ uniformly distributed over $0, 2\pi$. Use conditional expectations, with X the conditioning r.v., to show that, for any nonzero integer n,

$$\mathbf{E}[\cos(X + n\,\Theta)] = \mathbf{E}[\sin(X + n\,\Theta)] = 0.$$

70.2. A fair die is thrown. Let y be the number of dots obtained.

a) The die is then thrown y times and the number of dots obtained in these y throws is summed. What is the expected value of this sum? (Suggestion: Use Theorem 70.2.)

b) Repeat (a), but with the value y (i.e., the result of the initial throw) included in the summation.

70.3. What is the expected value of the sum of the numbers of dots obtained with a fair die that is thrown until the first 'six' appears? (See Example 34.3.)

70.4. Consider two r.v.'s X, Y whose joint distribution is concentrated on the line segment specified in Figure A.11, with probability 1 distributed in an arbitrary manner over this line segment. Simplify $\mathbf{E}\left[\dfrac{X}{Y}\Big|Y\right]$ and compare it to $\dfrac{\mathbf{E}[X|Y]}{\mathbf{E}[Y|Y]}$. (Suggestion: Use Theorem 70.3.)

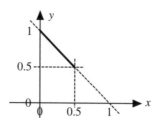

Figure A.11 Joint distribution over a line segment

70.5. Suppose two r.v.'s X, Y induce a uniform distribution over the unit square $\overline{0,1} \times \overline{0,1}$. Determine $\mathbf{E}[X - Y | X + Y]$.

70.6. Given two r.v.'s X, Y such that $\mathbf{E}[Y|X]$ exists. Show that $\mathbf{E}[Y|X]$ and $Y - \mathbf{E}[Y|X]$ are uncorrelated r.v.'s.

70.7. Let Y_1, Y_2 be two jointly Gaussian r.v.'s.

a) Use the result of Problem 69.8 to show that $\mathbf{E}[Y_2 | Y_1]$ is a linear function of Y_1.

b) Using the notation of Example 69.2, show that the angle of the regression line of Y_2 on Y_1 with respect to the y_1-axis is given by $\varphi - \arctan[(\sigma_2/\sigma_1)^2 \tan\varphi]$.

Section 71

71.1. The covariance matrix for the r.v. $X = (X_1, X_2, X_3)$ is

$$\mathbf{C}[X, X] = \begin{bmatrix} 1 & 0.2 & 0.1 \\ 0.2 & 2 & 0.2 \\ 0.1 & 0.2 & 1 \end{bmatrix}.$$

Find $\mathbf{C}[Y,Y]$ for the r.v. $Y = (Y_1, Y_2, Y_3)$ defined by

$$Y_1 = X_1 - X_2$$

$$Y_2 = 2X_1 + X_3 + 1$$

$$Y_3 = X_2 - 2X_3.$$

71.2. Which of the following matrices cannot be covariance matrices?

$$
c_1 = \begin{bmatrix} 1 & 0.5 & 0.1 \\ 0.2 & 1 & 0.2 \\ 0.1 & 0.2 & 0.2 \end{bmatrix} \quad c_2 = \begin{bmatrix} 1 & 0 & 0.5 \\ 0 & 1 & 0 \\ 0.5 & 0 & 1 \end{bmatrix}
$$

$$
c_3 = \begin{bmatrix} 10 & 1 & 0.1 \\ 1 & 1 & 1 \\ 0.1 & 1 & 10 \end{bmatrix} \quad c_4 = \begin{bmatrix} 1 & 1 & 0 \\ 1 & 1 & 0 \\ 0 & 0 & 3 \end{bmatrix}.
$$

71.3. Let $Y = (Y_1, \ldots, Y_n)$ be an n-dimensional Gaussian r.v. with p.d.f. (71.6). Show that if the components of Y are uncorrelated (i.e., $\rho[Y_i, Y_j] = 0$ for $i \neq j$, where $i, j = 1, \ldots, n$), then they are m.s.i. Thus, if $Y_1, \ldots, Y_n$ are pairwise s.i., they also are m.s.i.

71.4. Given an n–dimensional vector r.v. X with covariance matrix $\mathbf{C}[X,X]$. A new r.v. Y is obtained by multiplying each of the components of X by a constant, so that $Y_i = a_i X_i$, $i = 1, \ldots, n$. Show that $\mathbf{C}[Y,Y]$ is obtained from $\mathbf{C}[X,X]$ by multiplying the ith row and column of $\mathbf{C}[X,X]$ by a_i, for $i = 1, \ldots, n$.

71.5. Express the covariance matrix of a zero-mean bivariate Gaussian distribution, and also its inverse, in terms of the parameters α, β, γ in Equation (54.6).

71.6. Given a symmetric 2×2 matrix $c = \begin{bmatrix} c_{11} & c_{12} \\ c_{12} & c_{22} \end{bmatrix}$, with $c_{11}, c_{22} \geq 0$. Show that c is positive semi-definite if and only if $c_{12} \leq \sqrt{c_{11} c_{22}}$.

71.7. For the joint density (71.6), show that the marginal densities are zero-mean Gaussian. (Suggestion: Show that Y_n is zero-mean Gaussian and then provide an argument that allows you to conclude that all the Y_i's are Gaussian.)

71.8. A square matrix $c = [c_{ij}]$, where $i, j = 1, \ldots, n$, is called a *Toeplitz* matrix if the elements of the matrix satisfy $c_{ij} = c_{mn}$ if $|i - j| = |m - n|$. In other words, a Toeplitz matrix is of the form

$$
\begin{bmatrix}
a & b & c & d & \cdots & p & q \\
b & a & b & c & & & p \\
c & b & a & b & \ddots & & \vdots \\
d & c & b & a & & \ddots & \vdots \\
\vdots & & \ddots & & \ddots & & d \\
\vdots & & & \ddots & & \ddots & c \\
p & & & & \ddots & a & b \\
q & & \cdots & & d & c & b & a
\end{bmatrix}
$$

Let X, Y be n-dimensional r.v.'s whose covariance matrices are Toeplitz matrices. Also, let X and Y be uncorrelated; i.e., $\mathbf{C}[X_i, Y_j] = 0$ for all $i, j = 1, \ldots, n$, $i \neq j$. If $Z = X + Y$ (so that $Z_i = X_i + Y_i$, $i = 1, \ldots, n$), show that c_Z is also a Toeplitz matrix.

71.9. Three zero-mean Gaussian r.v.'s Y_1, Y_2, Y_3 have a joint p.d.f. as given by Equation (71.6), with

$$
c_Y^{-1} = \begin{bmatrix} 1.18 & -0.24 & 0 \\ -0.24 & 1.32 & 0 \\ 0 & 0 & 0.6 \end{bmatrix}.
$$

a) What are the variances of Y_1, Y_2, Y_3?
b) Determine an orthogonal transformation that will decorrelate the three r.v.'s; i.e., that will result in three r.v.'s that are uncorrelated (and therefore m.s.i.). Note: The inverse of a diagonal matrix is a diagonal matrix.

$$\text{Possible answer:} \begin{bmatrix} 0.8 & 0.6 & 0 \\ -0.6 & 0.8 & 0 \\ 0 & 0 & 1 \end{bmatrix}$$

71.10. A vector r.v. $X = \begin{bmatrix} X_1 \\ X_2 \end{bmatrix}$ induces probability uniformly over the region bounded by the parallelogram shown in Figure A.12.

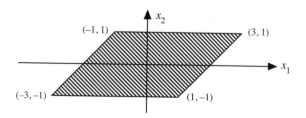

Figure A.12   Parallelogram

a) Determine the covariance matrix of X.
b) Determine a transformation matrix a that rotates X so that the resulting r.v. Y has a diagonal covariance matrix. What is the angle through which a rotates X?
c) By inspection, it can be seen that merely distorting the shaded parallelogram so as to make the sides vertical will also result in uncorrelated component r.v.'s. What is a transformation matrix that accomplishes this?

Section 72

72.1. Given n standardized m.s.i. Gaussian r.v.'s $X_1, \ldots, X_n$. Let Y and Z be linear combinations of $X_1, \ldots, X_n$:

$$Y = a_1 X_1 + \ldots + a_n X_n$$

$$Z = b_1 X_1 + \ldots + b_n X_n$$

where neither the coefficients $a_1, \ldots, a_n$ nor $b_1, \ldots, b_n$ are all zero. Show that Y and Z are statistically independent if and only if

$$\sum_{i=1}^{n} a_i b_i = 0.$$

72.2. Consider the optimum representation of a zero-mean r.v. Y in terms of n given zero-mean r.v.'s $X_1, \ldots, X_n$, with a remainder Z, in the manner of Equation (72.3). Prove that this requires Z to be orthogonal to the subspace of V defined by $X_1, \ldots, X_n$ (i.e., Z is orthogonal to $X_1, \ldots, X_n$).

72.3. Consider the situation specified in Query 72.2. For convenience, assume that X, Y, Z are standardized. What is the range of allowable values for ρ?

72.4. Given a probability space $(\mathsf{S}, \mathscr{A}, P)$ where $\mathsf{S} = \{\xi_1, \xi_2, \xi_3\}$ and the elementary sets $\{\xi_1\}, \{\xi_2\}, \{\xi_3\}$ have equal probability. What is the dimensionality of V?

72.5. On a given probability space a standardized r.v. X_s is defined. In the associated vector space V, X_s is a unit vector. Now let $Y = X_s^2$. What is the vector that represents Y_c, the centralized version of Y, relative to the unit vector that represents X_s, if:

   a) X_s is a standardized Gaussian r.v.
   b) X_s is a standardized exponential r.v.

Section 73

73.1. Consider the situation of Example 73.1, under the following conditions. In each case, what criterion of optimization might be most appropriate?

   a) You play the game only once. The stakes are high. Your loss is proportional to $|Q|$.
   b) You play the game a large number of times. The stakes are small in each individual play. Each time, your loss is proportional to $|Q|$.
   c) As in (b), but your loss is proportional to Q^2.

73.2. On the probability space associated with some p.e. E, a r.v. Y is defined that induces the triangular distribution whose p.d.f. is described in Figure A.13. Find the optimum estimate of the value $Y(\xi)$ the next time E is performed, for each of the four criteria of optimality used in Example 73.1.

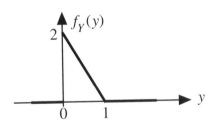

Figure A.13   A triangular p.d.f

If, instead, Y induces a Cauchy distribution (see Figure 39.1d), show that none of the four criteria of Example 73.1 can be used.

73.3. Two i.d.s.i. r.v.'s X, Y are unif. $\overline{0, 1}$. When the underlying p.e. gets performed, the values taken on by the two r.v.'s are not directly observable. Instead, only $X + Y$ is observed. What is the l.m.s. estimate of X in terms of $X + Y$? Of Y in terms of $X + Y$?

73.4. A p.e. has been formulated in which a random phenomenon is to be observed that is characterized by a standardized Gaussian r.v. X_s. However, it is in the nature of the p.e. that an observation of X_s is corrupted by 'multiplicative noise', so that the measurement carried out in the p.e. yields a value of the r.v. $Y = Z X_s$. Suppose that Z is binary, with

$$P_Z(\{z\}) = \begin{cases} 3/4, & z = 1 \\ 1/4, & z = -1. \end{cases}$$

a) What is the optimum mean-square estimate of X_s in terms of Y?
b) How much better is the estimate in (a) than the l.m.s. estimate?

73.5. Suppose X is to be estimated from n 'noisy' observations $Y_1, \ldots, Y_n$, where $Y_i = X + N_i$ $(i = 1, \ldots, n)$. Let $N_1, \ldots, N_n$ be i.d.s.i. zero-mean Gaussian r.v.'s. If $\mathbf{V}N_i = \mathbf{V}X$ $(i = 1, \ldots, n)$ and X is zero-mean, how large must n be so that the l.m.s. estimate of X yields an r.m.s. error no larger than

a) $0.1\sigma_X$?
b) $0.01\sigma_X$?

Section 74

74.1. Prove Theorem 74.1; that is, assuming that the various integrals exist, establish the following:

a) If $g(x) = c_1 g_1(x) + c_2 g_2(x)$, where c_1 and c_2 are constants, then

$$\int_a^b g(x)\, dF(x) = c_1 \int_a^b g_1(x)\, dF(x) + c_2 \int_a^b g_2(x)\, dF(x).$$

b) If $F(x) = k_1 F_1(x) + k_2 F_2(x)$, where k_1 and k_2 are constants, $k_1 + k_2 = 1$, and F_1, F_2 are c.d.f.'s, then

$$\int_a^b g(x)\, dF(x) = k_1 \int_a^b g(x)\, dF_1(x) + k_2 \int_a^b g(x)\, dF_2(x).$$

74.2. Prove Theorem 74.2.
74.3. Prove Theorem 74.3.
74.4. Let $F_X(x)$ be the c.d.f. specified in Example 74.2, and let $u(x)$ denote the right-continuous unit step function,

$$u(x) = \begin{cases} 0, & x < 0 \\ 1, & x \geq 0. \end{cases}$$

Now define $F_Y(y) = 0.5\, F_X(y) + 0.5\, u(y - 1/2)$.

a) Verify that F_Y is a c.d.f.
b) Determine the standard deviation of Y.

74.5. Consider the c.d.f. for a r.v. $Y = X_1 + X_2$ as expressed in Equation (50.3), and suppose that X_1, X_2 are statistically independent.

a) Show that Equation (50.3) can then be written

$$F_Y(y) = \int_{-\infty}^{\infty} F_{X_2}(y - x)\, dF_{X_1}(x).$$

b) Suppose X_2 is unif. $\overline{-1,1}$ and X_1 is singular with the c.d.f.

$$F_{X_1}(x) = \begin{cases} 0, & x < -1 \\ 1, & x \geq 1 \\ 1/2, & -1/2 \leq x < 1/2 \\ 1/4, & -7/8 \leq x < -5/8 \\ 3/4, & 5/8 \leq x < 7/8 \\ 1/8, & -31/32 \leq x < -29/32 \\ \quad \ldots \ldots \ldots \end{cases}$$

    i) Determine $F_Y(0)$.
   ii) Determine $F_Y(1)$.

74.6. In Example 56.1, the singular r.v. V_o distributes probability in a similar manner as the r.v. X in Example 74.2. Consider a generalization of V_o as follows. Let max(Y) and min(Y) be 1 and -1, respectively, as is the case for V_o. The symmetrical interval (centered at the origin) to which Y assigns 0 probability is $\overline{-w, w}$ where $0 < w < 1$. The distribution over the interval $\overline{-1, -w}$ is identical to the distribution over $\overline{w, 1}$, and is simply a horizontally scaled version of the distribution over $\overline{-1, 1}$, with half the weight.

a) Obtain the variance of Y as function of w.
b) Show that for the r.v. V_o of Example 56.1, the result of (a) gives $VV_o = (e-1)/(e+1)$.
c) Show that VX as found in Example 74.2 also agrees with the result of (a).

Part VI Problems

Section 75

75.1. Given an arbitrary complex r.v. Z_1 whose variance exists. Let $Z_2 = ae^{j\theta}Z_1$, where a and θ are real and $a > 0$. Thus, in C, P_{Z_2} is obtained by rotating P_{Z_1} counterclockwise through an angle θ around the origin, and applying the scale factor a.

  a) Show that VZ_2 also exists.
  b) Express $C[Z_1, Z_2]$ and $C[Z_2, Z_1]$ in terms of $V[Z_1]$.
  c) Let $Z = Z_1 + Z_2$. Express $V[Z]$ in terms of $V[Z_1]$.
  d) Is it possible for $V[Z]$ to equal $V[Z_1] + V[Z_2]$?

75.2. Given a complex vector r.v. $Z = \begin{bmatrix} Z_1 \\ \vdots \\ Z_n \end{bmatrix}$, show that c_Z is positive semi-definite. That is, show that for every column vector of complex numbers $s = [s_1 \ \dots \ s_n]^T$ it is true that $s^{*T} c_Z s$ is non-negative. (Note that $s^{*T} c_Z s$ is an Hermitian form and therefore real.)

75.3. Given two s.i. complex r.v.'s Z_1, Z_2. Each distributes all probability uniformly over the circumference of the unit circle in C. Show that the r.v. $Z_3 = Z_1/Z_2$ induces the same distribution.

75.4. Specify a transformation g to be applied to the die-throw r.v. G in Example 34.1 so that a r.v. Z is obtained that induces probability $1/6$ at six equally spaced points along the unit circle in C, if one of those points is to be:
  i) the point $1 + j0$
  ii) the point $0 + j1$.

75.5. If Z is a complex r.v. whose expectation exists, show that $E[|Z|] \geq |EZ|$.
  Suggestion: Express Z in exponential form and make use of Equation (75.2b).

75.6. $Z = X + jY$ is a zero-mean complex Gaussian r.v. with $VX = 1$, $VY = 0.5$, and $\rho[X,Y] = 0.2$. Also, let $Z' = e^{-j\varphi}Z = X' + jY'$. Determine the smallest positive value for the constant φ that will make X', Y' s.i., and find the resulting variances VX', VY'. (See Problem 69.6.)

Section 76

76.1. Obtain the characteristic function $\psi_X(\tau)$ if X is the indicator r.v. of an event occurring with probability p. Note that $\psi_X(\tau)$ is periodic. What is the period?

76.2. Let X be a Cauchy r.v. with p.d.f. $f_X(x) = [\pi(1 + x^2)]^{-1}$. Determine $\psi_X(\tau)$ and use the result to show that the first and higher moments of P_X do not exist.

76.3. Let $g(\tau)$ be any positive semi-definite complex function of the real variable τ. Show that for every integer $n > 1$ and every set of n real numbers $\{\tau_1, \dots, \tau_n\}$, the matrix

$$g(\tau_1, \dots, \tau_n) = \begin{bmatrix} g(0) & g(\tau_1 - \tau_2) & g(\tau_1 - \tau_3) & \cdots & g(\tau_1 - \tau_n) \\ g(\tau_2 - \tau_1) & g(0) & g(\tau_2 - \tau_3) & \cdots & g(\tau_2 - \tau_n) \\ \cdots & \cdots & \cdots & \cdots & \cdots \\ g(\tau_n - \tau_1) & g(\tau_n - \tau_2) & g(\tau_n - \tau_3) & \cdots & g(0) \end{bmatrix}$$

is a positive semi-definite matrix.

76.4. It is to be demonstrated that the function

$$\text{rect}(\tau) \equiv \begin{cases} 1, & -1/2 \leq \tau \leq 1/2 \\ 0, & \text{otherwise} \end{cases}$$

does not satisfy Theorem 76.2 and therefore cannot be the characteristic function of some r.v.

a) Show that for $n = 2$, no choice of parameters τ_1, τ_2, s_1, s_2 can be found such that Equation (76.14) is not satisfied.
b) On the other hand with $n = 3$, show that the following choice of parameters violates Equation (76.14): $\tau_1 = 0$, $\tau_2 = 1/3$, $\tau_3 = 2/3$; $s_1 = 1$, $s_2 = -1$, $s_3 = 1$.

76.5. For the complex r.v. $Z_\tau = e^{j\tau X}$ in Example 76.1, sketch the locus of $\mathbf{E}[Z_\tau]$ in $\mathbf{C}$ as a function of τ. Repeat with X changed to be unif. $\overline{-1/2, 1/2}$.

76.6. Express the correlation coefficient of two r.v.'s in terms of their joint characteristic function.

76.7. Let X_c be the centralized version of the r.v. X defined in Example 74.2. Expressing $\psi_{X_c}(\tau)$ as a Stieltjes integral and making use of Theorem 74.4, show that the characteristic function of X_c is

$$\psi_{X_c}(\tau) = \prod_{n=1}^{\infty} \cos \frac{\tau}{3^n}.$$

76.8. Let Y be unif. $\overline{-1/2, 1/2}$.

a) Obtain the characteristic function of Y by expressing $\psi_Y(\tau)$ as a Stieltjes integral and using the same approach as in Problem 76.7. In other words, change the integral into the sum of two integrals, in this case over $\overline{-1/2, 0}$ and $\overline{0, 1/2}$, respectively.
b) Use the result of (a) to prove that $\displaystyle\prod_{n=1}^{\infty} \cos \frac{x}{2^n} = \frac{\sin x}{x}$.

Section 77

77.1. a) Using the density function $f_X(x) = (1/2) [\delta(x) + \delta(x-1)]$ for the r.v. that expresses the number of heads observed upon tossing a coin, obtain $\psi_X(\tau)$.
b) Obtain an expression for the characteristic function of a *mixture* of n r.v.'s $X_1, \ldots, X_n$ (see Problem 51.4).
c) Apply the result in (b) to the mixture specified in Problem 51.4.

77.2. Make use of Theorem 77.1 to obtain the characteristic function of a binomial r.v.

77.3. a) Let X be an exponential r.v. with p.d.f. as given in Figure 39.1(b). Show that $\psi_X(\tau) = (1 - j\tau/\alpha)^{-1}$.
b) Use the result in (a) to obtain the characteristic function of the r.v. T_m that induces a Gamma distribution of order m and with parameter α, whose p.d.f. is given in Equation (60.5).

77.4. Given two s.i. Gaussian r.v.'s X, Y with means m_X, m_Y and standard deviations σ_X, σ_Y, let $V = X + Y$. Use characteristic functions to show that V is Gaussian.

77.5. Consider a discrete r.v. X that assigns positive probability only to points that are multiples of some nonzero real number a. Apply Equation (77.1) and Problem 77.1b to the result of Problem 76.1 to show that the characteristic function of X is periodic with period $2\pi/a$.

77.6. X_1, X_2, X_3 are i.d.s.i. r.v.'s, with first, second and third moments m_{1X}, m_{2X}, m_{3X}. Let $S = X_1 + X_2 + X_3$. Use characteristic functions to express m_{3S} in terms of m_{1X}, m_{2X}, and m_{3X}.

77.7. When two r.v.'s are given that are not statistically independent, it is sometimes possible to make use of conditional independence in order to be able to apply Theorem 77.1. Suppose the two-dimensional r.v. $X = (X_1, X_2)$ distributes probability uniformly over the region shown shaded in Figure A.14. Thus, X_1 and X_2 are not s.i. However, under the conditions $A = \{\xi: X_1(\xi) \geq 0\}$ and A^c, respectively, statistical independence holds. Let $V = X_1 + X_2$, determine $\psi_{V|A}(\tau)$ and $\psi_{V|A^c}(\tau)$, and then obtain $\psi_V(\tau)$.

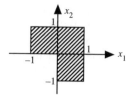

Figure A.14   Region where the r.v. X in Problem 77.7 distributes probability uniformly

77.8. Let $Y = X^2$, where X is a standardized Gaussian r.v. Then, Y has a χ^2 distribution with one degree of freedom (see also Problem 53.10).

a) Using the change of variable $z = \sqrt{1 - j2\tau} x$ in the integral, and integrating in the complex plane, show that

$$\psi_Y(\tau) = \frac{1}{\sqrt{1 - j2\tau}}.$$

b) Let $Y_n = X_1^2 + \ldots + X_n^2$, where the X_i are s.i. standardized Gaussian r.v.'s. Then, Y_n is χ^2 with n degrees of freedom. Obtain $\psi_{Y_n}(\tau)$.

c) What relationship between χ^2 and Gamma r.v.'s becomes apparent from a comparison with Problem 77.3?

Section 78

78.1. a) Prove Theorem 78.2.
  b) Prove Theorem 78.3.
  c) Prove Theorem 78.4.

78.2. Given $X = (X_1, X_2, X_3)$, where X_1, X_2 are i.d.s.i. r.v.'s with positive variance, and $X_3 = X_1 + X_2$. Express ψ_X in terms of ψ_{X_1}.

78.3. A four-dimensional r.v. $X = (X_1, X_2, X_3, X_4)$ is given, where X_1 and X_2 are s.i., $X_3 = aX_1$, and $X_4 = X_2 + b$, where a, b are real constants and $a \neq 0$.

a) Express $\psi_X(\tau_1, \tau_2, \tau_3, \tau_4)$ in terms of ψ_{X_1} and ψ_{X_2}.
b) Express $\psi_X(\tau_1, \tau_2, \tau_3, \tau_4)$ in terms of ψ_{X_3} and ψ_{X_4}.

78.4. Let $X = (X_1, \ldots, X_n)$ be a r.v. that distributes probability uniformly throughout the interior of the n-dimensional cube $\overline{-1,1} \times \overline{-1,1} \times \ \ldots \ \times \overline{-1,1}$. $Y = (Y_1, \ldots, Y_n)$ induces the same distribution but is statistically independent of X. What is the characteristic function of $X + Y$?

78.5. The r.v. $X = (X_1, \ X_2, \ X_3)$ distributes probability 0.5 uniformly over the cube $\overline{-1,1} \times \overline{-1,1} \times \overline{-1,1} \in \mathsf{R}^3$, and the remaining probability 0.5 uniformly over the line segment extending from $(-1, -1, -1)$ to $(1, 1, 1)$ in R^3. Determine $\psi_X(\tau_1, \tau_2, \tau_3)$.

78.6. An n-dimensional zero-mean Gaussian r.v. $X = (X_1, \ldots, X_n)$ has a diagonal covariance matrix,

$$
\mathbf{C}[X, X] = \begin{bmatrix} \sigma_1 & & & 0 \\ & \sigma_2 & & \\ & & \ddots & \\ 0 & & & \sigma_n \end{bmatrix}
$$

where $\sigma_i > 0$, $i = 1, \ldots, n$.

a) Determine ψ_X.
b) If $Y = aX$, where a is a nonsingular transformation matrix, what is ψ_Y?

78.7. Refer to the result obtained in Example 78.2.

a) Write the expression for $\psi_Y(\tau_1, \tau_2)$ in the case where $\varphi = \pi/4$.
b) Use part (a) to obtain $\psi_{Y_1}(\tau)$ for the case where $\varphi = \pi/4$. Then, verify that result by obtaining the characteristic function of Y_1 directly from f_{Y_1}, keeping in mind that a symmetrical triangular function can be expressed as the convolution of two identical rectangle functions.

Section 79

79.1. Given a generating function $G_X(z)$, then it is possible to recover the induced distribution P_X by expanding $G_X(z)$ in a power series. Of course, if $G_X(z)$ is a polynomial, then the power-series expansion is already at hand and P_X is given by the coefficients, as in the case of the binomial distribution. Another simple case is a generating function $G_X(z)$ that is a rational function of z and X induces zero probability on $\overline{-\infty, 0}$, because the series expansion can then usually be obtained by long division.

Thus, let an integer-valued r.v. X have $G_X(z) = (z - z^7)/6(1 - z)$. Determine $P_X(\{x\})$.

79.2. Let X be a discrete r.v. with generating function $G_X(z)$. Obtain an expression for m_{3X} in terms of derivatives of $G_X(z)$.

79.3. Obtain an expression for the generating function $G_M(z)$, where M is a Poisson r.v. with $P_M(\{k\}) = \lambda^k \, (k!)^{-1} e^{-\lambda}$, $k = 0, 1, 2, \ldots$. Let H be another Poisson r.v., statistically independent of M, with $P_H(\{k\}) = \gamma^k \, (k!)^{-1} e^{-\gamma}$, $k = 0, 1, 2, \ldots$. Show that $S = H + M$ is again a Poisson r.v., and obtain the distribution for this sum.

79.4. a) Given $G_X(z) = \alpha(1 - \beta z^2)^{-1}$, compute $P_X(\{x\})$ by long division, and determine the values of α and β.

b) Let Y be a geometric r.v. Show that $G_X(z)$ can be obtained from $G_Y(z)$ by applying scale change and shift to Y. Inversion of $G_X(z)$ is therefore also possible by converting it into a generating function whose inverse is already known.

79.5. A probabilistic experiment consists of a series of identical, independent trials, such as repeated rolls of a die. In each trial, an event A is of interest that occurs with probability $P(A) = p$. For $m = 1, 2, 3, \ldots$, let Y_m be the r.v. that expresses the number of trials until A occurs for the mth time. Each of the Y_m is a negative binomial r.v. that induces the distribution (see Problems 27.14 and 63.2)

$$P_{Y_m}(\{k\}) = \binom{k-1}{m-1} p^m (1-p)^{k-m}, \quad k = m, m+1, m+2, \ldots$$

a) Determine the generating function $G_{Y_m}(z)$.

b) For $m = 1, 2, 3, \ldots$, and $n = 1, 2, 3, \ldots$, let $X_{m,n}$ be the r.v. that expresses the number of trials, after the mth occurrence of A, until A occurs an additional n times. Explain why $X_{m,n}$ is also a negative binomial r.v. and is statistically independent of Y_m.

c) Using generating functions, show that $S_{m,n} = Y_m + X_{m,n}$ is again a negative binomial r.v.

d) In fact, $Y_m + X_{m,n} = Y_{m+n}$. However, this cannot be proven using the generating functions $G_{Y_m}(z)$, $G_{X_{m,n}}(z)$, $G_{S_{m,n}}(z)$. Why not?

79.6. By comparing the generating functions, show that a geometric distribution is related to a negative binomial distribution (with the same parameter p) in the same way as the binary distribution induced by an indicator r.v. is related to a binomial distribution (with the same parameter p).

79.7. *Mixture of Discrete Distributions.* Let $P_i(\{x\})$, $i = 1, \ldots, k$, be discrete induced distributions over the integers.

a) If the r.v. X induces the discrete distribution

$$P_X(\{x\}) = \sum_{i=1}^{k} \alpha_i P_i(\{x\}), \quad \text{where} \quad \sum_{i=1}^{k} \alpha_i = 1$$

show that $G_X(z) = \sum_i \alpha_i G_i(z)$, where for $i = 1, \ldots, k$, $G_i(z)$ is the generating function associated with $P_i(\{x\})$.

b) Let N be a geometric r.v. with parameter p as given in Example 34.3, and let X be a r.v. that induces the two-sided geometric distribution

$$P_X(\{x\}) = (1-p)^{|x|} p/(2-p), x = 0, \pm 1, \pm 2, \ldots$$

Using $G_N(z)$ and applying Part (a), show that

$$G_X(z) = \frac{pz}{1 - (1-p)z} \frac{p}{z - (1-p)}.$$

c) Consider the r.v. $Y = N + M$, where N is as in Part (b), and M, $-N$ are i.d.s.i. Use the result of Part (b) to show that Y induces the same distribution as does X in Part (b).

d) Express $\mathbf{V}X$ in terms of $G_N(z)$ and in this way determine the variance of X.

Section 80

80.1. Generalize the result (80.8) by letting $R_2 = \sqrt{X_1^2 + X_2^2}$, where X_1, X_2 are s.i. zero-mean Gaussian r.v.'s with variance σ^2.

80.2. Restate Theorem 80.4 for the case where the two Gaussian r.v.'s have arbitrary (but identical) variances.

80.3. Let $X_{1s}, \ldots, X_{ns}$ be s.i. standardized Gaussian r.v.'s, and $R_n = \sqrt{X_{1s}^2 + \ldots + X_{ns}^2}$.

a) Verify that the expressions obtained in this Section for f_{R_1}, f_{R_2} and f_{R_3} agree with the following formula:

$$f_{R_n}(r) = \frac{2r^{n-1}}{2^{n/2}\Gamma(n/2)} \exp\left[-\frac{1}{2}r^2\right], \quad r \geq 0. \qquad (*)$$

b) Use induction to show that (*) applies for all n.

Note: $\displaystyle\int_{-\pi/2}^{\pi/2} \cos^n\theta \, d\theta = \sqrt{\pi} \; \frac{\Gamma\left(\dfrac{n+1}{2}\right)}{\Gamma\left(\dfrac{n+2}{2}\right)}$, $\quad n = 1, 2, 3, \ldots$, and for n odd

$$\Gamma\left(\frac{n}{2}\right) = \sqrt{\frac{2}{\pi}} \frac{1 \cdot 3 \cdot 5 \cdot \ldots \cdot (n-2)}{2^{n/2}}.$$

c) Now let $R_n = (X_1^2 + \ldots + X_n^2)^{1/2}$ where the X_is are s.i. zero-mean Gaussian r.v.'s with identical variances σ^2. How does this change (*)?

80.4. Extend Theorem 80.4 to *three* normalized s.i. Gaussian r.v.'s with means b_1, b_2, b_3, where $\sqrt{b_1^2 + b_2^2 + b_3^2} = b$. (Make use of the fact that a rotation of coordinates that keeps the component r.v.'s s.i. but makes the x_1-axis coincide with the vector $\mathbf{b} = [b_1 \, b_2 \, b_3]^T$ causes X_2 and X_3 to be zero mean.) Verify that your result approaches Equation (80.9) as $b \to 0$.

Note: $\displaystyle\int_0^{\pi/2} \cos\theta \, I_0(a\cos\theta) \, d\theta = \frac{\sinh 2a}{2a}$.

([Gr2] Section 6.681 Equation (11); [AS1] Equations (10.2.13) and (10.2.14).)

80.5. *Decision Problem.* At a designated time t, a certain physical system can be in one of two states, call them A and B. It is known that the two states are governed by a random mechanism, causing state A to arise with a likelihood of p at time t. However, determining the actual state of the system requires a measurement. Suppose a measurement setup is designed to indicate $+1.0$ if the system is in state A, and -1.0 if it is in state B. Unfortunately, the measurement result is corrupted by additive Gaussian noise that is not influenced by the system state; that is, in state A, the measurement yields $1 + N$, and in state B it yields $-1 + N$, where N is a zero-mean Gaussian r.v. A *decision threshold* needs to get established, i.e., a real number τ, above which the measurement result will be interpreted as 'state A', and otherwise as 'state B'.

The decision threshold depends on the *decision criterion*. What is to be optimized? Suppose the decision criterion is to *minimize the probability of an error* (i.e., deciding state A if the system is actually in state B, or vice versa). Show that this results in a decision threshold that is given by $\tau = (\sigma_N^2/2) \ln (1-p)/p$.

80.6. Consider Problem 80.5 with the following modification: *Two* different measurements are made to identify the system state. Both of the measurements are designed to indicate $+1.0$ if the system is in state A, and -1.0 if it is in state B. Both measurements are corrupted by additive Gaussian noise. The noise contributions N_1 and N_2 affecting the two measurements are represented by zero-mean s.i. Gaussian r.v.'s with variances σ_1^2 and σ_2^2, respectively. The decision space is now $\mathbb{R}^2$, and in place of a decision threshold there is now a *decision boundary* that separates the two *decision regions*.

a) Determine the decision boundary for the minimum probability of error criterion.
b) Obtain an expression for the minimum probability of error as a function of p, σ_1, and σ_2.

Suggestion: Scale one of the axes so as to make the noise r.v.'s i.d.s.i. A rotation can then be applied in order to center the two conditional measurement distributions on the horizontal axis. This makes the problem one-dimensional.

Section 81

81.1. Prove Theorem 81.3.
81.2. Determine $f_{Y^{[4]}}$ in Example 81.2.
81.3. Obtain $f_{Y^{[1]}}$ and $f_{Y^{[2]}}$ in Example 81.2 for arbitrary α, β.
81.4. In Definition 81.1, let S be a binary r.v. that assigns probability p to $\{a\}$ and probability q to $\{b\}$, where $q = 1 - p$, and $a > b > 0$. Then $Y^{[n]}$ is a three-parameter family of s.s.v.r.v.'s such that each of its members is a mixture of two members of the Gaussian family of s.s.v.r.v.'s.

a) Obtain expressions for $f_Y[n]$, and $h_n(\gamma)$.
b) Verify that the components of $Y^{[2]}$ are not s.i. by showing that they do not satisfy Equation (46.1).

81.5. Let S be discrete, assigning probability 0.05 at $s = 0.1$ and probability 0.95 at $s = 1$. Sketch $f_{Y^{[1]}}$. Is the distribution 'Gaussian-like'?
81.6. Consider the circularly symmetric two-dimensional vector r.v. $Y = (Y_1, Y_2)$ whose radial component $R = \sqrt{Y_1^2 + Y_2^2}$ induces an exponential distribution with p.d.f.

$$f_R(r) = \begin{cases} e^{-r}, & r > 0 \\ 0, & \text{otherwise.} \end{cases}$$

Show that the marginal distributions P_{Y_1}, P_{Y_2} are 'K-distributions' described by the p.d.f. $f_Y(y) = \dfrac{1}{\pi} K_0(|y|)$, where

$$K_0(y) = \int_1^\pi \exp(-xy) \frac{dx}{\sqrt{x^2 - 1}}, \quad y > 0$$

is the zero-order modified Bessel function of the second type (cf. [AS1] Equation (9.6.23)). It should be noted that Y cannot be called an s.s.v.r.v. unless it is shown that Definition 81.1 is satisfied.

81.7. In Definition 81.1, let $f_S(s) = \begin{cases} s^{-2}, & s > 1 \\ 0, & \text{otherwise.} \end{cases}$

a) Obtain $f_{Y_{11}}(y)$. (Suggestion: Interchange the roles of f_S and f_X in Equation (81.2).)

b) Compare $f_{Y_{11}}(y)$ with the Cauchy p.d.f. $f_V(v) = (2\pi)^{-1/2} (\pi/2 + v^2)^{-1}$ by superimposing plots of the two functions.

c) Consider two i.d.s.i. Cauchy r.v.'s. centered at the origin. Plot a contour of constant density for their joint p.d.f., thus showing that the joint distribution does not have circular symmetry.

Section 82

82.1. *Encoding in terms of s.i. symmetrical binary r.v.'s.* In the discussion following Example 82.1 it is noted that the r.v. X in that Example, which has entropy $H_X = 2$ bits, can be encoded by two s.i. symmetrical binary r.v.'s. In general, however, the most efficient encoding of a r.v. X in terms of s.i. symmetrical binary r.v.'s requires that the encoding be applied to a product space consisting of k executions of the underlying experiment, and that a variable number of binary r.v.'s be used—where this number is governed by a r.v. N. It is then possible to say that $E[N/k] \rightarrow H_X$ as $k \rightarrow \infty$. Consider the following simple example.

A r.v. X assigns probabilities 1/4 and 3/4 to $\{-1\}$ and $\{1\}$, respectively. The result indicated by X is to be encoded by s.i. binary r.v.'s $B_1, B_2, B_3, \ldots$ each of which assigns to $\{0\}$ and $\{1\}$ probabilities as close to 1/2 as possible. Here, $H_X = 0.811278$ bits, but encoding of X by itself requires B_1, and $E[N/k] = 1/1 = 1$.

On the other hand, *two* executions of the underlying experiment result in a product space on which X is replaced by two i.d.s.i. r.v.'s X_1 and X_2. If X_1 and X_2 are each encoded individually, that encoding requires twice as many binary r.v.'s as before, and again $E[N/k] = 1$. Instead, consider the use of *three* binary r.v.'s, as follows:

If the values of X_1, X_2 are:	set B_1 to:	set B_2 to:	set B_3 to:
1, 1	1	don't use	don't use
1, −1	0	1	don't use
−1, 1	0	0	1
−1, −1	0	0	0

(The assumption here is that when B_2 and/or B_3 are not used in the encoding, they can serve to convey some other meaning.) Determine $E[N/k]$ for this encoding scheme.

82.2. Given an absolutely continuous n-dimensional r.v. $X = (X_1, \ldots, X_n)$ with entropy H_X. Show that result (82.4) also applies in this case.

82.3. The range of an absolutely continuous r.v. X is restricted to the unit interval, $\overline{0, 1}$. Use the approach of Example 82.3 to show that H_X is largest if X is unif. $\overline{0, 1}$, and determine the value of H_X for that case.

82.4. Given two s.i. absolutely continuous r.v.'s X, Y with entropies H_X, H_Y bits, respectively. Consider the mixture Z, where $f_Z(z) = af_X(z) + bf_Y(z)$, with $a, b \geq 0$ and $a + b = 1$. Furthermore, assume that the ranges of X and Y do not overlap; i.e., $f_X(z)f_Y(z) = 0$ for all z.

a) Express H_Z in terms of H_X and H_Y.

b) If $a = b = 1/2$, and $H_X = H_Y$, show that $H_Z = H_X + 1$ bits. In other words, added to the randomness associated with X (or Y) in this case is only the uncertainty of the equally likely binary choice between X and Y, which is 1 bit.

82.5. a) If X is a Gaussian r.v. with standard deviation σ_X, show that $H_X = \log_2 \sqrt{2\pi e}\,\sigma_X$.

b) If X, Y are s.i. Gaussian r.v.'s, show that $4^{H_{X+Y}} = 4^{H_X} + 4^{H_Y}$, where all the entropies are in bits.

82.6. a) Let N be the geometric r.v. defined in Example 34.3. Determine H_N as a function of p_0.

b) What value of p_0 minimizes H_N? What value of p_0 maximizes H_N?

c) Let $p_0 = 1/2$. Referring to Problem 82.1, what encoding of N (in terms of symmetrical binary r.v.'s) can you specify so that the expected number of binary r.v.'s needed equals H_N?

82.7. *Amplitude and scale change.* To simplify entropy calculations for absolutely continuous r.v.'s, it is useful to know how the integral in Equation (82.6) is affected when an amplitude and scale change is applied to $f_X(x)$. Given a p.d.f. $f(x)$ that yields an entropy of H_X bits, consider the function $g(x) = af(cx)$, where $a > 0$, $c \neq 0$ (which is a p.d.f. only if $|c| = a$), and show that

$$-\int_{-\infty}^{\infty} g(x) \log_2 g(x)\, dx = \frac{a}{|c|}(H_X - \log_2 a).$$

Using this result, express H_Y in terms of H_X if $Y = aX$ ($a \neq 0$).

82.8. Two discrete r.v.'s X, Y are given.

a) Show that $H_X - H_{X|Y} = H_Y - H_{Y|X} \geq 0$. (Suggestion: For the inequality, make use of the inequality $\ln x \leq x - 1$ examined in Problem 32.7.)

b) Determine $H_X - H_{X|Y}$ if

i) X, Y are s.i.;

ii) $Y = aX$, where a is a given real number.

c) Show that Part (a) also holds for two absolutely continuous r.v.'s X, Y.

82.9. Two r.v.'s X, Y jointly induce a uniform distribution over the parallelogram A identified in Figure A.15.

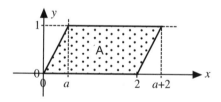

Figure A.15 Parallelogram over which the r.v.s X, Y of Problem 82.9 induce a uniform distribution

a) Determine $H_{(X,Y)}$, H_Y, H_X, $H_{Y|X}$, and $H_{X|Y}$ as a function of the parameter a, where $0 \leq a \leq 2$. (Suggestion: f_X can be viewed as a mixture of two triangular and a uniform density. In this way, the results of Problem 82.4 can be utilized.)

b) What is the relation between the quantity $H_Y - H_{Y|X}$ and $\rho[X,Y]$ in this case?

Section 83

83.1. Show that the joint c.d.f. $F_{X,Y}$ obtained in Example 83.1 satisfies Theorem 83.1.

83.2. If X is an absolutely continuous r.v. and $Y = aX + b$ where $a < 0$, show that over the unit square, $C_{X,Y}(u, v) = F_w(u, v)$ as defined in Theorem 83.1.

83.3. A copula C assigns no probability below the line $y = 1 - x$. Is the copula that coincides with F_w the only copula that satisfies this requirement?

83.4. Given the copula $C_{X,Y}(u, v)$ found in Example 83.4, and the marginal c.d.f.'s

$$F_X(x) = F_Y(x) = \begin{cases} 0, & x < 0 \\ 1.5x, & 0 \le x < 0.5 \\ 0.5 + 0.5x, & 0.5 \le x < 1 \\ 1, & x \ge 1. \end{cases}$$

a) Show that X, Y jointly induce the probability distribution described at the beginning of Example 83.4.

b) Show that the points $(3/8, 3/4)$ and $(3/4, 3/8)$ in the (u,v)-plane map into the points $(1/4, 1/2)$ and $(1/2, 1/4)$, respectively, in the (x,y)-plane. Thus, the four equally probable line segments in the u,v-plane can be thought of as mapping into four equally probable line segments in the (x,y)-plane.

83.5. Two r.v.'s X, Y jointly induce probability 1 uniformly over the right triangle with vertices at $(0, 0)$, $(1, 0)$ and $(0, 1)$. Determine their copula, $C_{X,Y}(u, v)$. Describe the joint distribution that this copula specifies for the r.v.'s U, V.

83.6. Determine the nature of the joint distribution of two standardized Gaussian r.v.'s X_g, Y_g whose coupling is defined by the copula $C_{X,Y}(u, v)$ found in Example 83.4.

Section 84

84.1. a) How much benefit, if any, does the casino in Example 84.2 derive from the fact that most players of the game will not be able to continue playing long enough to win \$1 000 000?

b) How does this compare with a lottery in which a million tickets are available at \$1 each, and there is only one prize of \$1 000 000?

84.2. In the continuation of Example 84.1, suppose that player A starts with \$1 and player B with \$2, and they continue to play till one player goes broke.

a) Draw a state-transition diagram.

b) For each player determine the probability of going broke.

c) What is the expected duration of this game?

84.3. In the continuation of Example 84.1, suppose that player A starts with \$2 and player B with \$3, and they continue to play till one player goes broke. Use a computer simulation to determine the expected duration of the game, and the probability of going broke for each player. Verify that the latter agrees with Equation (84.1).

84.4. You have a supply of 100 black and 100 white balls. Into an empty container you place b_0 black balls and $100 - b_0$ white balls, and mix up the contents of the container. You blindly withdraw a ball from the container, replace it with a ball of the opposite color and again mix up the contents of the container. This step gets repeated indefinitely. Let the number of black balls in the urn be represented by the sequence of r.v.'s $\{B_k\}$, where for $k = 1, 2, 3, \ldots$, the r.v. B_k expresses the number of black balls in the urn after the kth replacement.

a) Show that $\mathbf{E}B_1 = b_0$ if $b_0 = 50$; $\mathbf{E}B_1 < b_0$ if $b_0 > 50$; and $\mathbf{E}B_1 > b_0$ if $b_0 < 50$.

b) Show that the following relations hold for $k = 1, 2, 3, \ldots$:

   i) $\mathbf{E}[B_{k+1}|D_k] > \mathbf{E}[B_k|D_k]$, where $D_k = \{\xi: B_k(\xi) < 50\}$

   ii) $\mathbf{E}[B_{k+1}|E_k] = \mathbf{E}[B_k|E_k]$, where $E_k = \{\xi: B_k(\xi) = 50\}$

   iii) $\mathbf{E}[B_{k+1}|F_k] < \mathbf{E}[B_k|F_k]$, where $F_k = \{\xi: B_k(\xi) > 50\}$.

84.5. As in Problem 84.4, you have a container with a mixture of b_0 black balls ($0 < b_0 < 100$) and $100 - b_0$ white balls. You blindly withdraw a ball from the container, note its color, and return it to the container. Then you find in the container a ball of the opposite color and, if one can be found, replace it with a ball of the color that was drawn, and mix up the contents. The number of black balls in the urn is now expressed by the r.v. B_1. Repeating these actions indefinitely, the number of black balls in the urn is then, successively, $B_1, B_2, B_3, B_4, \ldots$. Consider the sequence of r.v.'s $\{B_k\}$.

a) Verify that $\mathbf{E}[B_{k+1}|B_k, B_{k-1}, \ldots, B_1] = \mathbf{E}[B_{k+1}|B_k]$.

b) Show that the following relations hold for $k = 1, 2, 3, \ldots$:

   i) $\mathbf{E}[B_{k+1}|C_k] = \mathbf{E}[B_k|C_k]$, where $C_k = \{\xi: B_k(\xi) = 0\}$

   ii) $\mathbf{E}[B_{k+1}|D_k] < \mathbf{E}[B_k|D_k]$, where $D_k = \{\xi: 1 \leq B_k(\xi) \leq 49\}$

   iii) $\mathbf{E}[B_{k+1}|E_k] = \mathbf{E}[B_k|E_k]$, where $E_k = \{\xi: B_k(\xi) = 50\}$

   iv) $\mathbf{E}[B_{k+1}|F_k] > \mathbf{E}[B_k|F_k]$, where $F_k = \{\xi: 51 \leq B_k(\xi) \leq 99\}$

   v) $\mathbf{E}[B_{k+1}|G_k] = \mathbf{E}[B_k|G_k]$, where $G_k = \{\xi: B_k(\xi) = 100\}$.

c) For $b_0 = 50$ compare $\{B_k\}$ to the sequence $\{A_k\}$ defined in the continuation of Example 84.1, but with starting capitals $c_A = c_B = 50$. What can be concluded about the probability that $B_k \to 0$ or 100?

84.6. Given an absolutely fair sequence $\{X_k\}$, let a sequence $\{W_k\}$ be defined by

$$W_1 = X_1 + c,$$

$$W_k = W_{k-1} + X_k. \tag{*}$$

Here, c is a constant that in a gambling situation, for instance, could represent an initial fee. The sequence $\{W_k\}$ defined in this way is called a *Martingale* [Fe2].

a) Is the sequence $\{B_k\}$ defined in Problem 84.4 a Martingale?

b) In the coin-tossing game of Example 84.1 suppose both players have unlimited funds. With $\{X_k\}$ as specified in that Example, let $\{W_k\}$ be a sequence defined by

$$W_1 = X_1,$$

$$W_k = \begin{cases} W_{k-1} + (0.5\,|W_{k-1}| + 1)X_k, & k \text{ odd} \\ W_{k-1} + (0.5|W_{k-1}| + 0.5)X_k, & k \text{ even} \end{cases}$$

Show that $\{W_k\}$ is a Martingale.

84.7. An experiment consists of repeated trials, where in each trial two coins are tossed simulta-neously—a nickel and a dime. Let $\{U_k\}$ be the sequence of indicator r.v.'s of 'heads' on the nickel, and $\{V_k\}$ the sequence of indicator r.v.'s of 'heads' on the dime. Furthermore, let $\{K_m\}$ be the sequence of order statistics of the event ' $+1$ ' on $\{U_k\}$, and let $\{L_m\}$ be the sequence of order statistics of the event ' $+1$ ' on $\{V_k\}$.

    a) If ξ is an element of the underlying probability space, what is the probability that $K_1(\xi) = L_1(\xi)$?

    b) What is the probability that $K_2(\xi) = L_2(\xi)$?

Section 85

85.1. Let $\{X_k\}$ be a sequence of i.d.m.s.i. r.v.'s that are Cauchy distributed, with probability density functions

$$f_{X_k}(x) = \frac{1}{\pi(1+x)^2}.$$

    Show that no finite constant m exists to which the sequence of partial averages $\{Y_n\}$ of $\{X_k\}$ converges in probability.

85.2. $\{X_k\}$ is a sequence of i.d.s.i. r.v.'s. Each of the r.v.'s X_k takes on integer values in the range from 1 to 10 with equal likelihood. For each integer m, where $1 \le m \le 10$, another sequence $\{Y_{n,m}\}$ is defined, where the r.v. $Y_{n,m}$ expresses the fraction of the r.v.'s $X_1, X_2, \ldots, X_n$ that take on the value m. Show that each of the sequences $\{Y_{n,m}\}$ converges to 0.1 with probability 1.

85.3. Given a sequence $\{X_k\}$ that converges to zero with probability 1. Prove that $\{X_{2k}\}$ also converges to zero with probability 1.

85.4. If $\{X_k\}$ satisfies the Strong Law of large numbers, show that the sequence of increments, $\{X_{k+1}\} - \{X_k\}$, also satisfies the Strong Law, and converges to zero.

85.5. Show that the sequence $\{X_k\}$ in Example 85.4 obeys the Strong Law of Large Numbers.

Section 86

86.1. Consider a sequence $\{X_n\}$ of m.s.i. r.v.'s X_n whose p.d.f.'s are given by

$$f_{X_n}(x) = \begin{cases} n\alpha\, e^{-n\alpha x}, & x \ge 0 \\ 0, & x < 0 \end{cases}$$

    for some fixed $\alpha > 0$. Another r.v. Z, s.i. of all the X_n's, is unif. $\overline{0, \beta}$ for some $\beta > 0$. Now let the sequence $\{Y_n\}$ be defined by $Y_n = X_n + Z$, $n = 1, 2, \ldots$.

    a) Obtain an expression for the correlation coefficient between Y_n and Y_{n+1}, for $n = 1, 2, \ldots$.

    b) Obtain an expression for the correlation coefficient between Y_n and Z, for $n = 1, 2, \ldots$.

    c) What kind of convergence is exhibited by the sequence $\{X_n\}$? The sequence $\{Y_n\}$?

86.2. Prove convergence in distribution for the sequence $\{X_k\}$ of Example 86.1. In other words, show that for every $\varepsilon > 0$, there is a k_0 such that for all $k > k_0$, $|F_{X_k} - F_X| < \varepsilon$ for all x.

86.3. *Random walk.* Given a sequence of i.d.m.s.i. binary r.v.'s $\{X_n\}$, each assigning probability 1/2 to $\{+1\}$ and $\{-1\}$. (These values can be taken as expressing the amount won in a coin-tossing game.) Let

$$Y_n = \sum_{i=1}^{n} X_i, \quad n = 1, 2, 3, \ldots$$

also, $Y_0 = 0$.

a) Use the Central Limit Theorem to determine the smallest n for which $P_{Y_n}(-\infty, -10) \geq 0.25$. (In other words, if a dollar is bet with each coin toss, how many coin tosses are needed so that the chance of losing $10 or more is at least 1/4?)

b) Let u be some positive or negative integer. What is the probability that the sample sequence $\{y_n\}$ includes the value u? (That is, what is the probability that the random walk reaches u in a finite number of steps?)

86.4. A r.v. B induces a binomial distribution that is symmetrical about its mean and assigns positive probability to $\{0\}, \{1\}, \{2\}, \ldots, \{10\}$ and to no other points. (B might express the number of times a 'head' is obtained in ten tosses of a fair coin.)

a) Explain how you can approximate $P_B(\{5\})$ from the standardized Gaussian c.d.f., and obtain that approximate value.

b) Compare your result with the value computed directly from the binomial distribution.

86.5. Consider a sequence of independent trials that are intended to yield an estimate of the probability of an event A.

a) Draw a plot similar to Figure 86.4, if in fact $P(A) = 0.1$. Sketch both $F_{Y_{1s}}$ and $F_{Y_{5s}}$.

b) Repeat for the case where $P(A) = 10^{-6}$.

86.6. Consider a sequence of i.d.m.s.i. r.v.'s $\{X_k\}$ with triangular p.d.f. as given in Example 50.2.

a) Obtain $f_{Y_{1s}}, f_{Y_{2s}}$ and $f_{Y_{3s}}$.

b) Plot a graph similar to Figure 86.3, in order to gain a visual impression of the trend of these unsymmetrical p.d.f.'s toward the Gaussian p.d.f.

Notation and Abbreviations

Standard mathematical notation is not included. Page numbers indicate where the item is first used. For this purpose, the problems of any given Section (in the Appendix) are considered as following that Section and thus precede the text of the next Section.

Plain lower case text letters

a, b, c, d, g, m, r, w, x, y   conceptual entities such as: a playing card, a ball of specified color, a die throw result (possibly with subscript)   76
a_i ith component of a k-tuple   126
c.d.f.   cumulative distribution function   173
c.g.   center of gravity   311
cerf   complementary error function   187
$\det(\boldsymbol{a})$ determinant of the square matrix $\boldsymbol{a}$ 369
erf   error function   187
exp   exponential function   186
i.d.   identically distributed   233
i.d.m.s.i.   identically distributed and mutually statistically independent   301, 549
i.d.s.i.   identically distributed and statistically independent   250
m.s.i.   mutually statistically independent   274
p.d.f.   probability density function   180
p.e.   probabilistic experiment   11
r   a relation   489
r.v.   random variable   171
rf(A)   relative frequency of an event A
s.i.   statistically independent   218
s.s.v.r.v.   spherically symmetric vector random variable   428
unif.$\overline{a, b}$ uniformly distributed (or, distributes probability uniformly) over the interval $\overline{a, b}$ 234
$\text{var}(x_1, \ldots, x_k)$ variance of a k-tuple of numbers   340
x, y   binary operations   490
$_c$ (subscript)   centralized version of a r.v.   350
^{c} (superscript)   complement of a set   26

Probability Concepts and Theory for Engineers, First Edition. Harry Schwarzlander.
© 2011 John Wiley & Sons, Ltd. Published 2011 by John Wiley & Sons, Ltd.

Plain upper case text letters

A, B   events in a probabilistic experiment   54

A, B, C, D, E, F, ...   conceptual entities such as playing cards, types of coins, players in a game, transmission routes   37, 483

I   the set of all integers   168

I_+ the set of positive integers   36

Im z imaginary component of the complex number z 397

MTBF   mean time between failures   558

Re z real component of the complex number z 397

$S_0, S_1, ..., S_3$ different states   137

Bold numerals and upper case text letters

0   zero column vector; zero element of a vector space   371–2

0, 1   binary digits   127

A   matrix of coefficients   546

I   unit matrix   371

Italic lower case text letters

a, b, c, d, e ...   numbers, parameters and coefficients   4, 485

a_p population average   337

a_T test sample average   337

a_{ij} (i, j)th coefficient in a set of linear equations   546

$c_{X,Y}$ covariance of X, Y 350

c_{ij} (i, j)th element of a covariance matrix   367

d distance   148

$d(A, B)$ distance between sets A and B   501

e base of natural logarithm   163

e_{ij} jth element of ith eigenvector   372

f, g, h functions   222

$f_X(x)$ p.d.f. of the r.v. X 180

$f_{X|A}(x)$ conditional p.d.f. of X given A   285

$f_X(\boldsymbol{x})$ two-dimensional or n-dimensional p.d.f.   210

$g(x)$ a real-valued function of x 178

h half-life   528

j imaginary unit   397

m_2 moment of inertia of a mass distribution   314

m_{1X} first moment of the distribution P_X 311

m_{2X} second moment of the distribution P_X 314

m_{kX} kth moment of the distribution P_X 314

p, q probabilities   122

q_i ith element of the vector $\boldsymbol{q}$ 368

r polar coordinate radial variable; regression function   182

r_g radius of gyration   314

t, u, v, w, x, y, z real variable or value   182

t time variable   280
t_0 observation time   280
t_x arbitrary time instant   555
$u(x)$ unit step function   573
$v(t)$ a time-varying voltage   398
$v_i(t)$ input voltage   279
$v_o(t)$ output voltage   279
v_p population variance   340
x_m median of a distribution P_X 312
z complex number or variable   397

Bold italic lower case text letters

a a transformation matrix   369
c a square matrix   368
c_X covariance matrix of X 367
e_i ith eigenvector   371
g two-dimensional or n-dimensional function   238
g^{-1} inverse of the function g 264
q a column vector of real numbers   368
$x, y, \ldots$ elements of a vector space   374

Italic upper case text letters

$A, B, C, D, \ldots$ random variables   157
A_T test sample average   338
$C_{X,Y}(u, v)$ copula of the r.v.'s X, Y 445
$F(x), G(x)$ function of a real variable x 173
$F_m(u, v)$ upper bound on the copula of two random variables   443
$F_w(u, v)$ lower bound on the copula of two random variables   443
$F_X, F_Y, \ldots$ c.d.f. for a r.v. X, Y, etc.   173
$F_{X,Y}$ two-dimensional c.d.f. for r.v.'s X, Y 202
$F_{X|A}$ conditional c.d.f. given the event A   284
F_X c.d.f. for a two-dimensional or n-dimensional r.v. X 202
G, H, N point function   33
G^{-1} inverse of the function G 34
$G^{-1}(A)$ inverse image of set A under G 159
$G(\omega)$ Fourier transform of $g(t)$ 402
$G_X(z)$ generating function of a discrete r.v. X 417
$G_{X|A}(z)$ conditional generating function of X given the event A   420
H additive set function   49
$H(p_1, p_2, \ldots, p_n)$ entropy associated with $p_1, p_2, \ldots, p_n$ 142
$H(\mathscr{P})$ entropy associated with a partition $\mathscr{P}$ 143
$H(\mathscr{P}_1|\mathscr{P}_2)$ conditional entropy   144
H_X entropy associated with a r.v. X 435
$H_{Y|X}$ conditional entropy of Y given X 436
$H_{\mathscr{E}}$ uncertainty experienced by experimenter   141

I_0 modified Bessel function of the first kind of order 0   426

$I_A, I_B, \ldots$ indicator functions or indicator r.v.'s   218, 491

I_1, I_2 indicator r.v.'s   323

J_g, J_g Jacobian for the transformation g, g 263

K, N set function   35

K_0 zero-order modified Bessel function of the second type   580

$\{K_m\}$ sequence of order statistics   459

L Lebesgue measure   168

$M, M_1, M_2, \ldots$ masses   311

P probability function or distribution   58

P_G, P_X, etc.   probability distribution induced by r.v. G, X, etc.   157

$P_{G,H}, P_{X,Y}$, etc.   joint probability distribution induced by the r.v.'s G and H, X and Y, etc.   192

$P_X(|A)$ conditional distribution induced by r.v. X, given the event A   284

$P_{Y|X}(|x)$ conditional distribution of Y given the value of X 293

P_X probability distribution induced by a two-dimensional or n-dimensional r.v. X 200

P_Z probability distribution induced by a complex r.v. Z onto the complex plane   398

$P(|A)$ conditional probability distribution given A   91

$P(B|A)$ conditional probability of B given A   87

Q estimation error r.v.   380

R polar coordinate r.v.   213

V_T test sample variance   341

X_s standardized version of the r.v. X 317

X_c, Y_c, etc.   centralized version of X, Y, etc.   350

X_d discrete r.v. approximating an absolutely continuous r.v. X 438

X_r a reference random variable   438

X^{-1} inverse of the function (r.v.) X 170

$X|B_a, X|B_d, X|B_s$ conditional r.v.'s   291

$\tilde{Y}$ estimate of Y 381

Z a complex r.v.   397

Bold italic upper case text letters

U, V, W, X, Y, Z two-dimensional and n-dimensional r.v.'s and vector r.v.'s   197

X_a absolutely continuous component of X 290

X_d discrete component of X 290

X_s singular component of X 290

$X^{(i)}$ X with ith component deleted   544

$Y^{[n]}$ n-dimensional spherically symmetric vector r.v.   428

Plain upper case block letters

A, B, C, D, $\ldots$ sets

A$'$ complement of set A 26, image of set A   286

C   unit cube; space of complex numbers   542

E   set of even positive integers   47

I_x an x-interval   236

K_y inverse image of $\{y\}$ 230

L_c ellipse of concentration   567
$N_1, N_2, N_3, \ldots$ time intervals   280
R the space of real numbers   156
$(R, \mathscr{A}_G, P_G)$ probability space induced by G 157
$(R, \mathscr{B}, P_X)$ probability space induced by a r.v. X 171
R^2 Cartesian plane, two-dimensional Euclidian space   192
$(R^2, \mathscr{A}_{G,H}, P_{G,H})$ probability space induced by G, H jointly   192
$(R^2, \mathscr{B}^2, P_X)$ probability space induced by a two-dimensional r.v. X 200
R^n n-dimensional Euclidean space   253
$(R^n, \mathscr{B}^n, P_X)$ probability space induced by an n-dimensional r.v. X 254
R_a the set of rational numbers   168
R_i set of all irrational numbers   523
S, T space of elements, or sample space   20
$(S, \mathscr{A}, P)$ probability space   62
U_y inverse image of $\overline{-\infty, y}$ 234
V vector space   374

Plain upper case enhanced block letters

C$[X,Y]$ covariance of X, Y 350
C$[X,X]$ covariance matrix of X 367
E$[X]$, **E**X expected value of the r.v. X 309
EX expectation of a vector r.v. X 367
E$[Y|X]$ conditional expectation of Y given X 363
E$[X|A]$ conditional expectation of X given (or with respect to) A 345
V$[X]$, **V**X variance of the r.v. X 324

Upper case script letters

$\mathscr{A}$ additive class   31
$\mathscr{A}_G$ additive class generated by $\mathscr{P}_G$ 157
$\mathscr{A}_{G,H}$ additive class generated by $\mathscr{P}_{G,H}$ 192
$\mathscr{A}(\mathscr{P})$ additive class generated by $\mathscr{P}$ 44
$\mathscr{A}(\mathscr{B} \times \mathscr{B})$ completely additive class generated by $\mathscr{B} \times \mathscr{B}$ 199
$\mathscr{B}$ class of Borel sets   169
$\mathscr{B}^2$ class of two-dimensional Borel sets   199
$\mathscr{B}^n$ class of n-dimensional Borel sets   254
$\mathscr{C}, \mathscr{D}, \mathscr{E}, \mathscr{Q}$ classes   23
$\mathscr{P}$ partition   29
$\mathscr{P}_o$ largest partition in a finite additive class   83
$\mathscr{V}$ class of semi-infinite intervals   169

Upper case gothic letters

$\mathfrak{E}, \mathfrak{F}$ probabilistic experiment   54
$\mathfrak{T}$ trial   133

Plain and italic lower case Greek letters

$\alpha, \beta, \gamma, \delta, \lambda, \pi_i, \sigma, \omega$ parameters, scale factors, expressions, coefficients   486

$\partial/\partial x$ partial derivative with respect to x 295

$\partial^2/\partial x \partial y$ second partial derivative with respect to x and y 210

$\delta(x)$ delta-function or unit impulse   183

$\in$ belongs to, is an element of   20

ε a small positive number   170

$\notin$ does not belong to, is not an element of   21

ϕ empty set   21

φ, ψ angle   268

$\lambda_1, \lambda_2, \ldots$ eigenvalues   371

μ_{kX} kth central moment of the distribution P_X 315

$\mu_A(\xi)$ membership function of a fuzzy set A   147

π_i population size of town i 148

$\psi_X(\tau)$ characteristic function of the r.v. X 402

$\psi_{X|A}(\tau)$ conditional characteristic function of the r.v. X, conditioned on A   409

$\rho_{X,Y}$ correlation coefficient of X, Y 351

σ-algebra   completely additive class   47

σ_X, σ_Y, etc.   standard deviation of the r.v. X, Y, etc.   324

τ time constant, a time duration, or a specified time instant; argument of a characteristic function   280

θ polar coordinate angular variable; angle   268

ω angular frequency variable   566

ξ, ζ element   20

ξ_o element representing the actual outcome, or a particular element   175, 489

ξ^* element that gets selected   158

Bold lower case Greek letters

$\boldsymbol{\rho}[X,Y]$ correlation coefficient of X, Y 351

$\boldsymbol{\tau}$ argument of a multidimensional characteristic function   414

Plain and italic upper case Greek letters

Δ symmetric difference (of sets); a small increment   30

Δx small interval at point x 189

$\Delta(x)$ difference between two functions of x 189

$\Phi(x)$ Gaussian integral   187

$\Gamma(x)$ Gamma function   302

Σ a probability   501

Θ polar coordinate r.v.; r.v. expressing the modulus (angle) of a complex r.v.   213

χ^2 chi-square distribution   545

Symbols and markings

$\binom{n}{k}$ binomial coefficient   45

$(a_1, a_2, \ldots, a_k)$ k-tuple   110
(X_1, X_2) two-dimensional r.v.   197
[P]   holds with probability 1   352
[Fe1]   pointer to a reference (see References)   9

$[X_1, X_2], \begin{bmatrix} X_1 \\ X_2 \end{bmatrix}$ two-dimensional vector r.v.'s   197

$[X_1, \ldots, X_n], \begin{bmatrix} X_1 \\ \vdots \\ X_n \end{bmatrix}$ n-dimensional vector r.v.'s   253

$\lfloor x \rfloor$ largest integer less than or equal to x 527
$\{\ldots\}$ the set whose elements are...   21
$\{X_k\}$ sequence of r.v.'s   454
$\langle x, y \rangle$ inner product of x and y 375
$'$ result of a modification 563 first derivative   321
$''$ second derivative   189
$|$ Scheffer stroke   491
$|A|$ size of the set A   36
$|z|$ magnitude of the complex number z 397
$\|x\|$ norm of x 375
$\equiv$ is defined as   173
$=$ equality of sets   22
$\doteq$ equals, except for rounding error   112
$\approx$ approximately equals, is approximated by   163
$\neq$ distinctness of sets   22
$:$ such that   21
$+$ union   26
$-$ directed difference (of sets)   30
$\times$ product (of experiments, spaces or sets)   114
$\times$ product (of classes of sets)   120
$\cdot$ product (of probability distributions)   121
$*$ any one of certain set operations; convolution; complex conjugate   252
$(*)$ identifies an equation in the Appendix within a particular problem statement   495
$\rightarrow$ converges to   461
$\nrightarrow$ does not converge to   554
$\leftrightarrow$ corresponds to   18
$\subset$ is a subset of, is contained in, is included in   20
$\not\subset$ is not a subset of, is not contained in   21
$\supset$ contains, includes   20
$\cup$ union   26
$\uplus$ disjoint union   49
$\cap$ intersection   26

$\bigcup\limits_{i=1}^{n}, \bigcup\limits_{i=1}^{\infty}, \bigcup\limits_{A \in C}$ finite union, countable union   40, 41

$\displaystyle\biguplus_{i=1}^{n}, \biguplus_{i=1}^{\infty}$ finite disjoint union, countable disjoint union   49, 51

$\displaystyle\bigcap_{i=1}^{n}, \bigcap_{i=1}^{\infty}, \bigcap_{A\in C}$ finite intersection, countable intersection   40, 41

$\overline{a,b}$ open interval   166

$\overline{a,b}$ closed interval   166

$\overline{a,b}, \overline{a,b}$ semi-closed intervals   166

References

AR1   J. H. Abbott and J. I. Rosenblatt, *Foundations and Methods of Probability Theory with Scientific Applications*, Course Notes for Statistics 527, Purdue University, West Lafayette, Ind. 1960.

AS1   M. Abramowitz and I. A. Stegun, eds., *Handbook of Mathematical Functions: With Formulas, Graphs, and Mathematical Tables*, National Bureau of Standards Applied Mathematics Series #55, U.S. Government Printing Office, Washington, D.C. 1964.

CL1   E. Conte and M. Longo, "Characterization of Radar Clutter as a Spherically Invariant Random Process," *IEE Proc. F: Communications, Radar, & Signal Processing*, **134**, 191–197 (1987).

Fe1   W. Feller, *An Introduction to Probability Theory and its Applications*, Volume I, John Wiley & Sons, London. 1950.

Fe2   W. Feller, *An Introduction to Probability Theory and its Applications*, Volume II, John Wiley & Sons, New York, N.Y. © 1966.

Fi1   T. L. Fine, *Theories of Probability*, Academic Press, New York, N.Y. 1973.

Fr1   Hans Freudenthal, ed., *The Concept and the Role of the Model in Mathematics and the Natural and Social Sciences*, Proceedings of a Colloquium organized at Utrecht, January 1960, Gordon and Breach, New York, N.Y. © 1961.

Fr2   A. A. Frankel, *Abstract Set Theory*, North-Holland Publishing Co., Amsterdam. 4th edn, 1976.

Gn1   B. V. Gnedenko, *Theory of Probability*, Chelsea Publishing Co., New York, N.Y. 2nd edn, 1963.

GR1   R. L. Graham, "On Partitions of a Finite Set," *Journal of Combinatorial Theory*, **1**, 215–223 (1966).

GR2   I. S. Gradshteyn and I. M. Ryzhik, *Table of Integrals, Series and Products*, Academic Press, San Diego, Calif. 6th edn, © 2000.

Ha1   P. R. Halmos, *Measure Theory*, D. VanNostrand Co., Inc., Princeton, N.J. 1950.

Hi1   F. B. Hildebrand, *Methods of Applied Mathematics*, Prentice-Hall, Inc., Englewood Cliffs, N.J. © 1952.

HJ1   R. A. Horn and C. R. Johnson, *Matrix Analysis*, Cambridge University Press, Cambridge. 1985.

HK1   S. Hayden and J. F. Kennison, *Zermelo-Fraenkel Set Theory*, Charles E. Merrill Publishing Co., Columbus, O. 1968.

JJ1   A. Juels, M. Jacobsson, E. Shriver, and B. K. Hillyer, "How to Turn Loaded Dice into Fair Coins," *IEEE Trans. Info. Th.*, **46**, 911–921 (2000).

KF1   G. J. Klir and T. A. Folger, *Fuzzy Sets, Uncertainty and Information*, Prentice-Hall, Inc., Englewood Cliffs, N.J. 1988.

Kh1   A. I. Khinchin, *Mathematical Foundations of Information Theory*, Dover Publications, Inc., New York, N.Y. 1957.

Ko1   A. N. Kolmogorov, *Grundbegriffe der Wahrscheinlichkeitsrechnung*, Springer-Verlag, Berlin. 1933.

Li1   M. J. Lighthill, *Fourier Analysis and Generalised Functions*, Cambridge University Press, Cambridge. 1962.

Ma1   J. I. Marcum, "A Statistical Theory of Target Detection by Pulsed Radar," *IRE Trans. Info. Th.*, **IT-6**, 145 ff. (1960).

Probability Concepts and Theory for Engineers, First Edition. Harry Schwarzlander.
© 2011 John Wiley & Sons, Ltd. Published 2011 by John Wiley & Sons, Ltd.

Mc1 K. McKean, "The Orderly Pursuit of Pure Disorder", *Discover Magazine,* Jan. 1987.

MM1 M. S. Morgan and M. Morrison, eds., *Models as Mediators: Perspectives on Natural and Social Science,* Cambridge University Press, Cambridge. 1999.

Ne1 R. B. Nelsen, *An Introduction to Copulas,* Springer Verlag, New York, N.Y. 2nd edn, 2006.

OS1 A. V. Oppenheim and R. W. Schafer, *Discrete-Time Signal Processing,* Prentice-Hall, Englewood Cliffs, N.J. 1989.

Pa1 E. Parzen, *Modern Probability Theory and its Applications,* John Wiley & Sons, Inc., New York, N.Y. 1960.

Pf1 P. E. Pfeiffer, *Concepts of Probability Theory,* McGraw-Hill Book Co., New York, N.Y. 1965.

Re1 A. Renyi, "On a New Axiomatic Theory of Probability", *Acta Math. Acad. Sci. Hung.* **6**, 285–335 (1955).

Re2 A. Renyi, *Wahrscheinlichkeitsrechnung mit einem Anhang über Informationstheorie,* VEB Deutscher Verlag der Wissenschaften, Berlin. 1966.

Re3 F. M. Reza, *An Introduction to Information Theory,* McGraw-Hill Book Co., New York, N.Y. 1961.

Ri1 J. Richardson, *Models of Reality,* Lomond Publications, Mt. Airy, Md. 1984.

Se1 S. Selvin, "A Problem in Probability," *American Statistician* **29**, 67 (1975) and: "On the Monty Hall Problem," *American Statistician* **29**, 134 (1975).

Za1 L. A. Zadeh, "Fuzzy Sets", *Information and Control,* **8**, 228–353 (1965).

Ze1 A. H. Zemanian, *Distribution Theory and Transform Analysis,* McGraw-Hill Book Co., New York, N.Y. 1965.

Subject Index

Absolutely fair sequence, 455
Actual outcome, 11, 14–15, 21, 37
Addition of two random variables, *see* Sum of two
    random variables
Additive class, 31–2, 44–7, 49–50, 55, 489
    generated by a partition, 44–6
    largest partition, 44–6, 493
Additive set function, 49–52, 494
Algebra of sets, 31
Almost certain convergence, 462
Annihilation law for directed difference and
    symmetric difference, 30
Assignment problem, 135–7
Associative law
    for directed difference and symmetric
        difference, 30
    for union and intersection, 27

Bar graph, 157
Bayes' rule, 95
Bernoulli trials, 112
Bessel function, 426, 580
Binomial coefficients, 45, 112–3
Binomial distribution, 113, 122, 134, 161–4
    Poisson approximation, 163–4, 521
    truncated, 522
Binomial random variable(s), 161, 236, 540
    characteristic function, 575
    expected value, 323
    generating function, 419
    mixture, 543
    sum of two statistically independent, 548
    variance, 329–30
Binomial theorem, 45

Birthdays, 509, 513
Bits, 141
Bivariate Gaussian distribution, *see* Gaussian
    distribution
Body of a distribution, 183
Boole's inequality, 494
Boolean algebra, 490
Borel function, 235, 239, 256
Borel set, 169–71
    n-dimensional, 253–6
    two-dimensional, 199–200, 239
Branches of a coordinate transformation, 266–7,
    288–9
Bridge game, 17, 111, 125, 483, 511, 514

C.d.f., *see* cumulative distribution function
Cardinality, 36
Cartesian plane, 192
Cartesian product, 115–7, 197–8, 510
Cauchy distribution, 181, 183, 528, 562
Cauchy random variable, 183, 527
Central Limit Theorem, 472–3
Central moments, 315–16
Centralized distribution, 312–13, 318
Centralized random variable, 312, 318
Certain event, 15, 55
Certainty required of a particular event, 86
Chain rule
    for conditional densities, 296
    for conditional probabilities, 95, 288
Chance selection of a real number, 175–6
Characteristic function, 402–3
    conditional, 409
    moment generating property, 405–6

Probability Concepts and Theory for Engineers, First Edition. Harry Schwarzlander.
© 2011 John Wiley & Sons, Ltd. Published 2011 by John Wiley & Sons, Ltd.

Characteristic function (*Continued*)
  multidimensional, 412–14
  standardized Gaussian random variable,
    404–5
  transformed random variable, 408–11
  two-dimensional, 405–6
Chebychev inequality, 332–4, 562–3
Chi-square distribution, 425, 546
  characteristic function, 576
  mean and variance, 561
Circular cone, 538
Circular symmetry, 242
Class (of sets), 23–5, 29, 31, 486–7
  empty, 23–4
  independent, 104–8, 216–7, 508
  partially ordered, 487
  product class, 120, 124
Closed interval, 165–6
Closure of a class, 507
Clusters, 517–18
Coin-toss experiment, 37, 157–8, 175
  with *n* tosses of one coin, 37, 112, 161
  with two coins, 88–9
Coin-toss sequence, 37–8
  Infinite, 37–8, 63, 175–6, 492, 496
  periodic, 492
Collections of objects, 17
Commutative law
  for directed difference and symmetric
    difference, 30
  for union and intersection, 27
Complementary error function, 187
Complementation, 26–8
Completely additive class, 47–8, 165, 169–70,
    493
  generated by a product set, 120
Complex Gaussian random variable, 399
Complex random variable, 397–401
Complex vector random variable, 400
Component experiments, 114–8
Component of a sequence of random
    variables, 454
Component random variables, 253, 255, 259–60,
    275
Concept, 3
Conceptual errors, 3
Conceptual model, 4, 75–6, 79
Conceptual world, 8–9
Conditional characteristic function, 409
Conditional cumulative distribution function
  given an event, 284–6, 288, 291
  given the value of a random variable, 294

Conditional density function, 285–9, 295–7
  given an event, 285–9, 409
Conditional entropy, 144, 436–7, 518
Conditional expectation, 345–7, 362–6
  given the value of a random variable, 362–6,
    381–2
Conditional generating function, 420–1
Conditional independence, 506, 576
Conditional probability, 87–91, 93–99
Conditional probability density function, *see*
    Conditional density function
Conditional probability distribution, 91–2, 97
  defined with probability one, 294
  given an event, 284–5
  given the value of a random variable, 293–4
Conditional probability function, 92–94
Conditional product experiment, 118, 123–6
Conditional random variable, 286, 291
Conditional relative frequency, 86–7
Conditional variance, 347–8
Conditionally independent events, 506
Contraction, 129–31, 133
  uniform, 129–31
Convergence
  almost certain, 462
  in distribution, 470
  in probability, 462–4
  of a sequence of random variables, 461–4
  pointwise, 462
  with probability one, 462
Convolution, 249–51
  discrete, 245
Convolution integral, 249
Coordinate transformation, 263–8, 288–9
  branches, 266–7, 288–9
Copula, 444–53
  statistically independent random variables,
    445–6
Correlation coefficient, 350–7, 360–2, 400
Covariance, 350–2, 357–8, 367, 399–400
Covariance matrix, 367–73, 400
  diagonalization, 370–3
  eigenvalues, 370–2
Crisp set, 147

Cumulative distribution function (c.d.f.), 173–7
  conditional, given an event, 284–6, 288, 291
  continuous, 176–7
  joint, 202–7
  multidimensional, 254–6
  two-dimensional, 202–7

Data transmission, 95–6
Decision problem, 347–9
Delta function, 183–4, 234–6, 291
  changing the scale, 234–5
DeMorgan's laws, 27–8, 41
Density function, *see* Probability density function
Deterministic parameters, 307–8
Diagonal sequence, 37
Die throwing experiment, 483–4
  with a loaded die, 68, 92–3, 498
  with one die, 7, 12, 14–16, 19, 34, 155, 159, 191,
    217, 229–30, 484
  with two dice, 11–12, 62–3, 84, 86–8, 100, 105–6,
    115–7, 194, 238, 244
Differential entropy, 438–9
Dilation, 129
  uniform, 129
Directed difference, 30–31, 489
Discrete convolution, 245
Discrete probability distribution, 157–8
Discrete random variable, 157–9
Disjoint class, 24–5, 487
Disjoint sets, 24–5, 27, 65
Disjoint union, 49, 65
Distance between sets, 22, 499
Distance measure, 499
Distinct sets, 22
Distribution, *see also* Probability distribution, 50, 58
  binomial, *see* Binomial distribution
  Cauchy, *see* Cauchy distribution
  equal-probability, 85
  exponential, *see* Exponential distribution
  gamma, *see* Gamma distribution
  Gaussian, *see* Gaussian distribution
  geometric, 123, 159, 556, 578
  hyperexponential, 287
  hypergeometric, 131, 559
  hypoexponential, 541, 554
  log-normal, 545
  multinomial, 133–4
  negative binomial, 512, 578
  normal, *see* Gaussian distribution
  Poisson, 164, 300
  Rayleigh, 242, 425, 537
  uniform, *see* Uniform distribution

Distribution function, *see* Cumulative distribution
    function
Distributive laws
  for directed difference and symmetric
    difference, 30
  for union and intersection, 27
Drawing, *see* Random selection

Eigenvalues, 371–3
Eigenvectors, 370–3
Elementary event, 15
Elementary probability problems, 79
Elementary set, 63
Elementary set operations, 26, 31
Elements, 18
Ellipse of concentration, 567
Empty set, 21–2, 27, 31
Encoding, 581
Endpoints of an interval, 165
Entropy, 142–5
  of a discrete random variable, 435
  conditional, 144, 436–7, 518
  differential, 438–9
  partial, 143–4
  relative, 438–9
Epoch, 299
Equal-probability distribution, 85
Equality of sets, 21–2
Error function, 187
Estimation
  error, 380–5
  of a probability, 474–6
  of a random variable, 379
  linear mean-square, 382–5
Estimator, 337–9, 341–2
  biased, 341
  unbiased, 338–9, 341
Euclidian space, 197, 208
  n-dimensional, 253, 259
Event, *see also* Events, 14–16
  certain, 15
  elementary, 15
  impossible, *see* Impossible event
  required to occur with certainty, 86
  zero-probability, 182
Events, *see also* Event
  conditionally independent, 506
  independent, 99–104, 216–8
  mutually exclusive, 25
  mutually independent, 104–7
  pairwise independent, 104–5
Exclusive 'or', 30

Expectation
  conditional, *see* Conditional expectation
  is a linear operation, 323
  of a binomial random variable, 323
  of a constant, 311
  of a function of a random variable, 320–5
  of a random variable, 308–9
  of a vector random variable, 312, 367
  of an expected value, 311
  of the product of two random variables, 325
  of the sum of two random variables, 322–3
Expected value, *see* Expectation
Experiment, 6
Experimental plan, 8, 11
Experimental procedure, 7, 53, 56
  execution, 7
Exponential distribution, 181, 183, 301, 524, 526,
    552–3
  characteristic function, 575
  moments, 315
  standard deviation, 558
  truncated, 557, 562
Exponential random variables
  maximum of two, 541
  minimum of two, 541
  statistically independent, 535, 540–1, 547,
    550, 557
  not statistically independent, 447

Fair game, 455–6
First moment, 311–12
Fourier transform, 402–3
Function, 33–5
  domain, 33–5
  indicator, 286, 491
  inverse, 34
  one-to-one, 34
  positive semi-definite, 406
  range, 33–5
  real-valued, 33–4, 49–50
  right-continuous, 173
  single-valued, 33
Function of a random variable, 229–30
  expectation of, 320–5
  variance of, 328–31
Functional dependence, 195
Fuzziness, 3, 8, 146–7
Fuzzy arithmetic, 522
Fuzzy set, 147–50

Gambling schemes, 455–8
Gamma distribution, 302

characteristic function, 575
  mean, 559
  mode, 557
  variance, 561
Gamma function, 302
Gaussian distribution, 181–3, 186–9
  bivariate, 269–72, 359–60
    standard ellipse, 270, 567
  conditional density of bivariate, 567
  degenerate multivariate, 370
  multivariate, 369–70, 422–4, 569–70
  n-dimensional, 369–70, 422–3
  standardized, 187–8
    moments, 558
    n-dimensional, 422–3
Gaussian integral, 187
  approximation, 189–90
Gaussian random variable(s), 186–8, 359–60, 534–5,
    541–2
  characteristic function, 404–5, 423
  complex, 399
  jointly Gaussian, 272, 535
  n-dimensional, 369–70, 422–3
  not jointly Gaussian, 272
  standardized, 187–8, 317–18
    characteristic function, 404–5
    n-dimensional, 422–3
  statistically independent, 224, 269–70, 534–5,
    546, 561
  sum of two, 542, 546
  two zero-mean, 359–60
Gaussian-like vector random variables, 428
Generating function, 417–21
  conditional, 420–1
Geometric random variable, 540
Greatest lower bound, 490

Half-life, 526
Hermitian matrix, 400
Higher moments, 314–18
Hyperexponential distribution, 287
Hypergeometric distribution, 131, 559
Hypergeometric random variable, 559
Hypoexponential distribution, 541, 554

Idempotent law for unions, intersections, 27
Identity function, 37
Impossible event, 15, 19, 22
Inclusion relation, 21, 23, 27
Index, 33–4, 36
Indicator function, 286, 491

Indicator random variable, 218, 294, 311, 313, 330, 533
  generating function, 418–19
Induced probability distribution, 157–8, 173–4, 176
  bounds, 333–5
Induced probability space, 170–1, 229, 254, 284
Infinite collection of random variables, 454
Infinite sequence of sets, 41–2, 51
Information Theory, 145
Inner product, 375–6
Integer pairs, 166–7
Intersection, 26–8, 31, 40
  infinite, 41
Interval, 165–6
  closed, 165–6
  endpoints, 165
  finite, 168
  infinite, 168
  open, 165–6
  semi-closed, 165–6
  semi-infinite, 168–70
  semi-open, 165–6
Inverse image, 159–60, 170, 232, 239–40, 256, 264
Irrational numbers, 166
Isosceles triangle, 537

Jacobian, 263–7
Joint cumulative distribution function, 202–7
Joint expectation of two random variables, 325, 329
Joint probability density function, 210–11, 213
Joint probability distribution, 192–3, 195
  rotation, 268–71, 357–60

K-distribution, 580
k-tuples, 109–11, 113, 115–17, 126, 128–9
  ordered, 115, 125

Laplace random variable, 534
Lattice, 490
  distributive, 490
Law of large numbers, 340, 464–5
  Strong Law, 464
  Weak Law, 465
Least upper bound, 490
Lebesgue measure, 168, 208
  in R*n*, 259
Leibnitz's rule, 234
Linear dependence, 352–3
Linear operation, 323, 352
Linear transformation, 357
Loaded die, 68, 92–3, 498
Low-pass filter, 279

Markoff inequality, 334–5
Martingale, 585
Mass distribution, 311
Matrix
  Hermetian, 400
  non-negative definite, 368
  orthogonal, 370
  positive semi-definite, 368
  Toeplitz, 569
Mean time between failures, 558
Mean-square error, 381, 384
Mean-square estimation, 381–5
  linear, 382–5
  nonlinear, 381–2
Measure, 58
Measurement of probabilities, 75
Median, 312, 380
Membership function, 147–51
Minimum property of the center, 559
Mixture of distributions, 287, 291, 543
Mixture of random variables, 543, 575
Mode, 308
Moment, 311–12, 314–15
  central, 315–16
  first, 311–12
  higher, 314–18
  product moment, 326, 350, 415
  second, 314–16
Moment generating function, 417–21
Moment generating property, 405–6
Moment of inertia, 314
Monty Hall paradox, 79
Multinomial distribution, 133–4
Mutually statistically independent random variables, 274–6

Nested sets, 507
Network theory, 4–5
Noise, 565
Nonlinear mean-square estimation, 381–2
Non-negative definite matrix, 368
Norm, 375
Normal distribution, *see* Gaussian distribution
Normal random variable, *see* Gaussian random variable
Normalized probability distribution, 318
Normalized random variable, 318
Numerical-valued possible outcomes, 155

Occurrence times, 299–300
Observable features, 89

One-element set, 47–8, 63
Open interval, 165–6
Optimization criterion, 379–81
Order statistics, 459–60
Ordered pair, 109–10, 115
Ordered sampling
   with replacement, 111–12, 114, 126
   without replacement, 126
Ordered triple, 126
Orthogonality principle, 383

P.d.f., *see* Probability density function
Pairs of integers, 166–7
Pairs of real numbers, 192
Pairwise independent events, 104–5
Parameters
   deterministic, 307–8
   probabilistic, 308
Partial entropy, 143–4
Partial ordering, 487
Partition, 23–5, 44–5
   additive class, largest partition, 44–6, 495
Payoff rule, 191–2, 195, 230
Periodic coin-toss sequence, 492
Permutations, 126
Point function, 33–6, 491
   real-valued, 33–4, 491
Point mass, 311, 314
Pointwise convergence, 462
Poisson distribution, 164, 300
Poisson random variable, 164, 577
Polar coordinate random variables, 213–14, 264–5,
   538–9
Polar coordinates, 213
   conversion from rectangular coordinates, 182,
    240–2, 538
   conversion to rectangular coordinates, 264–5
Population, 336–7
   average, 336–7, 339
   variance, 340–2
Positive probability points, 243
Positive semi-definite function, 406
Positive semi-definite matrix, 368
Possible outcomes, 11–18, 484–5
   equally likely, 81–5
   numerical-valued, 155
Probabilistic average, 308
Probabilistic parameters, 308
Probabilistic experiment(s), 6–11, 53–6, 62–4,
   75–9
   Identical, 53–6
   Independent, 54–56

probabilistically equivalent, 62–4
purpose, 6–8, 11–13
Probability, 54, 58–9
   conditional, *see* Conditional probability
   measurement, 75
Probability density function (p.d.f.), 180–5
   conditional, 285–9, 295–7
    given an event, 285–9, 409
   Fourier transform, 402–3
   joint, 210–11, 213
   marginal, 211, 260
   n-dimensional, 260
   two-dimensional, 210–11, 213
Probability distribution, *see also* Distribution, 58, 63,
   83–4
   absolutely continuous, 176, 208
   absolutely continuous n-dimensional, 259
   binomial, *see* Binomial distribution
   Cauchy, *see* Cauchy distribution
   centralized, 312–13, 318
   conditional, *see* Conditional probability
    distribution
   discrete, 157–8
   exponential, *see* Exponential distribution
   first moment, 312
   gamma, *see* Gamma distribution
   Gaussian, *see* Gaussian distribution
   induced, *see* Induced probability distribution
   joint, *see* Joint probability distribution
   mean, 312
   mode, 308
   n-dimensional, 254, 259
    absolutely continuous, 259
   normal, *see* Gaussian distribution
   normalized, 318
   resolved into components of pure type, 290
   second moment, 314–15
   singular, 212, 282–3
   standardized, 317–18
   two-dimensional, 192, 197–8
   uniform, *see* Uniform distribution
Probability function, 58–60, 65–7, 83, 92
   conditional, 92–4
   unconditional, 92, 94
Probability generating function, *see* Generating
   function
Probability measure, 58
Probability of failure, 102–3, 504
Probability one, 294
Probability problems, 75–80
Probability space, 62–3, 65, 75, 83–5, 120–2
   induced, 171, 192, 200

Product class, 120, 124
Product experiment, 114–18
Product moment, 326, 350, 415
Product probability distribution, 121
Product probability space, 121, 123
Product rule, 95, 126
Product set, 116–17
Product space, 115–18
Proper subset, 44

Q-function, 426
Queue, 137

R.v., *see* Random variable
Radioactive decay, 300–1
Radius of gyration, 314, 316
Random deck of playing cards, 163
Random digits, 539
Random errors, 186
Random integers, 59–60
Random mechanism, 77–8, 81–2, 92, 107, 140
    stationary, 299
Random number table, 539, 563
Random occurrences in time, 298–9
Random square-wave, 279–80
Random sums, 564
Random selection, 81, 92
    ball, 77–8, 81–2, 96–7
    card, 111–12
Random variable (r.v.), *see also* Random
        variables, 171
    absolutely continuous, 176–7, 208
    binary, 356–7, 360, 566
    binomial, 161, 236, 540
    centralized, 312, 318
    complex, 397–401
    conditional, 286, 291
    discrete, 157–9
    function, *see* Function of a random variable
    general definition, 171
    geometric, 540
    hypergeometric, 559
    indicator, *see* Indicator random variable
    mean-square value, 324–5
    mixed type, 178, 290
    most probable value, 308
    n-dimensional, 253–4, 259
        absolutely continuous, 259
    negative binomial, 559, 578
    normalized, 318
    pure type, 290
    Poisson, 164, 577

Properties, 307–9, 314
    range, 307
    scalar, 197, 200, 221
    singular, *see* Singular random variable
    standardized version, 317
    transformation, 229–37
    two-dimensional, 197, 200
        transformation, 238–42, 263–5, 268
    unary, 220
    vector, *see* Vector random variable
    vector-valued, 197
Random variables, *see also* Random variable
    addition of two, *see* Sum of two random variables
    component, 253, 255, 259–60, 275
    expectation of the product of two, 325
    expectation of the sum of two, 322–3
    Gaussian, *see* Gaussian Random variables
    identical, 209–10
    identically distributed, 209
    jointly absolutely continuous, 210, 259
    jointly Gaussian, 271–2, 535
    linearly dependent, *see* Linear dependence
    maximum of two, 245–6
    minimum of two, 541
    mutually statistically independent, 274–7
    n-dimensional, 253–4
    orthogonal, 376
    pairwise statistically independent, 274–6, 330, 561
    product of two, 541, 554
    sequence of, *see* Sequence of random variables
    several, 253–4
    statistically independent, *see* Statistically
            independent random variables
    sum of two, *see* Sum of two random variables
    two, 193, 197
    two binary, 356–7, 565–6
    uncorrelated, 353–4
Random walk, 516–17, 561
    with absorbing boundaries, 458
    with reflecting boundaries, 458
Randomness, 140–2
Ratio of two s.i. random variables, 541
Rational numbers, 166–7
Rayleigh distribution, 242, 425, 537
Real line, 165
Real numbers, 33, 156–7, 165–6
    chance selection of, 175–6
    pairs of, 192
    space of, 165
Real world, 3–4
Real-world problems, 4–5, 8–9
Rectangle set, 197–8

Reduction rule, 27
Refinement, 129–30
  uniform, 129, 145
Regression function, 363, 381
Relative entropy, 438–9
Relative frequency, 54, 58–9
  conditional, 87
Reliability, 558
Riemann integral, 386
Riemann–Stieltjes integral, 386–90
Rotation about the origin, 268–71
Rule of total probability, 94
Runs, 159–60, 518

Sample, 18
Sample average, 337–9
Sample points, 18, 23
Sample sequence, 454
Sample space, 18, 20–1, 37
  uncountable, 37
Sample variance, 339–42
Sampling, 336–8
  ordered, with replacement, 111–12, 114, 126
  ordered, without replacement, 126
  unordered, with replacement, 132, 342–3
  unordered, without replacement, 128, 336, 342
Scalar random variable, 197, 200, 221
Scale factor, 328, 332
Scheffer stroke, 26, 489
Schwarz inequality, 325–6, 559–60
Second moment, 314–16
Semi-closed interval, 165–6
Semi-infinite interval, 168–70
Semi-open interval, 165–6
Sequence of partial averages, 458
Sequence of partial sums, 456, 472
Sequence of probability distributions, 470
Sequence of random variables, 454–5
  absolutely fair, 455
  convergence, 461–4
Sequence of standardized partial averages, 471–2
Set, 18–19
  crisp, 147
  elementary, 63
  empty, 21–2, 27, 31
  finite, 36–7
  fuzzy, 147–50
  one-element, 47–8, 63
  uncountable, 37–8, 492
Set algebra, 31
Set function, 33–4, 49–51, 155, 159
  set-valued, 491

Set operations, 26–32, 487–90
  countable, 41, 47, 492
Set theory, 17–20, 33, 53, 58
Several random variables, 253–4
Shape of a function, 332
Sigma-algebra, 47
Signal-to-noise ratio, 354
Similarity transformation, 370
Simple random sample, 336
Single-valued function, 33
Singular probability distribution, 212, 282–3
Singular random variable, 212, 262, 279, 282–3
  in one dimension, 279, 282–3
  in two dimensions, 212
Sinusoidally varying voltage, 537–8, 566
Size of a set, 36–39
Space, 18
  uncountable, 37
Spherical symmetry, 422, 428–30, 433
Spherically invariant vector random variables,
  433–4
Spherically symmetric vector random variables,
  428–434
  family of, 428, 430–1, 433
Spinning pointer experiment, 177–8
Square-wave, random, 279–80
Standard deviation, 316–19, 324–5, 328, 567
Standard ellipse for bivariate Gaussian distribution,
  270, 567
Standardized probability distribution, 317
Standardized version of a random variable, 317
State diagram, 137–38
State-transition diagram, 457–8
Statistical average, 308
Statistical independence of random variables, 194–5,
  218, 220–4, 561
  conditional, 576
  mutual, 274–7
  pairwise, 274–6, 330, 561
Statistical regularity, 54–6, 468–9
Statistically independent random variables
  exponential, 535, 540–1, 547, 550, 556–7
  Gaussian, 224, 269–71, 546
  product of two, 541
  ratio of two, 541
  sum of two, 249, 553, 565
Statistics, 75, 82
Steady state, 137–8
Stieltjes integral, 386–90
  improper, 388
Stroke, *see* Scheffer stroke
Strong Law of Large Numbers, 464

Subset, 20–1
    proper, 44
Sum of two random variables, 243–5, 247–9
    expectation, 322–3
    variance, 329
Symmetric difference, 30–1

Tails of a distribution, 183
Test sample, 336
    average, 338–9
    variance, 339–40
Test sampling, 336
Throw of a die, *see* Die throwing experiment
Toeplitz matrix, 569
Toss of a coin, *see* Coin-toss experiment
Transformation of a random variable, 229–37
Transformation of a vector random variable, 238–42,
    263–4, 268, 368–9
    effect on statistical independence, 277
Tree diagram, 136–7
Trials, 111–12, 114
    independent, 112, 114, 117–18
Triangle inequality, 499, 561
Triangle, isosceles, 537
Triangular random variables, sum of two, 542
Triple, 62, 109
Tuples, 109–13
Two-dimensional probability distribution, 192,
    197–8
Two-dimensional random variable, *see* Random
    variable, two-dimensional

Unbiased estimator, 339
Uncertainty, 140–5
Uniform contraction, 129–31
Uniform distribution, 85, 181
    in two dimensions, 269
    standard deviation of, 316–17
Uniform dilation, 129
Uniform refinement, 129, 145
Union, 26–9

infinite, 41
Unit cube, 255, 259
Unit impulse, *see* Delta function
Universal lower bound, 490
Universal upper bound, 490
Unordered sampling with replacement, 132,
    342–3
Unordered sampling without replacement, 128,
    336, 342
Unsatisfactorily performed experiment, 12–13
Urn problems, 82

Variance, 316–18
    binomial random variable, 329–30
    complex random variable, 399–400
    conditional, 347–8
    function of a random variable, 328–31
    singular random variable, 391–2
    sum or difference of two random variables,
        329–30, 353
    k-tuple of numbers, 340
    sample, 340–1
    sample average, 340–2
Vector random variable, 197–8
    complex, 400
    expectation, 312, 367
    transformation, 238–42, 263–4, 268, 368–9
    n-dimensional, 253
Vector space, 374–5
    of random variables, 374–8
Vector-valued random variable, 197
Venn diagram, 20–1, 24

Waiting line, *see* Queue
Waiting time, 299, 301
Weak Law of Large Numbers, 465
Weather forecasting, 496
Well-defined collection, 17

Zero probability, 63–4, 182
z-transform, 417